D1037056

Mass Spectrometry in Biology & Medicine

Mass Spectrometry in Biology & Medicine

Edited by

A. L. Burlingame
University of California, San Francisco, CA

Steven A. Carr
SmithKline Beecham Pharmaceuticals, King of Prussia, PA

Michael A. Baldwin
University of California, San Francisco, CA

999 Riverview Drive, Suite 208
Totowa, New Jersey 07512

For additional copies, pricing for bulk purchases, and/or information about other Humana titles, contact Humana at the above address or at any of the following numbers: Tel: 973-256-1699; Fax: 973-256-8341; E-mail: humana@humanapr.com or visit our website at http://www.humanapress.com

This publication is printed on acid-free paper. ∞
ANSI Z39.48-1984 (American National Standards Institute)
Permanence of Paper for Printed Library Materials.

Printed in the United States of America. 10 9 8 7 6 5 4 3 2 1

Library of Congress Cataloging-in-Publication Data

Mass spectrometry in biology & medicine / edited by A. L. Burlingame, Steven A. Carr, Michael A. Baldwin
p. cm.
Includes bibliographical references and index.
ISBN 0-89603-799-1 (alk. paper)
1. Mass spectrometry. 2. Proteins–Analysis. 3. Macromolecules–Analysis. I. Burlingame, A. L. II. Carr, S. A. (Steven A.) III. Baldwin, Micahel A. IV. Title: Mass spectrometry in biology and medicine.
QP519.9.M3M358 2000 99-23633
572'.36–dc21 CIP

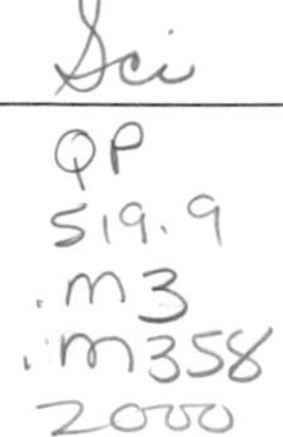

Dedicated to the memory of our dear colleague and friend

Wilhelm J. Richter

Table of Contents

Organizing Committee

Co-Chairs
A. L. Burlingame, University of California, San Francisco, CA
Steven A. Carr, SmithKline Beecham Pharmaceuticals, PA
Wilhelm J. Richter, Ludwig Institute for Cancer Research, UK

Conference Coordinator
Marilyn F. Schwartz, University of California, San Francisco, CA

Members
Robert Anderegg, Glaxo Wellcome Inc., NC
Robin T. Aplin, Oxford University, UK
Thomas A. Baillie, Merck, PA
Catherine E. Costello, Boston University, MA
Bhupesh C. Das, CNRS, France
Simon J. Gaskell, UMIST, UK
Emilio Gelpí, IIBB-CSIC, Spain
Michael L. Gross, Washington University, MO
Karl-Anders Karlsson, Gothenburg University, Sweden
Gennaro Marino, University of Naples, Italy
Takekiyo Matsuo, Osaka University, Japan
James A. McCloskey, University of Utah, UT
Robert A. Murphy, National Jewish Hospital, CO
Scott D. Patterson, Amgen Inc., CA
Jasna Peter-Katalinic, Universität Münster, Germany
Peter Roepstorff, Odense University, Denmark
Jan Sjövall, Karolinska Institutet, Sweden

Sponsors

Affymax Research Institute
Amgen
Astra Hässle AB
Baxter Healthcare Corporation
Bristol-Myers Squibb Co.
Finnigan Corporation
Genecor International, Inc.
Genentech, Inc.
Glaxo Wellcome Inc.
Institut Henri Beaufour
Isotec Inc.
Kratos Analytical Instruments
Merck & Co., Inc.
Micromass, Inc.
NeuroSearch A/S
Nippon Kayaku Co., Ltd.
Novartis Pharmaceuticals Corporation
Otsuka Pharmaceuticals Co., Ltd.
Parke-Davis Pharmaceutical Research
PE Biosystems
Pfizer, Inc.
Phyto Lierac
Proctor & Gamble Pharmaceuticals
Sankyo Co., Ltd.
SmithKline Beecham Pharmaceuticals
Sumitomo Chemical Co., Ltd.
Zeneca Ag Products
Zeneca Pharmaceuticals

We regret the passing of our esteemed colleague, Prof. Takekiyo Matsuo, in 1998.

Preface

As mass spectrometry enters its second century following the discoveries of the electron and ions by J. J. Thomson, no one could have anticipated the developments of the last decade that initiated structural studies of biomacromolecules, and even the detection and measurement of their mega-dalton noncovalent complexes and assemblages. This was not brought about by painstaking refinements of existing methods, but resulted from the discovery of new processes of ion formation from polar substances in liquid or solid solution that avoid chemical derivatization and/or destructive sample vaporization. New types of mass spectrometers are needed to exploit the inherent sensitivities of these high efficiency ion sources and permit routine realization of high mass resolution and accuracy of mass measurement. Of equal importance for sequencing and detailed structural characterization of components of biopolymers is the need to develop tandem instrumentation with optimized sensitivity able to produce high quality product ion spectra.

Considerable attention is currently focused on sample isolation and handling methodologies. Resolution of these issues is critical for successful work in the femto- and attomole ranges. Biological problems involving protein macromolecules, for example, usually require dealing with complex mixtures, such as immunoprecipitates, cell lysates, or molecular machines, that are in many cases well suited to SDS PAGE methods. These technologies mesh well with mass spectrometry and genomics. Hence, the digestion, extraction, and identification of peptides from gel spots using these mass spectrometric mapping and *de novo* peptide sequencing methods can be combined successfully with rapid protein identification by interrogation of protein, gene, and EST databases. However, concerns over how to analyze components of low abundance in the presence of many others with differing properties in widely varying amounts is an unsolved problem. IN the context of cell biology, the detection and identification of low copy number gene expression in a cell milieu is a daunting task. In addition, effective separation of membrane proteins remains a challenge, although amenable to electrospray measurement using suitable solvents. The detection and identification of DNA polymorphisms is readily tractable using chip-based robotics and mass spectrometric readout technologies.

To gain an understanding of the functional details of the machinery of cells, it is necessary to define the physiologically active forms of proteins and their modes of regulation. This requires a knowledge of their myriad posttranslational and xenobiotic modifications. The determination of the modifications of nucleic acids are of similar importance.

This volume contains a discussion of the new tools emerging to study molecular composition and the structural details required to elucidate function of macromolecules individually and as assemblages of communicating molecular machines. Individual contributors describe their strategies for solving problems that represent all the major topics of biomedical significance in drug discovery, lipid mediators, protein characterization and expression (proteomics or functional and structural genomics), glycobiology, and structural variation and modification of nucleic acids affecting function.

The participants and editors wish to acknowledge the outstanding organizational and managerial capabilities of Ms. Marilyn F. Schwartz for the success of the Fourth Symposium. The editors are indebted to the editorial talents of Ms. Candy Stoner for design and preparation of the final copy of this volume for publication. We also wish to acknowledge the financial support of the NIH NCRR Grant RR 01614 (to ALB).

A. L. Burlingame
Steven A. Carr
Michael A. Baldwin

A New Delayed Extraction MALDI-TOF MS-MS for Characterization of Protein Digests

Marvin Vestal, Peter Juhasz, Wade Hines and Stephen Martin
PerSeptive Biosystems, 500 Old Connecticut Path, Framingham, MA01701

Two-dimensional gel electrophoresis is commonly used to separate and visualize proteins present in complex mixtures such as cell lysates [1]. Digestion of selected spots by one or more endopeptidases followed by generation of peptide mass maps using MALDI-TOF MS is now widely accepted as the first step toward identifying and characterizing the proteins [2]. Following the development of delayed extraction techniques for MALDI [3-5], and improved sample preparation and clean-up methodology [6], this strategy is often successful at identifying proteins represented in a database [7] even when the proteins are present at low levels. In favorable cases protein identification is successful at the sub-femtomole level. When this approach fails, it is generally necessary to generate sequence data using either Edman degradation or MS-MS techniques. The MS-MS technique known as post-source decay (PSD) with MALDI-TOF [8] sometimes provides sufficient sequence information, but often the sensitivity and mass accuracy are inadequate and interpretation is difficult. Recent developments in electrospray ionization, such as nanospray [9] have dramatically improved the sensitivity of triple quadrupole and ion trap MS-MS, but these techniques are rather slow and tedious, and rapid, automated interpretation of data to provide reliable sequences is not yet routine. New instruments employing TOF analyzers in place of the third quadrupole in triple quadrupole MS-MS systems have very recently been described which provide improved resolution and mass accuracy in fragment ion measurements [10, 11].

The quadrupole and ion trap instruments employ multiple, low energy collisions to induce fragmentation. These approaches are very effective for low m/z precursor ions, and are particularly useful for multiply charged ions produced by electrospray or nanospray of tryptic fragments below mass 2000 u, but higher mass, singly charged ions, such as those produced by MALDI, are often not amenable to these techniques. Peptide sequencing, and structure determination in general for higher mass peptides has been successfully addressed by tandem double-focusing magnetic deflection mass spectrometers with high mass range [12]. Unfortunately, these large four-sector instruments are rather expensive, and sensitivity is limited. The addition of array detectors which allow portions of the spectrum to be acquired simultaneously has dramatically improved sensitivity [13], but even with these improvements more than 1 picomole of sample is generally required to obtain useful results.

Tandem TOF mass spectrometers employing either linear [14] or curved field reflectrons [15, 16] have been described by Cotter and coworkers, but both the sensitivity and resolution obtained in the early prototypes severely limited their applications. These approaches have apparently not been further developed.

The primary goal of the present work is to develop a new tandem TOF MS-MS technique with resolution and mass accuracy comparable to that of the four-sector magnetic instruments, while maintaining the sensitivity, speed, and mass range of MALDI-TOF. Ideally the sensitivity should be sufficient to obtain useful sequence information on the smallest peaks (ca. 1 femtomole) readily detected in a protein digest from a 2-D gel. A prototype of a new instrument aimed at achieving this goal has recently become operational, and the first experimental results are presented in this chapter. All of the performance goals have not yet been achieved, but the initial results are promising.

Design of the Instrument

The new instrument employs time-of-flight techniques for both primary ion selection and fragment ion determination. The heart of the system is a novel timed-ion-selector which allows an ion of a particular m/z, separated by a first time-of-flight analyzer, to be selectively transmitted while all others are rejected. This selector is designed so that the resolution of this selection process is determined primarily by the resolution of the first analyzer, and is not significantly degraded by the selector. This selector may be employed to select PSD ions or with a collision cell for high energy CID. To obtain high resolution in primary ion selection the extraction delay and extraction voltage applied to the DE source are adjusted to provide first order velocity focus at the timed-ion-selector. The second mass spectrometer includes an electrically isolated collision cell connected to a pulsed high voltage supply similar to that used in the MALDI-DE source. Thus, the second mass spectrometer also operates in DE mode, which

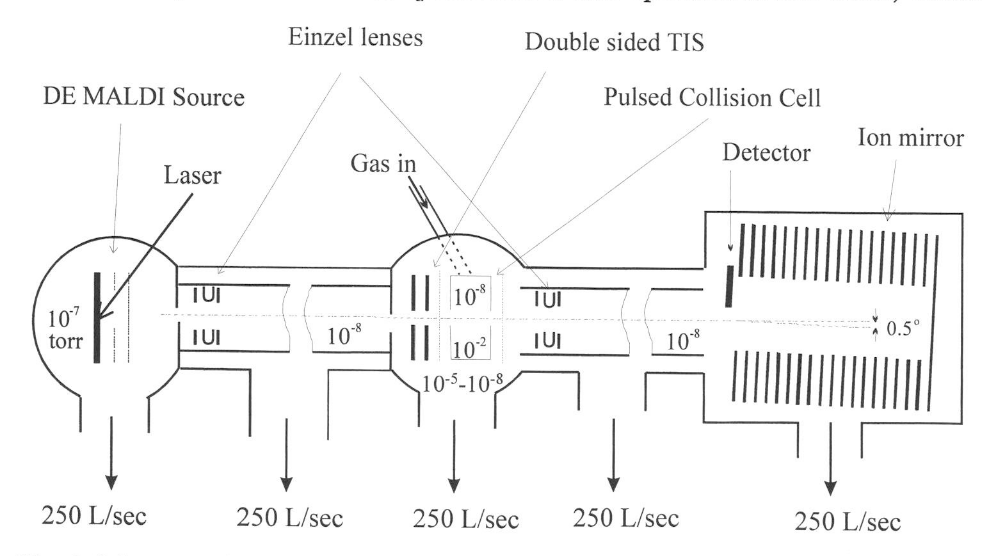

Fig. 1. Schematic diagram of T-Square MS-MS

allows the complete, high-resolution fragment spectrum produced by high energy CID to be obtained in a single acquisition.

A schematic diagram of the instrument is shown in Fig. 1. This instrument is based on the 6 m effective flight path instrument described earlier [5]. The total field-free path has been increased by about 1 m due to the addition of a 15 cm cube housing the collision cell and timed ion selector, and two additional differential pumping chambers. Three 250 L/sec turbomolecular pumps have been added, as indicated in Fig. 1, to provide pumping for the collision cell housing, and for the field-free regions connecting the ion source and the ion mirror to the collision cell. The 75 mm coaxial dual channel plate detector used in the original instrument has been replaced by a 40 mm hybrid detector mounted off axis, and the mirror is inclined by 0.5 degrees to the nominal axis of the instrument. By adjusting operating conditions, the instrument may be operated as a 7.5 m reflecting analyzer, a 5 m linear analyzer, or in MS-MS mode as a 1.5 m linear ion selector followed by a 6 m reflecting mass analyzer.

The dimensions of the instrument and a potential diagram for a typical MS-MS operating condition are shown in Fig. 2. In this example, ions are accelerated to approximately 18 kV in the MALDI DE source and the delay time is chosen to time-focus ions of a particular m/z at, or near, the timed-ion-selector. The collision cell is at 15 kV, so that ions are decelerated to 3 kV laboratory energy before they collide with neutral molecules (in this case air) to be excited and caused to fragment. Shortly after the fragment ions and any remaining precursor ions exit the collision cell, the collision cell potential is pulsed to 19.5 kV. The delay time of the pulse and the mirror voltage are adjusted so that the ions are time-focused at the detector located near the exit from the mirror. With this arrangement satisfactory resolution for the entire fragment spectrum can be obtained in a single acquisition, unlike conventional PSD which requires a

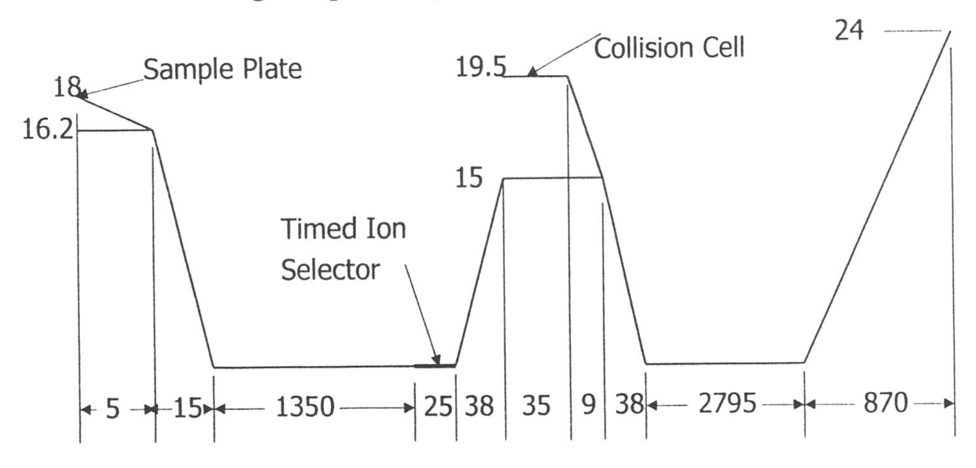

Fig. 2. Potential diagram for typical operating conditions corresponding to collisions occurring at 3 kV (laboratory).

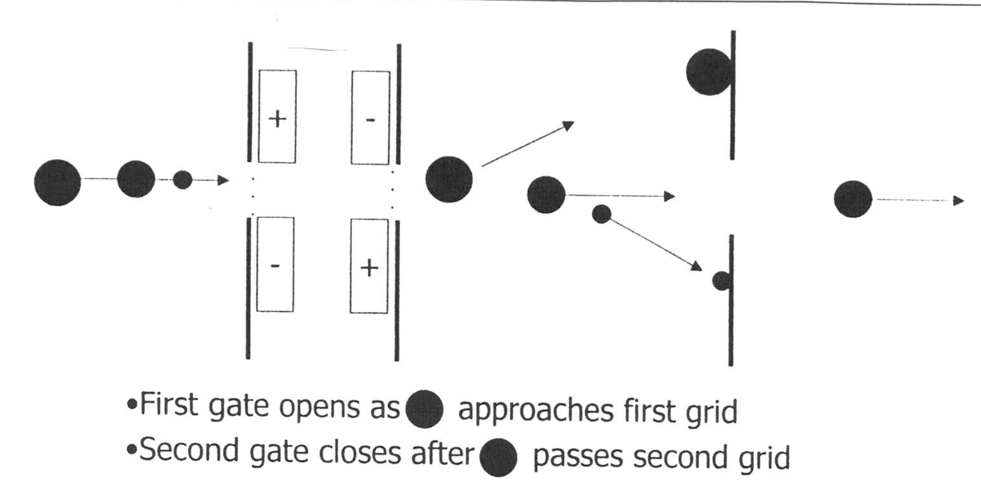

Fig. 3. Detailed schematic of timed ion selector.

series of spectra to be acquired at different mirror settings to obtain a complete fragment spectrum with adequate resolution.

The new timed-ion-selector employs two deflection gates in tandem as shown schematically in Fig. 3. The first gate is operated in "normally closed" mode, and the second is "normally open". When an ion of interest reaches the entrance to the first gate, the applied voltage is rapidly switched off, opening the gate, and allowing the ion to pass. As the selected ion passes the exit from the second gate, the deflection voltages are switched on and the gate is closed. Thus, the first gate rejects the lower mass ions, and the second gate rejects higher mass ions. The gate voltages are reset to their "normal" condition before ions from the next laser pulse can reach the selector. The resolving power of this selector is primarily determined by the width of the ion packet corresponding to a particular m/z, and the speed with which the gates can be switched on or off, and is nearly independent of the physical length of the gates.

Performance Evaluation

An example showing the performance of the new timed-ion-selector is given in Figure 4. In this example a mixture containing 1 picomole each of three peptide standards in α-cyano-4-hydroxycinnamic acid (CHCA) matrix was analyzed. The extraction delay was set to focus m/z 1570, MH^+ from glu-1-fibrinogen peptide, at the selector. Delays for the gate pulses were set so that the normally closed gate opened just before the ion of interest reached the gate, and the normally open gate closed just after the ion of interest passed the gate. The lower trace shows an expanded view of the spectrum when the gates are set so that only the ^{12}C isotope peak is transmitted. The middle trace shows the result of increasing the delay time for the second gate so that the first three isotope peaks are transmitted, and the top trace shows the delays set to transmit the second isotope peak with only a small contribution from the adjacent peaks. The highest resolution obtained with MS-1 is illustrated in Fig. 5, where a portion of

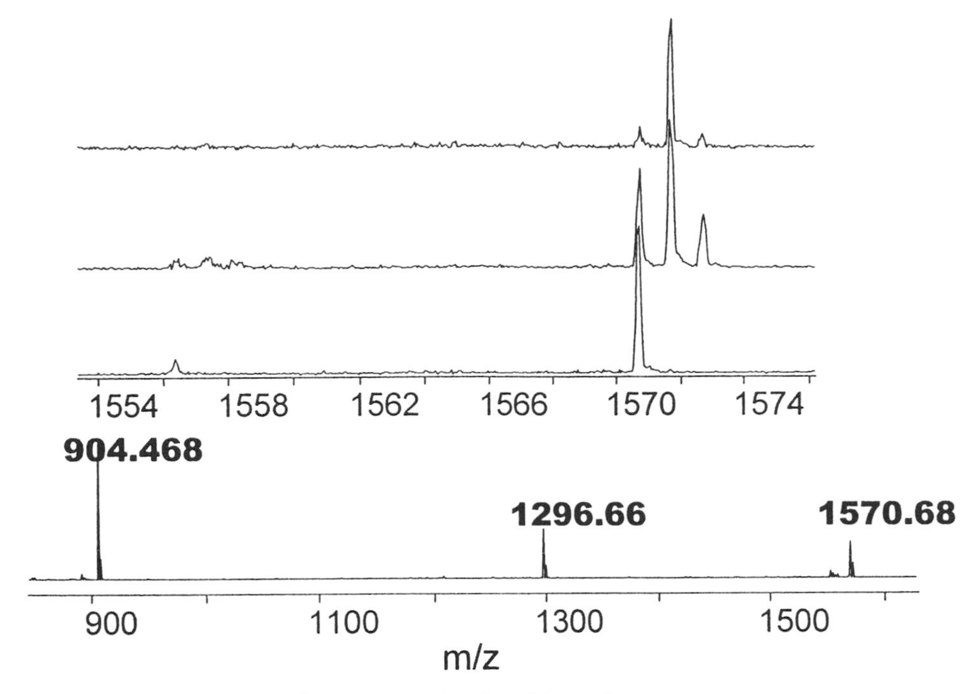

Fig. 4. Example of the performance of the timed ion selector.

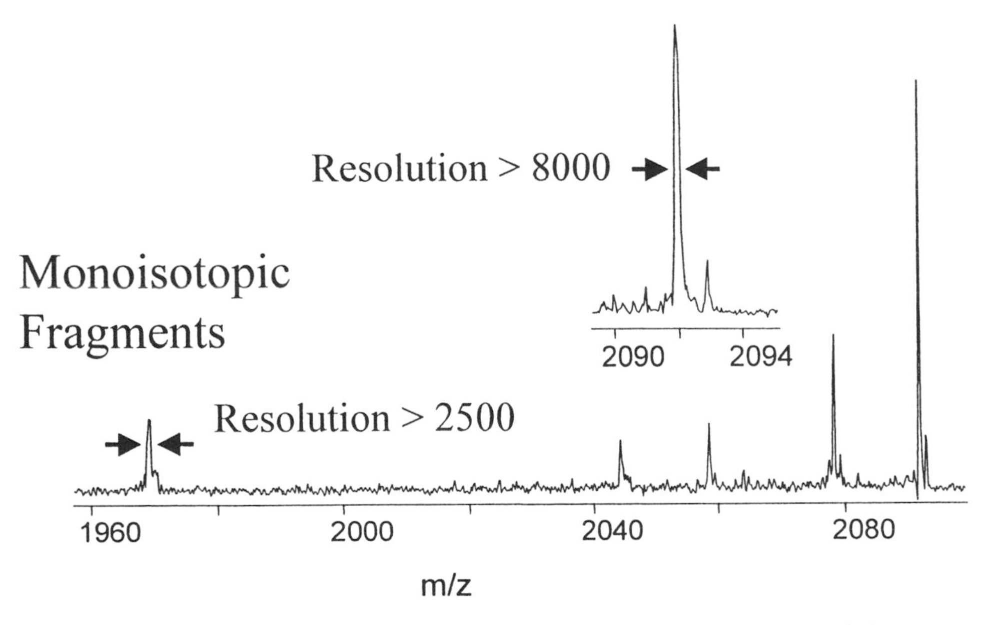

Fig. 5. Ultimate performance of timed ion selector showing selection of single isotope at m/z 2093.

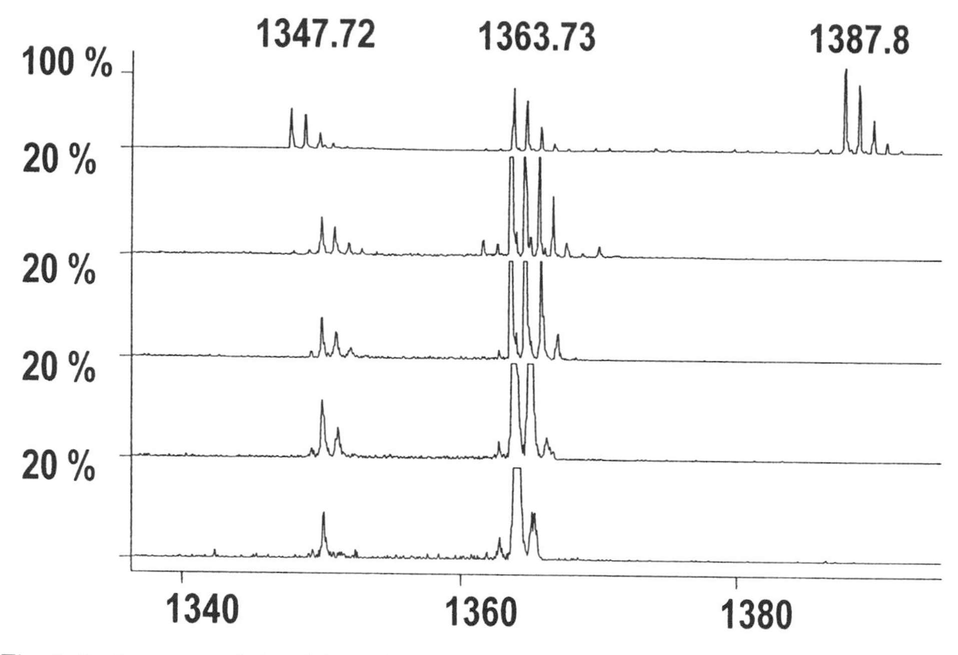

Fig. 6. Performance of timed ion selector at high laser intensity.

the MALDI spectrum for the peptide ACTH(1-17) is shown. The ^{12}C isotope peak at MH^+=2093 has been selected, and almost all of the higher mass isotopes are rejected. Note that the PSD fragment peaks in the spectrum are also monoisotopic. The resolving power of the first MS in this case is about 2000, using a 10% valley definition because the ratio of the ^{13}C isotope peak relative to the ^{12}C peak is reduced by about a factor of 10 compared to its normal value. This high resolution is accomplished with a sacrifice in sensitivity, because about 1/3 of the selected m/z 2093 ions are transmitted under these conditions. This resolving power is achieved with minimal distortion of the transmitted peaks as can be seen from the expanded view showing about 8000 resolving power on the transmitted peak. At masses below about m/z 1400 a single isotopic peak can be selected with negligible loss in transmission of the selected mass, provided that the source extraction delay is set so that the time dispersion of the ion packet for the selected m/z is minimum at the ion gate. At the higher laser intensities necessary to optimize sensitivity, significant broadening of the packet may occur. This is illustrated in Fig. 6 using a mixture of substance P and some analogs. The upper trace is an expanded view of the parent ion region with no selection, and the lower traces show the same region of the spectrum with m/z 1363 selected with successively narrower transmission windows. The vertical scale is expanded by a factor of 5 in the lower traces, where the transmission windows varies from about 8 u at the top to 4, 2, and 1 at the bottom. In this case some loss in transmission efficiency is observed for gate widths less than 4 u, along with noticeable broadening of the transmitted peaks.

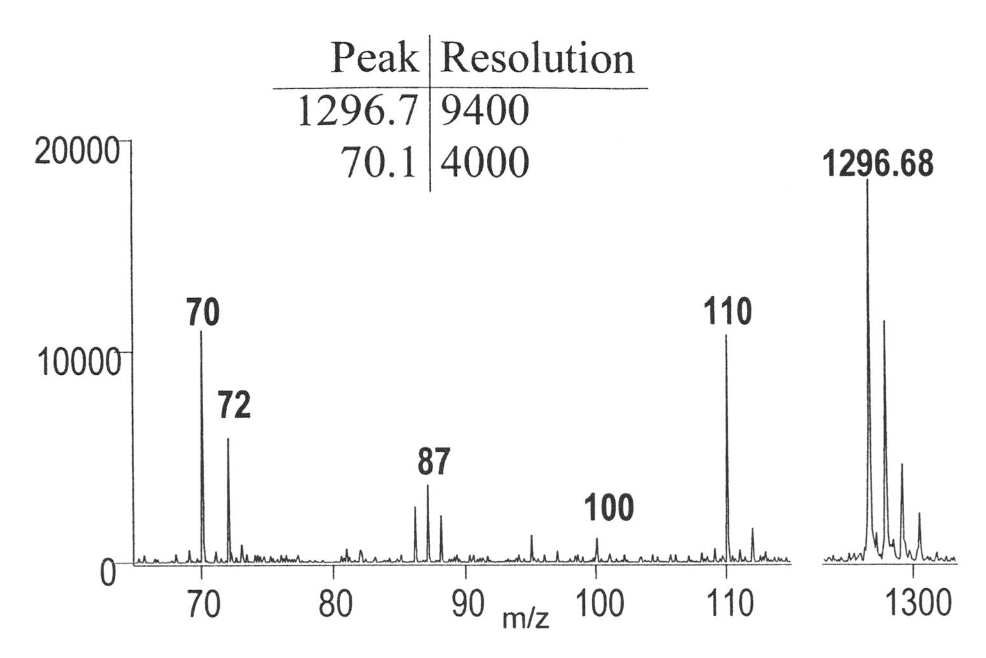

Fig. 7. Performance of MS-2 showing resolving power over entire fragment spectrum in single acquisition.

The performance of MS-2 with high energy CID is illustrated in Fig. 7, where expanded views of the low mass fragment ions and of the parent ion from Angiotensin I, obtained from a single acquisition are shown. The resolving power is greater than 4000 in the low-mass immonium ion region and approaches 10,000 at the precursor mass. The high energy CID fragment spectrum from substance P is shown in Fig. 8. This spectrum was obtained using 3 kV laboratory collision energy with air under single collision conditions. The operating conditions employed are illustrated in Fig. 2. Nearly complete sequences of a and d ions are observed, as found earlier in four-sector MS-MS measurements. This spectrum was obtained from 1 picomole of substance P present in the mixture illustrated in Fig. 6, with the ion transmission window set with a width of about 4 u centered on m/z 1348. Under these conditions the fragment spectra are dominated by the immonium ions at low m/z. This portion of the spectra obtained from substance P and two analogues is shown in Fig. 9. Except for the decrease in m/z 120 and small increase in m/z 136 due to substitution of Tyr for Phe at position 8, these spectra are virtually identical.

In comparing the high-energy CID spectra obtained here with data in the literature, obtained mostly from four-sector instruments, several similarities and differences are apparent. The fragment ions observed in the TOF-TOF spectra are in good agreement with those found from the four-sector measurements at comparable center-of-mass collision energies, but the relative intensities at low mass are much higher in the TOF measurements. This difference appears to be due to higher transmission efficiency for fragment ions in the TOF

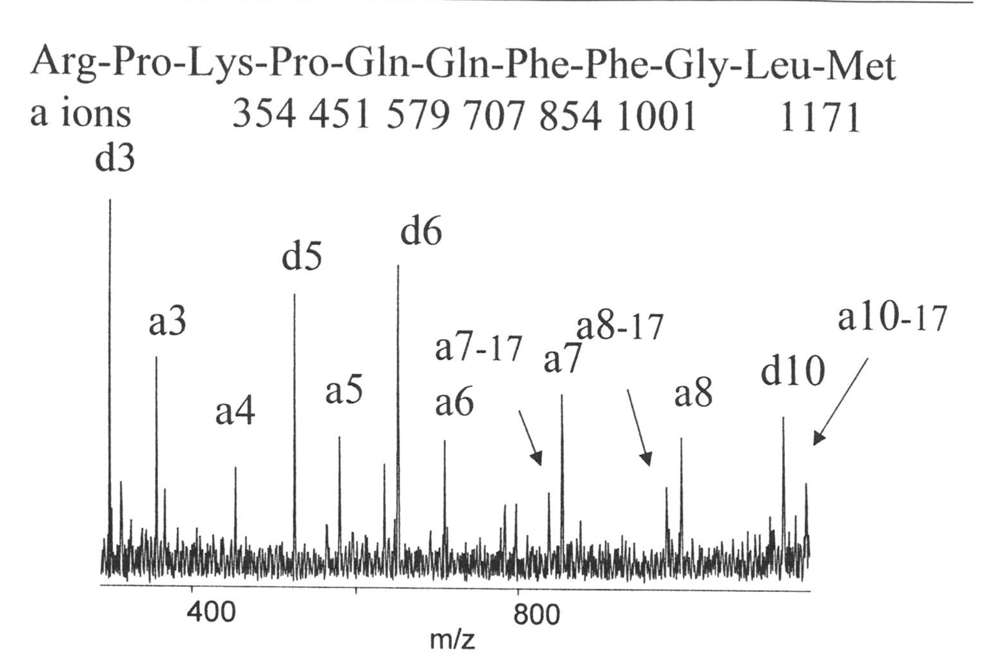

Fig. 8. High energy fragment spectrum from 1 picomole of substance P.

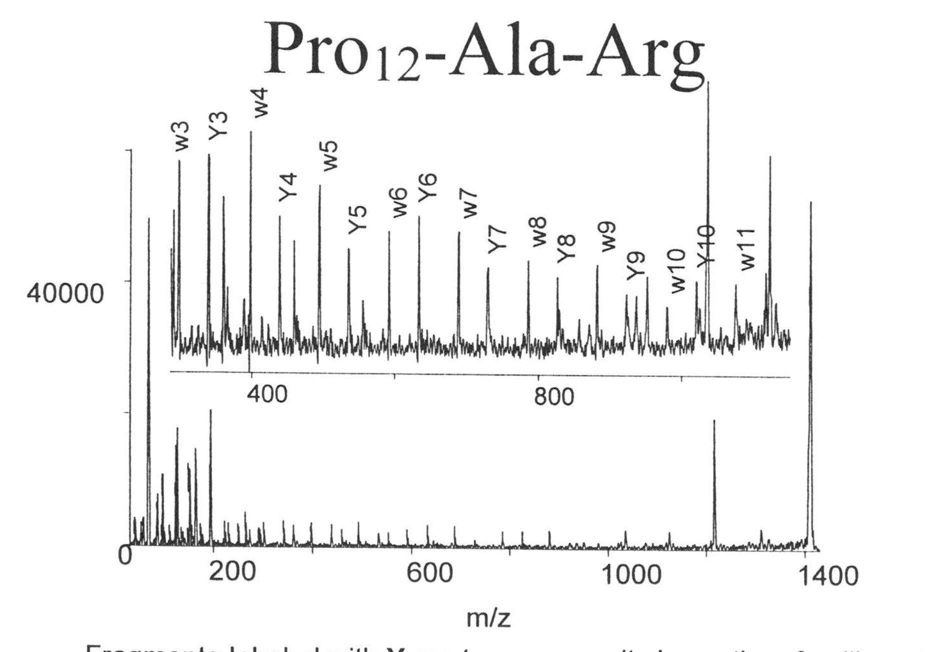

Fragments labeled with **Y** are two mass units lower than familiar **y** ion

Fig. 9. Comparison of immonium ion region from substance P and analogues.

	Relative Intensity (% of Precursor)		
m/z	**4-Sector(1)**	**Sector/TOF(2)**	**TOF-TOF**
650	0.4	5	1
226	0.3	10	8
70	0.2	40	100
Total Fragment	~10	~300	300
Reaction Time (usec)	1	15	1

(1) Reference [17]
(2) Reference [18]

Table 1. Comparison of ion detection efficiencies for substance P in different MS-MS instruments.

analyzer. Measured intensities for three fragment ions from the CID spectrum from substance P, relative to the precursor intensity, are given in Table 1 where they are compared to results at similar center-of-mass collision energies from a 4-sector instrument [17], and from a sector-TOF instrument [18]. An estimate of the total fragment intensity relative to precursor intensity is also given in Table 1, along with the average reaction time available for ions to fragment after collisional excitation. The total fragment intensity was estimated by summing peak heights compared to precursor ion peak height for the sector instruments, and for the TOF-TOF measurements the integrated fragment intensity was determined relative to the integrated intensity of the isotopic cluster of the precursor ion. In all cases the pressure in the collision cell was sufficient to reduce the precursor intensity to 20-30% of its value without gas in the cell. Clearly, both TOF fragment ion analyzers provide much higher transmission efficiency than the magnetic analyzer, particularly for low mass fragments. Also, these results indicate that processes producing the higher mass sequence ions are slower, and the sensitivity for detecting these fragments is enhanced when a longer reaction time is provided, as in the sector-TOF instrument.

A high energy CID spectrum from a synthetic peptide Pro_{12}-Ala-Arg is shown in Fig. 10. In this case a complete series of y-2 and w ions, corresponding to charge on the C-terminal arginine are observed. The y-2 ions (imine ions) are always particularly dominant for proline [12], and the w ions correspond to side chain cleavages generally specific to high energy fragmentation processes. This spectrum was used to calibrate the fragment spectra taking into account the initial velocity, which in this case was approximately 20,000 m/sec. The calibration used the parent ion and m/z 70 as standards. Results obtained from internal calibration on one spectrum and application of that calibration to external

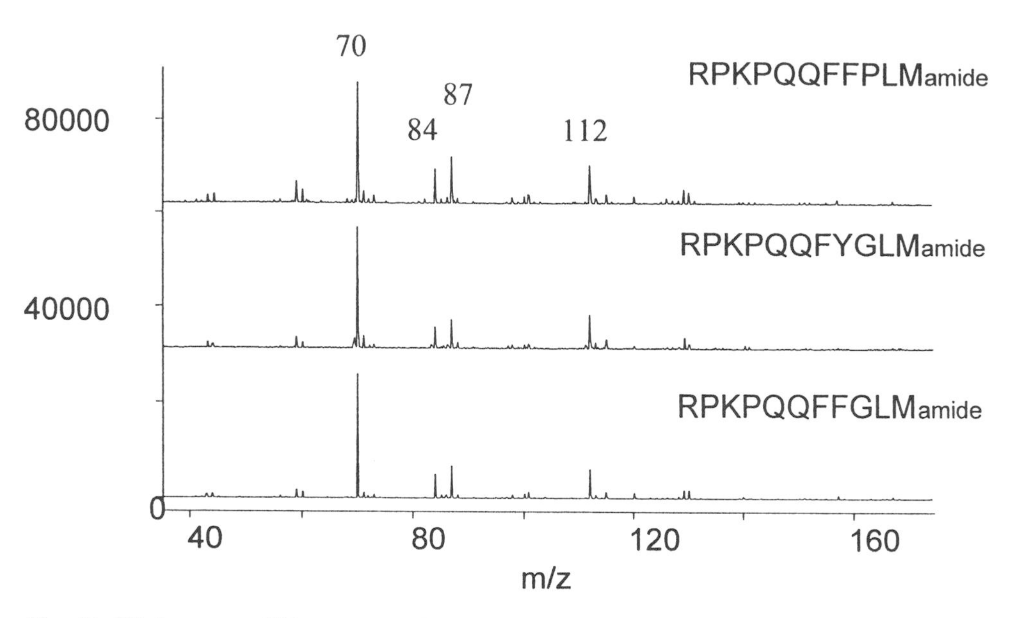

Fig. 10. High energy CID spectrum from 1 picomole of synthetic peptide Pro12-Ala-Arg.

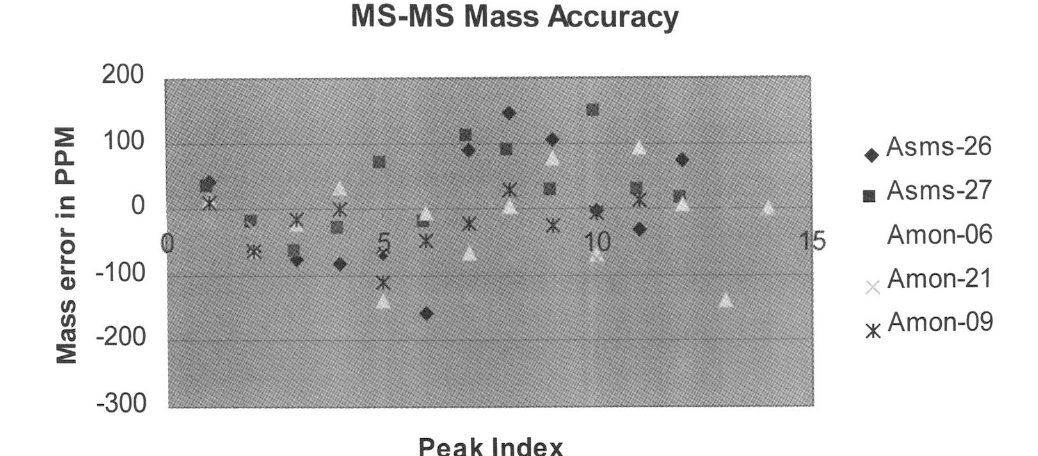

Fig. 11. Calibration and mass accuracy of fragment spectrum from synthetic peptide.

calibration of four additional spectra are summarized in Fig. 11. In this comparison, theoretical masses for fifteen peaks distributed across the spectrum were compared with experimental values for five acquisitions. The maximum observed error was 160 ppm, and for 90% of the peaks the error was less than 100 ppm. These errors appear to be primarily due to poor ion statistics from relatively weak peaks, and any systematic error is small compared to the statistical uncertainty.

The MALDI spectrum obtained from 1 picomole of a Glu-C digest of alpha Casein is shown in Fig. 12. The most intense peaks in this spectrum

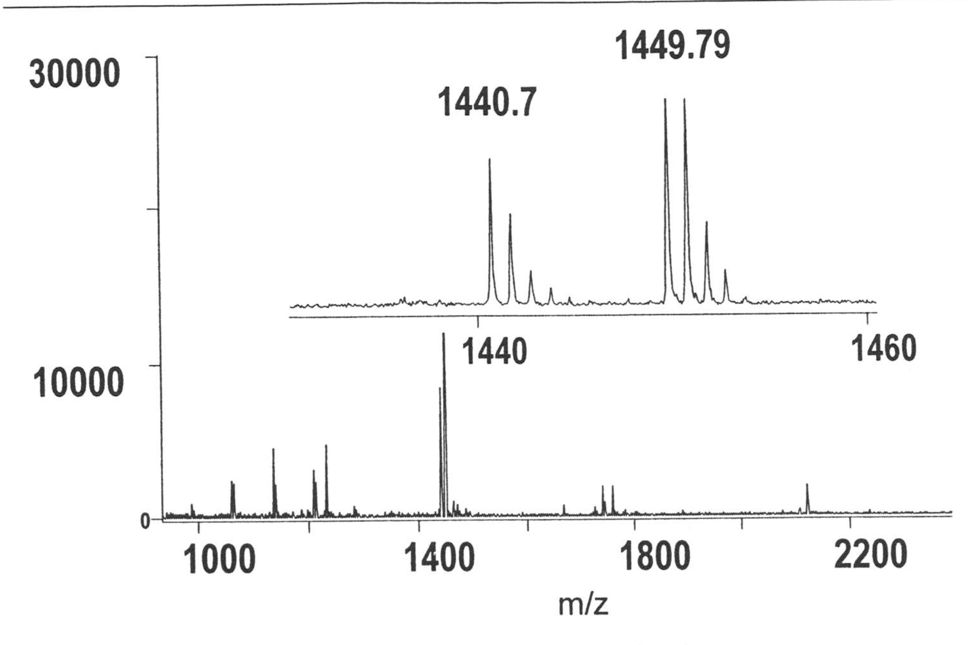

Fig. 12. MALDI mass spectrum of Glu-C digest of alpha Casein.

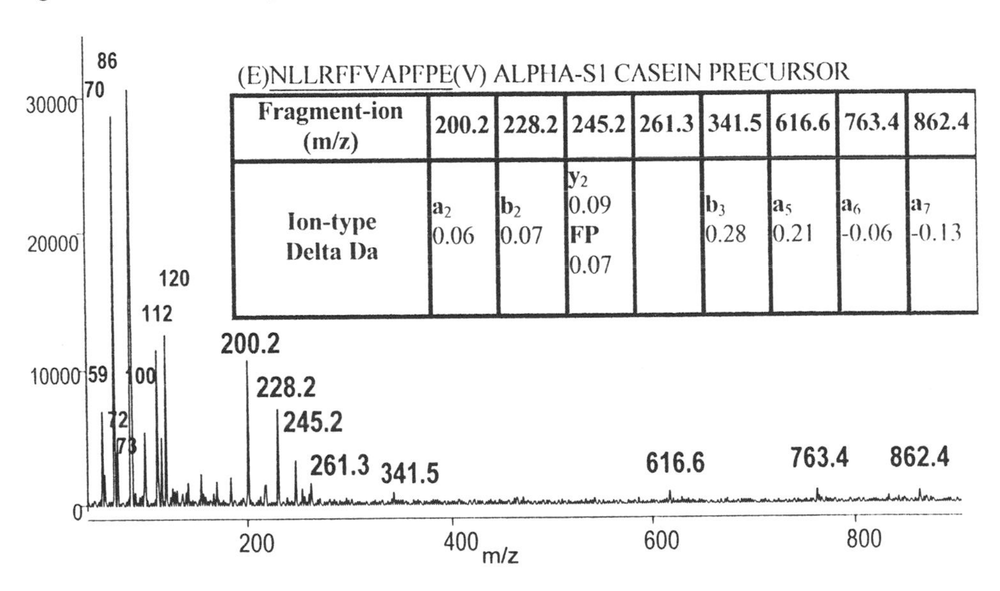

(E)NLLRFFVAPFPE(V) ALPHA-S1 CASEIN PRECURSOR

Fragment-ion (m/z)	200.2	228.2	245.2	261.3	341.5	616.6	763.4	862.4
Ion-type Delta Da	a_2 0.06	b_2 0.07	y_2 0.09 FP 0.07		b_3 0.28	a_5 0.21	a_6 -0.06	a_7 -0.13

Fig. 13. CID spectrum of component at m/z 1449.79 and result of MS-Tag search of database.

correspond to components at m/z 1440.70 and 1449.79. An expanded view of this portion of the spectrum is shown in the inset. A portion of the CID spectrum of the 1449.79 component is shown in Fig. 13. The inset Table shows the fragment ion masses entered into MS-Tag to search the database. The MS-TAG program was set to search release 35 of the SwissProt database via a local intranet

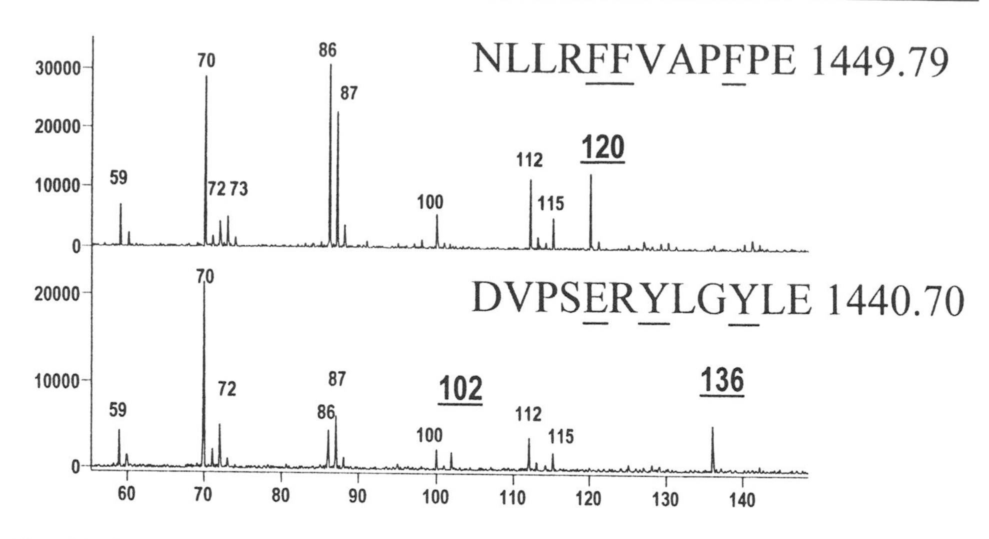

Fig. 14. Comparison of immonium ions from two component of Glu-C digest of alpha Casein.

connection on a Windows NT computer. A precursor mass of 1149.79 +/- 10ppm was specified, along with V8 digestion after glutamic acid residues. Fragment ion mass accuracy used the default option of 1500 ppm which was needlessly generous. Immonium ion data restricted fits to those which included the amino acids Leu or Ile, Val, Phe and Arg. Only a single match was found with 2 or less unidentified fragment masses. The peptide found in the database and the interpretation of the major fragment peaks is are included in the inset Table. The intense immonium ion region for the two peptides is compared in Fig. 14. The presence of Phe in m/z 1449.79 is clearly indicated by the m/z 120 peak, and Tyr and an internal Glu by 136 and 102, respectively, in m/z 1440.70.

Summary of Results

The feasibility of high performance MALDI-TOF MS-MS has been demonstrated, but some work remains to make the technique routinely applicable to determining primary sequence and structural details of peptides from protein digests. Resolution of MS-1 is sufficient to isolate the isotopic cluster of a peptide present in a mixture without sacrificing sensitivity, and a single monoisotopic peak can be selected up to ca. m/z 2000 with a modest loss in transmission efficiency. The resolution of MS-2 is sufficient to provide baseline resolution of adjacent masses in fragment spectra over the entire mass range with a single set of operating voltages and time delays. Measured resolving powers in the range 2000-10,000 (FWHM) are easily achieved. In the initial studies mass accuracies within 100 ppm of the theoretical values were generally obtained with external calibration, provided the initial velocity of the ions entering the collision cell was properly included. In a few cases errors approaching 200 ppm were observed, but these appear to be due to poor precision on very weak fragment peaks.

The major limitation, at present, is sensitivity for the higher mass ions providing sequence information. Protein digests loaded at a nominal level of 1 picomole with conventional sample preparation techniques for MALDI generally provide useful sequence ions from only the most intense peaks in the mixture. The major immonium ions can generally be detected at much lower levels. This sensitivity is typically obtained from summing the ions produced by 256 laser shots on a single region (ca. 100 micron in diameter) of the 1 picomole sample which is typically 2-3 mm in diameter, and often several such spectra can be obtained from a single 100 micron spot. Thus, only an infinitesimal portion of the sample loaded is actually consumed in generating a typical spectrum.

Conclusions and Future Work

The results presented here are from the first prototype of a new approach, based on delayed extraction technology, aimed at developing a high performance MALDI-TOF MS-MS system suitable for the routine characterization of protein digests. The resolution and mass accuracy achieved with this prototype are already adequate for many potential applications, but the sensitivity does not yet provide useful sequence information from peptides present at low concentrations in protein digests. Some improvement in sensitivity may be realized by improved ion optics which more efficiently transmit fragment ions, and by allowing more time for relatively slow, low-energy fragmentation processes to approach completion. A second generation system based on the principles described above is presently under construction, and is expected to provide more nearly optimum performance. However, it appears unlikely that these instrumental improvements will increase sensitivity by more than one order of magnitude. A more fruitful approach to enhanced sensitivity appears to be better sample utilization.

Recent developments in MALDI instrumentation and sample preparation techniques allow determination of accurate molecular weights on peptides present in proteins digests at sub-femtomole levels, and the sensitivity is generally limited by chemical noise rather than the efficiency of sample utilization. With MS-MS the chemical noise is drastically reduced by the primary ion selection process, but after fragmentation the ions initially present in a precursor peak are distributed among a large number of relatively small fragment peaks. Some fragment peaks which represent only 0.01% of the total ions produced may be vital for determining amino acid sequence. Sensitivity for detecting such peaks is primarily set by the total number of ions which are produced from the selected precursor. The recent development of high repetition rate UV lasers, such as diode-pumped, frequency-tripled Nd:YAG lasers, and high speed averaging transient digitizers appear to make it feasible to accumulate ions from a large number of laser shots in a reasonable total acquisition time. Thus, it should soon be possible to efficiently convert all of the sample to useful ion signals. Future work will focus on applying these new developments, along with the techniques described above, to achieving the overall performance goals required for routine analysis of protein digests by MALDI-TOF MS-MS.

REFERENCES

1. P. H. O'Farrell, *J. Biol. Chem.* **1975**, 250, 4007-4021.

2. W. J. Henzel, T. M. Billeci, J. T. Stults, S. C. Wong, C Grimley, and C. Watanabe, *Proc. Nat. Acad. Sci., USA* **1993**, 90, 5011-5015

3. R. S. Brown and J. J. Lennan, *Anal. Chem.* **1995**, 67, 1998-2003.

4. S. M. Colby, T. B. King, and J. P. Reilly, *Rapid Comm. Mass Spectrom.* **1994,** 8, 865-868.

5. M. L. Vestal, P. Juhasz, and S. A. Martin, *Rapid Comm. Mass Spectrom.* **1995,** 9, 1044-1050.

6. O. Vorm and M. Mann, *J Am. Soc. Mass Spectrom.* **1994**, 5, 955-958.

7. O. Vorm, P. Roepstorff and M. Mann, *Anal. Chem.* **1994**, 66, 3281-3287.

8. B. Spengler, D. Kirsch, R. Kaufmann and E. Jaeger, *Rapid Commun. Mass Spectrom.* **1992**, 6, 105-108.

9. M. S. Wilm and M. Mann, *Int. J. Mass Spectrom. Ion Proc.* **1994**, 136,167-180.

10. H. R. Morris, T. Paxton, A. Dell, J. Langhorne, M. Berg, R. S. Bordoli, J. Hoyes, and R. H. Bateman, *Rapid Commun. Mass Spectrom.* **1996**, 10, 889-896.

11. I. V. Chernushevich, A. A. Shevchenko, B. Thomson, E. W. Ens, M. Mann, and K. G. Standing, *Proc. 45th ASMS Conference* **1997**, p. 771.

12. K. Biemann, *Meth. Enzymol.* **1990**, 193, 455-479.

13. J. A. Hill, J. E. Biller, S. A. Martin, K Biemann, K. Yoshidome, and K. Sato, *Int. J. Mass Spectrom. Ion Processes* **1989**, 92, 211-230.

14. T. J. Cornish and R. J. Cotter, *Org. Mass Spectrom.* **1993**, 28, 1129-1134.

15. T. J. Cornish and R. J. Cotter, *Rapid Commun. Mass Spectrom.* **1993**, 7, 1037-1040.

16. R. J. Cotter, T. J. Cornish, and Marcelo Cordero, in Mass Spectrometry in the Biological Sciences, A. L. Burlingame and S. A. Carr, Eds.; Humana, Totawa, NJ, 1996, pp 5-23.

17. A. Falick, private communication.

18. K. F. Medzihradszky, G. W. Adams, A. L. Burlingame, R. H. Bateman, and Martin R. Green, *J. Am. Soc. Mass Spectrom*. **1995**, 7, 1-10.

Questions and Answers

Roland S. Annan (SmithKline Beecham Pharmaceuticals)

It seems a major drawback to this technique is its limitation to sequencing singly charged precursors, for which we know from our experience with FAB have a practical upper mass limit of 1800-2000 Da.

Answer. It is also applicable to multiply charged ions if they are made by the ionization technique employed. For larger peptides, doubly charged ions are often observed with α-cyano-4-hydroxy cinnamic acid matrix, but we have not yet looked at any of these.

A. L. Burlingame (UCSF)

What matrix was used? Have you tried "colder" matrices?

Answer. We used α-cyano-4-hydroxy cinnamic acid and no, we have not tried "colder" matrices.

Maria Person (UCSF)

How much of the increase in CID fragment ions in the 0-600 mass range with increased collision gas can be attributed to immonium ions or ions of mass less than 200?

Answer. For the limited number of peptides studied, it appears to be about 90% of the total.

Scott Patterson (Amgen, Inc.)

Do peptides that don't readily undergo PSD fragment in the TOF-TOF?

Answer. If high energy CID produces fragmentation, we will see it. We will certainly see the immonium ions, but it is difficult to generalize about the probability of seeing sequence ions.

Scott Patterson (Amgen, Inc.)

Can you select doubly charged peptide ions in the TOF-TOF and fragment them?

Answer. Yes, if they are produced.

B. Onisko (Zeneca Pharmaceuticals)

Is TOF-TOF compatible with ^{18}O methods to label C-terminal ends? Can parent precursor ion selection be adjusted to transmit ^{18}O and ^{16}O?

Answer. Yes. We have sufficient resolution to separate species containing one ^{18}O up to at least m/z 2000 without sacrificing sensitivity.

A. L. Burlingame (UCSF)

The point which has not yet been overtly made is that high quality CID for both MALDI and ESI are necessary. Samples that run well by one technique may get "lost" if ionization methods are changed just to obtain better quality CID information based on different instrumentation.

Answer. Amen.

Steven Carr (SmithKline Beecham Pharmaceuticals)

Have you considered putting electrospray onto this instrument and allowing for facile switching between these modes? Would this instrument have improved sensitivity in ESI MS/MS mode relative to current performance in MALDI MS/MS mode?

Answer. It appears that the general approach is also applicable to orthogonal injection TOF, but we have not yet done it. Ken Standing and coworkers at Manitoba have shown that both MALDI and ESI are compatible with orthogonal injection, so it should be possible in the future. It seems premature to speculate at this time about relative sensitivity of a technique at an early stage of development compared to one that has not yet been attempted.

Measurements of Protein Structure and Noncovalent Interactions by Time-of-Flight Mass Spectrometry With Orthogonal Ion Injection

A. N. Krutchinsky, I. V. Chernushevich, A. V. Loboda, W. Ens and K. G. Standing
Physics Department, University of Manitoba, Winnipeg MB R3T2N2, Canada

Genomic sequences are becoming available with increasing speed as a result of the advances in gene technology. Nevertheless, both covalent and noncovalent modifications can occur in the corresponding protein sequence before the protein is fully functional, so it is important to know the actual protein sequence to understand its function [1, 2]. It is equally important to determine the higher order protein structure, and its interaction with other biological components. Noncovalent interactions of this type play a key role in molecular recognition phenomena such as enzyme-substrate interaction, receptor-ligand binding, formation of oligomeric proteins, assembly of transcription factors, and formation of cellular structures themselves [3-4].

Mass spectrometry continues to become more important in such studies [1-4]. This is largely a result of the development of two new ionization methods, electrospray ionization (ESI) [5-7] and matrix-assisted laser desorption/ionization (MALDI) [8-10]. Using these techniques, intact molecular ions with molecular weights up to 100 kilodaltons or more can now be formed, placing increased demands on the corresponding methods of mass measurement. Because the ion beams produced by the two methods have very different characteristics, it has been customary to examine them in different types of mass analyzers.

MALDI ions are normally produced by a pulsed laser, so MALDI ion sources can be conveniently coupled to time-of-flight (TOF) mass spectrometers [11], which require a well-defined start time. Time-of-flight (TOF) mass analyzers have some significant advantages for analysis of biomolecules. They enable the full mass spectrum to be observed at the same time without scanning (parallel detection); in many cases this yields a large increase in sensitivity. TOF instruments also have an effectively unlimited m/z range, an important factor in the observation of large molecular ions or complexes. They have high sensitivity and rapid response. Early limitations on resolution have been largely removed by the use of electrostatic reflectors [12], and the rediscovery [13, 14] of the benefits of delayed extraction [15]. In addition, developments in fast electronics have removed earlier limitations in the recording of TOF spectra, and enable the rapid response of the instrument to be exploited more fully. Consequently, TOF instruments now provide in many cases an optimum combination of resolution, sensitivity, and fast response, particularly under conditions where the whole mass spectrum is required.

On the other hand, ESI produces a continuous beam of ions. Like other continuous ion sources, it is most compatible with mass spectrometers that themselves operate in a continuous fashion, such as quadrupole mass filters. The combination of an ESI source and a quadrupole mass filter has become a popular and satisfactory configuration. However, there are some problems. First, most commercial quadrupoles are restricted to m/z values below ~4000. Although this is usually not a difficulty when examining electrosprayed ions because of their high charge states, some entities, notably noncovalent complexes, may have m/z values up to 10,000 or more. Second, the quadrupole mass filter is a scanning device; it examines ion species in the spectrum one at a time, which implies a reciprocal relation between resolution and sensitivity. Thus an increase in sensitivity requires a decrease in resolution — a significant handicap in cases where the whole mass spectrum must be examined from a limited amount of sample.

The above remarks suggest that coupling an ESI source to a TOF spectrometer might overcome the problems just mentioned. However, there are difficulties in coupling any continuous source to a TOF instrument. The most straightforward technique is to inject ions along the spectrometer axis (the z axis), but this procedure results in an unacceptable loss of sensitivity, at least in its simplest form. However, TOF instruments can tolerate a relatively large spatial or velocity spread in a plane perpendicular to the spectrometer axis, as witnessed by the large sources (≤ 1 cm diameter) typically used in fission fragment desorption (PDMS) [16]. This tolerance can be exploited by injecting electrospray ions into the TOF instrument perpendicular to the z axis, i. e., "orthogonal injection". Such a geometry provides a high efficiency interface for transferring ions from a continuous beam to a pulsed mode. Another advantage is the small velocity spread in the z direction that is usually observed, making high resolution easier to obtain. Orthogonal injection into TOF instruments was first reported in 1964 [17], but the technique lay dormant until it was rediscovered by a number of workers some 20 years later. Dodonov *et al.* were the first to use the method to examine electrosprayed ions [18, 19].

Our laboratory has been active in the development of TOF methods and instruments, most recently using orthogonal injection. We have also been concerned with the application of these techniques to the study of proteins and other biomolecules, particularly noncovalent complexes. Some examples will be described in the following sections.

An ESI-TOF Spectrometer With Orthogonal Injection and Collisional Cooling

Figure 1 shows a schematic diagram of the Manitoba ESI-TOF mass spectrometer [20, 21] in its present configuration. Here electrosprayed ions are cooled in a quadrupole ion guide and enter a storage region in the TOF instrument with an energy of a few eV, moving perpendicular to the z axis. Pulses applied to the electrodes above and below the storage region at a frequency of a few kHz inject ions into an acceleration column. After acceleration the ions move

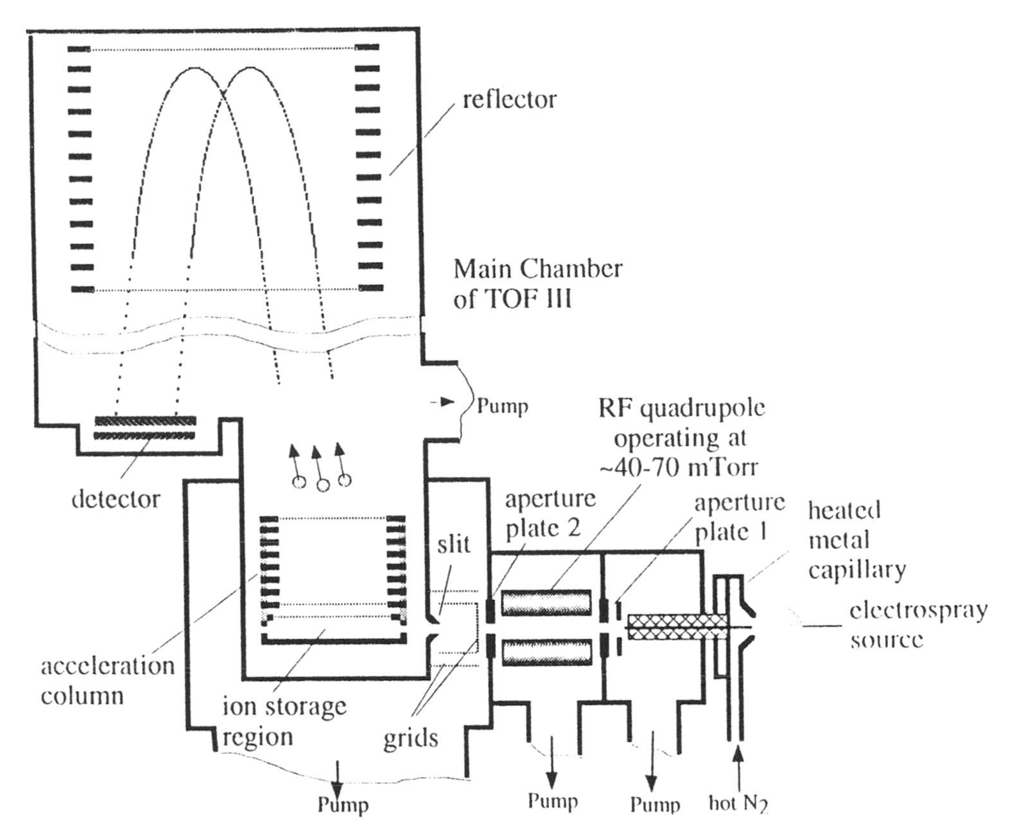

Fig. 1. Schematic diagram of the ESI-TOF instrument with a collisional damping interface.

almost parallel to the z axis with an energy ~4 keV and are time-focused onto a detector by a single-stage electrostatic mirror.

Properties of the original version of this instrument have been described previously [20, 21]. Since that time two major modifications have been made. First, the collisional damping region shown in Fig. 1 has been introduced [22]. Ions are cooled by collisions with the gas contained within the quadrupole ion guide (RFQ) (at pressures up to ~100 mTorr), during which time they are focused towards the quadrupole axis by the RF field. This gives improved resolution (up to $M/\Delta M_{FWHM}$ ~ 10,000 for bovine insulin) and removes mass discrimination [22]. Second, a separate electrostatic shield has been placed inside the flight tube so that the drift region is floated at high voltage [23], making it possible to inject ions into the instrument from a source close to ground potential.

STUDIES OF NONCOVALENT COMPLEXES

As mentioned above, noncovalent complexes are responsible for molecular recognition phenomena, and as such play a key role in biology. Because the complexes are held together only by noncovalent interactions, binding may be

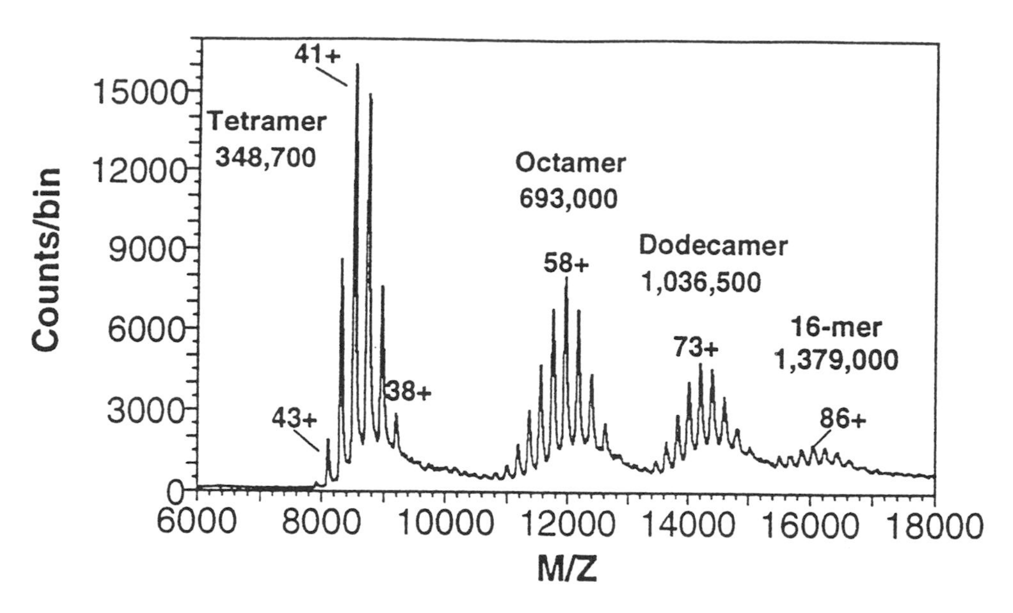

Fig. 2. The ESI mass spectrum of clusters of catalase HPII recorded using nanospray with 100V declustering potential.

weak. Thus the relatively gentle electrospray process is the best way to introduce the complex into the gas phase and to ionize it for examination by mass spectrometry. Many measurements of this kind have been carried out, usually in quadrupole mass filters [3, 4]. As remarked above such measurements in quadrupoles are possible only because of the high charge states normally produced by electrospray, as most commercial quadrupoles are not capable of measuring ions with m/z greater than ~4000. However, many complexes, particularly large ones, have significantly higher values of m/z because possible protonation sites on the constituent molecules are shielded by other parts of the complex. In such cases a large m/z range is necessary for observation of the complex, so TOF instruments are particularly suitable for the task [24]. An example is shown in Fig. 2 for the enzyme catalase HPII [24], where the tetrameric complex has m/z ~ 9,000, and clusters of tetramers, themselves of some interest, extend out to m/z ~ 16,000. Note that the 12- and 16-mers have masses exceeding one megadalton.

A similar situation is encountered in *E. coli* citrate synthase, the enzyme that initiates the citric acid cycle, which we have studied in some detail [25]. Under denaturing conditions (Fig. 3), a wide distribution of charge states corresponding to the monomer and centered around m/z ~1100 is observed, but when the pH is changed so as to be closer to physiological values, narrow charge distributions corresponding to dimers (around m/z~5300) and hexamers (around m/z~9000) appear (Fig. 4). The corresponding mass distributions, obtained by "deconvoluting" the m/z distributions [26], are also shown.

It has long been known that citrate synthases from gram-negative bacteria such as *E. coli* display allosteric inhibition by NADH (the reduced form

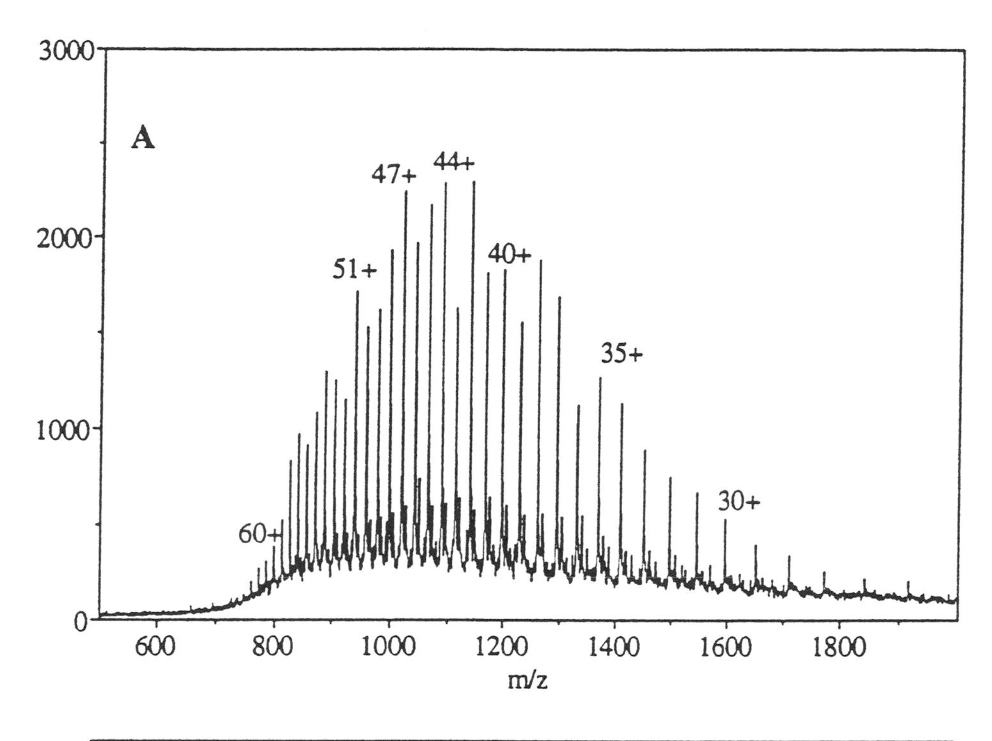

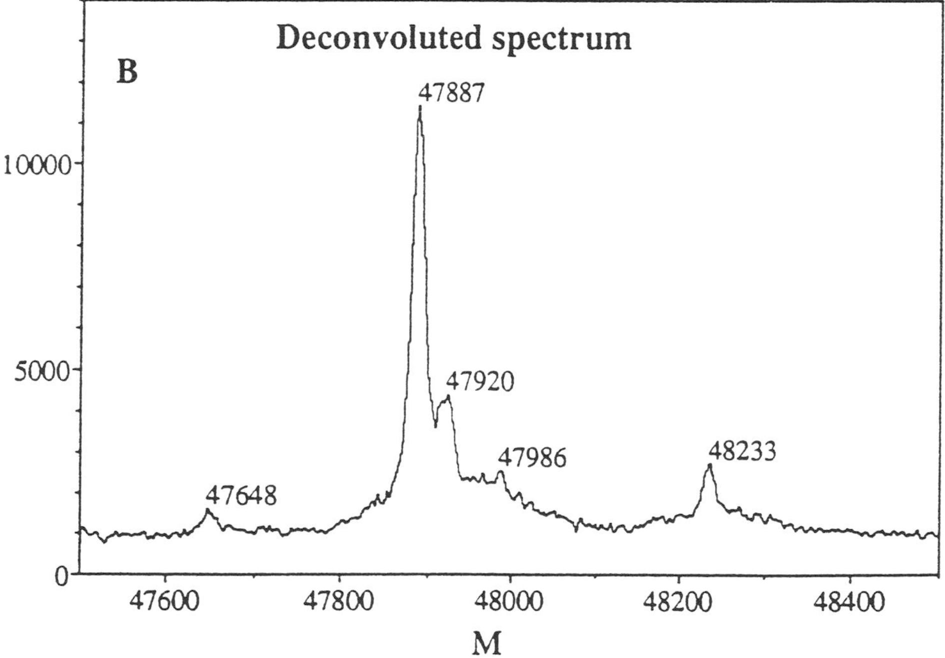

Fig. 3. (A) Electrospray spectrum of recombinant E. coli *citrate synthase obtained under denaturing conditions (water/methanol 1/1 v/v +5% acetic acid). (B) mass spectrum obtained by deconvolution of the electrospray spectrum.*

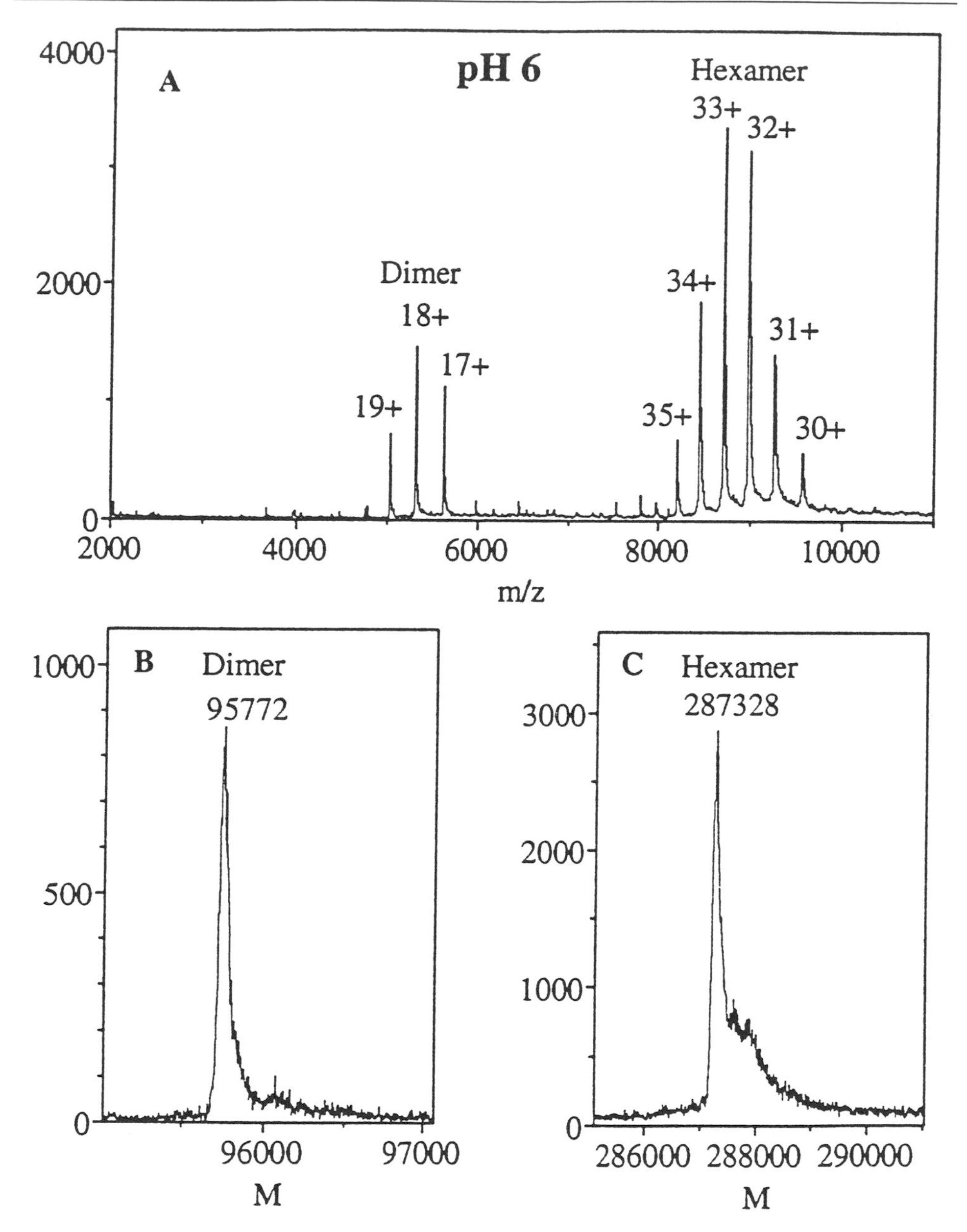

Fig. 4. (A) Electrospray spectrum of citrate synthase obtained from 5mM ammonium acetate buffer (pH 6). (B) and (C) mass spectra of the dimer and the hexamer of citrate synthase obtained by deconvolution of the m/z spectrum.

of nicotinamide adenine dinucleotide), but the mechanism for the inhibition was not understood. Our measurements have now clarified this question. Figure 5 shows m/z spectra and the corresponding mass spectra for citrate synthase in the presence of increasing concentrations of NADH. Binding of NADH to both dimers and hexamers is observed. As NADH is added, the hexameric part of the

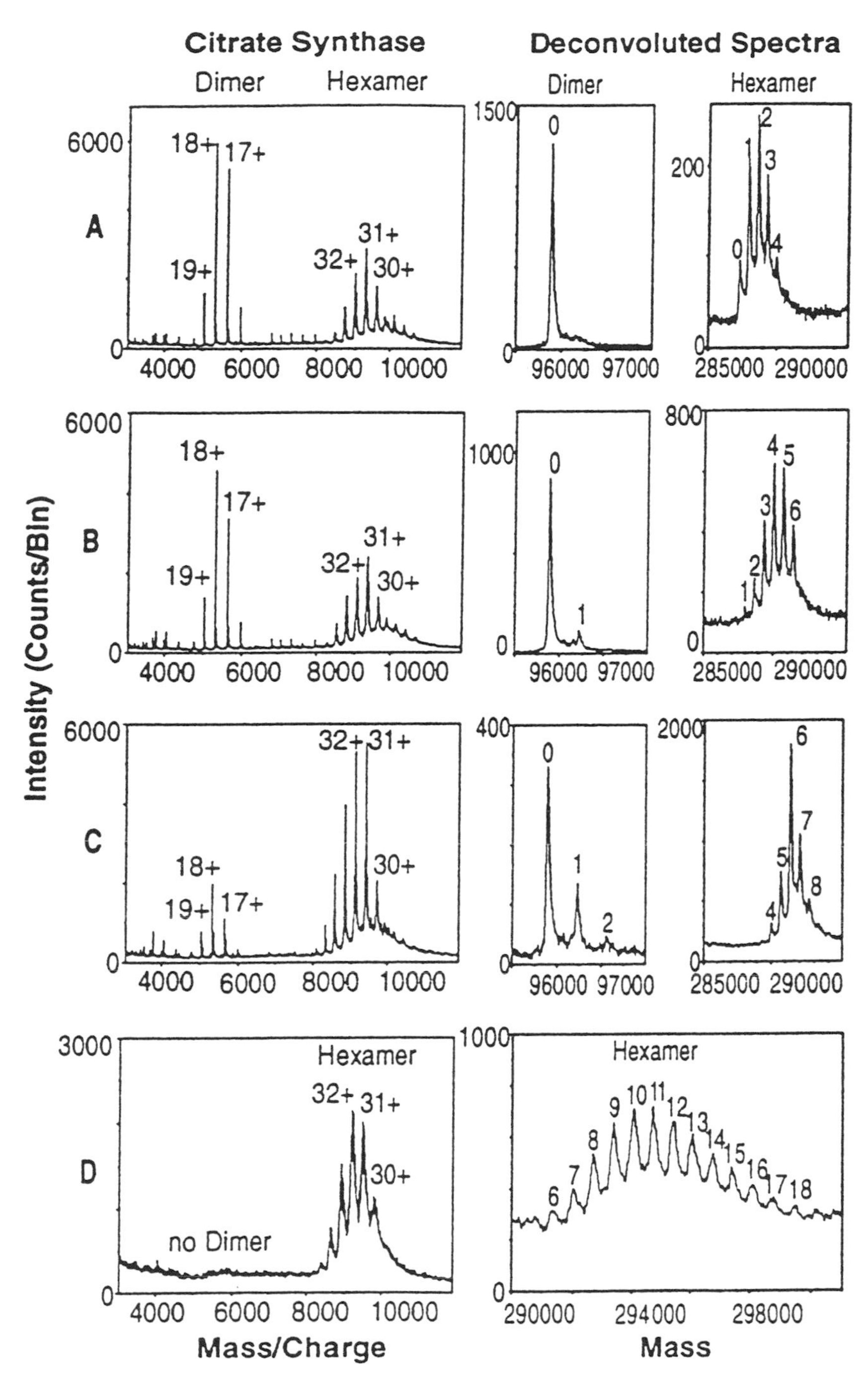

Fig. 5. Selected ESI-TOF m/z and corresponding mass spectra of CS (9 μM subunit concentration) in the presence of increasing concentrations of NADH in 5mM NH_4HCO_3 at ph 7.5. NADH concentrations were (a) 4.5 μM. (b) 9.0 μM. (c) 18 μM and (d) 108 μM. The digits labeling the deconvoluted spectra correspond to the number of NADH molecules bound.

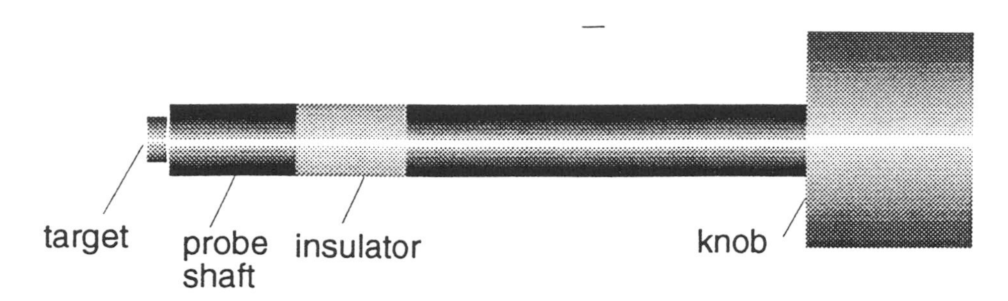

Fig. 6. The MALDI probe.

deconvoluted mass spectrum consists of a set of NADH-hexamer complexes, almost completely resolved from one another, and containing from one to as many as 18 NADH molecules. The hexamer binds NADH at very low concentrations, rapidly filling one site per subunit at about 10 μM NADH in a non-cooperative manner. At higher NADH concentrations, the number of bound molecules continues to increase, but more gradually. This behavior is in sharp contrast to that exhibited by the dimer, where the increase in the number of NADH molecules bound is gradual throughout the entire NADH concentration range, and the dimer disappears at high NADH concentration. This suggests that non-specific binding occurs to both dimers and hexamers, but that specific binding is exclusive to hexameric citrate synthase. It is evident that addition of NADH shifts the equilibrium away from the active dimeric and hexameric forms of the enzyme towards the inactive NADH-bound form of the hexamer, thus explaining its inhibiting effect.

ORTHOGONAL INJECTION OF MALDI IONS

As remarked above, MALDI is a technique that is well suited to TOF mass spectrometry in the usual axial injection geometry. However, some important benefits could be expected from the use of orthogonal injection. In particular, the latter mode of operation almost completely decouples the ion production from the mass measurement, increasing the freedom of choice for targets, matrices, etc. Direct injection of a MALDI beam in the orthogonal geometry suffers from very low efficiency because of the large velocity spread [27]. Therefore the use of a collisional damping interface appears to be essential in this application.

We have modified the ESI interface described above to permit orthogonal injection of MALDI ions [28]. The MALDI sample is applied to the end of the probe shown in Fig. 6. The probe fits through an inlet port, then through a valve into the quadrupole ion guide, where it is held in position by a supporting ring. The tip of the probe is electrically insulated from the rest of the shaft so the voltage on the target is determined by the voltage applied to the ring. The sample is irradiated by a pulsed N2-laser (Model VSL-337ND, Laser Science Inc.) operating at a repetition rate up to 20 Hz. The laser is focused onto the target surface by a lens so that the power density during the pulse is ~10^7 W/cm^2. Each laser shot generates a plume of neutral molecules and ions, which expands into

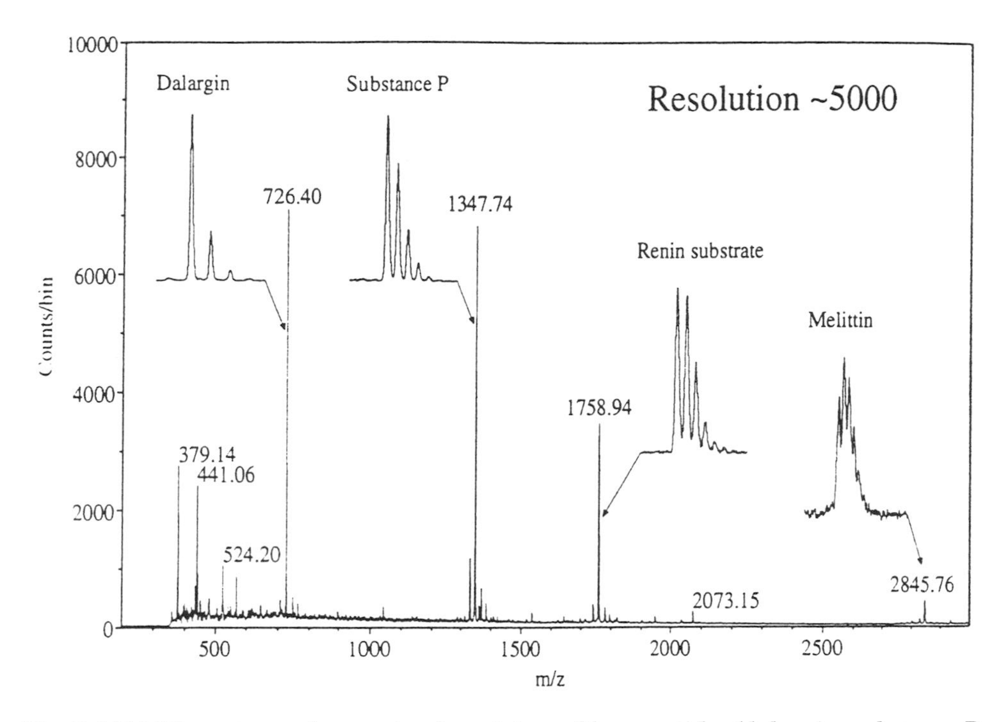

Fig. 7. MALDI spectrum of an equimolar mixture of four peptides (dalargin, substance P, renin substrate and melittin) from a matrix of α-cyano-4-hydroxy-cinnamic acid. 1 μL of a 10^{-6} M solution of each peptide was applied to the target of a MALDI probe shown. The repetition rate of the N_2-laser was 20 Hz and the spectrum was acquired for 1 min. Expanded views of the protonated ion peaks show a uniform resolution ($M / \Delta M_{FWHM}$) of 4000 to 5000. Substance P and melittin were used as internal calibrants, so the numbers over those peaks are theoretical values. Labels over other peaks indicate experimental masses.

the quadrupole ion guide operating at ~70 mTorr. Collisional cooling then produces a quasi-continuous ion beam of small cross section with a well defined kinetic energy, consistent with predictions of our computer simulation and measurements of transit times of ions through the ion guide [22].

Performance of the instrument with a MALDI ion source is similar to that obtained with the ESI source. A uniform mass resolution of about 5,000 (FWHM definition) is routinely obtained for molecular weights up to about 6,000 Da. Mass is determined with a simple two point calibration with mass accuracy around 30 ppm (using an external calibration). The sensitivity for peptides is in the low femtomole range. The mass range is currently limited by the low energy (5 keV) of the ions at the detector, but ions of cytochrome C (12,360 Da) have been detected at low intensity. Figure 7 shows the spectrum of an equimolar mixture of several peptides desorbed from an α-cyano-4-hydroxy cinnamic acid matrix. The spectrum was measured in a single 60 s run with the laser operating at 20 Hz.

A major advantage of the new interface is the ability to run the spectrometer with ESI as well as MALDI. The MALDI probe of Fig. 5 can be replaced quickly by the ESI probe shown in Fig. 8, which has a capillary (0.22 mm

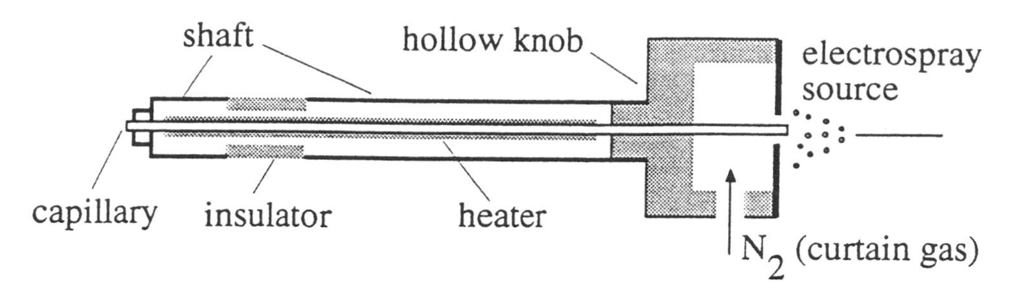

Fig. 8. The ESI probe.

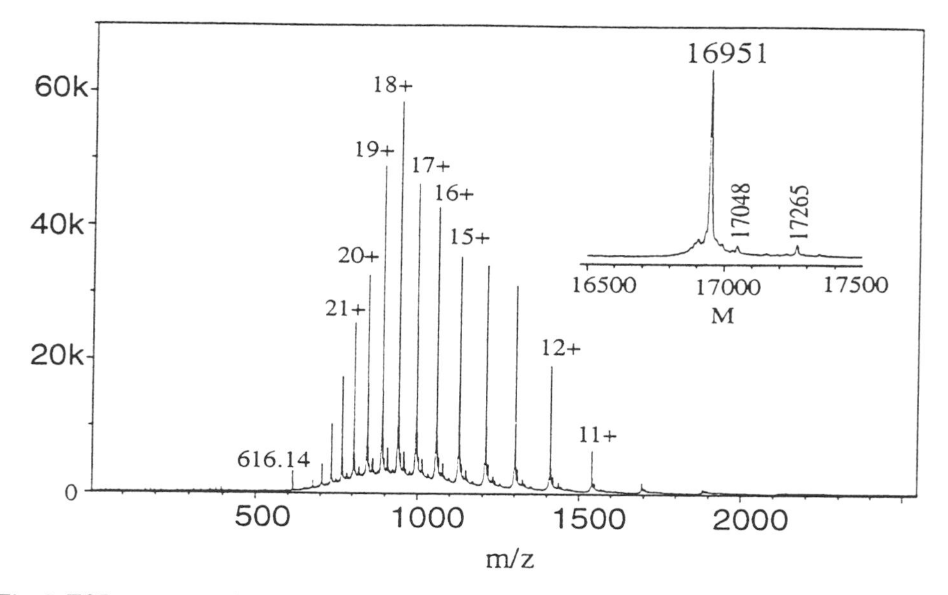

Fig. 9. ESI spectrum of myoglobin obtained by using the ESI probe inserted in the inlet port of the interface instead of the MALDI probe. Declustering of myoglobin ions was performed by setting a 40V potential difference between the first and second quadrupoles in the collisional ion guide.

i. d., 18.5 cm long) inside the probe shaft. Ions from a conventional electrospray source positioned at the end of the probe pass through the heated capillary, together with nitrogen used as a curtain gas. In this case the gas flow through the heated capillary determines the pressure inside the quadrupole ion guide. It is maintained between 100 and 180 mTorr, depending on the temperature of the capillary.

Myoglobin (from horse skeletal muscle, Sigma) is too large a molecule (16,951 Da) to be observed by MALDI with the present accelerating voltages, but it was examined by electrospray as a test of the ESI mode of instrument operation. Its m/z spectrum is shown in Fig. 9, with the deconvoluted mass spectrum as an inset. The ions were introduced directly from the atmosphere into vacuum through the capillary. Declustering of myoglobin was performed in the region between the two quadrupoles by setting a potential difference between them.

The MALDI probe has recently been connected also to our QqTOF instrument [23], and MS/MS experiments on MALDI ions are in progress.

ACKNOWLEDGMENTS

The measurements on citrate synthase were done in cooperation with A. Ayed, L. Donald and H. W. Duckworth at the University of Manitoba Chemistry Department. Our work is supported by grants from NSERC (Canada).

REFERENCES

1. P. Roepstorff, *Curr. Opinion Biotechnol.* **1997**, *8*, 6-13.

2. J. R Yates III, *J. Mass Spectrom.* **1998**, *33*, 1-19.

3. J. A. Loo, *Mass Spectrom. Rev.* **1997**, *16*, 1-23.

4. *New Methods for the Study of Biomolecular Complexes*, W. Ens, K. G. Standing and I. V. Chernushevich, Eds., NATO ASI Series, Kluwer Academic Press, Dordrecht 1998, pp. 354.

5. M. Yamashita and J. B. Fenn, *J. Chem. Phys.* **1984**, *88*, 4451-59.

6. M. L. Aleksandrov, L. N. Gall, N. V. Krasnov, V. I. Nikolayev, V. A. Pavlenko and V. A. Shkurov, *Dokl. Akad. Nauk* **1984,** *277*, 379-383 [*Dokl. Phys. Chem.* **1985**, *277*, 572-576].

7. *Electrospray Ionization Mass Spectrometry*, R. B. Cole, Ed.; John Wiley and Sons, New York, 1997.

8. M. Karas, D. Bachmann, U. Bahr and F. Hillenkamp, *Int. J. Mass Spectrom. Ion Proc.* **1987**, *78*, 53-68.

9. K. Tanaka, H. Waki, Y. Ido, S. Akita, Y. Yoshida and T. Yoshida, *Rapid Commun. Mass Spectrom.* **1988**, *2*, 151-153.

10. F. Hillenkamp, M. Karas, R.C. Beavis and B. T. Chait, *Anal. Chem.* **1991**, *63*, 1193A-1203A.

11. R. J. Cotter, *Time-of-Flight Mass Spectrometry*, Am. Chem. Soc., Washington, DC, 1997.

12. B. A. Mamyrin, V. I. Karataev, D. V. Shmikk and V. A. Zagulin, *Sov. Phys. JETP* **1973**, *37*, 45-48.

13. R. S. Brown and J. J. Lennon, *Anal. Chem.* **1995**, *67*, 1998-2003.

14. S. M. Colby, T. B. King and J. P. Reilly, *Rapid Commun. Mass Spectrom.* **1994**, 865-868.

15. W. C. Wiley and I. H. McLaren, *Rev. Sci. Instrum.* **1955**, *26*, 1150-1157.

16. R. D. Macfarlane and D. F. Torgerson, Science **1976**, *191*, 920.

17. G. J. O'Halloran, R. A. Fluegge, J. F. Betts and W. L. Everett, Technical Documentary Report, No. ASD-TDR-62-644, Part I & II, prepared under contracts Nos. AF33(616)-8374 and AF33(657)-11018 by the Bendix Corporation, Research Laboratory Division, Southfield, Michigan, 1964.

18. A. F. Dodonov, I. V. Chernushevich, T. F. Dodonova, V. V. Raznikov and V.L. Tal'roze, *USSR Patent #1681340A1* (Feb. 1987).

19. A. F. Dodonov, I. V. Chernushevich and V. V. Laiko, in *Time-of-Flight Mass Spectrometry*; R. J. Cotter, Ed. Am. Chem. Soc. Symposium Series 549, Washington, DC, 1994, pp. 108-123.

20. A. N. Verentchikov, W. Ens and K. G. Standing, *Anal. Chem.* **1994**, *66*, 126-133.

21. I. V. Chernushevich, W. Ens and K. G. Standing in *Electrospray Ionization Mass Spectrometry*, R. B. Cole, Ed.; John Wiley and Sons, New York, 1997, 203-234.

22. A. N. Krutchinsky, I. V. Chernushevich, V. L. Spicer, W. Ens and K. G. Standing, *J. Am. Soc. Mass Spectrom.* **1998**, *9*, 569-579.

23. A. Shevchenko, I. Chernushevich, W. Ens, K. G. Standing, B. Thomson, M. Wilm and M. Mann, *Rapid Commun. Mass Spectrom.* **1997**, *11*, 1015-1024.

24. I. V. Chernushevich, W. Ens and K. G. Standing in *New Methods for the Study of Biomolecular Complexes*, NATO ASI Series, Kluwer Academic Press, Dordrecht 1998, 101-116.

25. A. Ayed, A. N. Krutchinsky, W. Ens, K. G. Standing and H. W. Duckworth, *Rapid Commun. Mass Spectrom.* **1998**, *12*, 339-344.

26. M. Mann, C. K. Meng and J. B. Fenn, *Anal. Chem.* **1989**, *61*, 1702.

27. J. K. Olthoff, I. Lys and R. J. Cotter, *Rapid Commun. Mass Spectrom.* **1988**, *2*, 171-175.

28. A. N. Krutchinsky, A. V. Loboda, V. L. Spicer, R. Dworschak, W. Ens and K. G. Standing, *Rapid Commun. Mass Spectrom.* **1998**, *12*, 508-518.

Questions and Answers

A. L. Burlingame (UCSF)

How do you purify proteins too large for effective HPLC purification for ESI?

Answer. Purification may be carried out by chromatography and dialysis. For example, citrate synthase was first purified from cell extracts by diethylaminoethyl-cellulose chromatography followed by size exclusion chromatography through Sepharose 6B. Before ESI-TOF, the samples, originally in 20 mM TRIS-Cl, 1 mM EDTA, and 50 mM KCl, were washed 6 to 8 times with 2 mL aliquots of 20 mM ammonium bicarbonate in a Centricon 30 (30,000 molecular weight cutoff), and diluted to the appropriate protein concentration such that the buffer concentration was 5 mM.

A. L. Burlingame (UCSF)

Would you share "tricks" involved in getting non-covalent protein complexes airborne by ESI?

Answer. The usual problem is sample purity. This is a much more serious problem at physiological values of pH than in acidic solution. With inadequate purity the ESI peaks spread out to form a broad hump, which yields little information at best, and sometimes disappears in the chemical noise.

Michael O. Glocker (University of Konstanz)

K_D values for CS hexamer NADH dissociation were shown to range from nM to µM. What is the established limits for K_D-ranges that can be determined using the outlined approach and what is the actual protein concentration?

Answer. The citrate synthase subunit concentration used to obtain the spectra was 9 µM, a typical value required to obtain high quality spectra. The K_d values for NADH binding to citrate synthase varied from ~1 to ~150 µM. We estimate our observable Kd range at present as > 1 µM for conventional electrospray and > 0.1 µM for nanospray.

Joseph Loo (Parke-Davis Pharmaceuticals)

For your 11S REG protein example, where you showed a specific 3α4β heptameric complex, and a 3α3β hexameric complex (loss of β-subunit) from CAD experiments. Does this observation support any available crystal or NMR structure of the complex?

Answer. There is no crystallographic or NMR information available on the mixed αβ complexes of the 11S REG activator of the 20S proteasome. However, a crystal structure has recently been reported by Knowlton *et al.* (*Nature* **1997**, *390*, 639-643) for a complex composed of pure α subunits, which turns out to be a heptamer. The authors state: "As the biochemical properties of our human recombinant REGα closely resemble those of the mixed REGα/REGβ complex purified from red blood cells, it is likely that the heteromeric REGα/REGβ complex is also a heptamer ..., although others have concluded that the REGα/REGβ complex is a hexamer." Our observation of a 3α4β complex supports this point of view.

We do not really understand why the complex appears to break up into a hexamer and a β subunit, but we note that the 3α4β complex has obvious asymmetry, so it seems reasonable that it might spit out the extra β as an initial step in its breakup.

Carol Robinson (Oxford University)

How do you distinguish specific from non-specific binding of the NADH to citrate synthase? Is it possible to increase the energy in the ion source to remove non specific adducts?

Answer. We have fitted our data for the variation in numbers of bound NADH molecules vs the free NADH concentration to theoretical curves to obtain values for the dissociation constants K_d [25]. This is consistent with tight binding to 6 sites per hexamer (specific binding) and loose binding for the other NADH molecules attached (non-specific binding).

We have of course removed adducts of water, etc., by increasing the voltage between capillary and skimmer in the ion source. However, we have not observed any strong differentiation between specific and non-specific bound NADH in the present case, at least in the range of voltages available (~300 V).

Isotopic Amplification, H/D Exchange, and Other Mass Spectrometric Strategies for Characterization of Biomacromolecular Topology and Binding Sites

Alan G. Marshall[*†], *Mark R. Emmett*[*], *Michael A. Freitas*[*], *Christopher L. Hendrickson*[*], and *Zhongqi Zhang*[*‡]
*National High Magnetic Field Laboratory, Florida State University, Tallahassee, FL 32310;[†]Dept. of Chemistry, Florida State University, Tallahassee, FL 32310; [‡]Amgen Inc., Thousand Oaks, CA 91320

SUMMARY

Two recently developed techniques combine to extend by an order of magnitude the size of biomacromolecules whose surface exposure and contacts can be mapped by mass spectrometry. First, isotopic amplification of proteins, achieved by expression from growth media doubly-depleted in carbon-13 and nitrogen-15, narrows, amplifies, and shifts the isotopic mass distribution, to identify definitively the "monoisotopic" species (i. e., the only species with a unique elemental composition), for assignment of molecular weight to within 1 Da for biomolecules up to 80 kDa. Second, H/D exchange in solution, followed by peptic digestion and electrospray ionization Fourier transform ion cyclotron resonance mass spectrometry gives a profile of surface accessibility for backbone amide protons. The combination is being applied to individual protein structure and folding and also to topology of surface contacts in non-covalent complexes between two or more proteins. Moreover, a third technique, gas-phase H/D exchange, is shown to provide an extremely sensitive probe of accessibility to solvent of various labile protons in peptides and proteins.

INTRODUCTION

Multiple Charge States and Multiple Isotopes

The advantages of being able to measure protein molecular weight to within 1 Da are manifold: (a) count the number of disulfide bridges (–S–S– is 2 Da lighter than 2 –SH); (b) identify deamidation ($-NH_2$ is 1 Da lighter than –OH); (c) identify such post-translational modifications as phosphorylation and glycosylation; (d) resolve and identify adducts; (e) identify variant amino acid sequences; etc. Modern quadrupole mass analyzers achieve mass resolving power, $m/\Delta m_{50\%}$ (in which m is ion mass and $\Delta m_{50\%}$ is mass spectral peak full width at half-maximum peak height), of a few thousand for ions of mass-to-charge ratio, m/z (m in Dalton and z in multiples of the elementary charge), up to a few thousand. Modern time-of-flight mass analyzers can achieve mass resolving power of 10,000 or higher. Electrospray ionization can generate gaseous ions of proteins up to several hundred kDa in mass. Nevertheless,

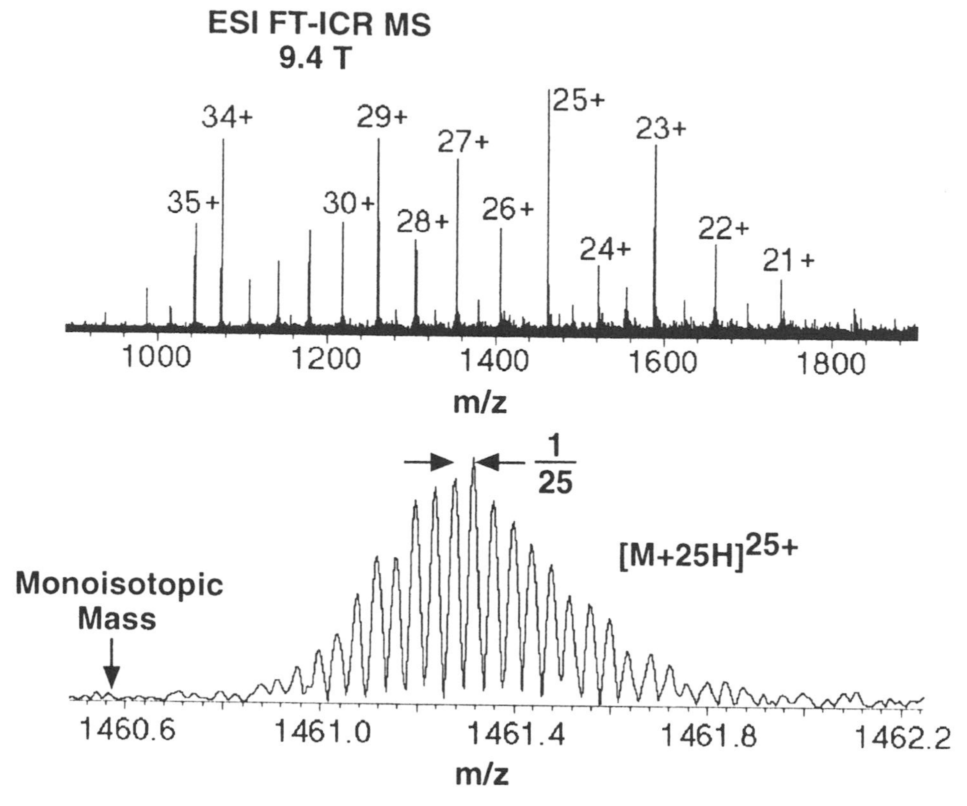

Fig. 1. Broadband electrospray ionization FT-ICR mass spectrum (top) of the alpha-subunit of RNA polymerase. The broadband spectrum shows several charge states, whereas a mass scale expansion showing a single charge state (bottom) reveals baseline-resolved individual isotopic peaks. The charge state, z, is determined as the reciprocal of the spacing between adjacent mass-to-charge ratio isotope peaks differing in mass by 1 Da. For online desalting, the sample was passed through a C8 microbore guard column (15 x 1 mm, Micro-Tech Scientific, Sunnyvale, CA), with consumption of ~100 pmol of protein (to allow for co-addition of multiple spectra). The sample solution flow rate was 40 µL / min.

determination of the mass of even a relatively small protein to within 1 Da with such instruments is rendered difficult by the presence of multiple isotopes and multiple charges.

In molecules with just a few atoms, the various "rare" isotopes (^{13}C, ^{15}N, ^{18}O, ^{34}S, etc.) are low in relative abundance. For example, carbon monoxide consists of ~99% $^{12}C^{16}O_2$ and ~1% $^{13}C^{16}O_2$. However, those "rare" isotopes become more prominent in larger molecules: e. g., for a fullerene, C_{70}, the probability that one of the carbons is ^{13}C rather than ^{12}C is ~70 x 1%, or a relative abundance of ~70% for $^{13}C^{12}C_{69}$ relative to $^{12}C_{70}$. Even a small protein, such as ubiquitin (8,560 Da), contains several hundred carbons, so that protein molecules containing several ^{13}C atoms are much more abundant than the all-^{12}C molecules.

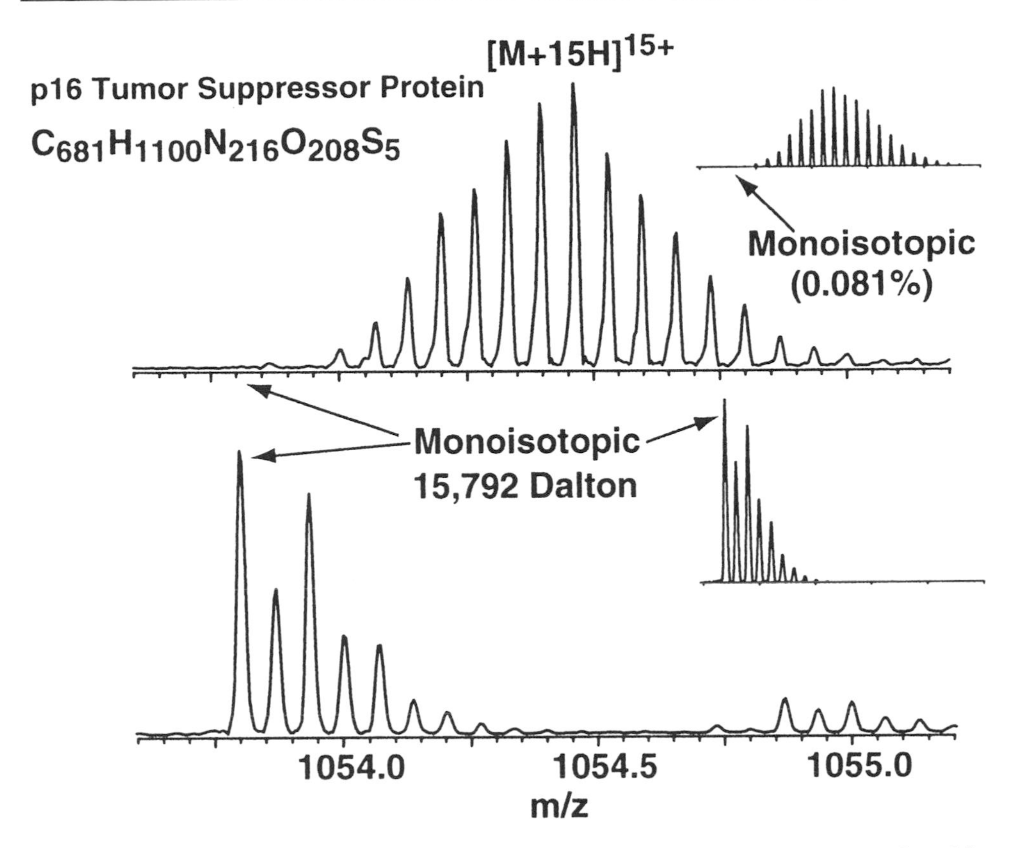

Fig. 2. ESI FT-ICR mass spectra (9.4 tesla) of p16 tumor suppressor protein, produced by direct infusion (2 μM, 50/50 methanol/water with 2.5% acetic acid, 0.5 μL/min). Top: Natural-abundance isotopic distribution (~98.89% ^{12}C; ~99.63% ^{14}N). Bottom: Isotopic distribution for the same protein grown on a medium with 99.95% ^{12}C and 99.99% ^{14}N. Insets: Isotopic distributions calculated (same vertical scale) from the chemical formula for natural-abundance (top) and ^{13}C, ^{15}N doubly-depleted (bottom) p16 tumor suppressor protein. Note the greatly enhanced monoisotopic ion relative abundance for the doubly-depleted protein.

Moreover, electrospray ionization forms multiply-charged ions by addition of (e. g.) protons [1, 2]. A typical protein can take on about 1 proton per kDa, by protonation of available basic residues (e. g., Fig. 1, top). Because mass analyzers (including Fourier transform ion cyclotron resonance, FT-ICR [3]) separate ions based on mass-to-charge *ratio*, m/z, the immediate problem becomes how to determine charge, independent from mass. Fortunately, if the isotopic envelope can be resolved to better than 1 u, then adjacent peaks differ by m/z = 1/z (Fig. 1, bottom); thus, the ion charge may be obtained simply as the reciprocal of the spacing between two adjacent peaks in the m/z spectrum [4]. Because only FT-ICR MS can resolve such multiplets for macromolecular ions of more than ~20 kDa [4], FT-ICR MS becomes the method of choice for mass analysis of such species.

A more severe problem is that even if m/z for each of the resolved isotopic peaks can be determined to ppm accuracy, the molecular weight can still be in error by a whole Dalton or more. One approach is to try to match the observed isotopic abundance distribution of a protein to that predicted for a protein of average amino acid composition [5]. However, if the experimental relative abundances are just a bit in error, or if the unknown protein differs in composition from the average protein in the database, then the estimated molecular weight can be in error by 1 or more Da.

Recently, we introduced a simple solution to the above problem, by expressing a recombinant protein from a minimal medium containing ^{13}C-depleted glucose and ^{15}N-depleted ammonium sulfate [6]. Compared to the natural-abundance protein (Fig. 2, top), a doubly-depleted protein yields a mass spectrum (Fig. 2, bottom) offering a host of advantages (see Results and Discussion).

Solvent Accessibility Determined from Hydrogen / Deuterium Exchange

Solution-phase hydrogen-to-deuterium exchange has become a powerful tool to probe the higher-order structure and dynamics of proteins [7, 8]. Although backbone amide hydrogen/deuterium (H/D) exchange rates for small soluble proteins can be determined by high-resolution multidimensional nuclear magnetic resonance (NMR) spectroscopy [9], mass spectrometry is becoming increasingly popular because (a) mass spectrometry can reveal multiple conformations of a protein (as reflected by multiple rates and/or extent of deuterium uptake); and (b) mass spectrometry can access proteins that aggregate at high concentration, or are too large (especially when non-covalently attached to other macromolecule(s)) for NMR analysis.

For example, Fig. 3 shows that native FK506-binding protein exhibits a *single* conformation (as judged from a single rate and extent of deuterium uptake and resulting mass increase). The same protein in 4.5 M urea consists of *two* comparably abundant protein conformations: one exhibits the same deuterium uptake extent (and is thus interpreted to have the same conformation) as the native protein, and a second (more complete) deuterium uptake extent (interpreted as a more unfolded conformation with more solvent-accessible backbone amide protons.

Typically, H/D exchange is initiated by suddenly diluting a dissolved protein in a large excess of D_2O buffer. Following a specified incubation period during which terminal amine and carboxyl protons, most side chain exchangeable protons, and exposed backbone amide protons are replaced by deuterium, further H/D exchange is quenched by changing the pH to ~2.5 and the temperature to ~0°C. Proteolysis with pepsin then yields numerous small peptide segments which may be electrosprayed (after on-line LC cleanup) for mass analysis, to determine the extent of deuterium uptake for each segment. From the deuterium uptake profile for a sufficiently large number of small peptide segments spanning most of the primary amino acid sequence of the protein, one can determine which segments of the protein backbone are solvent-accessible [10-13]. The ultrahigh mass resolving power of FT-ICR MS [3, 14-18] makes it optimal for H/D exchange analysis because isotopic distributions have been resolved in proteins up to 112,000 Da [4, 19-21].

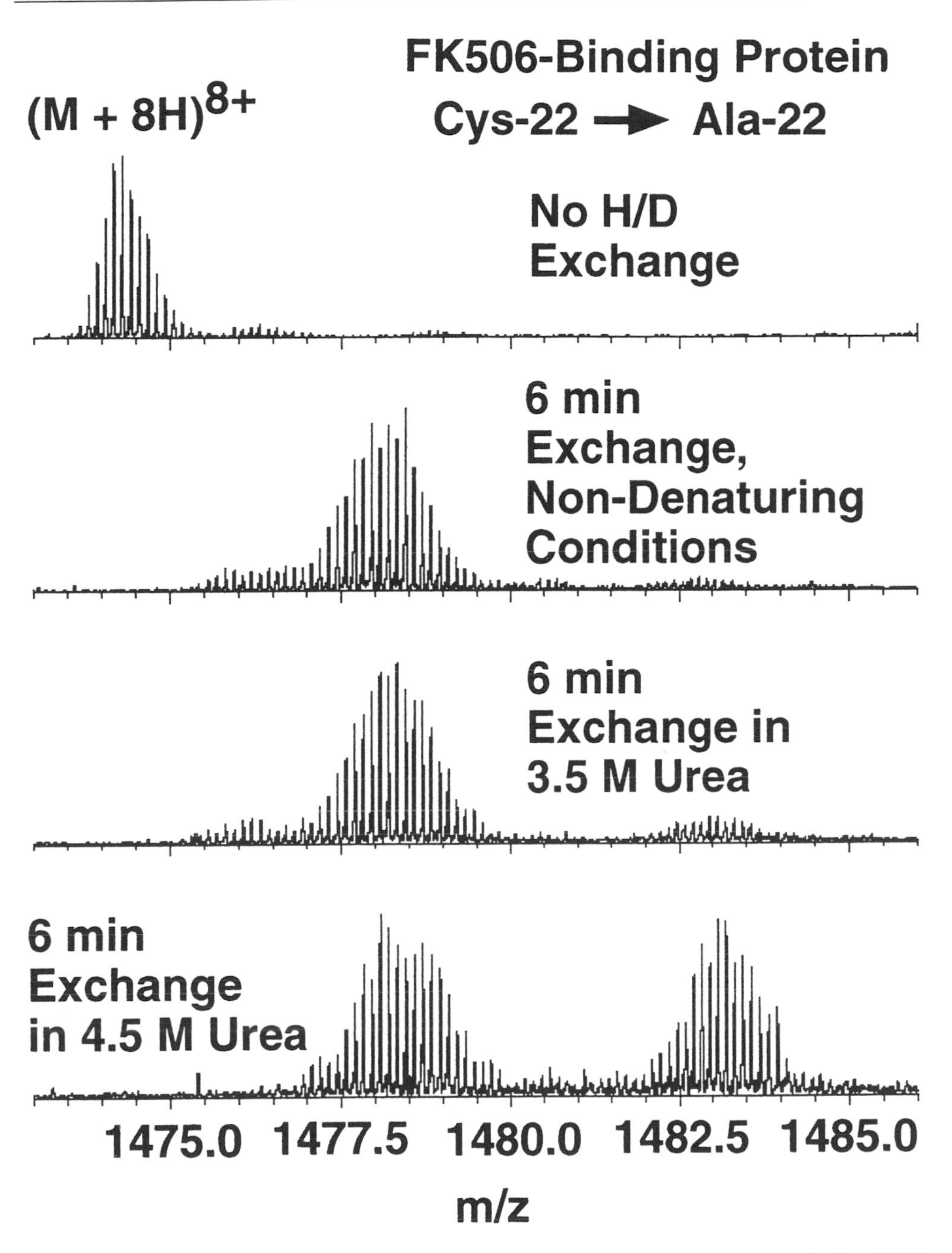

Fig. 3. FT-ICR high-resolution mass spectra of C22A-FKBP, $[M+8H]^{8+}$: Top: before H/D exchange. Second from top: after exchange under non-denaturing conditions for 6 min. Third from top: after exchange in 3.5 M urea for 6 min; Bottom: after exchange in 4.5 M urea for 6 min. (Reproduced with permission from [32].) Urea was removed by online desalting (1-2 min) with 10% water/acetonitrile, followed by LC-MS (~ 2.5 min) performed with a microbore C8 column at 40 μL/min, with a gradient from 10% to 90% H_2O/acetonitrile and 1% formic acid.

Gas-Phase Protein Conformation Revealed by Gas-phase H/D Exchange

Most H/D exchange studies of peptides and proteins have been conducted in *solution* [8, 10, 22-32]. However, interest is increasing in the *gas-phase* structure and conformations of peptides and proteins [33-42], because gas-phase peptide and protein ions typically contain no water molecules, thereby providing for more direct comparison to theoretically derived structures. Also, comparison of solution and gas-phase structures should help to clarify the role(s) of water in peptide and protein structure. Mass spectrometry is especially well suited for characterization of gas-phase H/D exchange of ionic peptides, because deuterium incorporation can be monitored directly from the time profile for mass increase of the peptide or protein ion following introduction of liquid or gaseous D_2O.

Measurement of the rate constants for gas-phase H/D exchange of singly charged amino acids or peptides with D_2O may require a long trapping period (minutes to hours) and/or high neutral reagent gas pressure [18], because the rate constants are small ($< 10^{-11}$ cm^3 molecule s^{-1}). Unfortunately, the same ion-neutral collisions required to produce H/D exchange also result in radial diffusion (and ultimate loss) of ions. Although radial expansion may be virtually eliminated by applying azimuthal quadrupolar excitation [43-48], the "axialization" process speeds up ions and increases their translational temperature, thereby perturbing ion-molecule reaction kinetics.

Fortunately, gas-phase H/D exchange benefits in several important ways by performing FT-ICR experiments at higher magnetic field [49]. First, the maximum number of ions that can be held in an ICR (Penning) ion trap (and hence the FT-ICR MS maximum signal-to-noise ratio and dynamic range) increases as B^2 [50]. Because the initial mass spectrum is broadened (and signal-to-noise ratio thereby reduced) by the deuterium uptake distribution, and because ions gradually diffuse radially beyond detectability, one needs (for example), an initial peptide ion signal-to-noise ratio of ~1,000:1 so that the signal-to-noise ratio is still 100:1 after an hour of exchange at 1×10^{-5} torr of D_2O [49]. Another advantage of high dynamic range is that experiments can be much faster and better controlled by simultaneous detection of several (as many as 10) different peptides (or nucleotides) at once, rather than having to compare experiments performed at different times on individual species. Second, because the maximum ion trapping period increases as B^2 [50], it becomes possible to measure much slower reaction rates. Third, because the cyclotron frequency separation between ions of adjacent m/z values increases directly with B, operation at higher magnetic field makes it easier to eject ions of nearby m/z values, making it easier and faster to isolate a particular isotopic species for unambiguous spectral analysis (see below). A nice consequence is that one can eject all but the monoisotopic isotopes initially, so that the subsequent deuterium uptake profile may be measured directly, without the need to deconvolve the initial natural abundance isotope distribution [51].

EXPERIMENTAL

Materials

C22A FK506-binding protein was expressed and isolated from *E. coli* grown on 99.95% glucose-$^{12}C_6$ and 99.99% ammonium sulfate-$^{14}N_2$ [52, 53] as previously described [6]. RNA polymerase α-subunit was provided by C. F. Meares (U. California, Davis); ^{13}C,^{15}N doubly-depleted p16 protein was kindly provided by T. Selby and M.-D. Tsai (The Ohio State University); and ^{13}C,^{15}N doubly-depleted cdc42 protein was a gift from E. Laue (Cambridge University). Bradykinin, des-Arg[1]-bradykinin, and des-Arg[9]-bradykinin were purchased from Sigma Chemical Company (St Louis, MO). Deuterium oxide (D_2O, ~99.9%D) was purchased from Aldrich Chemical Company (Milwaukee, WI). All reagents were used without further purification.

Solution-Phase H/D Exchange

Conditions for solution-phase H/D exchange are described in detail elsewhere [13, 32].

Gas-Phase H/D Exchange

Fig. 4 shows an experimental FT-ICR event sequence for gas-phase H/D exchange experiments on peptide cations. Ions electrosprayed from ~50 μM solution (50:50 MeOH:H_2O with 0.25% acetic acid) accumulate in a linear octopole ion trap (see Fig. 5) for a few seconds, and are then transmitted to the

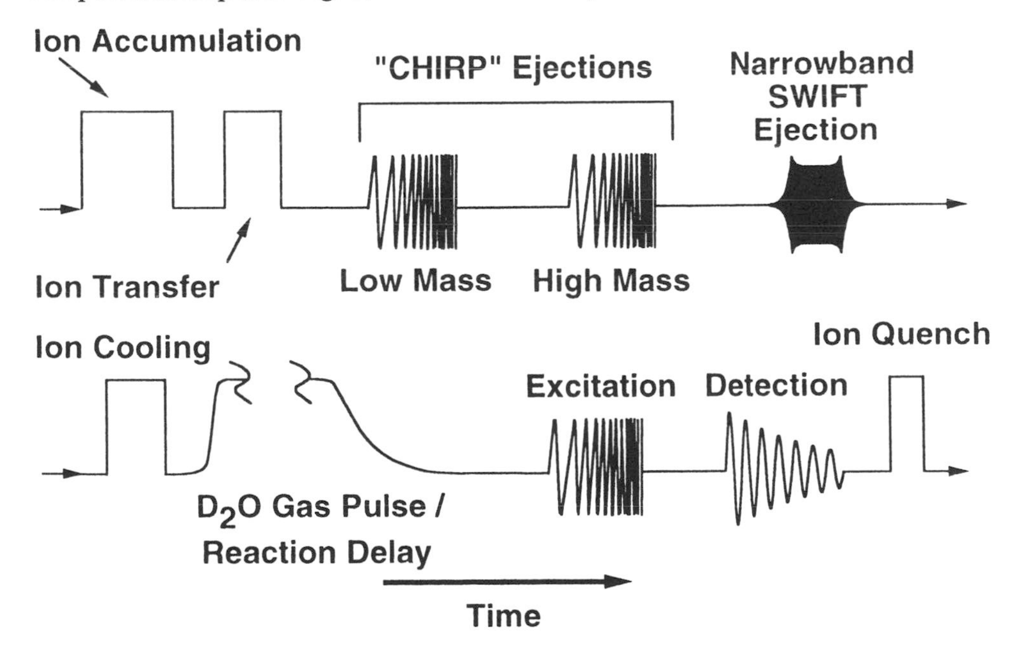

Fig. 4. Electrospray ionization FT-ICR MS experimental event sequence for slow gas-phase H/D exchange reactions. (Reproduced with permission from [49].)

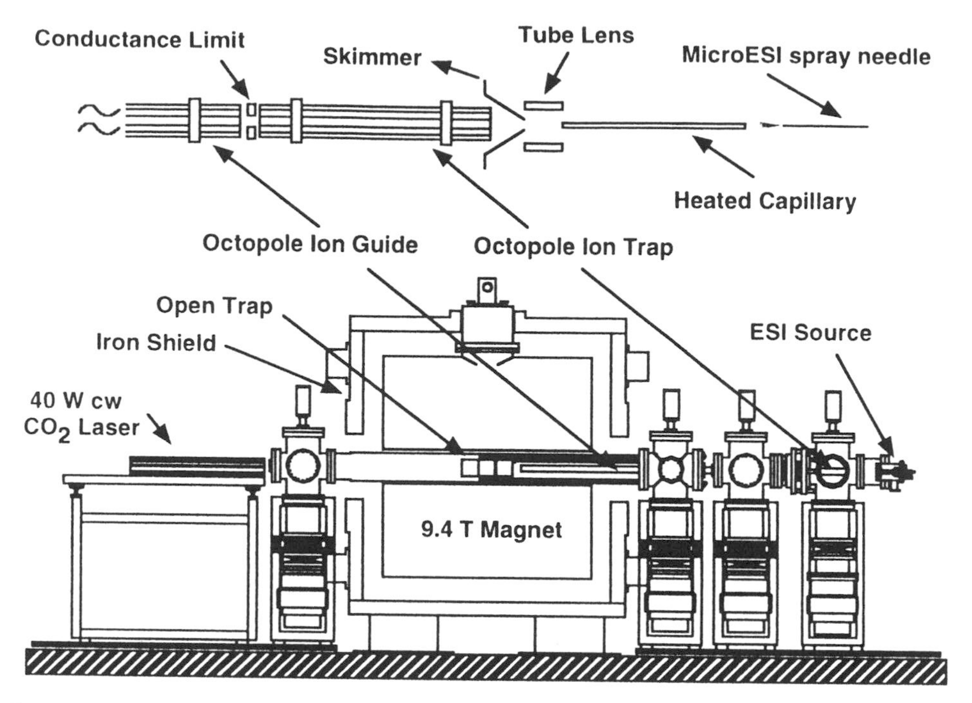

Fig. 5. Schematic diagram of the electrospray 9.4 T FT-ICR mass spectrometer at the National High Magnetic Field Laboratory in Tallahassee, FL.

Penning trap through a second octopole ion guide. The most abundant monoisotopic $[M + H]^+$ ions are isolated by a combination of frequency sweep [54, 55] and stored-waveform inverse Fourier transform (SWIFT) dipolar excitations [56-58] to eject radially ions of other mass-to-charge ratios. A pulse of helium gas and a 15 second delay is used to cool the initial ion translational energy. Parent ions then react with gaseous D_2O introduced into the vacuum system by a pulsed-valve in combination with a leak-valve [59]. The partial pressure of D_2O during the pulse typically rises to $\sim 1 \times 10^{-5}$ torr within 2 s and remains stable throughout the subsequent course of H/D exchange for time periods up to an hour. The exchange reagent pulse concludes with a 6 min pumpdown during which the pressure drops rapidly (~10 s) to $\sim 2 \times 10^{-7}$ torr, decreasing to a final pressure of $\sim 8 \times 10^{-8}$ torr. Immediately following hydrogen/deuterium exchange, the ions are dipole-excited and direct-mode broadband detected. Typical chamber base pressure is $\sim 1\text{-}5 \times 10^{-9}$ torr.

Electrospray FT-ICR Mass Spectrometry

ESI FT-ICR mass spectra were obtained with a homebuilt instrument (Fig. 5) incorporating a 9.4 T horizontal, 220 mm diameter bore, superconducting magnet[20], and featuring an octopole ion trap for external ion accumulation [60], a four-inch diameter three-section open cylindrical Penning trap, and either an Odyssey™ (Finnigan Corp., Madison, WI) or in-house modular ICR data system [61]. All experimental transients are subjected to baseline correction, Hanning apodization and one zero-fill prior to FFT followed by magnitude calculation.

RESULTS AND DISCUSSION

Isotopic Amplification

Figure 2 (bottom) shows an electrospray ionization FT-ICR mass spectrum of p16 tumor suppressor protein depleted in both ^{13}C (to leave 99.95% ^{12}C) and ^{15}N (to leave 99.99% ^{14}N). Compared to the same protein with a natural-abundance isotopic distribution (Fig. 2, top), both simulated and experimental electrospray FT-ICR mass spectra show that double-depletion of ^{13}C and ^{15}N effectively shifts the isotopic distribution to lower mass, so that the monoisotopic species is now prominent and easily identified [6]. (Of course, the observed isotopic distribution can be fitted to reveal the ^{13}C and ^{15}N relative abundances of the protein. In Figure 2, such a fit confirms that the expressed protein has essentially the same level of ^{13}C,^{15}N depletion as for the glucose/ammonium sulfate in the medium in which the bacteria were grown.) Isotopic double-depletion promises to extend the upper mass limit for protein mass spectrometry by about an order of magnitude.

Figure 2 suggests several additional advantages [6]. First, the monoisotopic species, present at only 0.08% at natural abundance, becomes the largest peak in the mass spectrum of the ^{13}C,^{15}N doubly-depleted protein! The monoisotopic molecular weight of the protein is thus determined immediately and unambiguously to high accuracy. For a similarly doubly-depleted protein of 80 kDa molecular weight, the monoisotopic peak should still be 1% abundant, and thus identifiable. Second, depletion of rare isotopes increases mass spectral signal-to-noise ratio (because the same number of ions now exhibit fewer isotopic variants); mass spectral sensitivity and detection limit thus improve accordingly. Third, space charge distortions are reduced, because a mass spectrum of a given peak-height-to-noise ratio requires fewer total ions. Fourth, MS^n experiments are improved, because the narrow m/z distribution makes it easier to isolate the desired parent ions and facilitates identification of fragments [e. g., one Da difference between loss of H_2O *vs.* NH_3 or glutamic acid (129 Da) vs. glutamine (128 Da)]. Fifth, depletion narrows all isotopic distributions, including any adducts (e. g., $[M+nH+mNa]^{(n+m)+}$), and thus dramatically increases the upper molecular weight limit before mass assignment is affected due to isotopic overlap of such impurities. Sixth, a narrower protein isotopic distribution makes it easier to observe and characterize non-covalent binding (protein:protein, protein:nucleic acid, enzyme:inhibitor, etc.). Seventh, identification of surface-accessible residues by H/D exchange is simpler, because of simpler deconvolution to yield the deuterium number distribution. Eighth, isotopically-depeleted proteins provide a good mass calibrant, because the monoisotopic mass is the most accurately assigned peak in the isotopic multiplet, and the depleted isotopic distribution is narrower than other natural-abundance proteins of similar molecular weight. Finally, isotopic amplification can be combined with H/D exchange for significant simplification and extension of that technique (see below).

Solution-Phase Hydrogen / Deuterium Exchange: Importance of Accurate Mass

Identification of peptide fragments, essential for analysis of H/D exchange experiments designed to locate the solvent-accessible amino acid residues in proteins [25], has proved a vexing problem in mass spectrometry. Recent prior approaches include: (a) digestion with a highly specific protease (e. g., trypsin) to reduce the number of possible fragments and the likelihood of misassignment [62]; (b) MS/MS to identify fragments with the same nominal mass [10, 11], and natural abundance *vs.* uniform ^{13}C, ^{15}N enrichment to aid in peptide identification [62, 63]. Fig. 6 illustrates a much more general approach, based on mass measurement accuracy to 5 ppm by FT-ICR mass spectrometry. The protocol is illustrative:

1. Peptides that fit pepsin's specificity (most frequent cleavage sites M-, F-, L-, E-, -Y, etc), and *at the same time* are the only peptides that fit the measured nominal mass [27, 64] are used for internal calibration (top spectrum): peptides 82-97, 1-29 and 30-50 (not labeled on the figure at m/z 853).

2. The spectrum is calibrated from the above three peptides. Peptides 4-29 (3+) and 30-53 (3+) are then assigned.

3. The spectrum is recalibrated, based on 4-29 (3+ and 4+), 30-50, 30-53 (3+ and 4+), 82-97 and 1-29. The average mass accuracy after calibration was ~5 ppm. One of the two remaining fragments is assigned as follows (bottom spectrum).

Assignment of 75-97: Four peptides fit the experimental nominal mass: of those, only two fit to within the measured accurate mass:

S39-R57, MW 2348.29, 43ppm
D41-W59, MW 2348.26, 30 ppm
L74-T96, MW 2348.18, 4 ppm
T75-L97, MW 2348.18, 4 ppm

Thus, accurate mass measurement immediately rules out S39-R57 and D41-W59. Although peptides L74-T96 and T75-L97 have exactly the same mass, both termini of T75-L97 match the proteolytic specificity of pepsin, whereas neither terminus of L74-T96 does. Ergo, **T75-L97 is the only possible internal fragment**.

This approach is equally applicable to peptide fragments produced by enzymatic cleavage in solution [25], gas-phase collision-induced dissociation [65], and gas-phase IR multi-photon dissociation [66], and eliminates the need for (more difficult and/or time- and sample-consuming) multiple-stage MS^n. In simple terms, experiments of the type described above lead to complete

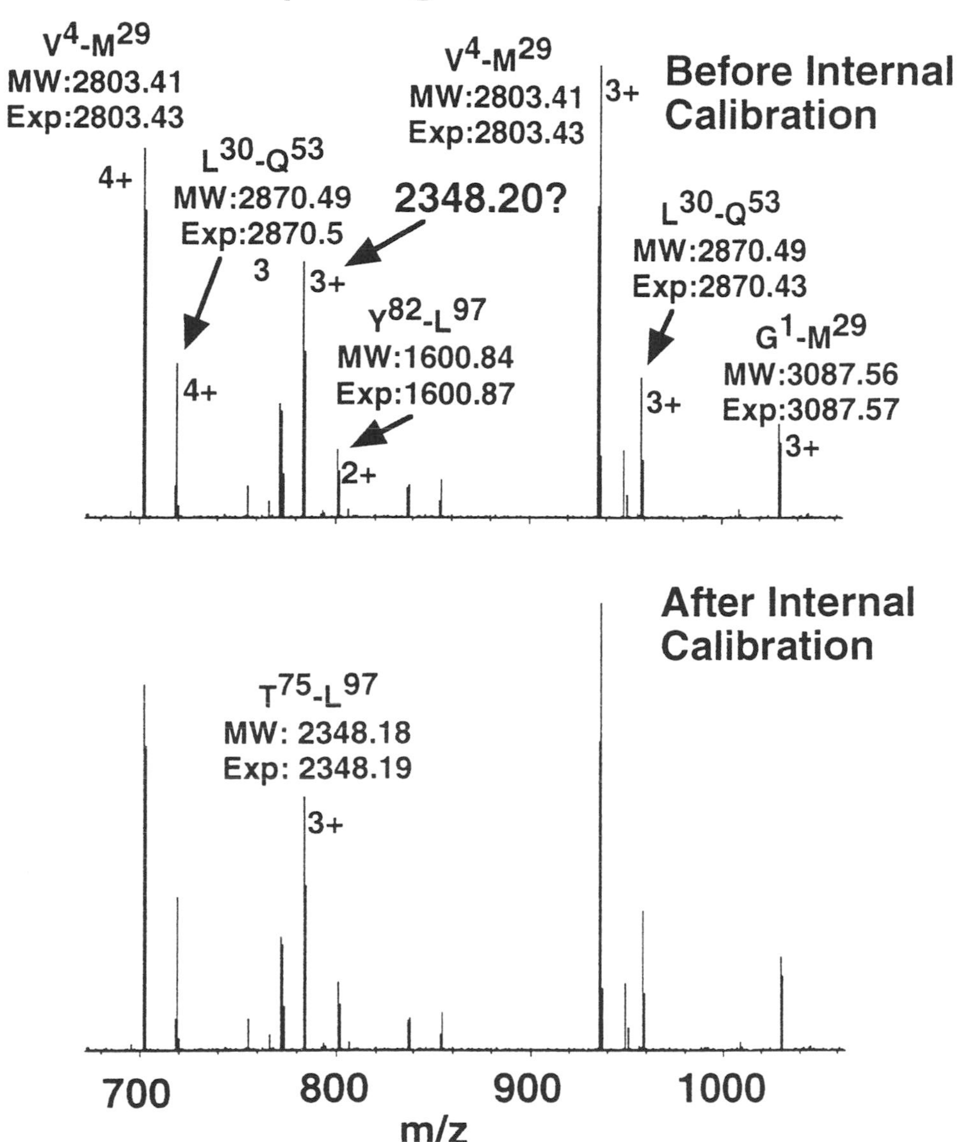

Fig. 6. Identification of ESI FT-ICR mass spectra of pepsin-cleaved fragments of C22A FK506-binding protein. Top: Peptides assignable from nominal mass accuracy and known specificity of pepsin. Bottom: Additional peptide assignment made possible (from four possibilities) by accurate mass measurement to 5 ppm by FT-ICR MS (see text). The sample was desalted online [microbore C8 column (50 x 0.3 mm, LC Packings)] for ~1 min, followed by a step increase to 90% acetonitrile / water at 40 μL / min.

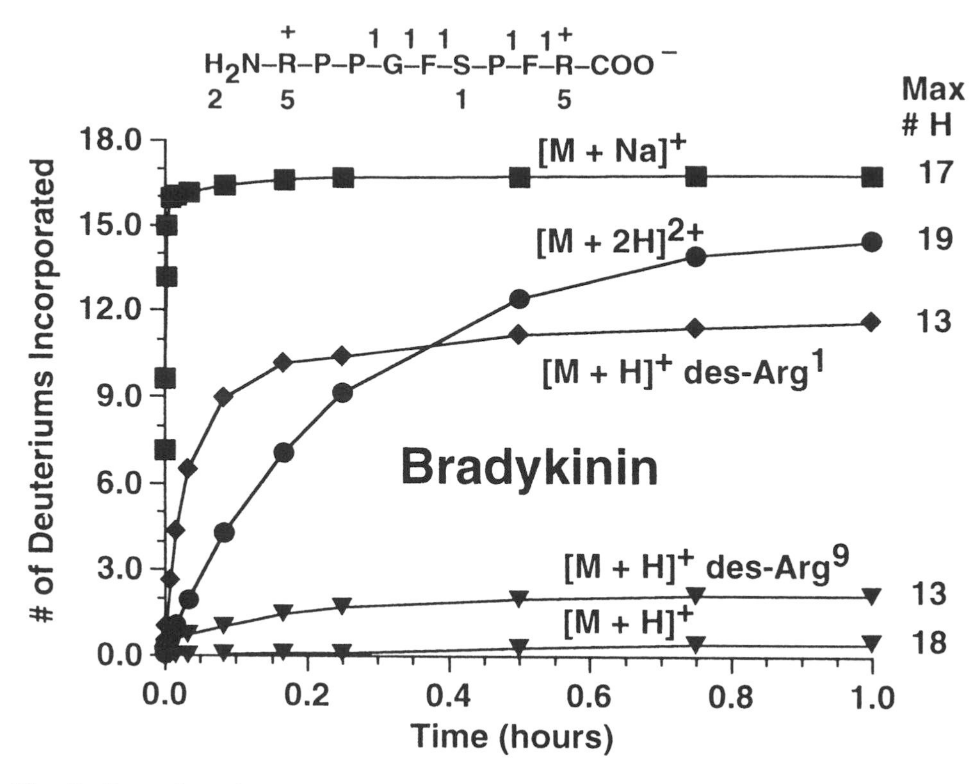

Fig. 7. Deuterium incorporation vs. time profiles for gas-phase simultaneous H/D exchange of singly protonated bradykinin, singly protonated des-Arg⁹-bradykinin, singly-protonated des-Arg¹-bradykinin, doubly protonated bradykinin, and singly sodiated bradykinin with D_2O (1 x 10^{-5} torr). The numbers above and below the bradykinin amino acid sequence denote the number of exchangeable hydrogens for each amide backbone proton, side chain proton(s), and N-terminal protons.

assignment of dozens of peptic fragments in a day, whereas extraction of similar information from a quadrupole or time-of-flight mass analyzer requires a more specific enzyme and/or MS^n for confirmation, and much more time. Of course, MS/MS or even MS^n is still available by FT-ICR (with accurate-mass measurement at each MS stage), if needed.

Gas-Phase Hydrogen/Deuterium Exchange

In the absence of quadrupolar axialization, bradykinin undergoes very slow exchange, whereas the removal of an arginine from either the amino- or carboxy-terminus increases the relative exchange rate dramatically (Fig. 7). Evidently bradykinin adopts a compact gas-phase conformation in which its exchangeable hydrogens are inaccessible to D_2O. Schnier *et al.* [67] have proposed that bradykinin adopts a zwitterionic form that may lead to a folded salt bridge-like conformation. Removal of either arginine residue results in increased exchange rate for the two des-Arg-bradykinins relative to bradykinin itself [49]. It is especially interesting to note that the bradykinins shown in Fig. 7 have similar collisional cross-sections [68], and thus do not differ significantly

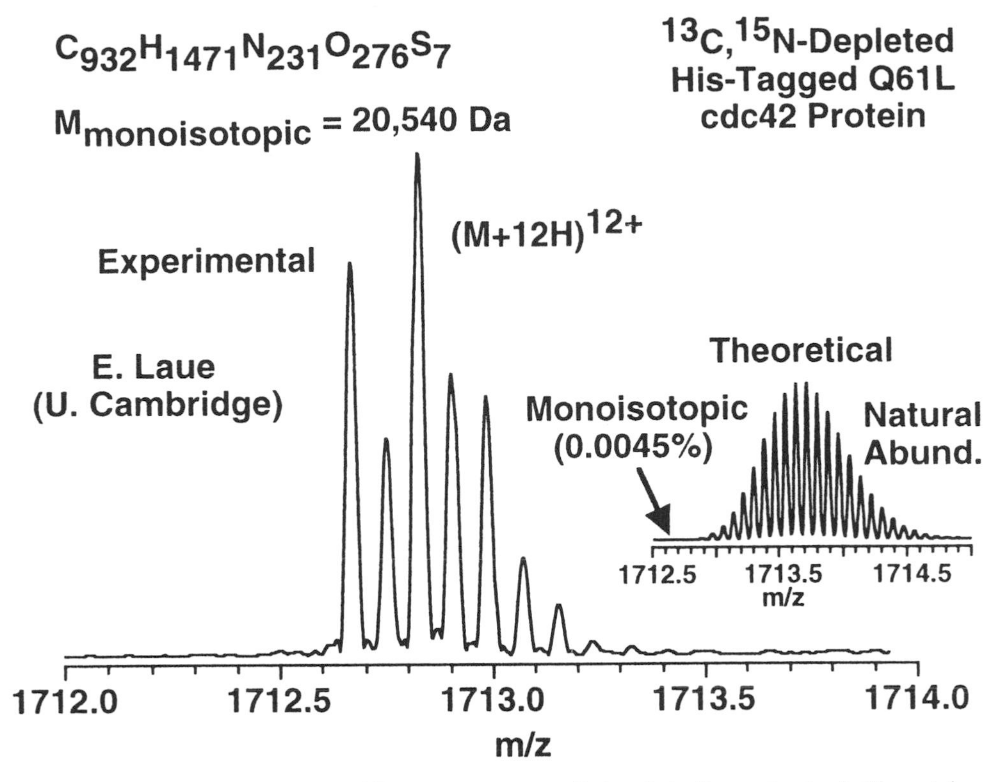

Fig. 8. Experimental ESI FT-ICR mass spectrum (9.4 tesla) of a mutant cdc42 protein, doubly-depleted as in Fig. 2. Inset: Isotopic distribution calculated for natural-abundance cdc42 protein. LC/MS was performed with a microbore C2 column at 30 μL min, with a gradient from 10/90 to 90/10 H_2O/acetonitrile containing 1% formic acid.

in overall size and shape. However, the H/D exchange rate and extent differ by several orders of magnitude: e. g., less than 1 proton exchanged for bradykinin (18 possible exchangeable protons) vs. *all* (17) possible exchangeable protons replaced by deuterium for sodiated bradykinin. Thus, it is abundantly clear that gas-phase H/D exchange offers an extraordinarily sensitive probe of solvent access to terminal, side chain, and backbone amide proton sites.

Note also that *all* of the possible exchangeable protons are accessed in the gas-phase H/D exchange experiment (Fig. 7), whereas NMR experiments in solution are typically restricted to the slowly-exchanging backbone amide protons. Stated another way, the gas-phase trapped-ion experiment extends back to the first ion-neutral collision, so it is as if the solution phase collision time scale (picoseconds) has been stretched into seconds or minutes. We are thus able to probe the very first sites at which H/D exchange occurs.

CONCLUSIONS AND FUTURE DIRECTIONS

Isotopic Amplification

An obvious future direction is to extend this technique to macromolecules of higher mass. Figure 8 shows an ESI FT-ICR mass spectrum of the highest-mass protein for which ^{13}C,^{15}N double-depletion has been achieved, namely a mutant cdc42 protein of more than 20 kDa in mass. Although the monoisotopic relative abundance for the natural-abundance protein is only 45 ppm relative to the most abundant isotopic peak, the isotopically depleted protein yields a prominent and easily recognized monoisotopic peak. Theoretical simulations suggest that the monoisotopic peak relative abundance should still be 1% (i. e., readily detectable) for doubly-depleted proteins up to 80 kDa [6]. Of course, extension of isotopic depletion to RNA and DNA is also feasible.

Solution-Phase H/D Exchange

A particularly exciting future direction for solution-phase H/D exchange is to perform the experiment with isotopically-depleted protein(s). For example, in mapping the contact surface between two non-covalently bound proteins, A:B, it will be very advantageous to prepare (e. g.) isotopically depleted proteins: (a) it will be much easier to identify the peptide fragments due to simpler, narrower, higher signal-to-noise ratio isotopic distributions for each charge state in the electrospray mass spectrum; (b) by depleting only one of the proteins in (e. g.) a binary complex, we can easily distinguish the depleted protein fragments from fragments of the other protein; and (c) we can accurately locate each peptide's monoisotopic peak (i. e., the *only* mass spectral peak with a unique isotopic composition and thus uniquely assignable molecular weight). In this way, it should be possible to determine surface contacts in binary and tertiary protein non-covalent complexes inaccessible to high-resolution NMR. Note that after H/D exchange, the final isotopic distribution represents a convolution of the initial (depleted or natural abundance) isotopic distribution and the deuterium uptake distribution. Therefore, as described in [32], one may deconvolve the final isotope distribution to leave just the deuterium uptake pattern.

Gas-Phase H/D Exchange

Gas-phase H/D exchange experiments for biomolecules are in their infancy. Nevertheless, it is already clear that different conformations of molecules as small as angiotensin II (L-Asp-Arg-Val-Tyr-Ile-His-Pro-Phe) can be distinguished by their different rate and extent of uptake of deuterium from gas-phase D_2O [49]. A dramatic example is provided by the 6+ and 7+ charge states of ubiquitin ($C_{378}H_{629}N_{105}O_{118}S_1$; monoisotopic mass = 8,559.616 Da) shown in Fig. 9. Several resolved deuterium uptake peaks are resolved, meaning that at least that number of protein conformations remain distinct without interconverting on the 30-min time scale of the experiment. In contrast, the 8+ charge state shows essentially a single deuterium uptake peak: those ions therefore either have conformations of near-identical H/D exchange access, or exhibit several rapidly interconverting conformations. Furthermore, in experiments not shown here, we have successfully isolated (by SWIFT [56, 58] ejection) individual species having a particular deuterium uptake (which may represent species of a particular conformation), without scrambling the deuterium distribution. Thus, we can look forward to experiments in which gas-phase H/D exchange

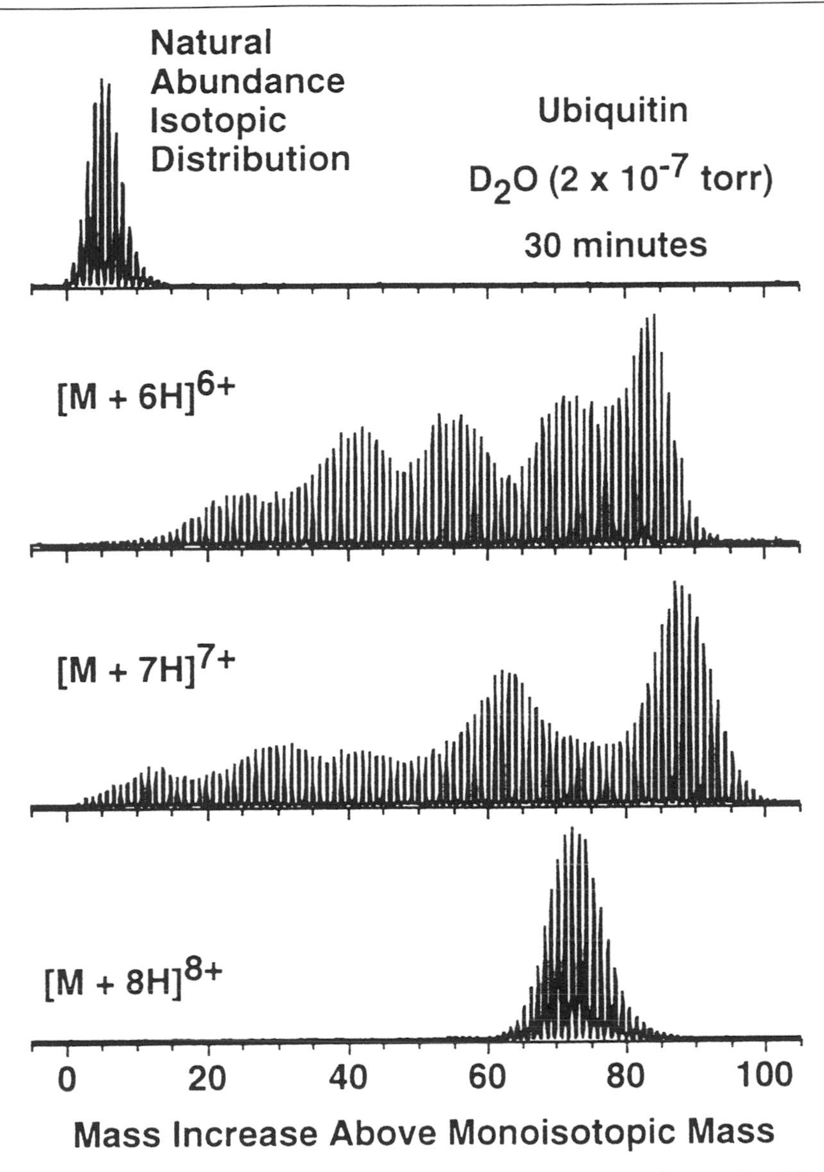

Fig. 9. ESI FT-ICR mass spectra following gas-phase H/D exchange for 30 min at $2x10^{-7}$ torr D_2O for the 6+ (top), 7+ (middle) and 8+ (bottom) charge states of bovine ubiquitin. The abscissa represents a convolution of the deuterium incorporation distribution after 1 h and the initial natural-abundance isotopic distribution of the fully protonated ion. Note the wide dispersion in deuterium uptake for the 6+ and 7+ charge states, corresponding to multiple slowly-interconverting gas-phase ubiquitin conformations. Experimental conditions: Ubiquitin (10 μM) in 50:50 (v/v) MeOH:H_2O with 0.25% acetic acid was infused into a tapered 50 μm i.d. fused silica micro-ESI needle at a rate of 200 nL/min (ESI needle voltage = 1.8 kV, heated capillary current = 2.5 A). Ions were accumulated in a linear octopole ion trap (1.5 MHz, 100 V_{p-p}) for 1.5 s and then transferred to the ICR cell through a second octopole ion guide (1.5 MHz, 100 V_{p-p}). Ions were cooled by a short pulse of He gas to $1x10^{-6}$ torr, and then allowed to react with D_2O ($2x10^{-7}$ torr) for 30 min. After a 5 min pumpdown to $5x10^{-8}$ torr (~10s), the ions were subjected to broadband frequency sweep excitation (50-300 kHz) and detection (300 kHz Nyquist bandwidth, 256 kWord data).

provides a means to separate and isolate a particular gas-phase protein conformation for subsequent chemical reaction or dissociation. By comparing different conformations, we may be able to decide which conformation (if any) corresponds to the solution conformation at a pH corresponding to that charge state.

Gas-phase H/D exchange is already being applied to nucleotide ions [51]. In those experiments, it was shown that the rate of H/D exchange may be "tuned" by choice of solvent acid. For example, uptake of deuterium from D_2O by $(5'GMP - H)^-$ is negligible even after incubation for 6 min at 5×10^{-8} torr. However, in the presence of D_2S at the same pressure, $(5'GMP - H)^-$ exchanges up to 5 hydrogens for deuterium [51], because H_2S is ~40 kcal/mole more acidic than H_2O in the gas phase [69].

ACKNOWLEDGEMENTS

The authors thank Timothy Logan for advice, assistance, and facilities for preparation of the ^{13}C,^{15}N doubly-depleted FK506-binding protein. We also thank C. F. Meares for providing a sample of RNA polymerase α-subunit, T. Selby and M.-D. Tsai for providing ^{13}C,^{15}N doubly-depleted p16 tumor suppressor protein, and E. Laue for providing ^{13}C,^{15}N doubly-depleted CDC42 protein. This work was supported by grants from NSF (CHE-93-22824), the NSF National High Field FT-ICR Mass Spectrometry Facility (CHE-94-13008), NIH (GM-31683), Florida State University, and the National High Magnetic Field Laboratory in Tallahassee, FL.

REFERENCES

1. J. B. Fenn, M. Mann, C. K. Meng and S. F. Wong, *Mass Spectrom. Rev.* **1990**, *9*, 37-70.

2. R. D. Smith, J. A. Loo, R. R. Ogorzalek Loo, M. Busman and H. R. Udseth, *Mass Spectrom. Rev.* **1991**, *10*, 359-451.

3. A. G. Marshall, C. L. Hendrickson and G. S. Jackson, *Mass Spectrom. Rev.* **1998**, *00*, 000-000.

4. F. W. McLafferty, *Acc. Chem. Res.* **1994**, *27*, 379-386.

5. M. W. Senko, S. C. Beu and F. W. McLafferty, *J. Am. Soc. Mass Spectrom.* **1995**, *6*, 229-233.

6. A. G. Marshall, M. W. Senko, W. Li, M. Li, S. Dillon, S. Guan and T. M. Logan, *J. Am. Chem. Soc.* **1997**, *119*, 433-434.

7. S. W. Englander and N. R. Kallenbach, *Quart. Rev. Biophys.* **1984**, *16*, 521-655.

8. R. B. Gregory and A. Rosenberg, in *Methods in Enzymology*; C. H. W. Hirs and S. N. Timasheff, Eds., Academic Press: Orlando, 1986; Vol. 131; pp 448-508.

9. S. W. Englander and L. Mayne, *Annu. Rev. Biophys. Biomol. Struct.* **1992**, *21*, 243-265.

10. Z. Zhang and D. L. Smith, *Protein Sci.* **1993**, *2*, 522-531.

11. R. S. Johnson and K. A. Walsh, *Protein Sci.* **1994**, *3*, 2411-2418.

12. F. Wang, J. S. Blanchard and X.-J. Tang, *Biochemistry* **1997**, *36*, 3755-3759.

13. Z. Zhang, W. Li, T. M. Logan, M. Li and A. G. Marshall, *Protein Sci.* **1997**, *6*, 2203-2217.

14. M. V. Buchanan and R. L. Hettich, *Anal. Chem.* **1993**, *65*, 245A-259A.

15. I. J. Amster, *J. Mass Spectrom.* **1996**, *31*, 1325-1337.

16. T. Dienes, S. J. Pastor, S. Schürch, J. R. Scott, J. Yao, S. Cui and C. L. Wilkins, *Mass Spectrom. Rev.* **1996**, *15*, 163-211.

17. D. A. Laude, E. Stevenson and J. M. Robinson, in *Electrospray Ionization Mass Spectrometry*, R. B. Cole, Ed.; John Wiley & Sons, Inc.: New York, 1997; pp 291-319.

18. M. K. Green and C. B. Lebrilla, *Mass Spectrom. Rev.* **1997**, *16*, 53-71.

19. Q. Wu, S. Van Orden, X. Cheng, R. Bakhtiar and R. D. Smith, *Anal. Chem.* **1995**, *67*, 2498-2509.

20. M. W. Senko, C. L. Hendrickson, L. Pasa-Tolic, J. A. Marto, F. M. White, S. Guan and A. G. Marshall, *Rapid Commun. Mass Spectrom.* **1996**, *10*, 1824-1828.

21. N. L. Kelleher, M. W. Senko, M. M. Siegel and F. W. McLafferty, *J. Am. Soc. Mass Spectrom.* **1997**, *8*, 380-383.

22. R. J. Anderegg, D. S. Wagner, C. L. Stevenson and R. T. Borchardt, *J. Am. Soc. Mass Spectrom.* **1994**, *5*, 425-433.

23. K. Dharmasiri and D. L. Smith, *Anal. Chem.* **1996**, *68*, 2340-2344.

24. S. W. Englander, T. R. Sosnick, J. J. Englander and L. Mayne, *Curr. Opin. Struct. Biol.* **1996**, *6*, 18-23.

25. D. L. Smith and Z. Zhang, *Mass Spectrom. Rev.* **1994**, *13*, 411-429.

26. R. D. Smith, J. E. Bruce, R. Chen, X. Cheng, B. L. Schuartz, G. A. Anderson, S. A. Hofstadler and D. C. Gale, in *Proc. 43rd ASMS Conference on Mass Spectrometry & Allied Topics*, Atlanta, Georgia, 1995; p. 682.

27. D. L. Smith, Y. Deng and Z. Zhang, *J. Mass Spectrom.* **1997**, *32*, 135-146.

28. D. S. Wagner and R. J. Anderegg, *Anal. Chem.* **1994**, *66*, 706-711.

29. B. E. Winger, K. J. Light-Wahl, A. L. Rockwood and R. D. Smith, *J. Am. Chem. Soc.* **1992**, *114*, 5897-5898.

30. Z. Zhang, C. B. Post and D. L. Smith, *Biochemistry* **1996**, *35*, 779-91.

31. Z. Zhang and D. L. Smith *Protein Sci.* **1996**, *5*, 1282-1289.

32. Z. Zhang, W. Li, M. Li, S. Guan and A. G. Marshall, in *Techniques in Protein Chemistry*, D. R. Marshak, Ed., Academic Press: New York, 1997, Vol. VIII; pp. 703-713.

33. S. Campbell, M. T. Rodgers, E. M. Marzluff and J. L. Beauchamp, *J. Am. Chem. Soc.* **1995**, *117*, 12840-12854.

34. C. J. Cassady and S. R. Carr, *J. Mass. Spectrom.* **1996**, *31*, 247-254.

35. X. Cheng and C. Fenselau, *Int. J. Mass Spectrom. Ion Processes* **1992**, *122*, 109-119.

36. E. Gard, D. Willard, J. Bregar, M. K. Green and C. Lebrilla, *Org. Mass Spectrom.* **1993**, *28*, 1632-1639.

37. E. Gard, M. K. Green, J. Bregar and C. Lebrilla, *J. Am. Soc. Mass Spectrom.* **1994**, *5*, 623-631.

38. M. K. Green, E. Gard, J. Bregar and C. B. Lebrilla, *J. Mass Spectrom.* **1995**, *30*, 1103-1110.

39. I. A. Kaltashov, V. M. Doroshenko and R. J. Cotter, *Proteins* **1997**, *28*, 53-58.

40. V. Katta and B. T. Chait, *J. Am. Chem. Soc.* **1993**, *115*, 6317-21.

41. S. J. Valentine and D. E. Clemmer, *J. Am. Chem. Soc.* **1997**, *119*, 3558-3566.

42. T. D. Wood, R. A. Chorush, F. M. I. Wampler, D. P. Little, P. B. O'Connor and F. W. McLafferty, *Proc. Natl. Acad. Sci. USA* **1995**, *92*, 2451-2454.

43. G. Bollen, R. B. Moore, G. Savard and H. Stolzenberg, *Appl. Phys.* **1990**, *68*, 4355-4374.

44. L. Schweikhard, S. Guan and A. G. Marshall, *Int. J. Mass Spectrom. Ion Processes* **1992**, *120*, 71-83.

45. S. Guan, M. C. Wahl and A. G. Marshall, *J. Chem. Phys.* **1994**, *100*, 6137-6140.

46. C. L. Hendrickson and D. A. Laude, Jr., *Anal. Chem.* **1995**, *67*, 1717-1721.

47. C. L. Hendrickson, J. J. Drader and D. A. Laude, Jr., *J. Am. Soc. Mass Spectrom.* **1995**, *6*, 448-452.

48. J. A. Marto, S. Guan and A. G. Marshall, *Rapid Commun. Mass Spectrom.* **1994**, *8*, 615-620.

49. M. A. Freitas, C. L. Hendrickson, M. R. Emmett and A. G. Marshall, *J. Am. Soc. Mass Spectrom.* **1998**, *9*, 0000-0000.

50. A. G. Marshall and S. Guan, *Rapid Commun. Mass Spectrom.* **1996**, *10*, 1819-1823.

51. M. A. Freitas, S. D.-H. Shi, C. L. Hendrickson and A. G. Marshall, *J. Am. Chem. Soc.* **1998**, *120*, 0000-0000.

52. D. Barettino, M. Feigenbutz, R. Valcarcel and H. G. Stunnenberg, *Nucl. Acids Res.* **1994**, *22*, 541-542.

53. T. L. Holzman, D. A. Egan, R. Edalji, R. L. Simmer, R. Helfrich, A. Taylor and N. S. Burres, *J. Biol. Chem.* **1991**, *226*, 2474-2479.

54. M. B. Comisarow and A. G. Marshall, *Chem. Phys. Lett.* **1974**, *26*, 489-490.

55. A. G. Marshall and D. C. Roe, *J. Chem. Phys.* **1980**, *73*, 1581-1590.

56. A. G. Marshall, T.-C. L. Wang and T. L. Ricca, *J. Am. Chem. Soc.* **1985**, *107*, 7893-7897.

57. A. G. Marshall, T.-C. L. Wang, L. Chen and T. L. Ricca, in *Amer. Chem. Soc. Symposium Series*; M. V. Buchanan, Ed.; American Chemical Society: Washington, D. C., 1987; Vol. 359; pp 21-33.

58. S. Guan and A. G. Marshall *Int. J. Mass Spectrom. Ion Processes* **1996**, *157/158*, 5-37.

59. C. Q. Jiao, D. R. A. Ranatunga, W. E. Vaughn and B. S. Freiser, *J. Am. Soc. Mass Spectrom.* **1996**, *7*, 118-122.

60. M. W. Senko, C. L. Hendrickson, M. R. Emmett, S. D.-H. Shi and A. G. Marshall, *J. Am. Soc. Mass Spectrom.* **1997**, *8*, 970-976.

61. M. W. Senko, J. D. Canterbury, S. Guan and A. G. Marshall, *Rapid Commun. Mass Spectrom.* **1996**, *10*, 1839-1844.

62. R. W. Kriwacki, J. Wu, G. Siuzdak and P. E. Wright, *J. Am. Chem. Soc.* **1996**, *118*, 5320-5321.

63. C. M. Dobson, *J. Am. Chem. Soc.* **1996**, *118*, 7402-7403.

64. J. C. Powers, A. D. Harley and D. V. Myers, *Adv. Exp. Med. Biol.* **1977**, *95*, 141-57.

65. M. W. Senko, J. P. Speir and F. W. McLafferty, *Anal. Chem.* **1994**, *66*, 2801-2808.

66. D. P. Little, J. P. Speir, M. W. Senko, P. B. O'Cornnor and F. W. McLafferty, *Anal. Chem.* **1994**, *66*, 2809-2815.

67. P. D. Schnier, W. D. Price, R. A. Jockusch and E. R. Williams, *J. Am. Chem. Soc.* **1996**, *118*, 7178-7189.

68. T. Wyttenbach, G. von Helden and M. T. Bowers, *J. Am. Chem. Soc.* **1996**, *118*, 8355-8364.

69. S. G. Lias, J. E. Bartmess, J. F. Liebman, J. L. Holmes, R. D. Levin and W. G. Mallard *J. Phys. Chem Ref. Data* **1988**, *17*, Supplement No. 1.

Questions and Answers

Julie Leary (University of California, Berkeley)

Could you address the viability of using ICR to perform high-resolution mass separation prior to MS/MS? Is there a dependence of magnet strength on the ability to effectively isolate the ion of interest and obtain reproducible MS/MS spectra; i. e., will off-resonance excitation be a problem in high-resolution MS/MS?

Answer. We (and others) have isolated precursor ions at a mass resolving power of ~20,000, either by dipolar ejection of ions of all but the desired m/z ratio or by quadrupolar axialization (i. e., retention) of ions of a single m/z ratio (while ions of other m/z passively diffuse radially outward and are lost). Resolution in both stages of an FT-ICR MS/MS experiments scales linearly with increasing magnetic field strength. Sustained off-resonance irradiation (SORI) is designed to modulate periodically the cyclotron radius (and thus linear speed) of a precursor ion, and will certainly affect nearby resonances (according to a "sinc" variation in excitation amplitude with offset frequency). However, if one isolates the precursor ion *before* SORI, then there are no other nearby resonances to worry about.

A. L. Burlingame (UCSF)

What is known about rearrangement during long observation periods?

Answer. The context of this question was gas-phase H/D exchange. The answer is that rearrangement is not a problem as long as ions are not excited (by collisions, UV/VIS/IR radiation, or cyclotron excitation). However, we have found that all three of those excitations can scramble the locations of deuteration of a gas-phase peptide or protein.

Robert Anderegg (Glaxo Wellcome, Inc.)

A followup to Al Burlingame's question: although the deuterons on the surface may scramble, the protons of interest are those tied up in H-bonds and those probably don't scramble.

Answer. That's what one would expect, but there isn't hard evidence yet. We are currently conducting experiments designed to answer that question.

Steven Carr (SmithKline Beecham Pharmaceuticals)

Coalescence of closely-spaced m/z signals is a common problem in FT-ICR, leading to problems in recognition of isotopes, ion relative abundance measurements, and accurate mass determination. Is this problem alleviated by going to higher magnetic field strengths? What other factors affect coalescence?

Answer. The tendency for closely-spaced ion cyclotron resonances to coalescence into a single resonance varies directly with ion mass and the number of trapped ions; and inversely with the mass separation between the two species, the magnetic field strength, and the ion cyclotron orbital radius. Thus, we minimize peak coalescence by use of a large-radius (4-inch diameter) cylindrical ion trap and by use of a high-field (9.4 tesla) magnet.

X. Christopher Yu (Chiron Corp.)

Is there a correlation between the amino acid solvent accessibility determined by NMR or X-ray diffraction and that determined by gas-phase hydrogen/deuterium exchange experiments?

Answer. Yes. We recently published [13] the first direct comparison between solution-phase H/D exchange determined from NMR and from ESI FT-ICR MS. NMR can yield the H/D exchange rate for each individual backbone amide proton (provided that its resonance can be resolved and identified), but does not directly reveal the number of H/D-distinguishable conformations. Mass spectrometry can (if a sufficiently large number of overlapping segments is available from pepsin cleavage) resolve segments from 1-20 backbone amide protons, but can also show the number of different conformations.

Michael O. Glocker (University of Konstanz)

Does the outlined approach for the characterization of non-covalent complexes of tyrosine phosphatase with vanadate [inhibitor], by use of H/D exchange with ESI FT-ICR MS give enough information to differentiate shielding effects that are due to direct contacts between the two complex consituents from those that might arise from conformational changes at a different location in the protein? If so, what would be the ranges of affinity/dissociation constant differences that can be differentiated with this methodology?

Answer. The figures I showed reveal only all-or-none reduction in H/D exchange rate on binding of inhibitor. However, it is roughly true that the greatest reduction in exchange rate was for the backbone amides covered by the flap drawn over the active site by binding of vanadate. We don't have enough data to generalize that result. However, in related experiments, David Smith (U. Nebraska) has recently used ESI MS to show that the first protein backbone segments to unfold in solution are those with highest H/D exchange rate. Therefore, it is probably reasonable to make qualitative (but probably not quantitative) distinctions between solvent accessibility based on H/D exchange rate.

Victor L. Talrose (Russian Academy of Sciences)

What are the smallest measurable rate constants for the second-order (H/D exchange) reactions you presented?

Answer. We can detect rates as slow as 1 deuterium reacting per ion after 1 hour at $\sim 10^{-5}$ torr, corresponding to a rate constant of $\sim 10^{-15}$ per molecule per cm^3 per second.

Probing the Nature of Amyloidogenic Proteins by Mass Spectrometry

Ewan J. Nettleton and Carol V. Robinson
Oxford Centre for Molecular Sciences, Oxford University, South Parks Road, Oxford OX1 3QT

Growing awareness that the underlying cause of many diseases involves the formation of amyloid plaques, from Alzheimer's through to the spongiform encephalopathies, has prompted the search for new methods to study the structural conversion of normally soluble protein to amyloid fibril. Recent evidence has suggested that the formation of amyloid fibrils may involve structural rearrangement of native monomeric protein through partially folded intermediates [1]. Mass spectrometry (MS) is playing an increasing role in characterizing the nature of proteins that are present in the fibril [2] and also in looking at intermediates that form during fibril formation [3, 4].

Hydrogen exchange (H-X), monitored by MS, can address the problem of measuring populations of folding intermediates in solution. One of the difficulties of studying amyloid forming proteins is their tendency to form fibrils or aggregate under the solution conditions required to make the measurement. The important characteristic of the MS method which render it particularly amenable to this type of study are its large dynamic range, enabling many different solution conditions from millimolar to nanomolar protein concentrations. Thus the effects of high protein concentration can be explored, while in the case of the tetrameric protein transthyretin (TTR) the concentration can be in the nanomolar range so that only the monomeric species is present in the exchanging solution.

H-X techniques exploit the fact that hydrogen atoms buried in the core of the protein or involved in hydrogen bonded secondary structure do not exchange readily with solvent deuterons. Those on the surface or undergoing fluctuations, which transiently expose the hydrophobic core to solvent, will exchange more readily. We have used these techniques to explore the folding and dynamics of variants of human lysoyme and in a detailed study of the transthyretin disassaembly and unfolding of the TTR tetramer. Both proteins are involved in a number of diseases and have very different 3D structures: lysozyme is a small globular protein while TTR is a non-covalently associated tetramer comprised of four identical subunits. In addition to monitoring H-X properties of these proteins, we use the ability of mass spectrometry to preserve non-covalent interactions to probe the subunit interactions in wild-type and variant TTRs.

Amyloidogenic Human Lysozymes

Two different amyloidogenic human lysozymes were identified relatively recently by Pepys and coworkers in patients suffering from non-neuropathic systemic amyloidosis [5]. X-Ray crystallographic analysis of the two

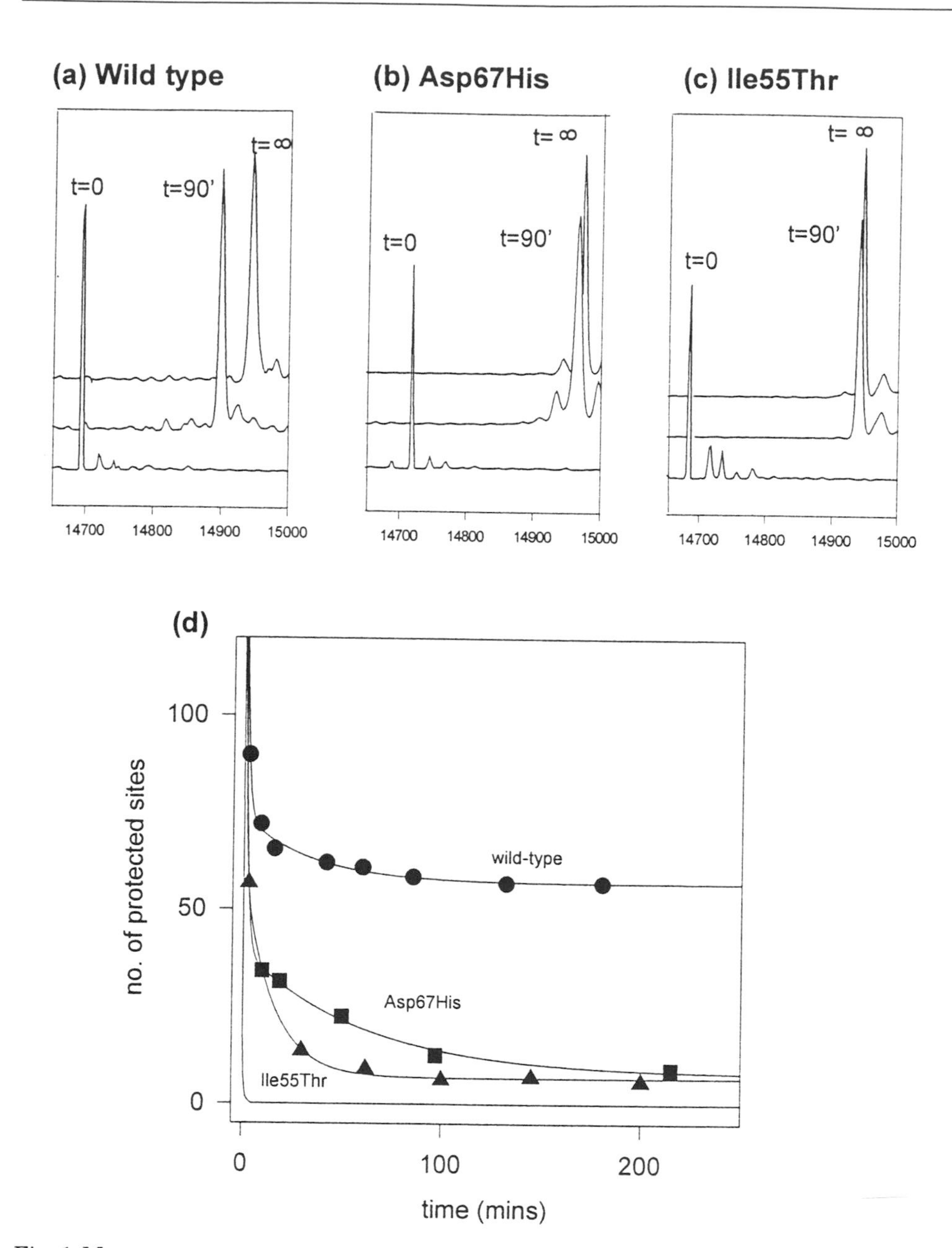

*Fig. 1. Mass spectra and H-X profiles of human lysozyme and variants at 37 °C and pH*5.0. Mass transformed ESI mass spectra of wild-type (a), Asp67His (b) and Ile55Thr (c) lysozymes showing the fully protonated protein (t=0), fully deuterated protein (t=∞) and partially exchanged protein after 90 min of H-X. Kinetic profile of H-X (d) of ● wild-type, ■ Asp67His, ▲ Ile56Thr human lysozymes and (—) random coil simulation of H-X from fully unstructured states of the proteins under the same solution conditions [22]. All pH measurements, represented by pH*, are glass electrode readings. No attempt was made to correct for differences in pH in H_2O and D_2O.*

amyloidogenic proteins shows that the native folds of the two variants are not affected by the mutations [3]. Given that the two variants and the wild-type protein have similar structures, H-X monitored by MS was employed to probe their conformational dynamics (Fig. 1).

Of the 262 exchange labile hydrogens in wild-type human lysozyme, 130 are located on backbone amides while 132 are in side chains and at the two termini. At t=0 the mass recorded is that of the fully protonated protein. At the endpoint of the experiment (t=∞) the spectrum for the fully exchanged protein, in which all labile protons are exchange for deuterium, is recorded. Because the time scale for complete exchange of wild-type human lysozyme is several months, this sample is prepared by heating to 70°C. After 90 minutes protons on the surface or exposed to D_2O will exchange. The recorded mass at this time point corresponds to wild-type protein with 205±2 sites exchanged for deuterium. This represents a stable core of 55±2 remaining protected from exchange under these conditions. By contrast, however, in the Asp67His and Ile56Thr variants, 250±2 protons have undergone H-X after 90 minutes leaving only 10±2 sites remaining protected. Comparison of hydrogen-exchange data over a four hour time period, Figure 1(d), confirms that the two variants are much less protective against hydrogen-exchange than the wild-type, the Ile56Thr variant being the least protective.

The results show that the hydrogen-exchange protection in the wild-type protein is dramatically different from that of the two variants. Given that X-ray analysis reveals that both variant proteins have a native-like fold [3], an increased amplitude and frequency of native state fluctuations explain the lack of protection in the variant proteins. These increased conformational dynamics allow solvent molecules to penetrate the interior of the protein leading to an increased hydrogen-exchange rate in the variants. This result, together with data from other biophysical techniques, suggests that a transient population of the amyloidogenic proteins in a partially-folded state is crucial for their structural conversion to the fibrillar form affected by this mutation [3].

Wild-Type and Variant Transthyretins

TTR is a tetramer containing four identical monomers consisting of 127 amino acid residues. The monomers are composed essentially of β-sheet and associate to form a tetramer with a large central channel, the binding sites for thyroid hormones [6]. The TTR tetramer with two thyroxine molecules bound is shown in Fig. 2.

Wild type TTR forms amyloid in a disease known as senile systemic amyloidosis (SSA), a widespread geriatric disease associated with deposition of fibrils in cardiac tissue [7]. Variant TTR molecules aggregate to form fibrils in familial amyloidotic polyneuropathy (FAP); over fifty different amyloidogenic variants have been identified, all of which are point mutants of the wild-type protein [8]. Autopsies reveal that TTR deposits occur throughout the peripheral and autonomic nervous systems and can accumulate in other organ systems. Many mutations in TTR appear to render the protein more amyloidogenic than

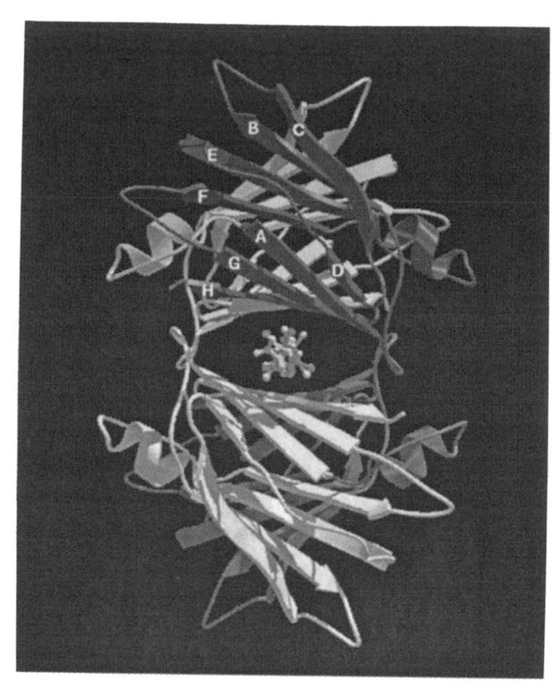

Fig. 2. The physiologically active form of TTR with two thyroxine molecules bound in the ligand-binding channel. A view of the TTR tetramer, along the axis of the ligand binding channel. The four monomers are illustrated in grey and thyroxine is in a ball and stick representation. Co-ordinates are from C. C. F. Blake et al. [6], and the figure was produced with the programme MOLSCRIPT adapted by R. Esnouf [23] and Raster3D [24].

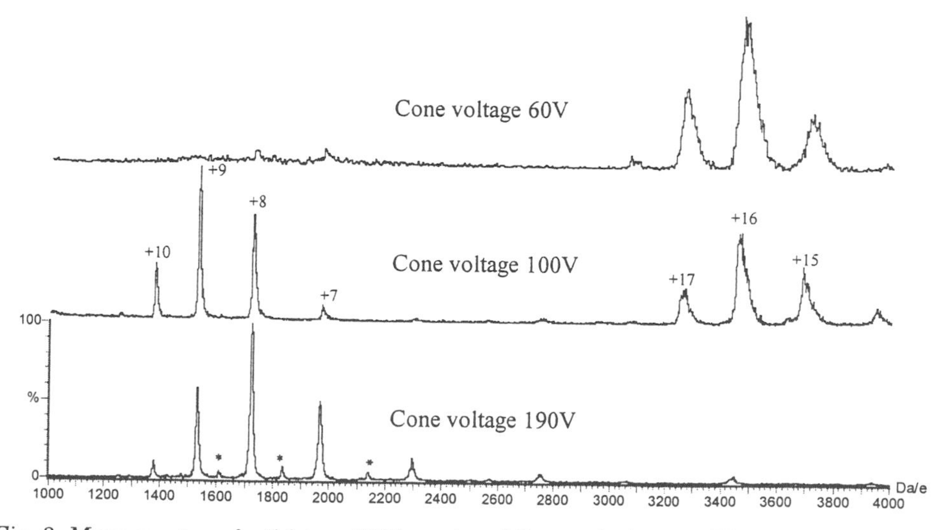

*Fig. 3. Mass spectra of wild-type TTR analyzed from solution at pH 4.0 and 0.18 mg/ml under different energy conditions. +7 to +10: charge states of monomer TTR; +15 to +17: charge states of tetramer TTR; * peaks corresponding to the y_{107} fragment of the TTR monomer that results from cleavage of the N-terminal peptide bond of Pro11. Reproduced with permission from [4].*

wild-type since the onset of FAP is relatively early, in some cases in the second decade of life, whereas SSA occurs only after the eighth decade [9, 10]. SSA affects some 25% of the population, whereas FAP affects approximately 1 in 100,000 and leads to death between two and twenty years after disease onset [11].

Mass Spectrometry of Wild-type Transthyretin

The mass spectra of wild-type TTR analyzed by nanoflow electrospray MS from aqueous solution containing 10mM NH_4OAc at pH 4.0 and physiological protein concentration (0.18 mg/ml) under different energy conditions are shown in Fig. 3. Under ESI-MS conditions at low energy, with a cone voltage of 60V, the energy of collisions in the atmospheric pressure region of the ESI source is minimized and a series of broad peaks is observed between m/z 3000 - 4000. Upon increasing the cone voltage to 100V, where collisions occur with higher energy, an additional well-defined series of peaks at lower m/z (1300 – 2000) is also seen. A further increase in the cone voltage to 190V leads to considerable reduction in the high m/z peak series; the low m/z species now dominate.

Analysis of the masses of the species giving rise to the two charge-state series confirms the identity of the low m/z series (+7 to +10) as monomeric TTR and the high m/z series (+15 to +17) as tetrameric TTR (Table 1). However, the measured mass of the tetramer is greater than that predicted from four times the monomer mass. This could be attributed to non-covalent binding of water molecules and buffer salts, forming a complex referred to hereafter as the solvent-bound tetramer. Such a species is also present in solution, because the assembly of four monomer units results in a central channel in the native tetramer, which is able to accommodate water molecules and buffer salts.

Species (cone voltage)	Charge states	Measured mass (Da)	Calculated mass (Da)	$w_{1/2}$ (Da/e) (charge state)
tetramer (60V)	+17 to +15	56 030 ± 12	55 044	57 ± 3 (+16)
tetramer (100V)	+17 to +15	55 262 ± 8.5	55 044	26 ± 3 (+16)
monomer (100V)	+7 to + 10	13 764 ± 4	13 761	8 ± 1 (+9)
monomer (190V)	+7 to +10	13 767 ± 3	13 761	10 ± 1 (+8)
monomer* (190V)	+6 to +8	12 836 ± 2	12 838	9 ± 2 (+7)

Table 1. Measured masses of wild-type TTR under different cone voltage conditions. Masses measured are those of the most abundant species present. The standard deviations of the measured masses arise from the number of charge states used to determine the mass of the species and do not reflect the broad nature of the peak.

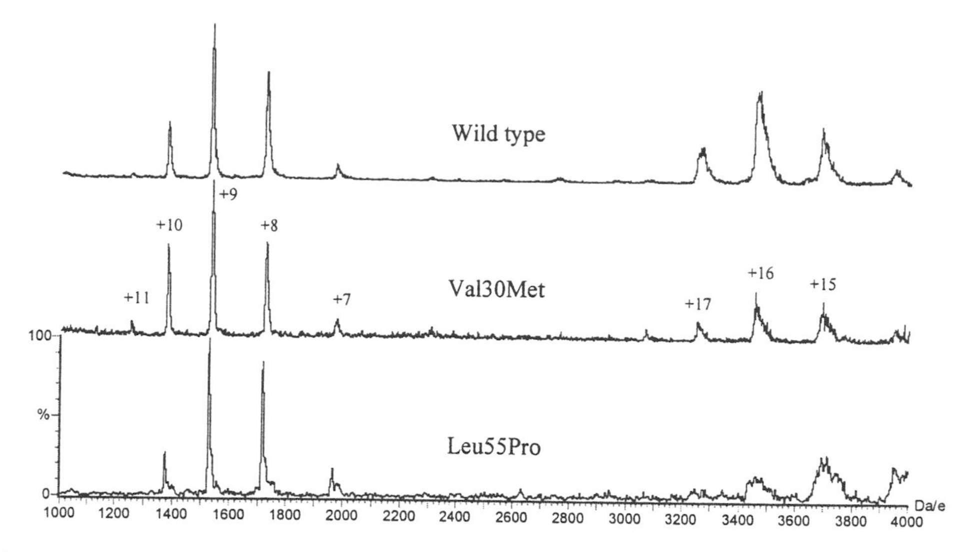

Fig. 4. Mass spectra of wild-type and variant TTRs analyzed from solution at pH 4.0 and 0.18 mg/ml under the same energy conditions (cone voltage 100V). +7 to +10: charge states of monomer TTR; +15 to +17: charge states of tetramer TTR. Reproduced with permission from [4].

It is remarkable that under all of the energy conditions employed (cone voltages between 50V and 200V), the only species detected by MS are the monomer and tetramer. In none of the spectra obtained are any intermediate or higher associated species present. The absence of such species would suggest that non-specific interactions between TTR molecules in the gas phase do not occur, and that dissociation of the tetramer does not result in stable, gas-phase dimer or trimer species. These results are in accord with solution data, where stable dimers and trimers have not been reported. They also support the suggestion that non-covalent interactions observed by MS closely mirror those occurring in solution.

The only peaks detected that have not been accounted for so far correspond to a species close in mass to monomeric TTR, denoted * (Fig. 3). This series is observed under the highest energy condition, and the peaks correspond to the y_{107} fragment that results from cleavage of the NH_2-terminal peptide bond of Pro11. Collision-induced dissociation at proline amide bonds is well-documented [12]. The fact that some cleavage of the peptide backbone occurs before complete disruption of the wild-type TTR tetramer suggests a remarkable stability of this species in the gas phase of the mass spectrometer. This result is in accord with the exceptional solution stability, the tetramer being resistant to chemical denaturation with 4M guanidine hydrochloride [13].

Difference in Stability of Variant TTR Tetramers

A comparison of wild-type TTR with Leu55Pro and Val30Met variant TTRs was carried out by ESI-MS under intermediate energy conditions (cone voltage of 100V) (Fig. 4). Similar charge state series are observed for all three

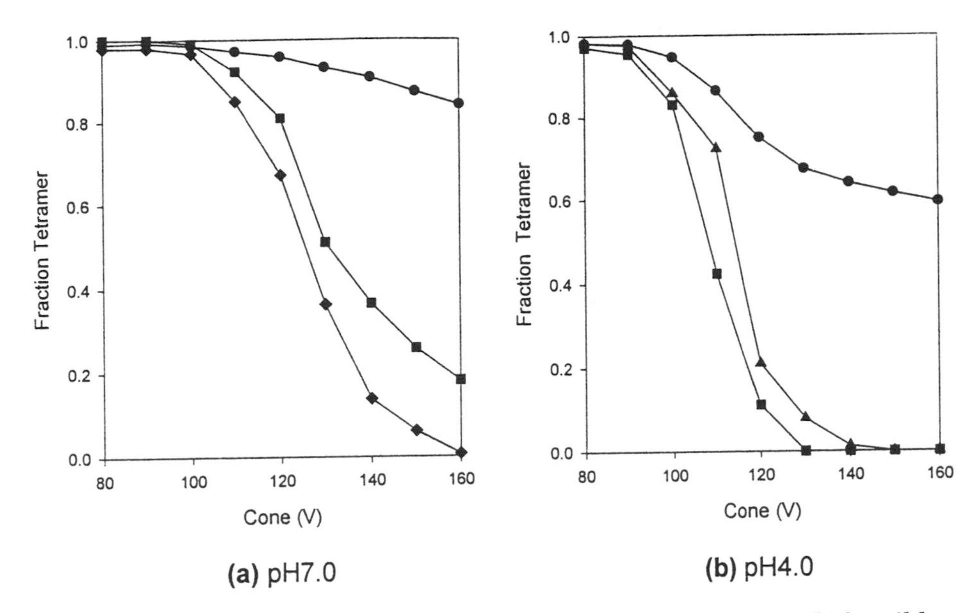

Fig. 5. Fraction of tetramer measured by ESI-MS at pH 7.0 (a) and pH 4.0 (b) for wild-type and variant TTRs as a function of cone voltage. Wild-type TTR: ●; Val30Met TTR: ■; Leu55Pro TTR: ◆. Reproduced with permission from [4].

TTRs, the measured masses being consistent with the presence of only monomeric and solvent-bound tetrameric species. However, the proportion of wild-type TTR tetramer is greater than that observed for Val30Met TTR, which in turn is greater than that observed for Pro55Leu TTR under identical MS conditions. These differences suggest a variation in gas phase stability of the three tetrameric species, which in turn may reflect their relative stabilities in solution. In particular, the results are in line with SDS PAGE analysis, which shows reduced tetramer concentrations in the variant protein solutions after incubation under fibril forming conditions [14, 15].

To understand further the effects of these mutations on the proportion of protein observed as a tetramer in the mass spectrum, a series of experiments was carried out. In these, the energy input into the gas phase tetramer was manipulated by varying the cone voltage at both pH 4.0 and pH 7.0 (Fig. 5). The wild-type TTR tetramer at pH 7.0 is stable showing very little dissociation across the range of cone voltages applied (Fig. 5a). At pH 4.0, however, the wild-type TTR does show a greater proportion of the monomeric species (Fig. 5b). The Val30Met and Leu55Pro variants clearly show more dissociation than wild-type TTR under all the conditions used, and the Leu55Pro TTR exhibits more dissociation than the Val30Met throughout the experiments.

It is interesting to compare the results obtained by ESI-MS with analytical ultracentrifugation data for wild-type TTR. Under similar solution conditions, at pH 7.0 and pH 4.1, molecular masses corresponding to 99% and 78% tetramer were measured [16]. At the mid-point of the cone voltage range, 120V, the ESI-MS results show 97% and 75% tetramer, in close agreement with ultracentrifugation data. However, under both low energy (80V) and high energy (180V) conditions the proportion of tetramer calculated from the ESI mass spectrum is markedly different from that in the solution sample [16]. These observations may be explained by consideration of the nanoflow-ESI process.

During the nanoflow electrospray process positively charged microdroplets are produced by electro-osmotic flow of protein-containing solution through a needle of internal diameter ~ 1 μm at a voltage of ~ 1kV. These charged microdroplets undergo evaporation and droplet fission processes until gas-phase protein ions are formed prior to analysis under vacuum. While evaporation is occurring, a balance appears to exist between desolvation of microdroplets and disruption of non-covalent forces maintaining complexes. It follows that during evaporation of microdroplets, changes in protein concentration may occur. In solution, changes in protein concentration have a marked effect on the tetramer to monomer equilibrium [15]. The observation of 100% tetramer for all three proteins under low energy conditions suggests that changes in local concentration are occurring and that assembly of the tetramer is possible during the shrinking of the microdroplets. Conversely, it might be expected that under high energy conditions, the energy imparted to the microdroplets causes disruption of non-covalent forces before solvent evaporation, resulting in the absence of tetrameric TTR.

Thus, for TTR, the balance between desolvation of microdroplets and disruption of non-covalent forces maintaining tetramer assembly appears to be obtained with a cone voltage of 120V. This cone voltage, at the midpoint of the range, effectively calibrates the MS measurements with the ultracentrifugation data and is used for comparison of the MS results from the two variant proteins where solution data are not available due to their propensity towards fibril formation. The variant proteins at 120V show 80% and 65% tetramer for Val30Met and Leu55Pro variants compared with 97% for wild-type TTR at pH 7.0. At pH 4.0 and 120V, only 20% and 10% tetramer are observed for the Val30Met and Leu55Pro variants, respectively, compared with 75% in the wild-type protein. SDS PAGE analysis of the Leu55Pro variant at pH 4.0, in the presence of detergent to prevent fibril formation, showed only 10% tetramer [14, 15], in agreement with the ESI-MS results.

Analysis of the Mass Distribution in the Spectra

One of the most remarkable features of the wild-type and variant TTR spectra, discussed in the previous two sections, is the increased peak width of charge states corresponding to the tetrameric form, relative to those of the monomer, as seen in Figs. 3 and 4. This broadening of the peak could arise from several factors, including heterogeneity of the protein subunits, formation of adduct ions from buffer salts or the presence of non-covalently-bound water molecules in the quaternary structure of TTR. To assess the contribution to peak width from buffer salts, spectra were acquired after extensive washing of the TTR proteins with pH 4.0 deionized water containing <1% HCl required for pH adjustment. These solutions are thought to be ostensibly devoid of buffer salts,

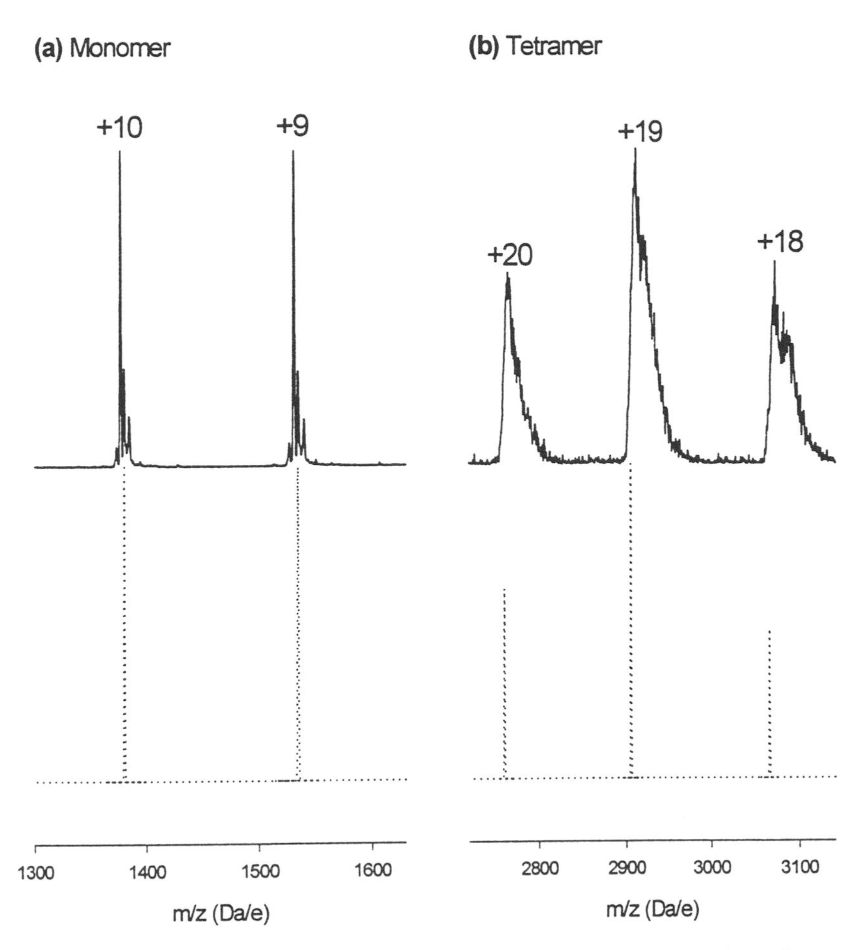

Fig. 6. Comparison of the experimental data obtained for the monomeric and tetrameric Val30Met TTR with a simulation of the natural abundance isotopes. —: experimental data for (a) monomer Val30Met TTR at cone 120V and (b) tetramer Val30Met TTR at cone 60V; ·····: simulated spectra calculated from a Gaussian distribution of the natural abundance isotopes for monomer and tetramer Val30Met TTR. Reproduced with permission from [4].

Species	Charge states	Measured mass (Da)	Calculated mass (Da)	Measured $w_{1/2}$ (Da)	Calculated $w_{1/2}$ (Da)
monomer	+8 to +11	13 792 ± 1	13 793	1.2 ± 1	0.9
tetramer	+17 to +21	55 320 ± 13	55 174	30 ± 5	3.2 (+19)

Table 2. Measured masses and peak width at half height ($w_{1/2}$) of monomeric and tetrameric Val30Met TTR in the absence of buffers and at pH 4.5. Spectra of monomeric Val30MetTTR were obtained at a cone voltage of 120V while those of the tetramer were obtained at 60V. Masses measured are those of the most abundant species present. The standard deviations of the measured masses arise from the number of charge states used to determine the mass of the species and do not reflect the broad nature of the peak.

such that any broadening of the tetramer charge states can be assigned to the binding of water molecules or the heterogeneity of the protein sample.

The spectra of monomer and tetramer of Val30Met TTR were acquired at 60V and 100V and compared with Gaussian models of their natural abundance isotopes (Fig. 6). Values of the peak width at half height for the peaks in the monomer spectrum show good agreement with those predicted by the Gaussian simulation (Fig. 6a and Table 2). In addition, it is clear from this spectrum of the fully-dissociated monomer that the small degree of heterogeneity in the Val30Met TTR sample is not sufficient to account for the broad nature of the peaks corresponding to the tetrameric form. Moreover, the substantial broadening of the tetramer peaks is also accompanied by a significant mass increase relative to the calculated mass (Fig. 6b and Table 2). This can only be accounted for by the non-covalent binding of water or other small molecules.

The mass increase for the most abundant species observed in the Val30Met tetramer spectrum (at the leading edge of the peaks), relative to the calculated value, corresponds to 6 ± 2 water molecules. In the X-ray structure, several water molecules are present and appear to be involved in the maintenance of the quaternary structure of the protein. Two tightly bound water molecules are observed between the two F strands in each dimer, and four internally bound water molecules satisfy the hydrogen bonding potential of the internal polar groups in each monomer [6]. Although it is not possible to determine the location in the structure of the 6 ± 2 water molecules observed in the mass spectrum, it seems likely that four of them are bound between the F and F′ strands in the monomer-monomer interface. Further support for this location comes from the observation that the same number of water molecules remains even when the channel is occupied (see below).

Because the binding affinity of water molecules will vary depending on their location, a distribution of water molecules is expected. Analysis of the peaks corresponding to the tetrameric species reveals that a maximum of 44 ± 10 water molecules is observed under salt-free conditions. Under the same ESI MS conditions, but in the presence of buffer, the peaks observed are significantly broader, suggesting an increase in the number of solvent molecules and adduct ions within the quaternary structure. This increase in mass corresponds to a maximum of 68 ± 10 NH_4^+ or H_2O molecules bound to the tetramer. This observation, coupled with the enhanced stability of the tetramer in the presence of excess buffer and solvent molecules, is consistent with a role for buffer and water molecules in the maintenance of non-covalent protein assemblies. The presence of buffer salts may stabilize the monomer-monomer interface while water molecules, occupying the central channel and surface of the TTR assembly, could be important for stabilization of the hydrophobic interactions at the dimer-dimer interface, essential for maintaining intact tetramer.

The Binding of Thyroxine

Analysis of of 3,5,3′,5′-tetraiodo-L-thyronine (thyroxine) binding to wild-type TTR by equilibrium dialysis has demonstrated binding of two molar

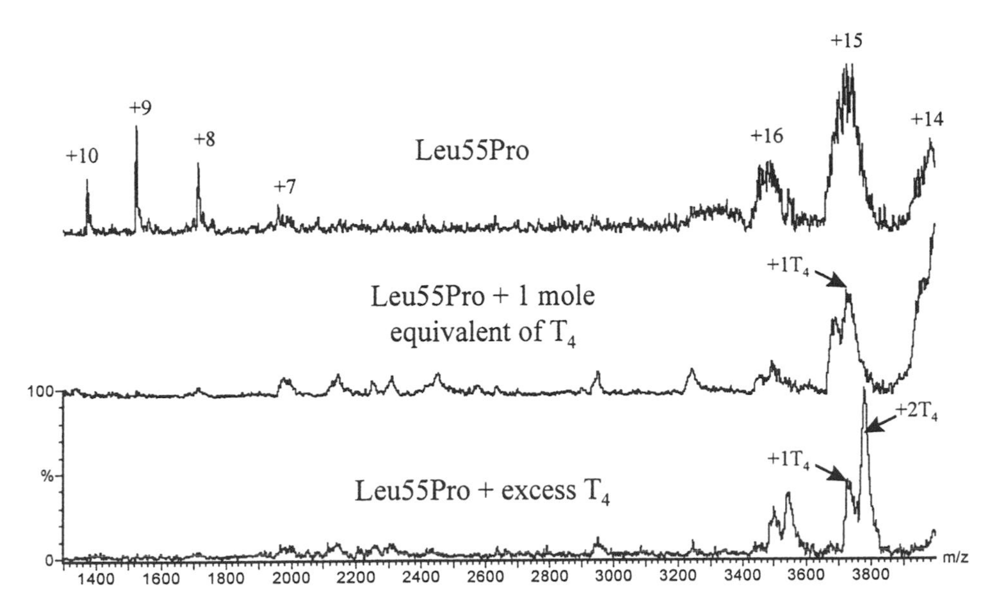

Fig. 7. The effects of thyroxine binding on the protein-protein interactions in Leu55Pro TTR followed by ESI-MS. Spectra were recorded for protein samples at pH 7.0 and 22°C, and identical MS conditions (cone voltage of 90V) were employed throughout. Reproduced with permission from [4].

equivalents of thyroxine under acidic conditions [17]. Interestingly, the binding of thyroxine has been shown to inhibit *in vitro* fibril formation by wild-type and variant TTRs. [17]. This inhibitory effect was attributed to increased stability of the wild-type TTR tetramer in the presence of thyroxine, revealed by sedimentation velocity analysis [17]. The variant proteins, unfortunately, could not be analyzed by this method due to their instability under the experimental conditions used. The effect of thyroxine binding on the protein-protein interactions in Leu55Pro TTR was investigated by recording ESI-MS spectra in the absence and presence of thyroxine molecules (Fig. 7). In the absence of thyroxine, two series of peaks are observed corresponding to monomeric and tetrameric TTR (Fig. 7, upper trace). In the presence of one mole equivalent of thyroxine per TTR tetramer, however, the monomeric species is not detected, and the peaks in the high m/z series appear as doublets corresponding to Leu55Pro TTR unliganded and bound to one thyroxine molecule (Fig. 7, middle trace). In the presence of a 40-fold excess of thyroxine, two high m/z series of charge states are observed corresponding in mass to binding of one and two molecules of thyroxine to the Leu55Pro TTR tetramer.

Further analysis of the masses of these two series shows that, in the presence of one thyroxine molecule, peaks in the charge state series remain broad and the masses are higher than predicted (Table 3). Binding of the second thyroxine molecule leads to a dramatic decrease in peak width and closer agreement with the mass calculated for two thyroxine molecules, one TTR tetramer and six water molecules (Table 3 and Fig. 7, lower trace). These

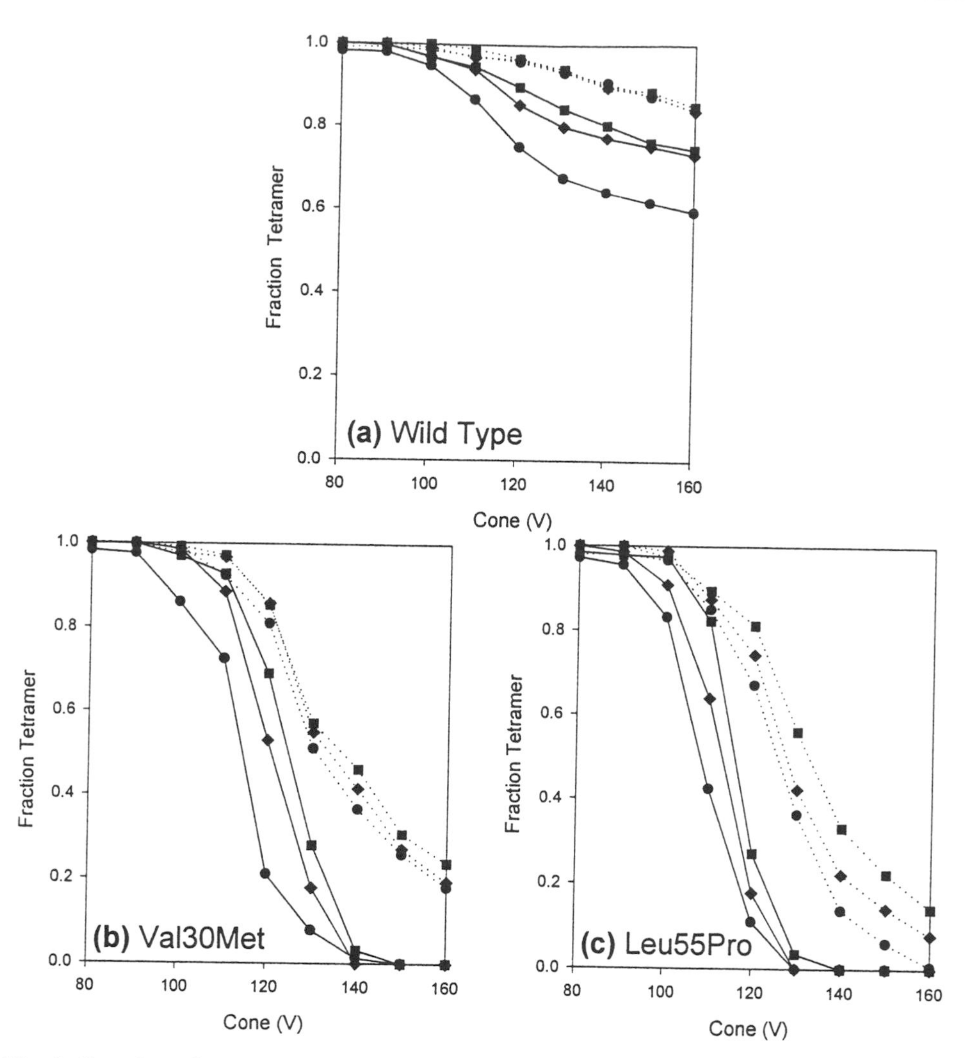

Fig. 8. Fraction of tetramer measured by ESI-MS at pH 7.0 and pH 4.0 for wild-type and variant TTRs with various amounts of T_4 present as a function of cone voltage. ●: no T_4; ■: one mole equivalent of T_4 per TTR tetramer; ◆: excess T_4; —: data at pH 4.0; ·····: data at pH 7.0. Reproduced with permission from [4].

observations could be explained by displacement of solvent molecules bound in the TTR channel by specific binding of thyroxine, providing further support for the location of the residual tightly bound water molecules at the monomer-monomer interfaces, which involve the F strands of the monomers.

To compare the effect of thyroxine binding on the stability of the TTR tetramers, the two variant proteins and wild-type TTR were analyzed at pH 4.0 and pH 7.0 under a series of different cone voltages (Fig. 8). The proportion of tetramer in the mass spectrum of the wild-type protein at pH 7.0 is not affected by thyroxine binding, presumably because the tetramer is stable under these

Moles T_4 per tetramer	Species observed	Measured mass (Da)	Calculated mass (Da)	Measured $w_{1/2}$ (Da/e) (charge state)
0	monomer	13 740 ± 3	13 745	5 ± 1 (+9)
0	tetramer	55 719 ± 13	54 982	73 ± 3 (+15)
1	tetramer + $1T_4$	55 979 ± 14	55 757	60 ± 3 (+15)
40	tetramer + $1T_4$	55 981 ± 5	55 757	20 ± 2 (+15)
40	tetramer + $2T_4$	56 657 ± 9	56 534	19 ± 3 (+15)

Table 3. Molecular masses and peak width at half height ($w_{1/2}$) of the Leu55Pro TTR monomer and tetramer species at pH 4.0 with 10mM NH4OAc and various amounts of thyroxine. Masses measured are those of the most abundant species present. The standard deviations of the measured masses arise from the number of charge states used to determine the mass of the species and do not reflect the broad nature of the peak.

conditions in the absence of thyroxine (Fig. 8a). For the amyloidogenic variants at pH 7.0, a small change in the proportion of tetramer in the mass spectrum was observed upon binding one molecule of thyroxine, and the amount of tetramer was seen to increase further upon binding the second thyroxine molecule (Figs. 8b and 8c). At pH 7.0, this increase in tetrameric species upon addition of thyroxine was most marked for the Leu55Pro variant.

The amyloidogenic variants at pH 4.0 show remarkable effects of thyroxine binding when analyzed by MS (Figs. 8b and 8c). For unliganded Val30Met at a cone voltage of 120V, the setting that calibrated MS data with those acquired using solution ultracentrifugation, only 20% tetramer is present, whereas in the presence of one and two molecules of thyroxine, 52% and 72% tetramer are observed, respectively. The effects of thyroxine binding on the proportion of tetramer in the ESI mass spectrum of Leu55Pro are also dramatic at pH 4.0, conditions under which this variant has proved difficult to study using solution techniques. Without thyroxine, only 10% tetramer is observed, while in the presence of one and two molecules of thyroxine, 17% and 26% of the tetramer are observed respectively.

This study clearly demonstrates increased tetramer stability in the presence of thyroxine, the effect being most dramatic upon occupancy of one thyroxine binding site although binding at the second thyroxine site leads to a further increase in tetramer stability. Furthermore, greater increases in tetramer stability are observed for the amyloidogenic variants upon thyroxine binding when compared with the wild-type protein. These findings are also in accord with the proposal that thyroxine-related ligands, which stabilize the tetramer of TTR, may act as therapeutic, anti-amyloid compounds [17].

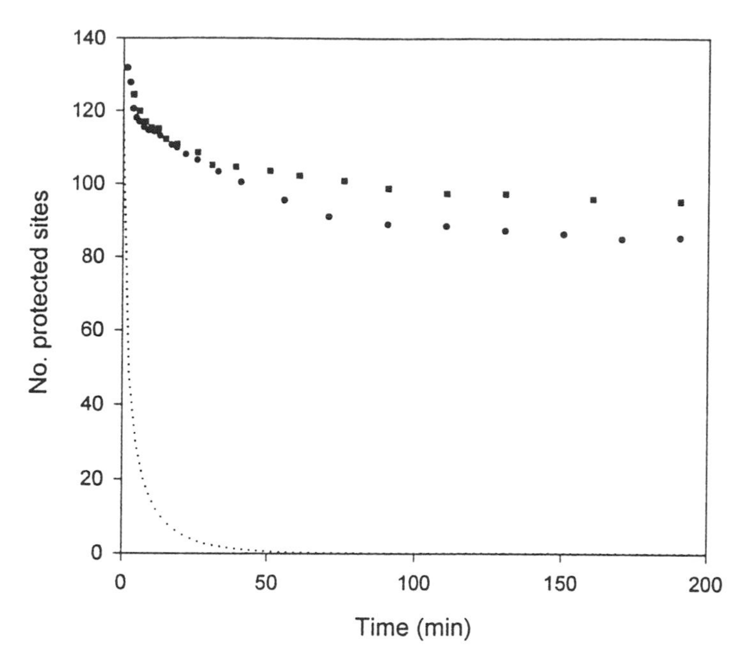

Fig. 9. H-X profiles obtained for wild-type TTR measured by MS at 0.018 mg / ml protein concentration, pH 4.0 both with and without 1 molar equivalent of thyroxine. ●: without thyroxine; ■: with 1 molar equivalent of thyroxine; ·····: data for the exchange profile of completely unstructured TTR at pH* 4.0, 22°C calculated from near-neighbor inductive effects on acid, base and water catalysis [22].*

H-X of Wild-type TTR

The H-X profiles obtained for wild-type TTR measured by MS at 0.18 mg/ml, the physiological protein concentration, pH* 4.0 both with and without 1 molar equivalent of thyroxine are shown in Fig. 9. In the absence of thyroxine, a core of 85 ± 2 protected sites remains at the end of the exchange, while introducing 1 molar equivalent of thyroxine results in an increase in protection, with 95 ± 2 sites remaining inaccessible to solvent deuterons.

Unstructured wild-type TTR has 210 sites that are labile to exchange, 118 from backbone amides, 89 from side-chains and three from the termini. The interpretation of H-X properties of proteins is complex but protection is known to arise, at least in part, from persistent hydrogen-bonded structure. From the crystallographic studies 44 hydrogen bonds are known to exist within the eight β-sheets and the helix in monomeric TTR [6]. Thus, the detection of 85 ± 2 protected sites in the wild-type protein at 0.18 mg/ml protein concentration, pH* 4.0 cannot be accounted for by persistent hydrogen-bonding within the monomer protein alone.

The data may be interpreted in the light of results obtained previously using analytical ultracentrifugation. [16]. These have demonstrated a concentration dependence for the monomer – tetramer equilibrium that exists for TTR in solution, and shown that wild-type TTR is extensively monomeric only at very low protein concentrations (below 0.09 μM). Thus, it seems likely that

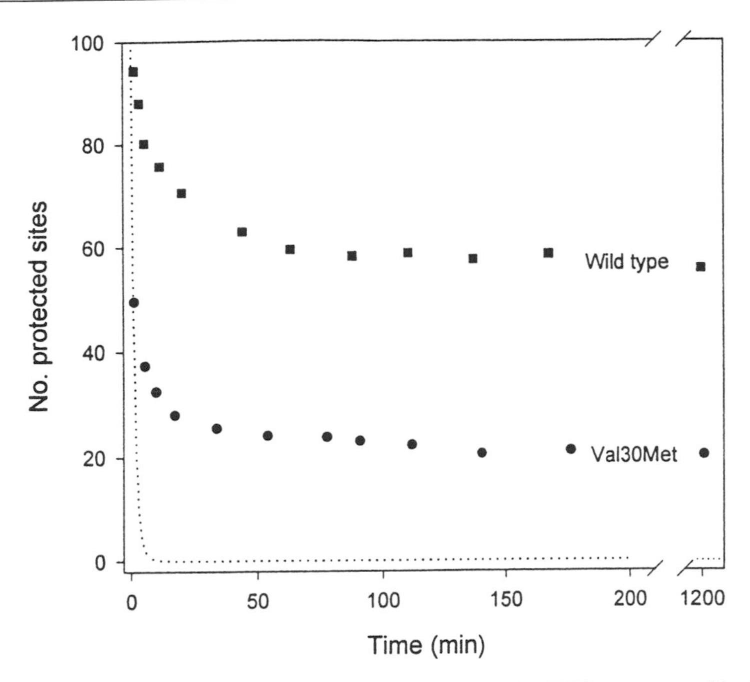

Figure 10. H-X profiles obtained for wild-type and Val30Met TTRs measured by MS at pH 4.5 and 37°C and at low protein concentrations (0.09 mM), where both are monomeric [13]. ●: Val30Met TTR; ■: Wild-type TTR; ····: data for the exchange profile of completely unstructured TTR at pH* 4.5, 37°C calculated from near-neighbor inductive effects on acid, base and water catalysis [22]. Reproduced with permission from [4].*

some H-X protection arises from interactions at the monomer – monomer and dimer – dimer interfaces in the tetramer. The presence of 1 molar equivalent of thyroxine in the solution of TTR undergoing H-X at physiological protein concentration, pH* 4.0, 22°C leads to an additional 7 ± 3 protons being protected. This presumably reflects added stability afforded the tetramer assembly upon the binding of one thyroxine molecule.

The insights gained by the H-X properties of wild type TTR at 0.18 mg/ml protein concentration, through the experiments detailed above, illustrate the difficulty in comparing the H-X kinetics of the variant and wild-type proteins. Clearly, the differences in stability of the wild-type and variant TTR tetramers complicates the H-X kinetics observed for these species at physiological protein concentrations.

H-X Wild-type and Val30Met TTR under Amyloid forming Conditions

The H-X properties of wild type and Val30Met TTR at pH* 4.5 and 37°C were examined at low concentrations of protein (0.09μM), where both are monomeric [16] (Fig. 10). These experimental conditions allow the conformational dynamics of the protein under amyloidogenic conditions to be assessed without the added complication of monomer – tetramer equilibria. Significant differences in the H-X properties are observed for the wild type and Val30Met TTR (Fig. 10). Of the 210 labile sites in wild-type TTR, an average of 55 ± 1 sites are protected from exchange with solvent after 20h at 37°C. This value is in reasonable agreement with the 44 protons shown to be hydrogen bonded within the eight β-sheets and the helix in monomeric TTR in the crystal structure [6]. The additional 11 ± 1 sites protected in the H-X experiment presumably arise from other hydrogen bonds involving side chains, or from other buried amides.

The extent of H-X protection within the Val30Met TTR monomer, however, is remarkably different with only 20 ± 1 sites remaining protected under identical solution and MS conditions. Using ESI-MS, it is not possible to assign to specific β-strands the loss of H-X protection without enzymatic digestion and assignment of the resulting fragments. The upper limits of the concentrations required to maintain both proteins in their monomeric form (0.09 μM) is prohibitively low for study directly by NMR spectroscopy, which could give site specific exchange protection data. However, the results presented here indicate that not all secondary structural elements in this subunit are stable. This suggests that the native fold of this variant may be in equilibrium with a partially-folded form which permits H-X of 24 ± 1 more backbone amides than predicted by the Val30Met crystal structure [18].

Computer modelling studies have indicated that such an intermediate could arise in the variant TTR from partial disruption of the β-sandwich structure, through disorder in the "edge strands" C and D [19]. Disorder of these strands, together with strand B that contains the Val30Met mutation, might be anticipated to lead to disruption of hydrogen bonding between strands E-B, C-B and D-A, G-A (Fig. 2). The results presented here are consistent with these modelling studies [20]. Stable hydrogen-bonded structure between the remaining β-strands G-H and E-F, together with the short region of helix between β-strands E and F, would give a total of 20 hydrogen-bonded amides. This number is in excellent agreement with the 20 ± 1 protected sites remaining in the Val30Met TTR after 20 hours at pH*4.5 at 37°C.

These H-X MS data have allowed a model for the amyloidogenic intermediate of Val30Met TTR to be proposed in which the native fold, observed by circular dichroism [16], is in equilibrium with a partially-folded form in which β-strands A, B, C and D are substantially disordered.

Conclusions

The present results are consistent with a monomeric TTR intermediate that undergoes transient unfolding in some β-sheet strands while retaining a stable β-sheet core which could provide a starting point for amyloid fibril formation in wild type TTR and its variants. The role of the partial folded amyloidogenic intermediate in converting normally soluble protein into the amyloid fibril is remarkably similar to the mechanism of fibril formation proposed for the immunoglobulin light chain [1]. Moreover, the two amyloidogenic lysozyme variants are thought to populate partially structured conformations, prone to self association [3]. Similarly, recent studies of a recombinant prion protein have indicated a relatively open conformation with a folded core [21]. The results from these studies not only demonstrate powerful strategies for studying protein interactions and conformational dynamics of amyloidogenic variants, but also reinforce the view that partially folded intermediates could underlie the conversion of normally soluble protein to amyloid fibril.

Acknowedgements

We thank Margaret Sunde and Christopher Dobson for helpful discussions and acknowledge with thanks collaborations with Mark Pepys and Jeffery Kelly. This is a contribution from the Oxford Centre for Molecular Sciences, which is funded by the BBSRC, EPSRC and MRC. EJN is grateful for financial support from Glaxo Wellcome and CVR thanks the Royal Society for support.

References

1. R. Wetzel, *Cell* **1996**, *86*, 699-702.

2. V. Bellotti, M. Stoppini, P. Mangione, M. Sunde, C. V. Robinson, L. Asti, D. Brancaccio and G. Ferri, *Eur. J. Biochem.*, in press.

3. D. Booth, M. Sunde, V. Bellotti, C. V. Robinson, W. L. Hutchinson, P. E. Fraser, P. N. Hawkins, C. M. Dobson, S. E. Radford, C. C. F. Blake and M. B. Pepys, *Nature* **1997**, *385*, 787-793.

4. E. Nettleton, M. Sunde, V. Lai, J. Kelly, C. Dobson and C. V. Robinson, *J. Mol. Biol.* **1998**, *281*, 553-564.

5. M. B. Pepys, P. N. Hawkins, D. R. Booth, D. M. Vigushin, G. A. Tennent, A. K. Soutar, N. Totty, O. Nguyen, C. C. F. Blake, C. J. Terry, T. G. Feest, A. M. Zalin and J. J. Hsuan, *Nature* **1993**, *362*, 553-557.

6. C. C. F. Blake, M. J. Geisow, S. J. Oatley, B. Rerat and C. Rerat, *J. Mol. Biol.* **1978**, *121*, 339-356.

7. A. Gustavsson, H. Jahr, R. Tobiassen, D. R. Jacobson, K. Sletten and P. Westermark, *Lab. Invest.* **1995**, *73*, 703-708.

8. M. Savaiva, *Human Mutat.* **1995**, *5*, 191-196.

9. M. D. Benson, *TIBS* **1989**, *12*, 88-92.

10. P. Westermark, K. Sletten, B. Johansson and G. G. Cornwell, *Proc. Natl. Acad. Sci. USA* **1990**, *87*, 2843-2845.

11. D. R. Jacobsen, D. E. McFarlin, I. Kane and J. N. Buxbaum, *Human Genet.* **1992**, *89*, 353-356.

12. J. A. Loo, C. G. Edmonds and R. D. Smith, *Anal. Chem.* **1993**, *65*, 425-438.

13. Z. Lai, J. McCulloch, A. Lahuel and J. W. Kelly, *Biochemistry* **1997**, *36*, 10230-10239.

14. S. L. McCutchen, W. Colon and J. W. Kelly, *Biochemistry* **1993**, *32*, 12119-12127.

15. W. Colon and J. W. Kelly, *Biochemistry* **1992**, *31*, 8654-8660.

16. Z. Lai, W. Colon and J. W. Kelly, *Biochemistry* **1996**, *35*, 6470-6482.

17. G. J. Miroy, Z. Lai, H. Lashuel, S. A. Peterson, C. Strang and J. W. Kelly, *Proc. Natl. Acad. Sci. USA* **1996**, *93*, 15051-15056.

18. J. A. Hamilton, L. K. Steinrauf, J. Liepnieks, M. D. Benson, G. Holmgren, O. Sandgren and L. Steen, *Biochim. Biophys. Acta.* **1992**, *1139*, 9-16.

19. J. W. Kelly and P. T. Lansbury, *Amyloid:Int .J. Exp. Clin. Invest.* **1994**, *1*, 186-205.

20. L. Serpell, G. Goldstein, I. Dacklin, E. Lundgren and C. C. Blake, *Amyloid: Int. J. Exp. Clin. Invest.* **1996**, *3*, 75-85.

21. H. Zhang, J. Stockel, I. Mehlhorn, D. Groth, M. A. Baldwin, S. B. Prusiner, T. L. James and F. E. Cohen, *Biochemistry* **1997**, *36*, 3543-3553.

22. Y. W. Bai, J. S. Milne, L. Mayne and S. W. Englander, *Proteins Struct. Funct. Genet.* **1993**, *17*, 75-86.

23. R. M. Esnouf, *J. Mol. Graphics Modelling* **1997**, *15*, 132.

24. E. Merritt and M. Murphy, *Acta Cryst.* **1994**, *D50*, 869-873.

QUESTIONS AND ANSWERS

J. E. Walker (MRC, Cambridge, UK)

Have you applied these techniques to more complex biological assemblies containing many different subunits, such as the mammalian proteasome or the ribosome?

Answer. Yes, we have preliminary spectra for the intact 30S and 50S subunits.

J. E. Walker (MRC, Cambridge, UK)

Were you able to interpret the spectra to obtain information about interactions between various subunits in the ribosome?

Answer. The spectra show that it is possible to maintain interactions in groups of proteins, particularly those in the stalk region which are known to form a stable complex in solution.

A. L. Burlingame (UCSF)

Is retinol binding protein acylated?

Answer. The chicken transthyretin retinol binding protein complex did not show any acetylation.

A. L. Burlingame (UCSF)

Have you studied any proteins that are N-terminally acylated? Would this modification dominate the noncovalent interaction?

Answer. I think it would be unlikely to affect the non-covalent interactions as the N-terminus is not involved in the binding.

Jasna Peter-Katalinic (University of Muenster, Germany)

You mentioned the process of disruption of non-covalent complexes by high energy collisions. Could you specify more precisely experimental conditions to induce these processes?

Answer. The work I have described was carried out on three different mass spectrometers. The transthyretin spectra were recorded on a quadrupole mass spectrometer and cone voltage was used to disrupt the complex. The transthyretin: retinol binding protein complex was obtained on an electrospray time-of-flight instrument and again was dissociated using increasing cone voltage. The Gro^{EL}14mer, however, was resistant to cone voltage, remaining intact at the highest cone voltages. We used high energy collisions in the collision cell of a Q-tof mass spectrometer to dissociate the complex.

X. Christopher Yu (Chiron Corporation)

Did you observe different amounts of solvent molecules and counter ions associated with protein complexes at different charge states?

Answer. Yes, there is clearly a difference in the amount of counter ions/ water molecules attached to different charge states. The lower charge states are generally much broader and have more small molecules attached than the higher charge states.

X. Christopher Yu (Chiron Corporation)

To what extent does the inability to measure molecular weights of large complexes impede the determination of the nature of these complexes?

Answer. The broad nature of the peaks will make the mass determination more difficult, but with the complexes I have described the masses of the individual subunits are sufficiently large such that the stoichiometry can be determined. However, looking at small ligands binding to large protein complexes is more difficult. It is difficult to discriminate between small molecules such as buffer salts or water molecules and potential ligands.

Kenneth Standing (University of Manitoba)

In the cases presented, the wild type was always more stable than the mutants. Do you think this might reflect the way the proteins have evolved over time?

Answer. It is interesting to speculate about this. In the cases I have described the wild-type is not associated with the same diseases as the variants. Whether this is an evolutionary process or the variants form as the result of a chance mutation is open to question.

Studying Noncovalent Small Molecule Interactions with Protein and RNA Targets by Mass Spectrometry

Joseph A. Loo, Venkataraman Thanabal, and Houng-Yau Mei
Chemistry Department, Parke-Davis Pharmaceutical Research, Division of Warner-Lambert Company, 2800 Plymouth Road, Ann Arbor, MI 48105

Developing methods for elucidating the mechanism of action of drugs and inhibitors on a structural level is an active area of biomedical and pharmaceutical research. To develop new therapies at an increasing pace, more new drugs need to be synthesized or discovered, more novel macromolecular targets need to be identified, and compounds need to be characterized, screened and evaluated at a faster rate. Advances and application of combinatorial chemistry address the need for more compounds [1, 2], whereas research endeavors such as genomics [3] and proteomics [4, 5] have shown great potential for uncovering novel targets. Methods for high-throughput screening using a variety of large-scale automated processes and robotics have been developed to further address some of these important needs [6]. However, biophysical techniques are still needed to fully elucidate the fundamental aspects of inhibitor action on its target.

Mass spectrometry (MS) [7] has played an increasingly important role in the overall drug discovery process, which includes the analytical characterization of candidate drug molecules, metabolite identification, and structural characterization of macromolecule targets. With the development of electrospray ionization (ESI) [8] and atmospheric pressure chemical ionization (APCI), commercial systems are available to allow non-expert analysts to operate the mass spectrometer and acquire data for their compounds. Medicinal and synthetic organic chemists successfully use these "walk-up" or "open access" mass spectrometry systems not only to analyze their intended final products, but also to characterize the intermediates of the reaction to investigate and optimize the reaction processes [9-11]. As more open access MS systems are being put into practice in the pharmaceutical lab, the throughput of the organic chemist increases.

In addition, methods using MS as a drug screening tool have been demonstrated for preclinical research (Fig. 1). Combinations of techniques such as size-exclusion chromatography, affinity binding [12-14], or ultrafiltration [15] and mass spectrometric detection are able to rapidly determine high-affinity ligands from complex mixtures of compounds from, for example, combinatorial chemistries (for a review, see [16]). In brief, a compound mixture is introduced to the macromolecular target and allowed to bind. The unbound compounds are either washed away or separated from the ligand-bound target. Subsequently, the bound ligands are released from the target, usually by a change in solution conditions, such as lowering the pH and/or increasing the organic solvent

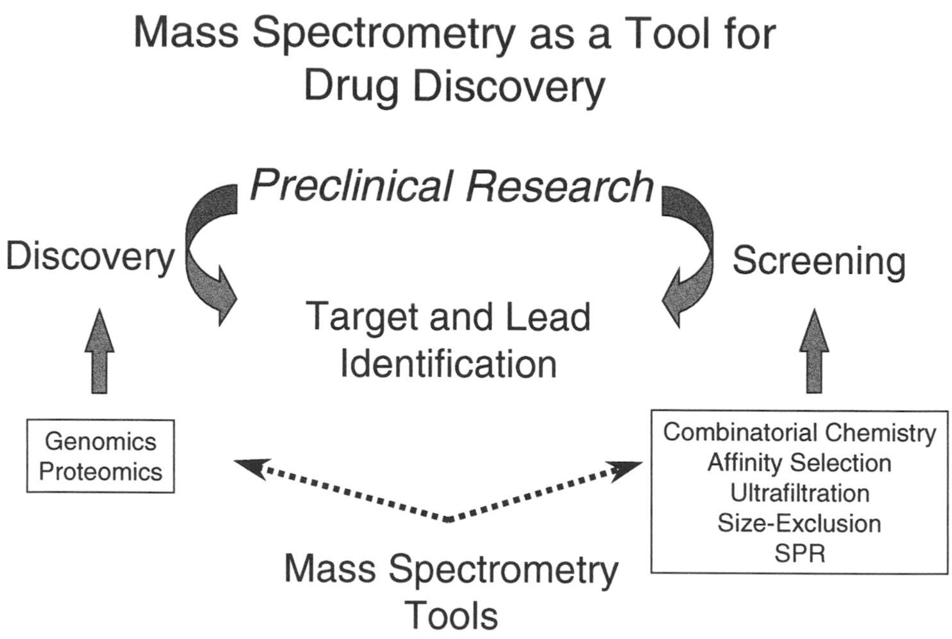

Mass Spectrometry as a Tool for Drug Discovery

Preclinical Research

Discovery

Screening

Target and Lead Identification

Genomics
Proteomics

Combinatorial Chemistry
Affinity Selection
Ultrafiltration
Size-Exclusion
SPR

Mass Spectrometry Tools

Fig. 1. Mass spectrometry plays an important role in the overall drug discovery process and in lead and target identification.

content to denature the complex [17, 18]. Those small molecular weight molecules that have a propensity to bind to the target are then measured and characterized by mass spectrometry. These methodologies offer the potential for high-throughput screening of large, complex mixtures and can be highly automatable.

In parallel to these applications, electrospray ionization with mass spectrometry has also developed as a biophysical tool for *directly* probing and detecting the molecular interactions between ligand and target, and monitoring noncovalently-bound complexes [19]. The gentleness of the ESI desorption/ionization process has been exploited in numerous studies of protein-protein, protein-oligonucleotide, and a variety of other molecular interactions. The gas phase-based mass measurement provides a direct read-out of the stoichiometry of the solution phase binding partners, as the molecular weights of each partner or ligand and the intact complex are measured.

Other methods such as nuclear magnetic resonance (NMR) and x-ray crystallography are superior for determining the precise three-dimensional structure of macromolecules and their complexes. Ultracentrifugation and surface plasmon resonance (SPR) can be applied for the study of molecular interactions and the determination of binding thermodynamics and kinetics [19]. ESI-MS has an important niche as a tool for the determination of binding stoichiometry. The high sensitivity and rapid analysis of mass spectrometry is an attractive feature for the research laboratory. Moreover, the resolution of MS allows for measuring the stoichiometry of small molecular weight ligands, such as drug molecules and metal ions, binding to much larger molecular weight targets, difficult endeavors for ultracentrifugation and SPR.

We have applied ESI-MS for the study of small molecule interactions with target RNAs and proteins for a number of drug discovery programs [20-22]. In some examples, information on the site(s) of interactions of these compounds on their targets can be confirmed. Often, this application of ESI-MS can provide information more easily than other biophysical methods. Moreover, the types of interactions that govern small molecule binding to biomolecule targets can be distinguished, as electrostatic forces are greatly strengthened in a solventless environment (and thus such complexes are extremely stable in the gas phase), whereas hydrophobic interactions are weakened in the gas phase. How these types of directed and focused studies can be used and fit into the overall drug discovery scheme will be discussed.

Experimental

Mass Spectrometry

ESI-MS was performed with a double focusing hybrid mass spectrometer (EBqQ geometry, Finnigan MAT 900Q, Bremen, Germany) with a mass-to-charge (m/z) range of 10,000 at 5 kV full acceleration potential. A position-and-time-resolved-ion-counting (PATRIC) scanning focal plane detector with an 8% m/z range of the m/z centered on the array detector was used. The ESI interface is based on a heated metal capillary inlet. The energetics of the gas phase collisions of the ions downstream from the heated metal capillary are controlled by adjustment of the voltage difference between the tube lens at the metallized exit of the glass capillary and the first skimmer element (DV_{TS}), and are used to augment the desolvation of the ESI-produced droplets and ions. A stream of sulfur hexafluoride coaxial to the spray suppressed corona discharges and is especially important for ESI of aqueous solutions. A low flow micro-ESI source was utilized, with control of the analyte flowrate (50-200 nL min^{-1}) accomplished

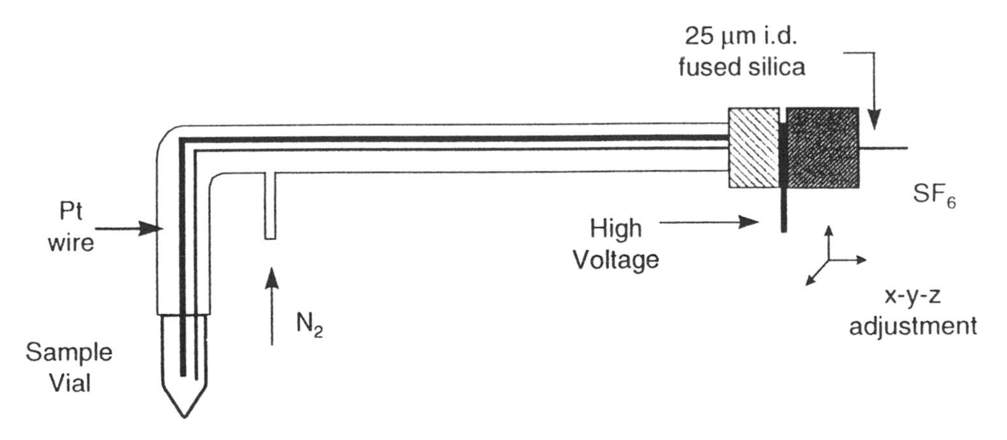

Fig. 2. Schematic of the low-flow micro-ESI source.

by adjustable pneumatic pressure. Approximately 10-20 μL of solution is loaded into small sample vials. The solution is pumped to the sprayer region by pressurizing the source with nitrogen gas (Fig. 2).

For most of the examples, protein or RNA samples (1-10 μM) were dissolved in 10 mM ammonium acetate (pH 6.9) or 10 mM ammonium bicarbonate (pH 7.5). The synthesis and purification of TAR RNA 31-mer have been previously reported [22, 23]. All TAR RNA solutions were ethanol precipitated three times to remove extraneous sodium salt. The addition of CDTA (1,2-diaminocyclohexane-*N,N,N',N'*-tetraacetic acid) to a concentration of 0.3 mM also helped reduce ubiquitous cations in solution for the RNA work. A 40 amino acid residue peptide (Tat peptide) derived from the C-terminal portion of Tat protein containing the basic regions important for TAR recognition was used for this study.

RESULTS AND DISCUSSION

The following examples from our laboratory are representative of the types of biochemical systems we have studied by ESI-MS. Noncovalent (reversible) drug binding to a protein or RNA target is a major interest in the pharmaceutical field. In some manner, the analytical laboratory represents a link between the chemist, the person who designed and/or synthesized the small molecule inhibitor, and the biochemist or molecular biologist, the scientist who selected the target. For all of the cases presented, the ESI-MS data confirmed the drug binding results obtained from other biophysical techniques.

HIV-1 Protease Dimer and Inhibitor Binding

The protease (PR) protein from the human immunodeficiency virus has been an attractive target to develop small molecule inhibitors. HIV PR is a 21.5 kDa homodimeric protein with a highly coordinated water molecule (water-301) in its binding pocket, as determined from the crystal structure [24]. Baca and Kent first reported the ESI mass spectrum of the noncovalent PR dimer bound with one molecule of synthetic inhibitor [25]. HIV protease is a member of the aspartyl protease family of proteins which include pepsin and renin. An inhibitor of aspartyl proteases and HIV protease is pepstatin A (isovaleryl-Val-Val-Sta-Ala-Sta, M_r 686) [26]; the ESI mass spectrum of the ternary PR dimer-pepstatin A complex from a pH 5.3 solution is shown in Fig. 3.

The stability of the gas phase PR dimer and dimer-inhibitor complexes is highly dependant on the solution pH and on the experimental variables affecting ion desolvation in the ESI atmosphere/vacuum interface. The solution phase stability of PR dimer maximizes at pH 5.5 and decreases with increasing pH [27]; the ESI spectrum from a pH 7.4 solution shows only ions for PR monomer, consistent with the solution phase characteristics (data not shown). The spectrum shown in Fig. 3 was acquired with a metal capillary temperature (T_{cap}) of 150°C and a DV_{TS} voltage difference of +10 V. Using a higher temperature capillary necessitates the use of lower DV_{TS} values to prevent dissociation of the noncovalent complex. The multiply charged ions exiting the metal capillary

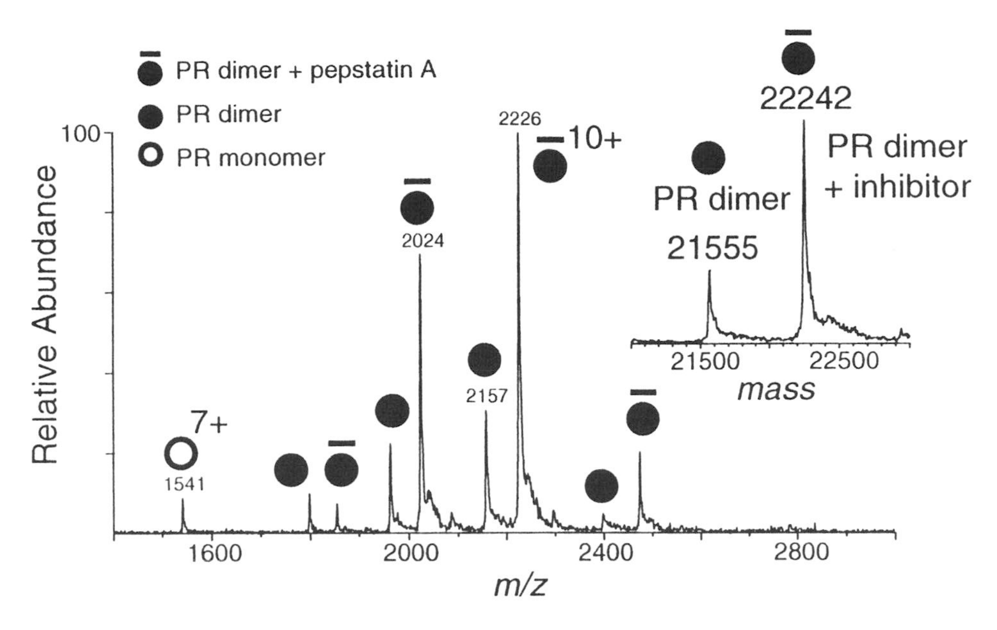

Fig. 3. Positive ESI mass spectrum of the HIV-1 protease dimer and protease dimer-pepstatin A complexes (10 mM ammonium acetate, pH 5.3, T_{cap} = 150°C, DV_{TS} = +10 V).

interface have variable amounts of bound solvent molecules, depending on the temperature (i. e., less solvation with higher temperatures). The effect of the energetic collisions on the solvated ions depends on the amount of solvation [19, 20]. Highly solvated ions (i. e., produced from lower capillary temperatures) require higher energy collisions to sufficiently desolvate the ions. However, less solvated ions exposed to the same high energy collisions undergo, not only desolvation, but also dissociation of the protein complex to the monomeric species. In effect, the solvent acts as a "buffer" or "cushion" against dissociation from collisional effects of the macromolecule complex.

This effect is illustrated in Fig. 4. At low T_{cap}, a higher DV_{TS} is needed to effectively desolvate the ions, although the peaks in Fig. 4 (top) are broad due to poorer desolvation efficiency. Higher DV_{TS} values result in dissociation of the gas phase complex. Too high a T_{cap} value (Fig. 4, bottom; T_{cap} = 210°C) results in dissociation of the complex, even at a DV_{TS} value of 0 V. There exists a fine balance of ESI interface tuning conditions for best results (especially T_{cap} and DV_{TS} in our ESI atmospheric pressure/vacuum interface). Unfortunately for the analyst, the optimum set of conditions for one system may be different from those measured for another biochemical system. For example, the gas phase streptavidin tetramer complex (13 kDa monomer, 52 kDa tetramer) [23] is highly stable at T_{cap} temperatures greater than 250°C with our interface. It is necessary to understand the effects of all the possible variables associated with each type of atmosphere/vacuum interface to achieve maximum performance. However, for the PR dimer example, the critical water-301 molecule was not "observed" by ESI-MS under any set of experimental conditions investigated. The lability of

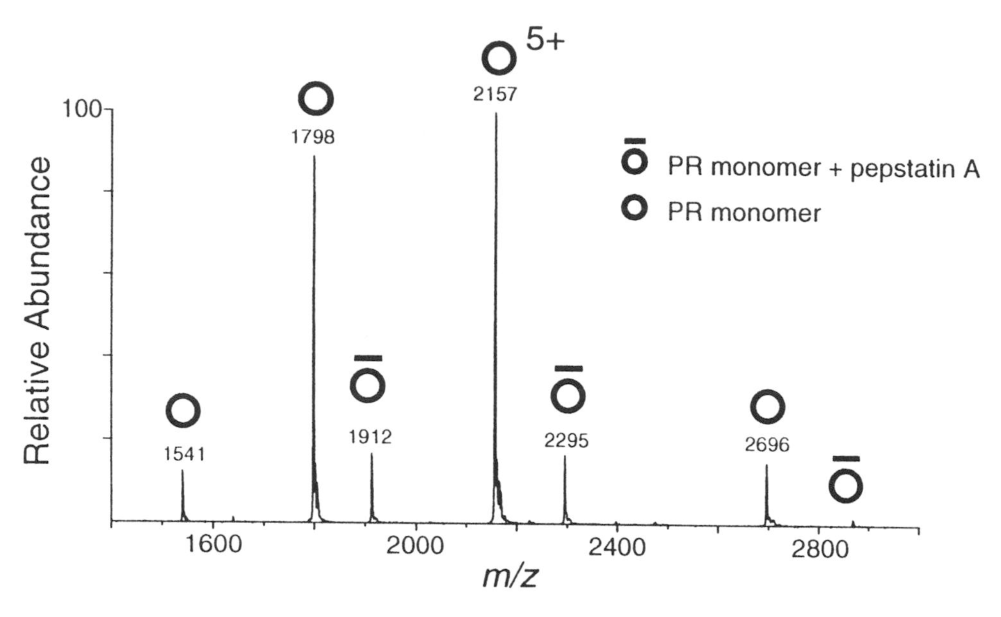

Fig. 4. ESI mass spectra of the HIV-1 protease-pepstatin A solution used for Fig. 3, but with different ESI interface conditions. The top spectrum was acquired with T_{cap} = 90°C and DV_{TS} = +65 V. The peaks are broader compared to the spectrum in Fig. 3, indicating more solvated ions. However, increasing DV_{TS} results in dissociation of the PR dimer and PR dimer-inhibitor complexes to the monomer species. The bottom spectrum was acquired with T_{cap} = 210°C and DV_{TS} = 0 V. More dissociation of the gas phase complexes is evident, even with a 0 V DV_{TS} difference. The symbols labeling the peaks are the same as those used in Fig. 3.

the interactions and/or short lifetime of this water molecule, which rapidly exchanges in solution, may preclude its observation by mass spectrometry.

The results from the dissociation of gas phase noncovalently-bound complexes may not represent the assembly of the complex in the solution phase. Increasing DV_{TS} from +10 to +100 V, with T_{cap} at 150°C, as for Fig. 3, results in total dissociation of the PR dimer ions to the monomeric state (Fig. 5). Although the dominant ions in the spectrum represent the 10.8 kDa protease monomer ions, a series of multiply charged ions consistent for the pepstatin A-bound PR monomer species is also observed. This monomer-inhibitor species is not known to exist in the solution phase and is most likely the result of the manner in which the gas phase complex is packed and assembled.

Metalloenzyme Stromelysin

Matrix metalloproteinases (MMPs) are a family of zinc-dependent enzymes involved in the selective degradation of various components of the extracellular matrix [28]. MMPs play a key role in diseases such as arthritis, heart failure, cancer, and atherosclerosis, and hence are important targets for therapeutic intervention. Numerous structural studies of the catalytic domains (CD's) of a variety of MMPs have been reported, in light of their identification as drug targets. In particular, the x-ray and NMR structures of the CD of

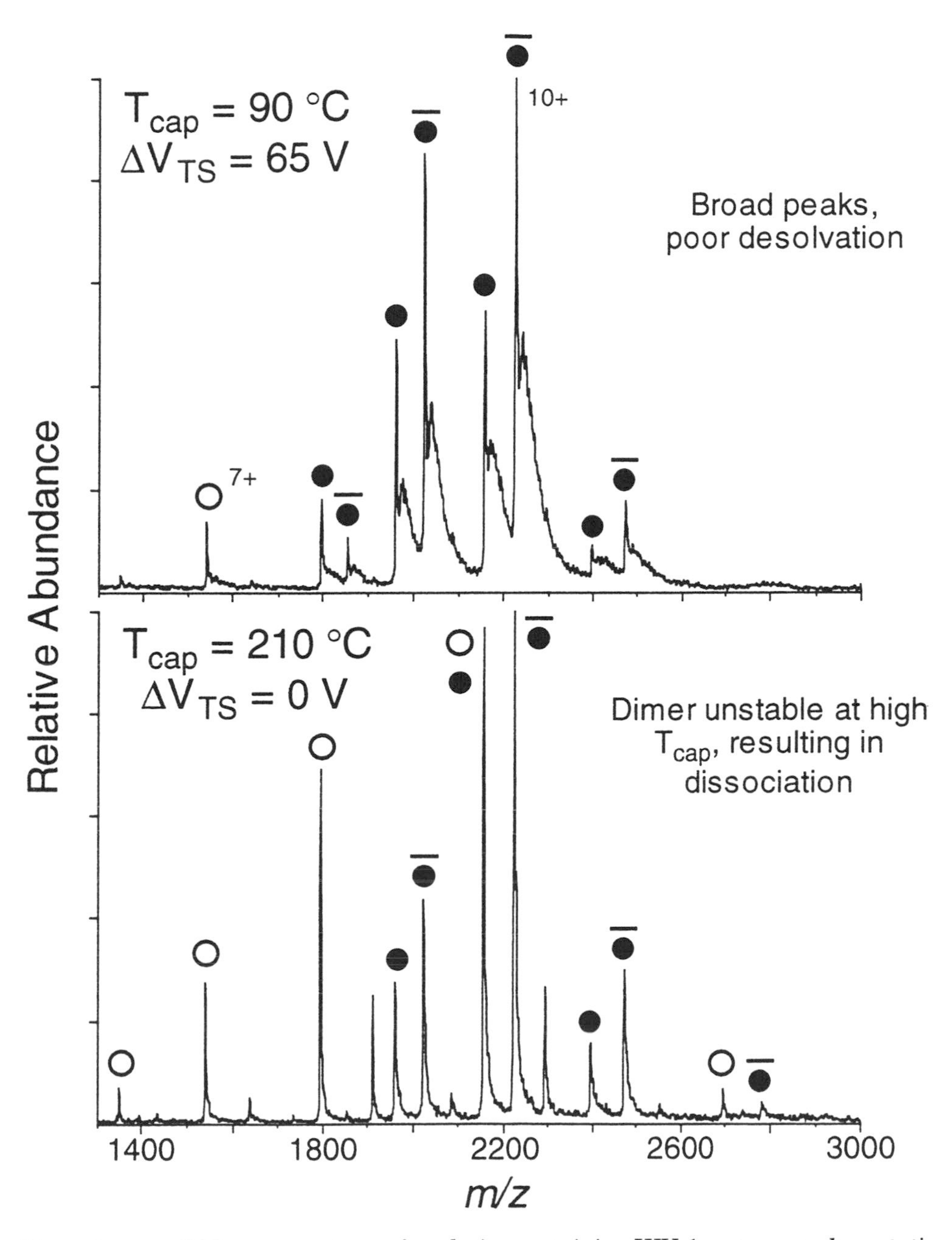

Fig. 5. Positive ESI mass spectrum of a solution containing HIV-1 protease and pepstatin A inhibitor (10 mM ammonium acetate, pH 5.3, T_{cap} = 150°C, DV_{TS} = +100 V). Higher DV_{TS} values (compared to those used for the spectrum in Figure 3) result in dissociation of the PR dimer complex to the monomer species.

stromelysin-1 (SCD, MMP-3) and bound inhibitors have been studied by a number of groups [29-32]. The catalytic domains of MMPs are 18-20 kDa proteins that typically bind 2 zinc ions and 2-3 calcium ions. The CD of

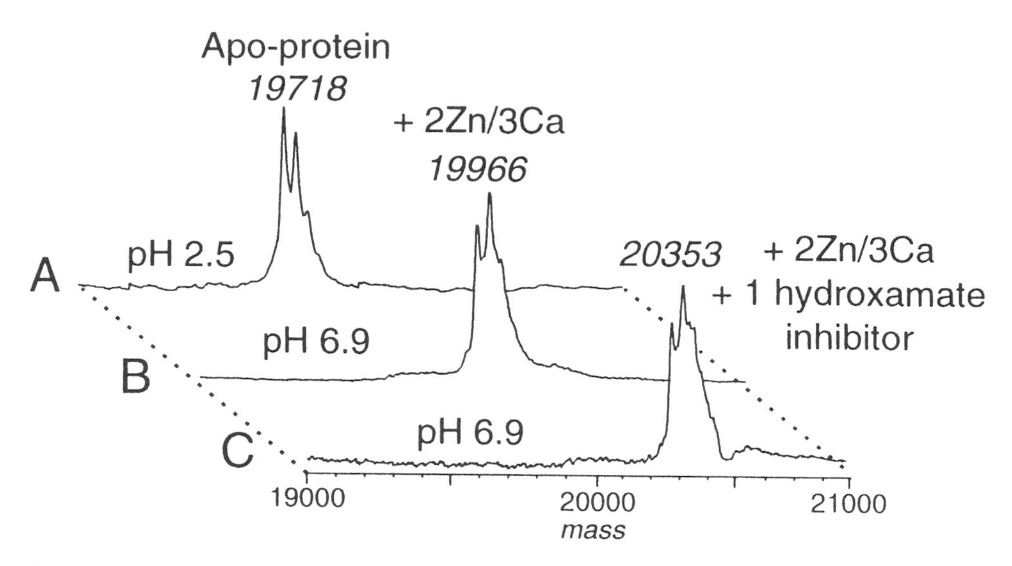

Fig. 6. ESI deconvoluted mass spectra (positive ion) of a solution containing MMP stromelysin-1 catalytic domain, zinc, calcium, and galardin-hydroxamate inhibitor. The solvent conditions were: (A) pH 2.5 with 2:1 acetonitrile / water and 5% (v / v) acetic acid and (B, C) aqueous 10 mM ammonium acetate. The mass spectra in (B) and (C) were acquired from the same solution, except that DV_{TS} was increased from (C) +40 V to (B) +90 V, causing dissociation of the inhibitor molecule from the metalloenzyme complex.

stromelysin-1 has two independent zinc binding sites: a catalytic site and a structural site. Many of the identified inhibitors of SCD are hydroxamate-based molecules and coordinate to a bound-zinc ion.

The results of ESI-MS studies to determine the stoichiometry of metal- and inhibitor binding to SCD is shown in Fig. 6. The application of ESI-MS for determining metal-binding stoichiometry of SCD was reported previously by our lab [33]. In addition, Feng *et al.* reported an ESI-MS study of another MMP catalytic domain, matrilysin, and its binding to zinc and calcium metal ions and inhibitors [34]. As are most other MMPs, SCD is highly unstable in solution; the enzyme rapidly undergoes autolytic cleavage. However, the enzyme is extremely stable in the presence of inhibitor. Figure 6 shows the deconvoluted ESI mass spectra of ^{13}C-, ^{15}N-labeled (for NMR structural studies) SCD in the presence of a hydroxamate inhibitor, galardin (GM 6001 or Ilomastat, M_r 388, IC_{50} ~ 3-10 nM) [35]. A solution of galardin inhibitor was added in excess to a solution containing SCD, zinc, and calcium during the purification process of the recombinant protein. Excess inhibitor and salts were removed by centrifugal ultrafiltration (10,000 molecular weight cutoff membranes, Microcon-10 microconcentrators, Amicon, Beverly, MA) and the sample was washed with 10 mM ammonium acetate (pH 6.9) solution. Any specific binding of inhibitors and metal ions should still be retained under these conditions. A portion of the washed solution was diluted with acetonitrile/water/acetic acid (pH 2.5) solution, and the spectrum shows only ions for the apo-protein form (Fig. 6A), as the noncovalent complexes are denatured under low pH and high organic solvent

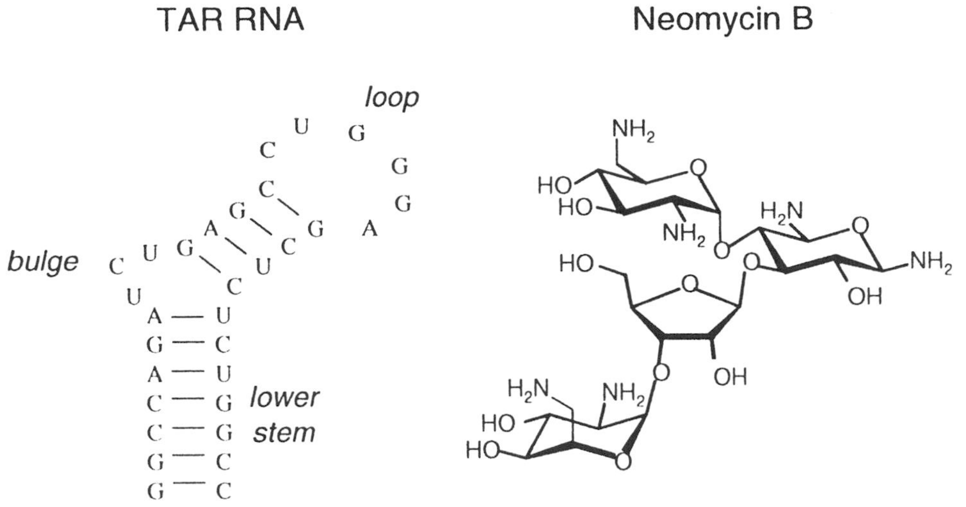

Fig. 7. Structures of TAR RNA (31-mer, M_r 9941) and neomycin B (M_r 615).

conditions. The ESI mass spectrum of the pH 6.9 solution reveals the holoprotein species with binding of 2 zinc ions, a distribution of 2 and 3 calcium ions, and only one molecule of inhibitor (Fig. 6C). Similar methods and conditions are being used in our lab to study other inhibitors and their binding to SCD and other matrix metalloproteinases.

Under gentle ESI interface conditions, inhibitor binding is observed. However, increasing the energy of collisions in the atmosphere/vacuum interface (by increasing DV_{TS} from +40 V to +90 V) results in dissociation of the inhibitor from the gas phase protein complex (Fig. 6B). Interestingly, dissociation of the metal ions is not observed under these energetic conditions, as the gas phase interaction is stronger for the metal ion(s)-protein complex than the inhibitor-protein species.

HIV Tat Peptide-TAR RNA Complex

Protein-RNA recognition is important for many biochemical systems. HIV gene expression is controlled by the binding of viral regulatory proteins to specific RNA target sequences. Tat protein is required to increase the rate of transcription from the HIV long terminal repeat (LTR) and its action is dependant on the region near the start of transcription in the viral LTR called the trans-activation responsive (TAR) element. A 3 nucleotide bulge region on TAR RNA represents a critical structural feature for Tat protein recognition (Fig. 7).

Our laboratory has identified a number of small molecule inhibitors that target the RNA structure of the Tat-TAR complex [36]. Compounds that target the RNA bulge region can effectively inhibit Tat-TAR complexation. However, small molecular weight compounds such as aminoglycoside antibiotics (e. g., neomycin B) which bind to regions removed from the bulge structure can also disrupt RNA-protein interactions [36-38]. TAR RNA undergoes a conformational change induced by small molecule interactions. The sites of these interac-

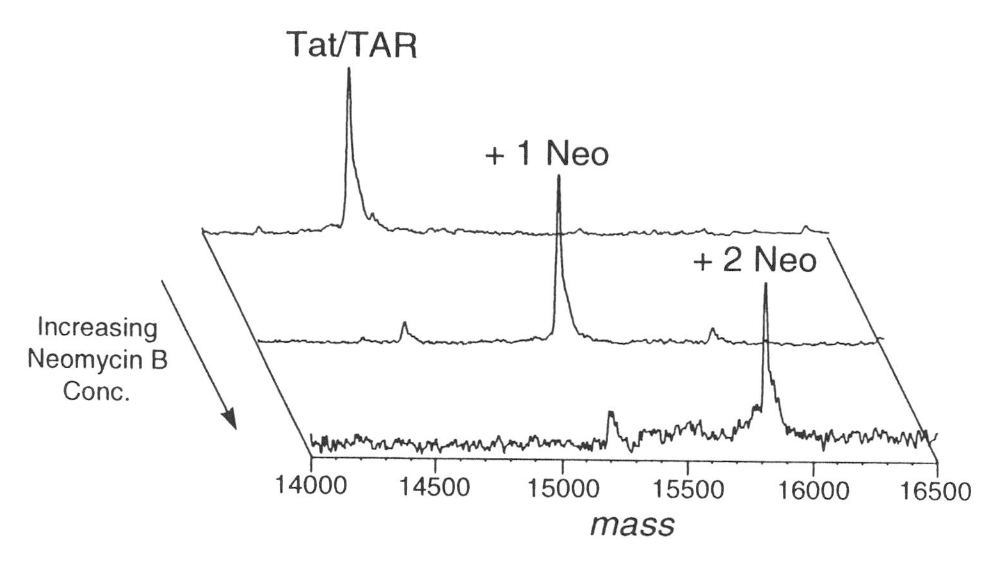

Fig. 8. ESI deconvoluted mass spectra (negative ion) of neomycin B binding to the Tat peptide / TAR RNA complex (Tat peptide, 40 amino acids, M_r 4644; Tat / TAR complex, M_r 14585) in 10 mM ammonium acetate, pH 6.9, and 0.3 mM CDTA. A maximum stoichiometry of 2 neomycin B molecules bind to Tat / TAR.

tions can be revealed from competition binding experiments with various TAR mutants, and the MS results confirm chemical and enzymatic footprinting studies [39].

By ESI-MS, a maximum of 3 neomycin B molecules was found to bind to TAR RNA, whereas 2 molecules bind to the Tat peptide/TAR complex (Fig. 8). This is consistent with a gel shift assay of the titration of neomycin to TAR and Tat/TAR [36-39]. However, the exact stoichiometries are difficult to interpret from the electrophoretic mobilities in a gel shift assay. The mass spectrum provides unambiguous stoichiometry measurements. From footprinting experiments, neomycin likely binds to the lower stem with high affinity, and the loop and bulge region of TAR with lower affinity (Fig. 7) [39], accounting for a total of 3 sites. Tat peptide and the "third" neomycin B molecule compete for the bulge region of TAR. With excess neomycin B in solution, Tat peptide is displaced and the peptide/RNA complex dissociates.

A class of compounds identified from screening our chemical compound library as TAR inhibitors, represented by the molecule 2,4,5,6-tetraaminoquinazoline (drug Y) [39], does not share the same binding site(s) with neomycin B, as the ternary TAR-Y-neomycin complex can be formed (Fig. 9). A maximum of only one molecule of drug Y binds to TAR. From the footprinting experiments, drug Y occupies the loop region of TAR, and does not compete for the same position in the loop region as neomycin B. A TAR-Y-neomycin$_3$ complex can also be formed and observed by ESI-MS (data not shown). For both the neomycin B and quinazoline cases, in experiments with only Tat peptide and TAR inhibitor in solution, no interactions with Tat peptide were observed.

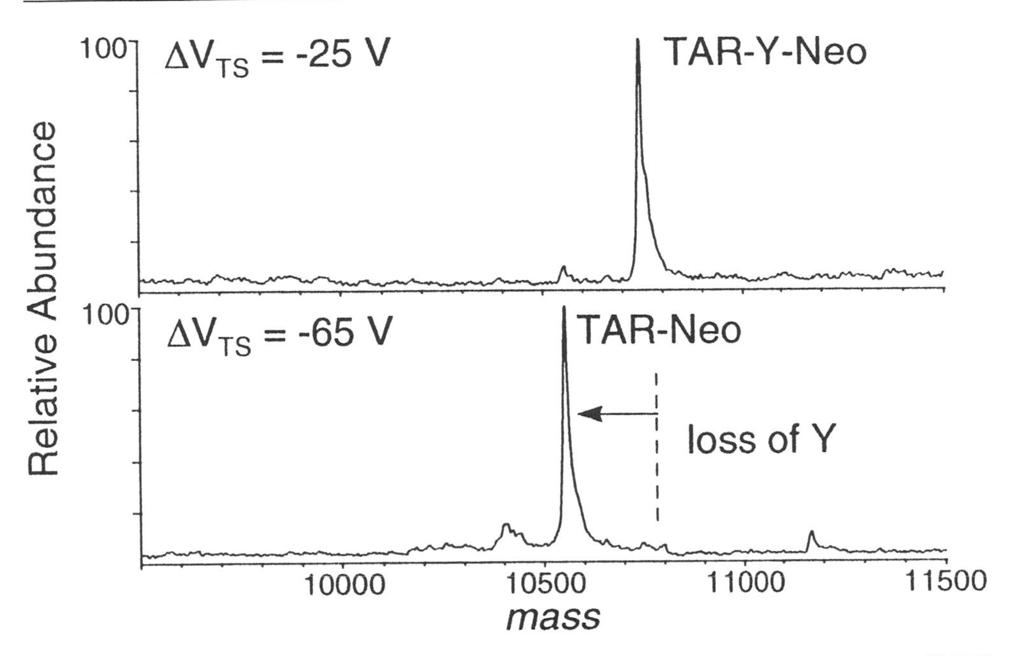

Fig. 9. ESI deconvoluted mass spectra (negative ion, 10 mM ammonium acetate, pH 6.9) of in-source CAD of the 1:1:1 TAR-drug Y-neomycin B complex. Drug Y is selectively dissociated from the gas phase complex, as neomycin B binds to TAR RNA through electrostatic interactions.

The types of interactions that govern small molecule binding to RNA and protein targets can be distinguished, as electrostatic forces are greatly strengthened in a solventless environment (and thus such complexes are extremely stable in the gas phase), whereas hydrophobic interactions are weakened in the gas phase. Electrostatic interactions in the gas phase are increased owing to the low dielectric of the surrounding vacuum media. The complex between TAR and drug Y is relatively unstable in the gas phase compared to TAR-neomycin B. Aminoglycosides are known to bind to RNAs through charge-charge interactions [40]. In-source CAD (high DV_{TS}) dissociates only drug Y from the ternary TAR-Y-neomycin complex (Fig. 9). Similarly, increasing the organic solvent content reduces the relative proportion of TAR-Y complexation, whereas the TAR-neomycin B complex remains unchanged.

CONCLUSIONS

The application of ESI-MS to these types of directed and focused studies can play an important role in the drug discovery process. ESI-MS of noncovalently-bound complexes can often reveal molecular details more readily than other biophysical methods. In many reported case studies, the interactions found in solution are mirrored by the results observed by the gas phase MS method [19]. This is an important point if the methodology is to be used to study previously unknown biochemical systems. As the ESI-MS method was developing, mostly well-studied systems, such as myoglobin-heme, DNA duplex, and quaternary protein complexes (e. g., streptavidin, hemoglobin) were used as test examples.

Encouragingly, more examples of unknown systems are being studied by ESI-MS.

However, as demonstrated by the examples discussed, the types of physical forces governing the ligand-target interaction can influence the results of the MS assay, as the stability of the gas phase noncovalent complexes may be different. Obviously, the results need to be carefully interpreted. For example, a mixture of potential ligands may contain those that bind predominantly through electrostatic forces and those where hydrophobic interactions dominate, depending on the functional groups and structure. Exposing this mixture to a macromolecule target and attempting to directly determine the higher affinity ligands through a competition binding ESI-MS experiment [21, 41] (e. g., ESI-MS of the noncovalent complexes formed between an enzyme target and a compound mixture synthesized by combinatorial chemistry) may be difficult. The mass spectra may reveal selectively those that bind through electrostatic interactions, even though the hydrophobic binding ligands may have higher affinities in solution [42]. In this case, the gas phase measurement may be misleading.

How does the study of noncovalent complexes and drug binding by mass spectrometry fit into a drug discovery program? Clearly, a complete three-dimensional structure from either NMR or x-ray crystallography techniques is useful, especially if coupled with rational drug design approaches. However, not every macromolecular target crystallizes well. Even obtaining a crystal of the apo-form does not guarantee that co-crystallization of protein and drug will be successful. NMR requires large quantities of protein and precipitation may occur at the concentrations needed for structure determination. The shorter analysis time and the small amount of sample required for the ESI-MS measurement should provide enough incentive and justification to use the method for determining drug binding stoichiometry. Often the question from the medicinal chemist is as simple as determining whether the drug binds to the protein target in a covalent or noncovalent manner. Mass spectrometry should be considered as a useful tool in the drug discovery toolbox. It can not solve every problem (as one would not use a hammer to cut a block of wood), but for the pharmaceutical researcher, it can be used to obtain very valuable structural information.

Acknowledgments

The authors wish to thank Patrick McConnell and W. Tom Mueller for providing the protease and stromelysin protein samples and Kristin A. Sannes-Lowery for making many of the Tat/TAR/neomycin measurements.

References

1. *A Practical Guide to Combinatorial Chemistry*, A. W. Czarnik and S. H. DeWitt, Eds.; American Chemical Society: Washington, DC, 1997.

2. A. W. Czarnik, *Anal. Chem.* **1998**, *70*, 378A-386A.

3. J. Lillie, *Drug Dev. Res.* **1997**, *41*, 160-172.

4. A. Shevchenko, O. N. Jensen, A. V. Podtelejnikov, F. Sagliocco, M. Wilm, O. Vorm, P. Mortensen, A. Shevchenko, H. Boucherie and M. Mann, *Proc. Natl. Acad. Sci. USA* **1996**, *93*, 14440-14445.

5. P. Roepstorff, *Curr. Opin. Biotechnol.* **1997**, *8*, 6-13.

6. L. Silverman, R. Campbell and J. R. Broach, *Curr. Opin. Chem. Biol.* **1998**, *2*, 397-403.

7. J. A. Loo, *Bioconjugate Chem.* **1995**, *6*, 644-665.

8. J. B. Fenn, M. Mann, C. K. Meng, S. F. Wong and C. M. Whitehouse, *Science* **1989**, *246*, 64-71.

9. F. S. Pullen, G. L. Perkins, K. I. Burton, R. S. Ware, M. S. Teague and J. P. Kiplinger, *J. Am. Soc. Mass Spectrom.* **1995**, *6*, 394-399.

10. L. C. E. Taylor, R. L. Johnson and R. Raso, *J. Am. Soc. Mass Spectrom.* **1995**, *6*, 387-393.

11. R. C. Spreen and L. M. Schaffter, *Anal. Chem.* **1996**, *68*, 414A-419A.

12. S. Kaur, L. McGuire, D. Tang, G. Dollinger and V. Huebner, *J. Protein Chem.* **1997**, *16*, 505-511.

13. M. A. Kelly, H. Liang, I.-I. Sytwu, I. Vlattas, N. L. Lyons, B. R. Bowen and L. P. Wennogle, *Biochemistry* **1996**, *35*, 11747-11755.

14. Y. F. Hsieh, N. Gordon, F. Regnier, N. Afeyan, S. A. Martin and G. J. Vella, *Mol. Diversity* **1997**, *2*, 189-196.

15. R. B. van Breemen, C.-R. Huang, D. Nikolic, C. P. Woodbury, Y.-Z. Zhao and D. L. Venton, *Anal. Chem.* **1997**, *69*, 2159-2164.

16. J. A. Loo, *Eur. Mass Spectrom.* **1997**, *3*, 93-104.

17. R. Wieboldt, J. Zweigenbaum and J. Henion, *Anal. Chem.* **1997**, *69*, 1683-1691.

18. M. M. Siegel, K. Tabei, G. A. Bebernitz and E. Z. Baum, *J. Mass Spectrom.* **1998**, *33*, 264-273.

19. J. A. Loo, *Mass Spectrom. Rev.* **1997**, *16*, 1-23.

20. J. A. Loo, K. A. Sannes-Lowery, P. Hu, D. P. Mack and H.-Y. Mei, in *New Methods for the Study of Biomolecular Complexes*, W. Ens, K. G. Standing, and I. V. Chernushevich, Eds.; Kluwer: Dordrecht, Netherlands, 1998; pp 83-99.

21. J. A. Loo, P. Hu, P. McConnell, W. T. Mueller, T. K. Sawyer and V. Thanabal, *J. Am. Soc. Mass Spectrom.* **1997**, *8*, 234-243.

22. K. A. Sannes-Lowery, P. Hu, D. P. Mack, H.-Y. Mei and J. A. Loo, *Anal. Chem.* **1997**, *69*, 5130-5135.

23. J. A. Loo and K. A. Sannes-Lowery, in *Mass Spectrometry of Biological Materials (2nd Ed.)*, B. S. Larsen and C. N. McEwen, Eds.; Dekker: New York, 1998; pp 345-367.

24. A. Wlodawer, M. Miller, M. Jaskolski, B. K. Sathyanarayana, E. Baldwin, I. T. Weber, L. M. Selk, L. Clawson, J. Schneider and S. B. H. Kent, *Science* **1989**, *245*, 616-621.

25. M. Baca and S. B. H. Kent, *J. Am. Chem. Soc.* **1992**, *114*, 3992-3993.

26. P. M. D. Fitzgerald, B. M. McKeever, J. F. VanMiddlesworth, J. P. Springer, J. C. Heimbach, C.-T. Leu, W. K. Herber, R. A. F. Dixon and P. L. Darke, *J. Biol. Chem.* **1990**, *265*, 14209-14219.

27. P. L. Darke, S. P. Jordan, D. L. Hall, J. A. Zugay, J. A. Shafer and L. C. Kuo, *Biochemistry* **1994**, *33*, 98-105.

28. Q.-Z. Ye, L. L. Johnson, D. J. Hupe and V. Baragi, *Biochemistry* **1992**, *31*, 11231-11235.

29. P. R. Gooley, J. F. O'Connell, A. I. Marcy, G. C. Cuca, S. P. Salowe, B. L. Bush, J. D. Hermes, C. K. Esser, W. K. Hagmann, J. P. Springer and B. A. Johnson, *Nature Struct. Biol.* **1994**, *1*, 111-118.

30. J. W. Becker, A. I. Marcy, L. L. Rokosz, M. G. Axel, J. J. Burbaum, P. M. D. Fitzgerald, P. M. Cameron, C. K. Esser, W. K. Hagmann, J. D. Hermes and J. P. Springer, *Protein Sci.* **1995**, *4*, 1966-1976.

31. S. R. Van Doren, A. V. Kurochkin, W. Hu, Q.-Z. Ye, L. L. Johnson, D. J. Hupe and E. R. P. Zuiderweg, *Protein Sci.* **1995**, *4*, 2487-2498.

32. V. Dhanaraj, Q.-Z. Ye, L. L. Johnson, D. J. Hupe, D. F. Ortwine, J. B. Dunbar, Jr., J. R. Rubin, A. Pavlovsky, C. Humblet and T. L. Blundell, *Structure (London)* **1996**, *4*, 375-386.

33. P. Hu, Q.-Z. Ye and J. A. Loo, *Anal. Chem.* **1994**, *66*, 4190-4194.

34. R. Feng, A. L. Castelhano, R. Billedeau and Z. Yuan, *J. Am. Soc. Mass Spectrom.* **1995**, *6*, 1105-1111.

35. D. Grobelny, L. Poncz and R. E. Galardy, *Biochemistry* **1992**, *31*, 7152-7154.

36. H.-Y. Mei, D. P. Mack, A. A. Galan, N. S. Halim, A. Heldsinger, J. A. Loo, D. W. Moreland, K. A. Sannes-Lowery, L. Sharmeen, H. N. Truong and A. W. Czarnik, *Bioorg. Med. Chem.* **1997**, *5*, 1173-1184.

37. H.-Y. Mei, A. A. Galan, N. S. Halim, D. P. Mack, D. W. Moreland, K. B. Sanders, H. N. Truong and A. W. Czarnik, *Bioorg. Med. Chem. Lett.* **1995**, *5*, 2755-2760.

38. S. Wang, P. W. Huber, M. Cui, A. W. Czarnik and H.-Y. Mei, *Biochemistry* **1998**, *37*, 5549-5557.

39. H.-Y. Mei, M. Cui, A. Heldsinger, S. M. Lemrow, J. A. Loo, K. A. Sannes-Lowery, L. Sharmeen and A. W. Czarnik, *Biochemistry*, in press.

40. H. Wang and Y. Tor, *J. Am. Chem. Soc.* **1997**, *119*, 8734-8735.

41. X. Cheng, R. D. Chen, J. E. Bruce, B. L. Schwartz, G. A. Anderson, S. A. Hofstadler, D. C. Gale, R. D. Smith, J. M. Gao, G. B. Sigal, M. Mammen and G. M. Whitesides, *J. Am. Chem. Soc.* **1995**, *117*, 8859-8860.

42. C. V. Robinson, E. W. Chung, B. B. Kragelund, J. Knudsen, R. T. Aplin, F. M. Poulsen and C. M. Dobson, *J. Am. Chem. Soc.* **1996**, *118*, 8646-8653.

Questions and Answers

Robert Anderegg (Glaxo Wellcome)

The data you show of the dissociation of ionic and hydrophobic interactions behaving differently is intuitively comfortable if you think about dissociating an ion pair in water *vs.* in the gas phase. It should serve as a warning about extrapolating gas-phase data to solutions.

My question is about the protease data. You showed a spectrum at low temperature that looked like a sharp spike followed by a broad lump formed by the non-specific water attachment. Was the mass of the sharp spike equal to the mass of protease dimer or protease dimer plus one water? Does that suggest that the binding of the "essential" water is no stronger than that of the non-specific waters?

Answer. The mass of the "sharp spike" peaks in the spectrum acquired at a low capillary temperature of 90°C is consistent for the protease dimer molecular weight. No experimental conditions were found to produce spectra showing the binding of the essential water-301 molecule of HIV protease. A large distribution of non-specific water molecules binding to protease dimer can be observed at low capillary temperatures and/or low ESI interface energies. You speculation may be correct. At least for the protease example, if the specific water-301 molecule is binding in the gas phase, its stability is no greater than the binding of non-specific water molecules.

Kenneth Standing (University of Manitoba)

Frequently we fail to see water in a complex, even when crystallography indicates that it is tightly bound (trp repressor-DNA complex in our case). Do you think this is a limitation of MS methods, or could the indicated presence of water be an artifact of crystallography?

Answer. Neither. I don't think the inability to observe specific water binding by ESI-MS is debilitating. It would certainly be wonderful if ESI-MS could reliably detect structurally significant waters. In fact, I believe Carol Robinson's group at Oxford has observed specific water binding to a different protein complex (but I don't know if it's been published yet). And at the past ASMS meeting in Orlando, FL, Catherine Fenselau's group presented a poster showing water molecules coordinated to the zinc metal ions of the insulin hexamer complex.

However, the measurement of binding stoichiometry of larger ligands is an important capability. Also, I would not want to speculate whether "bound" waters are an artifact of crystallography. I am not an expert in x-ray crystallography. The NMR folks tell me about "crystal patching effects" in a crystal structure that may explain subtle differences between x-ray and NMR structures. The bottom line is that each biophysical technique has its advantages, disadvantages, and mysteries.

Carol Robinson (Oxford University)

The observation of water molecules in non-covalent complexes is presumably related to the residence times. Do you know the residence time of the water molecule in HIV protease?

Answer. You may be correct. However, I've always viewed the electrospray mass spectrum as an instantaneous "snapshot" of the solution just prior to desorption. Regardless of the residence time or the on- or off-rates of binding, the mass spectrometry measurement should be able to observe some population of molecules binding to water. On the other hand, the "optimum" conditions for the ESI-MS experiment requires efficient ion desolution prior to ion detection. These gentle, yet still energetic conditions may dissociate the key water molecules, in addition to the bulk solvation sphere around the protein complex. To get back to your question, the water-301 molecule of HIV protease rapidly exchanges in solution. I believe the exchange rate is in the nanosecond range, but I would have to check my references to be sure.

Jonathan Weissman (UCSF)

How do you know complexes aren't formed during vaporization?

Answer. We can't be absolutely sure, but most of us who work in this area have done enough control experiments to convince ourselves that we are seeing specific binding from the solution phase as opposed to non-specific binding. For example, a few years ago we published a study of phosphopeptide inhibitors binding to src SH2 with a 1:1 stoichiometry. Binding of the unphosphorylated YEEIP peptide was not observed under the same conditions. If complexes are formed only during the vaporization or desorption step, then I would think that both PYEEIP and YEEIP should form a complex with the protein.

A. L. Burlingame (UCSF)

What solvent was used to study HIV protease dimer?

Answer. The protease samples were in 10 mM ammonium acetate and the pH of the solution was adjusted by the addition of either acetic acid or ammonium hydroxide.

Alan Marshall (Florida State University)

Steve Benner (University of Florida) exposes a receptor in solution to a combinatorial library of inhibitors, then washes off all but the tightly bound inhibitors, then knocks those off and electrosprays that mixture into a FT-ICR mass spectrometer. The relative abundances in the mass spectrum then directly reflect their relative binding strength in solution. Thus, one need not rely on gas-phase affinities only.

Answer. This affinity selection method also has the advantage of being automatable and should have high throughput. I am aware of efforts at Chiron Corporation and Novartis to use similar technologies to screen compound mixtures. However, there are other technologies available that make use of micro-chip arrays, picoliter-to-nanoliter sampling, and other non-MS detection methods. These ultra-high throughput screening schemes make extensive use of robotics and have the potential to screen over 100,000 compounds per day. I'm not sure the affinity-MS methods can approach this level of throughput. But the mass spectrometry-based methods, including the gas phase measurements I presented, may have an important role in more directed and focused research projects.

Steven Carr (SmithKline Beecham Pharmaceuticals)

Most of the drug targets in the pharmaceutical industry are not soluble receptors, but insoluble, multipass membrane proteins. This is a significant challenge to our community. Any thoughts?

Answer. You are absolutely correct. We've worked on several systems in which the protein target is insoluble under the conditions we need to use for ESI-MS. We need to develop new solvent conditions which are compatible with electrospray. Unfortunately, currently there are only a limited set of solution conditions to use when studying noncovalent conditions (e. g., ammonium acetate or ammonium bicarbonate buffers with only a small amount of organic solvent modifiers). Of course, for drug binding studies, keeping the drugs in solution is not an easy task. Many of the compounds we study are soluble in only DMSO or sometimes THF. It is difficult to spray solutions with concentration of DMSO without the addition of acid and/or methanol or acetonitrile, which would denature the complex.

Identification of Protein-Protein Interfaces by Amide Proton Exchange Coupled to MALDI-TOF Mass Spectrometry

Jeffrey G. Mandell, Arnold M. Falick[†] and Elizabeth A. Komives**
*Department of Chemistry and Biochemistry, University of California, San Diego, 9500 Gilman Dr. La Jolla, CA 92093-0601; [†]PerSeptive Biosystems 871 Dubuque Ave., South San Francisco, CA 94080

Abstract

Matrix assisted laser desorption/ ionization time-of-flight (MALDI/ TOF) mass spectrometry (MS) was used to determine amide proton exchange (H/D-X) rates. The method has broad application to the study of protein conformation and folding and to the study of protein-ligand interactions and requires no modifications of the instrument. Amide protons were allowed to exchange with deuterons in buffered D_2O, pD 7.25 at room temperature. Exchanged deuterons were "frozen" in the exchanged state by quenching at pH 2.5, 0°C and analyzed by MALDI-TOF MS using a matrix mixture consisting of 5 mg/ml α-cyano-4-hydroxycinnamic acid, acetonitrile, ethanol, and 0.1% TFA. To slow off-exchange of the amide deuterons, the matrix was chilled to 0°C and adjusted to pH 2.5, and the chilled MALDI target was rapidly dried. The unseparated peptic digest mixture was analyzed from a single spot to determine the deuterium content of all proteolytic fragments of the protein at the same time. The method was used to identify peptic fragments from protein complexes that retained deuterium under hydrogen exchange conditions due to decreased solvent accessibility in the protein-protein complex. The experimental scheme involved deuteration of protein amides, complex formation, off-exchange in H_2O, and detection of remaining deuterons by MALDI-TOF MS of the mixture of peptic fragments. Deuterons that did not off-exchange result in a higher mass of certain peptic fragments, and these identify the protein-protein interface. The method was used to identify the ATP and kinase inhibitor binding sites in the cyclic-AMP-dependent protein kinase. Three overlapping peptides identified the ATP-binding site, three overlapping peptides identified the glycine-rich loop, and two peptides identified the PKI(5-24) binding site. In another set of experiments, the surface of thrombin that interacts with a fragment of the anticoagulant protein, thrombomodulin [TMEGF(4-5)] was identified. The structure of this complex is not known, and five peptides from thrombin showed significantly decreased solvent accessibility in the complex. The peptides are consistent with mutagenesis data indicating that TMEGF(4-5) binds in the anion binding exosite, but also identify a new region that was not known to be involved in the thrombin-thrombomodulin interaction.

Introduction

Localization of protein sequences participating in protein-protein interactions is currently a time-consuming and difficult task, yet it is critical for advancing fields such as signal transduction and drug design. Thousands of proteins are being discovered by protein-protein interaction screens, and determination of which surface of each protein interacts with the other is frequently the next step in narrowing down the functionally important regions of these proteins. Amide hydrogens provide individualized probes along the entire protein sequence, and measurements of amide proton (H/D) exchange rates have been used to analyze protein conformational changes, protein folding, and protein-protein interactions [1-3]. NMR has been the method most used to study protein-protein interactions by amide H/D exchange since the first report by Paterson *et al.* in 1990 [4]. NMR studies of other protein-protein complexes have not yielded unambiguous identification of the interface because amide H/D exchange rate variations also resulted from conformational changes [5-11]. Re-examination of these data shows that the key to separating conformational changes from interface information is to consider only those amides with rapid exchange rates in the uncomplexed state. These amides are most likely to be solvent accessible and should give information about changes at the protein surface. When data reported by Mayne *et al.* [5] were re-tabulated according to the exchange rate of each amide in the uncomplexed state, amides with protection factors greater than ten and exchange rates in the uncomplexed state greater than five hr^{-1} were reliable indicators of the protein-protein interface. Amides with slower free exchange rates were not good indicators of the interface. Because NMR experiments require large amounts of protein and long data collection times, information on the rapidly exchanging surface amides is inaccessible by these methods. In contrast, mass spectrometry is an ideal method to acquire information about decreased solvent accessibility of rapidly exchanging amides [12].

We propose that the key to obtaining reliable localization of protein-protein interfaces is to consider only those amides with rapid exchange rates in the uncomplexed state because these amides should always be near the protein surface. An observed decrease in the exchange rate of a surface amide should only result from a surface accessibility change, and not from a conformational change. Experiments to measure solvent accessibility changes upon protein-protein complexation have been carried out on two different systems. The first was a complex of known structure, the catalytic subunit of murine cAMP-dependent protein kinase (PKA) complexed to the kinase inhibitor, PKI(5-24). A complex of unknown structure was also analyzed, human α-thrombin bound to an eighty-four amino acid fragment of human thrombomodulin [TM(4-5)]. The surfaces of PKA that interact with PKI(5-24) are consistent with the crystal structure of the complex. The results of the thrombin-TM(4-5) complex are consistent with the structure of the complex between thrombin and a 19 amino acid peptide from TM, and provide new information about where this important anticoagulant protein directly contacts thrombin.

Methods

Proteins and Reagents

PKA was prepared as described previously [13] and experiments were carried out in the presence of either ATP/Mg^{+2} or ATP/Mg^{+2} and a 2-fold molar excess of PKI(5-24). PKI(5-24), which has a K_d of 2 nM in the presence of 1 mM ATP and 5 mM Mg^{+2}, was prepared as described [14]. PKI(5-24) samples were dissolved with 6 μl D_2O containing 20 mM ATP, 100 mM $MgCl_2$, pD 7.4. Immobilized pepsin was obtained from Pierce Chemicals (Rockford, IL). Deuterium oxide (99.996%) was purchased from Cambridge Isotope Laboratories (Andover, MA). The matrix compound, α-cyano-4-hydroxycinnamic acid, was obtained from Aldrich Chemicals (Milwaukee, WI), and was re-crystallized once from ethanol. All other reagents were from Fisher Scientific.

Human α-thrombin (a generous gift of Dr. John Fenton) and was stored lyophilized at -20°C. Thrombin experiments were carried out with a 2.5-fold molar excess of TM(4-5). Upon addition of 6 μl D_2O to the thrombin sample (500 pmol), the buffer was 60 mM KH_2PO_4, 130 mM NaCl, pD 6.9, and the protein concentration was 80 μM. TM(4-5) was prepared as described [15], stored lyophilized at -20°C and dissolved by addition of 6 μl of 25 mM TRIS in D_20, pD 7.9.

MALDI MS

The matrix used in these experiments was 5 mg/ml α-cyano-4-hydroxycinnamic acid, in a solution containing 1:1:1 acetonitrile, ethanol, and 0.1 % TFA (pH 2.5). The resulting matrix was adjusted to pH 2.5 with 2% TFA, as determined by matching the color of the spot on pH paper to that of 0.1% TFA, pH 2.5. The matrix was incubated on ice and mixed by vortexing before use.

The MALDI targets were chilled on ice in a plastic case to prevent condensation of atmospheric H_2O prior to use. Frozen samples were quickly defrosted to 0°C. Small portions, typically 5 μl, were mixed with an equal volume of matrix solution, and 1 μl was spotted onto the chilled MALDI target. The target was immediately placed in a dessicator under a moderate vacuum such that the spots would dry in 1-2 minutes. Slow drying in moderate vacuum was found to improve sample analysis, presumably because of improved crystal growth. The chilled, dried plate was transferred as quickly as possible from the dessicator to the mass spectrometer.

Mass spectra were acquired on a PerSeptive Biosystems Voyager DE STR. Data were acquired at 2 GHz sampling rate, 250,000 data channels, with a 20,000 V accelerating voltage, 70% grid voltage, 0.01% guide wire voltage, and using delayed extraction with a 100 nsec pulse delay. Typically, 256 scans were averaged in approximately 3 minutes. All reported masses are monoisotopic MH^+ masses unless otherwise noted.

Identification of Protein-Protein Complex Interfaces

Interface localization was achieved by identification of those peptides in a single MALDI mass spectrum (256 scans) of a peptic digest of the protein of interest that retained excess deuterium after protein complexation. The amounts of each protein were adjusted for maximum signal and in accord with the K_d of the complex as follows: PKA (400 pmol) and PKI(5-24) (750 pmol) or thrombin (500 pmol) and TMEGF(4-5) (1200 pmol). A typical experiment required three samples of the protein of interest. One sample was deuterated for 10 min.; a second was deuterated for 10 min. and subsequently off-exchanged; a third was deuterated for 8 min. and complexed with ligand for 2 min. A short deuteration time was used so that only solvent accessible amides were labeled. Samples were off-exchanged for varying times by 1:10 dilution with H_2O, 25°C. The off-exchange reaction was quenched at 0°C to pH 2.5 by addition of 1% TFA. The complex was digested with a 1:1 molar ratio of immobilized pepsin for 10 min. at 0°C. A portion (10 µl of 132 µl total) was diluted 1:3 with cold 0.1% TFA pH 2.5, and 1 µl of this was mixed 1:1 with chilled matrix, spotted on a chilled MALDI target, and rapidly dried under vacuum. A single MALDI mass spectrum of the entire peptic digest mixture was immediately acquired. The centroid of the isotopic envelope of each PKA or thrombin peptide from the mass spectrum of the peptic digest mixture of the protein-protein complex was compared to the same peptide from control reactions in which only PKA or thrombin was present. The control reactions were off-exchanged for the same amounts of time as the complex. If the centroid of the peptide mass envelope had increased by more than 0.5 for all of the off-exchange times (1-10 min.), this peptide was taken to be at the interface in the complex.

Identification of Each Peptic Fragment Present in the Mass Spectrum of the Mixture

The analysis of each complex required the identification of the peptides produced when the protein of interest was cleaved with pepsin. This experiment was required because the pepsin cleavage sites can not be predicted in advance, although a particular protein will be cleaved reproducibly each time. Either PKA (400 pmol, 200 mM KCl, 40 mM KH_2PO_4, pH 6.85, at a protein concentration of 33 µM in 12 µl) or thrombin (500 pmol, 1 mM $CaCl_2$, 65 mM NaCl, 33 mM KH_2PO_4, pH 6.5, at a protein concentration of 40 µM in 12 µl) was diluted to 132 µl with 0.1% TFA, pH 2.5, and digested with a 1:1 molar ratio of immobilized pepsin for 10 min. at 0°C. Peptides were identified by a combination of sequence searching for accurate masses, post-source decay (PSD) sequencing and C-terminal ladder sequencing [13, 16].

Data Collection and Analysis.

The average number of deuterons present on each peptic peptide was determined by calculating the centroid of each isotopic peak cluster [13]. Briefly, spectra were calibrated according to known masses of two peptides internal to the mixture. After a two-point baseline correction had been applied, centroids were computed by integrating from the left-most edge of the monoisotopic peak to the right-most edge of the highest-mass observable isotopic peak in the envelope (26). A program that applies this calculation to data in the Galactic SPC

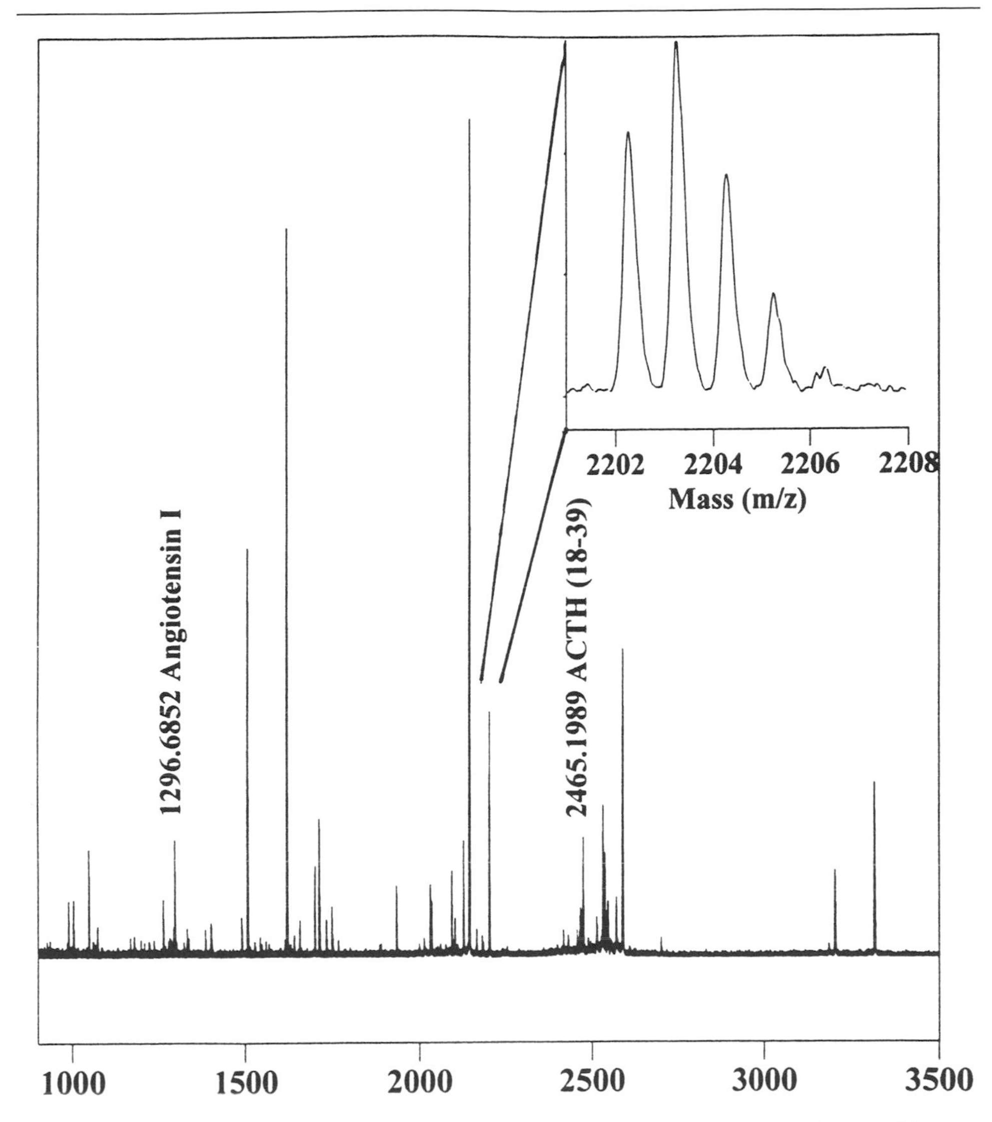

Fig. 1. MALDI mass spectrum of a peptic digest of thrombin. The inset shows a blow-up of the region surrounding the isotopic mass envelope for the peptide with mass 2202.2809 (MH^+, monoisotopic). All of the data required to completely identify the protein-protein interface can be obtained from four such spectra.

file format and that runs under Windows 95/NT (Microsoft, Redmond, WA) can be obtained from the authors. The GRASP [17] program was used to determine the solvent accessible surface area for the portion of either PKA or thrombin spanning a proteolytic fragment in the presence and absence of either PKI(5-24) or TMEGF(4-5). The change in solvent accessible surface area was defined as $SA_{unliganded} - SA_{liganded}$, where SA is the solvent accessible surface area. For the PKA-PKI(5-24) interaction, the structure with PDB ID 1ATP was used [14]. For the thrombin-thrombomodulin interaction, the structure with PDB ID 1HLT was used [18].

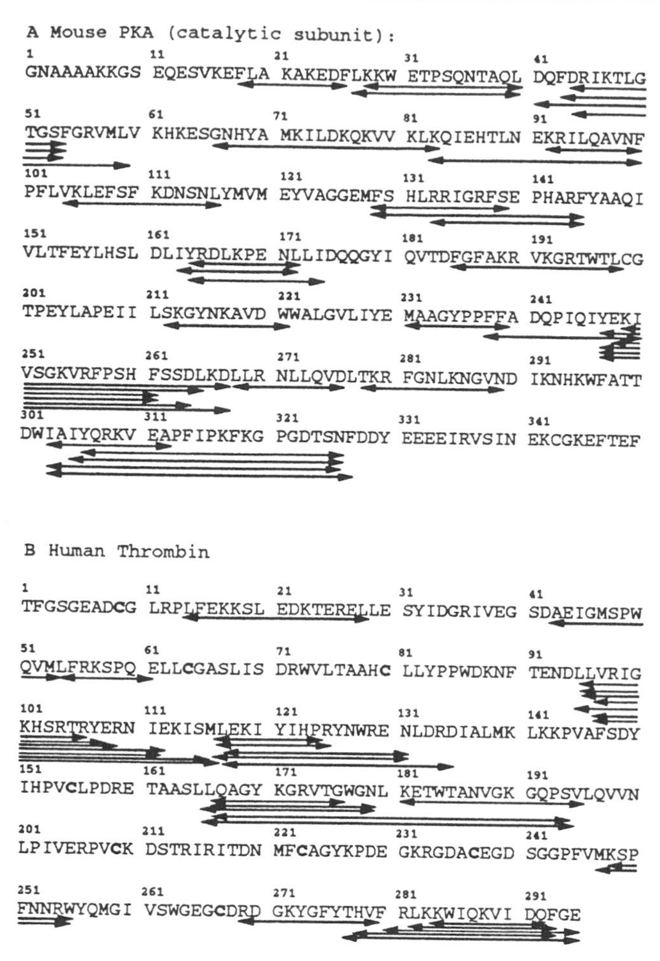

Fig. 2. (A) Sequence of PKA showing the 42 identified peptides that were observed in a single MALDI-TOF mass spectrum. The peptides cover 65% of the PKA sequence including two of the phosphorylation sites. (B) Sequence of human thrombin showing the 29 identified peptides that were observed in a single MALDI-TOF mass spectrum. The sequence is numbered sequentially, although in the text the chymotrypsin numbering system notation is given as well. The peptides cover 50% of the thrombin sequence. The cysteines that form the disulfide bonds between C9 and C155, between C64 and C80, between C209 and C223, and between C237 and C267 are shown in bold letters. The disulfide bonds were not reduced, and coverage of the sequence near the disulfide bonds appears to be inefficient.

Results

Measurement of Amide H/D Exchange by MALDI-TOF MS

We recently reported a method by which amide H/D exchange could be monitored by matrix assisted laser desorption ionization/time-of-flight (MALDI-TOF) MS [13]. All of the data for each protein analysis could be obtained from a single spot (256 scans) of the entire peptic digest of the protein (Fig. 1). For PKA, approximately 50 peptides were observed in the MALDI mass spectrum of 0.4

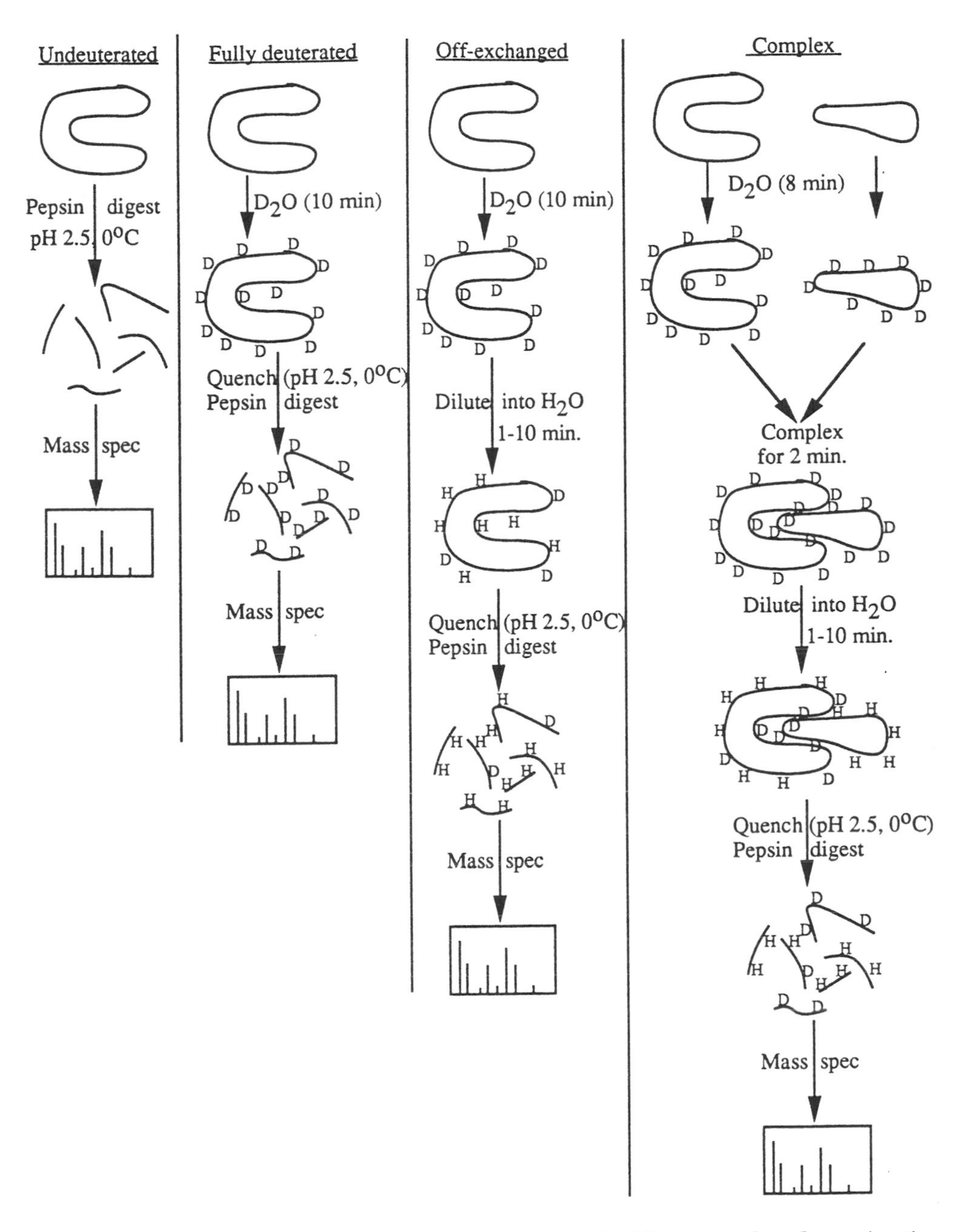

Fig. 3. Schematic diagram of the four experiments required for a complete determination of a protein-protein interface by MALDI-TOF MS. The "off-exchanged" control was important to determine the amount of deuteration remaining after 11-fold dilution. The peptides showing decreased solvent accessibility were identified from the "Complex" experiment as those that retained excess deuterium over the 9% that was present in all peptides after off-exchange.

pmol of the mixture of peptic fragments, and 42 of these were identified by a combination of sequence searching for accurate masses, post-source decay (PSD) sequencing and C-terminal ladder sequencing [13]. Thus, a single MALDI mass spectrum covered 65% of the PKA sequence and 80% of the solvent accessible protein surface (Fig. 2A). For thrombin, 29 peptides were identified out of the 40 in the spectrum, and these covered 50% of the sequence and 60% of the solvent accessible surface area (Fig. 2B). The lower level of coverage of the thrombin sequence is undoubtedly due to the fact that the four disulfide bonds were not reduced prior to pepsin cleavage resulting in fragments of higher masses.

The PKA-PKI(5-24) Interface

Amide H/D exchange rates were measured for the complex of the catalytic subunit of murine cAMP-dependent protein kinase (PKA) complexed to the 20 residue inhibitor PKI(5-24). The solvent accessibility of amide protons on PKA alone was first determined. The maximum deuteration for each peptide varied depending on its surface accessibility [13]. Analyses were then carried out on PKA complexed to ATP, and on PKA complexed to ATP and PKI(5-24) (Fig. 3). Figure 4 shows mass spectra expanded around the peptide of mass 1194.6463, one of the peptides from PKA that retained a significant number of deuterons in the PKI(5-24) complex compared to the controls, even after 10 min. of off-exchange. Much less deuteration was retained by this region of PKA in the presence of only Mg^{+2}/ATP. Of the 42 peptic fragments identified in the mass spectrum of PKA peptides, eight showed markedly reduced off-exchange rates in the complex (Table 1), while 34 others retained no excess deuterium (Fig. 5). Three overlapping peptides identified the ATP-binding site and catalytic loop (residues 164-172), including the essential residues K168, E170, and N171 [14]. Three overlapping peptides identified the glycine-rich loop (residues 44-54), including S53, F54, and G55 which anchor the ATP phosphates as well as L49, which binds the adenine ring [14]. One peptide corresponded to the substrate-binding shelf (residues 237-250), including residues 234-242 corresponding to the P11 pocket which binds PKI(5-24) [19, 20]. One peptide corresponded to a second contact with the PKI(5-24) (residues 133-145), including R133 which is essential for PKI(5-24) binding [20] (Fig. 6).

Variation of the off-exchange time allowed the rate of off-exchange of deuterons from the interfacial peptic fragments to be determined. Figure 7 shows representative plots of the rates of off-exchange of deuterons from peptides located in three different regions of PKA that contact either ATP or PKI(5-24) in the ternary complex. Kinetic data were fit to a bi-exponential because each peptide was expected to be only partly at the interface (only some of the amides would experience slowed exchange). The kinetic data gave four parameters: the number of deuterons that off-exchanged quickly, the corresponding rapid ensemble off-exchange rate, the number of deuterons experiencing slowed exchange, and the corresponding slow ensemble off-exchange rate [21]. Comparison of the kinetic plots for off-exchange of deuterons from surface amides of PKA alone, with those for the PKA-ATP complex and the ternary PKA-

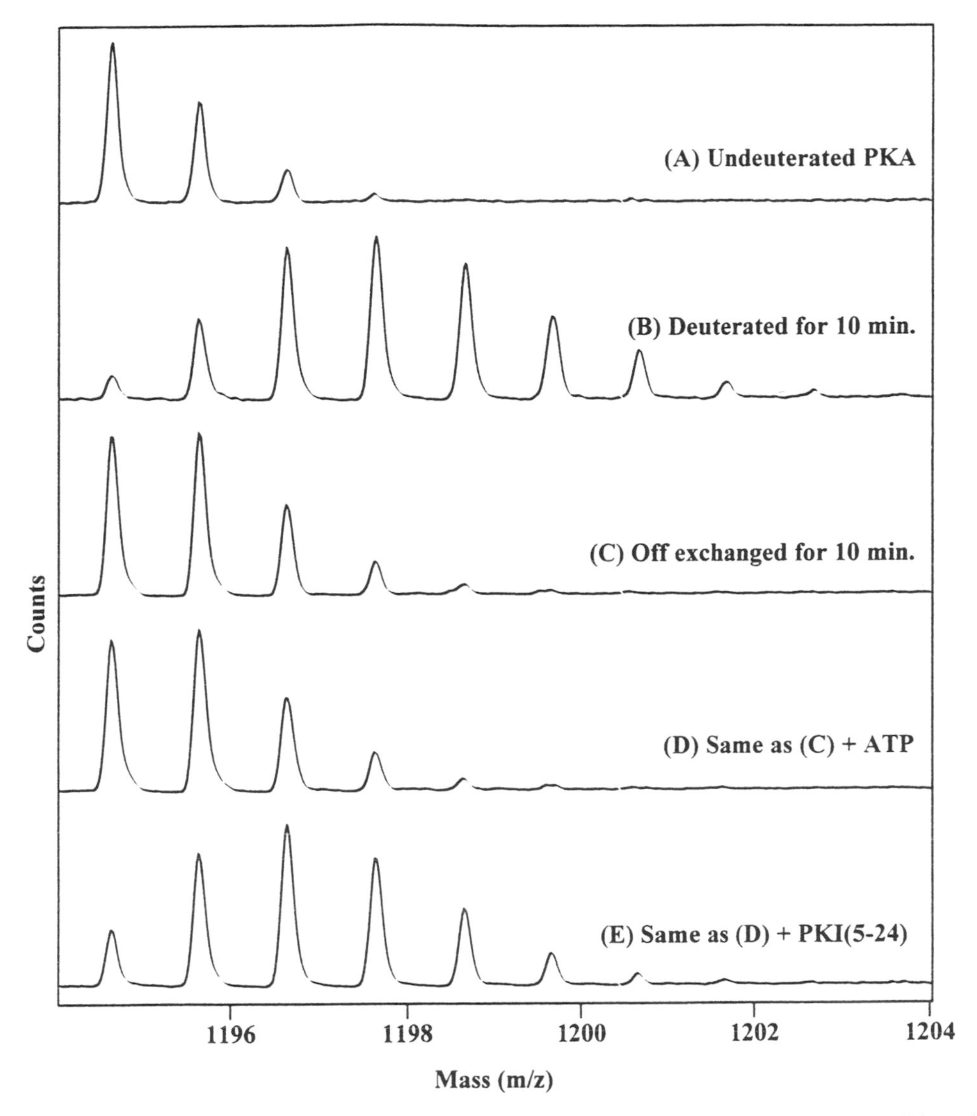

Fig. 4. MALDI-TOF mass spectra of one of the eight peptides from the PKA-PKI(5-24) analysis that experienced slowed exchange in the complex. The spectra are expanded so as to show the isotopic distribution for the ion of interest (MH^+ = 1194.6463). (A) The undeuterated peptide. The higher mass peaks in the envelope are due to naturally occurring isotopes. (B) The isotopic envelope for the same peptide obtained from the PKA sample that was fully deuterated before pepsin digestion. The mass shifts and additional peaks are due to the incorporation of deuterium in the peptide. (C) The peptide obtained from the deuterated PKA sample that was allowed to off-exchange for 10 min. (D) The peptide from the PKA-Mg^{+2}ATP complex that was allowed to off-exchange for 10 min. (E) The peptide from the PKA-Mg^{+2}ATP / PKI(5-24) complex that was allowed to off-exchange for 10 min.

ATP-PKI(5-24) complex show that some decrease in solvent accessibility is observed upon ATP binding, but a marked decrease in solvent accessibility is observed for the ternary complex. The lost solvent accessible surface area of each

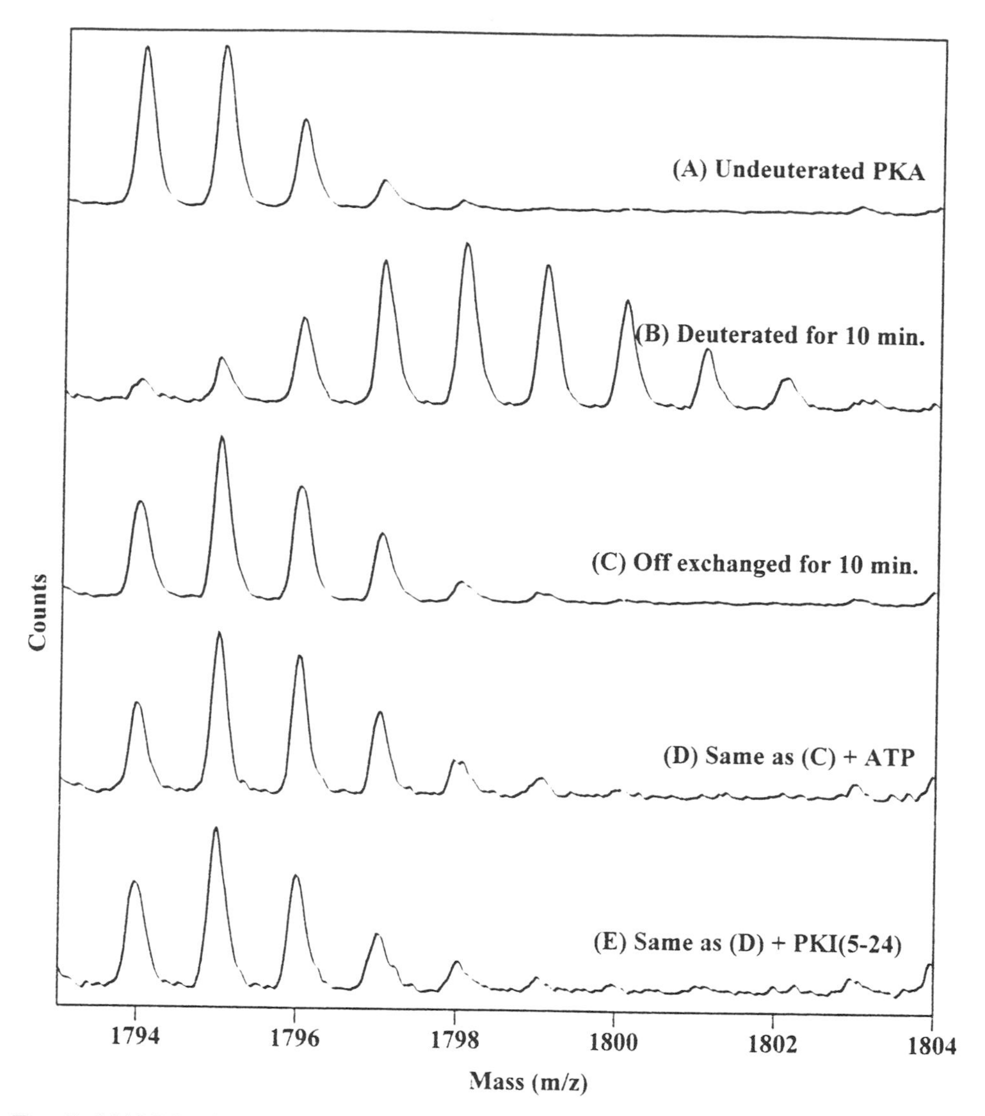

Fig. 5. MALDI-TOF mass spectra of one of the 34 peptides from the PKA-PKI(5-24) analysis that did not experience slowed exchange in the complex. The spectra are expanded so as to show the isotopic distribution for the ion of interest (MH^+ = 1793.9704). (A) through (E) are different regions of the same spectra shown in Fig. 4.

peptide upon formation of the ternary complex was found to correlate with the number of deuterons experiencing slowed exchange in the complex (r = 0.81) (Fig. 8).

The Thrombin-TMEGF(4-5) Interface

A complex of unknown structure was also analyzed, human α-thrombin bound to an eighty-four amino acid fragment of human thrombomodulin (TM(4-5)). This is the smallest fragment of TM that retains anticoagulant activity and

Peptide MH^+	residues	k(complex)[a] (min^{-1})	# of D's slowly exchanging[b]	Surface Area Protected $Å^2$ [c]
A. PKA COMPLEXED TO PKI(5-24)				
ATP-binding site				
1147.6052	164-172	0.004	1.3	148
1260.6933	163-172	0.028	1.6	148
1373.7837	164-174	0.007	1.4	180
Glycine-rich loop				
1194.6463	44-54	0.037	2.4	223
1341.7118	43-54	0.024	2.0	224
1584.8087	41-54	0.014	2.1	224
IP20 binding site				
1628.8925	133-145	0.016	1.0	80
1708.8816	237-250	0.0016	1.1	179
B. THROMBIN COMPLEXED TO TMEGF(4-5)				
Anion binding exosite I				
2127.2265	96-112	0.36	2.1	190[d]
2473.3970	97-116	0.13	3.7	223
2586.4805	97-117	0.23	3.2	224
Potential connection to active site				
2144.1661	117-132	0.74	2.4	7
2530.3606	118-136	ND[e]	ND	ND

[a]The value of the slowed off-exchange rate in the complex $k_{(PKA\text{-}IP20)}$ or $k_{(thrombin\text{-}TM)}$ was determined from the bi-exponential fit ($Ae^{-k_1t} + Be^{-k_2t}$) of the off-exchange rate data and is equal to k_2 in this equation.

[b]The number of deuterons experiencing slowed off exchange was determined from the bi-exponential fit of the off-exchange rate data and is equal to B in the equation above.

[c]The surface area protected was determined using the program, GRASP.

[d]These values are based on the structure of a 19 residue peptide from TM bound to thrombin. The structure of the complex studied here is not known.

[e] Although the 2530.3606 peak shifted significantly, the signal was weak and quantitative kinetics were not determined.

Table 1. Peptides that showed decreased solvent accessibility upon protein-protein complexation.

has a thrombin binding constant of approximately 120 nM [15]. Of the 29 peptides identified from the MALDI-TOF mass spectrum of the thrombin-TM(4-5) complex, five showed marked slowing of off-exchange compared to thrombin alone (Table 1). Mass spectral data from one of the peptides is shown in Fig. 9. Three overlapping peptides corresponded to anion-binding-exosite I (residues 97-117). This region contains residues R103 (73), T104 (74), R105 (75), Y106 (76) and R108 (77a) that have been shown by site-directed mutagenesis to be important for TM binding, or that make contacts to a 19 amino acid peptide from TM in the structure of the peptide-thrombin complex [22-24]. Two other overlapping peptides which showed decreased solvent accessibility upon binding of TMEGF(4-5) corresponded to residues 117-132 (Fig. 10). This is a region that was not previously known to interact with TM.

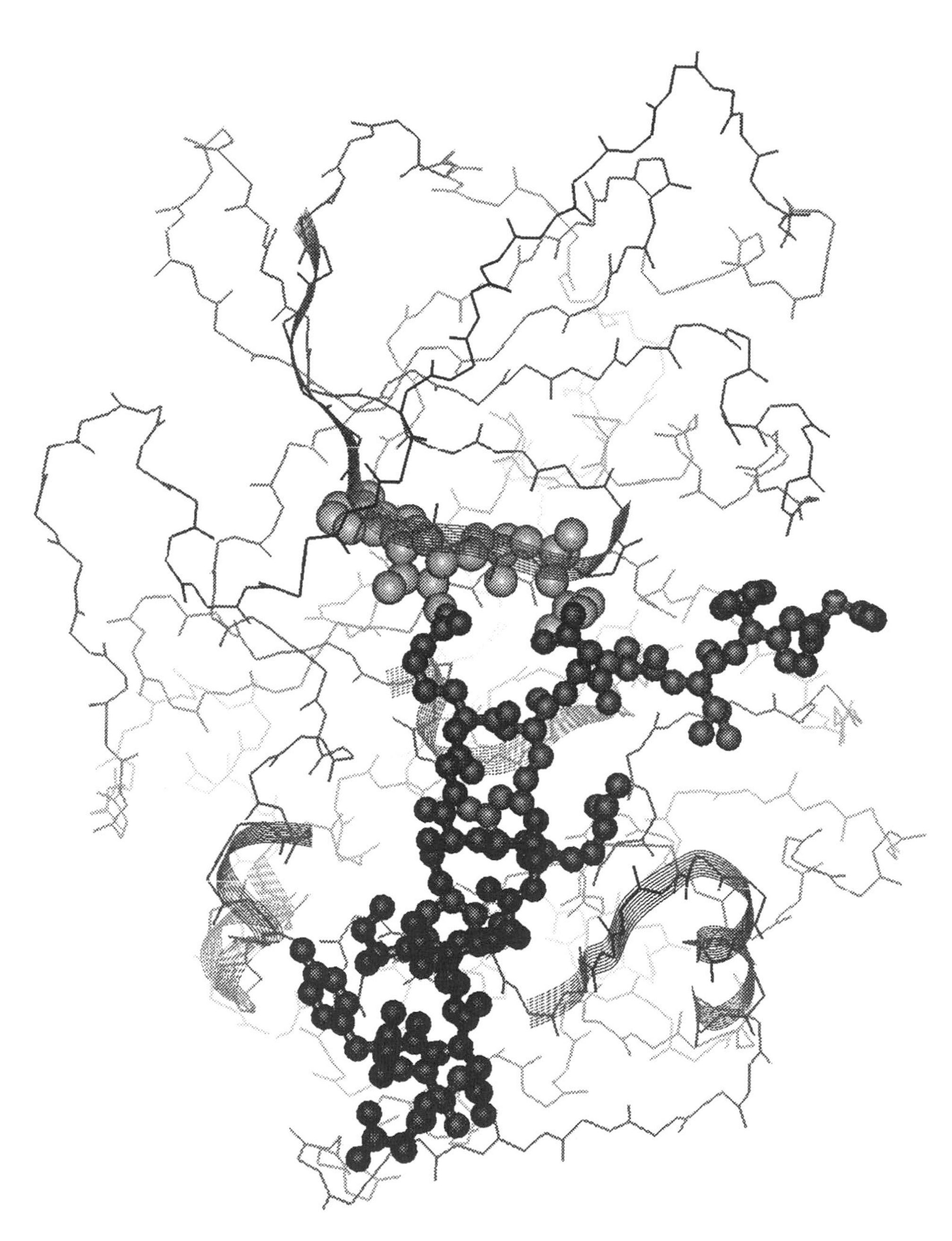

Fig. 6. Structure of the PKA-Mg^{+2}ATP/PKI(5-24) complex. Only the PKA backbone is displayed. PKI(5-24) is in stick and ATP in ball-and-stick (15). The regions of the PKA backbone that experienced slowed off-exchange in the complex are shown with ribbons. Three peptides identified the ATP-binding site and catalytic loop (residues 164-172). Three peptides identified the glycine-rich loop (residues 44-54). One peptide corresponded to the substrate-binding shelf (residues 237-250) and one peptide corresponded to a second contact with the PKI(5-24) (residues 133-145).

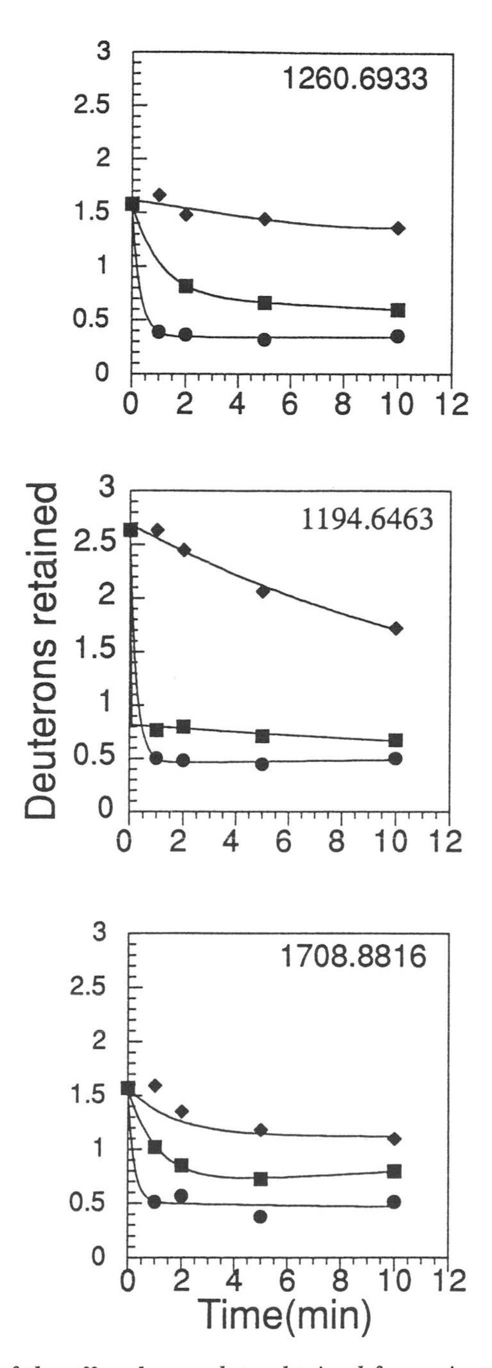

Fig. 7. Kinetic plots of the off-exchange data obtained for various peptides from pepsin digestion of PKA. (A) The peptide of mass 1260.6963, corresponding to part of the ATP-binding site. (B) The peptide of mass 1194.6463, corresponding to the glycine-rich loop. (C) The peptide of mass 1708.8816, corresponding to part of the PKI(5-24) binding site. In each graph, data for the PKA-Mg^{+2}ATP / PKI(5-24) complex (◆), the PKA-Mg^{+2}ATP complex (■) and the PKA alone (●) are shown.

Discussion

Rapid drying at 0°C preserves amide deuteration during MALDI-TOF analysis and allows rapid measurement of amide H/D exchange for many applications, including protein folding and ligand interactions. No modifications to commercially-available MALDI-TOF mass spectrometers are required, only small amounts of protein are used, and the experiments require very little analysis time. We have used MALDI-TOF detection to monitor rapidly exchanging surface amides to assess solvent accessibility changes upon protein-protein complexation. Data for an entire protein sample was obtained from a single spectrum of the mixture of peptides resulting from peptic digestion after H/D exchange. For PKA, this single spectrum gave information on 80% of the solvent

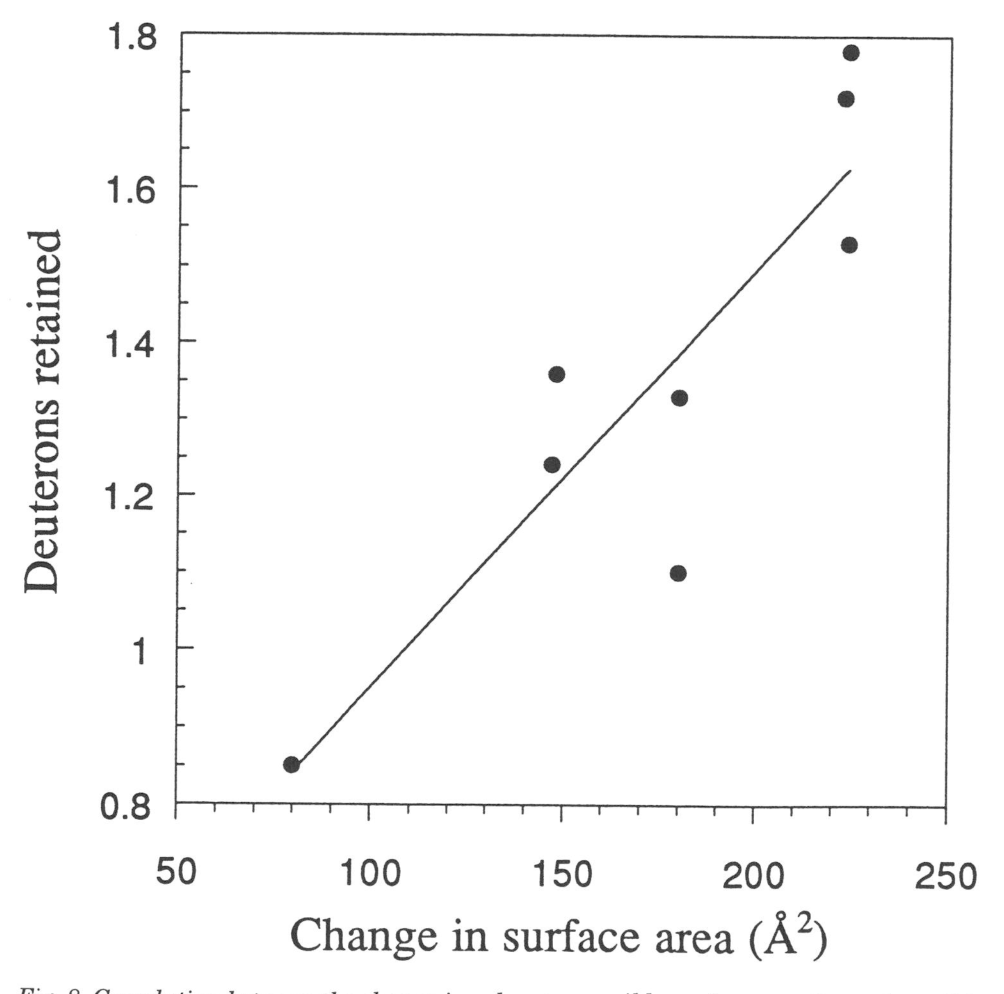

Fig. 8. Correlation between the change in solvent accessible surface area for each peptide found to contain amides that experienced slowed exchange upon binding of ATP and PKI(5-24) and the number of deuterons experiencing slowed exchange. The correlation coefficient for the least-squares line is 0.81.

accessible surface area. All of the regions represented by the peptic fragments that retained deuterium were indeed at the interface of the ATP or PKI(5-24) binding site, and together they comprehensively indicate the surface of interaction with PKA.

The results on the thrombin-TMEGF(4-5) complex show that H/D exchange can be used to identify protein-protein interfaces when no prior structural information is available. The data show that TMEGF(4-5) binds to anion binding exosite I, confirming site-directed mutagenesis and ligand competition studies. The results are also consistent with the structure of a 19 amino

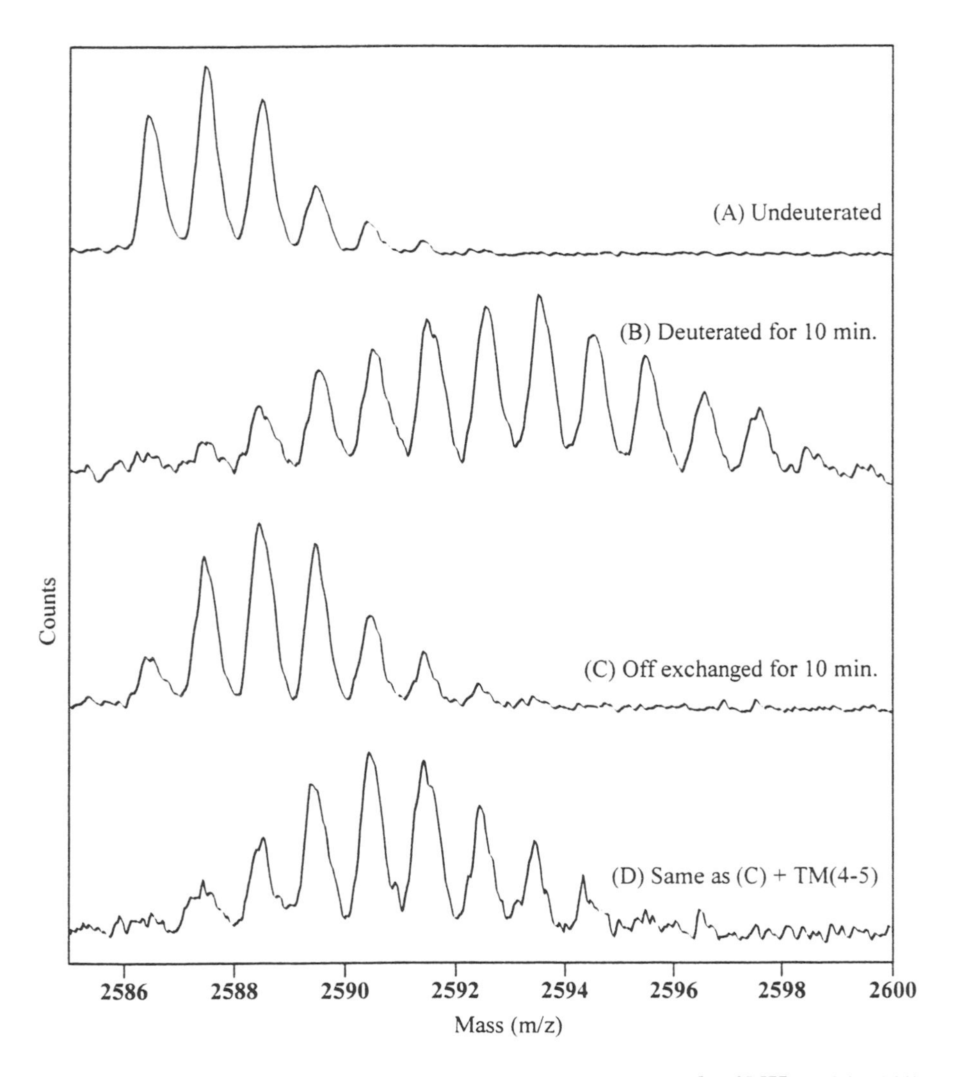

Fig. 9. Expansion of the isotopic envelope for the thrombin peptide of MH^+ 2586.4805, one of the four peptides that experienced slowed exchange in the thrombin:TM(4-5) complex. The order of spectra is the same as for PKA, but without the PKA:Mg^{+2}ATP complex.

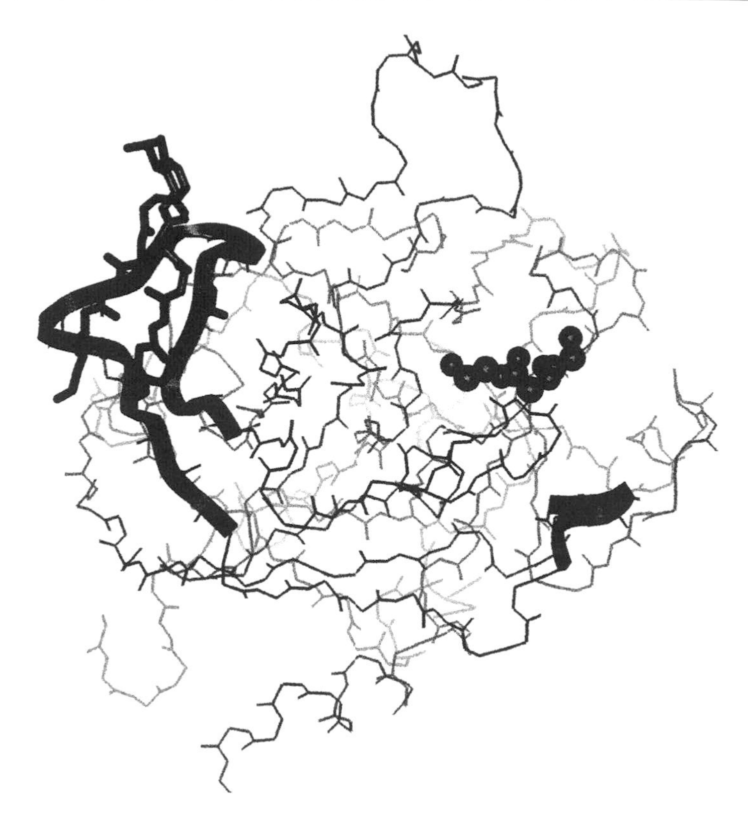

Fig. 10. Structure of human α-thrombin (backbone only) containing an active site inhibitor (ball-and-stick) complexed to a 19 residue fragment from thrombomodulin (stick) (22). The two regions of thrombin that experienced slowed off-exchange in the complex are shown in ribbon. The sequence that corresponds to anion binding exosite I (residues 97-117) lies directly under the 19 residue inhibitor and was expected from mutagenesis data. An additional sequence of decreased solvent accessibility upon TM (4-5) binding residues (126-132), is also shown.

acid loop peptide from thrombomodulin bound to thrombin (18). A new area of thrombin that interacts with TMEGF(4-5) was also discovered. This region corresponds to two β-strands connected by a loop that is within 5 Å of the active site, and its identification indicates a possible mechanism of TM alteration of thrombin substrate specificity. These results will direct future site-directed mutagenesis analysis toward this region of thrombin. It is also important that the binding affinity of TMEGF(4-5) for thrombin is > 100 nM. Protein-ligand interactions in this affinity range are traditionally difficult to analyze, but the amide H/D exchange experiments simply required increasing the molar ratio of TMEGF(4-5) to thrombin to 2.5:1.

For these two entirely different proteins, rapid deuteration of only the surface accessible amides gave reliable information about the protein-protein interface. No unexplainable retention of deuteration was observed, suggesting that rapid deuteration allows the measurement of only solvent accessibility changes without the added deuterium retention resulting from conformational changes. Further support for this contention is obtained by the good correlation (r = 0.81) that was obtained between the number of deuterons experiencing slowed exchange in each peptide and the change in solvent accessible surface area for that peptide. This result suggests that slowed off-exchange in the complex is indeed due to a decrease in solvent accessibility at the interface. Thus, measurement of decreases in H/D exchange rates of solvent-accessible amides can be utilized to monitor changes in solvent accessibility such as those occurring upon protein-protein complex formation.

ACKNOWLEDGEMENTS

This work was supported by NIH grant #HL47463. Jeffrey Mandell acknowledges NIH/NCI #T32 CA09523, and a fellowship from the La Jolla Interfaces in Science Program, funded by the Burroughs Wellcome Fund. We thank Susan Taylor for the gifts of PKA and PKI(5-24), as well as critical comments on the manuscript. We thank John Fenton for his generous gift of human α-thrombin.

REFERENCES

1. J.J. Englander, J. R. Rogero and S. W. Englander, *Anal. Biochem.* **1985**, *147*, 234-244.

2. D. L. Smith, Y. Deng and Z. Zhang, *J. Mass Spectrom.* **1997**, *32*, 135-146.

3. S. W. Englander, T. R. Sosnick, J. J. Englander and L. Mayne, *Curr. Opin. Struct. Biol.* **1996**, *6*, 18-23.
4. Y. Paterson, S. W. Englander and H. Roder, *Science* **1990**, *249*, 755-759.

5. L. Mayne, Y. Paterson, D. Cerasoli and S. W. Englander, *Biochemistry* **1992,** *31*, 10678-10685.

6. D. C. Benjamin, D. C. Williams, S. J. Smith-Gill and G. S. Rule, *Biochemistry* **1992,** *31,* 9539-9545.

7. J. Orban, P. Alexander and P. Bryan, *Biochemistry* **1994**, *33*, 5702-5710.

8. Q. Yi, J. E. Erman and J. D. Satterlee, *Biochemistry* **1994**, *33*, 12032-12041.

9. D. C. Williams Jr., D. C. Benjamin R. J. Poljak and G. S. Rule, *J. Mol. Biol.* **1996,** *257,* 866-876.

10. D. C. Williams, Jr, G. S. Rule, R. J. Poljak and D. C. Benjamin, *J. Mol. Biol.* **1997,** *270,* 751-62.

11. M. H. Werner amd D. E. Wemmer, *J. Mol. Biol.* **1992,** *225,* 873-889.

12. K. Dharmasiri, and D. L. Smith, *Anal. Chem.* **1996**, *68,* 2340-2344.

13. J. G. Mandell, A. M. Falick and E. A. Komives, *Anal. Chem.* **1998**, (in press).

14. J. Zheng, D. Knighton, L. Ten Eyck, R. Karlsson, N.-H. Xuong, S. S. Taylor and J. Sowadski, *Biochemistry* **1993**, *32,* 2154-2161.

15. C. E. White, M. J. Hunter, D. P. Meininger, L. R. White and E. A. Komives, *Protein Engineering* **1995,** *8,* 1177-1187.

16. D. H. Patterson, G. E. Tarr, F. E. Regnier and S. A. Martin, *Anal. Chem.* **1995**, *67*, 3971-8.

17. A. Nicholls, K. Sharp and B. Honig, *Proteins* **1991**, *11*, 281-296.

18. I. I. Mathews, K. P. Padmanabhan and A. Tulinsky, *Biochemistry* **1994,** *33*, 13547-13442.

19. N. Narayana, S. Cox, S. Shaltiel, S. S. Taylor and N.-H. Xuong, *Biochemistry* **1997**, *36,* 4438-4448.

20. N. Narayana, S. Cox, N-H. Xuong and L. F. Ten Eyck, *Structure* **1997,** *5,* 1-15.

21. K. A. Resing, and N. G. Ahn, *Biochemistry* **1998,** *37,* 463-475.

22. Q. Y. Wu, J. P. Sheehan, M. Tsiang, S. R. Lentz, J. J. Birktoft and J. E. Sadler, *Proc. Natl. Acad. Sci. USA* **1991,** *88,* 6775-6779.

23. E. Di Cera, unpublished data.

24. R. Hrabal, E. A. Komives and F. Ni, *Protein Sci.* **1996,** *5,* 195-203.

Questions and Answers

Michael O. Glocker (University of Konstanz)

When deuterium labels on peptic peptides are disappearing while the sample is introduced into the mass spectrometer, does this give rise to deuterated matrix ions?

Answer. We don't know where the deuterium is going. Most probably it is exchanging with carboxyl groups on the matrix or with waters of crystallization in the "dried" sample. An interesting suggestion is to prepare deuterated matrix to see if this stabilizes the sample. We plan to try these experiments soon.

Michael O. Glocker (University of Konstanz)

What would be the limits of on-/off-rates and/or K_D-values that are required to identify binding surfaces in non-covalent kinase-inhibitor complexes by the outlined combination of H/D-exchange and MALDI-MS analysis of the peptic fragments?

Answer. The conditions of the complexation reactions are such that the protein concentration is micromolar, so proteins are above the K_D for the complex. For complexes with K_D's in the 1-20nM range, there will be essentially no unbound components. For complexes with higher K_D's such as the thrombin-TMEGF(4-5) complex, $K_D \pm 120$nM, we used a two-fold molar excess to achieve 100% bound. The equation for calculating % bound is a quadratic:

$$x^2 - x(L+P+K_D) + L \bullet P = 0$$

where L = ligand concentration added; P = protein concentration added; x = complex concentration formed; and % bound can be determined as x/P x100.

Bradford W. Gibson (UCSF)

One would expect that the percent peptide coverage in a single MALDI spectrum of the protein after pepsin digestion would be progressively worse as one went up in mass. Can you comment on this? Have you considered some fast chromatography approaches to avoid this limitation?

Answer. You are correct, and the problem is largely due to overlap of mass envelopes of different peptic fragments. In our experience, overlap is not a big problem for proteins of 40 kDa, but above 60 kDa it begins to be a problem. We have tried a quick spin in a G25 column or a C_{18} column, and the results were as follows. The G25 column was effective to desalt, but signal-to-noise was not significantly improved. The C_{18} column allowed the separation of the mixture into four fractions, each containing a number of peptic fragments, so this shows promise.

Electron Capture Dissociation Produces Many More Protein Backbone Cleavages Than Collisional and *IR* Excitation

Roman A. Zubarev, Einar K. Fridriksson, David M. Horn, Neil L. Kelleher, Nathan A. Kruger, Mark A. Lewis, Barry K. Carpenter and Fred W. McLafferty
Department of Chemistry, Baker Laboratory, Cornell University, Ithaca, NY 14853-1301

Capture of free, low energy (<0.2 eV) electrons by electrospray-generated multiply-charged protein cations $(R–CO–NH–R' + nH)^{n+}$ produces the unusual cleavage products $[R–C(OH)=NH + (n-m)H]^{(n-m)+}$ (***c***) + $[\cdot R' + mH]^{(m-1)+\cdot}$ (***z·***), whereas conventional activation methods mainly cleave the amide bond R–C—/—NH–R' to yield ***b*** and ***y*** ions. In the postulated cleavage mechanism, electron capture to form $(M + nH)^{(n-1)+\cdot}$ reduced molecular ions is followed by the transfer of the labile H· atom to the site with the highest H· affinity (S-S bonds, Trp side chains, and the backbone carbonyls) with subsequent cleavage. The radical-site initiated fragmentation is facilitated by partial infusion of the charge recombination energy. The ***c, z·*** electron capture dissociation (ECD) is complementary to, and substantially less selective than, conventional ***b, y*** cleavage initiated either collisionally or by infrared irradiation. Information from a single ECD spectrum of multiply protonated ubiquitin (76 amino acids) provides, *de novo*, ~90% of its sequence, and ECD of smaller proteins and peptides generally provides 90%-100% of the sequence. However, this coverage decreases for proteins >10 kDa, and proteins >20 kDa only undergo multiple reduction steps without apparent fragmentation.

INTRODUCTION

Electrospray ionization [1] has revolutionized the application of mass spectrometry to biomolecules, both by forming gas phase ions from them with minimal dissociation, and by the fact that these are multiply-charged ions that can be readily fragmented by adding energy with collisionally activated dissociation (CAD) [2-4], infrared multiphoton dissociation (IRMPD) [5, 6], and other less-used methods [7-9]. The masses of fragments formed by backbone cleavages of linear biomolecules such as proteins [10-13] and DNA [14, 15] provide sequence information from knowledge of the masses of the component amino acids and bases, respectively. For example, the published CAD [4] and IRMPD [5] spectra of ubiquitin, 8.6 kDa, exhibit masses corresponding to amide bond cleavage of the $(RCO–NHR' + nH)^{n+}$ molecular ion to form the even-electron ***b*** (N-terminal) and ***y*** (C-terminal) ions shown in Fig. 1.

Such information is especially useful for structural problems of proteins expressed from DNA whose base sequence has been determined, as this also

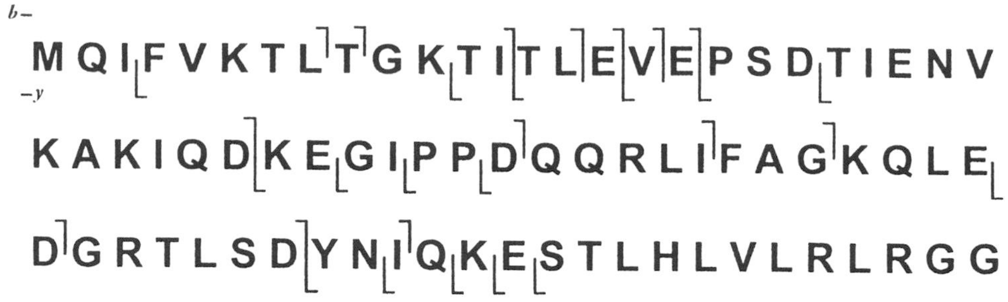

b–

M Q I F V K T L T G K T I T L E V E P S D T I E N V

–y

K A K I Q D K E G I P P D Q Q R L I F A G K Q L E

D G R T L S D Y N I Q K E S T L H L V L R L R G G

Fig. 1. Backbone dissociations of bovine ubiquitin ions, charge states 7+ through 12+, indicated by **b**, **y** *product ions from combined IRMPD and CAD data [4, 5].*

predicts the protein sequence. As a hypothetical example of using MS to confirm this, the relative molecular weight (M_r) value from the DNA prediction of ubiquitin (8.6 kDa) was found to be 14 Da lower than that in the ESI mass spectrum. However, many "complementary pairs" of ***b***, ***y*** ions are present in Fig. 1 (which shows the correct sequence) whose masses sum to the measured M_r value (the "top down" approach [16]). For example, the amide cleavage D^{32}-K^{33} yields a ***b*** ion of mass predicted by the hypothetical DNA sequence, but the ***y*** ion mass is 14 Da less than predicted, thus restricting the site causing this difference to the K^{33}-G^{76} region. Continuing this approach, the cleavages I^{44}-F^{45} and G^{47}-K^{48} yield ions whose mass agrees, and is 14 Da low, respectively. This isolates the erroneous sequence prediction to the three residues F^{45}, A^{46}, and G^{47}; the hypothetical DNA prediction was G^{46}. Other bonds can be cleaved by proteolysis; this combined approach has solved sequence problems for several larger proteins, such as identifying Cys-113 as the derivatized active site in the 379-residue enzyme thiaminase [17].

Such sample degradation outside the MS instrument greatly increases analysis time and sample requirements. Using only MS, the overall utility of this MS/MS approach (similar to that for 2D-NMR [18]) is basically limited by the proportion of backbone amide bonds that are cleaved [16]. Unfortunately, the several dissociation methods [2-9] give similar cleavages because they apparently add similar amounts of energy to the multiply charged protein ions; randomization of this energy over the whole ion causes dissociation of the weakest bonds. Blackbody infrared dissociation enhances the lowest energy dissociations, but these are mainly side chain losses (e. g., H_2O, NH_3) that provide no sequence information [6]. Single additions of much higher energy in the form of 193 nm (6.4 eV) photons also gave mainly the same ***b***, ***y*** product ions [19], except that sometimes ***c***, ***z*** product ions were observed. This anomaly was finally traced [20] to the interaction of the laser photons with the metal cell to produce low energy electrons; serendipitously, their capture produces many more backbone cleavages and ***c***, ***z*** ions through initial formation of an odd-electron molecular ion (Figs. 2 and 3). These electrons can be produced more simply with a classic "electron gun" (electrically heated filament) that is standard equipment on most mass spectrometers.

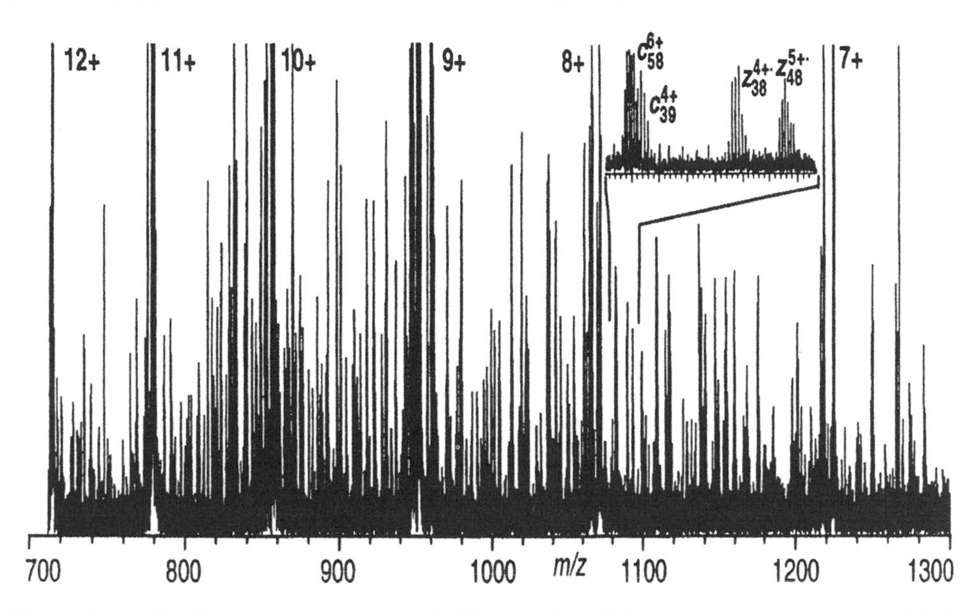

Fig. 2. Partial ECD spectrum (50 scans) of bovine ubiquitin, charge states 7+ through 12+.

ELECTRON CAPTURE DISSOCIATION (ECD)

This novel technique is based on recombination of multiply protonated molecular ions with free, low energy electrons [20]. For proteins, this yields an unusual backbone cleavage (eq 1).

$$[\text{R–CO–NH–R'} + \text{nH}]^{n+} + e^{-} \rightarrow \underset{\boldsymbol{c}}{[\text{R–C(OH)=NH} + (\text{n-m})\text{H}]^{(n-m)+}} + \underset{\boldsymbol{z\bullet}}{[\bullet\text{R'} + \text{mH}]^{(m-1)+\cdot}} \tag{1}$$

These ***c, z•*** ions are in contrast to the mainly ***b***, ***y*** product ions from CAD and IMRPD. A minor fragmentation channel, R–CO–NH-R' → R· (***a·***) + CO + NH_2–R' (***y***), is also observed, especially for peptides [21]. Mechanistically, ECD proceeds via formation of radical cations $(M + nH)^{(n-1)+\cdot}$ after capture of a single electron that is energetic as a result of this charge recombination. This reduced (odd-electron) molecular species can dissociate, yielding ***c*** and ***z·*** fragments. Although the reduced species are formed for proteins > ~ 20 kDa, they do not dissociate. However, for smaller proteins and peptides (Fig. 4) [21], ECD can cleave 80-100% of the backbone bonds. An exception is formation of ***c***, ***z·*** ions on the N-terminal side of Pro; this requires the cleavage of two bonds and is not observed. However, CAD/IRMPD formation of ***b***, ***y*** ions on this side of Pro is highly favored. In further contrast, secondary dissociation of the ***c, z•*** primary products to produce internal ions is negligible. ECD has proven to be a useful and convenient technique for peptides [21] (Fig. 4) and proteins as large as 17 kDa [20, 22-24] using electrospray ionization and Fourier transform mass spectrometry (FTMS) [25].

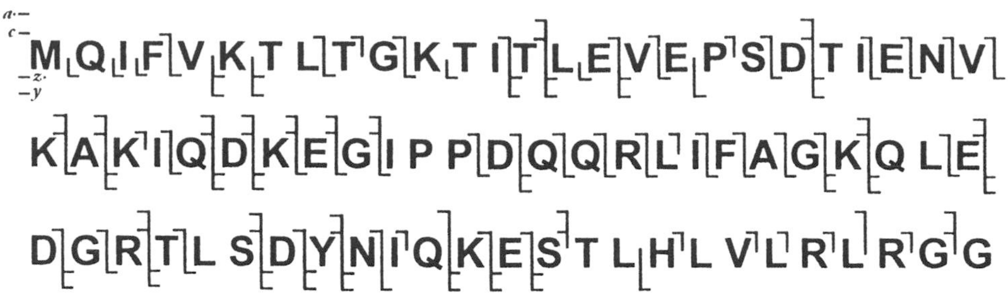

M Q I F V K T L T G K T I T L E V E P S D T I E N V
K A K I Q D K E G I P P D Q Q R L I F A G K Q L E
D G R T L S D Y N I Q K E S T L H L V L R L R G G

Fig. 3. Backbone dissociations of bovine ubiquitin indicated by the **c**, **z·**, **a·**, *and* **y** *products of the ECD data from Fig. 2. Tetrapeptide fragment ions are the smallest observed because the spectrum was not scanned below m/z 500; note that smaller ions would have provided no additional sequence information.*

ECD of 3+
Theor = 1305.65 Da
Exptl = 1305.65 Da
T S H Q A C P S L I L H
(M+3H)2+·
-63.02 Da
-18.05 Da
M-H2O
~ x4
(M+3H)3+
(M+3H)2+·
m/z

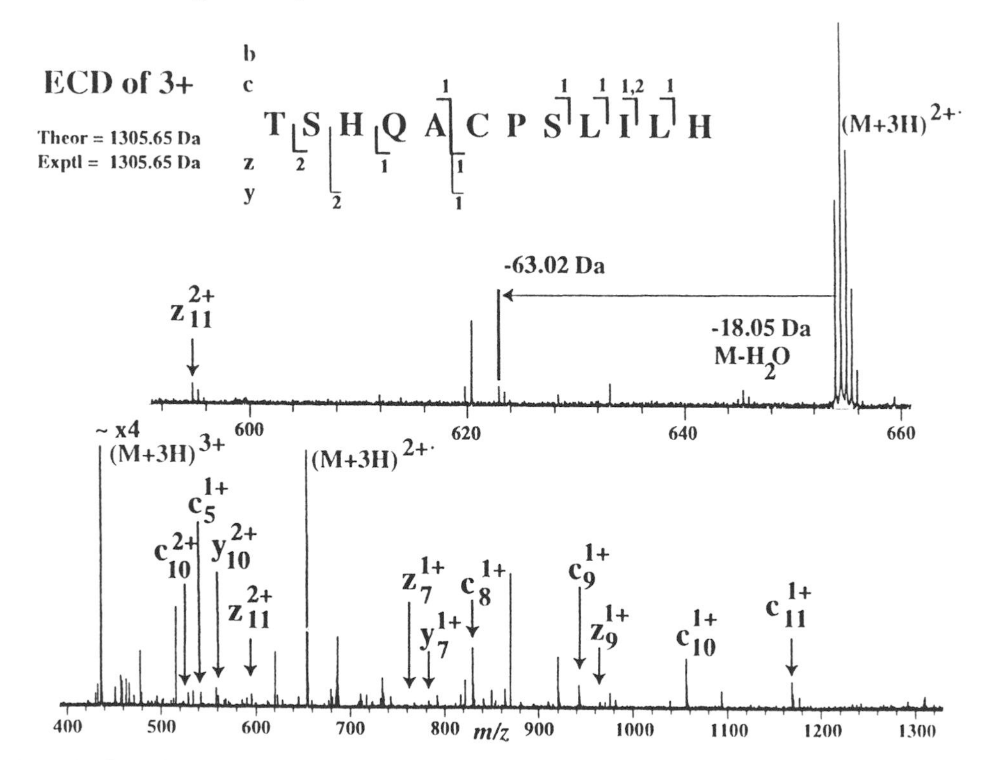

Fig. 4. The ECD spectrum of the (M + 3H)$^{3+}$ *ions of a 12-mer peptide. In the sequence above the spectrum, the ends of the vertical bars indicate the cleavages and resulting products, and the numerals indicate the product charge states.*

MECHANISMS OF ECD

The details of the electron capture and the experiments that led to the proposed ECD mechanism are described elsewhere [21-24]. Only low energy electrons (< 0.2 eV) were found effective in ECD, with the cross section decreasing by a factor of 10^3 for 1 eV electrons. After electron capture, the reduced odd-electron molecular ions $(M + nH)^{(n-1)+\cdot}$ that are undissociated can be fragmented using much less collisional activation than is needed to decompose the even-electron $(M + nH)^{n+}$ ions (and also the even-electron $(M + nH)^{(n-2)+}$ ions formed by capture of two electrons). CAD of the reduced $(M + nH)^{(n-1)+\cdot}$ ions also yields mainly ***c*** and ***z·*** ions, but many are from cleavages at different sites than in the direct ECD, consistent with migration of the unpaired electron before dissociation.

Such evidence indicates that direct ECD occurs too fast for randomization of the excess energy from electron capture; secondary dissociations do not occur because the remaining excess energy is randomized in the primary ***c, z•*** products. Such non-ergodic dissociation has never been observed in molecules anywhere near as large. Strong support for $\sim 10^{-12}$ s dissociation is provided by theory; B3LYP and RRKM calculations on the model system $CH_3C\bullet(O\text{–}H)NH\text{–}CH_3$ indicate that it is basically unstable, with H· loss favored over $\cdot CH_3$ loss at threshold; however, higher energies that shorten its lifetime to $<10^{-12}$ s favor $\cdot CH_3$ loss, a loss that corresponds to backbone protein cleavage [24]. Electron capture exhibits orders of magnitude larger cross sections than does collisional excitation, with the ECD cross section proportional to n^2, the ionic charge squared [22, 23]. The lack of secondary internal product ions is also consistent with the primary products' lower charge and cross section that makes the capture of another electron relatively unfavorable.

The original ECD mechanism postulated formation of the RC(OH)–NH–R' intermediate by e^- capture at a protonated carbonyl group. However, in the presence of a disulfide bond, the predominant channel is S–S bond cleavage: RS–SR' → RSH + ·SR' [25]. This cannot occur by direct e^- capture; the electron affinity of the uncharged –S–S– is far less than any protonated site. Further, the –S–S– group should not be protonated; the H^+ affinity of –S–S– is 1 eV less than that of an amide. However, the hydrogen atom affinity of –S–S– is 1 eV *higher* than that of an amide. Thus, ECD appears to occur through proton neutralization to release a hot H· that is subsequently captured by sites with high H· affinity; possibly the high H· atom polarizability keeps it trapped by the multiple cationic sites. Although H transfer across a substantial molecular distance has been noted in fatty acid ester mass spectra [26], this unusual mechanistic postulate requires more definitive mechanistic tests. The generally non-selective backbone bond rupture is consistent with *ab initio* calculations that predict closely similar H· affinity values for carbonyl groups of most amino acids. However, the H· affinity for the Trp indole side chain is similar to that of –S–S–; a subsequent study confirmed preferential Trp cleavage (although less than that for disulfides), consistent with indole radical attack at the C-terminal side of Trp [24].

ECD for Rapid Protein Structure Analysis

Using these spectra with ESI/FTMS provides far more sequence information than conventional CAD or IRMPD spectra. Figure 2 shows an ECD spectrum in which charge states 7+ through 12+ of bovine ubiquitin (76 residues, 8.6 kDa) were irradiated with low-energy electrons. The spectrum was obtained on the Cornell 6 T FTMS instrument and contains 50 integrated scans. Values of m/z < 500 were not recorded, so that small terminal fragment ions may also have been formed. The utility of the FTMS resolving power is demonstrated in the inset, where the resolved isotopic peaks allow overlapping ***c*** and ***z·*** fragments of different charge states to be identified. The cleavages are summarized in Fig. 3. This single spectrum contains 55 ***c*** and 50 ***z·***, as well as 18 ***a·*** and 24 ***y*** ions, with only 9 bonds remaining intact (12%). Of these, two bonds are on the N-side of prolines, and thus should not be cleaved by ECD.

The fragment masses give 42 complementary ***c, z·*** pairs (eq 1) and 11 complementary ***a·, y*** pairs. The former can be recognized because the sum of the ***c*** and ***z·*** fragment masses is equal to the molecular mass plus 1 Da (the neutralized H+), while the ***a·, y*** pair in addition yields a 27.995 Da mass sum deficit because this cleavage is accompanied by the loss of CO. Even more useful are the 23 ***c, z·*** or ***a·, y*** complementary pairs for which at least one fragment of the other type represents a cleavage between the same amino acid pair; for this the ***c*** ion is 44.014 Da heavier than the ***a·*** ion of the same charge value, and the ***z·*** ion is 16.019 Da lighter than the corresponding ***y*** ion of unit greater charge value. These then show the sequence direction, giving the partial sequence:

$$...K_{6}...T_{14}...A_{28}...D_{32}K_{33}E_{34}G_{35}...K_{48}...D_{52}...D_{33}Y_{59}...K_{63}E_{64}...$$

For this, only the exact mass differences between the amino acids are known; their positions (subscripts) are indicated for convenience. This 13-residue sequence tag [27, 28] is based on at least three detected fragment ions from cleavages on each side of the amino acid, and therefore is of a higher validity than ordinary mass spectrometric sequence information derived from one mass value on each side. This information is redundant for identification of the protein in the database, for which 5-7 residues usually suffice [27, 28]. This also gives higher confidence in extending the sequence information based on the mass difference between these and other fragment masses. With this, only the gaps TL, TI, TI, QL, LS, TL and LV remain after ordering IPP because N-terminal cleavage at Pro is not expected. It is noticeable that all resistant bonds involve hydrophobic residues, mostly on the C-terminal side of the bond, and in four cases the N-terminal residue is threonine. Of course these MS/MS data have not distinguished the isomers Leu and Ile. In contrast, IRMPD and CAD of the same ubiquitin ions produced together 14 ***b*** and 16 ***y*** fragments (Fig. 1) that confirmed the amino acid positions in the polypeptide chain found by ECD. Although the I_{36}-P_{37} cleavage gives positive evidence of the IPP ordering inferred by an absence of cleavage in the ECD spectrum, the CAD/IRMPD data do not order any of the seven doublets not ordered by ECD.

In a separate study [22, 23], individual charge states of ubiquitin molecular ions were isolated for ECD; the spectral signal/noise was enhanced by refilling the FTMS ion cell several times, each with the ejection of unwanted ions. These ECD spectra showed that product ion abundances can vary substantially with charge state; with these data, the complete amino acid ordering of ubiquitin except for three doublets was achieved.

Conclusions

The ECD mechanism appears to involve initial release of an excited H· atom that can explore many reactive sites within the folded protein ion before being captured to trigger a fast (non-ergodic) dissociation. This reaction, which appears to have little direct precedent, is under further investigation. The failure of ECD to dissociate protein ions >17 kDa could be due to the increasing difficulty of denaturing larger protein ions [29]; collisional or IR energizing of the ions during ("in beam") ECD results in more extensive ***c, z•*** cleavages [24].

From the analytical point of view, a present disadvantage is that ECD produces only ~30% yield of fragment ions, versus 90% for CAD and 80% for IRMPD. However, ECD yields far more extensive sequence information. A single ECD spectrum of a 76-residue protein provides the ordering of all but seven amino acid doublets. Disulfide bonds are resistant to CAD/IRMPD cleavage, but are readily cleaved by ECD, an advantage especially for –S–S– cyclized proteins such as insulin [24]. Thus, ECD gives promise for extensive *de novo* sequencing of much larger proteins by initial CAD/IRMPD or proteolytic dissociation into ~10 kDa protein fragments followed by their ECD sequencing. Conceivably, this could be applied routinely to 10^{-15} mol samples; MS injection of 10^{-17} mol of carbonic anhydrase (29 kDa) gave exact masses of nine CAD fragment ions [30], more than sufficient information for the correct retrieval of this protein from the 250 K protein database [27, 28].

References

1. J. B. Fenn, M. Mann, C. K. Meng, S. F. Wong and C. M. Whitehouse, *Science* **1989**, *246*, 64-71.

2. J. A. Loo, H. R. Udseth and R. D. Smith, *Rapid Commun. Mass Spectrom.* **1988**, *2*, 207-210.

3. J. W. Gauthier, T. R. Trautman and D. B. Jacobsen, *Anal. Chim. Acta* **1991**, *246*, 211-225.

4. M. W. Senko, J. P. Speir and F. W. McLafferty, *Anal. Chem.* **1994**, *66,* 2801-2808.

5. D. P. Little, J. P. Speir, M. W. Senko, P. B. O'Connor and F. W. McLafferty, *Anal. Chem.* **1994**, *66,* 2809-2815.

6. W. D. Price, P. D. Schnier and E. R. Williams, *Anal. Chem.* **1996**, *68,* 859-866.

7. M. E. Bier, J. C. Schwartz, K. L. Schey and R. G. Cooks, *Int. J. Mass Spectrom Ion Processes* **1990**, *103*, 1-19.

8. R. A. Chorush, D. P. Little, S. C. Beu, T. D. Wood and F. W. McLafferty, *Anal. Chem.* **1995**, *67*, 1042-1046.

9. B. Wang and F. W. McLafferty, *Org. Mass Spectrom.* **1990**, *25*, 554-556.

10. F. W. McLafferty, *Acc. Chem. Res.* **1994**, *27*, 379-386.

11. P. B. O'Connor, J. P. Speir, M. W. Senko, D. P. Little and F. W. McLafferty, *J. Mass Spectrom.*, **1995**, *30*, 88-93.

12. E. R. Williams, *Anal. Chem.* **1998**, *70*, 179A-185A

13. N. L. Kelleher, S. V. Taylor, D. Grannis, C. Kinsland, H.-J. Chiu, T. P. Begley and F. W. McLafferty, *Protein Sci.* **1998**, *7*, 1-6.

14. S. A. McLuckey and S. Habibi-Goudarzi, *J. Am. Chem. Soc.* **1993**, *115*, 12085-12095.

15. D. P. Little, D. J. Aaserud, G. A. Valskovic and F. W. McLafferty, *J. Am. Chem. Soc.* **1996**, *118*, 9352-9359.

16. N. L. Kelleher, H. Y. Lin, G. A. Valaskovic, D. J. Aaserud, E. K. Fridriksson and F. W. McLafferty, *J. Am. Chem. Soc.*, accepted.

17. N. L. Kelleher, R. B. Nicewonger, T. P. Begley and F. W. McLafferty, *J. Biol. Chem.* **1997**, *272*, 32215-32220.

18. F. W. McLafferty, N. L. Kelleher, T. P. Begley, E. K. Fridriksson, R. A. Zubarev and D. M. Horn, *Curr. Opin. Chem. Biol.* **1998**, *2*, 571-578.

19. Z. Guan, N. L. Kelleher, P. B. O'Connor, D. J. Aaserud, D. P. Little and F. W. McLafferty, *Int. J. Mass Spectrom. Ion Processes* **1996**, *157/158*, 357-364.

20. R. A. Zubarev, N. L. Kelleher and F. W. McLafferty, *J. Am. Chem. Soc.* **1998**, *120,* 3265-3266.

21. N. A. Kruger, R. A. Zubarev, D. M. Horn and F. W. McLafferty, *Int. J. Mass Spectrom.,* accepted.

22. N. A. Kruger, R. A. Zubarev, D. M. Horn and F. W. McLafferty, *Int. J. Mass Spectrom.,* accepted.

23. R. A. Zubarev, E. K. Fridriksson, D. M. Horn, N. L. Kelleher, N. A. Kruger, B. K. Carpenter and F. W. McLafferty, *to be submitted.*

24. R. A. Zubarev, N. A. Kruger, E. K. Fridriksson, M. A. Lewis, D. M. Horn, B. K. Carpenter and F. W. McLafferty, *submitted to J. Am. Chem. Soc., May 1998.*

25. A. G. Marshall, C. L. Hendrickson and G. S. Jackson, *Mass Spec. Rev.* **1998**, *17*, 1-35

26. F. W. McLafferty and F. Turecek, *Interpretation of Mass Spectra*, 4th Ed., University Science Books: Mill Valley, CA, 1993, p 197.

27. M. Mann and M. Wilm, *Anal. Chem.* **1994**, *66*, 4390-4399.

28. E. Mørtz, P. B. O'Connor, P. Roepstorff, N. L. Kelleher, T. D. Wood and F. W. McLafferty, *Proc. Natl. Acad. Sci. USA* **1996**, *93*, 8264-8267.

29. F. W. McLafferty, Z. Guan, U. Haupts, T. D. Wood and N. L Kelleher, *J. Am. Chem. Soc.* **1998**, *120*, 4732-4740.

30. G. A. Valaskovic, N. L. Kelleher and F. W. McLafferty, *Science* **1996**, *273*, 1199-1202.

Questions and Answers

Kenneth Standing (University of Manitoba)

For analyzing unknowns, is it really an advantage to increase the backbone fragmentation?

Answer. I think so. Since about 90% of the assignable peaks in ECD are ***c*** and ***z*** ions, the rest being ***a*** and ***y***, there are practically no secondary cleavages, so there is no danger of informational overflow.

Alan Marshall (Florida State University)

Electron capture will work best in a nested Penning trap, because both positive ions and electrons can be trapped simultaneously. In a quadrupole (Paul) trap, one can't trap electrons and positive ions simultaneously.

Answer. Not for a long time. Still, pulse introduction of electrons may be a feasible alternative.

Marvin Vestal (PerSeptive Biosystems)

Can you comment on whether this electron capture mechanism can account for the ***c*** and ***z*** ions observed in the prompt fragmentation of proteins in DE-MALDI?

Answer. So-called "in source decay" in MALDI seems to exhibit features closely resembling electron capture dissociation. There is plenty of secondary electrons in MALDI, and multiply charged ions (at least with z=2) are formed in reasonable amounts. S-S bond reduction observed after prolonged (hundreds of shots) laser irradiation is consistent with this view. It would be interesting to learn whether these effects are also observed in IR MALDI.

Protein Micro-Characterization by Mass Spectrometry: Sample Handling and Data Flow

Paul Tempst, Hediye Erdjument-Bromage, Matthew C. Posewitz, Scott Geromanos, Gordon Freckleton, Anita Grewal, Lynne Lacomis, Mary Lui and John Philip
Memorial Sloan-Kettering Cancer Center, New York, NY 10021

Mass spectrometry (MS) occupies center position in most current protein identification schemes. "Mass fingerprinting" techniques rely on composite mass patterns of proteolytic fragments, or dissociation products thereof, to query databases [1-6]. Two issues, however, need to be addressed at the present time. Are we ready for high throughput analyses in a reliable, efficient and timely manner, and to apply these tools towards rationally selected, meaningful research problems?

The second question is crucial, for biologists will be hard to convince to pursue research objectives with protein-based, MS techniques if answers can be obtained faster and more cost-effectively through nucleic acid-based approaches [e. g., measuring gene expression by comparative EST analysis, serial analysis of gene expression (SAGE), and hybridization to DNA-arrays], or with antibodies or two-hybrid systems. Translational control mechanisms and related abnormalities (with concomitant changes of protein levels in the cell) come to mind as potential "proteomics" study subjects. Instead, MS will likely feature a more immediate role in protein functional analyses, including spatial/temporal interactions and binding to specific nucleic acid sequences or small molecules (e. g., signaling molecules or drugs); studies that cannot be readily approached by "genomics". MS will also fill an important need in the analysis of posttranslational modifications, which, again, are not easily identified at the DNA level. Reversible phosphorylation, for example, is important for signal transmission in all living cells and its deregulation has been implicated in cancer. By studying substrates and sites of phosphorylation, diagnostic tools can be developed for certain tumors, and the modification process itself could be a target for therapeutic intervention. A major task will be precise comparative analysis of all the above in diseased versus healthy cells.

Four areas need further development to achieve these goals: protein micro-biochemistry; ultra-sensitive sequencing; structural chemistry bioinformatics; and overall technology merger. MS analysis of yet smaller amounts, to search rapidly expanding databases, will require higher sensitivity and more complete spectral information, both of which depend on proper sample handling, among other factors. Hence, "micro-biochemistry" is a term that collectively refers to manual micro-chromatographic techniques, handling and transfer (e. g., by infusion) into mass spectrometers. Furthermore, selective isolation of

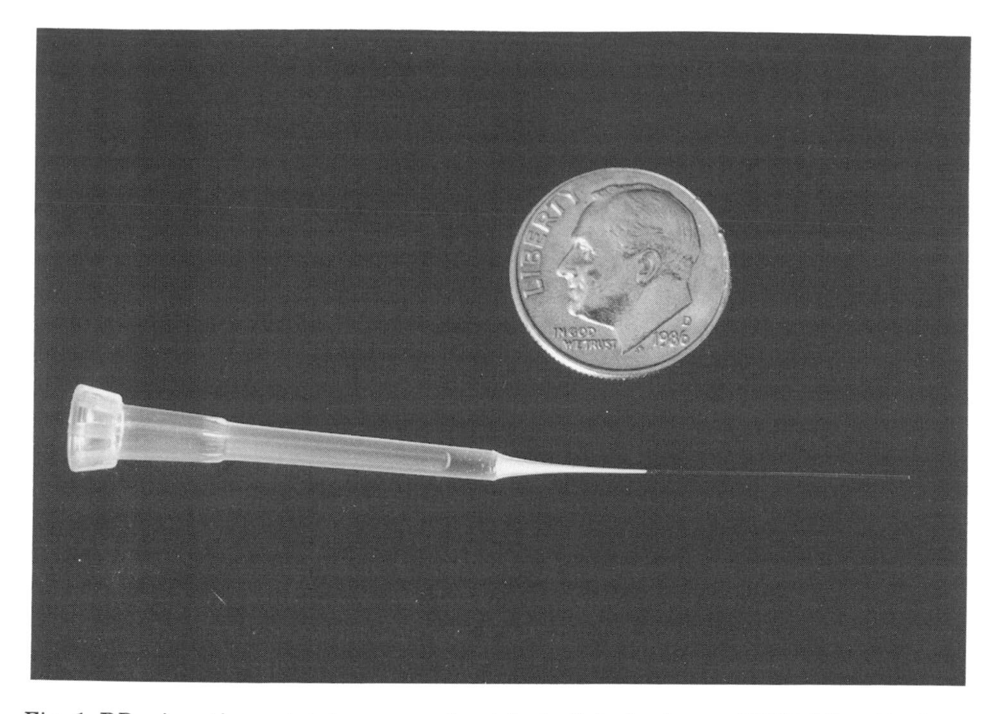

Fig. 1. RP micro-tip containing approximately 2 μL bed volume of POROS 50 R2 beads.

phospho-proteins and of proteolytic fragments bearing the modification are currently not very sophisticated and, in practice, not readily coupled with state-of-the-art micro-LC and MS. Performance evaluation of metal ion-derivatized perfusion beads for phospho-peptide isolation, in terms of both selectivity and recovery, is thus in order. The challenge is to subject small volume (in 1-4 μL) peptide mixtures to MS analysis with minimal sample losses, especially in the case of nanospray. We have constructed a continuous-flow nanoES source (termed "JaFIS") which utilizes a PCR tube as sample holder and can be operated at flows of 1-50 nL/min with detection at the 2-10 fmole/μL level. This technique can be automated and is uniquely suited for use in conjunction with micro-tip procedures. Finally, the multi-instrument analytical approach and a rapidly growing sample throughput require semi-automated interpretation, comparison, database searching and archival of mass spectrometric data, all tied into a streamlined project information management and reporting system.

Micro-Biochemistry

Preparation of Peptide Pools

Peptide masses can be quickly and accurately determined from unfractionated mixtures, and at concentrations below 100 femtomoles per μL, by either matrix-assisted laser desorption ionization (MALDI) or electrospray ionization (ESI) mass spectrometry (MS). Unfortunately, the quality of mass spectrometric data is severely compromised when samples are contaminated. In the case of MALDI and infusion-based ESI, signals are highly concentration-dependent as well. Therefore, both of these techniques benefit from having the sample available in the smallest possible volume. This issue was first addressed by Caprioli and coworkers [7] and then by Wilm and Mann, who used micro-columns filled with reversed phase (RP) chromatographic material to elute peptides directly into glass capillary spray needles by centrifugation [8]. Simplified, manually operated versions using eppendorf gel-loading pipet tips as the "column" or prepacked "micro-guard" columns, have subsequently been developed and shown to provide satisfactory cleanup and concentration of the sample, resulting in improved mass spectrometric data on minute amounts of gel-purified proteins [9-11]. Optimization of the pipet tip device, or "micro-tip" (shown in Fig. 1), to yield spectra of still better quality has awaited detailed investigation of several operational variables. Thus, we have studied quantitative and qualitative aspects of polypeptide extraction using these small manual devices [12] and shown that optimization of sample volume and additives, micro-tip bed volume, and eluant composition and volume, all contribute to effective recovery (~ 65-70%, on average).

We also explored the possibility of breaking up complex peptide mixtures into smaller pools by manual micro-tip chromatography, and how such partial fractionation might benefit subsequent mass spectrometric analysis in terms of total information content. A tryptic digest of yeast G6PD (*M*r ~ 55 kDa) was used as a model system. A total of 53 different peptides were found to be present in the G6PD digest mixture. Peptides (200 femtomoles) were thus adsorbed to an appropriately sized micro-tip and eluted, either in a single step (4 µL 30% MeCN/ 0.1% FA) or stepwise (10-30% in 4 equal increments), and the eluates analyzed by MALDI-TOF (Figs. 2A and 3) and JaFIS(ESI) MS (Fig. 2B). Single-step eluates resulted in largely overlapping peak patterns as depicted in Fig. 2. The majority of peptides mapped to the center portion of the HPLC trace; the thirty or so peptides (or, better, MS signals representing those peptides) missing from the spectra were either not recovered from the tip or their MALDI and ES spectra suppressed [13]. On the other hand, multi-step elution of a similar micro-tip yielded a combined set of MALDI-TOF MS signals (Fig. 3) exceeding those of the one-step elutions both in number (29 peaks) and overall quality (17 major signals). The additional MS signals were derived from either small, hydrophilic peptides or very hydrophobic ones. Multi-step elution may therefore offer a significant advantage by virtue of (i) additional signals when analyzing digests of either large proteins or simple mixtures (two to three proteins), or (ii) detecting unusually hydrophilic (e. g., phosphorylated) or hydrophobic polypeptides. In its most simple form, we found that a two-step elution is best done with MeCN concentrations of 16% and 30% (data not shown).

Micro-tip peptide pool extraction was tested using an "in-gel" tryptic digestion [14] of 100 femtomoles G6PD (amount loaded on the gel). We introduced an extra variable by supplementing the standard bicarbonate digest buffer (~ 10 µL) in one experiment with 0.1% Zwittergent 3-16. The rationale for this experiment was our earlier observation that the detergent, in addition to

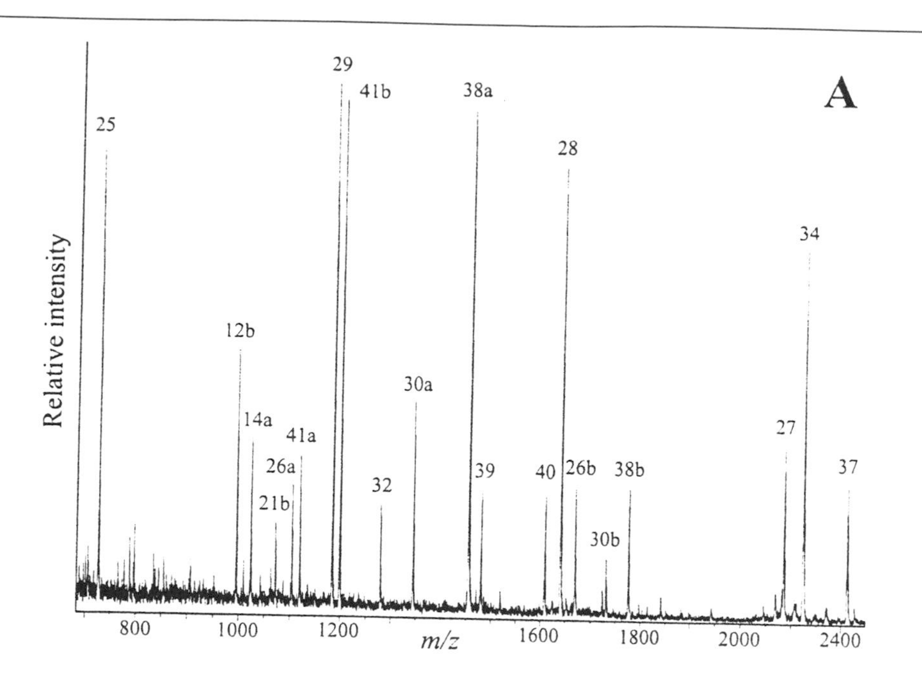

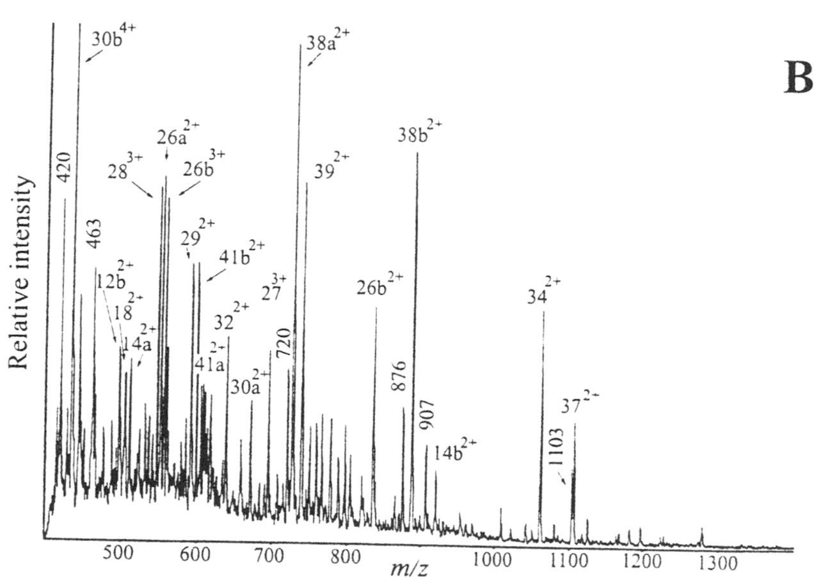

Fig. 2. Mass spectra of G6PD tryptic peptides after one-step RP micro-tip extraction. Two hundred femtomoles of G6PD tryptic digest were diluted in 2 μL of 0.5% Zwittergent 3-16 / 100 mM bicarbonate buffer, loaded onto a micro-tip (as shown in Fig. 1), washed three times with 20 μL 0.1% formic acid (FA), and eluted with 4 μL 30% MeCN / 0.1% FA. Eluates were then taken for ***(A)*** *MALDI-reTOF (0.5 μL deposited) and* ***(B)*** *continuous flow NanoESI ("JaFIS" source operated at 75 nL / min) quadrupole scan analysis. Numbering of the peaks corresponds to relative elution positions of the peptides from an RP-HPLC column (data not shown); peptide charge states are indicated on the JaFIS-Q1 spectrum as superscripts (panel B).*

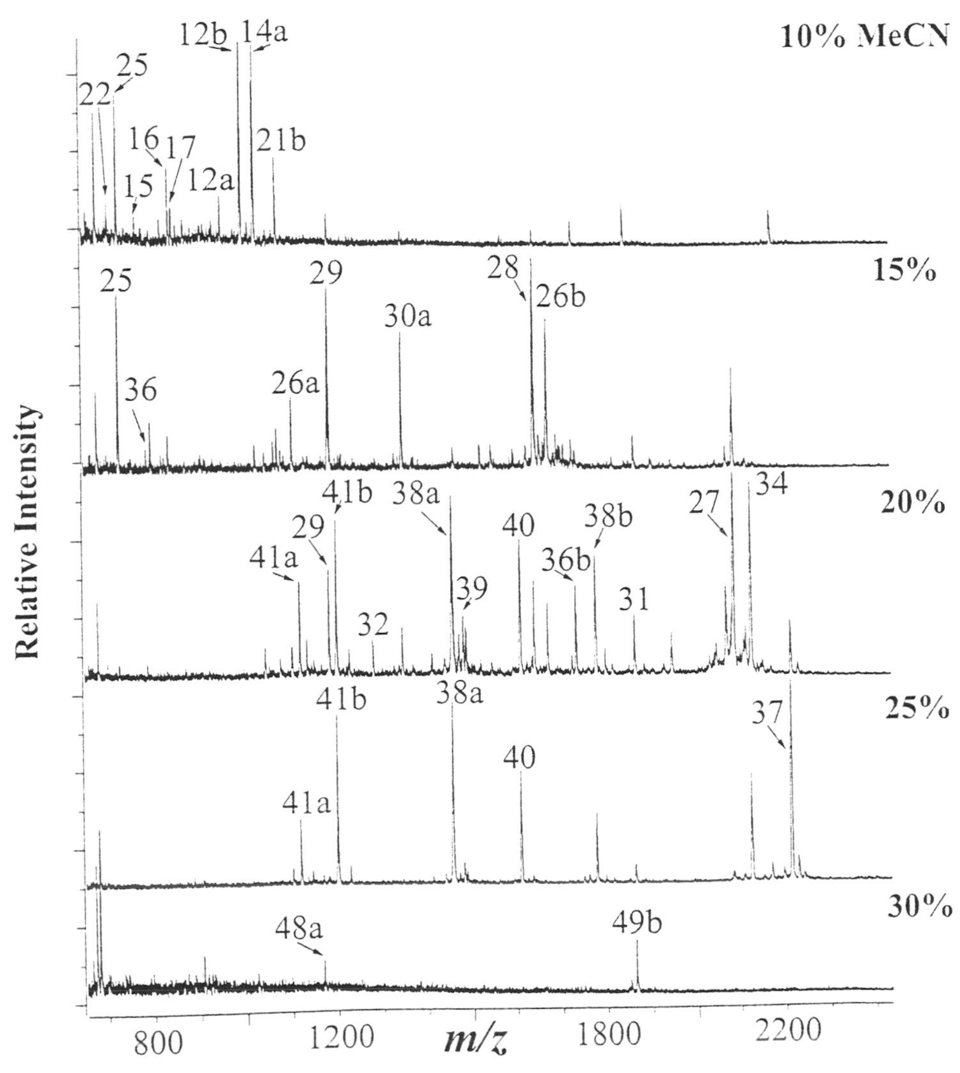

Fig. 3. MALDI-TOF mass spectra of G6PD tryptic peptides recovered from multi-step RP micro-tip extraction. Two hundred femtomoles of G6PD tryptic digest were prepared and used for micro-tip extraction as described under Fig. 2, except that step-wise elution was done with 4 μL 0.1% FA, containing increasing concentrations of acetonitrile (10-30% MeCN), as indicated on the panels. All five eluates were then taken for MALDI-reTOF analysis (0.5 μL deposited), which was done under the exact same conditions as used to generate the spectrum shown in Fig. 2A. Peak sizes in the different panels (and those in Fig. 2A) are to scale.

facilitating dissociation of polypeptides from membranes, helps prevent adsorptive losses to glass and plastic [15]. At the end of the procedure, the combined extraction volumes were dried , including the Zwittergent (a powder) when present, redissolved in 10 μL 0.1% formic acid, the entire volume loaded on a micro-tip (2 μL bed volume) and eluted in a single step. MALDI-TOF MS analysis of the eluates is pictured in Fig. 4. Two important conclusions could be drawn

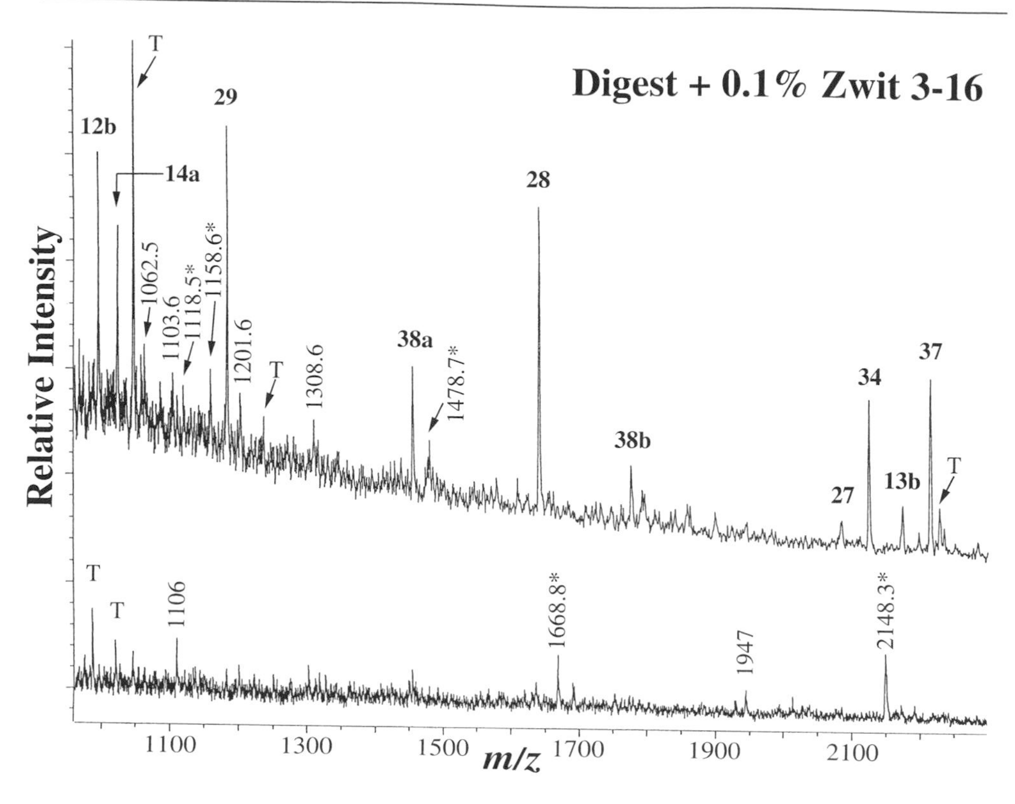

*Fig. 4. MALDI-TOF mass analysis of 100 femtomoles yeast G6PD, trypsinized in-gel and extracted over an RP micro-tip. Hundred femtomoles of G6PD (loaded on the gel) were digested with trypsin [14] in the presence **(top panel)** or absence **(bottom panel)** of 0.1% Zwittergent 3-16, extracted over an RP micro-tip and analyzed by MALDI-reTOF MS. Peaks labeled with bold numbers correspond to those marked in Figs. 2A and 3, those labeled with "T" are trypsin autolytic products, those labeled with an m/z value followed by the * mark are predicted G6PD derived peptides not observed in the earlier experiments, all others correspond to molecules of undefined nature.*

from these experiments. First, digests done in the presence of reduced Zwittergent concentrations yielded good quality, easily interpreted MALDI-TOF spectra (Fig. 4; top panel); from a total of 20 peaks, 14 are derived from G6PD peptides and 3 from trypsin autolytic products (the 3 others were of an undefined nature). The absence of detergent from the digest mixtures, on the other hand, resulted in spectra nearly devoid of any useful information; only a few small peaks could be produced after prolonged analysis (Fig. 4; bottom panel). It thus appeared that, in some instances, peptides derived from as little as 100 femtomoles of protein prepared by SDS-PAGE can be satisfactorily analyzed by mass spectrometry after RP micro-tip extraction. The importance of sample preparation, however, cannot be over-emphasized.

Isolation of Phosphopeptides

While adequate technology to map modified amino acids in peptides is already available, micro-preparative and selective isolation of the proteins, and

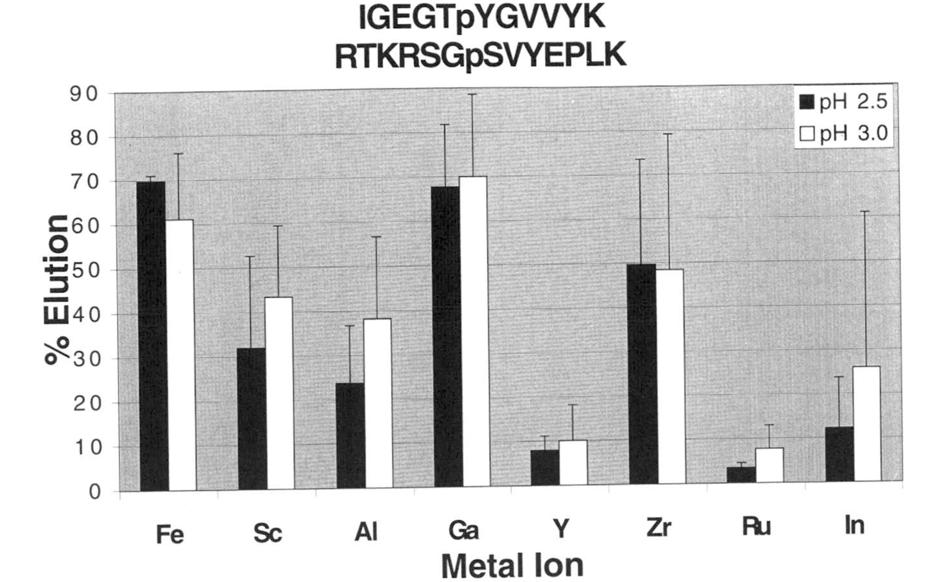

Fig. 5. Recovery of phosphopeptides from IMAC micro-columns: effects of metal ion. Two [^{32}P]-labeled phosphopeptides were separately loaded (at 2 different pHs, as indicated) onto POROS MC micro-columns to which different metal ions had been immobilized, extensively washed, eluted with 0.2M phosphate, and recoveries monitored by Cerenkov counting. Results are presented as average for the two peptides.

of the proteolytic fragment(s) bearing the modification(s), is considerably less sophisticated and, in practice, not as readily coupled with state-of-the-art micro-LC, chemical sequencing and MS. Isolation of phosphorylated polypeptides using immobilized metal ion affinity chromatography ("IMAC") has been reported [16-20]. We have extended these studies by systematically quantifying the ability of IMAC to retain and elute phosphorylated peptides. Pilot studies were initiated to quantitatively determine the capacity of several different metal ions to retain phosphopeptides using IMAC chromatography, and examined whether these IMAC columns could selectively retrieve phosphorylated peptides from tryptic protein digests under several different loading and elution conditions.

Eight water soluble, multi-valent metal ions (see Fig. 5) were selected for study, and immobilized on micro-tip packed (see above and [12]) POROS MC beads by equilibrating the column with 10 bed-volumes of the corresponding 50 mM chloride salt solutions. Two [^{32}P]-labeled phosphopeptides (one containing phosphotyrosine; the other phosphoserine) were separately loaded onto each of the micro-columns, in solutions of approximate pH of 2.5 and 3.0. Following extensive washing, the phosphopeptides were eluted with 0.2 M sodium phosphate. Binding and recovery were monitored by Cerenkov counting. The results presented in Fig. 5 show that of the eight metal ions, Fe(III), Ga(III) and Zr(IV)

A B-Gal + Src Peptide

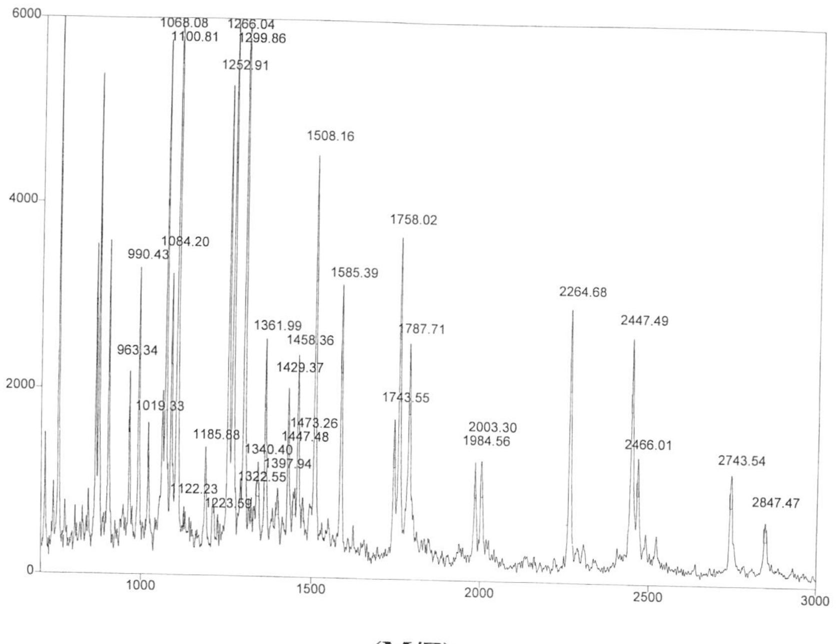

B Ga(III) IMAC/ pH 2.5

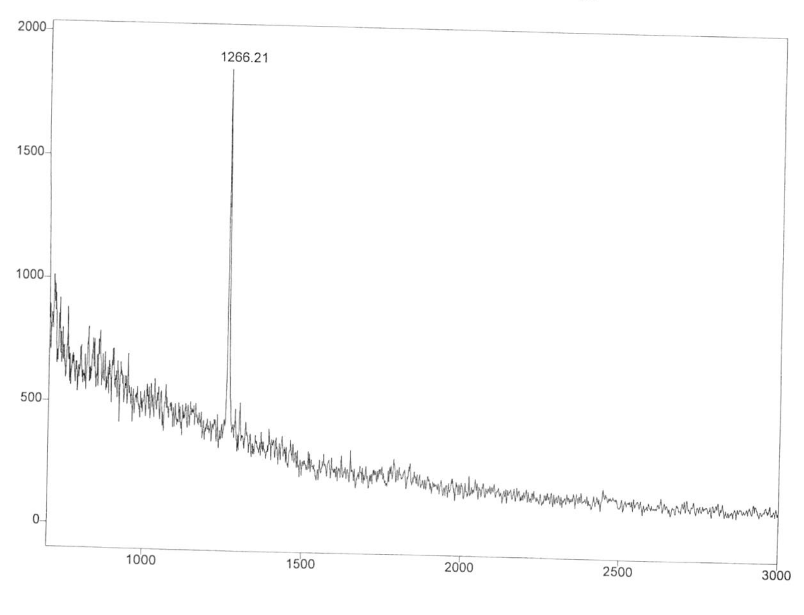

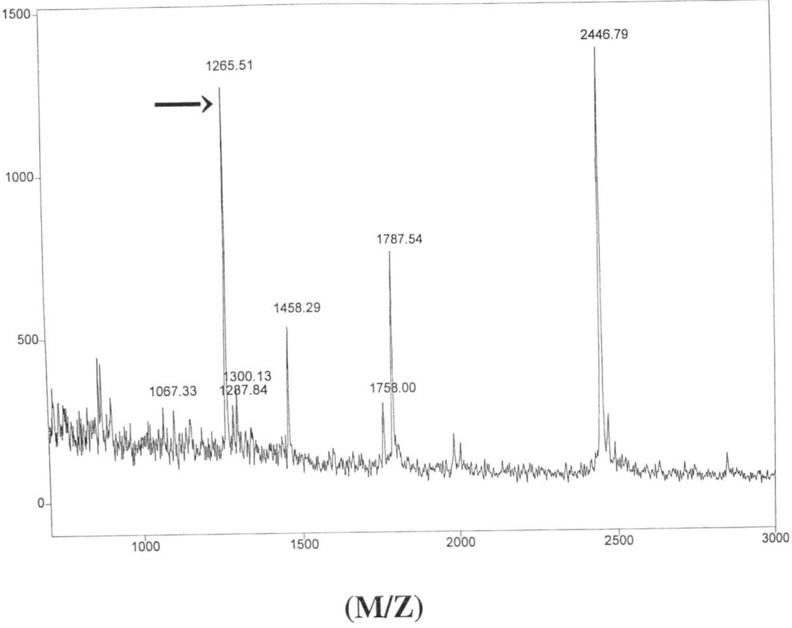

*Fig. 6. Selective IMAC isolation of phosphopeptide from a proteolytic digest. A phosphopeptide (sequence, see text) was mixed with β-galactosidase tryptic digest (1:2; phosphopeptide:digest) and analyzed by MALDI-TOF MS before (**panel A**) or after IMAC passage over an IMAC micro-column (**panels B and C**). Note that the eluted fractions were desalted over an RP micro-column before mass analysis. Phosphopeptide is marked with an arrow in panel C.*

gave the highest phosphopeptide recoveries. Al(III)-IMAC recoveries were lower, a rather unexpected finding in view of its popular use in many past studies on phosphopeptide characterization [18, 21]. It was also observed during these and later experiments that, for optimal performance, IMAC micro-columns must not be allowed to run dry, and that loading/washing/elution should be done slowly, using minimal argon pressure (2-3 psi).

To evaluate the capacity of Fe(III), Ga(III) and Zr(IV) ions for selective retention of phosphopeptides, a tryptic peptide (IGEGTpYGVVYK) was added to a tryptic digest of the 116 kDa protein β-galactosidase, in a 1:2 molar ratio (phophopeptide: β-galactosidase), and applied to and eluted from the IMAC columns. Eluates were desalted using RP micro-tips and then analyzed by MALDI-MS. The spectra shown in Fig. 6 indicate that Ga(III)-IMAC, while giving recovery comparable or better than the other metal ions (only Fe(III) is shown), retained the least β-galactosidase tryptic peptides. In fact, there were no contaminating peptides at all. To further probe general applicability, the experiment was repeated with two entirely different phosphopeptides (shown in Fig. 7), and again Ga(III) ions yielded superior results. Care should be taken,

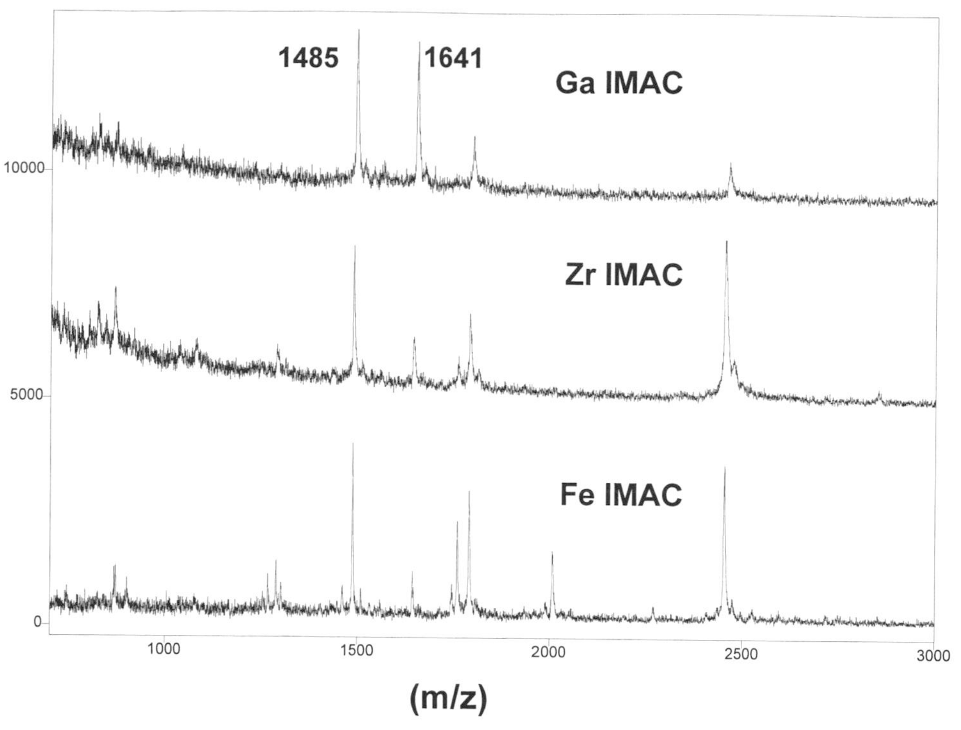

Fig. 7. Comparative IMAC micro-column isolation of phosphopeptides. Two phosphopeptides (RRLIEDAEpYAARG-amide; RLIEDAEpYAARG-amide) were added to a tryptic digest mixture, which was then divided into three equal parts and processed in parallel over IMAC micro-columns to which three different metal ions (as indicated) had been immobilized, desalted over an RP micro-column and analyzed by MALDI-TOF MS; all as described under Fig. 6.

however, not to load any peptide solutions of pH > 3.0 onto the column, as an increasingly larger number of unphosphorylated peptides will bind non-specifically.

A major drawback of IMAC-assisted phosphopeptide isolation as commonly done, and as described above, is the need for 0.2M phosphate in the eluant, which is incompatible with direct mass spectrometric analysis using either MALD or ES ionization (Posewitz *et. al.*; unpublished observations). Thus, we explored the possibility of eluting phosphopeptides with basic 0.075% ammonium hydroxide, pH 10.5. Under these conditions, and with the micro-column pre-equilibrated with 0.1% acetic acid (from washing), the collected fractions are mildly alkaline and essentially free of salt, and do not interfere with matrix/peptide crystallization (for laser desorption/ionization) or ESI. A radiolabeled phosphopeptide was stably adsorbed onto either a Fe(III), Ga(III) or Zr(IV) IMAC -column and eluted (in separate experiments) with either phosphate or base. The results in Fig. 8 indicate that base-promoted phosphopeptide elution was achieved only from the Ga(III)-affinity support, even slightly better, in fact,

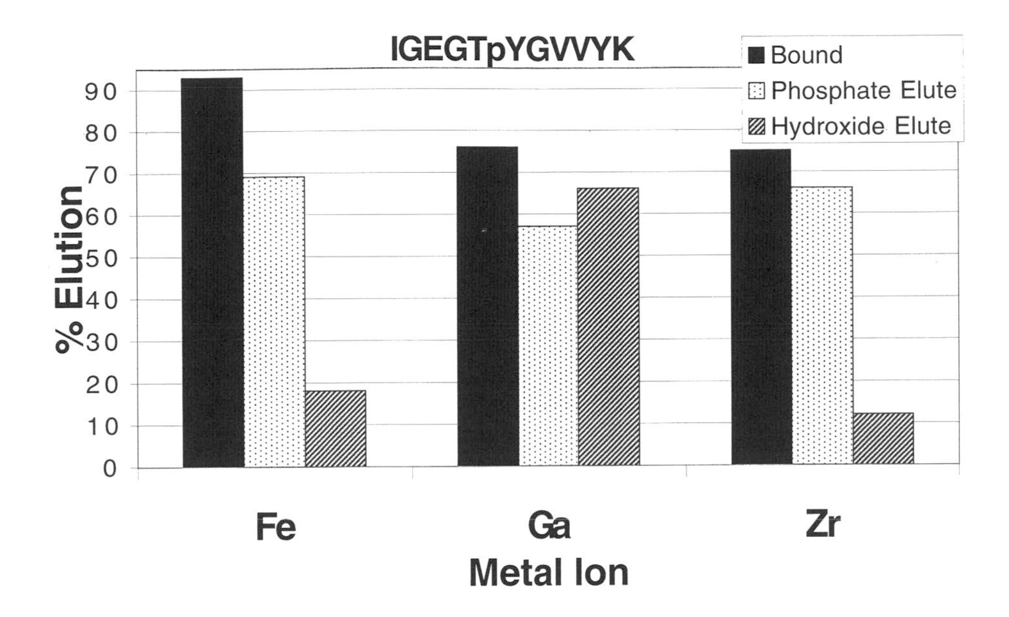

Fig. 8. Phosphate versus base elution of phosphopeptides from IMAC micro-columns. [^{32}P]-labeled phosphopeptide was loaded (at pH 2.5) onto a POROS MC (containing immobilized metal ions, as shown) micro-column, extensively washed, eluted with either 0.2M phosphate or 0.075% ammonium hydroxide pH 10.5, and recoveries monitored by Cerenkov counting. % bound was calculated (100-x %) by counting of the flow-through and wash solutions (combined x%).

than with phosphate. By contrast, poor (≤25%) recoveries from the Fe(III) and Zr(IV)-substituted resins were noted. These findings all but eliminate the use of the latter two metal ions for this purpose. Taken together, Ga(III)-IMAC binds phosphopeptides with the greatest selectivity and may be used to directly elute and analyze samples, thereby saving time and analyte. Moreover, mild base is favorable for directly coupled phosphopeptide analysis by infusion-ESI-MS in the negative ion mode [22].

Continuous Nano-Electrospray

Once prepared in a small volume of an ideal solvent as described in the preceding sections, complex peptide mixtures or purified phosphopeptides are then analyzed by MALDI-TOF and ESI-triple quad MS. In the case of the second mass spectrometric technique, capillary LC- or CZE-ESI-MS may offer advantages in terms of sensitivity [23, 24] but impose severe restrictions on analysis time for each peptide, essentially requiring full automation of MS/MS analysis. This experimental limitation can be alleviated by the use of ultra-low-flow infusion devices such as nano-electrospray ("NanoES") [8, 25]. Although applied superbly to solve a myriad of research problems [22, 26-28], the original NanoES assembly [8] does not allow for quality control of the spray needles, precise

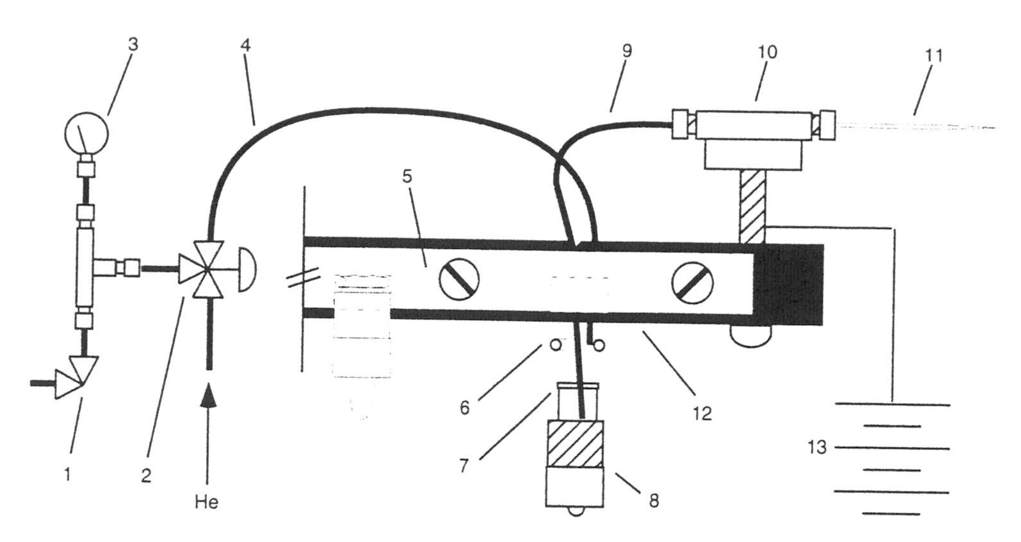

Fig. 9. Schematic diagram of an inJection adaptable Fine Ionization Source ("JaFIS"). A 0.2 mL PCR tube (sample reservoir; #7) is inserted into a nylon collar (#8), which fits into the threaded socket of a Teflon® sample holder (#5), also containing a O-ring (#6). Two 365 µm OD fused silica capillaries (FSC), a pickup line (#9) and pressure / vent line (#4), are inserted through the sample holder into the reservoir and sealed into place using adhesive epoxy. The 150 µm ID pressure / vent line connects to a pressure regulation station via a low dead volume union. The 25 µm ID pickup line is positioned with the inlet end just above the bottom of the reservoir. The outlet of the pickup line is connected to the ion spray body (ISB; #10) through a hybrid graphite / PTFE 0.4 mm ID ferrule with a 1 / 32 inch stainless steel bushing. The ISB consists of a 0.15 mm bore / 0.5 mm long stainless steel micro volume union. The ion spray needle (ISN; #11) is a 3 cm long, pencil-pointed piece of FSC which has been narrowed down at one end to an inner diameter of 1 to 5 micron using a laser puller. The ISN is connected to the ISB in the same way as the pickup line. Sample holder and ISB are attached to an 0.5 inch diameter Delrin® shaft (#12). The ion spray potential is supplied to the ISB from the ion source voltage connection (#13) on the PE-SCIEX API 300 instrument via a round head screw. The pressure control system consists of a regulator (#2) connected to an ultra-high purity helium source, and is monitored on a 0-60 psi gauge (#3), and vented via a purge valve (#1) connected to the gauge port of the reducer.

tuning of the flow, and any form of multi-sample analysis — features fairly important for high throughput and absolute reliability. Hence, we have constructed a *bona fide* NanoES source that incorporates these practical features: an in*J*ection *a*daptable *F*ine *I*onization *S*ource ("JaFIS").

A schematic view of the JaFIS assembly is given in Fig. 9. JaFIS is positioned through the use of three stepper-motor driven translation stages, controlled by a motion controller/driver and a Macintosh PowerBook G3 portable computer. The ICS was developed using LabView software. Monitoring of the ISN position and ionization characteristics, and of the liquid level in the sample reservoir, is done through a number of high resolution, magnifying CCD cameras mounted to a vibration-proof table. The video output is displayed on two monitors. Further technical details are beyond the scope of this chapter and can be found elsewhere [29].

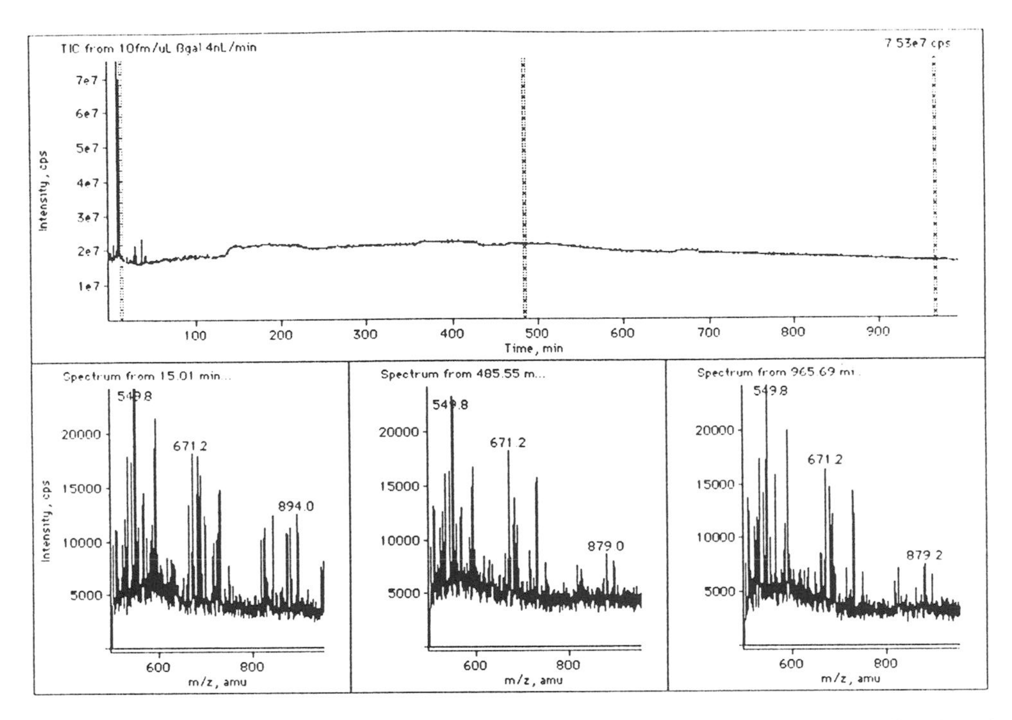

Fig. 10. Mass spectrometric data acquired during continuous operation of JaFIS using 4 μL of 10 fmoles / μL of β-galactosidase tryptic peptides. Panel A shows the total ion current over the full 17 hours of the experiment; indicating an average spray rate ≤ 4 nL / min. Lower panels are spectra averaged over 5 min (12 scans) and taken at time points (from left to right) 15 min, 8 hours and 16 hours, respectively. Selected ions are labeled to serve as reference points.

Previously described nanoelectrospray tips were gold-coated to provide ion spray potential to the analysis solution [8, 30, 31]. However, the gold burns off rather quickly with resultant loss of ion current [30, 32]. Because the ion spray potential in JaFIS is applied directly to the sample liquid through the ISB, the gold coating is no longer necessary and spray time is limited only by the sample volume. In a series of experiments, the system was allowed to operate continuously for increasingly longer periods of time while its performance was monitored. In one instance, 4 μL of control mixture (10 femtomoles of β-galactosidase tryptic peptides per μL) was loaded into the reservoir and Q1 scan data was acquired for 17 hours, as shown in Fig. 10. Stable TIC was sustained for the duration of the experiment. Moreover, the spectra shown in the three lower panels of Fig. 10 illustrate that relative ion intensity is remarkably constant, at least for the smaller hydrophilic peptides, despite being separated by over 15 hours of continuous spraying. More specifically, three ions, with *m / z* of 549.8, 671.2, and 879 (marked in Fig. 10; lower panels), were then arbitrarily selected and their relative intensities compared over time. Ion signals of larger, more hydrophobic peptides, such as the one marked "879", markedly decreased over time, the result of adsorption to the plastic vessel wall which is not uncommon

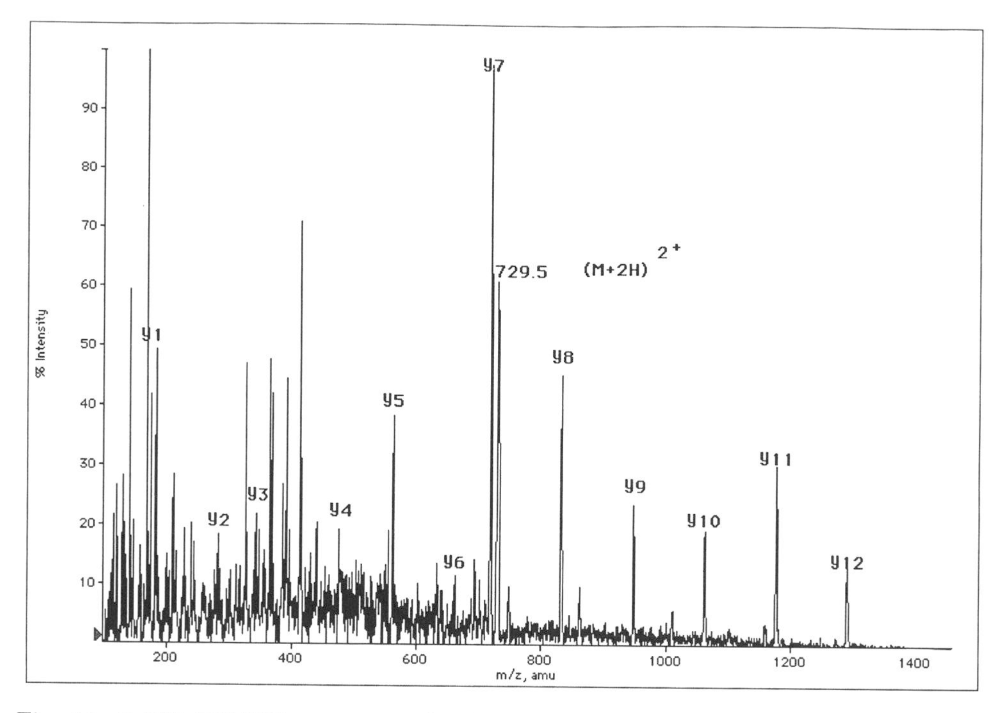

Fig. 11. JaFIS-MS/MS spectrum of 2 fmol/μL of a β-galactosidase tryptic peptide. Fragmentation of the doubly charged ion at m/z 729, selected from a Q1 spectrum similar to that shown in Fig. 10, is presented. The y"-ion series has been marked. Experimental: PE-SCIEX API300 triple-quadrupole instrument; 2.5 nL/min spray; 6.5 sec scans (1 amu stepsize, 5 msec dwell time); 700 scans averaged.

at room temperature and at such low concentrations. The above experiment clearly demonstrates that ample analysis time is available to dissect complex peptide mixtures, allowing MS/MS analysis not only of a large number of peptides but also for lengthy periods of time each. In turn, this enables optimization of instrument experimental settings for individual peptides and averaging of many spectra, hours on end if need be, thereby producing high quality spectra from low concentration analytes. An example of this approach is given in Fig. 11, depicting a product ion spectrum generated from a selected peptide during infusion of 2 femtomoles/μL of digest mixture (β-gal tryptic peptides) at 2.5 nL/min. Scans (1 amu stepsize, 5 msec dwell time; 6.5 sec each) were averaged for 76 min (totaling 700 scans), thereby consuming 175 nL (out of 2 μL in the PCR tube) of sample.

The extended ISN life time, combined with the ability to quickly and easily change samples, suggested that multiple samples could be analyzed consecutively through JaFIS with little cross-contamination. In one such "multi-sample" experiment, 100 fmol/μL alpha-casein tryptic peptide mixture (5 μL) was sprayed for 30 min from a previously optimized source. This sample was then replaced with a new reservoir containing 5 μL of BSA tryptic digest, which was sprayed for 30 min as well, and then again replaced with β-galactosidase

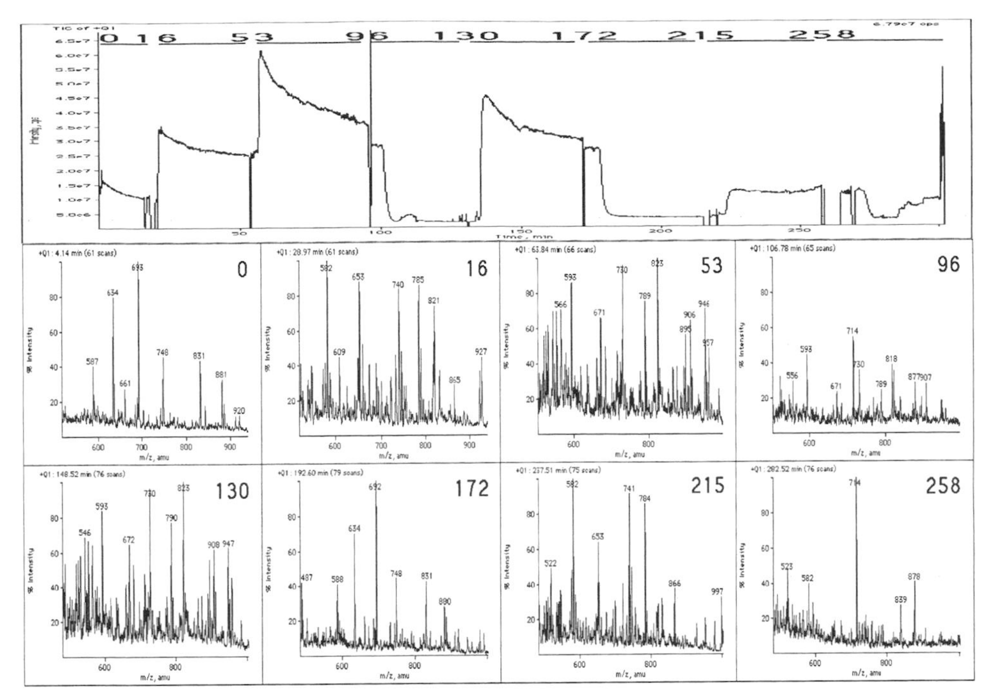

Fig. 12. Mass spectrometric data acquired during semi-continuous operation of JaFIS analyzing eight different samples consecutively. Samples are tryptic digests of various proteins (100 fmoles / μL). Top panel shows TIC over almost 5 hours of total analysis time. The sequence of events is depicted by horizontal bars (separated by gaps indicating time points — in min — where samples have been changed), representing a linear time line: 0, alpha-casein; 16, BSA; 53, β-galactosidase; 96, lysozyme; 130, β-galactosidase; 172, alpha-casein; 215, BSA; 258, lysozyme. Lower panels (marked with the corresponding time points) depict spectra averaged over 5 min (12 scans) to illustrate the lack of cross-contamination between the peptide mixtures being analyzed.

tryptic digest, and then with lysozyme derived tryptic peptides; this sequence of four sample exchanges was repeated once. The TIC was monitored until continuous ion current was fully restored and the Q1 scans indicated the perfect signature spectra of the four different tryptic maps. Figure 12 gives an overview of these experiments. As depicted in the top panel, TIC was lost as voltage was removed and one sample reservoir was replaced with the next, but returned when the voltage and spray position were restored. The differences in relative TIC between the 8 samples ("baseline" up and down in the top panel of Fig. 12) is related to the size of the respective proteins. Indeed, a tryptic digest of β-galactosidase (116 kDa) obviously generates more peptides than a similar digest of lysozyme (14 kDa), resulting in higher TIC. Note that there was no visible cross-contamination between peptides from the different digests. Using JaFIS under these low flow (~ 15-20 nL/min) operating conditions, the amount of sample lost while waiting for the cross-contamination to disappear can be considered negligible.

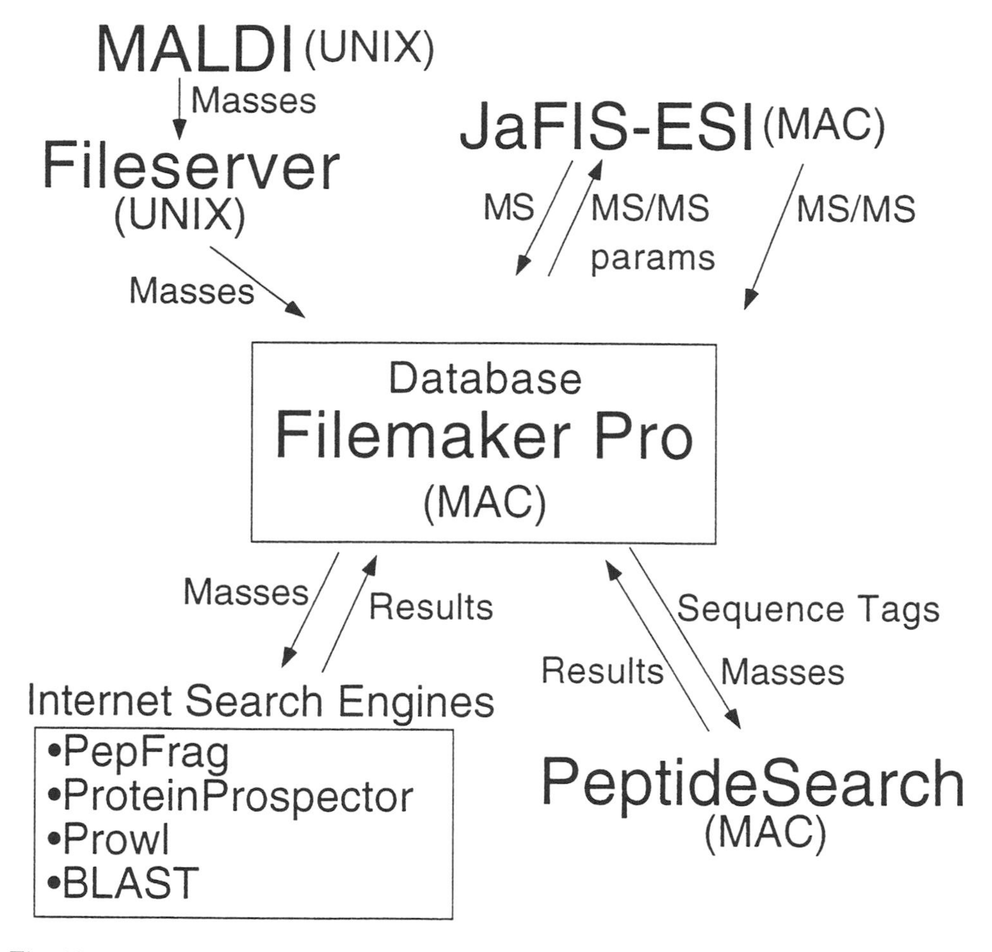

Fig. 13.

LIMS AND DATA FLOW

With the available techniques and strategies described in the previous sections, an applications lab should now be able to perform an increasing number of protein identifications. To handle the resulting stream of data, however, customized laboratory information management systems ("LIMS") have become essential. LIMS software is under development in our laboratory and consists of several FileMaker Pro databases and Applescripts. The LIMS tracks the project from start (administrative data) to finish (identification and report generation) and interacts with data from different mass spectrometers and search engines (Fig. 13).

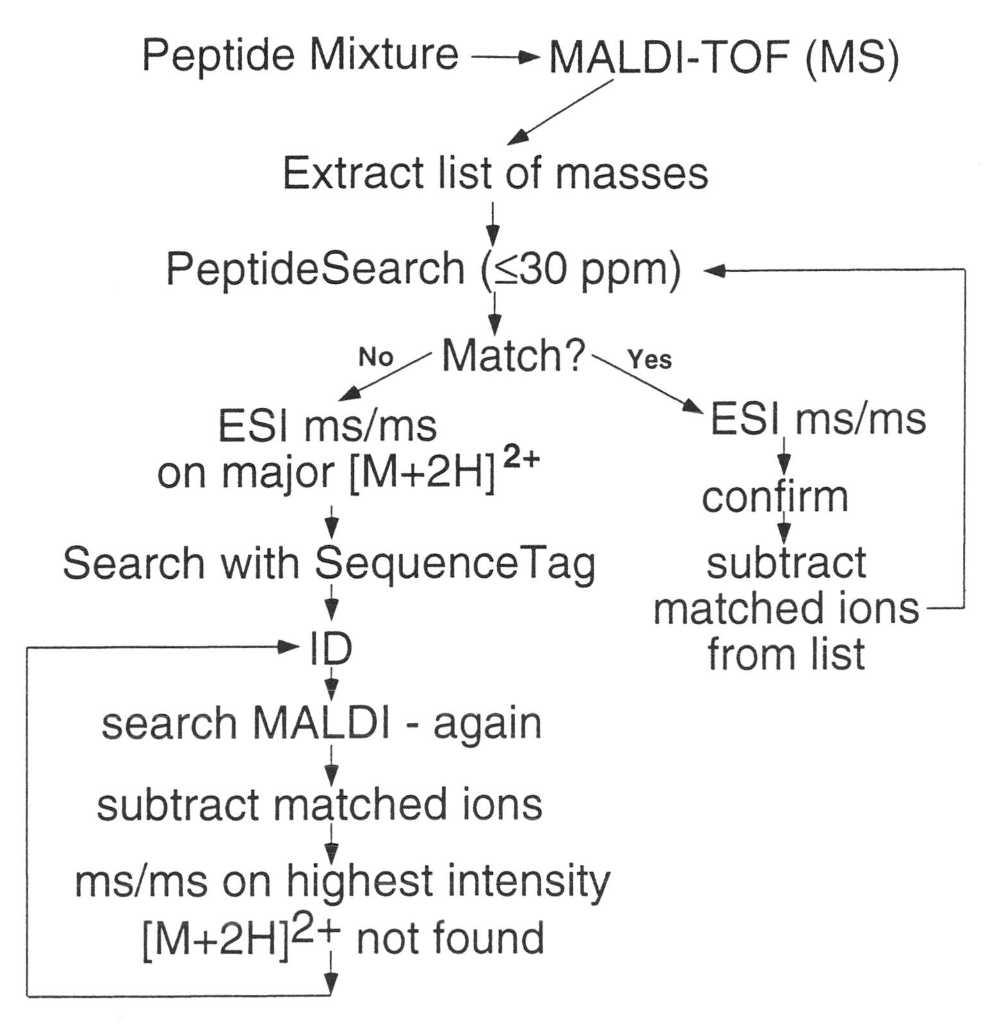

Fig. 14

Data from a MALDI-TOF (BRUKER Reflex III) is imported into the FileMaker Pro databases for data reduction and analysis. The signature MALDI data is screened against a nonredundant protein database (NRDB: ~ 330,000 entries) using PeptideSearch [2]. Identified masses are removed from the data set. To facilitate the identification of the remaining masses, the sample is loaded into the JaFIS and MS data (PE-SCIEX API300) is acquired for 5 minutes and then sent to the LIMS. After performing some calculations, it displays the ESI precursor masses for each of three possible charge states (1^+, 2^+, 3^+) plus relative intensities and recommends MS/MS parameters. The MS/MS fragmentation pattern can be interpreted manually or through automation. In the former case, the user (working in BioMultiview) selects a Sequence Tag and searches the tag against NRDB. In the automated mode, an Applescript queries the user for the number of ions, *m/z* of the precursor ion, the mass range for identification, whether or not to accept ions different by a mass of 17, precursor *m/z*, and the search stringency. These masses are then sent over the Internet to be searched using PepFrag at prowl.rockefeller.edu/PROWL/pepfragch.html [33]. All search results are stored in the LIMS (Figure 14).

SUMMARY

We have described ongoing efforts in our laboratory to put in place a well-structured, multi-step procedure for medium-volume (≤500 units / year) protein identification. The overriding priority in shaping this system has been to achieve the highest level of confidence, even when analyzing small amounts (≤200 fmoles) of protein. To that end, three areas were selected for fine-tuning: protein micro-biochemistry (sample handling and purification of phosphopeptides), ultra-sensitive "sequencing" (use of continuous flow nano-electrospray), and adequate management of the data flow. The final aim is integration of these efforts in a single package.

ACKNOWLEDGEMENTS

The work described herein was supported, in part, by Development Funds from National Cancer Institute Grant P30 CA08748.

REFERENCES

1. W. J. Henzel, T. M. Billeci, J. T. Stults, S. C. Wong, C. Grimley and C. Watanabe, *Proc. Natl. Acad. Sci. USA* **1993**, *90*, 5011-5015.

2. M. Mann, P. Højrup and P. Roepstorff, *Biol. Mass Spectrom.* **1993**, *22*, 338-345.

3. D. J. C. Pappin, P. Højrup and A. J. Bleasby, *Curr. Biol.* **1993**, *3*, 327-332.

4. J. R. Yates III, S. Speicher, P. R. Griffin and T. Hunkapiller, *Anal. Biochem.* **1993**, *214*, 397-408.

5. M. Mann and M. Wilm, *Anal. Chem.* **1994**, *66*, 4390-4399.

6. J. K. Eng, A. L. McCormack and J. R. Yates III, *J. Am. Soc. Mass Spectrom.* **1994**, *5*, 976-989.

7. H. Zhang, P. E. Andren and R. M. Caprioli, *J. Mass Spectrom.* **1995**, *30*, 1768-1771.

8. M. Wilm and M. Mann, *Anal. Chem.* **1996**, *68*, 1-8.

9. R. S. Annan, D. McNulty, M. J. Huddleston and S. A. Carr, *International Symposium ABRF'96: Biomolecular Techniques*, San Francisco, CA, 1996, Abstract S73.

10. P. L. Courchesne and S. D. Patterson, *BioTechniques* **1997**, *22*, 246-250.

11. F. Rusconi, J.-M. Schmitter, J. Rossier and M. le Maire, *Anal. Chem.* **1998**, *70*, 3046-3052.

12. H. Erdjument-Bromage, M. Lui, L. Lacomis, A. Grewal, R. S. Annan, D. E. McNulty, S. A. Carr and P. Tempst, *J. Chromatogr.*, in press.

13. T. M. Billeci and J. T. Stults, *Anal. Chem.* **1993**, *65*, 1709-1716.

14. U. Hellman, C. Wernstedt, J. Gonez and C.-H. Heldin, *Anal. Biochem.* **1995**, *224*, 451-455.

15. M. Lui, P. Tempst and H. Erdjument-Bromage, *Anal. Biochem.* **1996**, *241*, 156-166.

16. L. Andersson and J. Porath, *Anal. Biochem.* **1986**, *154*, 250-254.

17. G. Muszynska, L. Andersson and J. Porath, *Biochemistry* **1986**, *25*, 6850-6853.

18. L. Andersson, *J. Chromatogr.* **1991**, *539*, 327-334.

19. P. Scanff, M. Yvon and J. P. Pelissier, *J. Chromatogr.* **1991**, *539*, 425-432.

20. G. Muszynska, G. Dobrowolska, A. Medin, P. Ekman and J. Porath, *J. Chromatogr.* **1992**, *604*, 19-28.

21. M. T. Shoemaker and B. E. Haley, *Biochemistry* **1993**, *32*, 1883-1890.

22. S. A. Carr, M. J. Huddleston and R. S. Annan, *Anal. Biochem.* **1996**, *239*, 180-192.

23. J. M. M. den Haan, N. E. Sherman, E. Blokland, E. Huczko, F. Koning, J. W. Drijfhout, J. Skipper, J. Shabanowitz, D. F. Hunt, V. H. Engelhard and E. Goulmy, *Science* **1995**, *268*, 1476-1480.

24. D. Figeys, A. Ducret, J. R. Yates III and R. Aebersold, *Nature Biotechnol.* **1996**, *14*, 1579-1583.

25. M. S. Wilm and M. Mann, *Int. J. Mass Spectrom. Ion Processes* **1994**, *136*, 167-180.

26. M. Wilm, A. Shevchenko, T. Houthaeve, S. Breit, L. Schweigerer, T. Fotsis and M. Mann, *Nature* **1996**, *379*, 466-469.

27. M. Muzio, A. M. Chinnaiyan, F. C. Kischkel, K. O'Rourke, A. Shevchenko, J. Ni, C. Scaffidi, J. D. Bretz, M. Zhang, R. Gentz, M. Mann, P. H. Krammer, M. E. Peter and V. M. Dixit, *Cell* **1996**, *85*, 817-827.

28. J. Lingner, T.R. Hughes, A. Shevchenko, M. Mann, V. Lundblad and T.R. Cech, *Science* **1997**, *276*, 561-567.

29. S. Geromanos, J. Philip, G. Freckleton and P. Tempst, *Rapid Commun. Mass Spectrom.* **1998**, *12*, 551-556.

30. M. S. Kriger, K. D. Cook and R. S. Ramsey, *Anal. Chem.* **1995**, *67*, 385-389.

31. D. Figeys, I. van Oostveen, A. Ducret and R. Aebersold, *Anal. Chem.* **1996**, *68*, 1822-1828.

32. G. A. Valaskovic and F. W. McLafferty, *J. Am. Soc. Mass Spectrom.* **1996**, *7*, 1270-1272.

33. D. Fenyö, W. Zhang, B. T. Chait and R. C. Beavis, *Anal. Chem.* **1996**, *68*, 721A-726A.

Questions and Answers

John Stults (Genentech)

In the beginning of your talk you questioned the usefulness of *proteomics* showing a slide that stated "Why not (now)"? What will need to change to make *proteomics* useful in the future?

Answer. It all depends, of course, how you define *proteomics*. In the widest sense as "Simultaneous Analysis of All Proteins in the Cell", it seems rather ambitious at the present time. Especially, when approached by 2D-gel analysis of whole cell extracts. Many objectives that some are trying to pursue this way can more easily be realized using *genomics* technology. Moreover, there are a number of inherent problems with the comparative 2D-gel strategy that will almost certainly lead to incomplete information. On the other hand, there is an essentially unlimited list of relatively large-scale projects that can be done in the *proteomics* area, which may lead to better insights in basic biology and disease mechanisms in the short term. What comes to mind are dynamic complexes of proteins (e. g., in gene expression), the composition of which may vary upon signaling, activation and repression, by physiological or pharmacological concentrations of ligands and drugs, by incorporation of hybrid fusion proteins, or resulting from other genetically encoded abnormalities. All this in different cells, during different stages of development or differentiation. An enormous undertaking that will keep many labs occupied for years to come. Call it *targeted proteomics* if you will. If and when the time will come for a carpet-bombing style, "2D-gel-map-plus-mass-spec-ID" approach, major improvements in gel technology and automation will be required.

Steven Carr (SmithKline Beecham Pharmaceuticals)

What percentage of the *interesting* proteins are present at only very low copy number per cell (<10)? Is it possible / reasonable to use 2D-gel based approaches to detect these proteins?

Answer. I can't recall ever seeing exact or approximate numbers, even though they may be available somewhere. My guess is that a very large percentage falls in the "low abundant" category. This is based on reports (Zhang et al., *Science* **1997**, *276*, 1268-1272) that as many as 14,000 to 20,000 different transcripts can be found in a single cell population; yet, to my knowledge, only 500-1500 spots can be detected on the best 2D-gels, obviously representing only the top 5-10%. Although this doesn't put a precise numerical value on the distribution of abundance, it clearly answers the second question: by using 2D-gels, many proteins will be missed!

Steven Carr (SmithKline Beecham Pharmaceuticals)

How do you go about validating the results from chip-based or multi-hybrid genomic approaches?

Answer. By doing the biology, starting with northern blots and immunochemistry, all the way to functional assays and knock-outs. This is often a tremendous amount of work, but the only sure way. The same is true, of course, for mass-spec based protein identifications. Validation will take time. Which brings into focus two important issues. First, all talk about mega-efforts on protein identification should be tempered by the knowledge that in every such project an enormous bottleneck will present itself right thereafter. Second, identifications better be right for the validation process is long and costly. For these reasons, fully automated, absolutely no user-interaction type of identification schemes frighten me.

David Maltby (UCSF)

Why is the optimal flow rate of your tips lower (i. e., 2-3 nanoliters/min) than that shown by Arthur Moseley in yesterday's talk?

Answer. The differences in optimal flow rate between our two data sets is due to the difference in inner diameter of the spray tip. Tom Covey of SCIEX has previously shown that optimum sensitivity is found at different flow rates from different size tips. Our tip inner diameters are approximately one order of magnitude smaller than those of the Moseley group. So it is not surprising to see a similar difference in optimum flow rate.

Dan Kassel (Combichem)

Relating to the robustness of the extruded capillary with a 2 micron aperture opening, what is your experience with *real-world* proteins and the propensity for these capillaries to plug/clog?

Answer. There is actually no difference between standards and real-world proteins when it comes to using these spray needles, at least not the way we do it. All digest mixtures are first passed over a RP micro-tip before doing continuous flow nanospray. Besides providing sample concentration and clean-up, the beads in the micro-column also act as a filter to keep out dust and other small particles that otherwise might clog the spray needle. The best way to avoid plugging the needle with particulates from sources other than the sample is by working in a clean environment. Routinely, all needles are also pretested for possible spraying problems. If, despite all these precautions, the needle would still plug during mid-run, it can be quickly replaced without major sample losses (see [29]).

A. L. Burlingame (UCSF)

What is the basis for the selectivity of Ga(III) affinity for phosphopeptide isolation? Will other acidic components, such as sialyl glycosylated peptides, be bound as well?

Answer. We don't really know. As we showed, selection of Ga(III) metal ion for phosphopeptide isolation was the result of a comparative study and not based on any preconceived notions, except for the fact that it is multi-valent. While possibly a contributing factor, ionic interaction alone can not account for its specificity. We found that acidic peptides with net charges of -3 and -4 were only minimally (<1%) retained at pH ≤3.0. By contrast, these same peptides bound rather strongly to Fe-chelates at acidic pH.

Jasna Peter-Katalinic (University of Münster)

Would it be possible to detect glycosylated peptides using your approach with Ga(III) IMAC tips and base elution, monitoring in (-) ESI or MALDI-TOF MS?

Answer. We won't know that until we do the experiments. Our prediction is that they will probably not bind specifically. Chances are, though, that acidic glycopeptides can be retained non-specifically on Fe(III) IMAC columns, which could conceivably be used for enrichment purposes.

Towards an Integrated Analytical Technology for the Generation of Multidimensional Protein Expression Maps

Paul A. Haynes, David R. Goodlett, Steven P. Gygi, Julian D. Watts, Daniel Figeys and Reudi Aebersold
Department of Molecular Biotechnology, University of Washington, Seattle, WA 98195

A proteome has been defined as the protein complement expressed by the genome of an organism, or, in multicellular organisms, as the protein complement expressed by a tissue or differentiated cell [1]. In the most commonly used approach to proteome analysis the proteins extracted from the cell or tissue to be analyzed are separated by high resolution two dimensional gel electrophoresis (2DE), detected in the gel by staining or, if radiolabeled, by autoradiography, and identified by their amino acid sequence. The ease, sensitivity and speed with which gel separated proteins can be identified by the use of recently developed mass spectrometric techniques (MS) has led to a dramatic increase in interest in proteome studies (reviewed in [2]). One of the most attractive features of such analyses is that complex biological systems can potentially be studied as complete systems, rather than as a conglomerate of individual components.

Large-scale proteome characterization projects have been undertaken for a number of different organisms and cell types. Microbial proteome projects currently in progress include those studying, for example: *Saccharomyces cerevisiae* [3]; *Salmonella enterica* [4]; *Spiroplasma milliferum* [5]; *Mycobacterium tuberculosis* [6]; *Ochrobactrum anthropi* [7]; *Haemophilus influenzae* [8]; *Synechocystis spp.* [9]; *Escherichia coli* [10]; *Rhizobium leguminosarum* [11]; and *Dictyostelim discoideum* [12]. Proteome projects underway for tissues of more complex organisms include those for: human bladder squamous cell carcinomas [13]; human liver [14]; human plasma [14]; human keratinocytes [13]; human fibroblasts [13]; mouse kidney [13]; and rat serum [15].

In this chapter we discuss some of the challenges raised by proteome analysis and describe techniques which have been developed to meet these challenges.

Factors Complicating Proteome Analysis

In spite of the considerable effort spent on microbial and eukaryotic proteome projects, to date no entire proteome map has been completed. This is in contrast to genome sequences and quantitative mRNA expression maps which have been completed for prokaryotic and eukaryotic organisms [16-19]. There are a number of factors which complicate proteome analysis and have to date prevented the completion of proteome maps.

Proteins are Dynamically Modified and Processed

In the biologically active form many mature proteins are posttranslationally modified by, for example, glycosylation, phosphorylation, prenylation, acylation, ubiquitination, N- or C-terminal processing, or one or more of many other modifications [20]. Many proteins are only fully functional if specifically associated or complexed with other molecules, including DNA, RNA, other proteins, or organic and inorganic co-factors. Frequently, modifications are dynamic and reversible and may alter the precise three dimensional structure and the state of activity of a protein. The distribution of a constant amount of protein over several differentially modified isoforms further reduces the amount of each species available for analysis. Collectively, the states of modification of the proteins which constitute a biological system are important indicators for the state of the system. Complete analysis of posttranslationally modified proteins requires that the differentially modified forms be separated and analyzed [21], because the type of protein modification and the sites modified at a specific cellular state cannot generally be determined from the gene sequence alone. The complexity and dynamics of posttranslational protein editing thus significantly complicates proteome studies.

A further important consideration is that proteins cannot be amplified and therefore need to be isolated from their native source. It is possible to produce large amounts of a particular protein by over-expression in specific cell systems. However, because many proteins are dynamically posttranslationally modified, they cannot be easily amplified in the form in which they finally function in the biological system. It is frequently very difficult to purify from the native source sufficient amounts of a protein for analysis. From a technological point of view this translates into the need for highly sensitive analytical techniques.

Unlike Genomes, Proteomes are Inherently Dynamic

A single, essentially static, genome can give rise to many qualitatively and quantitatively different proteomes. Examples of cellular states which may be characterized by significantly different proteomes include specific stages of the cell cycle and states of differentiation, states induced by specific growth and nutrient conditions, temperature and stress, and pathological conditions. Unlike transcript profiling, proteome analysis, in principle, also reflects events which are under translational and posttranslational control. Therefore, it is expected that compared to mRNA transcript profiling, proteomics will be able to provide a more meaningful and detailed molecular description of the state of a cell or tissue, provided that the external conditions defining the state are carefully determined. The dynamic nature of proteomes also poses specific challenges in their analysis.

Proteins are difficult to maintain in solution and vary widely with respect to their solubility in commonly used solvents. Analytical techniques to be used must be able to solubilize all of the ever-changing components of a dynamic system. There are few, if any, solvent conditions which are compatible with protein analysis and in which all proteins are soluble. This makes the development of protein purification methods particularly difficult because both protein solubility and purification have to be achieved under the same conditions. Detergents, in particular sodium dodecyl sulfate (SDS), are frequently added to aqueous solvents to maintain protein solubility. The compatibility with SDS is a big advantage of SDS polyacrylamide gel electrophoresis (SDS-PAGE) over other protein separation techniques. Thus, SDS-PAGE and two-dimensional gel electrophoresis, which also uses SDS and other detergents, are the most general and preferred methods for purification of small amounts of proteins, provided that activity does not necessarily need to be maintained.

Collectively, these considerations indicate that a general proteome analysis technology needs to be highly sensitive, needs to be able to resolve and analyze differentially modified proteins, and needs to employ defined conditions which are suitable for the analysis of all proteins.

Current Technology for Proteome Analysis

A number of techniques for proteome analysis have been described [2]. They have in common that they combine high resolution protein and peptide analysis systems with analytical instruments for the structural characterization of the separated analytes. They differ in their performance and their specificity. Below we discuss methods which we have found particularly useful.

2D Electrophoresis-Mass Spectrometry

The most widely used implementation of proteome analysis technology is based on the separation of proteins by two-dimensional (IEF/SDS-PAGE) gel electrophoresis and their subsequent identification and analysis by mass spectrometry (MS) or tandem mass spectrometry (MS/MS). In 2DE proteins are first separated by isoelectric focusing (IEF), and then in the second perpendicular dimension by SDS-PAGE. Following separation, proteins are visualized at high sensitivity by either staining or autoradiography, producing two-dimensional arrays of proteins. Separation of thousands of proteins has been achieved in a single gel [22, 23], and differentially modified proteins are frequently separated. One reason the technique is so widely used is the compatibility of 2DE with high concentrations of detergents, protein denaturants and other additives promoting protein solubility.

The second step of this type of proteome analysis is the identification and analysis of separated proteins. Individual proteins from polyacrylamide gels have traditionally been identified using N-terminal sequencing [24, 25], internal peptide sequencing [26, 27], immunoblotting or comigration with known proteins [28]. The recent dramatic growth of large-scale genomic and expressed sequence tag (EST) sequence databases has resulted in a paradigm shift in the way proteins are identified. Rather than by the traditional methods described above, protein sequences are now more frequently determined by correlating mass spectral or tandem mass spectral data of peptides derived from proteins with the information contained in sequence databases [29-31].

Protein spots detected in a gel or after electroblotting on a membrane are enzymatically or chemically fragmented and the peptide fragments are isolated for analysis, most frequently by MS or MS/MS. There are many different protocols used for the generation of peptide fragments from gel-separated proteins. They can be grouped into two categories, digestion in the gel slice [26, 32] or digestion after electrotransfer out of the gel onto a suitable membrane ([27, 33-35] and reviewed in [2]). In most instances either technique is applicable and yields good results.

The analysis of MS or MS/MS data is an important step in the whole process because MS instruments can generate an enormous amount of information which cannot easily be managed manually. Recently, a number of groups have developed software systems dedicated to the use of peptide MS and MS/MS spectra for the identification of proteins. Proteins are identified by correlating the information contained in the MS spectra of protein digests of MS/MS spectra of individual peptides with data contained in DNA or protein sequence databases.

The systems we are currently using in our laboratory, which are described in detail below, are based on the separation of the peptides contained in protein digests by narrow bore or capillary liquid chromatography [21, 36] or capillary electrophoresis [37], the analysis of the separated peptides by electrospray ionization (ESI) MS/MS, and the correlation of the generated peptide spectra with sequence databases using the SEQUEST program [30, 31]. The individual systems described below differ in the way the peptides are introduced into the MS and therefore in their levels of sensitivity and automation.

Protein identification by MS is a complex multistep procedure. A typical example of how this procedure is employed when using a triple quadrupole MS is as follows. In the first part of the process, a particular peptide ion of a single mass-to-charge ratio is selected in the MS, collided with argon in a collision cell (collision induced dissociation, CID), and the masses of the resulting fragmentations are determined in the second stage of the tandem MS. In the second part of the process, the SEQUEST program is used to correlate this experimentally determined CID spectrum with the CID spectra predicted from all the peptides in a sequence database that are of essentially the same mass as the peptide selected for CID; this correlation matches the isolated peptide with a sequence segment in a database and thus identifies the protein from which the peptide was derived.

Protein Identification by LC-MS/MS, Capillary LC-MS/MS and CE-MS/MS

It has been demonstrated repeatedly that ESI-MS has very high intrinsic sensitivity. For the routine analysis of gel-separated proteins at high sensitivity, the most significant challenge is the handling of small amounts of sample. The crux of the problem is the extraction and transfer of peptide mixtures generated by the digestion of small amounts of protein, from gels or membranes into the MS/MS system without significant loss of sample or introduction of unwanted contaminants. We employ three different systems for introducing digests of gel purified proteins into a MS, depending on the level of sensitivity required. As an approximate guideline, for samples containing picomole amounts of peptides LC-MS/MS is most appropriate, for samples containing mid to high femtomole amounts we use capillary LC-MS/MS, and for samples containing femtomoles or less, CE-MS/MS is the method of choice.

The coupling of a HPLC system using 0.5 mm diameter or larger reverse phase (RP) column to an ESI-MS/MS instrument and its automated operation have been previously described in detail [38]. This system has several advantages if a large number of samples are to be analyzed and all are available in sufficient quantity. The LC-MS/MS and database searching program can be run in automated mode using an autosampler, thus maximizing sample throughput. The relatively large column is tolerant of high levels of impurities from either gel preparation or sample matrix. Lastly, if configured with a flow-splitter and micro-sprayer [36], analyses can be performed on a small fraction of the sample (less than 5%) while the remainder of the sample is recovered in very pure solvents. This feature is particularly useful when an orthogonal technique is also used to analyze peptide fractions, such as scintillation counting of an introduced radiolabel, and this data can be correlated with peptides identified by CID spectra.

An increase of sensitivity of approximately tenfold can be achieved by using a capillary LC system with a 100 µm i. d. column rather than a 0.5 mm i. d. column as referred to above. Because very low flow rates are required for such columns, most reports have used a pre-column flow splitting system for producing solvent gradients at very low flow rates (low nl-min) without the need for flow-splitting [39]. Using this capillary LC-MS/MS system we were able to identify gel separated proteins when high femtomole amounts were loaded onto the gel [36]. This system is not yet automated, and like all capillary LC systems, is prone to blockage of the columns by microparticulates when analyzing gel separated proteins.

The highest level of sensitivity for analyzing gel separated proteins can be achieved by using capillary electrophoresis-tandem mass spectrometry (CE-MS/MS). We have previously described a solid-phase extraction capillary electrophoresis (SPE-CE) system which was used with triple quadrupole and ion trap ESI-MS/MS systems for the identification of proteins at the low femtomole to subfemtomole sensitivity level [40, 41]. While this system is highly sensitive, it is also labor intensive and not yet automated. To devise an analytical system with both the sensitivity of a CE and the level of automation of LC, we have constructed micro-fabricated devices for the introduction of samples into ESI-MS for high sensitivity peptide analysis.

Microfabricated Devices for Sample Introduction into MS/MS Instruments

The basic version of the device is a piece of glass into which channels of 10-30 micrometers in depth and 50-70 micrometers in width are etched by using photolithography/etching techniques similar to the ones used in the semiconduc-

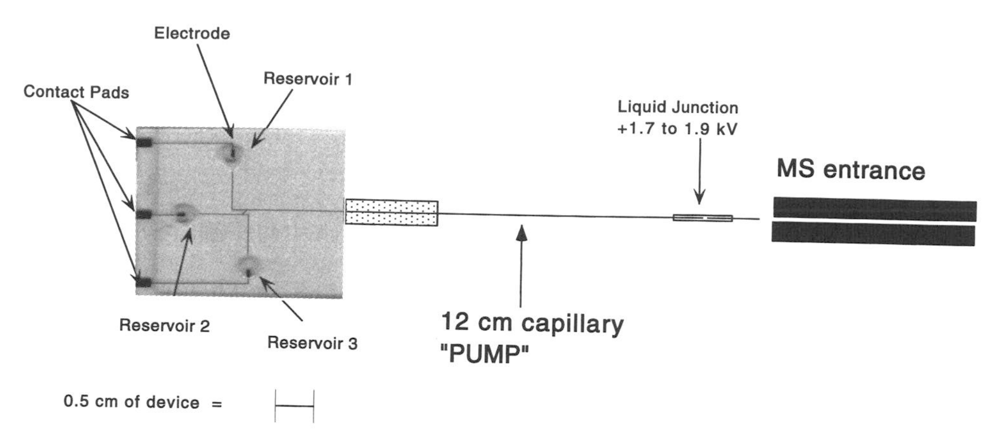

Fig. 1. Schematic illustration of a microfabricated analytical system for CE, consisting of a micromachined device, coated capillary electroosmotic pump and microelectrospray interface. The dimensions of the channels and reservoir are as indicated in the text. The channels on the device were graphically enhanced to make them more visible. Reproduced with permission from [42].

tor industry. A simple device is shown in Fig. 1. The channels are connected to an external high voltage power supply [42]. Samples are manipulated by applying different potentials to the reservoirs, which creates a solvent flow by electroosmotic pumping that can be redirected by changing the position of the electrodes or reversing the applied polarity. Therefore, flow can be redirected without the need for valves or gates and without any external pumping.

The type of data generated by the system is illustrated in Fig. 2. It shows the mass spectrum of a peptide sample representing the tryptic digest of carbonic anhydrase at 290 fmol/μl. Each numbered peak indicates a peptide successfully identified as being derived from carbonic anhydrase. The MS is programmed to automatically select each peak and subject the peptide to CID. The resulting CID spectra are then used to identify the protein by correlation with sequence databases. This system allows us to concurrently apply a number of protein digests onto the device, to sequentially mobilize the samples, to automatically generate CID spectra of selected peptide ions, and to search sequence databases for protein identification. These steps are performed automatically and proteins can be identified at a rate of approximately one protein per 15 minutes, at low femtomole level sensitivity. Subsequent samples are mobilized by switching of electrodes and the analysis cycle is repeated.

We have now produced a more advanced version of this device which has a 9-position reservoir system [43], allowing more samples to be loaded at once. The system can also be operated in an automated mode. The performance of this system, especially with respect to sensitivity of detection, is comparable to that of continuous-flow nanospray systems [32, 44].

In summary, the variety of techniques described in this section allow us to analyze and identify essentially any protein that can be detected in silver stained gels or membranes. The features of the various techniques are summarized in Table 1. These techniques are useful for the analysis of the more

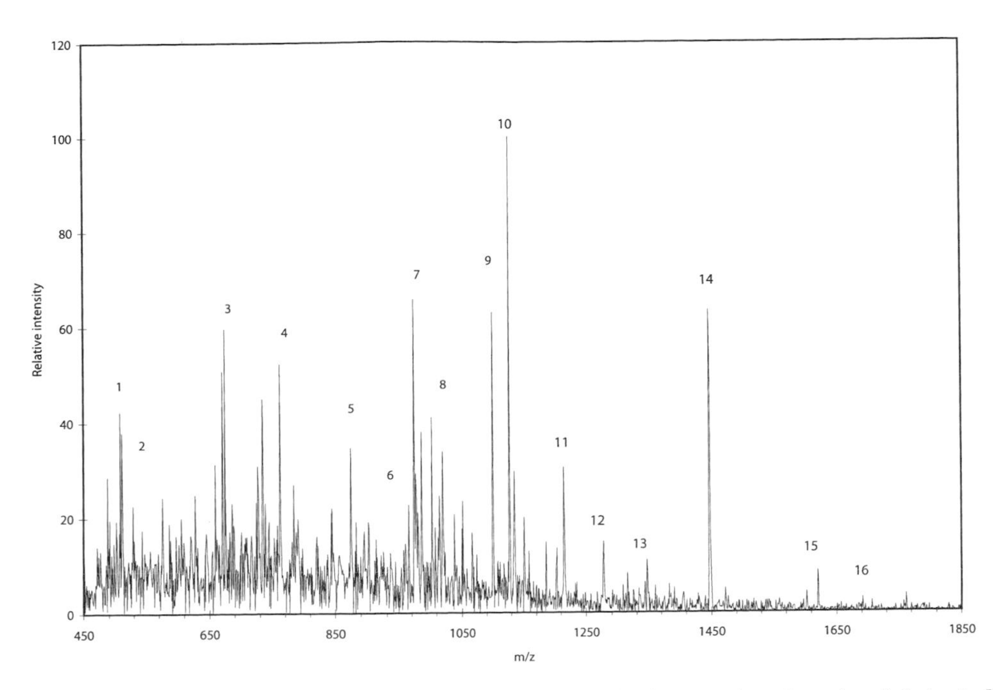

Fig. 2. MS spectrum of a tryptic digest of carbonic anhydrase using the microfabricated system shown in Fig. 3. 290 fmol / µl of carbonic anhydrase tryptic digest was infused into a Finnigan LCQ ion trap MS. Each peak was selected for CID, and those which were identified as containing peptides derived from carbonic anhydrase are numbered. Reproduced with permission from [42].

Technique	Routine Sensitivity	Automation	Sample Recovery	Reference
LC-MS/MS	picomoles and above	routine	>90% (with flow splitting)	[36]
capillary LC-MS/MS	mid femtomoles to picomoles	difficult	none	[36, 39]
SPE-CE-MS/MS	femtomoles	not yet available	none	[40]

Table 1. Features of techniques used for MS based protein identification.

abundant proteins in a cell, and thus allow many important biological problems to be addressed. However, they are still insufficient for the generation of truly comprehensive protein expression maps, as explained in the following sections.

SELECTIVE ENRICHMENT AND ANALYSIS OF STRUCTURALLY AND FUNCTIONALLY RELATED PROTEINS

The Challenge of Low Abundance Proteins

Using methods such as those described above, it is now possible to identify essentially any protein spot visible on a 2DE gel. If several cellular lysates from a given tissue or organism are separated, such analyses lay the foundation for proteome projects. In most cases, however, there is a large discrepancy between the number of protein spots observed and the number of potential gene products encoded for by a genome sequence.

To estimate to what extent this type of proteome analysis can perform in the identification of low abundance proteins, we have calculated the codon bias of the genes encoding the proteins identified from 2DE gels in which total yeast cell lysates were separated. Codon bias is a calculated measure of the degree of redundancy of triplet DNA codons used to produce each amino acid in a particular gene sequence. It has been shown to be a useful general indicator of the level of the protein product of a particular gene sequence present in a cell [45]. The general rule which applies is that the higher the value of the codon bias calculated for a gene, the more abundant the protein product of that gene becomes.

For a large number of yeast proteins identified from silver stained 2DE gels as part of an ongoing study [46], and from other literature reports, we found that nearly all have codon bias values of >0.2, indicating they are highly abundant in cells. In contrast, codon bias values calculated for the entire yeast genome show that the majority of expressed proteins have a codon bias of <0.2 and are thus of lower abundance that the proteins observed in 2DE separated unlabeled lysates.

This finding is of considerable importance in our assessment of the current status of proteome analysis technology. It is clear that even using highly sensitive analytical techniques, we are only able to visualize and identify the more abundant proteins. Because many important regulatory proteins are present only at low abundance, these would not be amenable to direct analysis without prior specific enrichment. This situation is expected to be exacerbated in the analysis of more complex proteomes than the approximately 6000 gene products present in yeast cells [16]. In the analysis of, for example, the proteome of any human cell types, 50,000 to 100,000 gene products are expected to be expressed [47]. Inherent limitations on the amount of protein that can be loaded on 2DE, and the number of components that can be resolved, indicate that only the most highly abundant fraction of the many gene products could be successfully analyzed. We believe this conclusion is extremely important in considering the future direction of proteomics research. As a result of this, we are exploring a number of ways to reduce the complexity of 2DE gel patterns to facilitate the identification of some of the less abundant proteins which would not be detected using current methodology without pre-enrichment.

The enrichment of proteins which physically interact to form a functional complex prior to the analysis of the individual complex components has been particularly successful for the analysis of low abundance proteins [48-52]. Another approach which we are exploring is to exploit the presence of a particular modification on a protein as a means of discriminating between that protein and all others in a complex mixture that do not contain the same modification. This can be done either by using pre-fractionation of lysate samples to enrich for the modified proteins, or by taking advantage of the power and versatility of a triple quadrupole mass spectrometer to distinguish the modified peptides, or, most promisingly, by a combination of both techniques. This type of approach necessarily results in the identification of all the proteins in a complex mixture that carry a particular type of modification. Some examples of our attempts to develop this type of strategy are presented in the following sections.

Direction and Identification of Phosphorylated Proteins in Complex Mixtures

Establishing the identity of unknown phosphoproteins, and the concomitant direct determination of the *in vivo* phosphorylation sites is an important area of research with broad applications. To date, such analyses have been performed protein-by-protein and not on a proteome-wide scale. Each analysis typically consists of the identification of the phosphoprotein, the isolation of the phosphorylated peptides from enzymatic digests of the phosphoprotein, followed by the identification of the sites of phosphorylation. An array of methods has been employed for such projects [53-55]. Several MS based techniques have been described which have the potential to identify phosphoproteins and analyze their state of modification in complex mixtures. They all have in common that they search for diagnostic fragment ions.

Neutral loss scanning in positive ion mode to monitor the loss of m/z 98 (H_3PO_4) neutral fragments has been demonstrated to be selective for phosphopeptides in mixtures [31]. This technique, however, is prone to false positives as specific peptide fragments of the same mass as the eliminated phosphate can be lost under the same conditions, and is only applicable when the charge state of the parent ion is already known. Precursor ion scanning to detect low mass phosphopeptide negative ion specific fragments (m/z 79 and 97) [56], and full range MS scanning which uses stepped voltage to detect both low mass and higher mass fragments in the same scan [57], have both been demonstrated to selectively detect phosphopeptides in complex mixtures. Both of these techniques use negative ion mode for the detection of fragments, and therefore cannot be used for the generation of desired sequence information from the peptide sequence. Also, they are not considered to be sufficiently sensitive to directly probe phosphorylation of low abundance proteins.

Current research in our laboratory includes the development of an MS based experimental approach for phosphopeptide identification in complex mixtures which is intended to be sufficiently sensitive for routine analysis of *in vivo* phosphorylation sites. The approach employs 50 μm i. d. capillary columns for separation of peptides selected out of complex mixtures on 500 μm i. d. iron immobilized metal-ion affinity chromatography (IMAC) columns. The mass spectrometer is then programmed to perform a repeating series of three consecutive scans in negative ion mode: 1) a multiple ion monitoring scan for m/z 63, 79 and 97 with in-source CID [58]; 2) a scan over the normal peptide range of m/z 300 to 1800 with in-source CID; and 3) a scan over the normal peptide range of m/z 300 to 1800 with no in-source CID. Collectively, these scans provide the following information: monitoring of m/z 63, 79 and 97 allows for three different

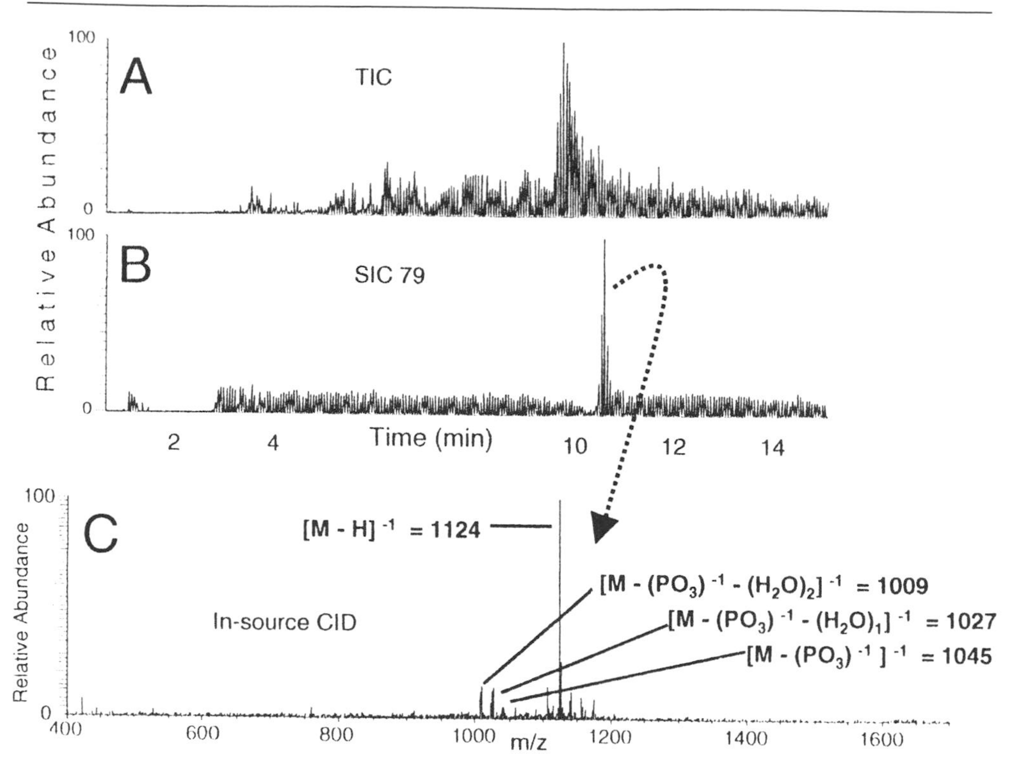

Fig. 3. Detection of a phosphopeptide using a multi-stage MS scan experiment. A standard containing a trypsin digested 32 kDa yeast lysate band plus 1 pmol of a phosphorylated angiotensin II peptide (DRVpYIHPF) was analyzed by capillary LC-MS using a 50 μm inner diameter C18 column. Panel A: total ion current chromatogram trace. Panel B: single ion current (m/z 79) chromatogram trace. Panel C: mass spectrum of source CID scan collected during peak indicated in panel B, with the mass of the phosphopeptide (m/z 1124) and the dephosphorylated peptide fragments indicated.

diagnostic fragment ions to be detected simultaneously, thus providing a high degree of confidence for the presence of a phosphopeptide at the ion source; the full scan with source CID provides the mass of the parent ion which has lost 79 ($H_2PO_4^-$) or 97 (PO_3^-); and the full scan without CID confirms the identity of the parent ion.

An example of this type of experiment is shown in Fig. 3. A complex peptide mixture was prepared by in-gel trypsin digestion of a 32 kDa silver stained band, containing several proteins, from a lysate for whole yeast cells. 1 pmol of a phosphorylated angiotensin II fragment (DRVpYIHPF) was added to this as a synthetic phosphopeptide standard and the mixture was analyzed by capillary LC-MS/MS using the multi-stage scan procedure described above. The total ion current chromatogram (Fig. 3A) shows that, as expected, numerous peptides were present in the sample. Analysis of another aliquot of the same sample using regular auto CID in positive ion mode LC-MS/MS confirmed this (data not shown). The multiple single ion monitoring scans revealed the presence of a single peak in the chromatogram where all three ions monitored showed a marked increase. The ion current chromatogram for m/z 79 is shown in Fig. 3B.

The m/z 63 and 97 chromatograms showed only a very small peak, because the phosphate group in this peptide is present in a phosphotyrosine residue. Due to the differences between their structures, phosphoserine and phosphothreonine give rise to all three fragment ions monitored, while phosphotyrosine produces mostly the m/z 79 fragment. A single mass spectrum using the in-source CID scan, collected during elution of the peak in Fig. 3B, is shown in Fig. 3C. This shows that the in-source fragmented parent ion which gave rise to the m/z 79 fragments was m/z 1124, and also indicates the presence of several dephosphorylated peptide ions as indicated. This result indicates that, using this procedure, the presence, elution time and mass of a phosphopeptide can be detected in a highly complex peptide mixture at the sub-picomole sensitivity level.

Detection and Identification of O-GlcNAc Modified Proteins

The O-linked N-acetyl-glucosamine (O-GlcNAc) modified intracellular proteins represent a second example for the possibilities for applying MS based technology to the identification and characterization of a class of modified proteins within a complex mixture. The class of glycoproteins was discovered relatively recently [59, 60]. It is now becoming apparent that the O-GlcNAc modification is not only widespread, but is dynamic and responsive to cellular stimuli [60, 61]. There appears to be a reciprocal relationship between phosphorylation and O-GlcNAc modification [57], and there is some evidence of O-GlcNAc involvement in signal transduction, oncogenesis and Alzheimers disease initiation [62-64]. Therefore, a clear need exists for high sensitivity analytical tools to study the presence and location of O-GlcNAc modifications at levels compatible with normal protein expression in cells. Analytical methodology currently used in this field involves radiolabeling O-GlcNAc groups with a galactosyl transferase [65], purifying the introduced label to apparent homogeneity, and only then identifying the underlying protein [66].

In parallel with the analysis of phosphoproteins described above, we are developing a MS based system for the *de novo* identification of O-GlcNAc modified proteins present in complex mixtures. The method is based on previous work in this field which has shown that sites of O-GlcNAc modification of peptides can be mapped using MS to detect the change in mass which occurs after the β-elimination of the glycosyl groups [53], and O-GlcNAc modified peptides can be selectively enriched by ricin affinity chromatography after galactosyl transferase labeling [67].

The labeling of a O-GlcNAc group with galactose using galactosyl transferase produces a disaccharide substituent with a diagnostic mass of 366 amu. The removal of this disaccharide from a glycopeptide by CID requires much less energy than the fragmentation energy by using a triple quadrupole mass spectrometer in a multi-stage experiment. Using a synthetic peptide, we have demonstrated that such labeled glycopeptides can be simultaneously detected and identified in a complex mixture. This is achieved by employing precursor ion scanning at a relatively low collision energy level to detect those species which produce a diagnostic m/z 366 fragment, followed by CID at higher energy of the

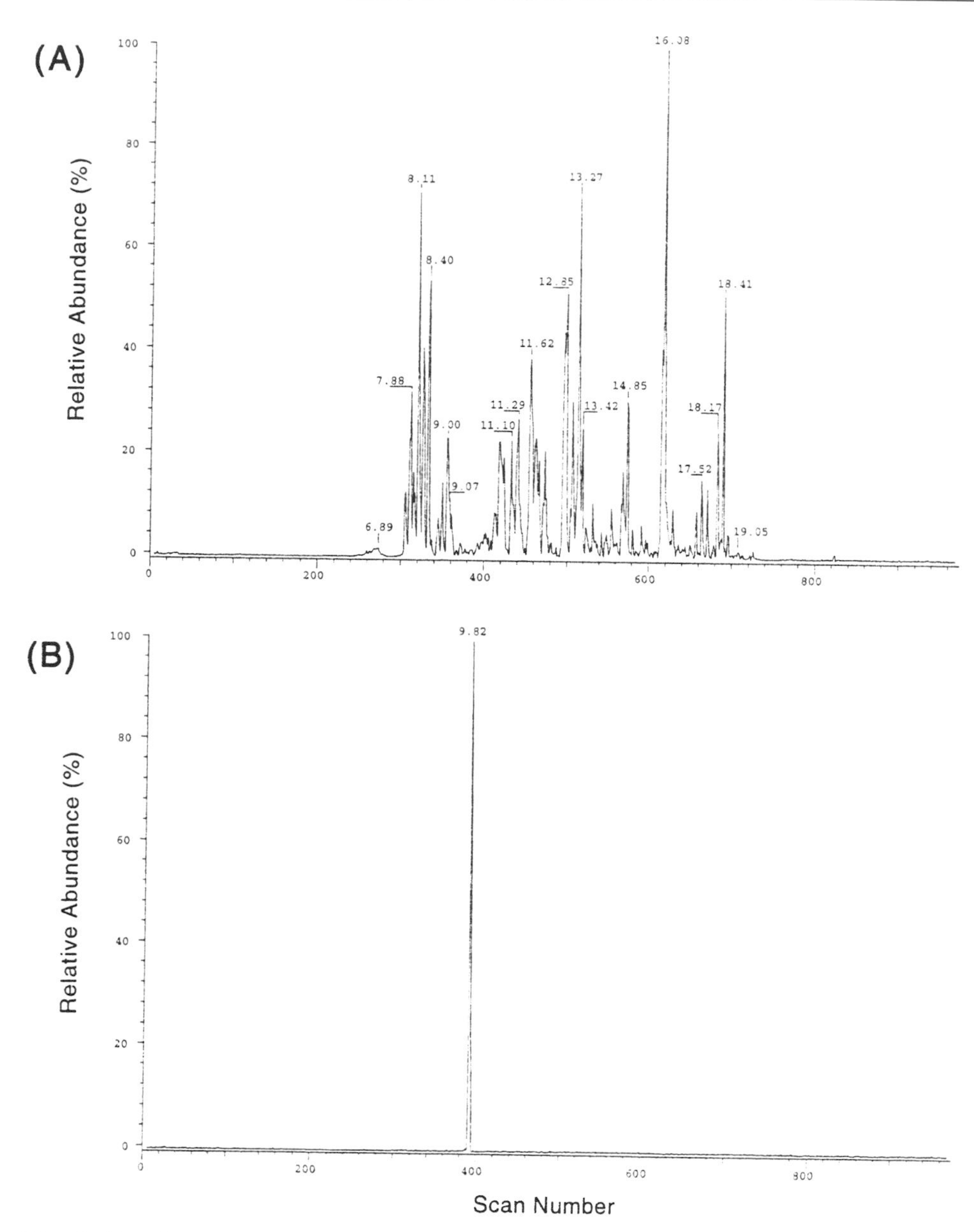

Fig. 4. Simultaneous detection and identification of an O-GlcNAc modified peptide. A standard containing 2.5 pmol each of trypsin digested carbonic anhydrase and bovine serum albumin plus 500 fmol of a galactosyl transferase labeled O-GlcNAc containing synthetic peptide (PSVPVS[GlcNAc]GSAPGR) was analyzed by capillary LC-MS/MS using a 100 µm inner diameter C18 column. Panel A: total ion current chromatogram trace in regular auto CID mode. Panel B: ion current chromatogram trace using m/z 366 precursor ion scanning/CID mode.

precursor ion to fragment the peptide backbone and thus identify the underlying peptide. This can be done in a single experiment using a procedure written in the Instrument Control Language of the Finnigan TSQ7000 mass spectrometer.

An example of this type of experiment is shown in Fig. 4. A sample consisting of 5 pmol of trypsin digested carbonic anhydrase and bovine serum albumin was spiked with 500 fmol of a synthetic O-GlcNAc containing peptide (PSVPVS[GlcNAc]GSAPGR) and analyzed by capillary LC-MS/MS in both regular CID mode and by m/z 366 precursor ion scanning/CID mode. The first stage of this experiment confirms that many different peptides are present in the mixture. The second stage detects a single glycopeptide of m/z 738 which is clearly not a major component of the sample and was not identified in the first stage. CID of this peptide produces a spectrum which was correctly identified by SEQUEST as the peptide sequence indicated above. The peptide is derived from a human cytomegalovirus protein (P100). We are currently applying this technique to the *de novo* detection and identification of O-GlcNAc modified peptides in nuclear lysates of Jurkat cells and mouse liver cells.

Concluding Remarks

In this report we have addressed several issues relevant to proteome analysis and discussed how these are shaping our future research directions. First, we discussed the reasons for studying proteomes and identified technical and conceptual challenges. Second, we assessed the technical feasibility of analyzing proteomes and described current proteome technology. Third, we described some of our current work involving the use of sample prefactionation and multiple stage MS experiments to specifically detect particular classes of modified peptides in complex mixtures.

It is becoming apparent that proteome analysis is an essential tool in the analysis of biological systems. Recently developed methods have enabled the identification of proteins at ever-increasing sensitivity levels, although a number of technical challenges remain. While it is currently possible to identify essentially any protein spot that can be visualized by common staining methods, it is clear that without prior enrichment only a relatively small and highly selected population of long lived, highly expressed proteins is observed. There are many more proteins in a given cell which are not visualized by such methods, and it is frequently low abundance proteins which execute key regulatory functions.

One way to successfully analyze more of the low abundance proteins in a given cellular preparation is to simply improve the sensitivity of detection. Because it is evident that proteins can be detected at much lower levels in dilutions of standard solutions than can be seen in even the faintest 2D gel spots, there must be considerable scope for improvement in the staining, in-gel digestion and sample handling stages. One other approach we are pursuing is the detection and identification of specific structural or functional classes of modified proteins as outlined in the report. We hope that this will have the dual advantages of enabling us to focus on some of the less abundant proteins, as well as providing an extra dimension of information concerning protein expression levels. We have demonstrated using peptide standards that phosphopeptides and O-GlcNAc modified peptides can be specifically detected and identified in the presence of complex mixtures. With the continued rapid progress in protein analysis technology, we anticipate that the field of proteome research will continue to produce not only more data, but also more informative data.

Acknowledgments

We would like to acknowledge funding for our work from the National Science Foundation Science and Technology Center for Molecular Biotechnology, from the NIH, and from Oxford Glycosciences. The human cytomegalovirus P100 synthetic peptide was a generous gift from Prof. Gerry Hart.

References

1. M. R. Wilkins, C. Pasquali, R. D. Appel, K. Ou, O. Golaz, J.-C. Sanchez, J. X. Yan, A. A. Gooley, G. Hughes, I. Humphery-Smith, K. L. Williams and D. Hochstrasser *Bio/Technology* **1996**, *14*, 61-65.

2. S. D. Patterson and R. A. Aebersold *Electrophoresis* **1995**, *16*, 1791-1814.

3. P. E. Hodges, W. E. Payne and J. I. Garrels *Nucl. Acids Res.* **1998**, *26*, 68-72.

4. C. D. O'Connor, M. Farris, R. Fowler and S. Y. Qi *Electrophoresis* **1997**, *18*, 1483-1490.

5. S. J. Cordwell, D. J. Basseal and I. Humphery-Smith *Electrophoresis* **1997**, *18*, 1335-1346.

6. B. L. Urquhart, T. E. Atsalos, D. Roach, D. J. Basseal, B. Bjellqvist, W. L. Britton and I. Humphery-Smith *Electrophoresis* **1997**, *18*, 1384-1392.

7. V. C. Wasinger, B. Bjellqvist and I. Humphery-Smith *Electrophoresis* **1997**, *18*, 1373-1383.

8. A. J. Link, L. G. Hays, E. B. Carmack and J. R. Yates, III *Electrophoresis* **1997**, *18*, 1314-1334.

9. T. Sazuka and O. Ohara *Electrophoresis* **1997**, *18*, 1252-1258.

10. R. A. Van Bogelen, K. Z. Abshire, B. Moldover, E. R. Olson and F. C. Neidhardt *Electrophoresis* **1997**, *18*, 1243-1251.

11. N. Guerreiro, J. W. Redmond, B. G. Rolfe and M. A. Djordjevic *Mol. Plant Microbe Interact.* **1997**, *10*, 506-516.

12. J. X. Yan, L. Tonella, J. C. Sanchez, M. R. Wilkins, N. H. Packer, A. A. Gooley, D. F. Hochstrasser and K. L. Williams *Electrophoresis* **1997**, *18*, 491-497.

13. J. Celis, P. Gromov, M. Ostergaard, P. Madsen, B. Honore, K. Dejgaard, E. Olsen, H. Vorum, D. B. Kristensen, I. Gromova, *et al.*, *FEBS Lett.* **1996**, *398*, 129-134.

14. R. D. Appel, J. C. Sanchez, A. Bairoch, O. Golaz, M. Miu, J. R. Vargas and D. F. Hochstrasser *Electrophoresis* **1993**, *14*, 1232-1238.

15. P. Haynes, I. Miller, R. Aebersold, M. Gemeiner, I. Eberini, M. Rosa Lovati, C. Manzoni, M. Vignati and E. Gianazza *Electrophoresis* **1998**, *19*, 1484-1492.

16. A. Goffeau, B. G. Barrell, H. Bussey, R. W. Davis, B. Dujon, H. Feldmann, F. Galibert, J. D. Hoheisel, C. Jacq, M. Johnston, *et al.*, *Science* **1996**, *274*, 546.

17. R. D. Fleischmann, M. D. Adams, O. White, R. A. Clayton, E. F. Kirkness, A. R. Kerlavage, C. J. Bult, J.-F. Tomb, B. A. Dougherty, J. M. Merrick, *et al.*, *Science* **1995**, *269*, 496-512.

18. C. M. Fraser, S. Casjens, W. M. Huang, G. G. Sutton, R. Clayton, R. Lathigra, O. White, K. A. Ketchum, R. Dodson, E. K. Hickey, *et al.*, *Nature* **1997**, *390*, 580-586.

19. V. E. Velculescu, L. Zhang, W. Zhou, J. Vogelstein, M. A. Basrai, D. E. J. Bassett, P. Hieter, B. Vogelstein and K. W. Kinzler *Cell* **1997**, *88*, 243-251.

20. R. G. Krishna and F. Wold *Adv. Enzymol. Relat. Areas Mol. Biol.* **1993**, *67*, 265-298.

21. A. Ducret, C. Foyn Brunn, E. J. Bures, G. Marhaug, G. Husby and A. R. *Electrophoresis* **1996**, *17*, 866-876.

22. A. Gorg, W. Postel and S. Gunther *Electrophoresis* **1988**, *9*, 531-546.

23. J. Klose and U. Kobalz *Electrophoresis* **1995**, *16*, 1034-1059.

24. P. Matsudaira *J. Biol. Chem.* **1987**, *262*, 10035-10038.

25. R. H. Aebersold, D. B. Teplow, L. E. Hood and S. B. Kent *J. Biol. Chem.* **1986**, *261*, 4229-4238.

26. J. Rosenfeld, J. Capdevielle, J. C. Guillemot and P. Ferrara *Anal. Biochem.* **1992**, *203*, 173-179.

27. R. H. Aebersold, J. Leavitt, R. A. Saavedra, L. E. Hood and S. B. Kent *Proc. Natl. Acad. Sci. USA* **1987**, *84*, 6970-6974.

28. B. Honore, H. Leffers, P. Madsen and J. E. Celis *Eur. J. Biochem.* **1993**, *218*, 421-430.

29. M. Mann and M. Wilm *Anal. Chem.* **1994**, *66*, 4390-4399.

30. J. Eng, A. L. McCormack and J. R. Yates, III *J. Am. Soc. Mass Spectrom.* **1994**, *5*, 976-989.

31. J. R. Yates, III, J. K. Eng, A. L. McCormack and D. Schieltz *Anal. Chem.* **1995**, *67*, 1426-1436.

32. A. Shevchenko, M. Wilm, O. Vorm and M. Mann *Anal. Chem.* **1996**, *68*, 850-858.

33. D. Hess, T. C. Covey, R. Winz, R. W. Brownsey and R. Aebersold *Protein Sci.* **1993**, *2*, 1342-1351.

34. I. van Oostveen, A. Ducret and R. Aebersold *Anal. Biochem.* **1997**, *247*, 310-318.

35. M. Lui, P. Tempst and H. Erdjument-Bromage *Anal. Biochem.* **1996**, *241*, 156-166.

36. P. A. Haynes, N. Fripp and R. Aebersold *Electrophoresis* **1998**, *19*, 939-945.

37. D. Figeys, I. van Oostveen, A. Ducret and R. Aebersold *Anal. Chem.* **1996**, *68*, 1822-1828.

38. A. Ducret, I. van Oostveen, J. K. Eng, J. R. Yates, III and R. Aebersold *Prot. Sci.* **1997**, *7*, 706-719.

39. A. Ducret, N. Bartone, P. A. Haynes, A. Blanchard, and R. Aebersold *Anal. Biochem.*, in press.

40. D. Figeys, A. Ducret, J. R. Yates, III and R. Aebersold *Nature Biotech.* **1996**, *14*, 1579-1583.

41. D. Figeys and R. Aebersold *Electrophoresis* **1997**, *18*, 360-368.

42. D. Figeys, Y. Ning and R. Aebersold *Anal. Chem.* **1997**, *69*, 3153-3160.

43. D. Figeys, S. P. Gygi, G. McKinnon and R. Aebersold *Anal. Chem.*, in press.

44. M. Wilm, A. Shevchenko, T. Houthaeve, S. Breit, L. Schweigerer, T. Fotsis and M. Mann *Nature* **1996**, *379*, 466-469.

45. J. I. Garrels, C. S. McLaughlin, J. R. Warner, B. Futcher, G. I. Latter, R. Kobayashi, B. Schwender, T. Volpe, D. S. Anderson, R. Mesquita-Fuentes, *et al.*, *Electrophoresis* **1997**, *18*, 1347-1360.

46. S. P. Gygi, Y. Rochon, B. R. Franza, and R. Aebersold **1998**, submitted.

47. G. D. Schuler, M. S. Boguski, E. A. Stewart, L. D. Stein, G. Gyapay, K. Rice, R. E. White, P. Rodriguez-Tome, A. Aggarwal, E. Bajorek, *et al.*, *Science* **1996**, *274*, 540-546.

48. X. Y. Fu, C. Schindler, T. Improta, R. Aebersold and J. E. Darnell, Jr. *Proc. Natl. Acad. Sci. USA* **1992**, *89*, 7840-7843.

49. S. A. Veals, C. Schindler, D. Leonard, X. Y. Fu, R. Aebersold, J. E. Darnell, Jr. and D. E. Levy *Mol. Cell Biol.* **1992**, *12*, 3315-3324.

50. J. R. Carter, M. A. Franden, R. Aebersold and C. S. McHenry *J. Bacteriol.* **1992**, *174*, 7013-7025.

51. J. R. Carter, M. A. Franden, R. Aebersold and C. S. McHenry *J. Bacteriol.* **1993**, *175*, 5604-5610.

52. G. Neubauer, A. King, J. Rappsilber, C. Calvio, M. Watson, P. Ajuh, J. Sleeman, A. Lamond and M. Mann *Nat. Genet.* **1998**, *20*, 45-50.

53. J. D. Watts, M. Affolter, D. L. Krebs, R. L. Wange, L. E. Samelson and R. Aebersold in *Biochemical and Biotechnological Applications of Electrospray Ionization Mass Spectrometry*, A. P. Snyder, Ed.; American Chemical Society: Washington, DC, 1996; pp 381-407.

54. J. D. Watts, M. Affolter, D. L. Krebs, L. E. Samelson and R. Aebersold *J. Biol. Chem.* **1994**, *269*, 29520-29529.

55. W. J. Boyle, P. Van Der Geer and T. Hunter *Meth. Enzymology* **1991**, *201*, 110-149.

56. S. A. Carr, M. J. Huddleston and R. S. Annan *Anal. Biochem.* **1996**, *239*, 180-192.

57. M. J. Huddleston, R. S. Annan, M. F. Bean and S. A. Carr *J. Am. Soc. Mass Spectrom.* **1993**, *4*, 710-717.

58. A. P. Hunter and D. E. Games *Rapid Commun. Mass Spectrom.* **1994**, *8*, 559-570.

59. C.-R. Torres and G. W. Hart *J. Biol. Chem.* **1984**, *259*, 3308-3317.

60. G. D. Holt, C. M. Snow, A. Senior, R. S. Haltiwanger, L. Gerace and G. W. Hart *J. Cell Biol.* **1987**, *104*, 1157-1164.

61. R. S. Haltiwanger, W. G. Kelly, E. P. Roquemore, M. A. Blomberg, L.-Y. D. Dong, L. Kreppel, T.-Y. Chou and G. W. Hart *Biochem. Soc. Trans.* **1992**, *20*, 264-269.

62. C. S. Arnold, G. W. V. Johnson, R. N. Cole, D. L.-Y. Dong, M. Lee and G. W. Hart *J. Biol. Chem.* **1996**, *271*, 28741-28744.

63. T.-Y. Chou, C. V. Dang and G. W. Hart *Proc. Natl. Acad. Sci. USA* **1995**, *92*, 4417-4421.

64. T.-Y. Chou, G. W. Hart and C. V. Dang *J. Biol. Chem.* **1996**, *270*, 18962-18965.

65. E. P. Roquemore, T.-Y. Chou and G. W. Hart *Meth. Enzymology* **1994**, *230*, 443-460.

66. K. D. Greis, W. Gibson and G. W. Hart *J. Virol.* **1994**, *68*, 8339-8349.

67. B. K. Hayes, K. D. Greis and G. W. Hart *Anal. Biochem.* **1995**, *228*, 115-122.

Questions and Answers

Tim Wehr (Bay Bioanalytical Laboratory)

In your micro-fabricated sample introduction device, electro and osmotic pumping is used to transport peptides to the MS. As this requires charge on the channel wall, does this cause selective loss of basic peptides which could compromise quantitative accuracy?

Answer. It is possible that the charged surface of the pumping capillary (and the chip) may selectively absorb basic peptides. This may have consequences for our ability to identify proteins and may limit peptide coverage. However, it is expected to be inconsequential for quantitation for the following reasons: 1) Protein quantitation is based on isotopic ratios of derivatized peptides. We do not expect that differentially isotopically labeled residues will be selectively absorbed (i. e., the ratio will remain constant, even if the signal dips). 2) Protein quantitation, in general, will be based on more than one peptide. Therefore, the relative quantities obtained with different peptides can be averaged.

Michael O. Glocker (University of Konstanz)

Where would you expect are the limits with regard to mixture complexity that could be resolved by the protein profiling approach you described as compared to the 2D-gel separation technique?

Answer. We have not determined this limit. However, we expect that the mixture complexity and the ability to identify low abundance proteins will compare favorably with 2DE for the following reasons: 1) We can carry out an experiment with essentially any amount of protein, whereas 2DE has limited sample loading capacity. 2) The complexity of the peptide mixture generated by co-digestion of proteins is reduced by the affinity relation of specific peptides. 3) We have developed electrophoretic peak parking techniques (not discussed in this chapter) which allow the analysis of very complex peptide mixtures.

We also expect that classes of proteins that are underrepresented in 2D gels (e. g., membrane proteins, very basic, very large proteins, etc.) will be efficiently analyzed by the method.

Pierre Thibault (National Research Council)

What are the major selling features of the chip microfabricated devices coupled to mass spectrometry compared to sample flow injection using nanoelectrospray?

Answer. The advantages are several-fold. 1) The system can be automated and sequential samples can be automatically applied without manual intervention. 2) Once the sprayer is optimally placed in front of the MS, the position does not need to be changed for all the samples analyzed in sequence. 3) The system provides a high degree of control over flow, etc. However, it is fair to say that with the current system the performance is not better than with suitable capillary-based systems. We see the advantage in the micro-fabricated approach in the possibility to build modular, integrated systems performing complete analyses in an automated and integrated fashion. This would be attractive for reasons of process control, cleanliness and avoidance of sample loss due to absorption.

A. L. Burlingame (UCSF)

What fraction of mammalian proteins have free thiols?

Answer. The fraction of proteins for mammalian species is not known. For yeast *S. cerevisiae*, 92% of proteins contain cysteines.

A. L. Burlingame (UCSF)

Is the amino modification strategy practical?

Answer. Amino-reactive reagents for protein quantitation have several potential uses, in addition to protein profiling. These include comprehensive analysis of immunoprecipitates, and phosphoproteins.

Robert Anderegg (Glaxo Wellcome)

In the comparison of mRNA and protein level, you looked at a single time point. Do you think that the mRNA and protein expression might follow different time courses (e. g., mRNA comes up first, then protein later), and that by looking a single time point you may be biasing the experiment to *not* see a correlation?

Answer. This is an important point. Our measurements were done in cells grown at a steady state. Therefore, the ratios represent averages over many cells, which may be in different stages of the cell cycle. The data do not make any prediction about correlations of the time course and magnitude of induced changes, e. g., by shifting cells into different growth media. These experiments are currently underway.

Sequencing the Primordial Soup

Jeffrey Shabanowitz, Robert E. Settlage*, Jarrod A. Marto*,*
Robert E. Christian, Forest M. White*, Paul S. Russo*, Susan E. Martin**
and Donald F. Hunt,†*
Departments of *Chemistry and †Pathology, University of Virginia,
Charlottesville, VA, 22901

Abstract

Proteomics is the brute force attempt to systematically identify and map the total protein complement expressed by a genome or tissue. The intracellular protein differences between whole cells (diseased vs. healthy) or between other complex biological fluids are important to understanding protein function. Once the structure and interactions of all proteins are characterized — in physiological and pathological situations — therapeutic strategies can be more easily designed. Typically, these analyses are accomplished by 2D-gel electrophoresis for separation and quantitation, followed by Edman degradation and/or mass spectrometry (MS) for protein identification. Matrix assisted laser desorption/ionization time-of-flight (MALDI-TOF) and/or nanospray ionization-MS are routinely used to identify tryptic maps of individual gel spots. Interfacing HPLC with electrospray ionization (ESI) on a tandem mass spectrometer (MS/MS) offers an attractive alternative. When coupled with automated sample handling techniques, data-dependent scanning, and database searching, it can be used with or without preliminary protein separation (1D or 2D gels). Combining these automated capabilities and adapting the sensitivity of nanospray to an on-line nanoflow-HPLC/micro-ESI assembly interfaced with ion trap MS has resulted in an even more attractive and sensitive method for protein identification and characterization. The methodology developed can be used to analyze complex biological mixtures containing hundreds of proteins, covering a wide dynamic range of concentrations. Detection limits of approximately 50 attomoles make possible the identification of several co-migrating proteins present in a single Coomassie- or silver-stained gel spot.

Introduction

Mass spectrometry has experienced tremendous growth and success as applied to various areas of biological research in recent years. One has to look no further than the biennial reviews to document these applications [1]. Klaus Biemann extols the contributions of mass spectrometry as they relate to peptide and protein chemistry over the past 30 years [2]. Indeed, the future of mass spectrometry appears to be just as promising in the post-genomic era [3-5].

Our efforts to develop methodology for sequencing peptides and proteins by tandem mass spectrometry began in 1978. At that time, a newly constructed triple quadrupole mass spectrometer was employed to sequence permethylated peptides in a mixture by the combination of chemical ionization and collision activated dissociation. In 1981, shortly after the introduction of fast atom bombardment, we published a paper that outlined a strategy for sequencing proteins [6]. The paper concluded with the statement; "We envision that this will entail degradation of the protein to peptides, the optimal length of which will be determined by the mass range of the mass spectrometer (1,000-3,000 Da) rather than by the volatility of the mixture components. The mixture of peptides will then be fractionated by high performance liquid chromatography and each fraction will be analyzed directly, without further purification, by the combination of secondary ion/collision activated dissociation mass spectrometry on a multi-analyzer instrument. Generation and analysis of the peptides could be completed in a matter of hours." The above methodology was described in somewhat greater detail in a second paper published in the *Proceedings of the National Academy Sciences* in 1986 [7].

Reports from our laboratory detailing the structural characterization of unknown peptide and protein sequences by this approach were first published in 1984 and 1987, respectively. In 1988, we were the first group to elucidate the structure of a test peptide (provided by the Protein Society) by mass spectrometry alone. In 1990, the Hunt group was again the only team to successfully characterize unknown test peptides synthesized for both the ASMS and Protein Society meetings. Introduction of electrospray ionization by John Fenn made it possible to interface liquid chromatography (LC) to mass spectrometers [8, 9], thus extending the power of tandem mass spectrometry to the analysis of peptides in complex mixtures. The Hunt group in 1991 [10] introduced the use of electrospray ionization with microcapillary HPLC for protein sequence analysis. Capillary electrophoresis (CE) was interfaced to the instrument two years earlier and reported in 1992 [11]. Sequence analysis of peptides in mixtures at the low femtomole level was achieved with both of these techniques. The combination of microcapillary HPLC, electrospray ionization, and tandem mass spectrometry made it possible to sequence peptides presented to the immune system in association with class I and class II MHC molecules, and has had a major impact on the field of immunology [12-14]. Twenty-eight antigens, including 8 from melanoma, have been identified with this technology. One or more melanoma antigens are presently in clinical trials as cancer vaccines/therapeutics at 7 centers around the world.

Over the past three years, we have achieved another breakthrough in technology that could dramatically impact protein characterization in the post-genome era. After considerable effort, we have successfully interfaced a Finnigan LCQ ion trap mass spectrometer to both nanoflow-HPLC (5-200 nL/min) and CE via low-flow electrospray ionization. Sequence analysis of peptides in mixtures can now be achieved at the 10-50 attomole level [15, 16]. A novel approach to variable-flow HPLC (5-200 nL/min) (patent pending) has also been implemented. This innovation allows us to sample a 10 second wide chromatographic peak for more than several minutes, with little loss of ion current, and thus to acquire MS/MS spectra on a large number of peptides co-eluting into the mass spectrometer [16]. Both the chromatography and data acquisition have been automated and are controlled by the mass spectrometer's data system. The same technology has also been implemented on our home-built Fourier transform mass spectrometer (FTMS). This instrument allows us to obtain accurate mass measurement data at high resolution on complex mixtures of peptides on sample levels as low as 10 amol. This instrument provides an intriguing alternative as it has, in principle, the ability both to provide accurate molecular mass information and to perform MS/MS experiments, thus combining the strengths of MALDI-TOF and quadrupole ion trap instruments. Although the FTMS is not yet capable of automated data acquisition, we have nonetheless obtained MS tryptic mass maps at the 10 amol level and performed peptide sequence analysis at the low fmol level.

The above methodology allows us to analyze mixtures containing several hundred proteins without prior separation. Samples are treated with trypsin, and the resulting mixture of several thousand peptides is then injected directly into the nanoflow-HPLC/MS system for automated analysis. We present results that demonstrate the feasibility of using this technology to characterize proteins involved in a number of key cellular functions, signaling pathways, and disease states.

Methods

Sample Preparation

Protein standards, including beta lactoglobulin (BLG), beta casein (BC), bovine serum albumin (BSA), carbonic anhydrase II (CAH), cytochrome C (CC), and glyceraldehyde-3-phosphate dehydrogenase (GAPDH) were obtained from Sigma (St. Louis, MO) and used without further purification. Sequencing grade, modified trypsin was obtained from Promega (Madison, WI). Tryptic digests were prepared as follows: Protein and trypsin were added to ammonium acetate buffer (25 mM, pH = 8.0, 100 μL total vol.) in a substrate:enzyme mass ratio of ~25:1. The solution was incubated at 37°C overnight, after which protease activity was quenched by addition of 1 μL acetic acid. Prior to MS analysis, the stock solutions were diluted further with 0.1% acetic acid to give working analyte solutions. Appropriate aliquots of digest solutions were loaded onto the microcapillary HPLC column and gradient eluted directly into the mass spectrometer (described below).

In-gel digests were performed by excising 1D or 2D Coomassie- or silver-stained gel spots and treating them according to procedures previously described [17].

HPLC/Ion Trap Mass Spectrometry

Peptide separation and analysis were performed on a nanoflow-HPLC column connected to a fused silica micro-ESI spray tip (Fig. 1) on-line with an LCQ ion trap mass spectrometer (Finnigan MAT, San Jose, CA). The HPLC column consisted of a "pre-column" and a "resolving" column. The pre-column

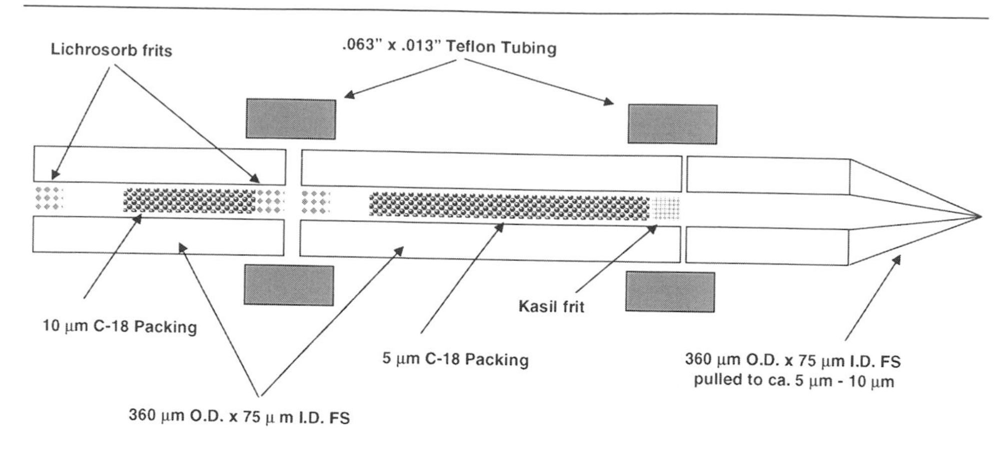

Fig. 1. Nanoflow-HPLC column and micro-ESI tip construction.

was constructed from 360 μm o.d. x 75 μm i.d. fused silica (Polymicro, Phoenix, AZ) packed with 3-4 cm of 10 μm C18 material (YMC, Wilmington, NC), while the resolving column was made from 360 μm o.d. x 75 μm i.d. fused silica (FS) packed with 7-9 cm of 5 μm C18 material (YMC, Wilmington, NC). A Kasil (The PQ Corporation, Valley Forge, PA) frit was used at the front of the resolving column because of its smaller and more uniform pore size, which prevents beads leaking into and clogging the spray tip. All samples were first loaded onto the pre-column, offline and separated from the resolving column and spray tip. This allowed for rapid loading (>10 μL/min) and thorough desalting of the sample before analysis. All samples were gradient eluted with 0 to 100% B (A = 0.1M acetic acid in water, B = 0.1M acetic acid in 70% acetonitrile) over times ranging from 10-60 minutes. The mobile phase flow rate was 200 nL/min at the beginning of the gradient and was subsequently reduced to <20 nL/min, either under computer control (described below) or manually. The ESI spray tip (5-10 μm diameter) was pulled from 360 μm o.d. x 75 μm i.d. FS on a P-2000 laser puller (Sutter, Novato, CA).

Mass spectrometric data were acquired using data dependent software provided with the LCQ. The experimental method was designed to perform one MS scan (300-2000 m/z) followed by MS/MS scans of the 5 most abundant ions present in the original MS scan. Employing dynamic and isotope exclusion functions increased the total number of unique peptides analyzed. These features decrease MS/MS redundancy by ensuring that each ion is selected for collisional activation no more than twice in a selected time frame and corresponding isotopes are not analyzed. MS/MS data were searched against the Owl database (NCBI) using SEQUEST [18] with peptide precursor mass tolerances of 1.4 Da and fragment ion mass tolerances of 0.4 Da. The above methodology is described in more detail elsewhere [16].

HPLC / Fourier Transform Mass Spectrometer

Our home-built Fourier transform mass spectrometer has been modified from previous descriptions [19, 20] and is described in detail elsewhere [21]. Ions produced externally in a modified Finnigan electrospray ion source are transported to an open, cylindrical Penning trap through a series of RF-only ion guides, consisting of two octapoles followed by two quadrupoles. External ion accumulation [22] in the first octapole was employed to maintain chromatographic resolution and to achieve high ion sampling efficiency. Using this approach, high resolution, full mass spectra may be acquired every 1-2 seconds, depending on the external ion accumulation period. For MS/MS experiments a 45-W, continuous-wave CO_2 laser (Synrad, Bothell, WA) was used to dissociate isolated precursor ions. Generally ten MS/MS spectra were summed for a given peptide species; data were subsequently analyzed using SEQUEST against the Owl (NCBI) protein database, with peptide precursor and fragment mass tolerances set to 0.03 Da.

HPLC Gradient / Flow Control

We have recently developed a simple and robust alternative implementation of the "peak parking" strategy first introduced by Davis and Lee [23]; the apparatus is described in detail elsewhere [16]. Briefly, a stainless steel tee (Swagelok, Solon, OH, 0.013" i.d.) serves as a pre-column splitter. The gradient is delivered through one arm of the tee by an ABI 140B (Applied Biosystems, Foster City, CA) HPLC pump. The column/needle assembly is connected to the opposite arm, while the 90° port goes to waste. Flow from the syringe pump (200 μL/min) is split at the head of the HPLC column such that 99.9% flows to waste, with the remainder entering the HPLC column. Flow rate through the column (and thus chromatographic peak width) is determined by the use of restrictors on the waste port. For example, MS-only scans are performed in "high-flow" mode (150-300 nL/min), while MS/MS experiments are performed in "peak-park" mode, with a flow rate through the column <20 nL/min.

Results and Discussion

Nanoflow-HPLC and Micro-ESI on an Ion Trap Mass Spectrometer

We have successfully interfaced the Finnigan LCQ ion trap mass spectrometer to both nanoflow-HPLC (5-200 nL/min) [16] and CE [15] via micro-electrospray ionization. Chemical noise has been reduced by a factor of 3, microspray needles can now be constructed reproducibly with a laser capillary puller, and column breakage and clogging has been largely eliminated. As a result, sequence analysis of peptides in mixtures can now be achieved at the 10-50 attomole level on this instrument.

To further extend our sequencing capabilities, we have also implemented a novel approach to variable-flow HPLC. This approach (described above) offers the following advantages; (a) it requires only a single HPLC pump and is compatible with any commercial HPLC unit; (b) it does not employ a preformed gradient; (c) multiple flow rates can be easily programmed; (d) peaks can be parked for a much longer time scale than previously reported; and (e) peptide sequence analysis at the low amol level can be accomplished on multiple components with this new innovation.

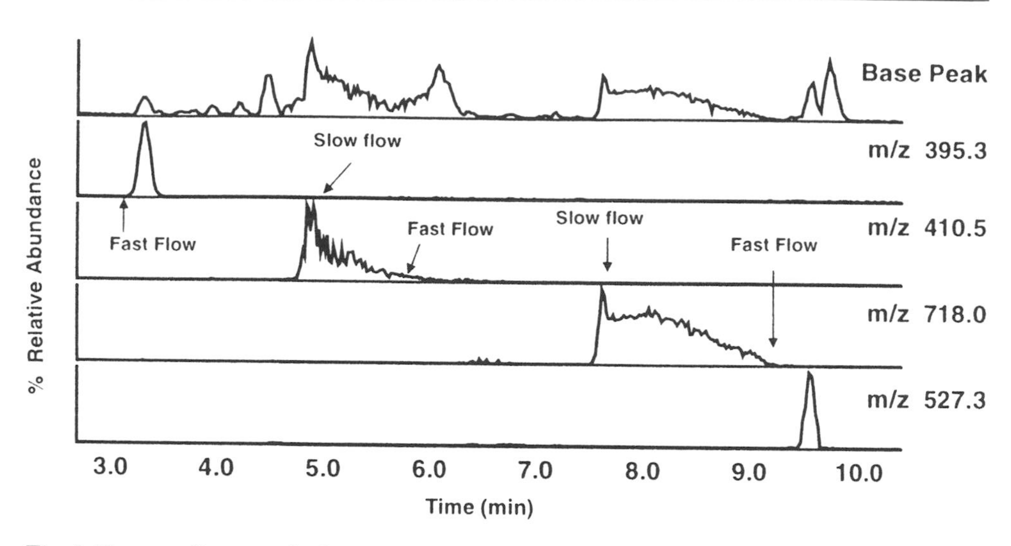

Fig. 2. Fast gradient analysis on 100 fmol tryptic digest of GAPDH with peak-parking (500 nL / min to 50 nL / min).

With a flow rate of 200 nL/min, low femtomole level samples elute with a peak width of 12-15 sec. Because MS and MS/MS spectra over the mass range, 100-2,000, can be recorded in 1.5 sec, up to 10 spectra can be acquired during the elution time of any particular chromatographic peak. To handle more complex mixtures, the HPLC has been placed under control of the LCQ data system, which allows the mass spectrometer to control the chromatographic flow rate [16].

When a mixture of peptides elutes into the mass spectrometer, the instrument detects the signal and reduces the chromatographic flow rate from several hundred nL to <50 nL/min. This happens in less than 2 seconds. The result is that the original 15 second chromatographic peak can elute over several minutes (Fig. 2). Even though sample elution now has been spread out over a much larger time window, the actual ion current obtained in any one scan is seldom less than what was recorded on the original 15 sec peak. This is the expected result, because the efficiency of ion generation and introduction into the instrument both increase as the flow rate through the fused silica micro-ESI tip decreases. In the low-flow mode, the first mass spectrum is acquired to determine the molecular weight of the components eluting in the chromatographic peak. The next five scans record MS/MS spectra on the 5 most abundant species in the mixture. These masses are then placed on an exclusion list along with their associated isotope peaks. The next scan is used to remeasure the molecular weights of the components and the next five scans record MS/MS spectra on the 5 most abundant species not already on the exclusion list (Fig. 3). MS/MS spectra can be recorded on as many as 100 different co-eluting peptides using this protocol. Once the list of candidates to be fragmented is exhausted, the flow rate is changed back to 200 nL/min until the next group of peptides is detected. The above process is then repeated. At the end of the chromatographic run, proteins

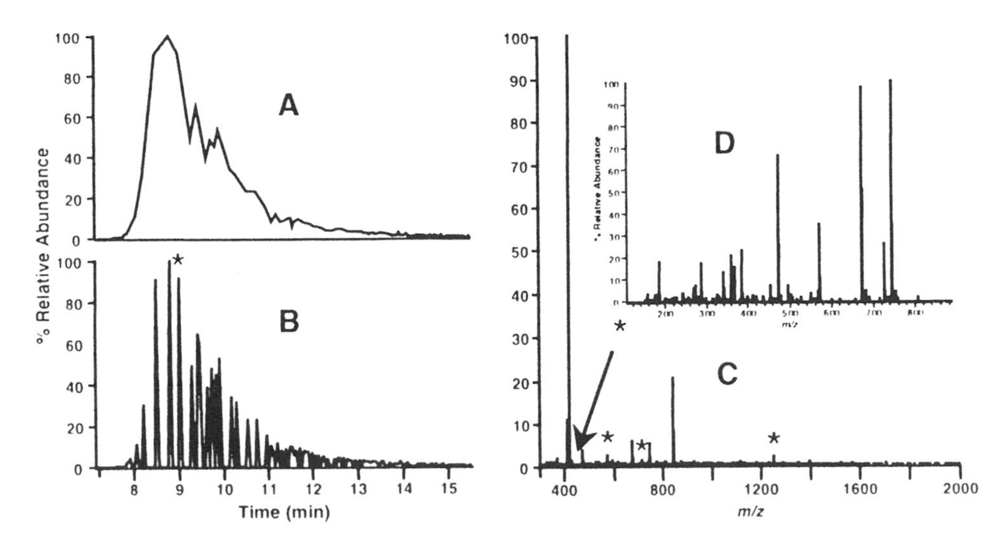

*Fig. 3. Data dependent analysis illustrating dynamic exclusion and dynamic range. (A) Total ion chromatogram of base peak MS only. (B) Total ion chromatograph of base peak including all scans. (C) MS scan of * peak in panel B. Asterisks indicate other ions selected for MS / MS analysis after dynamic exclusion. (D) MS / MS of 428^{+2} as indicated by arrow in panel C. The 428^{+2} ion is the 15th most abundant ion at ca. 1% of the base peak.*

in the original mixture are identified by processing the complete set of several thousand acquired MS/MS spectra against the protein and nucleic acid databases using the SEQUEST software program. Spectra with no match have to be interpreted manually.

Although we continue to develop the above protocol, it already has proven to be exceptionally powerful for characterizing peptides and proteins in complex mixtures. Results from several experiments are described here. In the first example, cytochrome C was digested with trypsin and an aliquot corresponding to 1.5 fmol was injected into the HPLC-LCQ system operating at a flow rate of 200 nL/min (no peak-parking). Total time for the experiment was 10 minutes. When SEQUEST searched the database with the acquired MS/MS spectra, 8 tryptic peptides were found to match those of cytochrome C. In short, 67% of the sequence was confirmed without operator intervention at any point in the experiment.

In the second example, a mixture of six proteins was digested with trypsin and an aliquot corresponding to 500 fmol of casein, 100 fmol of bovine serum albumin, 20 fmol of glyceraldehyde phosphate dehydrogenase, 7.5 fmol of carbonic anhydrase, 2 fmol of beta lactoglobulin and 500 amol of cytochrome C was injected into the HPLC-LCQ instrument operated in the automated peak-parking mode. MS/MS spectra were searched against the entire protein database with SEQUEST and 32 tryptic peptides from the above proteins were identified. This included 2 each from BLG and cytochrome C. Thus even when the mixture contains proteins present at a concentration dynamic range of 1,000:1, the protocol is still capable of correctly identifying all components.

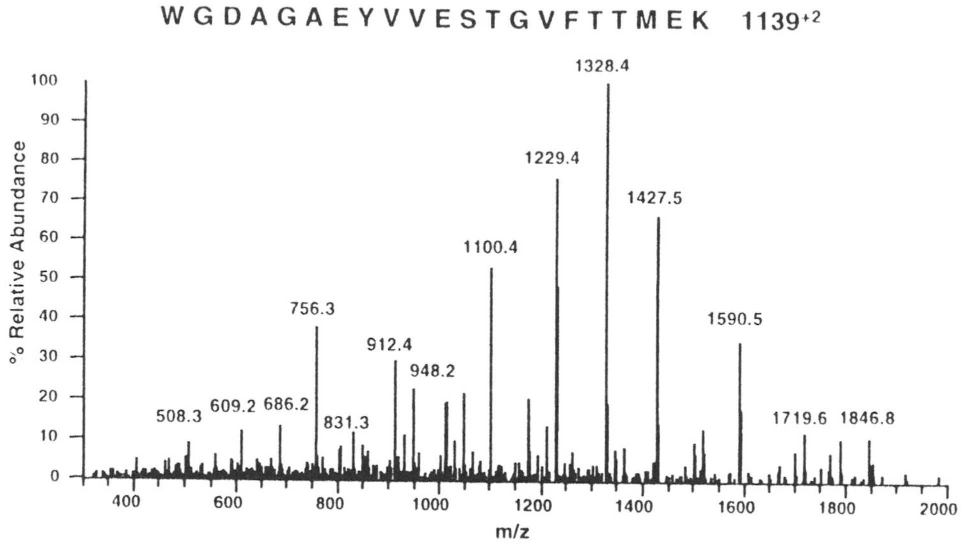

Fig. 4. MS/MS of 1139+2 taken from the analysis of PP1/PP2A. Tryptic fragment was identified as being from GAPDH by SEQUEST.

The third example is taken from a collaboration with Dr. Timothy Haystead in the Department of Pharmacology at the University of Virginia. Dr. Haystead is attempting to characterize all of the proteins that associate with and regulate the activity of two human phosphatases, PP1 and PP2A [24-26]. The experiment involves passing cellular extracts over a column containing co-valently attached recombinant phosphatase. After washing the column with salt solutions to remove non-specific binders, proteins associated with the phos-phatase are eluted with high salt and detergent and then fractionated on a 1D SDS-PAGE gel. More than 100 bands are observed with silver staining. Dr. Haystead has characterized approximately 15 of the most abundant proteins by Edman sequencing. We took the unfractionated sample, treated it with trypsin, and injected an aliquot of the resulting mixture containing more than 1,000 peptides into the HPLC-LCQ instrument operated in the automated peak-parking and data dependent scanning mode. Matches for 42 proteins in the databases by SEQUEST, including the 15 already sequenced by Edman were obtained in this single experiment. An example of a typical MS/MS spectrum matched by SEQUEST is given in Fig. 4. Many more MS/MS spectra failed to match with any entries in the database and presumably correspond to proteins of unknown sequence. Since the odds of finding a human protein sequence in the database are presently about one in four, we conclude that the sample probably contains 160-200 proteins. This conclusion is similar to that based on the number of bands observed on the 1D gel. Unmatched MS/MS spectra are sequenced manually and checked against alternate databases.

In a final example, Dr. Shu Man Fu from the Department of Pathology at the University of Virginia is interested in finding new proteins associated with CD40. Engagement to the CD40 ligand in B cells causes a variety of cellular changes and some of these associated proteins may be important to signal

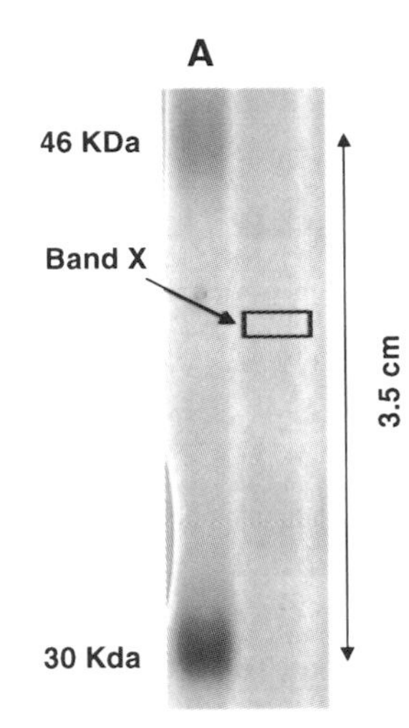

B

Protein in Band X	*Molecular Weight (Da)*	*Peptides Found*
GAPDH	35835	14
Translation initiation factor	35940	9
NKFA	22106	7
Alpha complex protein 1	37488	5
Actin	41671	5
K+ channel beta subunit	46555	4
DNA-lyase	35622	4

Fig. 5. (A) 30 - 46 kDa segment of a Coomassie blue stained SDS-PAGE gel (12.5% T) containing CD40 binding proteins. Band X represents an "in gel" digestion of a 1.6 mm x 10 mm gel slice. (B) List of proteins identified in a 40 minute nanoflow-HPLC micro-ESI MS analysis of 20% of the "in gel" digest of Band X.

transduction [27, 28]. Immunoprecipitation was used to select and enrich the accessory proteins. Whole cell lysates were prepared and added to precleared sepahrose beads containing a CD40 specific monoclonal antibody. The beads and sample were placed in a column, washed, and proteins eluted with 100 mM glycine at pH 2.5. The proteins were then concentrated on a 3 kDa molecular weight cut off filter. In one experiment, seven different proteins were identified from a 1.6 mm x 10mm (HxW) gel slice at faint Coomassie stain levels in a single 40-minute analysis (Fig. 5). This example illustrates the high sensitivity of our methodology and highlights the lack of resolution in 1D SDS-PAGE gels when working with complex mixtures.

Nanoflow-HPLC and Micro-ESI on a Fourier Transform Mass Spectrometer

During the past 12 months, we have interfaced our home-built, 7-Tesla Fourier transform mass spectrometer [21] with an electrospray ion source and implemented nanoflow-HPLC and microspray ionization. The instrument is operated in external ion accumulation mode, whereby ions are accumulated in an rf-only octapole guide at all times except ion injection; because ion flight times constitute a small fraction of total scan time (~2 ms out of ~1-2 s), ion sampling efficiency is quite high (> 99%). The FTMS has also been equipped with a CO_2 laser for infrared multiphoton dissociation (IRMPD) of peptide ions. In addition, a variable-flow HPLC gradient delivery system, which has proven successful on the LCQ, has been adapted to our FTMS as well. Due to the lack of automated data acquisition software available for FTMS instruments, our ability to control chromatographic peak width is critical for performing MS/MS experiments; peak-parking allows the operator to manually adjust ion isolation and dissociation parameters. Our preliminary data illustrates the strengths and potential of FTMS for peptide analysis.

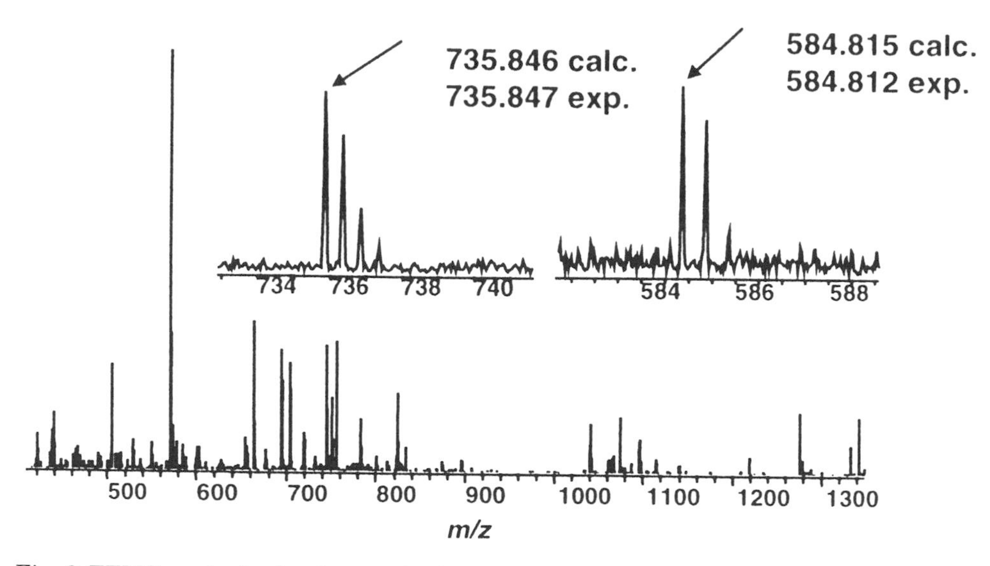

Fig. 6. FTMS analysis of a six-protein digest mixture. Proteins are present in a concentration dynamic range of 1000:1, with cytochrome C at 10 amol. Insets show individual CC peptides detected with high mass accuracy and mass resolution.

To test the utility of LC-FTMS for tryptic mapping, an aliquot containing 10 fmol BC, 2 fmol BSA, 400 amol GAPDH, 150 amol CAH, 40 amol BLG, and 10 amol CC was subjected to MS analysis. A total of 72 tryptic masses were detected and entered into PeptideSearch, a database search algorithm developed and maintained at the European Molecular Biology Laboratory. Typical search parameters were as follows: mass tolerance = 0.01 Da; protein mass range 10 – 75 kDa; 5 peptides required for a protein match; allow up to 2 missed cleavages per peptide. Masses assigned to the top scoring protein were removed prior to subsequent searches. This strategy correctly identified all six proteins in the mixture. Figure 6 (bottom) shows the average of some 200 mass spectra collected during the gradient. The insets show two different, *single scan* mass spectra extracted from the TIC, in which cytochrome C peptides were detected. Note that the high resolution inherent in FTMS allows easy identification of charge state (and thus mass), while the excellent mass accuracy allows one to specify narrow mass tolerances in database searches. In addition, the MS detection limits for FTMS are outstanding; note that a total of 5 CC tryptic peptides were detected, representing ~43% sequence coverage.

The MS/MS capabilities of our FTMS were evaluated using an aliquot (100 fmol BC, 20 fmol BSA, 4 fmol GAPDH, 1.5 fmol CAH, 400 amol BLG, and 100 amol CC) from the same standard protein digest; in this experiment we sought to obtain at least one MS/MS spectrum from each protein in the mixture. The experimental protocol was similar to that described above; initially, mass spectra were acquired in "high-flow" mode until peptides began to elute. The remainder of time was spent in "low-flow" to provide time for (manual) optimization of ion isolation and activation conditions. Figure 7 shows an MS/MS spectrum from 4 fmol of a GAPDH tryptic peptide. To date our general observa-

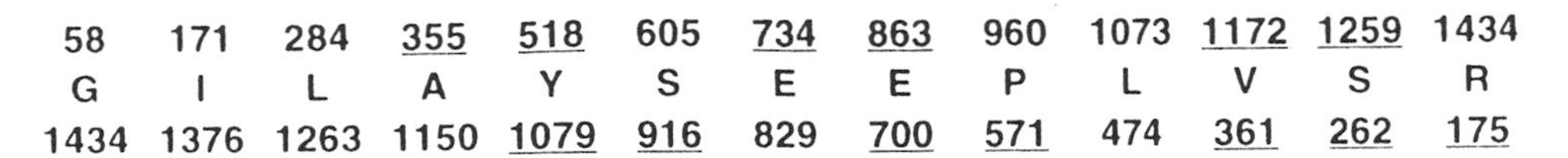

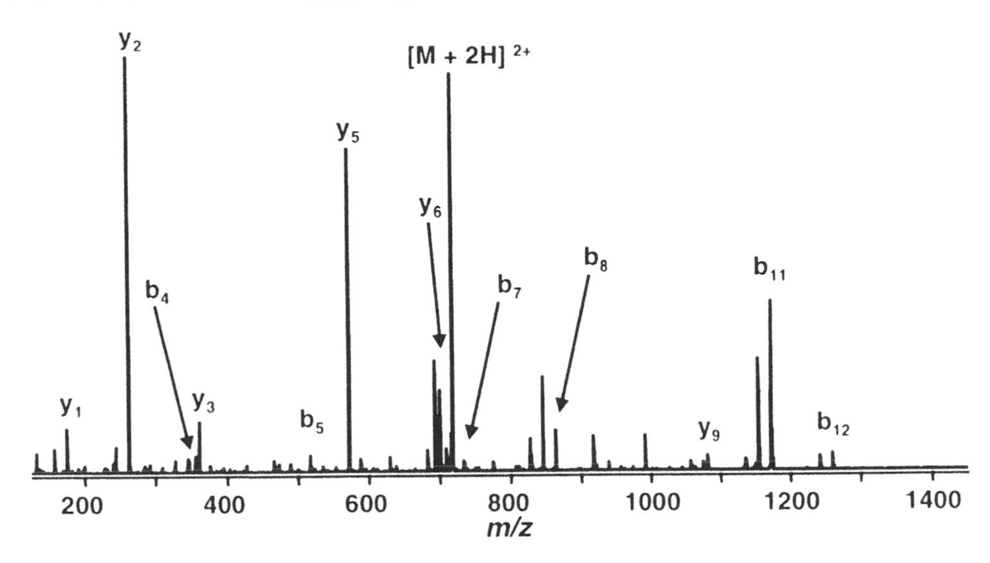

Fig. 7. FTMS / MS analysis (via IRMPD)of 4 fmol GAPDH tryptic peptide, loaded from a six-protein digest mixture.

tion is that photodissociation generates qualitatively similar sequence specific ions as compared to collisional activation. Peptides from each protein, including CAH (400 amol) were dissociated and successfully matched by SEQUEST. Fragment ion mass accuracy is generally < 0.01 Da; this allows for identification of glutamine vs. lysine and phenylalanine vs. oxidized methionine. A peptide from CC (100 amol) was subjected to laser activation but failed to generate useful sequence information.

At present our variable-flow HPLC apparatus is readily interfaced to both FTMS and ion trap mass spectrometers, with somewhat complementary results. FTMS is the preferred instrument for MS analysis due to its lower detection limit (10 vs. 100 amol), superior mass accuracy (0.005 Da vs. 0.3 Da), and higher mass resolution (several thousand vs. unit). In addition, we observe far less chemical noise in MS mode on the FTMS. At present we believe that the bulk of chemical noise is derived from solvent clusters which are not stable under the relatively long, external ion accumulation conditions used on the FTMS. Finally, the larger physical size of a Penning trap, relative to a quadrupole ion trap, provides greater dynamic range, meaning that low abundance species may be readily detected in the presence of high abundance ions. For MS/MS analysis, the LCQ is the superior instrument. In addition to its ability to acquire high quality MS/MS spectra on as little at 10-50 amol of peptide, it can be configured to operate automatically, generating hundreds of spectra per hour, with little or no operator input. Lastly, the LCQ is truly an "affordable" benchtop instrument; its footprint and cost are a small fraction of those compared to FTMS.

Conclusions

The nanoflow-HPLC micro-ESI MS source described is capable of identifying and sequencing proteins at low Coomassie- or silver-stain levels from gels. The high sensitivity of the methodology developed highlights the lack of presumed resolution in 1D or 2D gels when working with very complex mixtures. The low-flow capability of the nanoflow-HPLC micro-ESI source allows separation of peptides for an hour or more, making it suitable for the analysis of these highly complex protein/peptide mixtures with or without prior separation.

The combination of automation, data dependent software, and fast gradient HPLC also makes this system suitable for rapid protein identification. A single 13.5 minute data dependent run on 25 fmol of a tryptic digest of GAPDH (36.5 kDa) revealed 51% of the protein's sequence. When automated, analysis of protein digests could result in as many as 96 proteins/day being identified. This number could increase to >400 proteins/day if simple mixtures (3-5 proteins) are analyzed. Data dependent software complements the low-flow analysis as >1500 MS/MS spectra can be collected in 1 hour at protein concentration dynamic ranges of >1000:1.

Fully automated mass spectrometric methods for the rapid characterization of known and unknown proteins present in complex mixtures at the low femtomole or attomole level has been demonstrated. Development of the above technology will make it possible to monitor proteins in biological fluids and on the cell surface and thus play an important role in developing new tests for early disease diagnosis and new targets for disease treatment. The technology should also become the method of choice for unraveling protein function and expression in the post genome era.

References

1. A. L. Burlingame, R. K. Boyd and S. J. Gaskell, *Anal. Chem.* **1998**, *70*, 647-716.

2. K. Biemann, *Protein Sci.* **1995**, *4*, 1920-1927.

3. E. S. Landers, *Science* **1996**, *274*, 536-539.

4. P. Roepstorff, *Curr. Opin. Biotechnol.* **1997**, *8*, 6-13.

5. J. R. Yates, 3rd, *J. Mass Spectrom.* **1998**, *33*, 1-19.

6. D. F. Hunt, W. M. Bone, J. Shabanowitz, G. Rhodes and J. M. Ballard, *Anal. Chem.* **1981**, *53*, 1704-1706.

7. D. F. Hunt, J. R. Yates, 3rd, J. Shabanowitz, S. Winston and C. R. Hauer, *Proc. Natl. Acad. Sci. USA* **1986**, *83*, 6233-6237.

8. C. M. Whitehouse, R. N. Dreyer, M. Yamashita and J. B. Fenn, *Anal. Chem.* **1985**, *57*, 675-679.

9. J. B. Fenn, M. Mann, C. K. Meng, S. F. Wong and C. M. Whitehouse, *Science* **1989**, *246*, 64-71.

10. D. F. Hunt, J. E. Alexander, A. L. McCormack, P. A. Martino, H. Michel, J. Shabanowitz, N. Sherman, M. A. Moseley, J. W. Jorgenson and K. B. Tomer, in *Techniques in Protein Chemistry II*, J. J. Villafranca, Ed.; Academic Press: New York, **1991**; pp 441-454.

11. M. A. Moseley, D. F. Hunt, J. W. Jorgenson, J. Shabanowitz and K. B. Tomer *J. Am. Soc. Mass Spectrom.* **1992**, *3*, 289-300.

12. D. F. Hunt, R. A. Henderson, J. Shabanowitz, K. Sakaguchi, H. Michel, N. Sevilir, A. L. Cox, E. Appella and V. H. Engelhard, *Science* **1992**, *255*, 1261-1263.

13. A. L. Cox, J. Skipper, Y. Chen, R. A. Henderson, T. L. Darrow, J. Shabanowitz, V. H. Engelhard, D. F. Hunt and C. L. Slingluff, Jr., *Science* **1994**, *264*, 716-719.

14. D. F. Hunt, J. Shabanowitz, H. Michel, A. L. Cox, T. Dickinson, T. Davis, W. Bodnar, R. A. Henderson, N. Sevilir, V. H. Engelhard, K. Sakaguchi, E. Appella, H. M. Grey and A. Sette, in *Methods in Protein Analysis*, K. Imahori, and F. Sakiyama, Eds.; Plenum Press: New York, **1993**; pp 127-133.

15. R. E. Settlage, P. S. Russo, J. Shabanowitz, and D. F. Hunt, *J. Microcolumn Sep.* **1998**, *10*, 281-285.

16. R. E. Settlage, R. E. Christian, J. Shabanowitz and D. F. Hunt, *Anal. Chem.* **1998**, submitted.

17. A. Shevchenko, M. Wilm, O. Vorm and M. Mann, *Anal. Chem.* **1996**, *68*, 850-858.

18. J. Eng, A. L. McCormack and J. R. Yates, *J. Am. Soc. Mass Spectrom.* **1994**, *5*, 976-989.

19. D. F. Hunt, J. Shabanowitz, R. T. McIver, Jr., R. L. Hunter and J. E. P. Syka, *Anal. Chem.* **1985**, *57*, 765-771.

20. K. D. Henry, E. R. Williams, B. H. Wang, F. W. McLafferty, J. Shabanowitz and D. F. Hunt, *Proc. Natl. Acad. Sci. USA* **1989**, *86*, 9075-9078.

21. S. E. Martin, F. M. White, R. E. Settlage, R. E. Christian, J. Shabanowitz, D. F. Hunt and J. A. Marto, *Anal. Chem.* **1998**, submitted.

22. M. W. Senko, C. L. Hendrickson, M. R. Emmett, S. D. -H. Shi, and A. G. Marshall, *J. Am. Soc. Mass Spectrom.* **1997**, *8*, 970-976.

23. M. T. Davis and T. D. Lee, *J. Am. Soc. Mass Spectrom.* **1997,** *8*, 1059-1069.

24. T. S. Ingebritsen and P. Cohen, *Science* **1983**, *221*, 331-338.

25. T. A. J. Haystead, A. T. R. Sim, D. Carling, R. C. Honner, Y. Tsukitani, P. Cohen and D. G. Hardie, *Nature* **1988**, *337*, 78-81.

26. P. Cohen, *Ann. Rev. Biochem.* **1989**, *58*, 453-508.

27. S. Lederman, A. M. Cleary, M. J. Yellin, D. M. Frank, M. Karpusas, D. W. Thomas and L. Chess, *Curr. Opin. Hematol.* **1996**, *3*, 77-86.

28. M. Faris, F. Gaskin, J. T. Parson and S. M. Fu, *J. Exp. Med.* **1994**, *179*, 1923-1931.

QUESTIONS AND ANSWERS

Lowell H. Ericsson (University of Washington)

Have you seen evidence of keratins in solvents, reagents, gels, etc. in addition to keratin from laboratory dust? How do you minimize keratin contamination?

Answer. Yes, keratins seem to be everywhere. All of the reagents, sample handling (pipetman tips, tubes, gels) or manipulating equipment is kept separate and dust free. We have even received contaminated shipments of sequencing grade trypsin. It is reasonable and recommended that we start using "clean room" or HEPA-filter possible pressure hoods to handle samples. We also have to educate collaborators in handling their samples.

We are developing "exclusion" software in which we can parse out MS/MS spectra of known contaminants (i. e., keratins) or just subtract out a "control" sample from the sample of interest.

Liang Tang (Affymax Research Institute)

Have you used smaller particles (<5μm) for packing capillary LC?

Answer. Yes, we have gone down to 3 micron particles, but we have not tried the 1.5 micron packings yet. The smaller particles do give better chromatographic resolution and necessitate less packing to achieve the same results.

John Stults (Genentech, Inc.)

How do you explain the detection limit differences between FTMS and the LCQ for MS spectra?

Answer. Many more ions may be stored in the Penning trap of the FT-ICR. We are achieving an effective dynamic range on the LCQ by mass exclusion, whereas we can easily see several orders of magnitude between ion intensities in the FT-ICR. The FT-ICR is also trapping ions in the injection rf-only octapole to improve duty cycle and increase signal dynamic range. The LCQ is limited to the number of ions that can be put in the trap at a time. Also, the baseline chemical noise (solvent-associated clusters) appears to be lower in the FT-ICR, probably because of the octapole trapping. This also increases the signal-to-noise observed for the FT-ICR.

Dan Kassel (CombiChem, Inc.)

To facilitate your peak-parking experiments, are you incorporating an automated switching valve? Have you applied this approach to glycopeptide and phosphopeptide identification?

Answer. Yes, a computer controlled switching valve can rapidly select different restrictors. The flow can be reduced an order of magnitude in a couple of seconds without appreciable loss in sensitivity.

The technique could be used to select for any preprogrammed situations, i. e., ion current above a certain level, a particular mass/charge, ions above a certain m/z, or whatever you like. We have tried it for glycopeptides, but have not tried it with phosphopeptides, although it should work fine.

Coaxial Nanospray Coupled with a Hybrid Quadrupole/Time-of-Flight Tandem Mass Spectrometer for Proteome Studies

M. A. Moseley
Glaxo Wellcome, Research Triangle Park, NC 27709

Genome sequencing projects have dramatically changed the approach to the study of biology, and genomics will have a similar impact on the way in which pharmaceutical companies discover and develop drugs. These projects have two principal goals — the identification of the best disease genes/targets on which to initiate drug discovery programs, and the identification of patients who will respond favorably to drugs based on their genotype. To accomplish the first of these goals one must determine the interrelated expression of thousands of genes, and how these patterns of expression define an organism's phenotype. Two different products of gene expression can be studied — nucleic acids and proteins. Significant progress has been made in the application of differential displays of mRNA, using PCR with primers to generate high-density expression arrays of cDNA species. Study of the other product of gene expression, proteins, can provide information complementary to that obtained from nucleic acids. Disease mechanisms will not be defined in molecular terms by studying only nucleic acids. Proteomics, the study of the final products of gene expression (proteins), can provide much information which the study of nucleic acids cannot, including [1]:

1) if, when, and where final gene products are translated
2) the relative concentration of final gene products
3) the extent of post-translational modifications
4) the effects of gene knock-out or over-expression
5) the phenotype of multigenic phenomena, including: disease; drug administration; cell-cycle state; ontogeny; aging; and stress.

An excellent review of additional applications of proteomics has recently been published [2], and they include: the confirmation of open reading frames in DNA sequences; the assignment of function to novel genes by quantitatively monitoring changes in the proteome after overexpression or gene-knockout; determination of subcellular localization of proteins to monitor changes in the localization of intracellular signaling proteins; and the identification of the components of complex protein assemblies involved in sub-cellular activities, including vesicle transport, transcription, mRNA processing, and transport across nuclear membranes.

It is in the determination of the phenotype of multigenic phenomenon that proteomics may have the greatest impact in the discovery of medicines.

Proteomics has significant potential to assist and complement genomic differential gene expression technology for the identification of targets associated with disease, the tissue distribution of these targets, and their linkages in metabolic and signaling pathways, as well as identifying proteins other than the specific target(s) whose expression levels are affected by drug treatment (such as toxicological effects). The identification of proteins, which are potential targets as a result of misregulation in the disease state or their location in biochemically interesting pathways, is crucial. The rapid identification of these potential targets and advancement of knowledge regarding their regulation and interactions with signaling pathways in disease will significantly contribute to the identification of pharmacologically interesting targets.

In 1995 the genome (total DNA sequence) for a self-replicating organism (*Haemophilus influenzae*) was obtained for the first time [3]. It is expected that within the next few years the complete genomic sequence will be available for 30 to 40 microbes. These databases of genetic information have given key information needed to assess the ability of proteomic technology to detect and quantify the complete proteome of an organism. The ultimate goal of obtaining the total human genome will be finished within 5 to 7 years, at which time full power of genomics and proteomics can be applied to drug discovery.

Within the past several years significant advances have been made in three key fields required for proteomic studies: separation methods for proteins and peptides; mass spectrometry technology for the rapid, high throughput detection and identification of proteins and peptides (MALDI-TOF/MS and nanospray/MS/MS); and bioinformatics (both hardware and software) to correlate genes and these gene products. The nanospray/MS/MS technique developed by Wilm and Mann [4, 5], along with their sequence tag method of protein identification [6], has proven to be an extraordinarily useful technique for proteomic studies. The very low detection limits (tens of femtomoles) and long analysis times (approximately 1 hour/μL of sample) obtained via nanospray has permitted the identification of proteins which could not be identified from peptide maps obtained by MALDI-TOF/MS. However, nanospray is a manual, labor intensive process, and the high sample throughput required for proteome characterization will greatly benefit from the automation of MS/MS data acquisition. One option for automation of nanospray has been developed in our labs — a coaxial nanospray interface [7, 8]. We report the details on the fabrication of the interface, and its application to the identification of proteins using both flow-injection analysis and on-line nanoscale capillary LC/MS/MS, using both a triple quadrupole mass spectrometer and a hybrid quadrupole-TOF mass spectrometer.

The coaxial nanospray interface has several notable differences from the nanospray interface of Wilm and Mann. Their interface uses a metallic sputter coating on the needles to make electrical contact between the source probe and the analyte solution. The electrical potential in their nanospray interface serves a dual function — to generate the electrospray and an electroosmotic flow of analyte solution to the tip of the needle. Due to this coupling of spray and flow one cannot directly control the flow rate of the analyte solution. In addition, the metallic coating is quite fragile, and these needles cannot be used reliably for multiple analysis. The design of Wilm and Mann uses one needle for each sample, minimizing problems with the stability of the metallic coating, and precluding problems with sample-to-sample cross contamination. However, this approach requires manual loading of the sample into the needle, mounting of the needle into the probe, and initiating flow by manually touching off the needle tip on the source probe. Although this approach works very well, it is slow and labor intensive.

The coaxial nanospray interface was designed to use a pressure induced flow of analyte to the spray tip at flow rates of 20 nL/min., permitting automated flow-injection analysis of multiple samples without manipulation of the spray needle or the interface. Electrical contact is made between the probe and spray tip through the makeup fluid, which flows coaxially around the sample delivery column to the spray tip. This precludes the need to sputter coat the spray needles, and makes the interface quite rugged — single spray tips have been used for weeks of operation. Note that it is not necessary for the makeup fluid to be flowing during analysis — best concentration sensitivity is obtained in this manner. For nanospray-type infusion experiments the makeup fluid flow is useful for flushing residual sample from the tip between samples. When used with nanoscale capillary LC columns, the make-up fluid minimizes chromatographic band broadening by sweeping the analytes from the capillary end into the Taylor cone. In addition, the ability to use a makeup fluid adds flexibility to the nanoscale capillary LC analysis by permitting post-column modifications to the spray solution. For example, post-column modification of pH permits independent optimization of the pH conditions for the chromatographic separation and for nanospray analysis. Importantly, the use of a high percentage of organic modifier in the makeup fluid stabilizes the ion beam across the chromatographic gradient by minimizing the changes in surface tension of the sprayed fluid. Finally, "exotic" mobile phase modifiers such as "TFA Fix" can be easily added to the analytes post column, and mass reference standards can be introduced for accurate mass measurements.

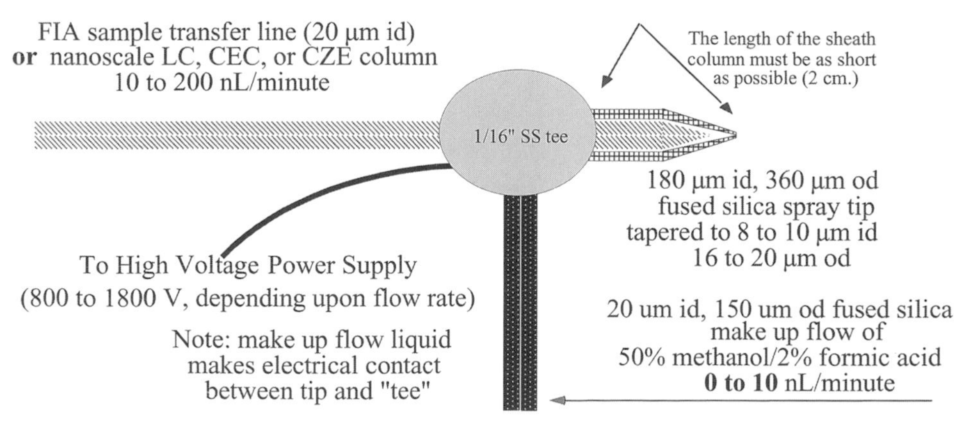

Fig. 1. Schematic of the interface used for infusion/coaxial nanospray/MS/MS analysis and nanoscale capillary LC/coaxial nanospray/MS/MS.

Experimental

Chemicals

The peptides and proteins used in this work were acquired from Sigma (St. Louis, MO), with the exception of trypsin, which was acquired from Promega (Madison, WI). Trifluoracetic acid was acquired from Aldrich (Milwaukee, WI). Acetonitrile (HPLC Grade) was acquired from Burdick and Jackson (Muskegon, MI). All solutions were prepared from 18 Mohm water (Hydro PicoPure Water System, RTP, NC).

Coaxial Nanospray Interface

The design of the coaxial nanospray interface is shown in Fig. 1. The interface uses a 1/16" tee (Valco, Houston, TX) which has been drilled to permit the 150 µm OD fused silica column (Polymicro Technologies, Phoenix, AZ) to pass completely through the tee and into the fused silica tip. Fused silica tips were drawn from 180 µm ID/360 µm OD fused silica capillaries to give an ID of 8 to 12 µm. The tips were sealed in the tee with a short length of 0.020" ID, 1/16" OD PEEK tubing (Upchurch Scientific, Oak Harbor, WA) using a stainless steel ferrule and nut. Note that for use on the Sciex API-III triple quadrupole mass spectrometer (Thornhill, Ontario) the head of this nut was machined down in both thickness and outer diameter to allow the coaxial interface to be placed as close as possible to the mass spectrometer sampling orifice. This modification was not required for use on the Z-Spray source of the Micromass Q-Tof (Manchester, UK). The length of the fused silica tips (approximately 2 cm. total length) has proven to be important for low nL/min. flow operations. As with nanospray tips, application of the electrospray voltage to the tee will generate an electroosmotic flow within this coaxial tip, and the magnitude of this flow has been observed to be proportional to the length of the tip. The sample is delivered to the tip via the 150 µm OD fused silica capillary. For infusion studies a 20 µm ID, 150 µm OD, 60 cm long fused silica capillary has been used. The coaxial makeup solution was delivered to the tee by a fused silica column with dimensions identical to those of the analytical column. This column was sealed in the tee with a short length of teflon tubing 0.010" ID, 1/16" OD (Upchurch Scientific, Oak Harbor, WA) using a PEEK ferrule and fingertight PEEK nut (Upchurch Scientific, Oak Harbor, WA).

To accurately deliver sample and makeup solutions to the coaxial interface in a pulse free manner at low nL/min. flow rates, stainless steel pressure vessels were used (Fig. 2), the design of which has been previously described [9]. Typical pressures required for the sample delivery in the nL/min. range were 25 to 200 psi, dependant on sample viscosity. Absolute flow rates (5 to 500 nL/min.) were measured as described in the Results and Discussion section.

Nanoscale Capillary LC

Nanoscale capillary LC columns were prepared by a modification of a method previously reported [9]. Fused silica capillary columns (75 µm ID, 150 µm

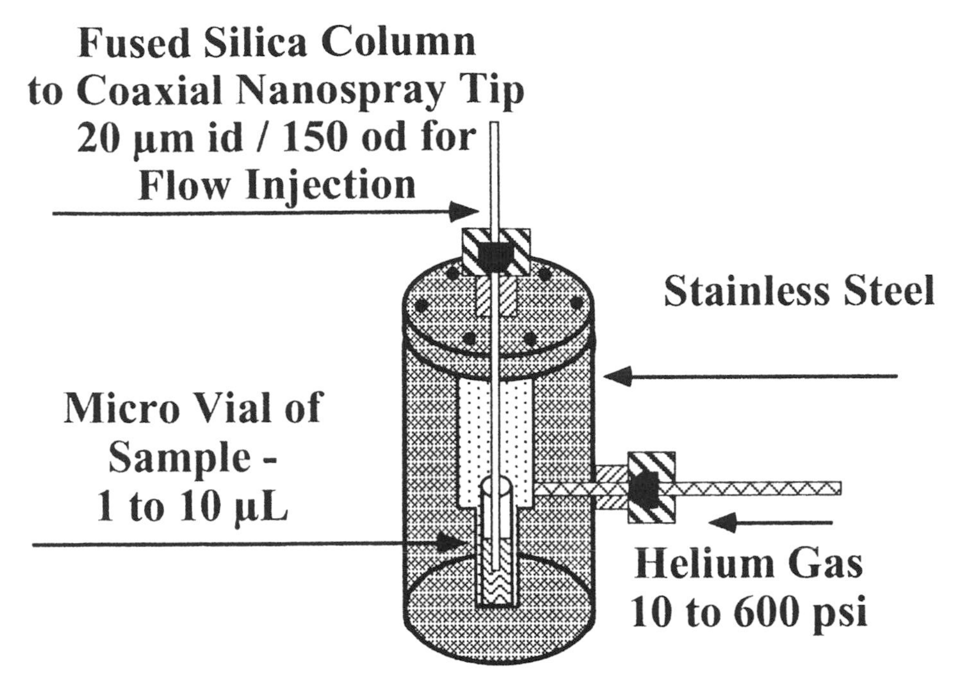

Current Sample Introduction System
Gas Pressurized Liquid "Pump"
10 to 1,000 nL/minute

SAMPLES CHANGED BY CHANGING MICROVIALS

Fig. 2. Stainless steel pressure vessel used with the coaxial nanospray interface. Two separate vessels were used for infusion/coaxial nanospray/MS/MS analysis — one to deliver the sample solution and one to deliver the makeup solution. Typical pressures used with these vessels were 10 to 200 psi, although they were designed for use with pressures up to 10,000 psi.

OD, Polymicro Technologies, Phoenix, AZ) were cut to 1 m lengths. Approximately 0.5 cm of the polyimide coating was removed from the end of the capillary using an electric arc. The column was then rinsed with 2-propanol and dried with a stream of helium. The frit was prepared by tapping the end of the column from which the polyimide was removed into a vial containing 5 μm glass microspheres (Duke Scientific Corp., Palo Alto, CA) until a 200 μm length of the column was filled. The particles were scintered into place using the electric arc. The integrity of the frit was tested prior to packing the column by infusing 2-propanol at 1,500 psi. A 30:1 (milliliter:gram) slurry of stationary phase (5 μm Hypersil BDS-C18, Shandon, Astmoor, UK) in 33% acetone/67% hexane was sonicated for 5 minutes and then loaded into a stainless steel slurry packing reservoir. A syringe pump

(ABI-140A, Applied Biosystems, Foster City, CA) was used to pressurize the slurry reservoir to a pressure of 4,000 psi to pack a 25 cm length of the stationary phase into the capillary.

Sample injections were made using either the pressure vessel or a 0.5 μL internal loop valve (Rheodyne, Cotati, CA). The pressure vessel proved to be useful for injecting large volumes of dilute samples, with sample preconcentration on the head of the stationary phase. Gradient elution with acetonitrile:water solutions containing 0.05% trifluroracetic acid was performed at column flow rates of 200 nL/min. using syringe pumps (ABI-140A) operated at a flow rate of 50 μL/minute. The flow from the pumps was split precolumn (250:1) using a splitter fabricated in-house.

Mass Spectrometry

Two mass spectrometers were used in this work — a Sciex API-III triple quadrupole (Toronto, Ontario, Canada) and a Micromass Q-Tof (Manchester, UK). The Sciex API-III triple quadrupole mass spectrometer was equipped with a nanospray interface (Protana A/S, Odense, Denmark). As with conventional nanospray, the best signal is obtained with coaxial nanospray when the spray tip is almost in the plane of the sampling orifice. The coaxial nanospray interface was mounted onto the Protana nanospray interface using an adapter fabricated

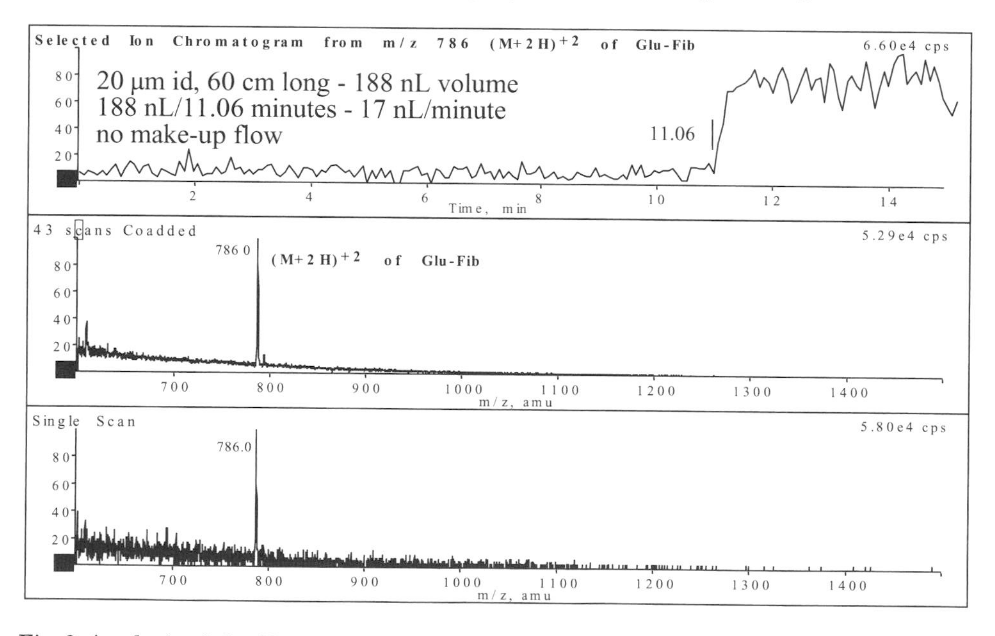

Fig. 3. Analysis of glu-fibrinopeptide (5 μM in 70% methanol / 5% formic acid) by infusion / coaxial nanospray / MS on the API-III at a scan rate of 5 sec / scan. The transfer line was initially filled with blank solvent, and switched to the sample solution at the start of data acquisition. The elution time of the sample (11.06 min.) divided by the volume of the column (188 nL) gives a measured flow rate of 17 nL / min. Top: selected ion chromatogram of the doubly charged ion showing an elution time of 11.07 min; middle: co-addition of 43 scans; bottom: single scan.

in-house, which permitted the coaxial nanospray tip to be in the plane of the sampling orifice. The Micromass Q-Tof was equipped with the Micromass Z-Spray source, and the coaxial nanospray interface was mounted directly on the Z-Spray probe stage.

Results and Discussion

Coaxial Nanospray

Analysis of peptides with the coaxial nanospray interface is simple and straightforward. The results from a typical coaxial nanospray analysis on the API-III are shown in Fig. 3. Prior to analysis the flow from the makeup reservoir (50% methanol/5% formic acid in water) was started to fill the line and tee. Pressures of 250 to 500 psi can be used to speed up the initial filling, and then reduced to the normal operating pressure of 0 to 20 psi. Also prior to analysis the sample transfer line is flushed with blank sample buffer (typically 70% methanol/5% formic acid in water) by placing a vial of solvent into the sample pressure vessel, inserting the column, and pressurizing the vessel. The sample was introduced by exchanging the vial containing the blank buffer with a vial containing the sample and repressurizing the vessel. The normal operating voltage for the coaxial nanospray probe was 1200-1600V. The top trace in Fig. 3 shows the selected ion trace for the double charged ion of the analyte, glu-fibrinopeptide (5 μM). The volume of a 20 μm ID column is 3.14 nL/cm, thus a 60 cm length has a volume of 188 nL, and the time required for the sample to flow through the column was measured to be 11.06 min., corresponding to a flow rate

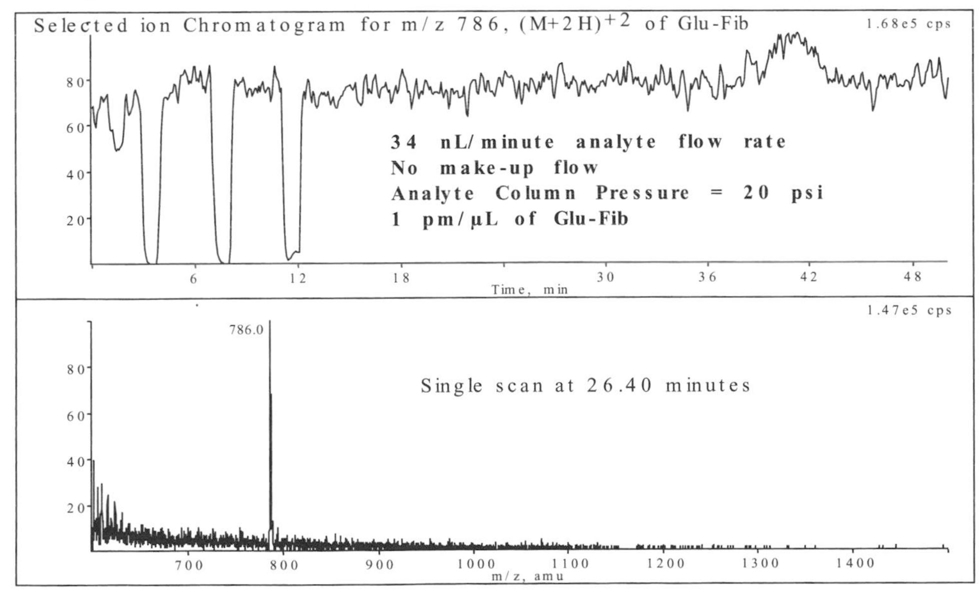

Fig. 4. Demonstration of the dependence of the analyte signal on applied pressure rather than applied voltage. The sample vessel was depressurized for one minute three times during the first 15 minutes of the 50 minute data acquisition on the Sciex API-III.

of 17 nL/min. The second and third traces in Fig. 3 contain spectra representing the sum of 43 scans and a single scan, respectively. A 10X higher pressure (200 psi) can be used to speed the sample introduction process, which is reduced to the normal operating pressure of 20 to 50 psi immediately prior to the sample entering the spray tip.

The analyte flow in the coaxial nanospray interface is pressure dependent and not voltage dependent, as demonstrated in Fig. 4. In this analysis glu-fibrinopeptide was delivered to the spray tip at a measured flow rate of 34 nL/minute, and the sample vessel was depressurized three times for one minute each during the first 15 minutes of the 50 minute data acquisition. The selected ion chromatogram in Fig. 4 shows the signal from the doubly charged ion of glu-fibrinopeptide drops out when the sample vessel is depressurized, and quickly returns when the vessel is repressurized.

The flow rate of the makeup fluid has an impact on the signal-to-noise ratio of the resultant mass spectrum because this flow dilutes the sample. The effect of varying the flow rate of the makeup fluid is shown in Fig. 5. The analyte was delivered to the tip at a measured flow rate of 44 nL/min., and the makeup fluid flow rate was varied from 0 nL/min. (top chromatogram) to 10 nL/min. (middle chromatogram) to 40 nL/min. (bottom chromatogram). Note that the ion counts per second were observed to be inversely proportional to the flow rate of the makeup fluid, with the highest ion counts obtained with the makeup fluid flow rate at 0 nL/min. In routine use the makeup flow was used only between samples to flush residual sample from the coaxial nanospray tip.

The stability of the ion beam from the coaxial nanospray interface was evaluated by analysis of a 4 μL sample placed in a microvial in the sample pressure vessel. A stable signal from the doubly charged ion of glu-fibrinopeptide

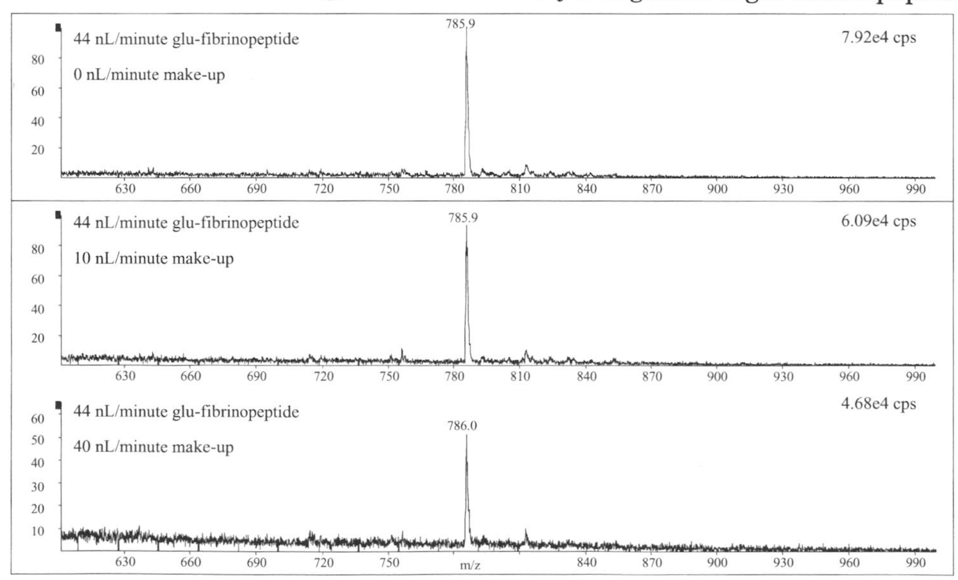

Fig. 5. Effect of makeup fluid flow rate on the signal-to-noise ratio of the selected ion trace for glu-fibrinopeptide on the Sciex API-III.

was obtained for 209 minutes (Fig. 6), giving a measured flow rate of 19 nL/min. No make-up flow was used in this analysis. The spike at the end of the analysis was due to the analyte solution in the sample capillary being replaced with helium gas when the sample was depleted, which increases the flow rate of the analyte solution. This indicated that the signal intensity was not completely independent of flow rate. The dependence of signal intensity on flow rate was quantitatively evaluated by acquiring data at 3, 6, 12, 23, 46, and 92 nL/min. The mass spectra and selected ion chromatogram acquired during this analysis (Fig. 7) show that while the ion beam was stable over the entire flow rate range, the signal intensity increased by approximately an order of magnitude in going from 3 to 46 nL/min., and did not increase further in going from 46 to 92 nL/min. Direct comparisons of coaxial nanospray with nanospray have shown that similar signal-to-noise ratios are obtained with both systems (Fig. 8).

The coaxial nanospray interface has been successfully used for the identification of liver tissue proteins separated by 2D gel electrophoresis. After an in-gel digestion of the protein, the resultant peptides were analyzed by coaxial nanospray MS (Fig. 9) and MS/MS (Fig. 10). A sequence tag search [6] of this MS/MS spectrum identified the peptide as T17 of rat ATP synthase beta chain. Three additional MS/MS spectra obtained in this experiment confirmed the protein identification, and six additional peptides gave molecular weights matching tryptic peptides of this protein but were not interrogated by MS/MS.

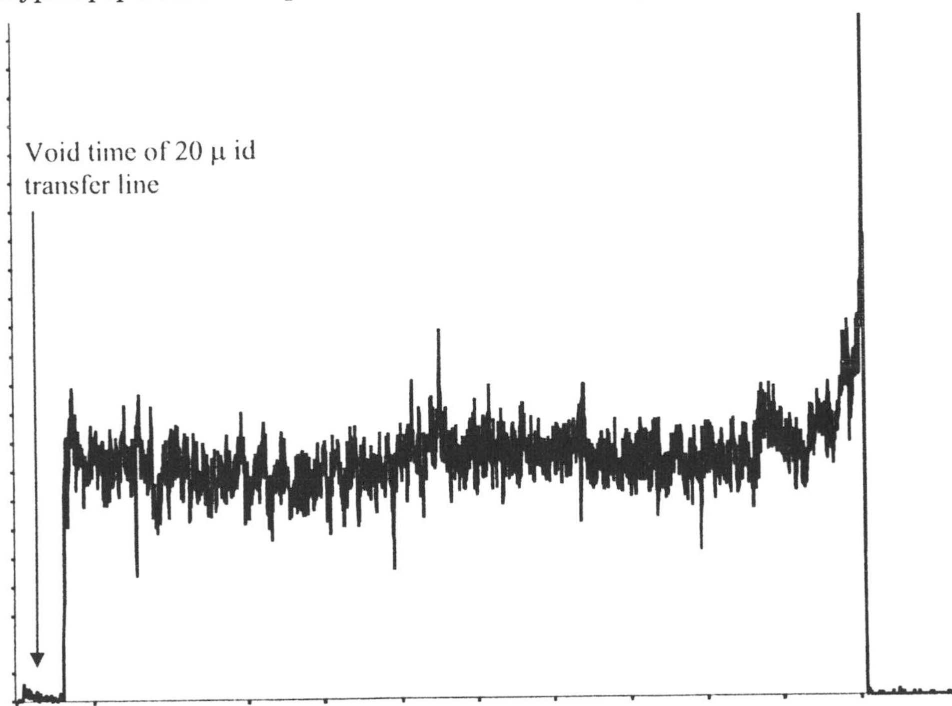

Fig. 6. Analysis of a 4 μL aliquot of glu-fibrinopeptide with the coaxial nanospray interface on the Sciex API-III. A stable ion beam was observed for 209 minutes, giving a measured flow rate of 19 nL/min.

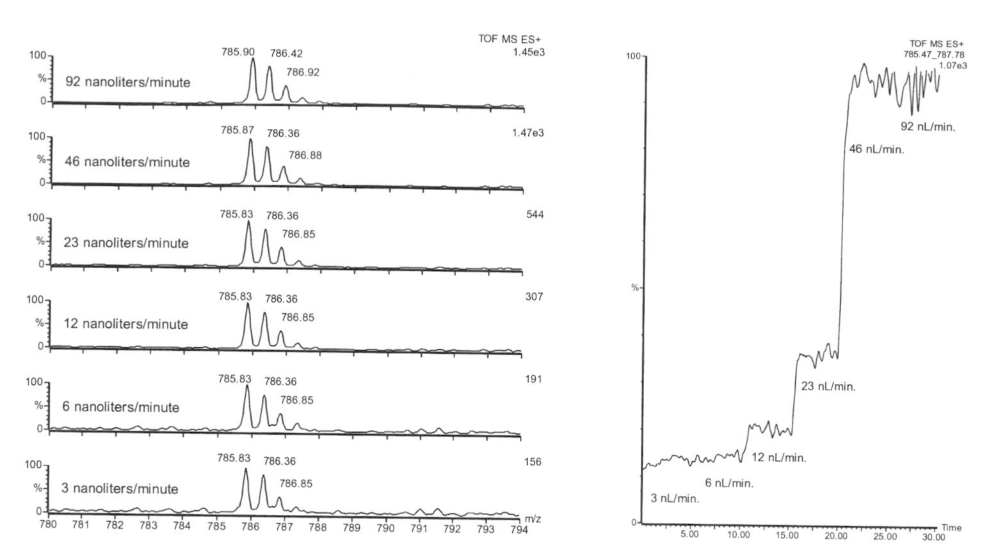

Fig. 7. Signal intensity from the coaxial nanospray interface as a function of flow rate on the Micromass Q-Tof. The optimum flow rate for this 11 μm ID spray tip was observed to be 46 nL / min.

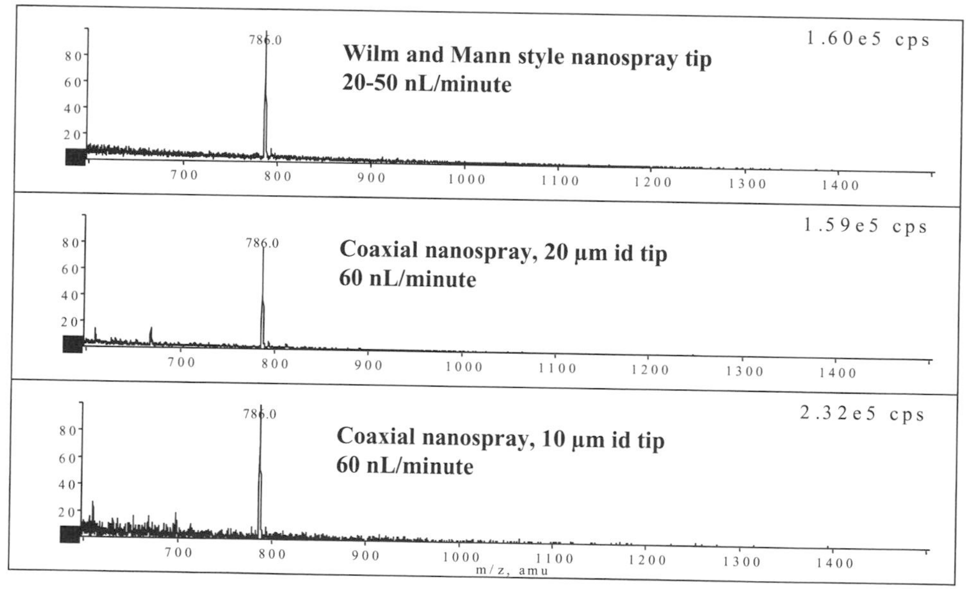

Fig. 8. Comparison of nanospray and coaxial nanospray interfaces for the analysis of glu-fibrinopeptide (5 μM) on the Sciex API-III.

Nanoscale Capillary LC / Coaxial Nanospray

The coupling of nanoscale capillary LC with mass spectrometry was first reported in 1991, using continuous-flow fast atom bombardment [9] and electrospray ionization [10] interfaces. This combination of techniques has proven useful for the analysis of peptides and proteins, and several reviews have

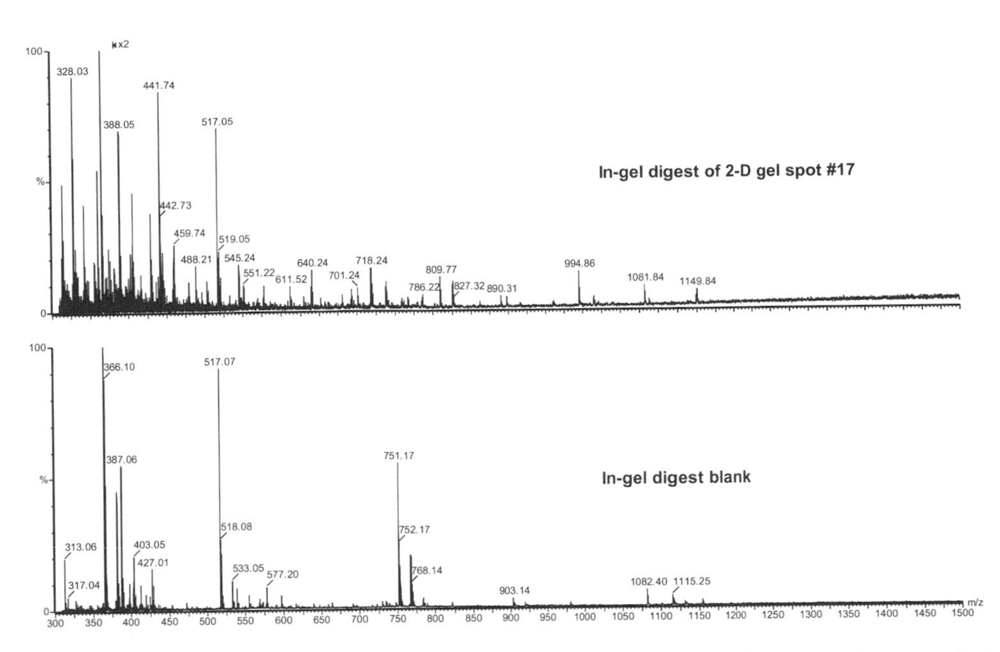

Fig. 9. Coaxial nanospray/MS analysis of an in-gel digest of a rat liver protein separated by 2D gel electrophoresis and a digest of a "blank" portion of the gel using the Micromass Q-Tof.

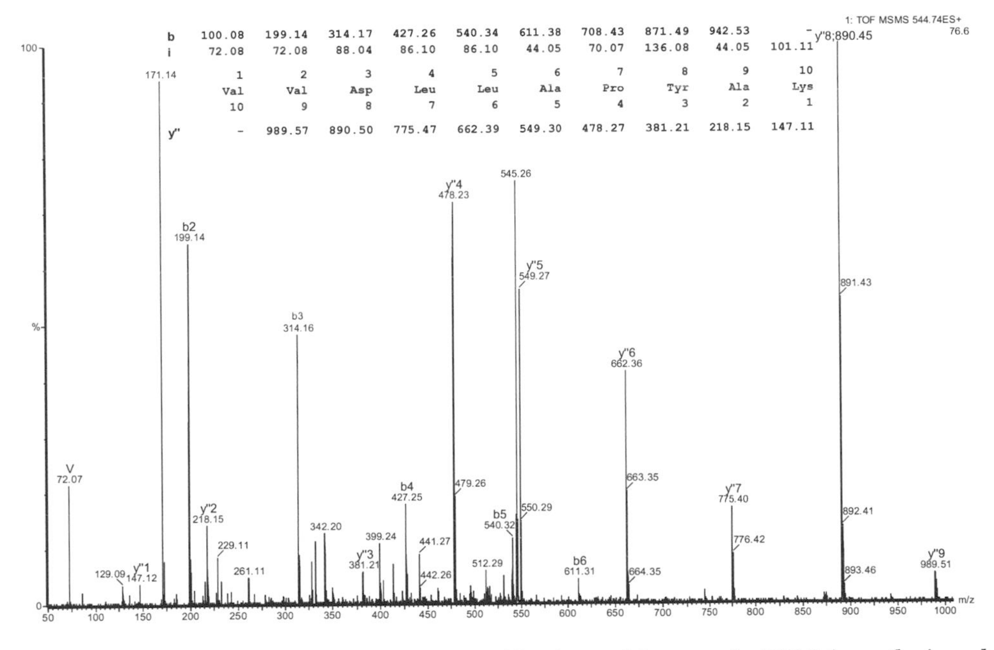

Fig. 10. Coaxial nanospray/MS/MS of doubly charged ion at m/z 545.2 from the in-gel digested protein using the Micromass Q-Tof. A sequence tag database search identified this peptide as the tryptic peptide #17 from ATP synthase beta chain (rat).

been published [11, 12]. A comparison of the figures of merit for the analysis of peptides by nanospray and nanoscale capillary LC leads to several interesting conclusions. Consider the following analysis conditions:

- 1 µL sample of a 1 pm/µL protein digest containing 50 tryptic peptides
- nanospray flow rate of 20 nL/min.
    - 50 minute analysis time
    - 1 minute analysis of each peptide
- nanoscale capillary LC flow rate of 200 nL/min.
    - 50 minute gradient elution
    - chromatographic peak width of 30 seconds at base

Under these conditions, the concentration of each peptide in the nanospray analysis would be 1 pm/µL, whereas the concentration in the nanoscale capillary LC experiment would be 10 pm/µL (1 pm of analyte eluting in 100 nL of mobile phase). Thus the nanoscale capillary LC system would give a 10-fold improvement in analyte concentration, which would yield a commensurate increase in the concentration-dependant electrospray response. In a nanospray analysis all analytes are flowing into the mass spectrometer simultaneously, so the duty cycle for the analysis of each peptide by MS/MS is actually quite low. In this example the nanospray analysis would interrogate each of 50 peptides for 1 minute of the 50 minute analysis, even though each peptide is constantly flowing into the mass spectrometer, giving a duty cycle for each peptide of 2% (1 min./50 min.). For a peptide separated by nanoscale capillary LC the duty cycle of its MS/MS analysis would be 100%, giving a 50 fold improvement in duty cycle over nanospray in this example. Not all peptides will be chromatographically separated, but the improvement in overall duty cycle is apparent. In addition, nanoscale capillary LC offers automation advantages over nanospray, including on-line desalting and concentration of samples, automated sample introduction, and automated data acquisition (data dependant MS to MS/MS switching). These advantages allow the unattended analysis of samples 24 hours/day, maximizing the sample throughput on the mass spectrometer and giving a higher return on the capital investment in the mass spectrometer. Nanospray does offer its own set of advantages, notably the ability to perform precursor and product ion scans from a single sample.

For the advantages of nanoscale capillary LC/MS/MS to be realized requires a tandem mass spectrometer capable of fast scanning (at least 1 second/scan), and with automation features including data dependant scanning (automated MS to MS/MS switching). The Micromass Q-Tof hybrid quadrupole/time-of-flight mass spectrometer has been found to be well suited for coupling with nanoscale capillary LC. Studies in our lab have shown that decreasing the scan rate (actually sample integration time) from 5 seconds to 1 second does reduce the overall signal intensity, but equivalent signal-to-noise ratios are obtained. Note that the nature of the TOF/MS/MS data acquisition precludes the spectral skewing which can be observed with fast data acquisition rates on scanning type spectrometers (such as quadrupole and sector instruments).

The coaxial nanospray interface has been successfully coupled with nanoscale capillary LC columns and applied to the analysis of protein digests using a Micromass Q-Tof hybrid spectrometer equipped with a Z-Spray source. Columns were fabricated from fused silica capillaries (75 µm ID, 150 µm OD) with a packing bed length of 25 cm. Gradient elution was performed at 200 nL/min. with an acetonitrile/water gradient of 0.5%/min. Trifluoroacetic acid was added to both buffers at a concentration of 0.05%. Formic acid is a preferred mobile phase modifier for mass spectral analysis of peptides because formate ions do not ion-pair with peptides and give better signal-to-noise ratios than those obtained with TFA. However, TFA provides superior chromatographic peak shape due to its ion-pairing effects, and one goal of this experiment was the evaluation of the extent of band broadening in the coaxial interface. The coaxial nanospray interface was changed from infusion analysis mode to chromatographic analysis mode by replacing the flow injection capillary with the packed capillary LC column such that the nanoscale capillary LC column passes through the tee and into the spray tip. Note that the packing bed terminates within a few hundred microns of the spray tip. This precludes the chromatographic band broadening that can occur when chromatographic unions and transfer lines are used post-column to deliver the chromatographic eluant to the spray tip. The makeup flow [20 to 50 nL/min. of 50% methanol/(5% formic acid in water)] minimizes chromatographic band broadening in the interface by sweeping the eluant from the end of the column into the spray tip.

This system was used to analyze a tryptic digest of bovine serum albumin (reduced and alkylated). The sample was first analyzed in MS mode using a data acquisition rate of 1 second/integration. At low analyte levels the total ion chromatogram can be dominated by chemical background noise, so the data set may be best visualized in the form of selected ion chromatograms (SICs). The SICs of 36 tryptic peptides from an analysis of 150 femtomoles of digest (on-column) are shown in Figs. 11-14. Peak width for individual peptides was typically 25-30 seconds at base. Good chromatographic peak shape was observed, along with high signal-to-noise ratios (typically >25:1), and a 61% sequence coverage of the protein was obtained. Data dependent scanning (automated MS to MS/MS switching) was also used to analyze this digest, again at the 150 fm level. Using a single collision energy of 28V, a total of 21 MS/MS spectra were acquired, 10 of which gave significant sequence information with good signal-to-noise ratios (>10:1) (Figs. 15-17). The remaining 11 MS/MS spectra were not useable due to the collision energy being either too low or too high for the m/z of the peptide interrogated. Note that the current release of MassLynx software (3.1) permits three different collision energies to be used for each precursor ion, and the magnitude of these collision energies can be varied as a function of the m/z of the precursor. Initial experiments with this software in our laboratory have shown extensive sequence information may be obtained from >80% of the peptides analyzed at low femtomole levels by data dependant scanning with collision energy switching.

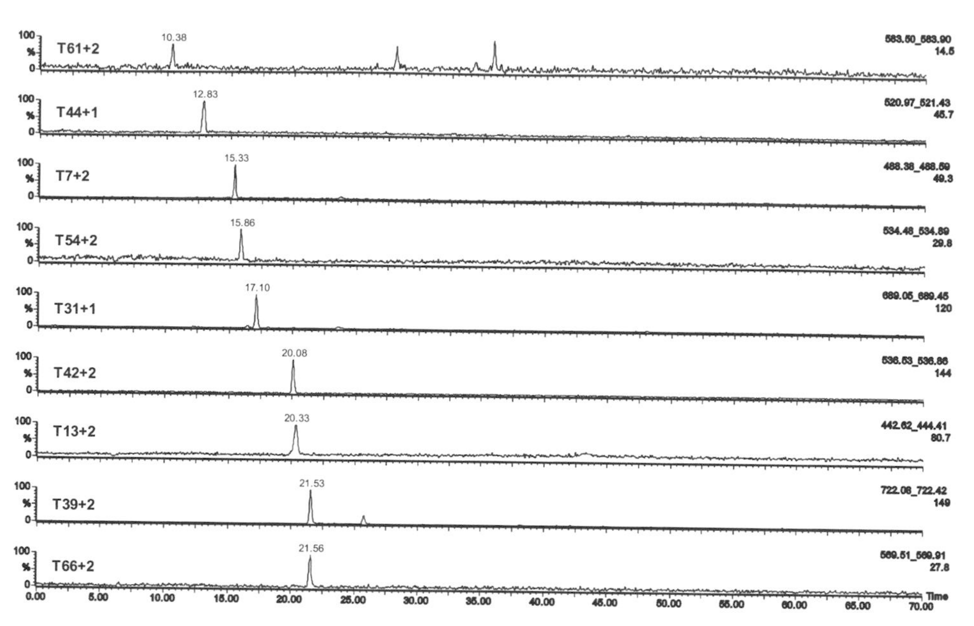

Fig. 11. First set of selected ion chromatograms from the nanoscale capillary LC / coaxial nanospray / MS analysis of 150 femtomoles (on-column) of a tryptic digest of bovine serum albumin (reduced and alkylated) on the Micromass Q-Tof. An acetonitrile:water gradient (0.05% TFA in each) was used at a flow rate of 200 nL / min. with a 75 μm ID, 25 cm long fused silica capillary LC column.

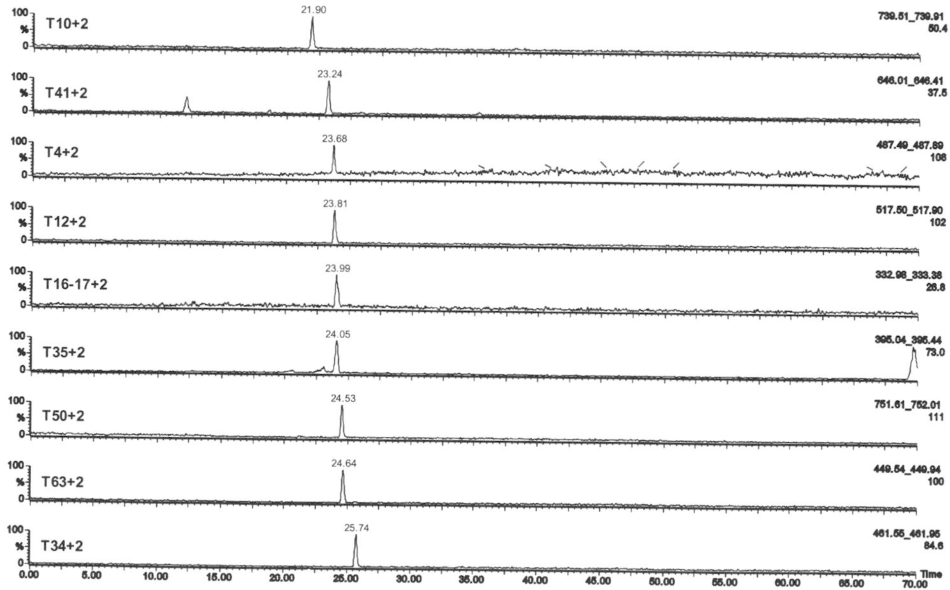

Fig. 12. Second set of selected ion chromatograms from the nanoscale capillary LC / coaxial nanospray / MS analysis of 150 femtomoles (on-column) of a tryptic digest of bovine serum albumin (reduced and alkylated) on the Micromass Q-Tof. An acetonitrile:water gradient (0.05% TFA in each) was used at a flow rate of 200 nL / min. with a 75 μm ID, 25 cm long fused silica capillary LC column.

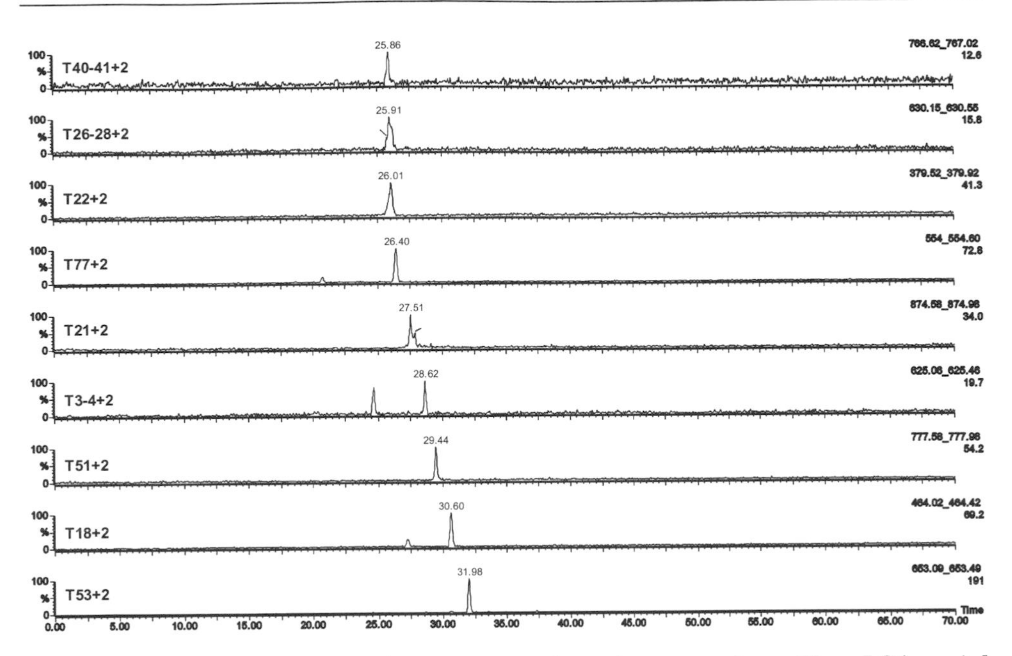

Fig. 13. Third set of selected ion chromatograms from the nanoscale capillary LC / coaxial nanospray / MS analysis of 150 femtomoles (on-column) of a tryptic digest of bovine serum albumin (reduced and alkylated) on the Micromass Q-Tof. An acetonitrile:water gradient (0.05% TFA in each) was used at a flow rate of 200 nL / min. with a 75 µm ID, 25 cm long fused silica capillary LC column.

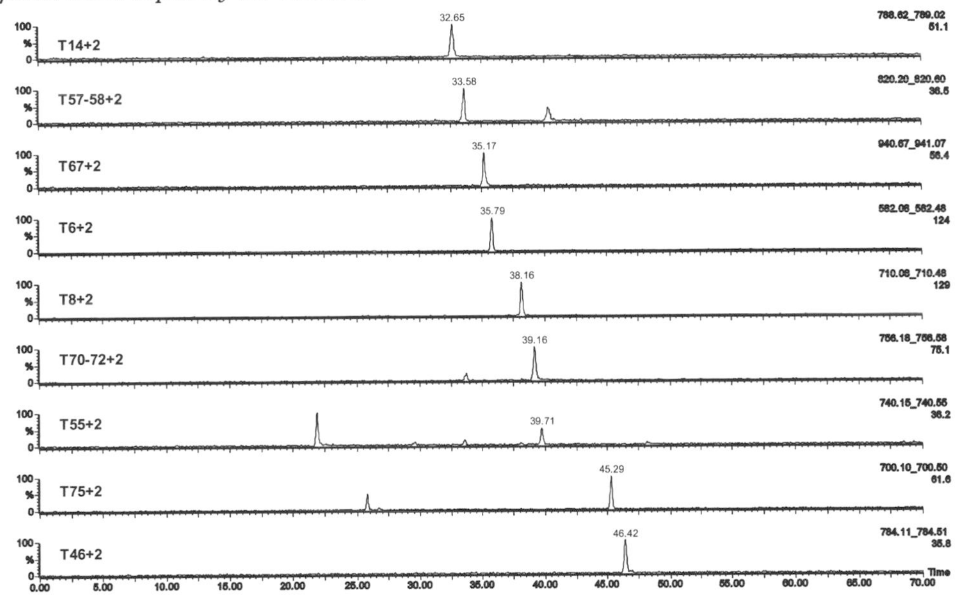

Fig. 14. Fourth set of selected ion chromatograms from the nanoscale capillary LC / coaxial nanospray / MS analysis of 150 femtomoles (on-column) of a tryptic digest of bovine serum albumin (reduced and alkylated) on the Micromass Q-Tof. An acetonitrile:water gradient (0.05% TFA in each) was used at a flow rate of 200 nL / min. with a 75 µm ID, 25 cm long fused silica capillary LC column.

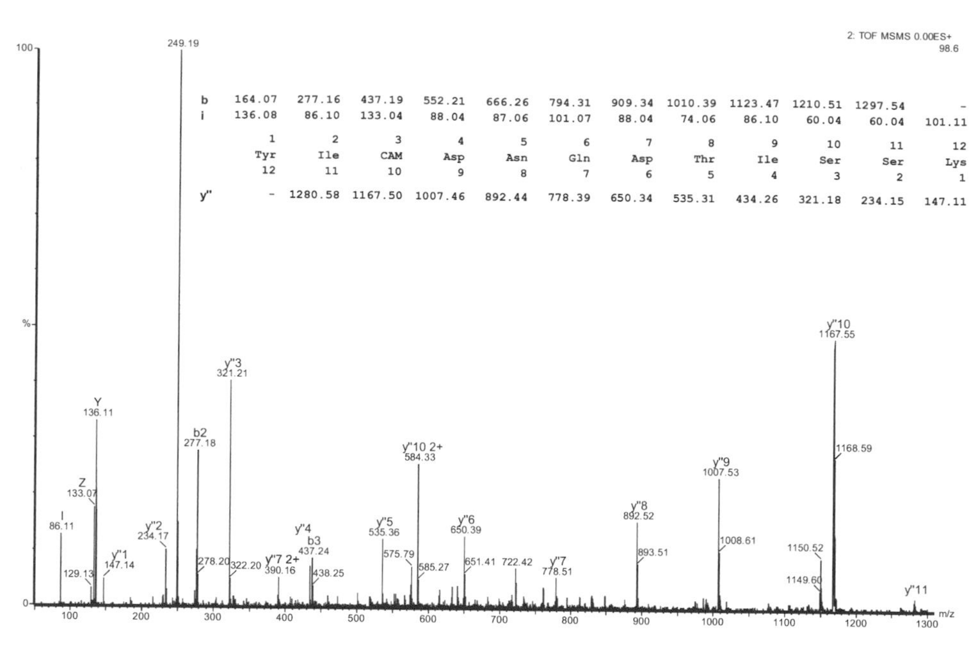

Fig. 15. Nanoscale capillary LC/coaxial nanospray/MS/MS spectrum of tryptic peptide T39 of bovine serum albumin. Data was acquired using automated data dependant scanning from a 150 femtomole injection of the digest using the Micromass Q-Tof. An acetonitrile:water gradient (0.05% TFA in each) was used at a flow rate of 200 nL/min. with a 75 µm ID, 25 cm long fused silica capillary LC column.

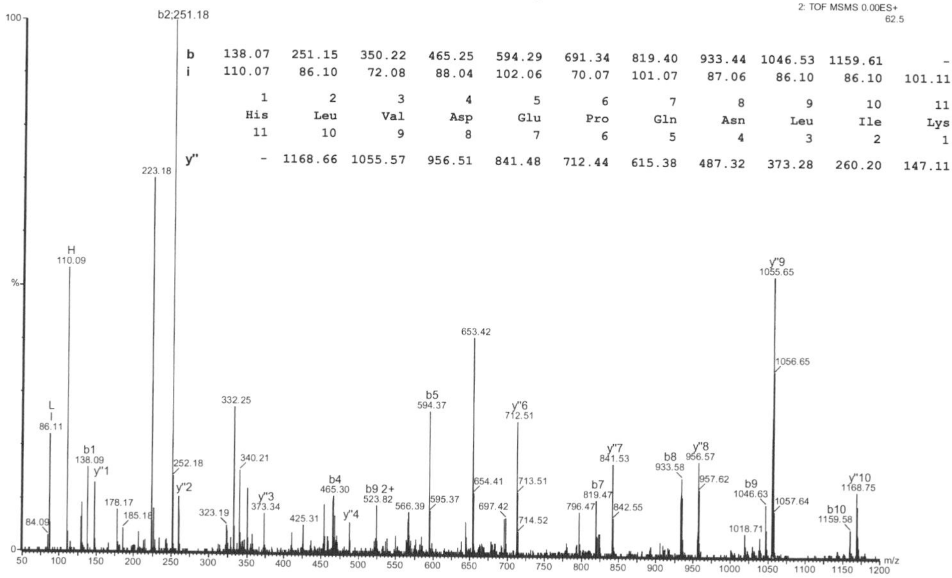

Fig. 16. Nanoscale capillary LC/coaxial nanospray/MS/MS spectrum of tryptic peptide T53 of bovine serum albumin. Data was acquired using automated data dependant scanning from a 150 femtomole injection of the digest using the Micromass Q-Tof. An acetonitrile:water gradient (0.05% TFA in each) was used at a flow rate of 200 nL/min. with a 75 µm ID, 25 cm long fused silica capillary LC column.

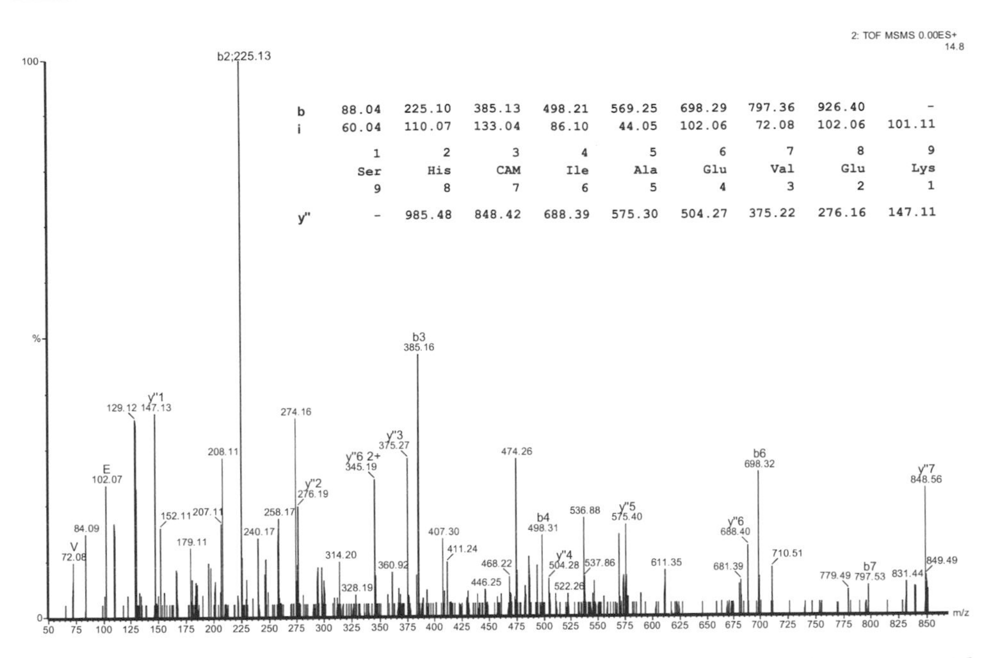

Fig. 17. Nanoscale capillary LC / coaxial nanospray / MS / MS spectrum of tryptic peptide T42 of bovine serum albumin. Data was acquired using automated data dependant scanning from a 150 femtomole injection of the digest using the Micromass Q-Tof. An acetonitrile:water gradient (0.05% TFA in each) was used at a flow rate of 200 nL / min. with a 75 μm ID, 25 cm long fused silica capillary LC column.

Conclusions

The coaxial nanospray interface has been used for the analysis of peptides under infusion and chromatographic conditions on both triple quadrupole and hybrid quadrupole/time-of-flight mass spectrometers. The combined systems have been found to be reliable, rugged, and easy to use. With infusion analysis at low nL/min. flow rates, coaxial nanospray has given signal-to-noise ratios equivalent to those obtained with the nanospray technique of Wilm and Mann. Sample manipulation was found to be simple, as changing samples requires only exchanging sample vials in the pressure vessel — no manipulation of the probe or the spray needle was required. The method of its operation permits the automation of sample introduction using a low-flow autosampler operated in flow injection mode. More significant improvements for automated MS/MS analysis of protein digests were found by using the coaxial nanospray interface with nanoscale capillary LC columns. Theoretically, nanoscale capillary LC offers at least a 10-fold improvement in analyte concentration delivered to the spray tip (and thus in electrospray response), and a similar overall improvement in the duty cycle of MS/MS data acquisition. When used with data dependant scanning, MS and MS/MS data acquisition was fully automated. The use of an autosampler with such a system permits unattended sample analysis 24 hours per day. High throughput proteomic studies will benefit from the automation advantages of nanoscale capillary LC. The coaxial nanospray interface offers advantages to nanoscale capillary LC/MS analyses by permitting modification of the pH and surface tension of the eluant post-column.

The use of nanoscale capillary LC technology, however, does add to the complexity of the experiment, and requires a significant investment in hardware, including an autosampler, LC pumps, UV detector (for off-line column evaluations), flow splitter, and column packing equipment. In addition, time must be invested in column packing and column evaluation. Care must be taken in using these columns in an automated manner, since they are prone to plugging. Nanospray does offer its own set of unique advantages, most notably the ability to acquire both precursor and product ion data, and this facilitates targeted analysis of specific peptides in the sample. Both nanospray and nanoscale capillary LC will play significant roles in proteomic studies in the future.

REFERENCES

1. I. Humphery-Smith, S. J. Cordwell and W. P. Blackstock, *Electrophoresis* **1997**, *18*, 1217-1242.

2. S. R. Pennington, M. R. Wilkins, D. F. Hochstrasser and M. J. Dunn, *Trends Cell Biol.* **1997**, *7*, 168-173.

3. R. D. Fleischmann, M. D. Adams, O. White, R. A. Clayton, E. F. Kirkness, A. R. Kerlavage, C. J. Bult, J. Tomb, B. S. Dougherty, J. M. Merrick, *et. al.*, *Science* **1995**, *269*, 496-512.

4. M. S. Wilm and M. Mann, *Int. J. Mass Spectrom. Ion Processes* **1994**, *136*, 167-180.

5. M. S. Wilm and M. Mann, *Anal. Chem.* **1996**, *68*, 1-8.

6. M. Mann and M. S. Wilm, *Anal. Chem.* **1994**, *66*, 4390-4399.

7. M. A. Moseley, K. C. Lewis, G. J. Opiteck, S. Ramirez, J. W. Jorgenson, and R. J. Anderegg, *Proc. 45th ASMS Conference on Mass Spectrometry and Allied Topics*, June, 1997, Palm Springs, CA.

8. M. A. Moseley, *Proc. 46th ASMS Conference on Mass Spectrometry and Allied Topics*, June, 1998, Orlando, FL.

9. M. A. Moseley, L. J. Deterding, K. B. Tomer and J. W. Jorgenson, *Anal. Chem.* **1991**, *63*, 1467-1473.

10. D. F. Hunt, J. E. Alexander, A. L. McCormack, P. A. Martino, H. Michel, J. Shabanowitz, N. Sherman, M. A. Moseley, J. W. Jorgenson and K. B. Tomer, in *Techniques in Protein Chemistry II*, J.J. Villafranca ed., Academic Press, San Diego, California, 1991; pp 441-465.

11. K.B. Tomer, M. A. Moseley, L.J. Deterding and C. E. Parker, *Mass Spectrom. Revs.* **1995**, *13*, 431-457.

12. J. T. Stults, B. L. Gillece-Castro, W. J. Henzel, J. H. Bourell, K. L. O'Connell and L. M. Nuwaysir, in *New Methods for Peptide Mapping for the Characterization of Proteins*, W. S. Hancock, Ed., CRC, Boca Raton, FL, 1996 119-141.

QUESTIONS AND ANSWERS

Kenneth Standing (University of Manitoba)

Regarding the sheath liquid, what is the relative flow compared to analyte? What is the composition of the sheath flow? What is the extent of mixing with the analyte?

Answer. The relative flow is 0-50%. The composition is 50% methanol/ (5% formic acid in H_2O). The extent of mixing can be varied, but overall the mixing is poorly quantitated.

Steve Barnes (University of Alabama, Birmingham)

Can the Q-TOF decide "on the fly" which molecular ions to select for MS/ MS analysis? This is vital for critical one-shot analyses.

Answer. Yes, the MassLynx automatic switching set-up allows you to determine which peptide to interrogate "on the fly" on the basis of the m/z signal rising above a user-set threshold. You have the option of having MS/MS data acquired from any signal which rises above the threshold, or only on the signals for which the user has specified.

Steve Barnes (University of Alabama, Birmingham)

With regard to interpreting tryptic digest MS and MS/MS data, is it yet possible to take into account unpredicted cleavages, i. e., not at Lys and Arg residues (and not posttranslational modifications)?

Answer. The SEQUEST database search method of Eng and Yates permits database searches with or without regard to enzyme specificity.

A. L. Burlingame (UCSF)

What stain was employed to detect spot 17 on your 2D gel.

Answer. The gel was stained with Coomassie blue.

Surinder Kaur (Chiron Corporation)

What type of increase do you see in sample throughput using the LC-nanospray? Presumably, for low level samples, your system would require controls to insure that there is no cross-contamination from different samples.

Answer. We do not have a sufficient database of sample analysis results to allow us to make this comparison.

John Stults (Genentech, Inc.)

Your data were all obtained with TFA as the organic modifier for HPLC. Do you always prefer TFA over formic or acetic acids for low-level samples?

Answer. No. TFA was used in these examples to obtain the best chromatographic peak shape. This facilitates the comparison of band broadening in the two interfaces used in this study — the coaxial nanospray interface and the nebulized nanoflow interface. We have observed formic acid to give a 4-5 fold improvement in sensitivity over TFA, so low-level samples benefit from the use of formic acid rather than TFA.

Roland Annan (SmithKline Beecham Pharmaceuticals)

Your data do not seem to support your hypothesis that LC-MS/MS is superior to nano-ES MS/MS. In spite of the fact that a spectrum in the LC-MS/MS experiment was recorded in 6 secs., you only managed to record ten useful spectra. Would you elaborate further on your conclusions?

Answer. The experiment generating 10 useful MS/MS spectra (good fragmentation and s/n) was an earlier version of MassLynx software which did not permit collision energy switching. The experiment gave approximately 50% success rate in obtaining useful MS/MS data. The subsequent experiment I showed on ATP synthase beta chain gave 8 MS/MS spectra from ATP beta chain, and about 25 additional MS/MS spectra that were useful. The success rate for obtaining useful MS/MS data in this experiment was 90%, because collision energy switching was used.

The most significant difference between the nanospray data and the nanoscale capillary LC data was in the observed s/n ratios, where capillary LC gave significantly better s/n in both MS and MS/MS data acquisition modes, even though at least 2x more sample was used with nanospray. We will be evaluating the relative merits of these techniques in more detail in the near future.

Also, a more detailed study by Al Burlingame's group at UCSF (reported at ASMS this year) found that studying the exact same digests by nanospray and by capillary LC found an average of 21 peptides detected by nanospray, whereas the capillary LC experiment found an average of 56 peptides, showing a 2x improvement in peptide detection using capillary LC over nanospray.

Life Without Databases: *De Novo* Sequencing of Small Gene Products and Complete Characterization of Posttranslational Modifications

Roland S. Annan, Michael J. Huddleston*, Dean E. McNulty*[†] *and Steven A. Carr**

*Department of Physical & Structural Chemistry and [†]Protein Biochemistry, SmithKline Beecham Pharmaceuticals, King of Prussia, PA 19406

The current revolution in modern biology has been largely fueled by advances in DNA technology, including but not limited to the development of restriction mapping, cloning, DNA sequencing, and PCR. Indeed, advances in nucleic acid techniques have led skeptics to predict that protein chemistry would go the way of the dinosaurs. Ironically, it is the most recent fruits of the DNA revolution, i. e., genomic sequence databases, that have breathed new life into protein science. Small amounts of previously undiscovered, biologically significant proteins, purified in many cases by SDS-PAGE, can now be rapidly identified by searching translated genomic or EST databases with limited amounts of amino acid sequence [1, 2]. Thus the previously difficult and time consuming task of protein sequencing has now in many cases become largely a matter of protein identification. However, a major remaining hurdle in this regard is sensitivity. While improvements in the sensitivity of Edman sequencing have been modest over the last decade (10-fold increase), the levels at which mass spectrometry can obtain a partial peptide sequence have decreased by 4-5 orders of magnitude over the same time period. Low femtomole amounts of peptides produced from enzymatic digestion of SDS-PAGE gel derived proteins are sufficient to identify proteins which are represented in a protein or EST database [3]. If the mass of the fragment ions used to produce the sequence information and the mass of the peptide from which the fragment ions were derived can be measured with accuracies on the order of 10-20 ppm, then even 2 residues of contiguous sequence data (together with the accurate molecular mass of the peptide) are often sufficient to unambiguously identify a known sequence in the database [4]. In many cases it is possible to make a positive identification without even interpreting the fragment ion spectrum [2]. Using this powerful mass spectrometry based approach to protein identification in combination with gene tagging techniques and one or two step affinity purification protocols, it is now possible to identify members of protein complexes[5], suggest functions for novel genes by identifying known interaction partners [6], find ligands for orphan receptors [7], and identify enzyme substrates [8]. When used in combination with high resolution 2D gel electrophoresis [9], mass spectrometry can be used to catalog changes in protein expression after cellular stimulation or insult, and to establish phenotypes at the subcellular level after gene knock-out.

However, not all amino acid sequences are available through databases. Complete genomes are available for only a relatively few organisms, largely bacteria and viruses; a notable exception being that of the yeast, *Saccharomyces cerevisiae*. In the meanwhile, the biological community eagerly awaits the completion of the *C. elegans* genome in late 1998, as the first organism with a central nervous system. For mammalian genomes, the picture is far from complete. The human genome is only about 2-3% finished, and the mouse genome is less than 1% completed. The NCBI nonredundant protein database contains approximately 50,000 human protein entries, but because most of these are mutants and partial sequences, this number only accounts for about 8,000-10,000 different protein sequences. The number of mouse proteins is estimated to be about half of this number, and for the rat, half as many again.

On the other hand the human genome is represented by a database of over 1 million (public domain) expressed sequence tags (ESTs), derived from small pieces of DNA which are produced from mRNA libraries of expressed genes. But although 60-90% of all human genes may be represented by these small pieces of DNA sequence, the level of coverage of any given gene is highly variable. Certainly the sequence of actin is well represented by ESTs, but very low copy number genes and very large genes are poorly represented, especially at the 5' end. The mouse genome on the other hand is represented by only about 300,000 ESTs, and the rat genome is represented by less than 21,000 ESTs. Thus the probability of finding a human protein as an EST are good, but variable, and the probability increases with the number of peptide sequences available to search. But for mouse proteins the likelihood of identifying a protein as an EST are greatly diminished, and for rat proteins it is quite poor. For most organisms, there is little or no DNA or protein sequence data available.

Small gene products (less than 3-4 kDa) are even less likely to be well represented in genomic or EST databases. However, the peptide products of these small genes are critical components in cellular communication, being classified as neuropeptides, peptide hormones, or local chemical mediators. These molecules exert their effects by binding to the extracellular domains of transmembrane receptors. As such they are excellent candidates for drug intervention. The sequence of the peptide ligand can be used for the design of an assay for antagonists or agonists, and the three-dimensional structure affords a useful starting point for the design of non-peptide molecules by the medicinal chemists. Because many of these peptides are differentially expressed, and occur for only short periods or at specific times in the cell cycle, their representation in mRNA libraries is likely to be quite low.

Peptides not present in databases must be fully characterized at the amino acid level. These sequences represent a significant challenge for mass spectrometry. Because the peptides are not generated by *in vitro* enzymatic cleavage, they are not normally ideal for tandem MS. In addition, these peptides are often posttranslationally modified. Therefore, all of the tools in the laboratory arsenal and a large measure of creative thinking need to be brought to bear on the problem. Fortunately, some of the same advances in mass spectrometry which have facilitated protein identification, have made *de novo* sequencing of limited amounts of endogenous peptide messengers a reality. Here we will describe strategies we have employed to solve difficult structures including orexin-A [10], a peptide which was subsequently shown to control feeding behavior in rats.

Another important challenge facing protein biochemistry is understanding the function and characterizing the activity of the thousands of new gene products identified as a result of genomic databases. In many cases this will involve the identification and localization of functionally significant posttranslational modifications, a problem to which MS is ideally suited. Genomic databases will be of little use in the analysis of posttranslational modifications, however. In cases where the consensus-site sequence for the modification is known (e. g., the consensus site for attachment of N-linked carbohydrate is Asn-X-Ser/Thr where X is any amino acid except Pro), the presence of the consensus sequence is necessary but it is not sufficient for recognition by the transferase, etc., responsible for the modification. Second, consensus site sequence rules are made to be broken; exceptions exist to most of these "rules". Third, not all consensus-sites are known or understood sufficiently to be used in a predicative manner. Many of the kinases fall into this latter category. For several years we have been working to develop sensitive and selective techniques for mapping phosphorylation sites in proteins [11-13]. Using orthogonal scanning techniques which are based on a common diagnostic fragment ion, we have successfully mapped *in vivo* phosphorylation sites, where the level of incorporation is less than 5% at a given site. These approaches are illustrated by the phosphopeptide mapping of *in vitro* and *in vivo* phosphorylated Sic1p, an S-phase Cdk inhibitor in budding yeast [14], a collaboration with Dr. Raymond Deshaies and Dr. Rati Verma, Division of Biology, California Institute of Technology.

Experimental

Electrospray mass spectra were acquired on a modified Perkin-Elmer Sciex API-III atmospheric pressure ionization triple quadrupole tandem mass spectrometer. This instrument is equipped with a high pressure collision cell [15], and either the standard Sciex ionspray source, or an articulated nanoelectrospray interface developed by Wilm and Mann [16] and built at the European Molecular Biology Laboratory in Heidelberg, Germany. The operation of the API-III in the ionspray mode for the detection of phosphopeptide marker ions [11] or the nanoelectrospray mode for precursor ion scanning of phosphopeptides [12] and peptide sequencing [12, 17] has been described in detail elsewhere. We first reported on the use of precursor-ion scanning for selective detection of glycopeptides in complex mixtures [18, 19].

MALDI mass spectra were recorded on a Micromass TofSpec SE, a reflectron, time-of-flight mass spectrometer equipped with a time-lag-focusing source. Samples were mixed with an equal volume of matrix solution containing two internal mass standards at a concentration of 200 fmol/µL. A 0.5 uL aliquot of this sample/matrix solution is applied to the MALDI target. The matrix solution was 10 µg/mL alpha-Cyano-4-hydroxycinnamic acid in 50:50 ethanol:acetonitrile.

Orexin-A was purified from 200 g of frozen rat whole brain as described previously [10] . After reduction and alkylation with DTT and iodoacetic acid, respectively, the peptide was digested with Lys-C and analyzed by Edman sequencing, MALDI, and nanoES mass spectrometry.

For nanoES mass spectrometry, the orexin Lys-C digest was desalted and concentrated on a disposable micro-cleanup column [20] which consists of Poros RII resin (40-60 μm particles) packed into an Eppendorf Geloader pipette tip. The peptides were eluted into a nanospray needle using 2.0 μL of 70% methanol, 5.0% formic acid. Molecular weights and precursor ion masses for peptides in the unfractionated digest were determined by scanning the first quadrupole (Q1) with the resolution adjusted such that singly, doubly and triply charged ions can be distinguished from one another. Tandem mass spectra were generated by causing selected precursor ions to undergo collision induced dissociation (CID) in the second quadrupole (Q2) and mass analyzing the resulting CID product ions by scanning the third quadrupole (Q3). Second generation tandem mass spectra (MS^3) were acquired by fragmentation in Q2 of a product ion which had been generated using in-source CID.

$Sic1p^{HAHis6}$ and MBP-$Sic1p^{mycHis6}$ were purified from cdc34 cells containing 19% yeast extract and BL21 cells, respectively, as described previously [14]. Purified Sic1p or *in vitro* phosphorylated purified MBP-Sic1p were digested with trypsin in the presence of 2M urea. The peptide digests were subjected to two dimensional phosphopeptide mapping using a recently developed method, which employs electrospray mass spectrometry in both dimensions, relying on orthogonal scanning techniques for selectivity [13] .

The first dimension of the phosphopeptide mapping experiment uses LC-MS, operating in the negative ion mode, with the mass spectrometer optimized to produce and detect an in-source CID generated m/z 79 product ion (PO_3-), which is highly specific for Ser, Thr, or Tyr phosphorylation [11] . The MS is operated in a single ion monitoring mode for enhanced sensitivity. The UV trace is taken into the MS data system, so that the UV and MS chromatograms can be easily aligned. Reversed phase HPLC was performed using a Michrom UMA. Protein digests were loaded onto a Michrom 1mm Peptide Trap Cartridge which had been installed in place of the sample loop on a Rheodyne model 8125 injector. After washing with 200 μL of 5% solvent B, the cartridge was rotated into the inject position. The peptides are chromatographed on a 0.5 mm x 15 cm Reliasil C_{18} column using water:acetonitrile:TFA gradients at 50 μL/min. After the UV detector, the column eluent was split 10:1 with 45 μL/min going to a fraction collector taking one minute fractions and 5 μL/min sent to the mass spectrometer. HPLC fractions which are shown by the first dimension MS response to contain phosphorylated peptides are analyzed in the second dimension.

The second dimension utilizes negative ion nanoelectrospray to perform precursor ion scans of m/z 79 [12] on targeted fractions. The precursor scan produces molecular ions only for those peptides which can produce the phosphopeptide specific marker ion. Therefore, the molecular weight of phosphorylated peptides can be determined in this dimension, even though other, more abundant peptides are present in the same fraction, and even though the signal in the conventional molecular weight scan is indistinguishable from background. After the molecular weights of the phosphopeptides have been determined from the m/z 79 precursor ion scan, if sufficient signal is present, the mass spectrometer is switched into the positive ion mode, and the phosphopeptides are sequenced from the same sample loading by acquiring a full scan CID product ion spectrum.

Results

De Novo Sequencing of Small Gene Products: Orexin

Many small gene products (<3000-4000 Da) are involved in some aspect of cellular communication. Whether they are peptide hormones, neuropeptides or local chemical mediators, all known signalling peptides exert their influence by binding to members of the seven-transmembrane, G protein-coupled cell surface receptor (GPCRs) family. Over the past 15 years nearly 350 drugs that act at GPCRs have been marketed, with examples including antihypertensives that act at the alpha- and beta adrenoreceptors, anti-anxiolytics and antidepressants that act at the serotonin receptors, and analgesics that interact with opiate receptors. Recently, large scale genomic sequencing efforts have identified numerous cDNA sequences that, based on low (<40%) but significant amino acid sequence similarity, encode novel GPCRs. These receptors are called "orphans" because their categorization as GPCRs is only putative, and their cognate natural ligands are unknown. SmithKline Beecham has embarked on a systematic program to define tissue localization of these orphan receptors and to isolate and structurally characterize the endogenous "orphan" ligands [21]. In collaboration with Professor M. Yanagisawa and coworkers at Howard Hughes Medical Institute, Dallas, screening experiments employing cell-based fluorescent reporter system assays led to the isolation of two novel neuropeptides, termed orexin-A and orexin-B that bind and activate two novel, closely related GPCRs, designated OX1 and OX2 receptors [10]. The name "orexin" is derived from the Greek word orexis, which means appetite. The mRNA for the precursor of these peptides is abundantly and specifically expressed in the lateral hypothalamus and adjacent areas, a region classically implicated in the central regulation of feeding behavior and energy homeostasis. Called orexins (A and B), these peptides stimulate appetite when administered centrally, and are themselves regulated by the nutritional state of the animal. Here we describe the detailed structural characterization of orexin-A, a 33 residue peptide having pyroglutamic acid at the N-terminus, a C-terminal amide, and four Cys residues in disulfide linkage.

Orexin-A was isolated from 200 g of rat whole brain as a biologically active, high-resolution HPLC fraction, specific for an orphan GPCR originally termed HFGAN72 [10]. Isolation and purification of the orexins involved a combination of ion exchange chromatography and four steps of HPLC. Biological

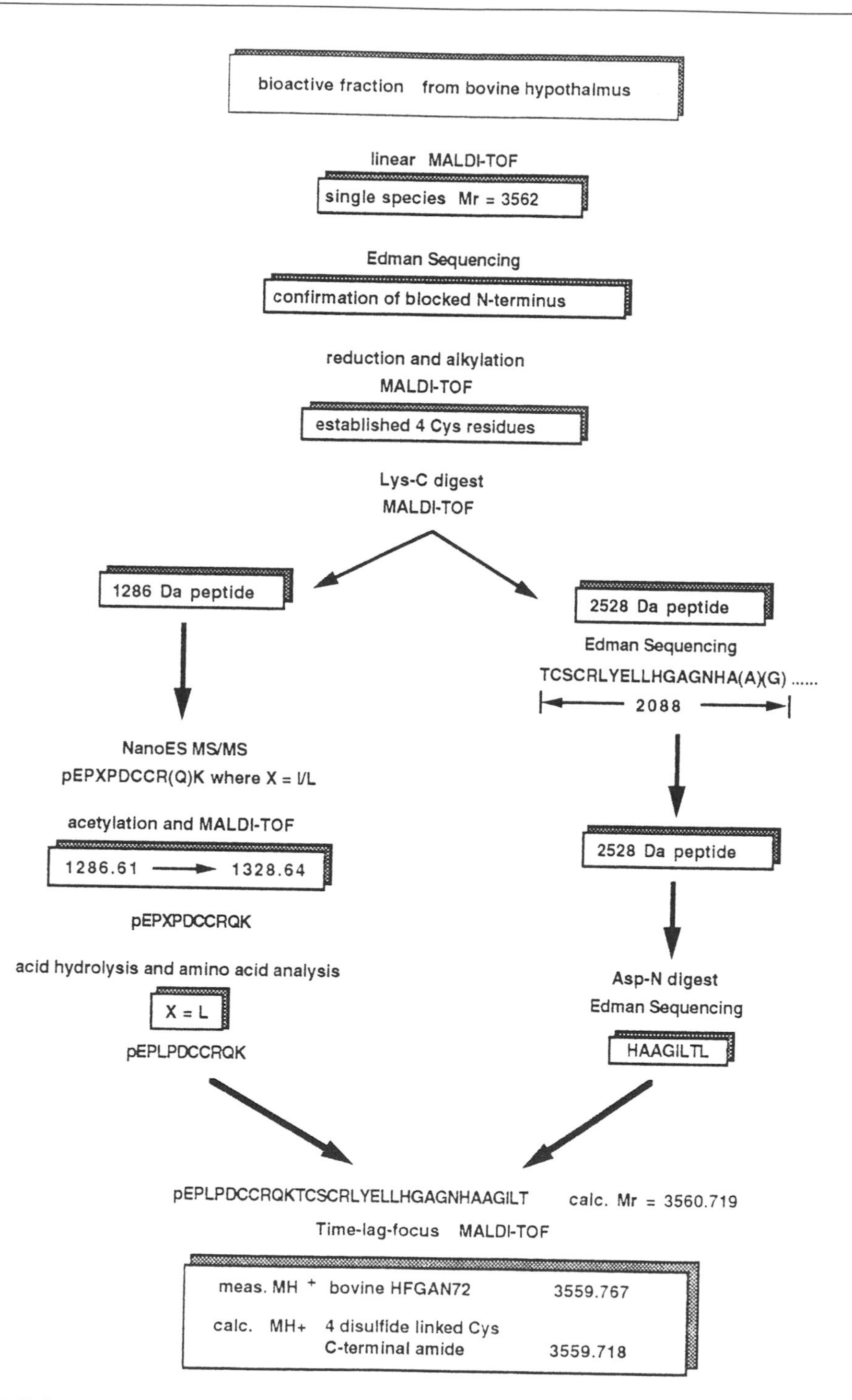

Fig. 1. Schematic diagram of the multidimensional approach used to complete the de novo *sequence of orexin-A.*

activity was monitored by assaying for [Ca^{2+}]i transients in OX1R-expressing HEK293 cells [10]. The purified active fractions from the final RP-HPLC separation step were analyzed by MALDI-MS, Edman degradation, and nanoES-MS (Fig. 1). The isolation and assay procedure is shown schematically in Fig. 1. After four steps of HPLC, the active fraction was analyzed by MALDI-MS, Edman degradation, and nanoES-MS.

MALDI-MS of the active fraction showed that it contained a single major species with a molecular mass of 3558.7 Da. Direct N-terminal sequencing of orexin-A was unsuccessful, suggesting a blocked terminus. Reduction and alkylation of the fraction with DTT and iodoacetic acid, respectively, showed the addition of 232 Da after re-measurement of the molecular mass, suggesting the presence of four cysteine residues involved in two intramolecular disulfide bonds. At this point an aliquot of the sample was digested with endoproteinase Lys-C. MALDI-MS showed that the orexin-A peptide had been cleaved into two peptides with molecular masses of 1286 and 2528 Da. Half of the unfractionated Lys-C digest was subjected to Edman sequencing and 17 residues of the C-terminal fragment were read off (the N-terminal peptide being blocked, yielded no sequence). These 17 residues, TCSCRLYELLHGAGNHA, account for 2088 Da of the mass of the C-terminal peptide, and indicated that between 3-7 amino acids were still unaccounted for. The determined partial sequence was searched against all of the existing protein and DNA databases and proved to be novel.

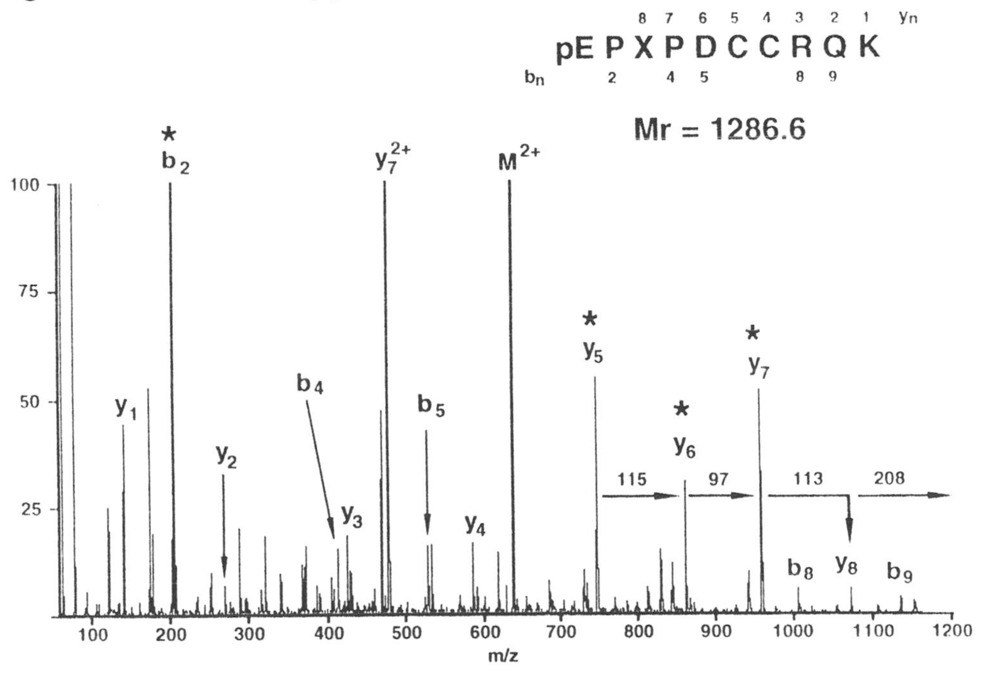

Fig. 2. Collision-induced dissociation product ion spectrum of m/z 644.3, the [M+2H]$^{2+}$ ion for the N-terminal Lys-C peptide from orexin-A. Peaks marked with an asterisk were subjected to a further stage of tandem MS as described in the text. Numbers above and below the sequence correspond to b_n and y_n ions actually observed in the spectrum.

The remainder of the unfractionated Lys-C digest was concentrated and analyzed by nanoelectrospray MS. We sequenced the N-terminal peptide by collision induced dissociation (CID) of the doubly charged precursor at m/z 643.8 (the spectrum is shown in Fig. 2). In addition, we performed MS^3 on selected product ions (marked with an * in Fig. 2). From the initial MS/MS spectrum we determined the partial sequence (208)XPD——K, where m/z 208 was believed to be the residue mass of the first two N-terminal amino acids (predicting a b_2 ion at m/z 209), and X was either Leu or Ile. The C-terminal Lys residue was deduced from the presence of a m/z 147 ion (the y_1 ion for peptides ending in Lys) and from the specificity of the enzyme, endoproteinase Lys-C. We were able to complete the sequence by interpreting the various MS^3 spectra in combination with the primary MS/MS spectrum as described below.

Interestingly, there were very few y_n fragment ions observed in the MS^3 spectra, even though the precursors were themselves y_n ions. For instance, the product ion spectrum of the y_7 ion (m/z 965, Fig. 3A) had a prominent y_5 ion from the loss of Asp and Pro from the precursor, but no other y_n ions. There were, however, consecutive losses of 128 Da from the precursor, suggesting y_n cleavage from the N-terminus. However, we were fairly certain that the N-terminus of this fragment was Pro-Asp, and so we were unable initially to explain the dual loss of 128 Da from the precursor. When we examined the MS^3 spectra from

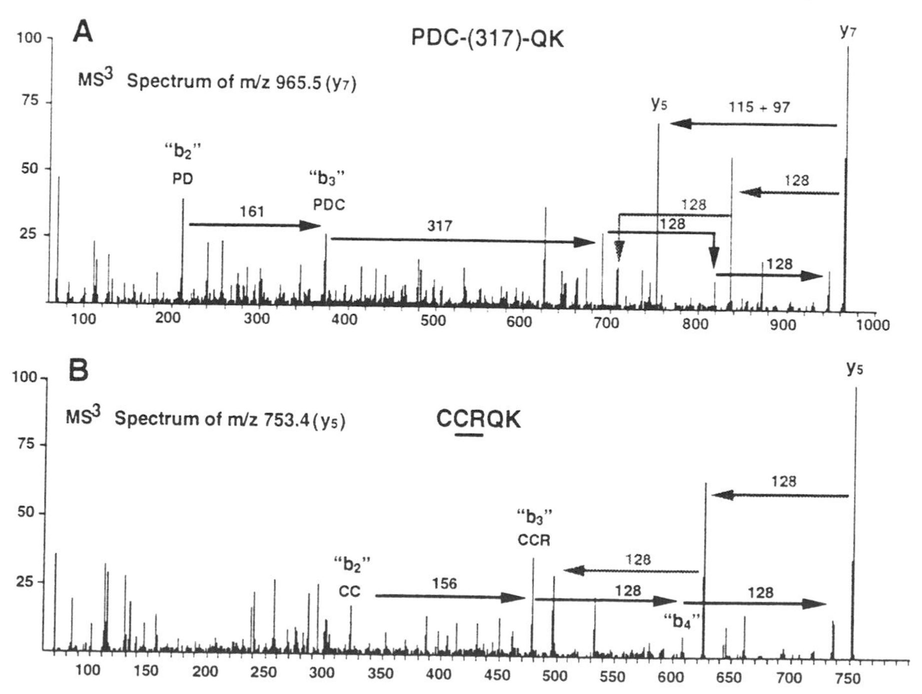

Fig. 3. Collision-induced dissociation MS^3 product ion spectra of y_n ions (see Fig. 2) from the orexin-A, N-terminal Lys-C peptide. The precursor y_n ions were produced by in-source CID. A) m/z 965.5; B) m/z 753.4. Ions marked "b_2", "b_3" or "b_4" are secondary product ions resulting from CID of a y_n ion. These do not have the same structure as b_2, b_3, or b_4 ions from the primary MS/MS spectrum shown in Fig. 2.

succeeding y_n ions (for instance, Fig. 3B) they also showed the dual loss of 128 Da, thus ruling out the possibility that these losses were from the N-terminus (which was changing with each successive y_n ion), and suggesting that an alternative explanation was necessary. A closer inspection of the MS^3 spectra for each successive y_n ion showed that an informative "b_n" ion series existed in each case, and that it terminated with consecutive additions of 128 Da to give the highest "b_n" ion. This suggested that the adjacent 128 Da amino acids were in fact the last two C-terminal residues and that a probable explanation for the unexpected successive losses of 128 Da from the precursor was that these ions are of the type b_n+18, as described by Gaskell [22]. Following this logic we arrived at the extended partial sequence, (208)XDP——QK, where the Q could in fact be either Gln or Lys.

The mass difference of 161 Da between the "b_2" ion (PD at m/z 213) and a prominent ion at m/z 374 in the spectrum of the m/z 965 fragment (Fig. 3A) suggested the next N-terminal residue was carboxymethylcysteine (Cmc). This was supported by the presence of the corresponding "b_2" ion in the MS^3 spectrum of the next lower y_n ion (data not shown), which matched the composition DC. There was also a prominent ion at m/z 592.4 (marked y_4 in Fig. 2) in the primary MS/MS spectrum which fit y_5-161. The probable inclusion of this cysteine gave us the partial sequence (208)XPDC(317)QK.

From the primary MS/MS spectrum, we then determined the next residue to be either Arg or Cmc. We determined that Cmc was the correct assignment from the MS^3 spectrum of the y_5 ion (C-317-QK), which had a "b_2" ion at m/z 323 that fit only CC (see Fig. 3B). The remaining 156 Da we assigned as Arg based on the absence of any intervening ions in either the primary MS/MS spectrum or any of the MS^3 spectra to support the combination G,V which has the same incremental mass as Arg. At this point we had determined the sequence to be (208)XPDCCRQK, and needed only to identify the first two residues and to distinguish between Gln and Lys at residue 8.

Because the N-terminus was believed to be blocked, the residue mass of the b_2 ion (208 Da) needed to fit a combination of residues, the first of which had to have a modified alpha-amino group. Our initial suspicions fell upon pyroglutamic acid (a common N-terminal modification) and proline. This preliminary assignment was confirmed by the MS^3 spectrum of the suspected b_2 product ion at m/z 209, which showed strong immonium ions for both proline and pyroglutamic acid (data not shown).

We distinguished between Gln and Lys at position 9 (both amino acids have the same residue mass) by acetylation of the peptide and re-measurement of the molecular weight. The MH+ ion for the N-terminal peptide shifted by 42 Da from 1286.6 to 1328.6, indicating the addition of only one acetate group (calc. 1328.6). Because Gln residues cannot be acetylated, and the N-terminus is blocked, the addition of only one acetate group is consistent with the C-terminal sequence of QK, not KK.

The sequence of the N-terminal orexin peptide as determined from the MS data was shown to be **pEPXPDCCRQK** (calc. MH+ = 1286.624), where pE is pyroglutamic acid, and X is either leucine or isoleucine. The X was determined by amino acid analysis to be Leu. At this point, the final residues of the C-terminal peptide were determined by Edman sequencing of an Asn-N derived peptide from the intact orexin ligand, which had the sequence NHAAGILTL.

The MH+ ion for the intact orexin ligand was measured by MALDI and found to be 3559.767. This mass is consistent (within 15 ppm of the calculated value, MH+ = 3559.718) with the determined sequence **pEPLPDCCRQ KTCSCRLYELLHGAGNHAAGILTL-NH2** provided all four Cys are in disulfide linkages, and that the C-terminus is an amide rather than a free acid. The topology of the disulfide bonds in orexin-A was established by synthesizing all three possible disulfide isomers of orexin-A and then comparing their retention with the natural product on RP-HPLC under isocratic conditions (data not shown). These results established that the disulfide bonds in orexin-A are formed between Cys6-Cys12 and Cys7-Cys14.

Using highly degenerate primers based on a part of the orexin-A sequence, a cDNA fragment was obtained from rat brain mRNA by RT-PCR. This led to isolation of the full length cDNA which was shown to encode the entire orexin-A sequence within a 130-residue polypeptide, prepro-orexin. The first 33 residues were predicted to be a signal sequence, with cleavage between Ala^{32} and Gln^{33}. The first residue of the orexin sequence is Gln^{33}, which is presumably cyclized enzymatically into the N-terminal pyroglutamic acid.

In addition to the major activity characterized above, a second minor activity described as orexin-B was characterized by MALDI-MS and Edman sequencing and found to be a 28 amino acid linear peptide, **RSGPP GLQGRLQRLLQANGNHAAGILTM** (mass 2937 Da), amidated at the C-terminus. This sequence was also contained within the orexin gene and began 3 residues down stream of orexin-A.

Mapping Posttranslational Modifications: Phosphopeptide Mapping of Sic1p

The mature form of nearly every protein represents a posttranslationally processed form of the gene product. In fact, many genes are represented by several protein products, differing primarily in the nature of their posttranslational modifications (PTM). It is now clear that phosphorylation is the most common and physiologically relevant posttranslational modification of proteins. Among the thousands of proteins expressed in a typical mammalian cell, as many as one third are thought to be phosphorylated. Phosphorylation and dephosphorylation is governed by a complex integrated network of kinases and phosphatases which are the enzymes responsible for transfer and removal of phosphate from specific target protein. This exquisite regulatory mechanism governs many aspects of cell growth, metabolism, division, motility and differentiation, and it plays a key role in signal transduction.

In the cell cycle of budding yeast, exit from the G_1 phase and the initiation of DNA synthesis are dependent on the G_1 cyclin-dependent kinase (Cdk)-triggered destruction of Sic1p, an S-phase Ckd inhibitor [23]. A key question to be answered was whether the G_1 Cdk activity of Cln2p/Cdc28 modulates the activity of the Sic1p degradation machinery or phosphorylates Sic1p directly, thereby allowing it to be recognized as a substrate for proteolysis.

To test whether Siclp serves as an efficient substrate for Cln2p/Cdc28p, a purified Siclp, maltose-binding protein (MBP) chimera, which had been expressed in *E. coli*, was phosphorylated *in vitro* by anti-HA immune complexes from Sf9 cell lysates containing both Cln2p and Cdc28p^{HA}. An observed decrease in mobility upon SDS-PAGE showed that the MBP–Siclp was quantitatively phosphorylated. There are eight S/T P candidate Cln2p/Cdc28 phosphoacceptor sites in the MBP-Siclp chimera. To map which sites were phosphorylated, we employed a multidimensional electrospray mass spectrometry (ESMS) approach developed in our laboratory. *In vitro* phosphorylation of MBP itself, showed that it was not phosphorylated by Cln2p/Cdc28p^{HA} complexes, despite the presence of two potential Cdk phosphorylation sites within MBP.

Figure 4 shows the result from the first dimension analysis of phosphorylated MBP-Siclp after digestion with trypsin. LC-ESMS in the negative-ion mode, using single-ion monitoring of the phosphopeptide specific marker ion, m/z 79 (gray trace in Fig. 4) allows us to identify areas of the HPLC chromatogram (white trace in Fig. 5) which contain phosphopeptides. Fractions are collected simultaneous with the MS analysis and those containing phosphopeptides undergo second dimension analysis by precursor ion scanning for m/z 79, to determine the molecular weights of the phosphopeptides. For the MBP-Siclp tryptic peptides, we found that in each case peptide molecular weight alone was insufficient to determine the specific site of the phosphorylation, or in some cases to even deduce the correct amino acid sequence. Thus we found it necessary to perform tandem MS on each phosphopeptide precursor.

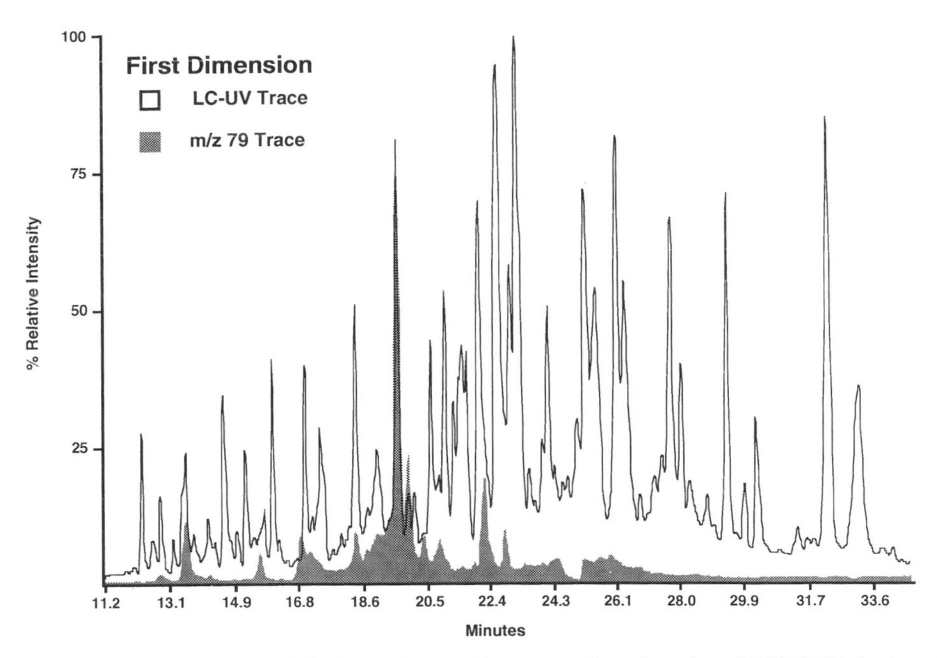

Fig. 4. First dimension LC-MS analysis of in vitro *phosphorylated MBP-Siclp tryptic digest. The grey trace represents SIM-MS of the phosphopeptide specific marker ion m/z 79 [PO$_3$]$^-$.*

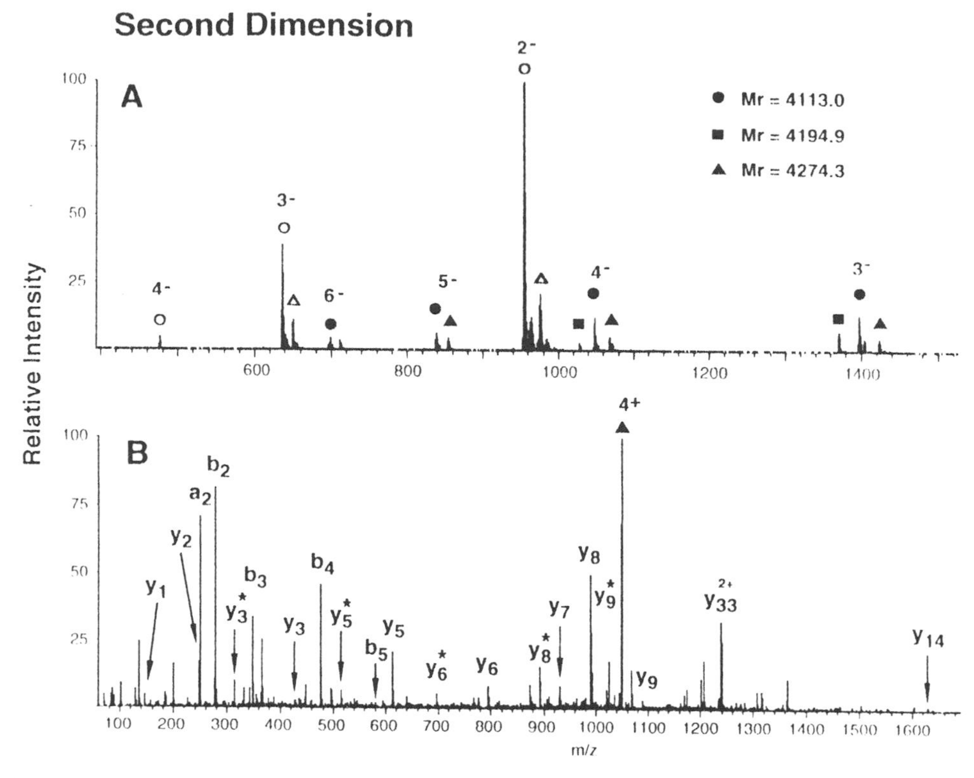

Fig. 5. Second dimension nanoES analysis of fraction 19 (19-20 minutes in Fig. 4) from the LC-MS analysis of in vitro *phosphorylated MBP-Sic1p tryptic digest. A) Negative ion, precursor ion scan for m/z 79. Ion series for five phosphopeptides are shown. For the sake of clarity only the molecular weights of the peptides described in the text are listed; B) Collision-induced dissociation product ion spectrum of m/z 1069, the $[M+4H]^{4+}$ ion for Mr 4274 (s). Ions marked y_n* have the structure y_n-H_3PO_4.*

Figure 5A shows the negative ion, m/z 79 precursor ion scan analysis of fraction 19. Five phosphopeptide molecular weights can be determined from the various charge series present in the spectrum. Three of these peptides differ in mass by 80 Da (4113.0, 4194.9, 4274.3), suggesting incremental phosphorylation of a single amino acid sequence. However, taking into account differing amounts of phosphate incorporation, the masses of the three peptides fit two possible amino acid sequences as shown in Fig. 6. *A priori*, one would expect the sequence 52-85 to be the correct peptide, because it contains three Cdk consensus sites, and the peptide molecular masses fit the 52-85 amino acid sequence if we allow for the addition of 1-3 moles of phosphate. Nevertheless, the MS/MS spectrum of the quadrupley charged precursor for the 4274.3 Da peptide (Figure 5B) shows that the sequence is in fact amino acid residues 14-51 containing 3-5 moles of phosphate. Interpretation of this and other MS/MS spectra revealed that in addition to quantitative phosphorylation at the two consensus Cdk sites within residues 14-51, phosphate was also incorporated in varying degrees at three nonconsensus sites.

14–51 + 3–5 HPO_3

Y L A Q P S G N T S S S A L M Q G Q K T P Q K P S Q N L V P V T P S T T K

52–85 + 1–3 HPO_3

S F K N A P L L A P P N S N M G M T S P F N G L T S P Q R S P F P K S S V K

SP or TP consensus sequence for Cdk/cyclin kinase

Fig. 6. Sequence possibilities for the putative phosphopeptide series 4113.0, 4194.9, 4274.4.

All eight potential Cdk consensus sites within the Siclp sequence of the MBP-Siclp fusion protein were quantitatively phosphorylated *in vitro*. In addition, we confirmed phosphorylation at four nonconsensus sites. This phosphorylation at non-preferred sites is probably due to the long incubation of the fusion protein with the kinase. In spite of this, no phosphorylated residues were found in the MBP domain.

After demonstrating that Siclp was a substrate for Cln2/Cdc28 *in vitro*, we showed that phosphorylation of Siclp *in vivo* was dependant upon Cln2/Cdc28 expression. Siclp purified from cells expressing active kinase complex, contained 12 times more phosphate than Siclp isolated from cells depleted of the kinase complex [14]. To identify which sites on Siclp were phosphorylated *in vivo*, Siclp was purified from cdc34 cells, digested with trypsin, and analyzed by two dimensional phosphopeptide mapping using the ESMS approach described above. We found that peptides encompassing six of the consensus sites are modified to some extent, with the three major *in vivo* phosphorylation sites being determined as Thr^{5} (100%) Thr^{33} (28%) and Ser^{76} (23%). This analysis agreed well with data derived from the molecular mass measurement of intact, phosphorylated Sic1P from cdc34 cells (data not shown). The mass spectrum of the intact protein showed a distribution of phosphoforms ranging from 1-6 moles of phosphate incorporated.

To test their role in the destruction of Siclp, the *in vivo* phosphorylated sites determined by ESMS were mutated, either singly or in various combinations and subjected to an *in vitro* ubiquitination assay. Whereas all of the single mutants were ubiquitinated as efficiently as wild type Siclp, various combinations of triple mutants, including one containing the three major sites determined by ESMS, all had a severe ubiquitination defect [14]. The data thus supports the notion that Siclp must be phosphorylated on a subset of sites by G_1 Cdk complexes to be recognized by the ubiquitination machinery, but that no single site is either absolutely necessary or sufficient to target Siclp for destruction.

Conclusion

A great variety of problems remain to be solved in biology and biochemistry which can benefit from the application of mass spectrometry. Complete sequence determination or complete definition of the modification state of a protein, including site-specific stoichiometry, is a far more complicated analytical challenge than identification of a protein in a database. Posttranslational modification analysis requires coverage of the entire sequence of a protein, not simply defining 3-4 residues of sequence from a randomly selected tryptic peptide. The challenge lies with developing ever more selective and sensitive methods (preferably non-radioactive!) for recognition of key posttranslational modifications, and techniques for sequencing these modified peptides at low femtomole levels and below. We have demonstrated several approaches to this problem in this report, but even higher sensitivity will be required to address many of the critical problems in biology. Mass spectrometry itself is no longer the major barrier to achieving this goal. Rather, sample handling methods, which include targeted sample enrichment, high resolution micro-separation and biochemical manipulation, are not sufficiently well developed to permit routine analysis of low femtomole amounts of modified proteins from biological sources by mass spectrometry without loss of most or all of the sample. We are confident that expertise gained from years of practice, coupled with the creativity and insight of the spectroscopists and the biologists will ultimately ensure that full advantage is taken of the power of modern instrumentation.

Acknowledgements

The authors wish to gratefully acknowledge our collaborators, Dr. Masashi Yanagisawa from the Howard Hughes Medical Institute at the University of Texas, Southwestern Medical Center, and Drs. Ray Deshaies and Rati Verma from the Division of Biology at the California Institute of Technology. We are also grateful for the many contributions from our various colleagues at SmithKline, especially Drs. D. Bergsma and W-S. Liu.

References

1. M. Mann and M. Wilm, *Anal. Chem.* **1994,** *66*, 4390-4399.

2. J. Eng, A.L. McCormick and J.R. Yates III, *J. Am. Soc. Mass Spectrom.* **1994**, *5*, 976-989.

3. A. Shevchenko, M. Wilm, O. Vorm and M. Mann, *Anal. Chem.* **1996**, *68*, 850-858.

4. A. Shevchenko, I. Chernushevich, W. Ens, K. G. Standing, B. Thomson, M. Wilm and M. Mann, *Rapid Commun. Mass Spectrom.* **1997**, *11*, 1015-1024.

5. C.L. Peterson, Y. Zhao and B.T. Chait, *J. Biol. Chem.* **1998**, *11*, 23641-23644.

6. J. L. Joyal, R.S. Annan, M.J. Hudldleston, S.A. Carr, M.J. Hart and D.B. Sacks, *J. Biol. Chem.* **1997**, *272*, 15419-15425.

7. H. Walczak, M. A. Degli-Esposti, R. S. Johnson, P. J. Smolak, J. Y. Waugh, N. Boiani, M. S. Timour, M. J. Gerhart, K. A. Schooley, C. A. Smith, R. G. Goodwin and C. T. Rauch, *EMBO J.* **1997**, *16*, 5386-5397.

8. M. Muzio, A. M. Chinnaiyan, F. C. Kischkel, K. O'Rourke, A.Shevchenko, J. Ni, C. Scaffidi, J. D. Bretz, M. Zhang, R. Gentz, M. Mann, P. H. Krammer, M. E. Peter, V. M. Dixit, *Cell* **1996**, *85*, 817-827.

9. J. Klose and U. Kobalz, *Electrophoresis* **1995**, *16*, 1034-1059.

10. T. Sakurai, A. Amemiya, M. Ishii, I. Matsuzaki, R. I. Chemelli, H. Tanaka, S. C. Williams J. A. Richardson, G. P. Kozlowski, S. Wilson, *et al.*, *Cell* **1998**, *92*, 573-585.

11. M.J. Huddleston, R.S. Annan, M. F. Bean and S. A. Carr, *J. Amer. Soc. Mass Spectrom.* **1993**, *4*, 710 - 717.

12. S. A. Carr, M. J. Huddleston and R. S. Annan, *Anal. Biochem.* **1996**, *239*, 180-192.

13. R. S. Annan and S. A. Carr, *J. Protein Chem.* **1997**, *16*, 391-401.

14. R. Verma, R. S. Annan, M. J. Huddleston, S. A. Carr, G. Reynard and R. J. Deshaies, *Science* **1997**, *278*, 455-560.

15. B. A. Thomson, D. J. Douglas, J. J. Corr, J. W. Hager and C. L. Jollife, *Anal. Chem.* **1995**, *67*, 1696-1704.

16. M. Wilm and M. Mann, *Int. J. Mass Spectrom. Ion Processes* **1994**, *136*, 167-180.

17. M. Wilm and M. Mann, *Anal. Chem.* **1996**, *68*, 1-8.

18. S. A. Carr, M. J. Huddleston and M.F. Bean, *Protein Sci.* **1993**, *2*, 183-196.

19 M. J. Huddleston, M.F. Bean and S. A. Carr, *Proc. 39th ASMS Conf. on Mass Spectrometry and Allied Topics* **1991**, pp. 280-281.

20. H. Erdjument-Bromage, M. Lui, L. Lacomis, A. Grewal, R. S. Annan, D. E. McNulty, S. A. Carr and P. Tempst, *J. Chromatog.* **1998**, in press.

21. J. M. Stadel, S. Wilson and D. J. Bergsma, *Trends Pharmacological Sci.* **1997**, *18*, 430-437.

22. G. C. Thorne, K. D. Ballard and S. J. Gaskell, *J. Am. Soc. Mass Spectrom.* **1990**, *1*, 249-257.

23. R. Verma, R. Feldman and R. J. Deshaies, *Mol. Biol. Cell* **1997**, *8*, 1427-1436.

QUESTIONS AND ANSWERS

Michael O. Glocker (University of Konstanz, Germany)

Have you tried to apply the experimental data you obtained from small peptides such as neuropeptides to search the databases for (unknown) precursor proteins in, for instance, EST databases?

Answer. Yes, with the orexin sequence, but it was not in any database.

A. L. Burlingame (UCSF)

How much *in vivo* protein was required to obtain the molecular weight of the phosphoprotein Sic 1p?

Answer. We're not really sure, but we estimate that around 10 pmol was used. This was a tremendous effort to get these spectra. The proteins were first purified by 0.5 mm column reversed phase HPLC, and then analyzed by nanoelectrospray. In addition to the differential phosphorylation, we noticed oxidation, covalent addition of urea, and Na ion adducts. The amount of any individual protein species was quite small.

A. L. Burlingame (UCSF)

What phosphorylation site stoichiometry required for biological activity?

Answer. That is a very difficult question to answer. We know that in this case, phosphorylation at residues T5, T33 and S76 is essential, but that the stoichiometry at the latter two sites is only 58% and 68%, respectively.

Ruedi Aebersold (University of Washington)

Concerning the small bioactive peptides, are the genomic structures known? Specifically, I am interested in whether the peptides are synthesized as such or cut out from precursor peptides?

Answer. The rat gene which codes for the 33 residue orexin ligand is a 130 amino acid polypeptide, containing a 32 residue signal sequence, followed by the orexin-A and orexin-B sequences. The gene for the l-factor peptide has not been found yet, but the a-factor peptide in *S. cerevisiae* comes from a somewhat longer peptide, which loses 3 residues from the C-terminus and 21 residues from the N-terminus to form the mature a-factor amino acid sequence.

Daniel Kassel (CombiChem, Inc.)

Would you comment on the "meaningfulness" of the orphan receptor peptide structure that you elucidated in the context of drug discovery and the design of small molecule antagonists?

Answer. Elucidating the structure of the orexin-A ligand helped us determine it's function. The posttranslational modifications present in the orexin structure are common to neuropeptides. Northern blot analysis of rat tissue showed that *prepro-orexin* mRNA is found exclusively in the brain. Further study with *prepro-orexin* cDNA probes showed that orexin-containing neurons were localized exclusively in the hypothalamus, an area of the brain known to contain neuropeptides which regulate feeding behavior. We have recently completed the NMR structure of the ligand and it is being used to aid in the development of small molecule drug candidates.

Investigation of Intact Subunit Polypeptide Composition of the 20S Proteasome Complex from Rat Liver Using Mass Spectrometry

Lan Huang, C. C. Wang and A. L. Burlingame
Department of Pharmaceutical Chemistry, University of California, San Francisco, CA 94143-0446

Proteasomes are multicatalytic proteolytic complexes found in almost all living cells that are responsible for protein degradation in both the cytosol and nucleus. They are involved in many important biological processes, including the removal of abnormal, misfolded or improperly assembled proteins, stress response, cell differentiation, metabolic adaption, and cellular immune response [1, 2]. The 20S proteasome is the catalytic core of the larger, ATP-dependent 26S complex that is responsible for degradation of ubiquitin-conjugated proteins. With a molecular weight of approximately 750 kDa, the 20S proteasome complex has a cylindrical structure consisting of four stacked rings, each of which is organized from seven α and β subunits, assembled in the order αββα(1-4). This overall structure is conserved from archebacteria to eukaryotes, which has been shown clearly in the crystal structures of the 20S proteasomes from *Thermoplasma acidophilum* and *Saccharomyces cerevisiae* [5, 6]. The proteasome complex from archeabacterium *Thermoplasma acidophilum* contains only two different but related subunits, α and β, while it is known that the eukaryotic proteasome complex is composed of at least 14 subunits with molecular masses of 21 to 34 kDa, and different charges (pI 3-10). These subunits can be divided into α- and β-type based on their polypeptide sequence homology with the *T. acidophilum*

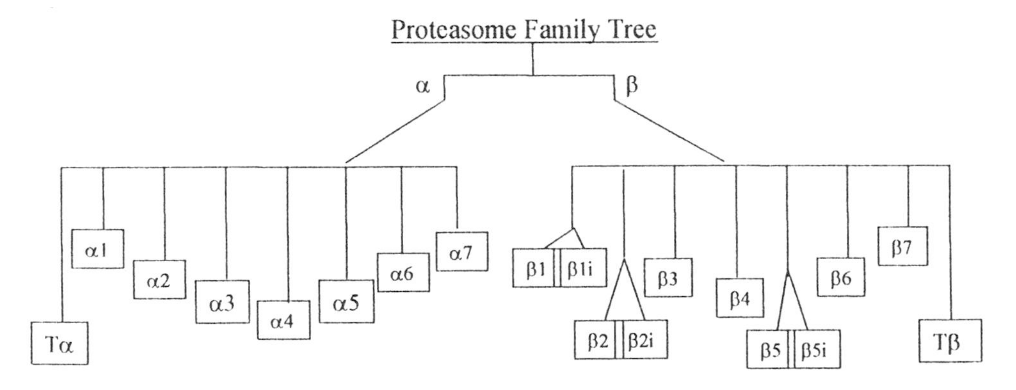

Fig. 1. The dendrogram of proteasome subunits into α - and β-type subunits. Tα and Tβ represent Thermoplasma acidophilum *α and β subunits. β-type subunits can be divided into two subgroups as "active" and "inactive", which refer to the presence or absence of Thr-1 as the N-terminal nucleophile, which is believed to be essential for the peptidase activities.*

α- or β-subunit. The primary structure of these subunits shows high inter-subunit homology within species, and high evolutionary conservation in various eukaroytes, suggesting that they constitute a multi-gene family and may have originated from a common ancestral gene [4].

Figure 1 gives the dendrogram showing the classification of 20S proteasome subunits into α- and β-type subunits. The α-type subunits do not appear to have catalytic acitivity, but they are responsible for the assembly and interactions of the proteasomes with regulatory subunits [2]. In comparison, β-type subunits can be divided into two subgroups as "active" or "inactive", based on the presence or absence of Thr-1 as the N-terminal nucleophile, which is believed essential for the peptidase activities. Three of the seven β subunits, β_1(X), β_5(Y), β_2(Z) appear to be functional in the mature enzyme complex [6]. In mammalian cells, these three β subunits can be replaced by three non-essential subunits, β_{1i}(LMP2), β_{5i}(LMP7) and β_{2i}(MECL-1), upon induction by the T-cell-derived antiviral cytokine interferon-γ [7-9]. This subunit replacement process generates "immunoproteasomes" responsible for production of MHC class I ligands, through proteolytic processing of intracellar antigens with altered proteolytic cleavage specificities [10]. Therefore, 20S proteasomes may exist in cells as heterogeneous populations with functional diversity, and the total number of β-type subunits may be increased in multi-cellular organisms during evolution for adaption to environmental stresses [11]. Analysis of the subunit composition of highly purified proteasomes by 2-D PAGE reveals more than 20 protein spots [12-16]. Most of the β-type subunits are synthesized as proproteins,

New Systematic Name[a]	Yeast Gene	Human Gene	Rat Gene
α1	PRS2	PROS-27(Iota)	rIota
α2	Y7	HC3	RC3
α3	Y13	HC9	RC9
α4	PRE6	XAPC-7-S[c] XAPC-7-L[c]	RC6-I-S[c] RC6-I-L[c]
α5	PUP2	Zeta	rZeta
α6	PRE5	HC2[c] PROS-30[c]	RC2
α7	PRS1	HC8	RC8
β1	PRE3	Y(Delta)	rDelta
β1i	-	LMP2	rLMP2
β2	PUP1	Z	unknown[b]
β2i	-	MECL-1	unknown[b]
β3	PUP3	HC10-II	RC10-II
β4	PRE1	HC7-I	RC7-I
β5	PRE2	X(MB1,ε)	rX
β5i	-	LMP7-E1[c] LMP7-E2[c]	RC1
β6	PRS3	HC5	RC5
β7	PRE4	HN3	RN3

[a] The new systematic name was determined based on yeast crystal structure(6).
[b] Rat subunits Z and MECL-1 are not present in protein databases, they are newly identified subunits by mass spectrometry.
[c] These four pairs of subunit s(RC6-I-S and RC6-I-L, HC2 and PROS-30, XAPC7-S and XPAC7-L, LMP-E1 and LMP-E2) are almost identical. Therefore their mRNAs may arise by alternative splicing or use of different transcription initiation sites(2).

Table 1. Nomenclature of proteasome subunits.

which undergo posttranslational processing during proteasome maturation [17]. This process is thought to generate further subunit composition complexity. To understand the 20S proteasome subunit composition and characterize their putative posttranslational modifications and physiologically functional forms, we have first undertaken a comprehensive characterization of 20S proteasomes from rat liver using advanced methods of mass spectrometry. To clarify the nature of each subunit, the nomenclatures of 20S proteasome subunits from yeast, human and rat are summarized in Table 1. The new systematic names are based on the location of subunits in yeast crystal structure.

Strategy #1 for Protein Characterization and Identification

Figure 2 illustrates the procedure for characterization and identification of 20 S proteasome complex from rat liver. The purified 20S proteasome complex from rat liver was first separated by 2-D gel electrophoresis and visualized by either Coomassie blue or silver staining. The gel spots of interest were excised and in-gel digested with trypsin overnight. The resulting tryptic peptides were subsequently extracted as unseparated digests. The peptide mass values were determined either by MALDI-DE-TOF measurements on unseparated digests and/or capillary LC-ES-orthogonal acceleration-TOF MS measurements of eluant peptides. Matrix-assisted laser desorption mass spectrometry (MALDI-TOF) with delayed extraction [18] has been widely employed in identification

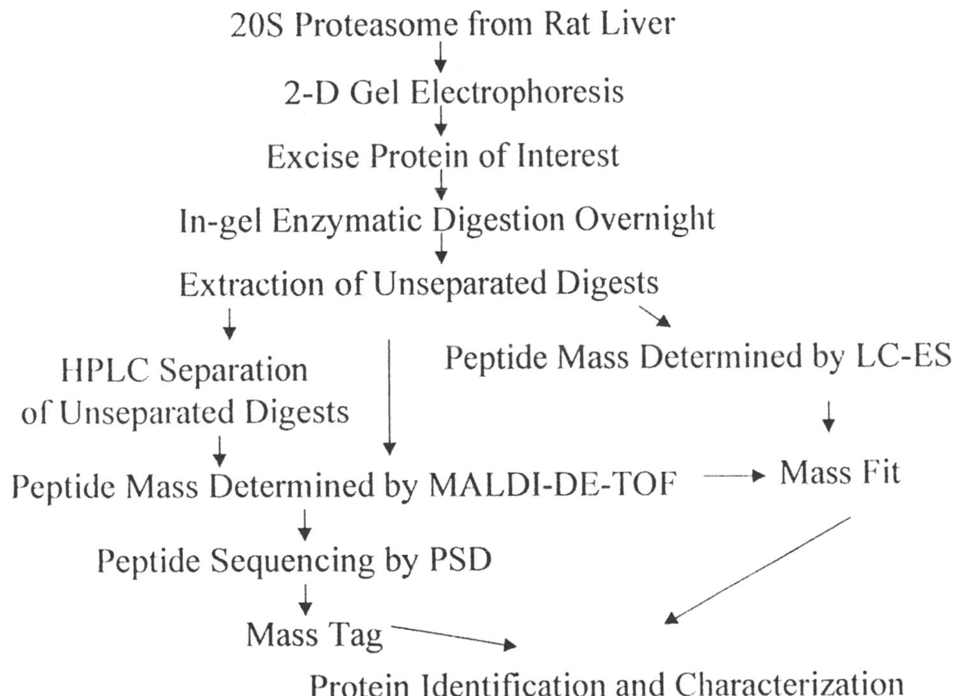

Fig. 2. Strategy #1 for protein characterization and identification.

and characterization of proteins isolated by 1-D or 2-D gel electrophoresis [16, 19-23]. This mass spectrometric technique has shown extremely low detection limits, high speed, high mass resolution and ease of dealing with simple protein mixtures. Capillary LC-ES has also demonstrated its powerful ability to measure peptide masses for protein identification separated by gel electrophoresis. The high sensitivity due to sample clean-up and concentration by HPLC results in better peptide mass mapping coverage in general (data not shown). All the measured tryptic peptide masses for each spot were submitted to the MS-Fit program developed by Karl Clauser and Peter Baker in our laboratory [24]. Basically, this program was designed to compare the experimental mass values with theoretical values from protein and gene databases, calculated by applying the enzyme cleavage rules. By using an appropriate scoring algorithm, the closest match or matches can be identified. Although peptide mass mapping is fast and simple, its success can be compromised by the purity of the protein spot in one gel spot, errors in the sequence database, and the observation of too few peptides in the MS map from a given protein. Therefore, partial amino acid sequencing should be employed to confirm or complete unambiguous protein identifications [19].

In this study, MALDI-DE-PSD (post source decay) was employed as a major tool to characterize peptide structures. This technique has been proven to be convenient, fast and sensitive for peptide sequence identification at the sub-picomole level [16, 21, 23, 25]. After screening of the MALDI mass spectra of tryptic digests of each spot, the peptides with the highest pseudomolecular ion abundance were subjected to post source decay (PSD) analysis for partial amino acid sequence determination. Some unseparated tryptic digests were separated subsequently by HPLC. The fractions were collected, concentrated and analyzed to obtain better peptide composition coverage, and especially enhanced quality PSD spectra. PSD spectra were analyzed by the MS-Tag program developed also by Clauser, et al. [26], which was designed to match the experimental fragment-ion derived partial sequence data to a peptide sequence in an existing database. Ideally, one would prefer to obtain a tandem mass spectrum with sufficient fragment ions from which a complete peptide sequence can be determined [19]. In practice, peptides from in-gel digestion usually yield only a limited number of fragment ions by PSD, which prevents unambiguous manual interpretation of these particular mass spectra. However, each fragment ion is characteristic of the sequence of the peptide under analysis and adds a constraint to database searching. Therefore, all ions present provide very high discriminating power for database searching. MS-Tag integrates all of these constraints by considering each fragment ion and parent ion mass independently to match a single peptide sequence in a genomic database. Combining both MS-Fit and MS-Tag searching results, proteins in each spot on a 2-D gel were identified with certainty. Occasionally, unknown peptides were found in gel spots. Therefore, de novo sequencing of unknown peptides by MALDI-DE-PSD was also used to design the oligonucleotide probes for cDNA cloning. The detailed experimental procedures have been described elsewhere [16].

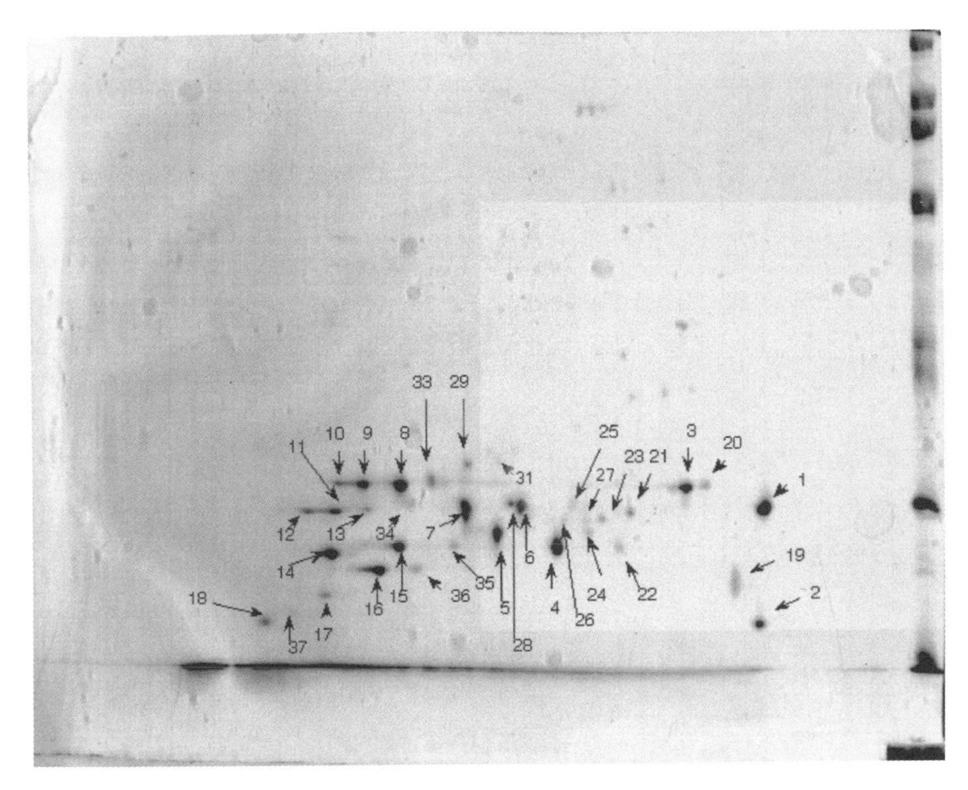

Fig. 3. 2-D gel picture of the 20S proteasome complex from rat liver.

IDENTIFICATION AND CHARACTERIZATION OF KNOWN 20S PROTEASOME SUBUNITS FROM RAT LIVER

The 20S proteasomes from rat liver isolated by 2-D gel electrophoresis were characterized and identified using MALDI-TOF and -PSD [16, 27]. Analysis of purified rat proteasomes by 2-D gel electrophoresis reveals more than 30 protein spots (Fig. 3). This suggests more subunits present than required to form any particular 20S proteasome complex. The amount of protein present in each spot varies as observed from the intensity of Coommassie blue stain on the gel, which might indicate multiple gene copies during protein maturation. The pattern of the major spots on two-dimensional gels is highly reproducible from batch to batch, but the relative abundance of minor spots varies with different proteasome preparations. Seven α subunits, C2, C3, C6-1, C8, C9, Iota and Zeta, and eight β subunits, C1(LMP7), C5, RN3, X, Y (Delta), C10-II(Theta), C7-I, Ring12 (LMP2) were identified using peptide mass mapping and peptide partial sequencing [16]. In addition, peptides with sequences identical to human and/or mouse related proteasome subunits were also found in some spots, where the corresponding rat proteasome subunits identified contain homologous sequences. Furthermore, subunits C8, RN3, C9, C3, C7-I, C1(Lmp 7), C2, C6-1 were present in more than one spot, thus suggesting the compositional heterogeneity of the 20S proteasomes from rat liver.

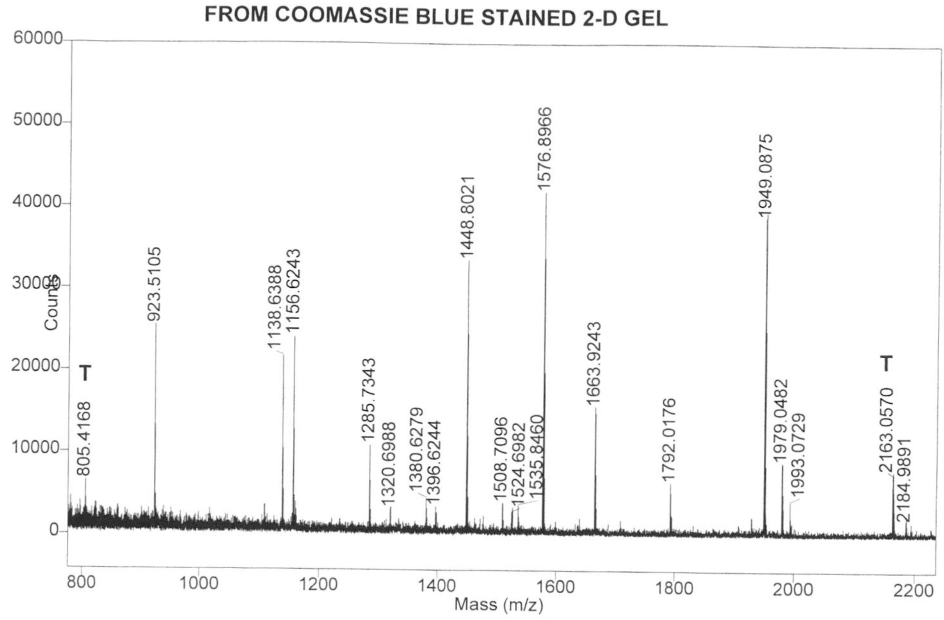

Fig. 4. MALDI-DE-TOF spectrum of unseparated digest of spot 6 from Coomassie blue stained 2-D-gel.

Identification and Characterization of New 20S Proteasome Subunits from Rat Liver

In addition to all the rat 20S proteasome subunits with cDNA sequences present in current databases (as shown in Table 1) that have been identified, two new subunits were found [16]. The MALDI-DE-TOF spectrum of unseparated digest of spot 6 from Coomassie blue stained 2-D gel is shown in Fig. 4. The mass values of two bovine trypsin autolysis products (805.4168, 2163.0570) were used as internal mass calibration standards. All the peptide mass values were submitted to the MS-Fit program, but only four peptides (1380.6279, 1396.6244, 1508.7096, 1524.6982) matched mouse proteasome subunit Z(PSMB7), whereas the dominant peaks could not be interpreted. To identify the protein in spot 6, the unseparated digest was further separated by HPLC. Each fraction was analyzed and all the dominant peaks were sequenced by MALDI-DE-PSD. PSD spectra of peptides with MH^+ at 1448.8021 and 1576.8966 from HPLC fraction #20 are displayed in Fig. 5. These two peptides differ only by mass of 128, indicating a lysine residue, which is confirmed from MS-Tag homology searching results shown in Fig. 6. The sequences are determined as LDFLRPYSVPNK and LDFLRPYSVPNKK, respectively, because the peptide with a sequence of LDFLRPY was also found in spot 6 [16, 27]. The sequences are homologous to human and mouse proteasome subunits Z with one mutation. The rest of unmatched peptides in the MS-Fit result have been identified using the same approach and have sequences homologous to human and/or mouse proteasome

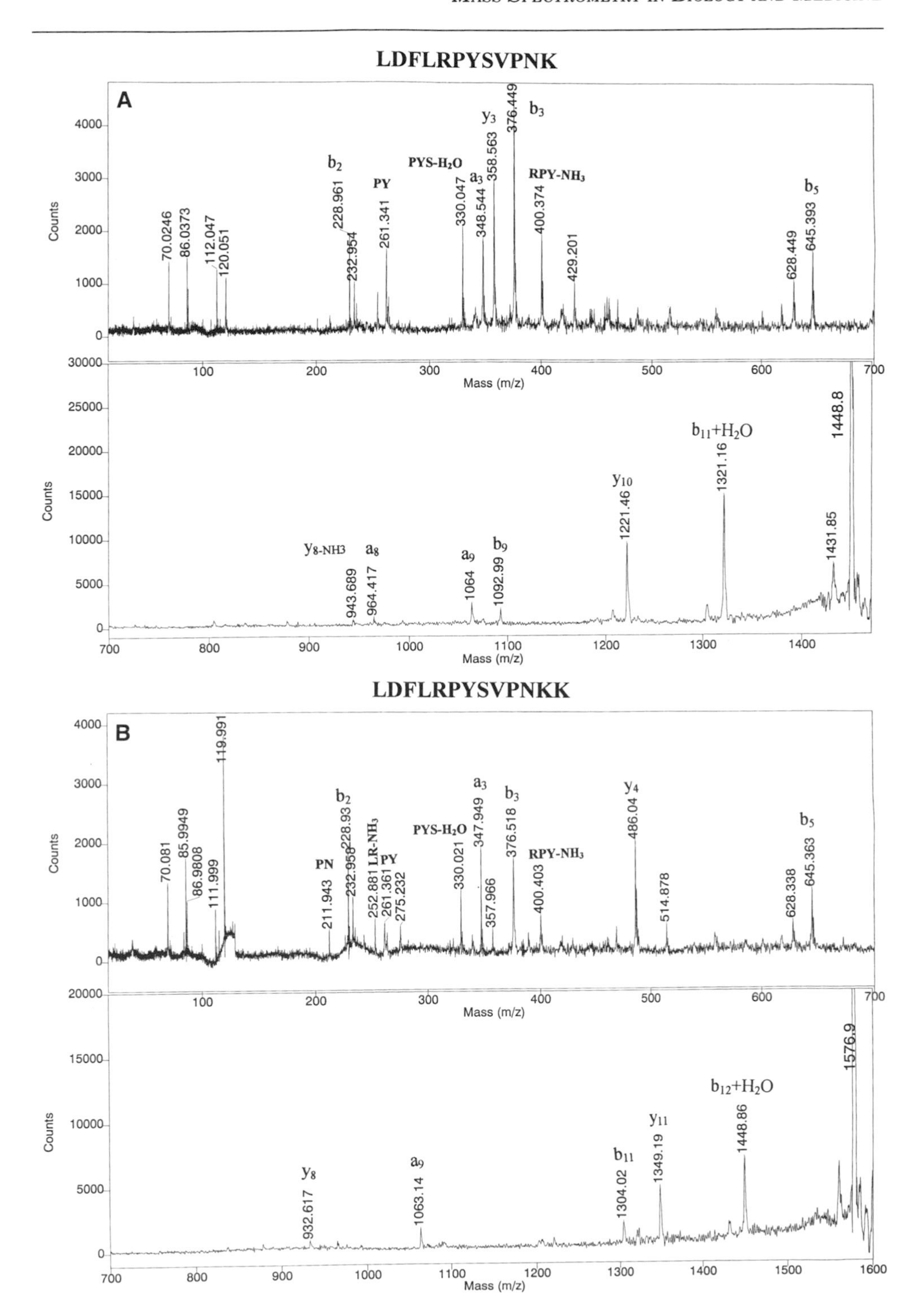

Fig. 5 (a) MALDI-DE-PSD spectrum of peptide with MH+ at 1448.8021 from HPLC fraction #20 of spot 6 digest; (b) MALDI-DE-PSD spectrum of peptide with MH+ at 1576.8966 from HPLC fraction #20 of spot 6 digest.

MS-Tag Search Results

Press stop on your browser if you wish to abort this MS-Tag search prematurely.

Sample ID (comment): **Rat proteaspme, spot6, fr.20, 1448.8021PSD**
Database searched: **NCBInr.10.04.98**
Molecular weight search (**1000 - 100000 Da**) selects **313061** entries.
Full pI range: **328782** entries.
Combined molecular weight and pI searches select **313061** entries.
Number of sequences passing through parent mass filter: **635661**
MS-Tag search selects **4** entries.

Parameters used in Search

Parent mass: **1448.8021** (+/- **100.0000 ppm**)
Fragment Ions used in search: **212.0, 229.0, 253.4, 261.3, 330.0, 348.5, 358.6, 376.4, 400.4, 645.4, 1064.0, 1093.0, 1321.2** (+/- **1500.00 ppm**)
Parent mass shift: +/- **45.1 Da**
Composition Ions present: **[RP][LI]F**

Result Summary

Rank	# Unmatched Ions	Modification	Sequence	MH⁺ Calculated (Da)	MH⁺ Error (Da)	Protein MW (Da)/pI	Species	NCBInr.10.04.98 Accession #	Protein Name
1	0/13	F7->Y	(K)LDFLRPYSVPNK(K)	1432.7953	16.0068	29891.6 / 8.14	MUS MUSCULUS	1632755	(D85570) PSMB7
1	0/13	P6->L	(K)LDFLRLFSVPNK(K)	1432.7953	16.0068	29891.6 / 8.14	MUS MUSCULUS	1632755	(D85570) PSMB7
1	0/13	P6->I	(K)LDFLRIFSVPNK(K)	1432.7953	16.0068	29891.6 / 8.14	MUS MUSCULUS	1632755	(D85570) PSMB7
1	0/13	T8->S	(K)LDFLRPYSVPNK(K)	1462.8058	-14.0037	29965.6 / 7.58	HOMO SAPIENS	1531533	(D38048) proteasome subunit z

Detailed Results

Rank	# Unmatched Ions	Modification	Sequence	MH⁺ Calculated (Da)	MH⁺ Error (Da)	Protein MW (Da)/pI	Species	NCBInr.10.04.98 Accession #	MS-Digest Index #	Protein Name
1	0/13	F7->Y	(K)LDFLRPYSVPNK(K)	1432.7953	16.0068	29891.6 / 8.14	MUS MUSCULUS	1632755	203346	(D85570) PSMB7
1	0/13	P6->L	(K)LDFLRLFSVPNK(K)	1432.7953	16.0068	29891.6 / 8.14	MUS MUSCULUS	1632755	203346	(D85570) PSMB7
1	0/13	P6->I	(K)LDFLRIFSVPNK(K)	1432.7953	16.0068	29891.6 / 8.14	MUS MUSCULUS	1632755	203346	(D85570) PSMB7

Fragment-ion (m/z)	212.0	229.0	253.4	261.3	330.0	348.5	358.6	376.4	400.4	645.4	1064.0	1093.0	1321.2
Ion-type Delta Da	PN -0.23	b_2 -0.26	RI-NH_3 0.07	y_2 -0.00 IF -0.05	IFS-H_2O -0.41	a_3 0.08 IFS 0.08 DFL-28 0.08	y_3 0.18	b_3 -0.03 DFL -0.03	RIF-NH_3 -0.10	b_5 -0.38 DFLRI -0.38	a_9 -0.32	b_9 0.67	b_{11}+H_2O -0.37 y_{11}-NH_3 1.65

Rank	# Unmatched Ions	Modification	Sequence	MH⁺ Calculated (Da)	MH⁺ Error (Da)	Protein MW (Da)/pI	Species	NCBInr.10.04.98 Accession #	MS-Digest Index #	Protein Name
1	0/13	T8->S	(K)LDFLRPYSVPNK(K)	1462.8058	-14.0037	29965.6 / 7.58	HOMO SAPIENS	1531533	197735	(D38048) proteasome subunit z

Fragment-ion (m/z)	212.0	229.0	253.4	261.3	330.0	348.5	358.6	376.4	400.4	645.4	1064.0	1093.0	1321.2
Ion-type Delta Da	PN -0.23	b_2 -0.26	LR-NH_3 0.07	y_2 -0.00 PY -0.00	PYS-H_2O -0.36	a_3 0.08 PYS 0.12 DFL-28 0.08	y_3 0.18	b_3 -0.03 DFL -0.03	RPY-NH_3 -0.06	b_5 -0.38	a_9 -0.28	b_9 0.71	b_{11}+H_2O -0.32 y_{11}-NH_3 1.69

Parameters Used in Search

Ion Types Considered: **a b B y n h I**

Search Mode	Max. # Unmatched Ions	Peptide Masses are	Digest Used	Max. # Missed Cleavages	Cysteines Modified by	Peptide N terminus	Peptide C terminus
homology	**0**	**Par(mi)Frag(av)**	**Trypsin**	**2**	**acrylamide**	**Hydrogen (H)**	**Free Acid (O H)**

MS-Tag 4.1.1, ProteinProspector 3.1.1

Fig. 6. MS-Tag search results of PSD spectra in Figure (a) 5a; (b) 5b.

MS-Tag Search Results

Press stop on your browser if you wish to abort this MS-Tag search prematurely.

Sample ID (comment): **Rat proteasome, spot6, fr.20, 1576.8966 PSD**
Database searched: **NCBInr.10.04.98** — Parameters used in Search
Molecular weight search (**1000 - 100000 Da**) selects **313061** entries.
Full pI range: **328782** entries.
Combined molecular weight and pI searches select **313061** entries.
Number of sequences passing through parent mass filter: **605093**
MS-Tag search selects **4** entries.

Parent mass: **1576.8966 (+/- 100.0000 ppm)**
Fragment Ions used in search: **212.0, 229.0, 253.4, 261.3, 330.0, 347.9, 376.5, 400.4, 486.0, 645.4, 932.6, 1063.2, 1304.1, 1448.9 (+/- 1500.00 ppm)**
Parent mass shift: **+/- 45.1 Da**
Composition Ions present: **[RP][LI]F**

Result Summary

Rank	# Unmatched Ions	Modification	Sequence	MH^+ Calculated (Da)	MH^+ Error (Da)	Protein MW (Da)/pI	Species	NCBInr.10.04.98 Accession #	Protein Name
1	0/14	F7->Y	(K)LDFLRPYSVPNKK(G)	1560.8902	16.0064	29891.6 / 8.14	MUS MUSCULUS	1632755	(D85570) PSMB7
1	0/14	P6->L	(K)LDFLRLFSVPNKK(G)	1560.8902	16.0064	29891.6 / 8.14	MUS MUSCULUS	1632755	(D85570) PSMB7
1	0/14	P6->I	(K)LDFLRIFSVPNKK(G)	1560.8902	16.0064	29891.6 / 8.14	MUS MUSCULUS	1632755	(D85570) PSMB7
1	0/14	T8->S	(K)LDFLRPYSVPNKK(G)	1590.9008	-14.0042	29965.6 / 7.58	HOMO SAPIENS	1531533	(D38048) proteasome subunit z

Detailed Results

Rank	# Unmatched Ions	Modification	Sequence	MH^+ Calculated (Da)	MH^+ Error (Da)	Protein MW (Da)/pI	Species	NCBInr.10.04.98 Accession #	MS-Digest Index #	Protein Name
1	0/14	F7->Y	(K)LDFLRPYSVPNKK(G)	1560.8902	16.0064	29891.6 / 8.14	MUS MUSCULUS	1632755	203346	(D85570) PSMB7
1	0/14	P6->L	(K)LDFLRLFSVPNKK(G)	1560.8902	16.0064	29891.6 / 8.14	MUS MUSCULUS	1632755	203346	(D85570) PSMB7
1	0/14	P6->I	(K)LDFLRIFSVPNKK(G)	1560.8902	16.0064	29891.6 / 8.14	MUS MUSCULUS	1632755	203346	(D85570) PSMB7

Fragment-ion (m/z)	212.0	229.0	253.4	261.3	330.0	347.9	376.5	400.4	486.0	645.4	932.6	1063.2	1304.1	1448.9
Ion-type Delta Da	PN -0.23	b_2 -0.26	**RI**-NH_3 0.07	IF -0.05	**IFS**-H_2O -0.41	a_3 -0.52 **IFS** -0.52 **DFL**-28 -0.52	b_3 0.07 **DFL** 0.07	**RIF**-NH_3 -0.10	y_4 -0.59	b_5 -0.38 **DFLRI** -0.38	y_8 -0.54	a_9 -1.12	b_{11} 0.55	b_{12}+H_2O -0.84 y_{12}-NH_3 1.17

Rank	# Unmatched Ions	Modification	Sequence	MH^+ Calculated (Da)	MH^+ Error (Da)	Protein MW (Da)/pI	Species	NCBInr.10.04.98 Accession #	MS-Digest Index #	Protein Name
1	0/14	T8->S	(K)LDFLRPYSVPNKK(G)	1590.9008	-14.0042	29965.6 / 7.58	HOMO SAPIENS	1531533	197735	(D38048) proteasome subunit z

Fragment-ion (m/z)	212.0	229.0	253.4	261.3	330.0	347.9	376.5	400.4	486.0	645.4	932.6	1063.2	1304.1	1448.9
Ion-type Delta Da	PN -0.23	b_2 -0.26	**LR**-NH_3 0.07	PY -0.00	**PYS**-H_2O -0.36	a_3 -0.52 **PYS** -0.48 **DFL**-28 -0.52	b_3 0.07 **DFL** 0.07	**RPY**-NH_3 -0.06	y_4 -0.59	b_5 -0.38	y_8 -0.50	a_9 -1.08	b_{11} 0.59	b_{12}+H_2O -0.80 y_{12}-NH_3 1.22

Parameters Used in Search

Ion Types Considered: **a b B y n h I**

Search Mode	Max. # Unmatched Ions	Peptide Masses are	Digest Used	Max. # Missed Cleavages	Cysteines Modified by	Peptide N terminus	Peptide C terminus
homology	**0**	**Par(mi)Frag(av)**	**Trypsin**	**2**	**acrylamide**	**Hydrogen (H)**	**Free Acid (O H)**

MS-Tag 4.1.1, ProteinProspector 3.1.1

Fig. 6b.

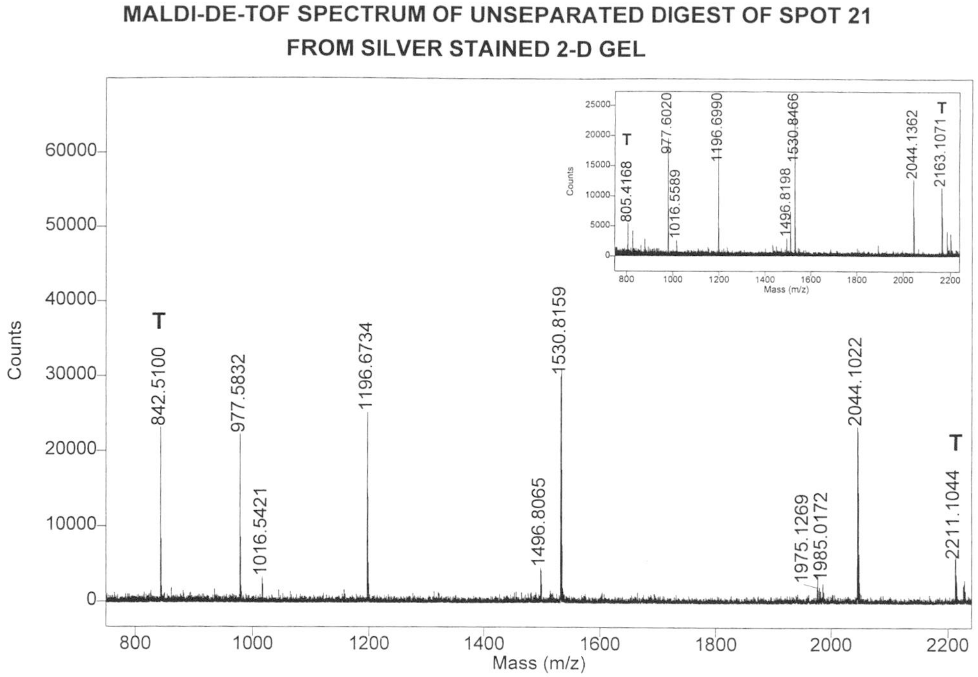

MALDI-DE-TOF SPECTRUM OF UNSEPARATED DIGEST OF SPOT 21 FROM SILVER STAINED 2-D GEL

Fig. 7. MALDI-DE-TOF spectrum of unseparated digest of spot 21 from silver-stained 2-D gel. The inset is the spectrum obtained from spot 2 on Coomassie blue stained 2-D gel.

subunit Z, with one or more mutations [16]. As a result, the new subunit in spot 6 is identified as rat proteasome subunit Z, a homolog of human and mouse subunit Z.

Similarly, another new subunit was found in spot 21. Figure 7 gives the MALDI-DE-TOF spectrum of unseparated digest of spot 21 from silver stained 2-D gel. The inset shows the MALDI-DE-TOF spectrum obtained from Coomassie blue stained gel. The pattern of spectra are almost identical, even though the type of trypsin used in both cases is different. Actually, both types of trypsin work well for either silver or Coomassie blue stained gels. Combining peptide mass mapping results with peptide partial sequencing information, we have identified the protein in spot 21 as a homolog of human and mouse subunit MECL-1, which is the substitute for subunit Z upon interferon-γ induction. In addition to the peptides with sequences identical to human and/or mouse subunit MECL-1, new peptides were also found. The PSD spectrum of peptide with MH^+ at 1530.8159 is shown in Fig. 8. The fragment tag ions in PSD spectrum did not match any existing sequences in databases using the MS-Tag program. Therefore, de novo sequencing of this peptide using the PSD spectrum was attempted, and the sequence was proposed as [FA]PGTTP[PV][Q/K]T[Q/K]EVR. To further refine this sequence, Qq-TOF tandem mass spectrometry equipped with a nanoelectrospray source was also used, because fragment ion peaks have much higher mass resolution and mass accuracy than those obtained in the PSD

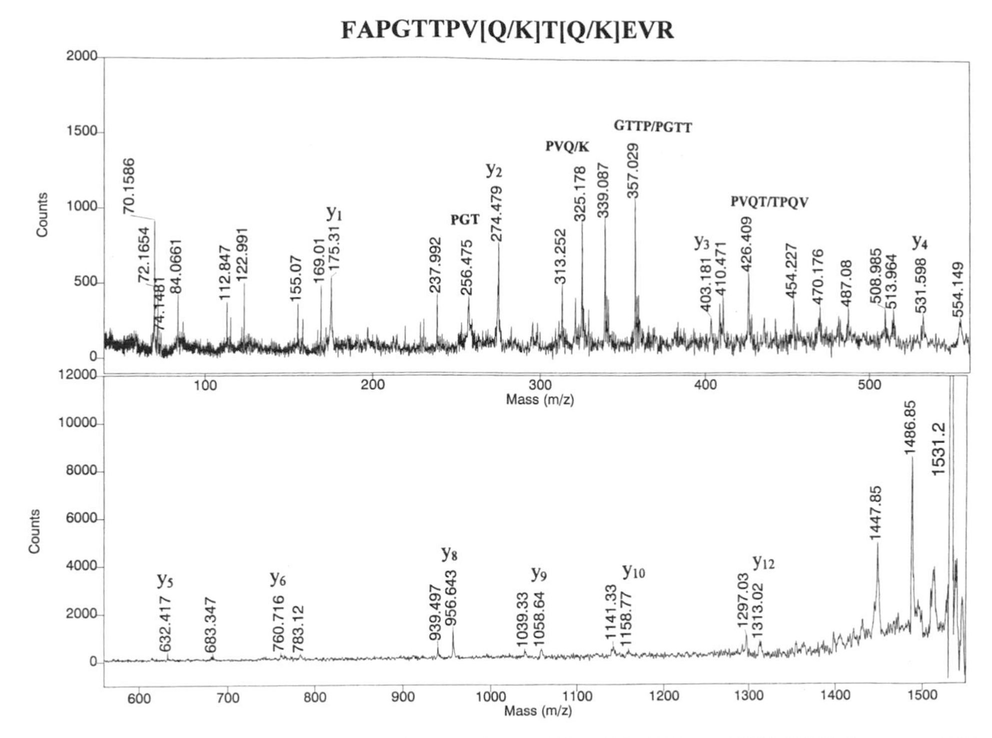

Fig. 8. MADL-DE-PSD spectrum of a tryptic peptide with MH⁺ at 1530.8159 from spot 21.

spectrum [28] (Fig. 9). The final sequence was suggested as FAPGTTPPVQTQEVR, which has homology to mouse proteasome subunit MECL-1 with 2 mutations, and to human MECL-1 with 5 mutations (Fig. 10). Another tryptic peptide with MH^+ at 1196.6734 was also de novo sequenced using PSD spectra, and has homology to both human and mouse subunits MECL-1 [16].

Interestingly, these two new subunits also have isoforms present at different spot coordinates [16]. The reason for this is not understood currently. However, it suggests that subunit compositional heterogeneity may be due to multiple functional diversities, because they represent catalytically active subunits responsible for enzymatic activities.

Furthermore, new peptides were observed with sequences identical to human and/or mouse proteasome subunit sequences, but not identical to rat subunit identified in these spots [16]. These sequence discrepancies found in the rat proteasome cDNA have been confirmed by molecular weight measurement of each individual intact proteasome subunit using LC-MS [29].

Strategy #2 for Characterization of Proteasome Subunits from Rat Liver

These investigations have revealed that the composition of 20S proteasomes from rat liver are much more complicated than expected from the results of previous proteasome subunit identification. The heterogeneity of the

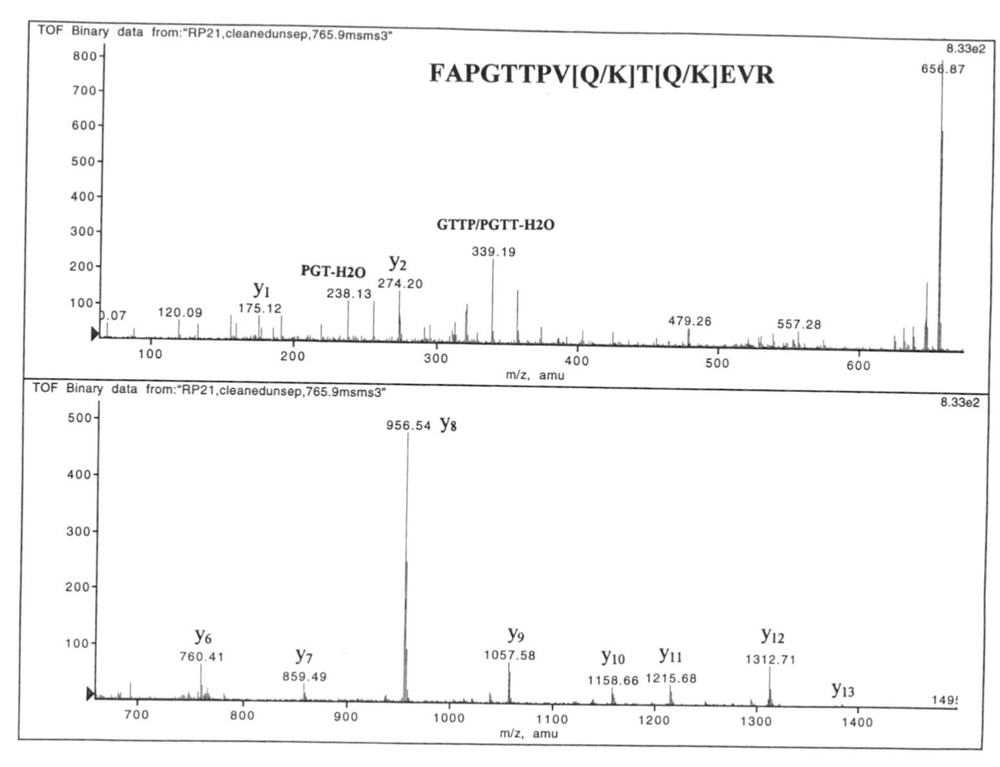

Fig. 9. Q-TOF tandem mass spectrum of a tryptic peptide with MH+ at 1530.8159 from spot 21.

MS-Edman Search Results

Press stop on your browser if you wish to abort this MS-Edman search prematurely.

Sample ID (comment): **Rat proteasome, spot 21, 1530.8**
Database searched: **NCBInr.10.04.98.sub** Parameters used in Search
Molecular weight search (**1000 - 100000 Da**) selects **313061** entries.
Full pI range: **328782** entries.
Combined molecular weight and pI searches select **313061** entries.
MS-Edman search selects **5** entries.

	Number of Substitutions	Matching Sequence	Peptide M+H	Protein MW (Da)/pI	Species	MS-Digest Index #	Accession #	Protein Name
1	2	(R)FAPGTTPV**L**T**R**EVR(P)	1543.8597	29105.5/6.40	MUS MUSCULUS	229530	2062109	(Y10875) proteasome subnuit MECL-1
2	2	(R)FAPGTTPV**L**T**R**EVR(P)	1543.8597	29063.4/6.40	MUS MUSCULUS	242807	2252790	(D85562) proteasome subunit MECL1
3	2	(R)FAPGTTPV**L**T**R**EVR(P)	1543.8597	9342.8/4.71	MUS SPRETUS	259460	2547072	(D85572) proteasome subunit MECL1
4	5	(H)F**V**PGTT**AVL**TQ**T**V**K**(P)	1461.8317	28936.5/7.69	HOMO SAPIENS	78221	730376	(Y13640) proteasome subunit MECl-1
5	5	(I)**V**APGT**FE**VQ**IE**EVR(Q)	1573.8226	44697.5/8.26	HOMO SAPIENS	71381	1082855	(U10323) NF45 protein

Parameters Used in Search

Peptide Masses are	Cysteines Modified by	Peptide N terminus	Peptide C terminus	Combinatorial Output	Max # AA Substitutions	Search Type	Regular Expression
monoisotopic	**acrylamide**	**Hydrogen (H)**	**Free Acid (O H)**	**off**	**5**	**Sequence Only**	**FAPGTTPV[Q/K]T[Q/K]EVR**

MS-Edman 2.1.1, ProteinProspector 3.1.1

Fig. 10. MS string-search (Edman) result of a tryptic peptide at MH+ 1530.8159 from spot 21.

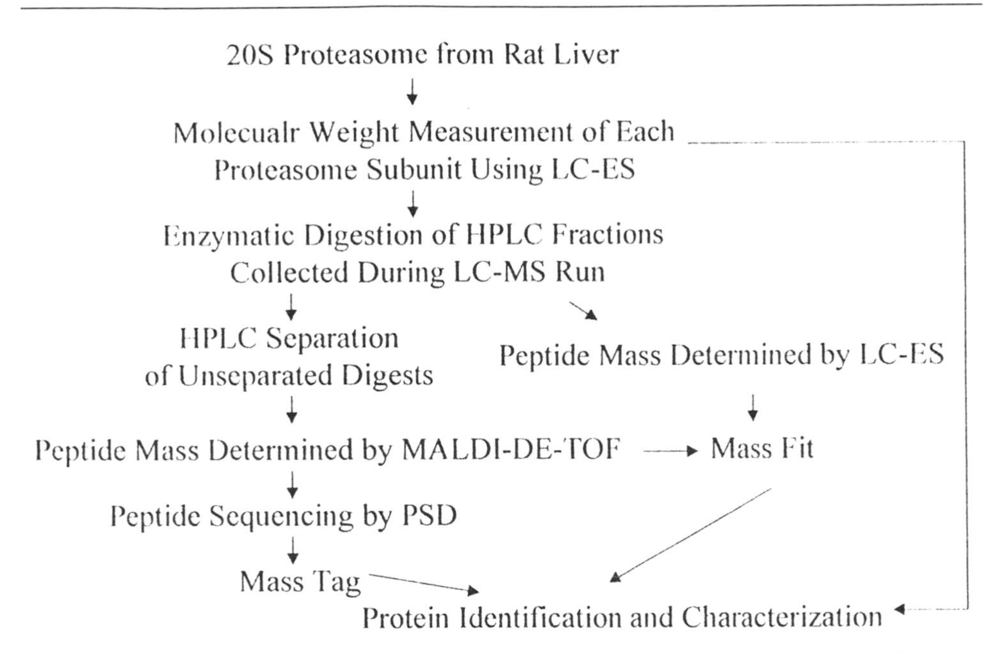

Fig. 11. Strategy #2 for identification and characterization of 20S proteasomes from rat liver.

subunit isoform composition suggests that multiple functional forms of particular proteasomes may be present in the cells. It has been suspected that the proteasome subunits are posttranslationally modified, possibly by phosphorylation [30, 31]. However, thus far no evidence has been found in our studies that show such modifications in rat proteasome subunits. To further investigate the proteasome complex composition heterogeneity, we have developed the second strategy for further characterization of 20S proteasomes from rat liver (Fig. 11). In this approach, we first measured the intact protein molecular weight of each individual proteasome subunit from rat liver using LC-ES-orthogonol acceleration-TOF MS [29]. Each fraction during the LC-MS run was collected, lyophylized and digested. The tryptic peptide masses in each fraction were either measured by LC-ES or MALDI-DE-TOF for peptide mapping. Some peptides were further sequenced by MALDI-DE-PSD. The proteins in each fraction were identified by combining both MS-Fit and MS-Tag search results. The molecular weights calculated from cDNA sequences of those identified proteasome subunits were compared with experimental values obtained by LC-ES. In those cases when the calculated molecular weights match well with anticipated experimental values, the protein sequences deduced from cDNA are correct; otherwise either the cDNA sequences have errors or the subunits have been modified.

Thirteen major peaks were observed on both total ion current and HPLC chromatograms, when rat 20S proteasome complex mixture was injected onto a microbore C4 column [29]. Twenty-six molecular weights were measured, which revealed again directly the compositional heterogeneity present in rat 20S proteasomes. The number of individual polypeptides (molecular weights)

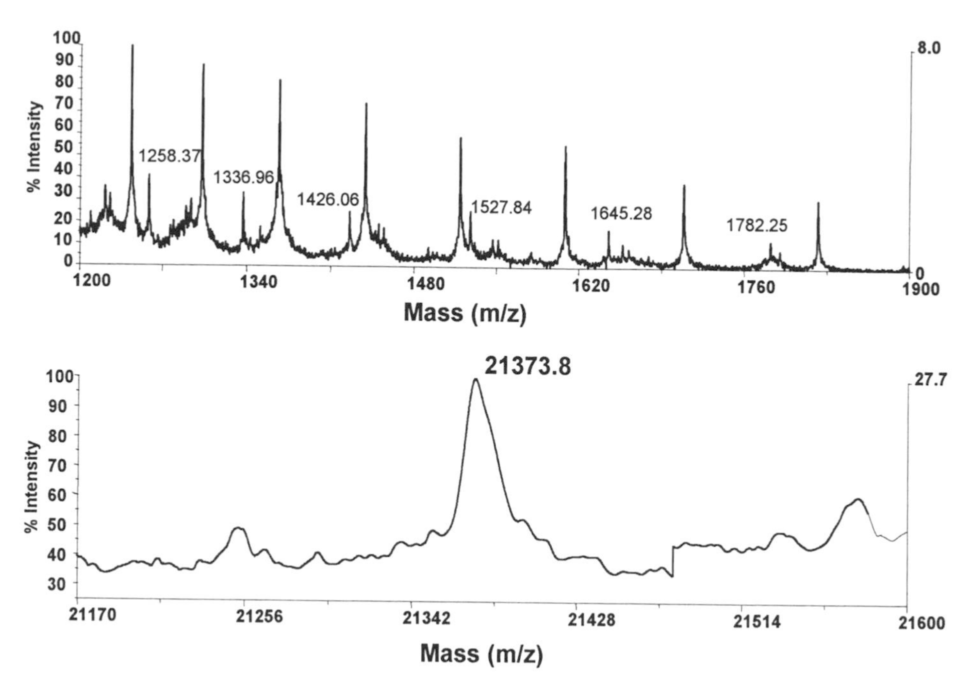

Fig. 12. ESI spectrum of peak #15.

obtained was much larger than expected for the known proteasome composition, i. e., 17 for mammalian proteasomes, with 14 essential and 3 non-essential subunits. Only three of the measured molecular weights matched well to the calculated molecular weights of proteasome subunits identified in HPLC fractions. The ESI spectra of peak #15 in total ion current chromatogram [29] is shown in Fig. 12. One of the molecular weights measured in this fraction is 21,373.8, which is close to the expected value of one of the subunits present in this fraction, i. e., mature form of Lmp2 subunit (MW ~21,354.3). The experimental and calculated values differ by 19u, which is much larger than experimental error. From peptide mass mapping of rat Lmp2 subunit, we have found two new tryptic peptides with sequences of (K)Y^{110}REDLLAHLMVAGWDQ**R**^{126}(E), and (**R**^{126})EGGQVYGTMGGMLIR^{151}(Q), respectively, which are homologous to rat Lmp2 subunit with one mutation at location 126. The residue was supposed to be histidine, not arginine, according to the cDNA sequence of rat Lmp2 subunit in the databases. Comparing rat Lmp2 sequence with human and mouse proteasome Lmp2 subunit sequences using multiple sequence alignment (Fig. 13), we have noticed that arginine was conserved at location 126 in both human and mouse subunits. The cDNA codons of arginine are C**A**U or C**A**C, and those of histidine are C**G**U or C**G**C. With one difference in the condons, reading errors could easily occur in the cDNA sequence, which has been found in several cases [29]. Therefore, it is reasonable to correct the sequence and replace histidine by arginine. The recalculated molecular weight of mature rat Lmp2 subunit with corrected sequence (H ➡ R) is 21,373.4, which is identical to the experimental

```
              Multiple Alignment of Mouse, Rat, and Human Lmp2 Subunits

              1                                ↓mature form                       50
M(lmp2)       MLRAGAPTAG SFRTEEVHTG TTIMAVEFDG GVVVGSDSRV SAGTAVVNRV
R(ring12)     MLQAGAPTAG SFRTGEVHTG TTIMAVEFDG GVVVGSDSRV SAGAAVVNRV
H(lmp2)       ~~~~~~~~~~ MLRAGEVHTG TTIMAVEFDG GVVMGSDSRV SAGEAVVNRV

              51                                                                 100
M(lmp2)       FDKLSPLHQR IFCALSGSAA DAQAIADMAA YQLELHGLEL EEPPLVLAAA
R(ring12)     FDKLSPLHQR IYCALSGSAA DAQAIADMAA YQLELHGLEL EEPPLVLAAA
H(lmp2)       FDKLSPLHER IYCALSGSAA DAQAVADMAA YQLELHGIEL EEPPLVLAAA

              101                              ↓sequence error                   150
M(lmp2)       NVVKNISYKY REDLLAHLIV AGWDQREGGQ VYGTMGGMLI RQPFTIGGSG
R(ring12)     NIVKNISYKY REDLLAHLMV AGWDQHEGGQ VYGTMGGMLI RQPFAIGGSG
H(lmp2)       NVVRNISYKY REDLSAHLMV AGWDQREGGQ VYGTLGGMLT RQPFAIGGSG

              151                                                                200
M(lmp2)       SSYIYGYVDA AYKPGMTPEE CRRFTTDAIT LAMNRDGSSG GVIYLVTITA
R(ring12)     STYIYGYVDA AYKPGMTPEE CRRFTTDAIT LAMNRDGSSG GVIYLVTITA
H(lmp2)       STFIYGYVDA AYKPGMSPEE CRRFTTDAIA LAMSRDGSSG GVIYLVTITA

              201                 219      MWcal (corrected mature form)
M(lmp2)       AGVDHRVILG DELPKFYDE                 = 21373.4
R(ring12)     DGVDHRVILG DELPKFYDE         MWExp
H(lmp2)       AGVDHRVILG NELPKFYDE                 = 21373.8±1.4(n=10)
```

Fig. 13. Multiple alignment of mouse, rat and human LMP2 subunits.

value observed. As a result, a sequence error was found in rat proteasome subunit Lmp2, and no posttranslational modifications of this particular subunit were observed. Similarly, after correcting the errors of cDNA sequences of several rat proteasome subunits in the databases, 10 additional measured molecular weights fit well with expected values. Although the remaining thirteen subunit molecular weight values cannot be explained at present, most likely errors in the cDNA sequences will eventually account for these discrepancies.

CONCLUSIONS AND FUTURE WORK

We have successfully identified and characterized rat 20S proteasome subunits using MALDI-DE-TOF and -PSD [16, 27]. All of the known rat proteasome subunits in the databases (Table 1), seven α and eight β subunits, have been identified and characterized. In addition, two new subunits and different isoforms of most of the proteasome subunits were revealed. The pattern of 20S proteasome from rat liver obtained on 2-D gels reported here is generally similar to that observed by other groups [15, 31], and also similar to that of 20S proteasomes from human and mouse [8, 12, 31-34]. Furthermore, the pattern of the identified rat proteasome subunits in major spots matched well with that obtained using immunoblotting [31]. Intact molecular weight measurement of individual proteasome subunits from rat liver using LC-ES has revealed not only the heterogeneity of proteasome subunit composition, but also the cDNA sequence errors present in the databases. The cDNA sequence errors can be an obstacle to successful characterization of proteins, because erroneous conclusions may be drawn in terms of chemical modification of proteins based on the differences between experimental molecular weight and calculated molecular weight from cDNA sequences. Extra precautions are needed in the complete characterization of proteins.

The rat 20S proteasome subunit composition appears to be more complicated than expected, because most of the proteasome subunits in rat occur in more than one form. These different isoforms may reflect multiple forms of the proteinase for diverse functions, possibly from different cell types or subcellular localization, or at different stages of differentiation [15]. Morphometric studies showed that proteasomes are localized in the nucleus, the cytoplasm and closely associated with the ER, and exist in several different molecular forms [35]. However, the precise distribution and the mechanisms of interconversion between different molecular forms of proteasomes are not yet clear. It has been suggested that the proteasome multi-subunit complex acquires functional diversity by change in molecular composition in response to environmental stimuli [36]. The subunit changes of proteasomes in various physiological conditions may be one reason why the proteasome has such a complex composition [9]. In higher eukaryotes, the subunit composition of the 20S proteasome can vary in a given species and is subject to precise regulation. Furthermore, how the isoforms of proteasome subunits have been processed still remains unknown. Further studies will be focused on clarifying the structural and functional relationships among different subunit isoforms.

Acknowledgements

Financial support was provided by the National Institutes of Health, National Center for Research Resources grant RR01614 (to ALB).

References

1. J. J. Monaco and D. Nandi, *Annu. Rev. Genetics* **1995**, *29*, 729-754.

2. O. Coux, K. Tanaka and A. L. Goldberg, *Annu. Rev. Biochem.* **1996**, *65*, 801-47.

3. W. L. H. Gerards, W. W. de John, W. Boelens and H. Bloemendal, *Cell. Mol. Life Sci.* **1998**, *54*, 253-262.

4. K. Tanaka, *Biochem. Biophys. Res. Commun.* **1998**, *247*, 537-541.

5. J. Lowe, D. Stock, B. Jap, P. Zwickl, W. Baumeister and R. Huber, *Science* **1995**, *268*, 533-539.

6. M. Groll, L. Doitzel, J. Lowe, D. Stock, M. Bochtler, H. Bartunik and R. Huber, *Nature* **1997**, *386*, 463-471.

7. M. Kasahara, M. Hayashi, K. Tanaka, et al. *Proc. Natl. Acad. Sci. USA* **1996**, *92*, 5072-5076.

8. D. Nandi, H. Jiang and J. J. Monaco, *J. Immunol.* **1996**, *156*, 2361-2364.

9. H. Hisamatsu, N. Shimbara, Y. Saito, P. Kristensen, K. B. Hendil, T. Fujiwara, E. Takahashi, N. Tanahashi, T. Tamura, A. Ichihara and K. Tanaka, *J. Exp. Med.* **1996**, *183*, 1807-1816.

10. K. Tanaka and M. Kasahara, *Immuno. Rev.*, in press.

11. K. Tanaka, *Mol. Biol. Reports* **1995**, *21*, 21-26.

12. P. Kristensen, H. Johnsen, W. Uerkvitz, K. Tanaka and K. B. Hendil, *Biochem. Biophys. Res. Commun.* **1994**, *205*, 1785-1789.

13. A. J. Rivett, *Biochem. J.* **1993**, *291*, 1-10.

14. C. Haass and P. M. Kloetzel, *Exp. Cell Res.* **1989**, *180*, 243-252.

15. A. J. Rivett and S. T. Sweeney, *Biochem. J.* **1991**, *278*, 171-177.

16. L. Huang, W. To, C. C. Wang and A. L. Burlingame, *J. Biol. Chem.*, submitted.

17. G. Schmidtke, R. Kraft, S. Kostka, P. Henklein, C. Frommel, J. Lowe, R. Huber, P. M. Kloetzel and M. Schmid, *EMBO J.* **1996**, *15*, 6887-6898.

18. M. L. Vestal, P. Juhasz and S. A. Martin, *Rapid Commun. Mass Spectrom.* **1995**, *9*, 1044-1050.

19. K. Clauser, S. C. Hall, D. M. Smith, J. W. Webb, L. E. Andrews, H. M. Tran, L. B. Epstein and A. L. Burlingame, *Proc. Natl. Acad. Sci. USA* **1995**, *92*, 5072-5076.

20. A. Shevchenko, M. Wilm, O. Vorm and M. Mann, M. *Anal. Chem.* **1996**, *68*, 850-858.

21. A. Schevchenko, O. N. Jensen, et al. *Proc. Natl. Acad. Sci. USA* **1996**, *93*, 14440-14445.

22. M. Wilm, A. Shevchenko, et al. *Nature* **1996**, *379*, 466-469.

23. Y. Qiu, L. Z. Benet and A. L. Burlingame, *J. Biol. Chem.* **1998**, *273*, 17940-17953.

24. K. Clauser, P. Baker and A. L. Burlingame, *Anal. Chem.*, in press.

25. O. N. Jensen, A. Podtelejnikov and M. Mann, *Rapid Commun. Mass Spectrom.* **1996**, *10*, 1371-1378.

26. K. R. Clauser, P. Baker and A. L. Burlingame, *Proc. 44th ASMS Conference on Mass Spectrometry and Allied Topics*, Portland, OR, 1996, p. 365.

27. L. Huang, W. To, C. C. Wang and A. L. Burlingame, *Proc. 45th ASMS Conference on Mass Spectrometry and Allied Topics*, Palm Springs, CA, 1997, p. 283.

28. A. Shevchenko, I. Chernushevich, W. Ens, K. G. Standing, B. Thomson, M. Wilm and M. Mann, *Rapid Commun. Mass Spectrom.* **1997**, *11*, 1015-1024.

29. L. Huang, R. Whittal, Y. Yao, C. C. Wang, A. L. Burlingame, in preparation.

30. G. G. Mason, K. B. Hendil and A. J. Rivett, *Eur. J. Biochem.* **1996**, *238*, 453-462.

31. A. Wehren, H. E. Meyer, A. Sobek, P-M. Kloetzel and B. Dahlmann, *Biol. Chem.* **1996**, *377*, 497-503.

32. R. Stohwasser, S. Standera, I. Peters and M. Groettup, *Eur. J. Immunol.* **1997**, *27*, 1182-1187.

33. K. Akiyama, S. Kagawa, T. Tamura, N. Shimbara, M. Takashina, P. Kristensen, K. B. Hendil, K. Tanaka and A. Ichihara, *FEBS Lett.* **1994**, *343*, 85-88.

34. M. Groettrup, T. Ruppert, L. Kuehn, M. Seeger, S. Standera, U. Koszinowski and P. M. Kloetzel, *J. Biol. Chem.* **1995**, *270*, 23808-23815.

35. A. J. Rivett, *Curr. Opin. Immunol.* **1998**, *10*, 110-114.

36. R. Ni, Y. Tomita, F. Tokunaga, T. J. Liang, C. Noda, A. Ichihara and K. Tanaka, *Biochim. Biophys. Acta* **1995**, *1264*, 45-32.

QUESTIONS AND ANSWERS

Scott Patterson (Amgen Inc.)

What percentage of peptides yielded usable for MALDI-PSD spectra?

Answer. It depends on the amount of peptides, ranging from 20% to 80%.

Scott Patterson (Amgen Inc.)

For those peptides for which de novo sequences were determined, did you feel you needed synthesize the peptide for confirmation?

Answer. No.

Micheal O. Glocker (University of Konstanz)

To match your experimental results to the protein sequence of the C5 subunit from the database, you suggested that a sequence error (E->G) could explain the observed deviation. Have you looked at the DNA sequence to determine what kind of nucleotide base exchange(s) was needed to match the suggested amino acid exchange?

Answer. Yes. Only the middle base in the codon has to be changed from A->G to match the suggested amino acid exchange.

Maria Person (UCSF)

How did % protein coverage compare between MALDI and HPLC-ESI?

Answer. HPLC-ESI mass mapping usually gives better protein coverage, due to sample cleanup and concentration. However, MALDI mass mapping can get as good coverage as HPLC-ESI mass mapping when the amount of proteins is more than 10 pmol.

Connie R. Jimenez (UCSF)

You found C-terminally trimmed forms (with lysine cleaved off) of the rat C6-1 subunit. Is that an artifact of your isolation procedure or does this truncation occur *in vivo*?

Answer. I am not sure.

Deciphering Functionally Important Multiprotein Complexes by Mass Spectrometry

Andrej Shevchenko and Matthias Mann†*
*European Molecular Biology Laboratory (EMBL), Meyerhofstraße, 1 69012 Heidelberg Germany; †Center for Experimental BioInformatics (CEBI) Odense University, Stærmosegaardsvej, 16 DK-5230 Odense M Denmark

Simple chemical reactions, such as the hydrolysis of a peptide bond or the attachment of a phosphate moiety, are catalyzed by individual enzymes. Often the enzymatic activity can be reproduced in an *in vitro* experiment and the chemical mechanisms of the reaction can be elucidated in detail [1]. However if the enzyme is involved in regulation of a cellular activity it seldom acts by itself. Enzymatic activity is controlled and targeted to specific substrates localized in specific cellular compartments by recruitment of other proteins acting as receptors, ligands, adaptors, etc. [2]. Thus complex multiprotein assemblies constitute important functional units of the cell regulation pathways and are involved in numerous cellular processes such as cell division, signal transduction, mRNA splicing and nuclear transport [2-5]. Study of the structure, regulation and relationship of protein complexes provides clues to understanding in molecular detail how cellular activities are carried out. Dissection of the protein complex places its individual subunits in a context of the specific cellular function carried out by the entire assembly. Taken together with specific motifs recognized in the sequence of an individual subunit, this often suggests a possible role of the protein in the concerted functioning of the whole assembly and allows the design of further suitable functional experiments.

Recent advances in gene-manipulating technology suggest the feasibility of large scale screening for protein interaction partners using two-hybrid and three-hybrid systems [6]. However in many instances biochemical isolation of protein complexes followed by characterization of their individual subunits would be preferred. Protein complexes are isolated from living cells and comprise proteins adequately modified to allow full functionality of the complex. Often physical interaction of proteins depends on the presence of non-protein constituents of the complex like RNA molecules. Another advantage of the biochemical approach over the genetic approach is that the protein complex is isolated as a whole functional unit, containing core subunits as well as loosely interacting proteins. Often, clever biochemical strategies allow isolation of large protein complexes containing dozens of interacting subunits along with the non-protein constituents in a single biochemical purification.

The generic approach to the isolation of multi-protein complexes requires targeting of one of the already known subunits (not necessarily proteins) by immunoprecipitation or by affinity chromatography, resulting in co-isolation of the associated protein subunits from the whole cell lysate. Individual protein subunits are then separated by one dimensional or two dimensional gel electrophoresis, and in principle can be subsequently identified by immunostaining, Edman degradation or mass spectrometry [7]. However, many interesting protein complexes are available only in minute amounts. Furthermore, protein subunits are held together by non-covalent interactions and therefore it is difficult to preserve the integrity of the complex during the purification process. Some of the complexes contain proteins which are associated with the core subunits only transiently and therefore are co-purified in sub-stoichiometric amounts. Often only low picomole or femtomole amounts of proteins are available even if large scale biochemical preparation was undertaken. It is therefore not surprising that until recently immunological methods like Western blotting were the most widely used approach for the detection of co-isolated protein components. Western blotting has the principal disadvantage that it only allows confirmation of the presence of expected proteins in the complex provided the corresponding antibodies are available, and thereby does not lend itself to identification of new protein subunits. Other disadvantages are that antibodies are not available for many proteins and cross reactions are common. The immunoprecipitated proteins could be sequenced by Edman degradation, however the sequencing requires large and seldom available amounts of protein starting material (typically more than 50 pmols of the protein loaded on a gel). Any efficient biochemical strategy for characterization of protein complexes should be supported by an analytical technique which enables unambiguous protein identification at extremely low levels of protein starting material, while retaining relatively high throughput of the analysis.

Recent developments in mass spectrometric technology have made extensive characterization of protein complexes feasible [7, 8]. Implementation of delayed extraction in MALDI instruments [9-11] allow part-per-million mass accuracy even on peptide mixtures produced by *in-gel* enzymatic digestion. Very high mass accuracy boosts the specificity of database searching with MALDI peptide maps [12], thus allowing identification of proteins which can be visualized on a gel only by silver staining [13]. Higher search specificity also enables proteins to be discerned in mixtures solely by MALDI peptide mass mapping [14].

Tandem mass spectrometry using electrospray as the ionization method is an alternative method to peptide mass mapping for the identification of proteins. In our laboratory the nanoelectrospray ion source [15] combined with a triple quadrupole mass spectrometer has been used very successfully for the tandem mass spectrometric sequencing of individual peptides from unseparated peptide mixtures recovered after tryptic *in-gel* digestion of gel separated proteins [16]. Taking advantage of the peptide sequence tag database searching algorithm [17], nanoelectrospray tandem mass spectrometry allows identification of gel separated proteins at the femtomole level in comprehensive protein and EST databases. If the protein of interest is not present in a database, tryptic peptides are sequenced "*de novo*" by nanoelectrospray tandem mass spectrometry with less than one picomole of protein starting material on a gel [16, 18].

Based on the sequence stretches obtained, oligonucleotide probes are designed and the protein is subsequently cloned by PCR based approaches. High mass accuracy MALDI peptide mass mapping and nanoelectrospray tandem mass spectrometric sequencing have been combined in a layered approach allowing protein identification at the femtomole/low picomole level with high certainty and throughput [19, 20].

In the present paper we discuss the application of mass spectrometry to deciphering protein complexes involved in cell cycle control in budding yeast [21, 22]. Several new proteins were identified by mass spectrometry after protein complexes were immunoprecipitated and individual subunits were separated by one dimensional gel electrophoresis. Protein subunits of all complexes were only available in femtomole amounts and were detected as faint silver stained bands. It is important to note that biochemical purification of these complexes was launched after previous attempts to identify additional protein subunits by several kinds of genetic screens had failed. One of the proteins, Cdc20p, was found to be a "missing link" between protein complexes which control different stages of cell cycle progression [23, 24]. Taken together, the data elucidate the relations between protein complexes involved in the regulation of the cell cycle machinery and show the power of mass spectrometry to solve problems previously only analyzable by genetic methods.

Outline of the General Approach

In recent years more than ten multiprotein complexes of different complexity, functionality, abundance and species of origin have been isolated and successfully characterized by our laboratory in collaboration with various molecular and cell biology research groups. The general approach for purification of protein complexes and identification of their protein components (outlined in Fig. 1) is fairly straightforward. However, there is a large variety of protein purification techniques appropriate to complexes of different abundance, molecular weight and physical properties which have been successfully interfaced to mass spectrometric identification, and which we feel are worth being discussed.

Identification of a Target Component for Subsequent Tagging of the Gene

At the initial stage of the investigation one or several already known subunits of the complex are selected for subsequent gene tagging. This preliminary information is often obtained by genetic screening for mutants in which the characteristic phenotype is observed. Alternatively, the candidate gene can be identified by biochemical investigation of various cellular phenomena or by "database mining" of the data produced by genomic sequencing projects.

Gene Tagging

Using molecular biology and genetic methods, the selected target gene can be tagged with different types of epitopes (myc-, HA-) [25, 26], or a fusion construct with protein A or gluthatione-S-transferase (GST) can be made. In this

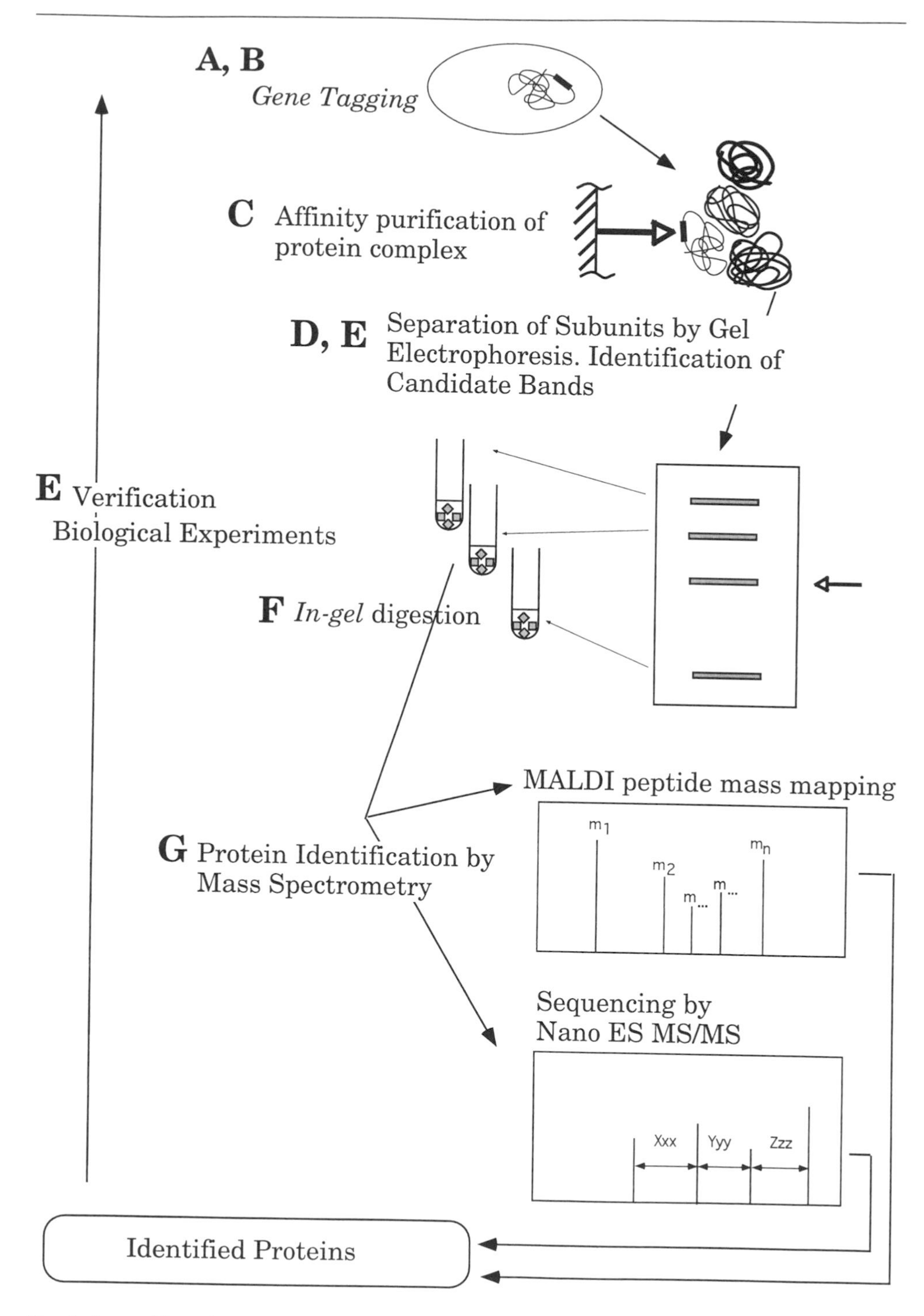

Fig. 1. An outline of a general approach for the characterization of multiprotein complexes. Steps A and B are applied if tagging of the genes of interest is feasible. For more explanation, see text.

experiment the endogenous gene is replaced with a gene containing the affinity tag at its C- or N- terminus. Alternatively, the tagged gene can be expressed from a plasmid. The gene tagging approach is becoming increasingly popular in projects dealing with organisms like yeast or bacteria, where straightforward genetic manipulation is feasible. All known subunits of the complex, in principle, can be easily tagged. In a case when the complex under study comprises many proteins, immunoprecipitation (IP) of a single tagged subunit results in co-isolation only of a limited subset of the complex. Thus tagging of more than one subunit becomes essential to elucidate the components of the entire complex [21]. Tagging of genes with different epitopes also allows optimization of IP conditions to increase the yield of protein complex components, along with moderate amount of co-precipitated background proteins.

Affinity Purification of Protein Complexes

Protein complexes can be pooled from cell lysate by taking advantage of any specific affinity of their individual subunits, such as specific binding to a protein [27], binding to a stretch of DNA [28] or RNA [29] sequence, etc. It is beyond the scope of the present paper to discuss the affinity purification stage in detail. The main issue affecting subsequent identification of interacting partners is the amount of co-isolated background proteins. Sometimes it is advantageous to perform IP from partially purified fractions of total cell lysate [30], although loosely associated subunits might be lost. A TEV proteinase cleavage site can be incorporated into the fusion construct together with protein A [31]. The protein complex is then trapped by beads containing crosslinked immunoglobulins. Subsequently, instead of direct elution of adsorbed proteins by salt gradient or acidification, the protein complex is cleaved off by virus proteinase and is released in solution virtually intact. Other non-tagged proteins non-specifically bound by the beads still remain adsorbed. Although the amount of the recovered protein subunits can be compromised, the observed background can be remarkably clean [32, 33].

Separation of Subunits by Gel Electrophoresis

The protein complex retained by affinity sorbent is eluted and the eluate is subsequently analyzed by one dimensional (1D) or two dimensional polyacrylamide (2D) gel electrophoresis. 2D electrophoresis offers better separation specificity, but may result in considerable losses of protein material, especially of poorly soluble membrane proteins. Separation of complex protein mixtures by 1D electrophoresis often results in co-migration of two or more proteins in a single band. However, the latter is now a matter of less concern because nanoelectrospray tandem mass spectrometry [16, 34] and to a lesser extent MALDI mass spectrometry [14] can efficiently discern proteins in mixtures. How the protein complex has been treated prior to electrophoresis might influence the results of the analysis. If the complex was purified by IP it is very likely that the sample is contaminated with antibodies "leaking" from the beads. Upon reduction with β-mercaptoethanol, immunoglobulin chains separate and migrate as intense although diffuse bands at about 30 kDa. Therefore sequencing of proteins having molecular weight close to 30 kDa will likely be hampered by contamination with immunoglobulin chains. It can be an advantage to omit the reduction step in the separate experiment, thus making low molecular weight proteins accessible for sequencing. Without prior reduction immunoglobulins migrate at around 150 kDa or do not enter the separation gel at all.

Identifying the Candidate Bands for Sequencing

Protein bands are visualized by Coomassie or silver staining and excised from the gel. In our experience, purification of protein complexes, especially from whole cell lysates is often accompanied by co-isolation of large amounts of proteins nonspecifically bound to antibodies. Most of these background proteins can be recognized as such by comparison of the gel pattern of immunoprecipitate with the pattern of the blank immunoprecipitate. However, it is our experience that at very low levels the difference between experiment and blank patterns is not straightforward. Inevitably, along with identification of true subunits of the complex, obviously unrelated contaminating proteins like abundant ribosomal proteins, metabolic enzymes or heat shock proteins are also identified. In 1D SDS PAGE the band corresponding to the bait protein should also be sequenced because it may be obscuring a subunit of the protein complex.

In-gel Digestion of Excised Bands

Excised protein bands are digested *in-gel* with trypsin [34]. We found it essential to perform *in-gel* reduction of the proteins with DTT followed by alkylation of cysteine residues with iodoacetamide, even if proteins had been reduced in Laemmli buffer prior to electrophoresis.

Mass Spectrometric Identification of Individual Subunits

Protein digests are analyzed by MALDI peptide mass mapping and nanoelectrospray MS/MS and interacting proteins are identified. Application of the mass spectrometric techniques is the focus of the present paper and will be discussed below in detail.

Verification by Biological Experiments

Identification of related protein bands by mass spectrometry is usually unambiguous. However, identified proteins still have to be regarded only as plausible candidates for components of the complex because they could also be artifacts of purification. Folding of tagged proteins or fusion proteins might be disturbed, thus facilitating their interaction with heat shock proteins and other chaperones. Chaperones interacting with the tagged proteins may then bind other proteins from cell lysate. Loosely associated but genuine and functionally important components are frequently co-isolated in even smaller quantities than unspecifically bound but abundant proteins. Therefore biological experiments have to be designed to provide functional evidence that identified proteins are true components of the complex. Functional experiments can be performed in different ways, depending on the biological function of the studied complex. Novel subunits can be fused to green fluorescent protein (GFP) and co-localized with already known components of the complex. The genes encoding novel subunits can be knocked out and the phenotype of the resulting mutant observed. Antibodies can be raised against novel subunits and then a subsequent stage of immunoprecipitation experiments is expected to yield some of the already identified subunits.

Mass Spectrometric Sequencing of Components of Protein Complexes

Layered Analytical Strategy — an Outline.

Let us summarize what has been discussed above regarding the requirements for the mass spectrometric technique for identification of the components of protein complexes. It is essential that the applied technique meet four criteria:

- femtomole sensitivity (protein applied to a gel)
- high throughput
- ability to identify proteins in mixtures
- the technique should be "expandable" for "*de novo*" sequencing of proteins not found in the database.

To tackle the problem we have developed a layered strategy for sequencing of gel separated proteins [19, 20] (Fig. 2). Two mass spectrometric techniques — high mass accuracy matrix-assisted laser desorption/ionization mass spectrometric peptide mapping (MALDI MS) and nanoelectrospray tandem mass spectrometric sequencing (nano ES MS/MS) — are combined in a single analytical approach. Proteins of interest are digested *in-gel* with trypsin using a protocol [34, 35] vastly simplified from previous approaches which can also be automated with a "digestion robot" [36]. The digests are analyzed by MALDI MS and then, if conclusive identification has not been reached, by nano ES MS/MS. Searching the databases with both nano ES MS/MS and MALDI MS data is performed by the same stand alone software, the PeptideSearch program developed at EMBL.

In-gel digestion of proteins results in the release of tryptic peptides from the gel matrix into the buffer solution surrounding the gel pieces. A small aliquot (ca 0.25 μL which constitutes 3-5% of the total digest volume) is taken out and analyzed directly using a delayed extraction MALDI TOF mass spectrometer. The analysis produces a peptide map with high mass accuracy. Searching a sequence database with the list of peptide masses will usually retrieve the corresponding protein, if it is present full-length in the database.

If no proteins have been retrieved or if the identification is not conclusive, the gel pieces are extracted, and extracts are pooled and dried down in a vacuum centrifuge. Importantly, the MALDI MS analysis already allows a rough estimate of the amount of peptide material actually available for sequencing. If the protein has not been identified by MALDI MS because very few and low intensity peptide ions were observed in the peptide map, in our experience it will

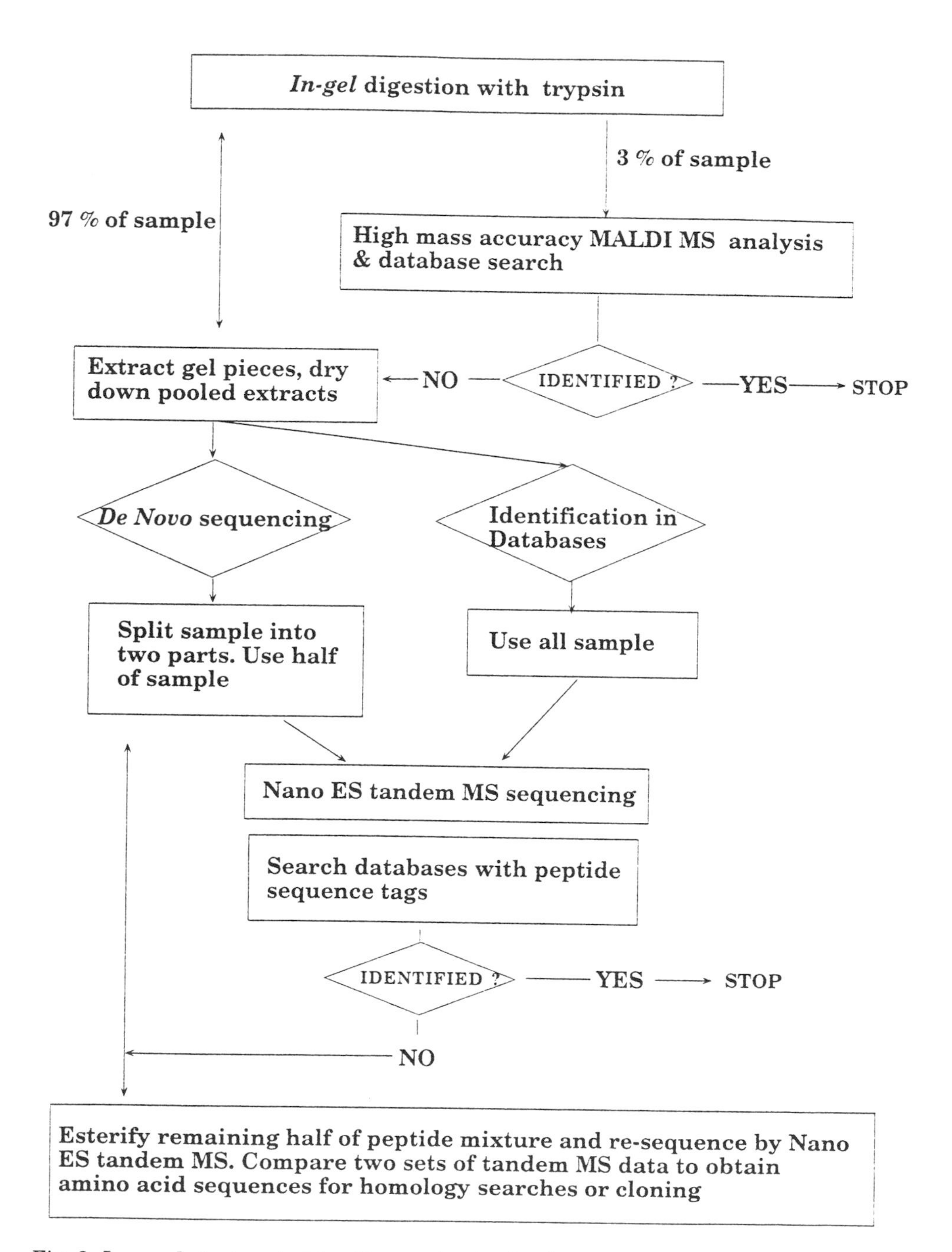

Fig. 2. Layered strategy to identify proteins in databases or to obtain stretches of amino acid sequence for cloning.

rarely be possible to sequence the corresponding peptides *'de novo'* by subsequent nano ES MS/MS analysis. In this case nano ES MS/MS can only be expected to provide the identification of the protein in the protein or EST sequence databases, and therefore the entire peptide material extracted from the gel matrix will be used in a single nanoelectrospray analysis. If intense peptide ions were observed in the MALDI peptide map of the digest then the most likely reason why conclusive protein identification has not been reached is because the sample constitutes a mixture of two or more proteins or it contains at least one unknown protein (i. e., a protein not present in a database). In this case the recovered peptide material is split into two parts and only one part is used for direct nano ES MS/MS sequencing (the rest is kept for an additional experiment, see below).

After a single desalting/concentration step using a pulled glass capillary containing 50-100 nL of POROS R2 reversed phase sorbent [15, 37], the unseparated peptide mixture is subjected to nano ES MS/MS. Short stretches of amino acid sequence are deduced from tandem spectra of individual peptide ions. A short (two to four amino acid residue) sequence stretch is combined with the molecular weight of the intact peptide and the masses of the corresponding fragment ions to give a peptide sequence tag [17]. Peptide sequence tags are then used for searching protein and EST databases with the PeptideSearch software. After the protein match has been found, all ion series of the fragment spectrum are used to verify the match against the calculated fragment masses for the retrieved peptide sequence. Usually more than one peptide ion is fragmented. If the sequence tags deduced from the fragment spectra of different peptide ions independently map to the same protein upon searching a database, this additionally verifies the protein identification.

If database searches produce no hits in either protein or EST databases, then longer sequence stretches have to be generated by complete interpretation of tandem mass spectrometric data and subsequently used to design oligonucleotide probes.

Characterization of gel separated proteins by MALDI MS and nano ES MS/MS has already been discussed in the literature [20]. In this paper we will focus only on certain features of the techniques which have proven to be especially valuable in deciphering protein complexes.

First Stage of the Layered Approach: Protein Identification by MALDI Mass Spectrometry

MALDI mass spectrometry has two features that make it the method of choice for initial screening of peptide mixtures generated by *in-gel* tryptic digestion. First, sample preparation is very simple, especially if the "fast evaporation method" is used to place a matrix layer on the target [38, 39]. Addition of nitrocellulose in the matrix solution [12, 34] leads to better representation of larger peptides and higher sensitivity. The extensive washing of the samples deposited on a matrix layer prepared by fast evaporation technique strongly reduces the intensity of salt adducts. The sensitivity of the MALDI reflector TOF machine for peptides in the mass range 800-3000 Da (typical range

Protein: P10592 HS72_YEAST HEAT SHOCK PROTEIN SSA2

Supernatant

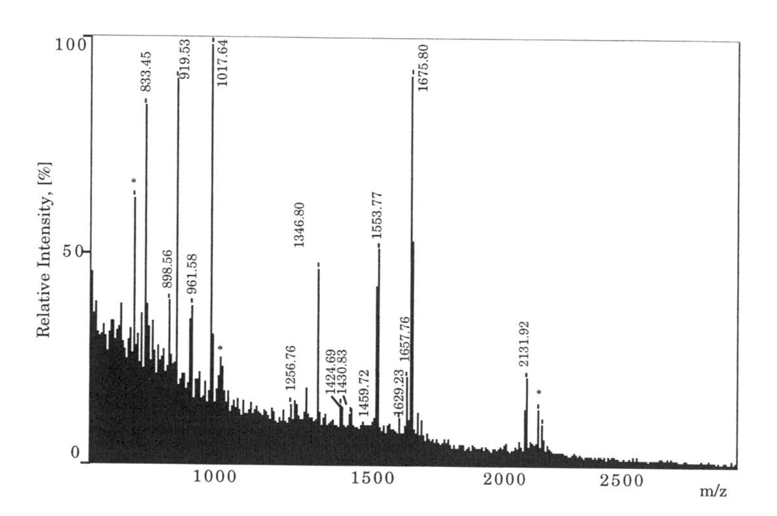

Extract

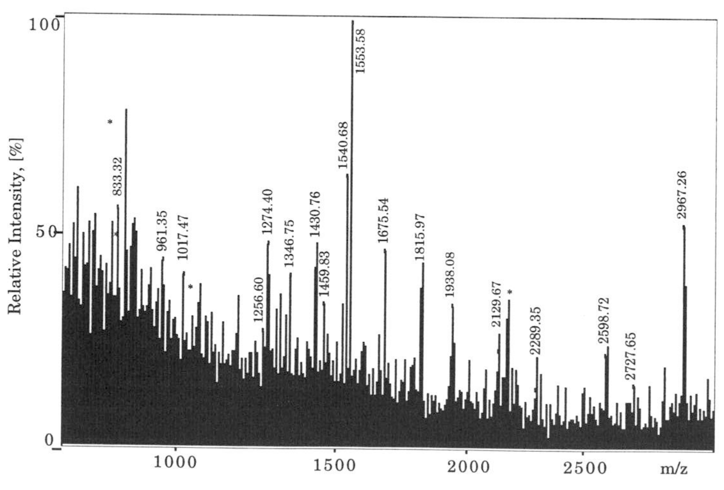

Fig. 3. MALDI mass spectrum acquired by direct analysis of an aliquot of the supernatant of in-gel digest of yeast heat shock protein Ssa2p (upper panel). Gel pieces were then extracted, extracts pooled together, dried down, redissolved in 5% formic acid, and an aliquot of the same volume was taken up and analyzed by MALDI (lower panel).

of masses of tryptic peptides) such as are encountered after *in-gel* digests is normally at the low femtomole level. Therefore when attempting to analyze Coomassie or even silver stained bands (spots) with MALDI MS, it is sufficient to load only a small aliquot of the digest (< 0.5 μL) to obtain the peptide map with good signal-to-noise ratio.

To assess the relative quality of the peptide maps obtained from *in-gel* digestion before and after peptide extraction the following experiment was performed. A 0.3 μL aliquot was taken directly from an *in-gel* digest of Coomassie stained protein band and analyzed by MALDI TOF (Fig. 3). The gel pieces were then extracted with 50 mM ammonium bicarbonate/acetonitrile and 5% formic acid/acetonitrile, the extracts combined and dried down in a vacuum centrifuge. The same extraction procedure is used routinely to prepare samples for sequencing by nano ES MS/MS. The extract was taken up in 10 μL of 5% formic acid (a volume similar to the volume of the buffer in the digest) and a 0.3 μL aliquot was analyzed by MALDI MS. Comparison of two peptide maps (Fig. 3) shows that direct analysis of an aliquot of *in-gel* digest provides a spectrum with better signal-to-noise ratio compared with the spectrum obtained from the peptide extract. Therefore for most of the samples containing known proteins, labor and time consuming extraction steps can be omitted thus increasing the throughput. Large protein complexes containing tens of subunits have been quickly dissected solely by this MALDI MS mapping approach [40].

The second remarkable feature of MALDI MS is its ability to identify proteins in mixtures [14]. Figure 4 shows an example of MALDI MS identification of two proteins co-migrating in a Coomassie stained band isolated by 1D electrophoresis. The ability to dissect mixtures rests on the high mass accuracy of MALDI MS. Usually five to six masses of peptide ions measured with mass accuracy better than 50 ppm are sufficient to identify the protein upon searching a comprehensive database. Once the matching protein sequence has been retrieved from the database, the spectrum is subjected to a "second pass search", or detailed comparison between measured peptide masses and possible peptide masses derived from the protein sequence. Peptide masses are calculated taking into account common protein modifications such as oxidation of methionine residues and acrylamidation of cysteine residues. The "ragged ends patterns" produced by incomplete tryptic cleavage of the sequence patterns consisting of two or more consecutive arginine and/or lysine residues [41] are also accounted for and used to validate the identification. Then the list of masses matching the already identified sequence in a second pass search is subtracted from the full list of masses measured in the spectrum. The database search is repeated with the list of remaining masses and if new matching database entries are found, the second pass search procedure is repeated again until all prominent peaks in the peptide map are assigned. To discern proteins in mixtures, high quality MALDI peptide maps are required. These peptide maps are not always available when analyzing proteins at low levels. MALDI MS peptide mapping offers high throughput protein identification, and therefore is applied as a first "screening" layer in the layered analytical approach. It is important that the screening technique should at least provide evidence that the particular sample is in fact

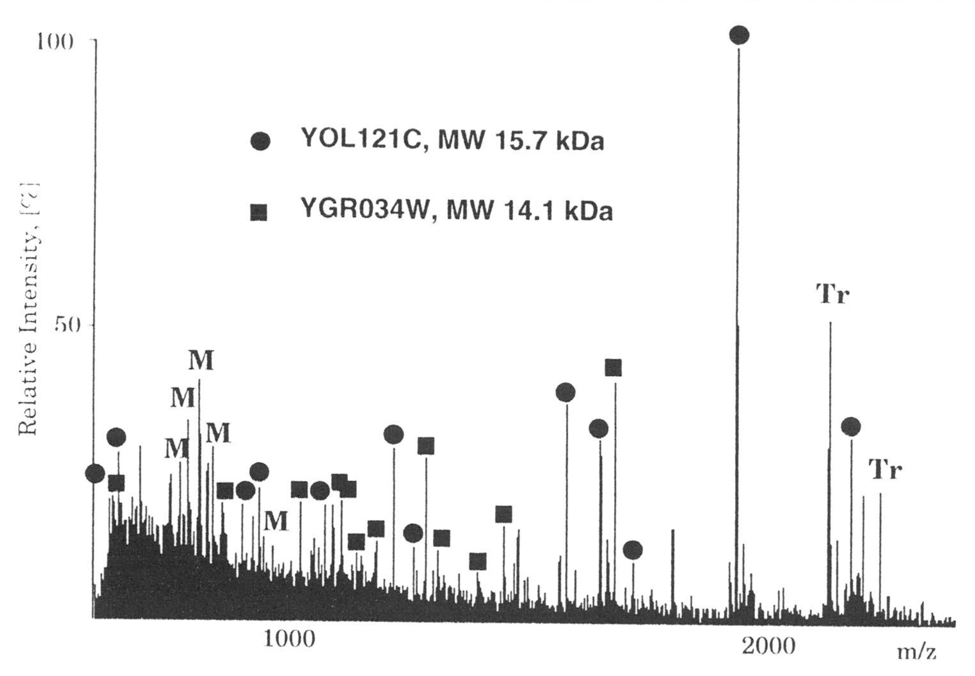

Fig. 4. Identification of proteins in a mixture by MALDI peptide mapping. A Coomassie stained band having apparent molecular weight 15 kDa was excised from a 1D polyacrylamide gel, in-gel digested with trypsin and 0.3 μL aliquot was analyzed by MALDI. Two proteins having similar molecular weights were identified in the mixture. Ions of peptides originating from the 40S ribosomal protein (gene YOL121c) and from the 60S ribosomal protein (gene YGR034w) are designated with filled circles and filled squares, respectively. Thirteen peptides match the sequence of YGR034w with mass accuracy better than 75 ppm, covering more than 45% of protein sequence. 11 peptides match YOLL121c with the same mass accuracy of 75 ppm, covering 68% of the sequence.

a mixture of proteins. Even if it would not be possible to identify the second component of the mixture solely by MALDI MS the results will "sound an alarm". Later this sample can be analyzed by nano ES MS/MS and the second component almost certainly will be identified (see discussion below).

Despite its advantages, protein identification by MALDI MS peptide mapping does also have some inherent limitations. Detection of at least 4-5 peptides from an *in-gel* digest of a protein is necessary to achieve unambiguous identification. However, this is not always possible, in particular in the case of small (less than 20 kDa) proteins, basic proteins or very hydrophobic proteins of any amount. If the protein amount is in the subpicomole range (silver stained protein bands) then the efficiency of protein digestion declines and very few peptides may be detectable in a MALDI peptide map. The resulting low sequence coverage may only allow tentative identification. Furthermore, it is not possible to search EST databases using MALDI peptide maps, because the vast majority of EST sequences are short and of poor quality. Although the feasibility of "*de novo*" peptide sequencing by post-source decay (PSD) analysis has been claimed [42], this application is very far from being routine for subpicomole levels of gel separated proteins.

Second Stage of the Layered Approach: Protein Identification by Nanoelectrospray Tandem Mass Spectrometry

Nano ES MS/MS as a second layer in the strategy efficiently complements MALDI peptide mapping. Protein identification by tandem mass spectrometry is based on matching the fragment spectrum of the peptide ion to the corresponding peptide sequence in a database [17, 43]. Thus one or a few peptides are sufficient for unambiguous identification. In our experience, nano ES MS/MS sequencing is the method of choice for protein identification when only femtomole amounts of gel separated protein are available and only few peptides can be recovered after *in-gel* digestion.

Practice has shown that the limit of detection is not ultimately determined by the sensitivity of the mass spectrometer. Peptide ion signals can still be present in the spectrum at "sequenceable" levels when they are already obscured by chemical noise. In this regard we found parent ion scanning — a feature of the triple quadrupole tandem mass spectrometer — to be vitally important for successful protein sequencing at the femtomole level [20, 44]. When operating in the parent ion scanning mode only those ions are recorded which produce upon their collisional fragmentation a daughter ion at m/z 86 — the immonium ion of leucine or isoleucine. This fragment ion is not produced by chemical background but is specific for peptides containing these abundant amino acid residues. The following case study (Fig. 5) demonstrates the utility of parent ion scanning for femtomole sequencing of protein mixtures. A silver stained band observed on a 1D gel after immunoprecipitation of a protein complex was *in-gel* digested. MALDI MS mapping, however, did not conclusively identify the protein. Nano ES MS/MS analysis revealed a number of peptide ions (peaks labeled in panel A) along with intense chemical background (unlabeled peaks). Upon sequencing and subsequent database searching, peptides (labeled Ig) were identified as tryptic peptides originating from immunoglobulins. The explanation for these peptides was an intense spot of reduced immunoglobulin chains which smeared along the electrophoretic lane and contaminated higher molecular weight bands. Sequencing of the doubly charged ion of peptide T_3 identified yeast protein Rrp4p (panel C). Application of the parent ion scan effectively eliminated chemical noise from the low m/z region of the spectrum (panel B). Along with a number of peptide ions originating from human cytokeratin (ubiquitous protein contamination at low levels), a peptide ion with m/z 521.8 was detected and subsequently fragmented. A peptide sequence tag assembled using the tandem mass spectrum conclusively identified another yeast protein, Rrp43p.

The case discussed above serves as a good example to demonstrate the specific difficulties encountered in low level sequencing. Protein mixtures are a common occurrence and often the target protein is not a major component of the mixture. Human and sheep keratins are by far the most common and ubiquitous contaminants at low levels. Even if all operations during the sample preparation have been done very carefully, it is our experience that a certain background level of keratin peptides always occurs and will likely affect the analysis (Fig. 6).

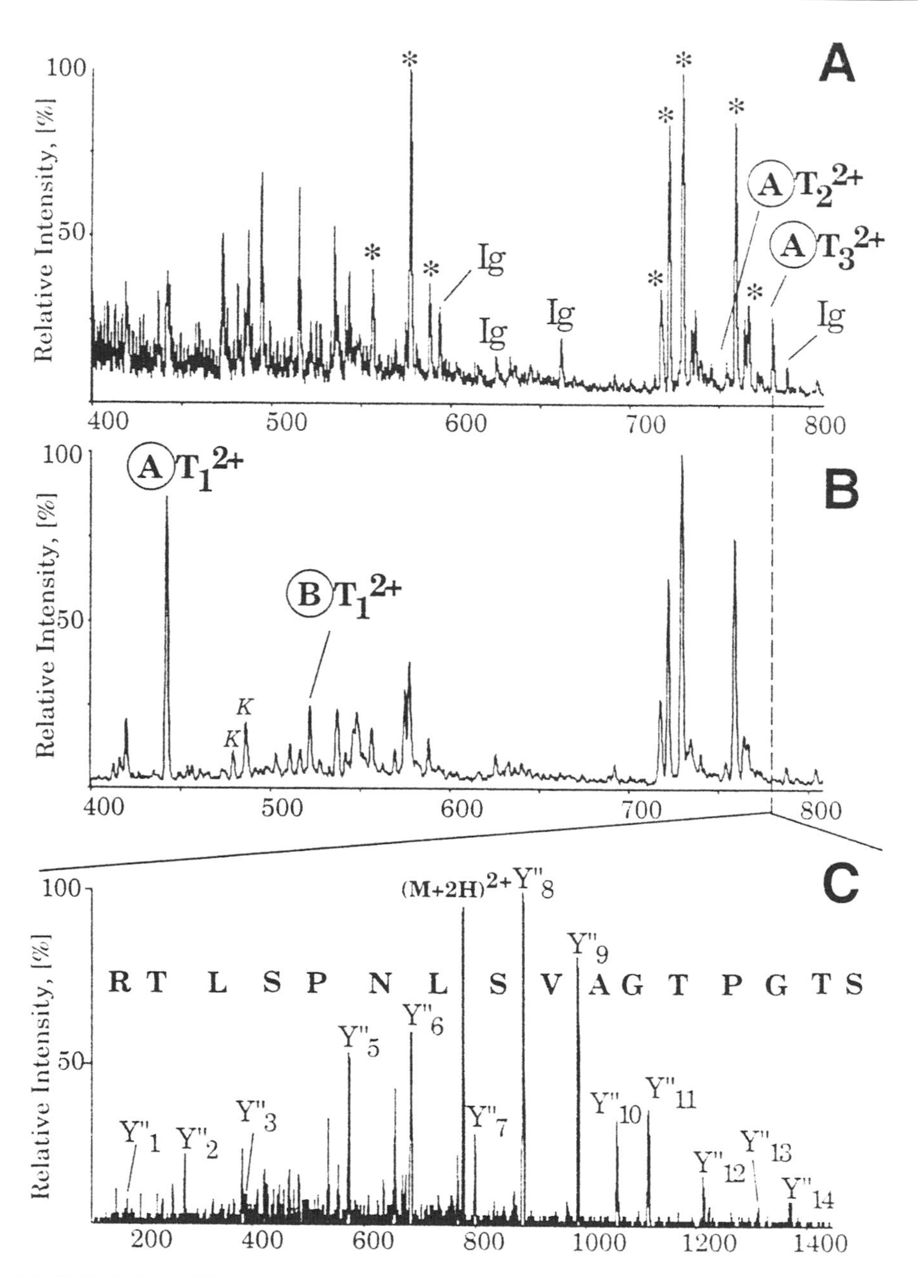

Fig. 5A, B & C. Identification of subunits of the yeast Rrp - complex. A - mass spectrum of an in-gel digest of a silver stained band with apparent molecular weight ca. 40 kDa. Peaks designated with asterisks are trypsin autolysis products. Peaks designated with Ig originate from antibodies used in immunoprecipitation of the Rrp-complex. Peptides designated with circled A originate from yeast protein Rrp4p. The tandem mass spectrum of the doubly charged ion T_3 is shown in panel C. A parent ion scan spectrum (scanning for precursor ions of daughter ion m/z 86) of the digest is shown in panel B. Along with cytokeratin peptides (K), and another peptide from Rrp4p (labeled with circled A) the ion with m/z 521.8 (designated with circled B) was sequenced (panel D) and identified as protein Rrp43p - another component of the Rrp-complex.

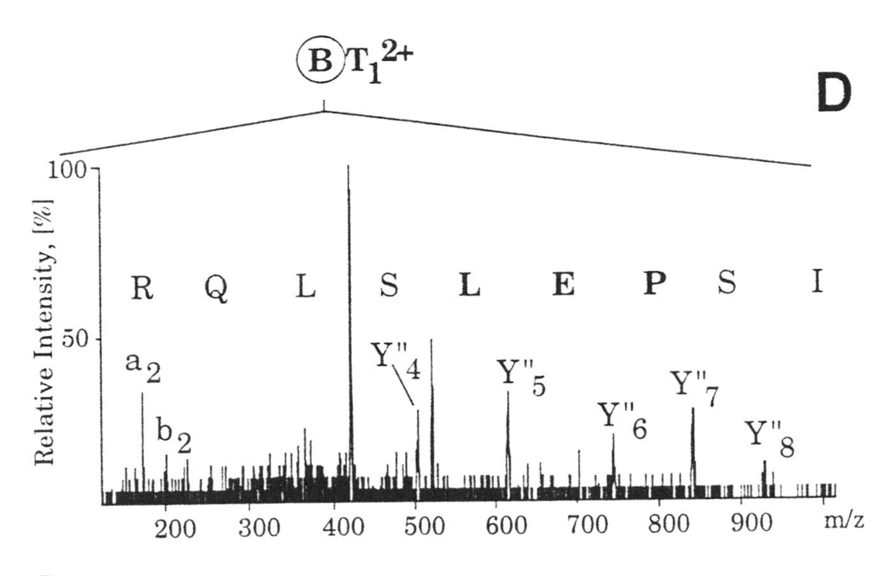

Fig. 5D.

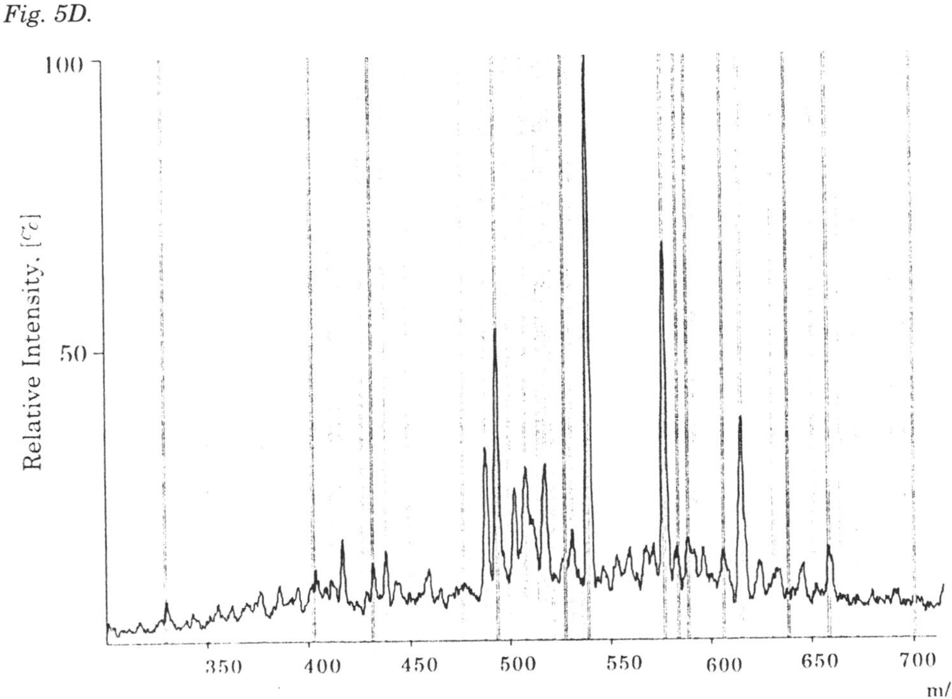

Fig. 6. Parent ion scan (daughter ion m/z 86) spectrum of a blank sample demonstrating the pattern of ubiquitous contaminants encountered in sequencing at low levels. Trypsin autolysis products and peptides from human and sheep keratins are highlighted.

Keratin background peaks are real peptides and thus cannot be eliminated by any type of a selective scanning. Therefore, when sequencing low amounts of protein it is beneficial to fragment as many ions detected by parent ion scanning as possible. Later during the interpretation of the acquired data, keratin

Full scanning region fragment spectrum, **9 spectra accumulated in 45 sec**

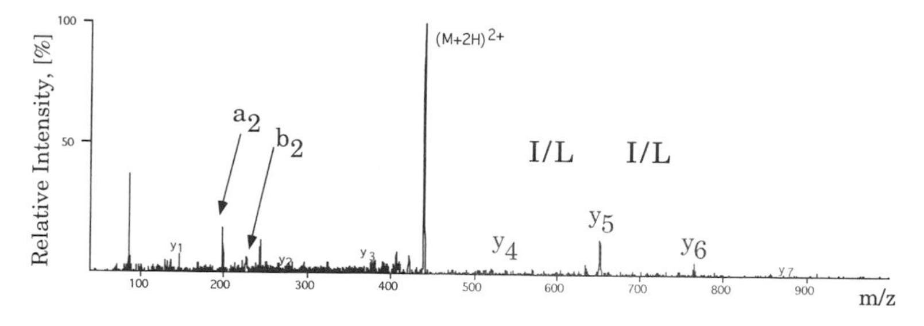

Narrow scanning region fragment spectrum, **31 spectra accumulated in 70 sec**

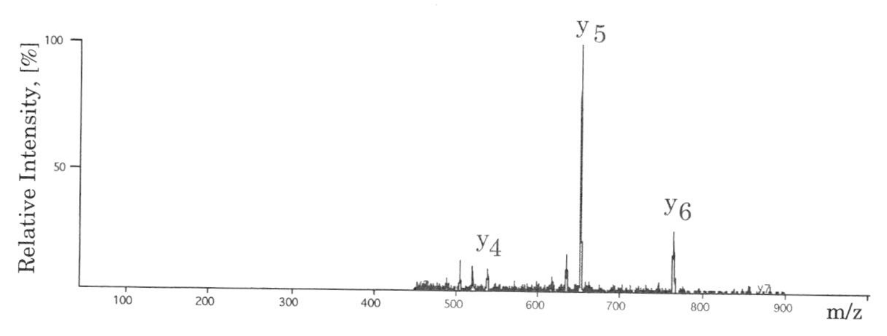

Fig. 7. Partitioning of the acquisition time for sequencing of low abundance peptide ions using a triple quadrupole mass spectrometer.

peptides and minor trypsin autolysis products can be sorted quickly. Occasionally some of the peptides of the protein to be identified will be found at the same m/z position as known keratin or trypsin contaminants (see example below).

Parent ion scanning helps to suppress chemical background, thus revealing the genuine multiply charged peptide ions amenable to subsequent sequencing. However when the peptide ion is isolated by the first quadrupole of the triple quadrupole mass spectrometer, large amounts of background ions are also co-isolated. Fortunately, most of the ions of the chemical background are singly charged and they are not fragmented at the collision energies sufficient to fragment multiply charged ions. Therefore, the intensity of the "precursor ion" observed in the fragment spectrum (of which genuine peptide ions constitute only a small part) should not be used as a guide to optimize the collision energy. If higher collision energies are applied, singly charged background ions produce singly charged fragment ions populating the fragment spectrum ***below*** the m/z of the multiply charged precursor ion. The part of the spectrum ***above*** the precursor ion remains relatively clean (Fig. 7). Therefore, for protein identification at low levels it is advantageous to acquire two tandem mass spectra from each precursor ion. The first tandem mass spectrum covers the entire m/z range, and only a small number of scans is accumulated. In the second spectrum, only a relatively narrow range of m/z above the parent ion is scanned and a large number of scans is accumulated, allowing the acquisition of a spectrum with very good signal-to-noise ratio. Then these two spectra are overlaid. The tandem mass spectrum of the region above the precursor ion allows the construction of reliable peptide sequence tag even if fragment ions have been recorded with poor ion statistics. When the matching peptide is found upon searching a database, the full tandem spectrum containing the complete ion series (along with fragments of background ions) is used to confirm the match. Rational partitioning of the acquisition time is important if a triple quadrupole mass spectrometer (a machine with a scanning mass analyzer) is used.

Peptide sequence tags can be applied equally for searching protein and EST databases. The identification of the components of the human signalsome, a complex of kinases involved in phosphorylation of the inhibitor (IκB) of NFκB transcription factor, is discussed here as a case study [27] (Fig. 8). A Coomassie stained 87 kDa band was analyzed by nanoelectrospray tandem mass spectrometric sequencing. Tandem mass spectra of peaks labeled with A quickly identified conserved helix-loop-helix ubiquitous kinase in a protein database (later called IKK-1) (GenBank acc. number U22512). The spectrum of the digest also contained a number of prominent peaks (labeled with B in Fig. 8) tandem mass spectra of which did not match any peptide sequence in a database. However, when a comprehensive EST database was searched, three EST clones were identified (Fig. 8B). When the full length sequence of the protein (later named IKK-2) was obtained using the relevant EST clones, it emerged that IKK-2, like IKK-1, is a helix-loop-helix protein containing a kinase sequence motif. IKK-2 shares only 59% identity with IKK-1 and both kinases are components of NFκB-activating protein complex. Recent findings suggest that these kinases are also differently regulated [45].

The EST database provides an important resource for identification of yet unknown biologically interesting genes [46]. The large multisubunit protein complex of human spliceosome-associated proteins, comprising more than fifteen novel proteins, has been deciphered using mass spectrometric identification of relevant EST clones [47].

Third Stage of the Layered Approach: "De Novo" Sequencing by Nanoelectrospray Tandem Mass Spectrometry

If the protein of interest has not been identified in a protein or EST database, nano ES MS/MS provides a tool to sequence tryptic peptides "*de novo*". The remaining part of the peptide material recovered after *in-gel* digestion (see above) is esterified with 2 M HCl in anhydrous methanol [48], and the unseparated digest mixture is sequenced again by nano ES MS/MS. Upon esterification, methyl groups are attached to side chain carboxyl group of aspartic and glutamic acid residues and to the carboxyl group of the C-terminal amino acid residues of peptides. Therefore the m/z of peptide ions is shifted upon esterification by 14(n+1)/z, where n is the number of aspartic and glutamic residues in the peptide and z is the charge of the peptide ion (Fig. 9). The same peptide which was sequenced from an underivatized digest has to be identified in the mixture after derivatization, and fragmented. Sequences of individual peptides are deduced by

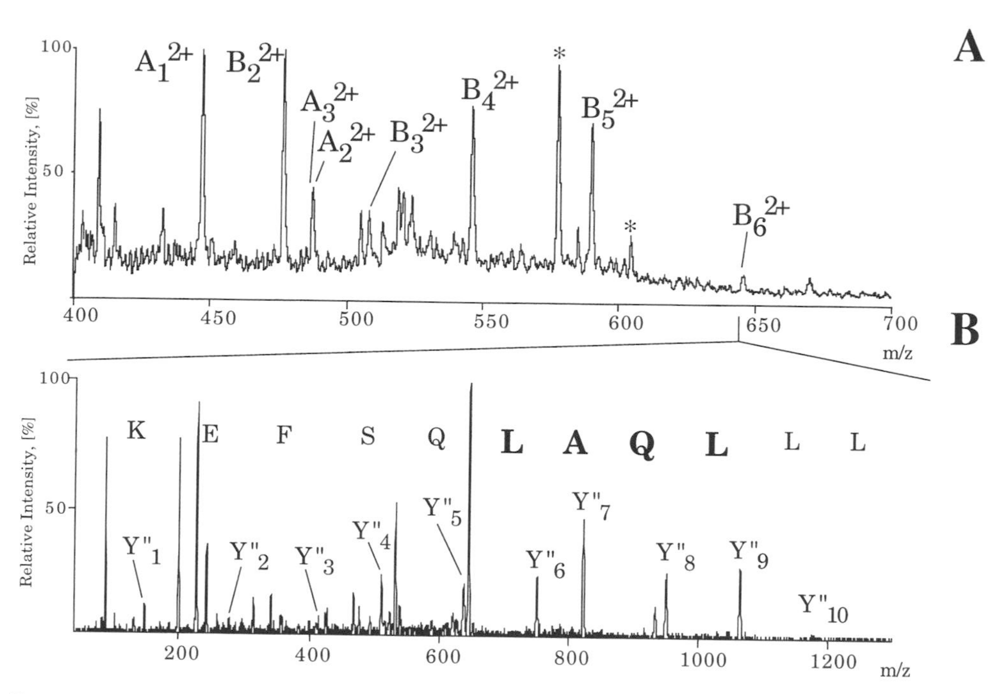

Fig. 8A & B. Sequencing of a 87 kDa protein with IκB kinase activity (Panel A). Peptide peaks designated with A originate from U22512 helix-loop-helix ubiquitous kinase (IKK-1). Tandem mass spectra acquired from the ions designated with B (panel B) did not identify any protein upon searching a comprehensive protein database (Panels A and B). However searches in an EST database identified three relevant EST clones (panel C). The full length sequence of novel protein (IKK-2) comprises all peptides which were sequenced by tandem mass spectrometry (in bold and underlined).

considering both the precise mass difference between adjacent fragment ions in the series ***and*** the characteristic mass shift induced by esterification (Fig. 9B). This approach is time consuming and requires experience in the interpretation of tandem mass spectra. However, it produces long and accurate peptide sequences. Even if only low to sub pmol amounts of protein are available on a gel, enough sequence information can be generated to allow successful cloning of the cognate gene. More than 10 proteins have been cloned by a PCR approach using degenerate oligonucleotide probes designed exclusively from the peptide sequences determined by nano ES MS/MS in our laboratory [28, 29, 49- 51].

As an example, less than ten picomoles of p123 catalytic subunit of telomerase holoenzyme was purified from ciliated protozoan *Euplotes aediculatus* [52] and separated by 1D electrophoresis. Half of the Coomassie stained band was subjected to nano ES MS/MS. sequencing. Altogether 24 peptide ions were fragmented. Complete sequences of 14 tryptic peptides were determined. The peptides consisted of 6 to 16 amino acid residues with an average length of 10.5 residues, and covered altogether more than 150 amino acid residues. Additionally, two peptides were partially sequenced (sequence stretches > 5 amino acid residues were obtained) and sequence tags were obtained for the remaining 8 peptide ions. The protein was cloned by PCR using degenerate primers designed

C

AA326115, Frame #3

B6 **B2**

PMGxKQGGTLDDLEEQARELYRRLREKPRDQRTEGDSQEMVR**LLLQAIQSFEK**KVR**VIYTQ**
LSKTxVCKQK**ALELLPK**VEEVVSLMNEDEKTVVRLQEKRQKELWNLLKIA

B1

AA480228, Frame #1

B1

KQK**ALELLPK**VEEVVSLMNEDEKTVVRLQEKRQKELWNLLKIACSKVRGPVSGSPDSMNASRL
SQPGQLMSQPSTASNSLPEPAKKSEELVAEAHNLCTLLENAIQDTVREQDQSFTVTACVRLLRF
HVLSFYGKIEEKMEMQSGIILNL

R06591, Frame # 1

GGCIxPGLxPQPxLPRPGAHPLPRVPADRVSTTSV*AGHLP*QRRHVGLGNERA
LGTGGFGNVIR WHNQADPPQCGxLPRCPLGDSELLAAN*LALLWHx

B4

IKK - 2, full length sequence

MSWSPSLTTQTCGAWEMKER**LGTGGFGNVIR**WHNQETGEQIAIKQCRQELSPRNRERWCLEI
QIMRRLTHPNVVAARDVPEGMQNLAPNDLPLLAMEYCQGGDLRKYLNQFENCCGLREGAILTLL
SDIASALRYLHENRIIHR**DLKPENIVLQQGEQR**LIHKIIDLGYAKELDQGSLCTSFVGTLQY
LAPELLEQQKYTVTVDYWSFGTLAFECITGFRPFLPNWQPVQWHSKVRQKSEVDIVVSEDLNGT
VKFSSSLPYPNNLNSVLAERLEKWLQLMLMWHPRQRGTDPTYGPNGCFK**ALDDILNLK**LVHIL
NMVTGTIHTYPVTEDESLQSLKARIQQDTGIPEEDQELLQEAGLALIPDKPATQCISDGKLNEG
HTLDMDLVFLFDNSKITYETQISPRPQPESVSCILQEPKRNLAFFHLRKVWGQVWHSIQTLKED
CNRLQQGQRAAMMNLLRNNSCLSKMKNSMASMSQQLKAKLDFFKTSIQIDLEKYSEQTEFGITS
DKLLLAWREMEQAVELCGRENEVKLLVERMMALQTDIVDLQRSPMGRKQGGTLDDLEEQARELY
RRLREKPRDQRTEGDSQEMVR**LLLQAIQSFEK**KVR**VIYTQLSK**TVVCKQK**ALELLPK**VEE
VVSLMNEDEKTVVRLQEKRQKELWNLLKIACSKVRGPVSGSPDSMNASRLSQPGQLMSQPSTAS
NSLPEPAKKSEELVAEAHNLCTLLENAIQDTVREQDQSFTALDWSWLQTEEEEHSCLEQAS*

Fig. 8C.

on the basis of the peptide sequences obtained. After the full length sequence of the protein was obtained only two discrepancies between the peptide sequences and corresponding sequences deduced from the cDNA were observed. All partial sequences and peptide sequence tags were retrospectively matched to the full length protein sequence, thus confirming the identity of the discovered gene [29].

Although sensitive and accurate, the approach employing esterification and multiple tandem mass spectrometric analysis does not lend itself to characterization of protein complexes because of the limited throughput. An alterna-

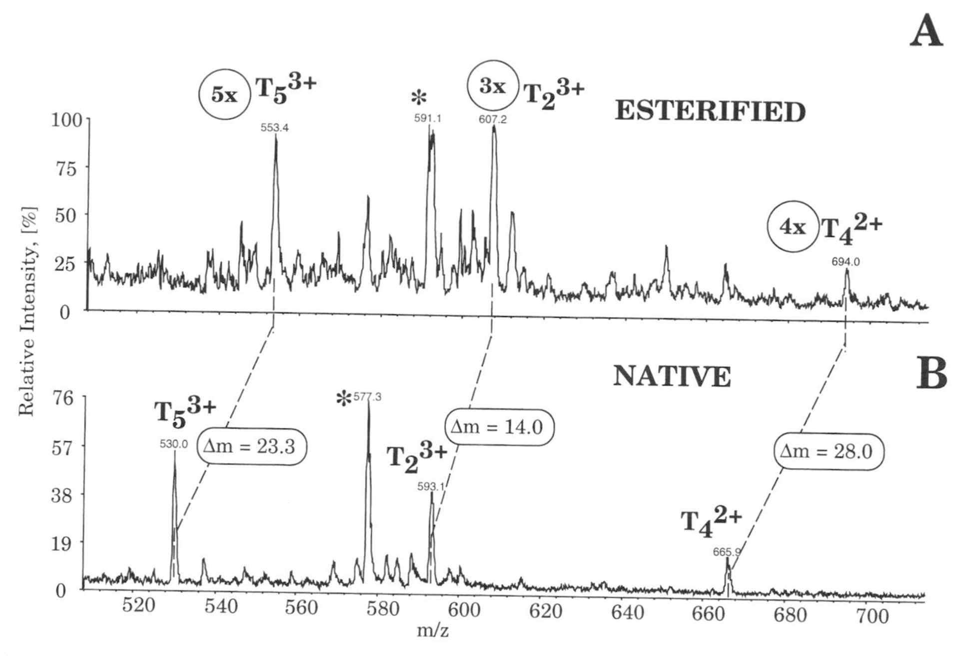

Fig. 9 (A & B). Sequencing of unknown protein by nano ES MS / MS. Part of the spectrum of the unseparated in-gel digest of the protein before (panel B) and after esterification with methanol (panel A). Ions of the same peptide before and after esterification are matched considering the shift in m / z, the charge and the fact that the attachment of one methyl group shifts the peptide mass by 14 Da. Numbers of attached methyl groups are circled. Tandem mass spectra of matching ions are compared and the amino acid sequence is deduced with software-assisted interpretation.

tive strategy for "*de novo*" peptide sequencing based on a combination of peptide isotopic labeling followed by sequencing on a hybrid quadrupole time-of-flight mass spectrometer [53] has been suggested [54]. Briefly, proteins are digested *in-gel* with trypsin in a buffer containing 50% of $H_2{}^{16}O$ and 50% $H_2{}^{18}O$ (v/v). Upon hydrolysis, C-terminal carboxyl groups of peptides incorporate ^{18}O atoms in the same proportion as ^{16}O atoms. The entire isotopic cluster of the peptide precursor ion is selected by a quadrupole mass analyzer and fragmented in the collision cell. Fragment ions which include the C-terminus of the peptide are detected as characteristic isotopic features — doublets split by 2 Da. The other fragment ions which do not contain the C-terminal carboxyl group have the normal isotopic pattern (Fig. 10). The peptide sequence can be quickly "read out" by considering precise mass differences between the adjacent characteristic isotopic multiplets corresponding exclusively to Y" ions [54].

A remarkable feature of the quadrupole time-of-flight mass spectrometer is that high resolution (> 8000 FWHM) and high mass accuracy (better than 20 ppm) are achieved without compromising sensitivity. High mass accuracy allows sequence ambiguities between amino acid residues having the same nominal mass to be resolved, such as phenylalanine and methionine-sulfoxide (mass difference 0.033 Da), as well as glutamine and lysine (mass difference 0.037 Da), thus gaining additional certainty of sequence assignment [55].

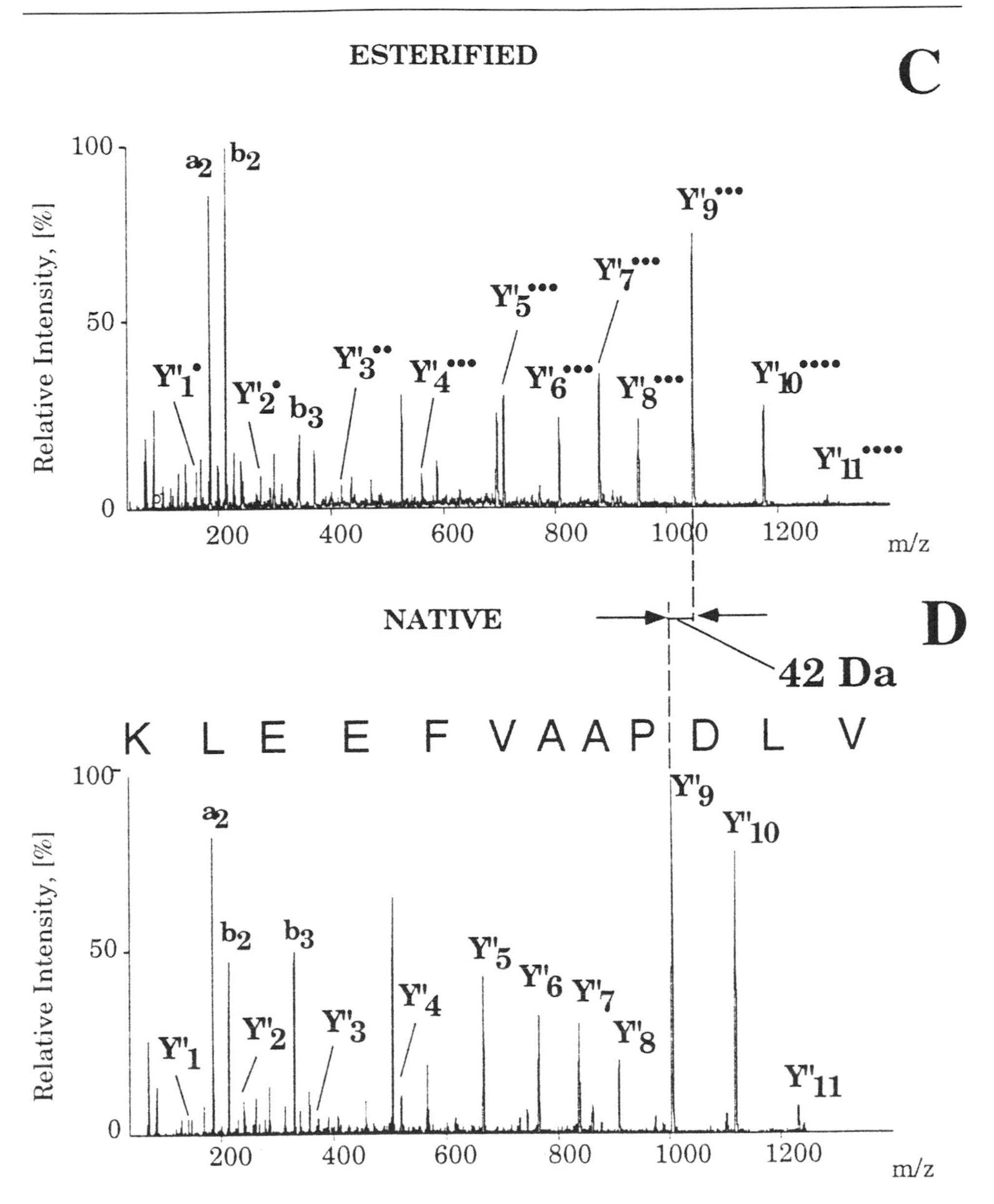

Fig 9 (C&D). Tandem mass spectra of esterified and native forms of peptide T_4 are shown in panels C and D. The precise mass difference between adjacent Y" ions is considered, as well as the characteristic mass shift resulting from esterification (14 Da per each glutamic acid and aspartic acid residue plus 14 Da because of the C-terminal carboxyl group). The number of methyl group attached to Y"-ions is indicated by filled circles.

MASS SPECTROMETRIC ANALYSIS OF COMPLEXES IMPLICATED IN CELL CYCLE CONTROL IN BUDDING YEAST

Budding yeast *S. cerevisiae* provide an exciting opportunity to study important cellular processes by biochemical rather than genetic approaches. The

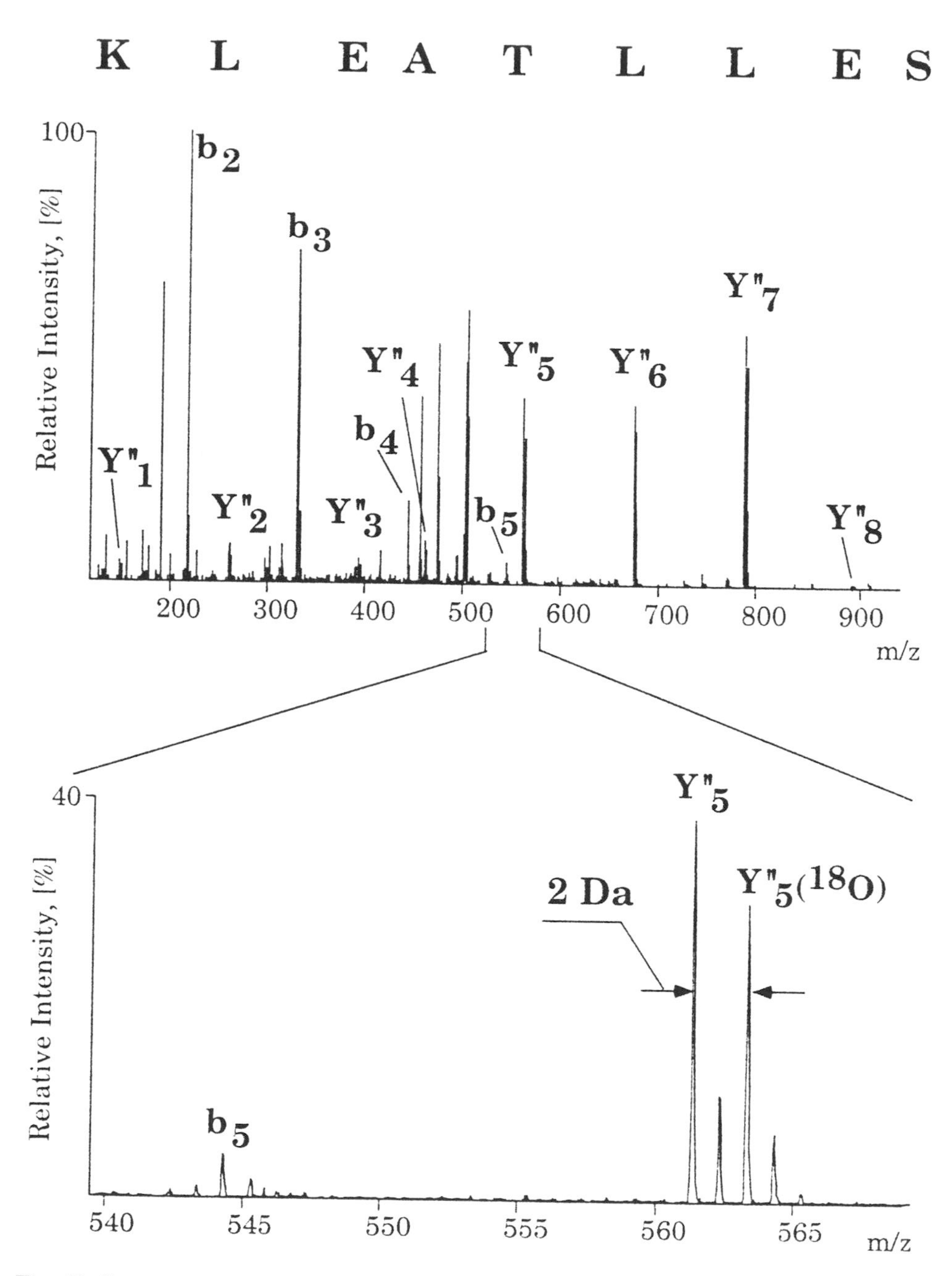

Fig. 10. Sequencing of isotopically labeled peptide using a hybrid quadrupole-TOF mass spectrometer (QqTOF). Peptide fragments containing the C-terminus (mostly Y"-ions in the case of a tryptic peptide) are detected as a doublet split by 2 Da, whereas ions not containing the C-terminal carboxyl group (in particular b-ions) have the normal isotopic pattern. Therefore peptide sequence could be quickly read out considering the precise mass difference between isotopically labeled Y"-ions.

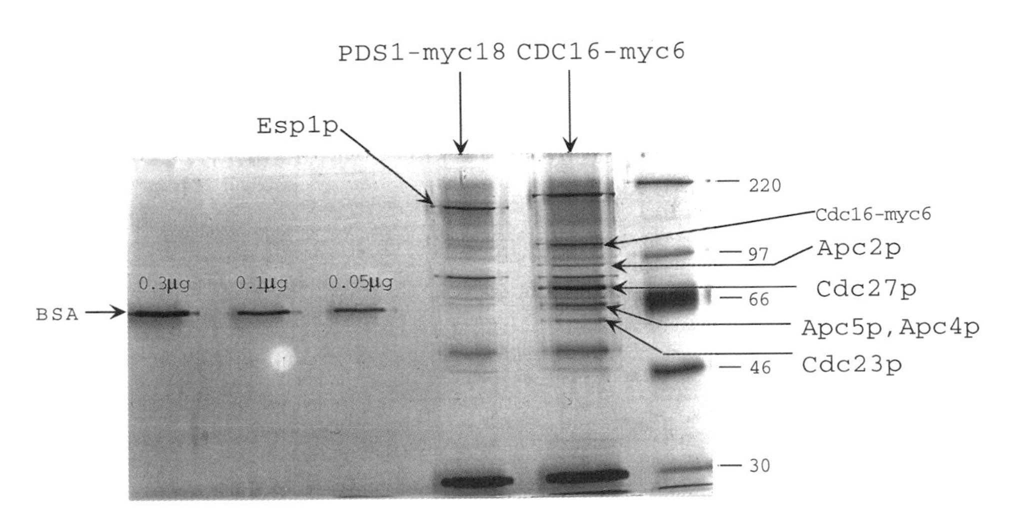

Fig. 11. APC subunits are immunoprecipitated and separated by 1D gel electrophoresis (lane at the right). Proteins were visualized by silver staining. The first three lanes from the left contain BSA in indicated amounts as a reference. Protein co-immunoprecipitated with tagged Pds1p was identified by mass spectrometry as Esp1p (middle lane).

sequence of the entire yeast genome [56] and recent advances in mass spectrometry allow rapid identification of interesting gene products even if they are available in low nanogram amounts in polyacrylamide gels. Thus it is often more straightforward to identify biologically important genes via isolation and characterization of protein complexes than by sophisticated genetic screening.

Here we applied the strategy outlined above to characterize protein complexes implicated in the regulation of the cell division machinery in budding yeast.

Identification of the Components of Anaphase Promoting Complex (APC)

Cell cycle progression in eukaryotes depends on the degradation of mitotic cyclins, which is in turn controlled by ubiquitin-mediated proteolysis [3]. Ubiquitin-mediated proteolysis of proteins containing a specific sequence motif – the "destruction box" — depends on a multisubunit protein assembly called Anaphase Promoting Complex (APC) or cyclosome, which functions as a cell cycle regulated ubiquitin-protein ligase. First subunits of APC from budding yeast, Cdc16p, Cdc23p and Cdc27p [57, 58], and later two additional subunits, Apc1p and Cdc26p [59], were discovered by genetic screening for mutants defective in cyclin degradation. However it was clear that yeast 32S APC comprises more proteins. Therefore the genes encoding for the known subunits were Myc-tagged and the complex was immunoprecipitated from a whole cell lysate. APC is not abundant in budding yeast, therefore only 10-20 ng of gel

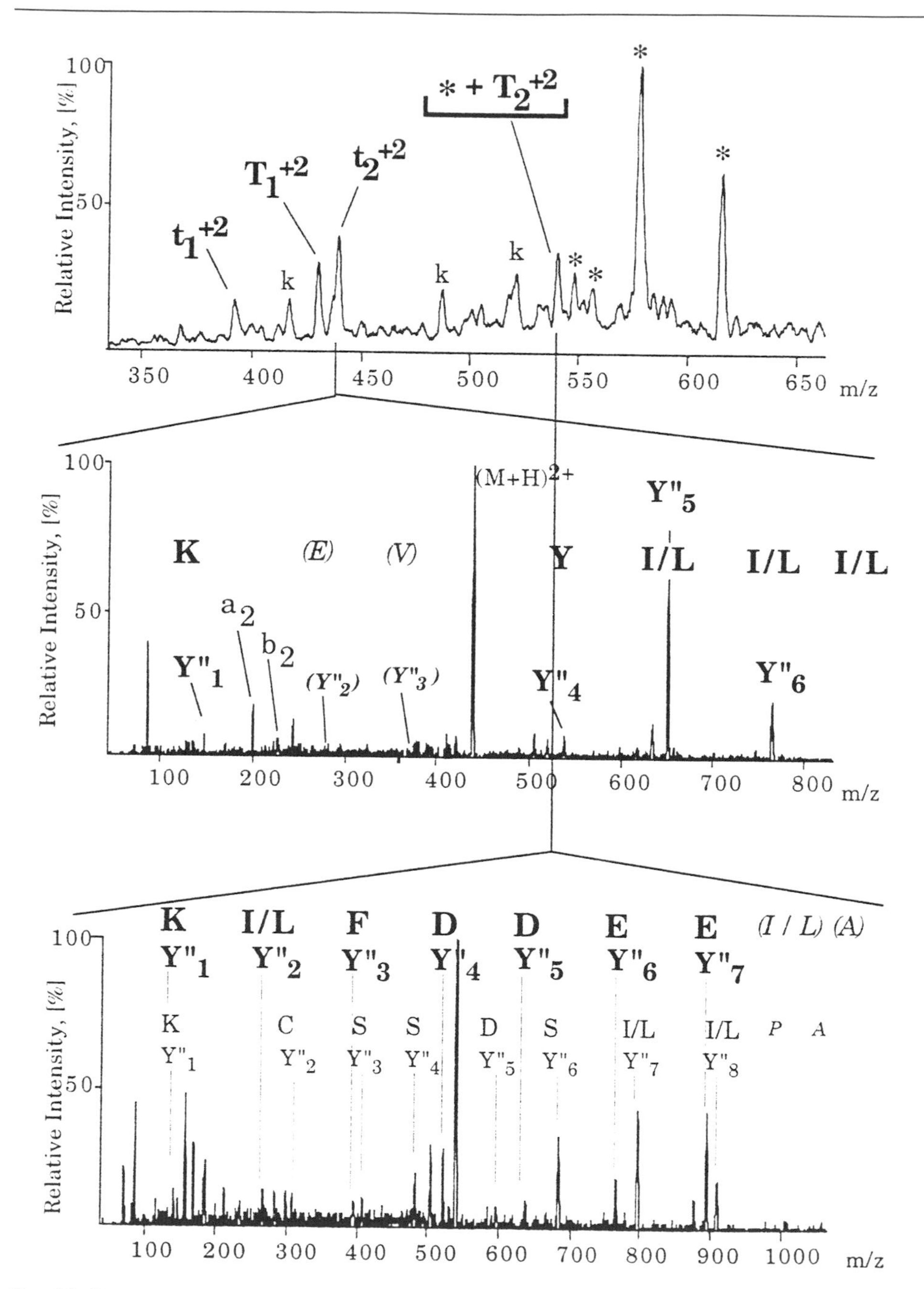

Fig. 12. Sequencing of APC 80 kDa APC subunit (Fig. 11). The upper panel shows a parent ion scan spectrum (daughter ion m/z 86) of the digest of the 80 kDa band. Ions designated with asterisks are trypsin autolysis products. The middle panel shows the tandem mass spectrum of the doubly charged t_2 ion which identified gene YDR118w (Apc4p). Interpretation of the tandem mass spectrum in the lower panel acquired from T_2 ion revealed that it contains two non-overlapping series of Y" ions. One ion series identified the trypsin autolysis product APLLSDSSCK (plain letters), whereas the other series identified gene YOR249c (Apc5p).

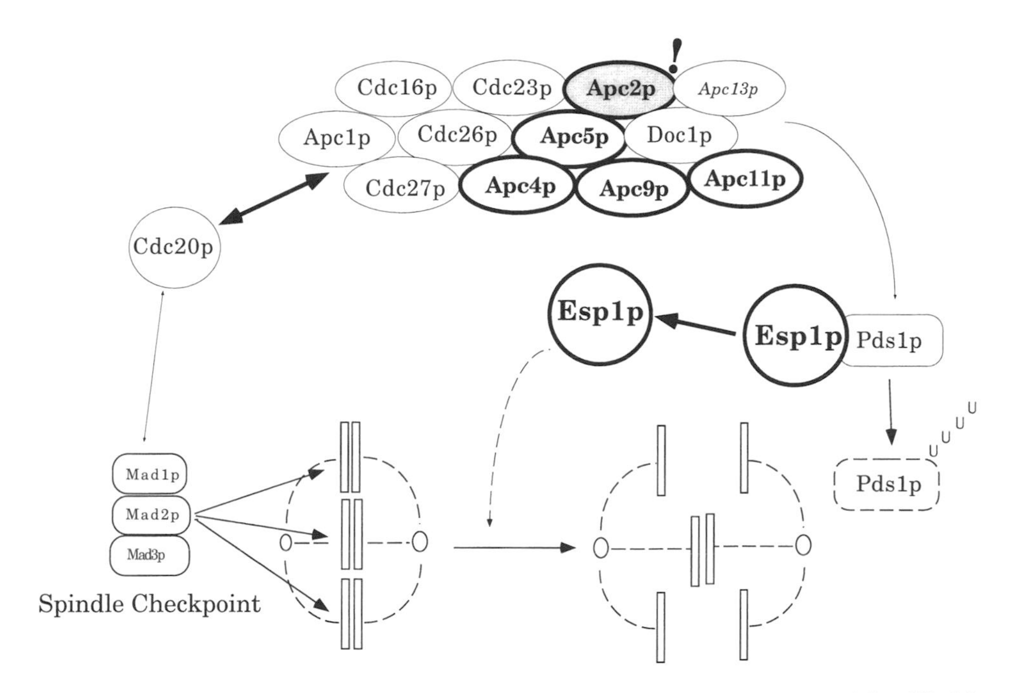

Fig. 13. A scheme of novel protein subunits of APC and protein interactions identified by mass spectrometry (in bold). Known proteins and interactions are in plain text. Apc2p is highlighted because it shares substantial homology to cullins.

separated protein subunits were obtained as a result of a large scale purification started from 5 x 10^{10} cells (Fig. 11). Protein bands were identified by nano ES MS/MS (Fig. 12). Because of the very low amount of protein material available, no peptide ions apart from the known trypsin autolysis products were observed in the spectrum of the digest (data not shown). Parent ion scanning for m/z 86 revealed a number of low intensity peptide ions which were sequenced in turn by tandem mass spectrometry. To obtain better ion statistics for very weak signals of fragment ions, partitioning of the acquisition time was applied. The tandem mass spectrum of the t_2 ion identified the gene YDR118w (later called APC4). Interpretation of the fragment spectrum acquired from T_2 ion (m/z 539.8) was less straightforward. Two non-overlapping ion series were deduced from the high m/z region of the spectrum. A peptide sequence tag assembled from one of the series of fragment ions identified a trypsin autolysis product, whereas the sequence tag assembled from the other fragment series identified gene YOR249c (later called APC5). Peptide APILSDSSCK (molecular weight 1076.5) (C stands here for cysteine-S-acetamide) and peptide ALEEDDFLK (molecular weight 1078.5) were both detected as doubly charged ions and their peaks were not resolved by parent ion scanning. The reason for this is that for parent ion scanning the first (scanning) quadrupole of the triple quadrupole machine was operated under low resolution settings to allow maximal transmission of parent ions to increase sensitivity.

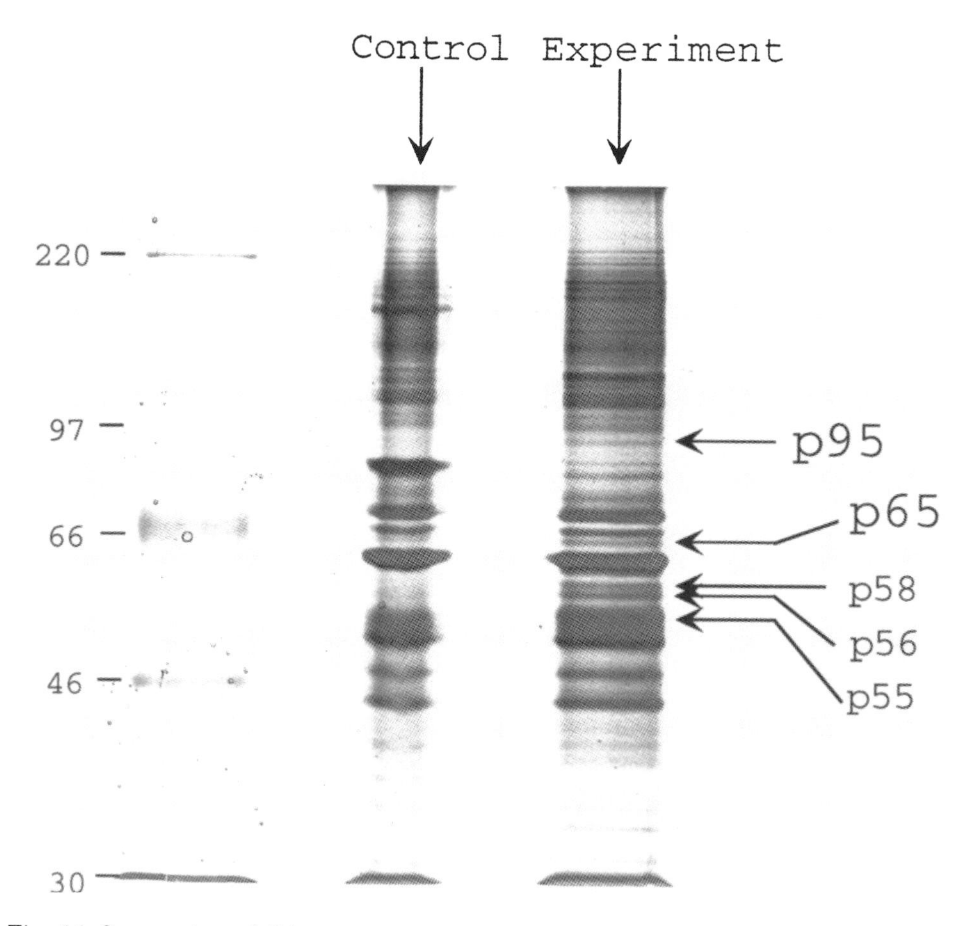

Fig. 14. Sequencing of Cdc20p interacting proteins identified as known APC subunits. Bands p95 and p65 were sequenced by nano ES MS / MS and identified as known members of the APC complex Cdc16p and Apc5p, respectively.

The other bands containing APC subunits were sequenced in a similar fashion. Altogether five new components of budding yeast APC were identified by mass spectrometry (Fig. 12).

APC Promotes Separation of Sister Chromatids via Release of Esp1p

Loss of cohesion between sister chromatids depends on the proteolysis of Pds1p mediated by APC [3]. However it was shown that Pds1p is not physically associated either with chromosomes or with cohesins (chromosomal proteins held together by sister chromatids), and therefore must fulfill its separating function in an indirect way via the interaction with other yet unknown protein(s).

To identify the interaction partners of Pds1p, its gene was tagged and the protein was immunoprecipitated from whole cell lysate. The protein in the silver stained band (Fig. 11) was sequenced by nano ES MS/MS and identified as a 180 kDa protein, Esplp [22]. Further biological experiments showed that Pdslp inhibits sister chromatids separation, whereas Esplp promotes it. Thus it has been concluded that APC promotes separation of chromatids by initiating the destruction of Pdslp, acting as an inhibitor of a specialized sister-separating factor Esplp (Fig. 13).

Cdc20p Links Spindle Checkpoint with APC

The spindle checkpoint improves the fidelity of chromosome segregation by delaying anaphase until all chromosomes are correctly aligned on the mitotic spindle [60]. It was shown that a 67 kDa protein, Cdc20p, associates with the components of the checkpoint Madlp, Mad2p, Mad3p [24]. To identify other interaction partners of Cdc20p its gene was tagged and the protein was immunoprecipitated from whole cell lysate (Fig. 14). Proteins p95 and p65 were sequenced by nano ES MS/MS (Fig. 15).

All prominent peptide ions observed in the parent ion scan spectrum of p65 digest were sequenced by nano ES MS/MS. The peptide ions designated as T_a identified a 60 kDa component of chaperonin-containing T-complex (TriC). Peptide ion T_b identified Pablp, a 64 kDa poly(A)-binding protein. Obviously these two abundant proteins were contaminations, co-purified with the tagged protein . From our experience in sequencing at low levels we knew that peptide ions with m/z 417.0, m/z 438.2 and m/z 487.6 most likely are tryptic peptides originating from human cytokeratins. However, they were also sequenced following the rule of thumb that all prominent peaks in the spectrum should be fragmented if possible. The peptide ion with m/z 417.0 was indeed identified as a doubly charged ion of human keratin peptide SISISVAR. But the fragment spectrum of peptide ion with m/z 517.4 contained two non-overlapping series of Y"-ions, identifying peptide TAEVSSSLLK from the known APC component, Apc5p, and a peptide from human keratin, TLLEGEESR (Fig. 14). Similarly, the fragment spectrum of the peptide ion with m/z 438.2 contained the other Apc5p peptide, LINQYVK, together with human keratin peptide SLVNLGGSK. Thus unambiguous identification of Apc5p has been achieved in spite of the fact that the protein band contained the two other proteins of similar molecular weight, and that it was furthermore contaminated with human keratins.

Similarly, protein p95 was identified as Cdc16p, also known to be a component of APC.

Thus we conclude that Cdc20p physically interacts with APC and constitutes a link between the spindle checkpoint and APC (Fig. 13).

Conclusions and Outlook

As shown in this chapter, robust and sensitive methods now exist for the unambiguous identification of gel separated proteins. We have focused here on application of these techniques to the identification of the components of functional protein complexes. As has been shown, biochemical purification now provides a powerful supplement and in some cases replacement for genetic

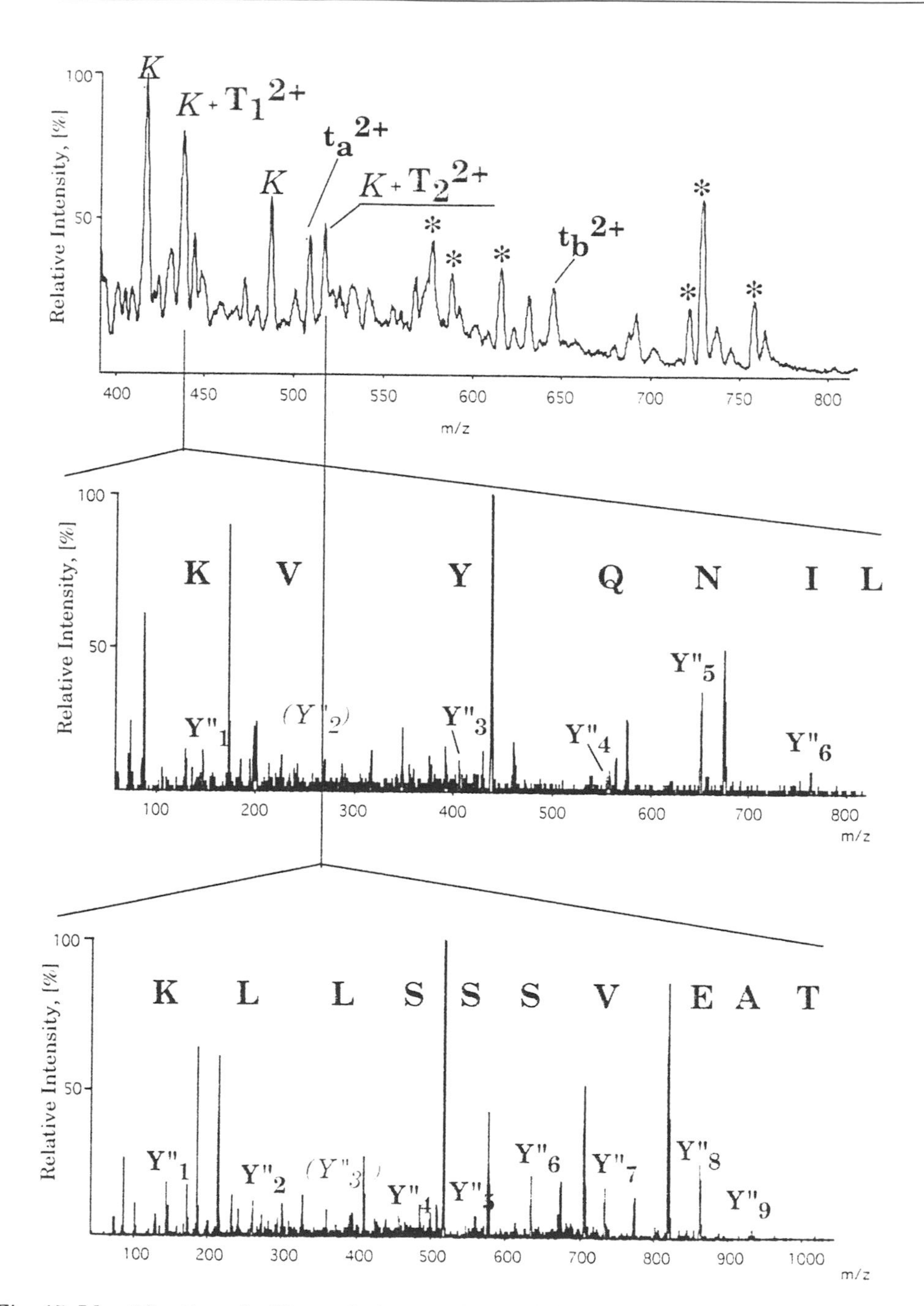

Fig. 15. Identification of p65 protein interacting with Cdc20p. The upper panel shows the parent ion scan spectrum (daughter ion m/z 86) of the unseparated digest. All prominent peaks in the spectrum were sequenced. Peptide ions t_a and t_b identified the 60 kDa chaperonine and poly(A)-binding protein, respectively. Peptide ions T_1 and T_2 (middle and lower panel) both identified Apc5p (series of Y"-ions labeled in the spectra). Ion series which are not labeled belong to human cytokeratin peptides.

screening techniques. As sophisticated mass spectrometric techniques become available to most biologists over the next few years, we expect an explosion of biological insight based on biochemical techniques coupled to mass spectrometric "read out".

Acknowledgements

We are grateful to members of the laboratories headed by Dr. Kim Nasmyth (Research Institute for Molecular Pathology, Vienna), Dr. D. Tollervey (EMBL, Heidelberg) and Dr. A. Manning (Signal Pharmaceuticals Inc., San Diego) for excellent collaboration in biological projects. Peptide sequencing on a QqTOF mass spectrometer was performed in collaboration with Dr. I. Chernushevich (PE Sciex, Canada). We thank members of the Peptide and Protein Group in EMBL (Heidelberg) and members of the Center for Experimental Bioinformatics (CEBI) (Odense) for fruitful discussions. Matthias Mann's laboratory at Odense University is supported by a grant by the Danish Foundation for Fundamental Research.

References

1. B. Alberts, *Cell* **1998**, *92*, 291-294.

2. T. Pawson and J. D. Scott, *Science* **1997**, *278*, 2075-2080.

3. R. W. King, R. J. Deshaies, J. M. Peters and M.W. Kirschner, *Science* **1996**, *274*, 1652-1659.

4. J. P. Staley and C. Guthrie, *Cell* **1998**, *92*, 315-326.

5. E. Fabre and E. Hurt, *Annu. Rev. Genet.* **1997**, *31*, 277-313.

6. M. Fromont-Racine, J. C. Rain and P. Legrain, *Nat. Genet.* **1997**, *16*, 277-282.

7. A. Lamond and M. Mann, *Trends Cell. Biol.* **1997**, *7*, 139-142.

8. J. R. Yates, *J. Mass Spectrom.* **1998**, *33*, 1-19.

9. R. S. Brown and J. Lennon, *Anal. Chem.* **1995**, *67*, 1998-2003.

10. M. L. Vestal, P. Juhasz and S. A. Martin, *Rapid Commun. Mass Spectrom.* **1995**, *9*, 1044-1050.

11. R. M. Whittal and L. Li, *Anal. Chem.* **1995**, *67*, 1950-1954.

12. O. N. Jensen, A. Podtelejnikov and M. Mann, *Rapid Commun. Mass Spectrom.* **1996**, *10*, 1371-1378.

13. A. V. Podtelezhnikov, O. N. Jensen and M. Mann, *Proc. 45th ASMS Conf. on Mass Spectrom. and Allied Topics* **1997**, Palm Springs, 269.

14. O. N. Jensen, A. V. Podtelejnikov and M. Mann, *Anal. Chem.* **1997**, *69*, 4741-4750.

15. M. Wilm and M. Mann, *Anal. Chem.* **1996**, *66*, 1-8.

16. M. Wilm, A. Shevchenko, T. Houthaeve, S. Breit, L. Schweigerer, T. Fotsis and M. Mann, *Nature* **1996**, *379*, 466-469.

17. M. Mann and M. Wilm, *Anal. Chem.* **1994**, *66*, 4390-4399.

18. A. Shevchenko, M. Wilm and M. Mann, *J. Protein Chem.* **1997**, *16*, 481-490.

19. A. Shevchenko, M. Wilm, O. Vorm, O. N. Jensen, A. V. Podtelejnikov, G. Neubauer, A. Shevchenko, P. Mortensen and M. Mann, *Biochem. Soc. Trans.* **1996**, *28*, 893-896.

20. A. Shevchenko, O. N. Jensen, A. V. Podtelejnikov, F. Sagliocco, M. Wilm, O. Vorm, P. Mortensen, A. Shevchenko, H. Boucherie and M. Mann, *Proc. Natl. Acad. Sci. USA* **1996**, *93*, 14440-14445.

21. W. Zachariae, A. Shevchenko, P. D. Andrews, R. Ciosk, M. Galova, M. J. Stark, M. Mann and M. K. Nasmyth, *Science* **1998**, *279*, 1216-1219.

22. R. Ciosk, W. Zachariae, C. Michaells, A. Shevchenko, M. Mann and K. Nasmyth, *Cell* **1998**, *93*, 1067-1076.

23. M. Shirayama, W. Zachariae, R. Ciosk and R. K. Nasmyth, *EMBO J.* **1998**, *17*, 1336-1349.

24. L. H. Hwang, L. Lau, D. L. Smith, C. A. Mistrot, K. G. Hardwick, E. S. Hwang, A. Amon and A. W. Murray, *Science* **1998**, 279, 1041-1044.

25. P. A. Kolodziej and R. A. Young, *Methods Enzymol.* **1991**, *194,* 508-519.

26. A. Cravchik and A. Matus, *Gene* **1993**, *137*, 139-143.

27. F. Mercurio, H. Zhu, H.; B. W. Murray, A. Shevchenko, B. L. Bennet, J. Li, D. B. Young, M. Barbosa, M. Mann, A. Manning and A. Rao, *Science* **1997**, *278*, 860-866.

28. T. Kawata, A. Shevchenko, M. Fukuzawa, K. A. Jermyn, N. F. Totty, N. V. Zhukovskaya, A. Sterling, M. Mann and J. C. Williams, *Cell* **1997**, *89*, 909-916.

29. J. Lingner, T. R. Hughes, A. Shevchenko, M. Mann, V. Lundblad and T. R. Cech, *Science* **1997**, *276*, 561-567.

30. P. Mitchell, E. Petfalski, A. Shevchenko, M. Mann and D. Tollervey, *Cell* **1997**, *91*, 457-466.

31. B. Senger, G. Simos, F. R. Bischoff, A. Podtelejnikov, M. Mann and E. Hurt, *EMBO J.* **1998**, *17*, 2196-2207.

32. B. Seraphin, *private communication*.

33. K. Siegers, M. R. Leroux, T. Waldmann, W. Schiebel, A. Shevchenko, F. U. Hartl and E. Schiebel, **1998**, submitted.

34. A. Shevchenko, M. Wilm, O. Vorm and M. Mann, *Anal. Chem.* **1996**, *68*, 850-858.

35. O. N. Jensen, A. Shevchenko and M. Mann, in *Protein Structure – A Practical Approach, 2nd Edition*; Chreighton, T. E., Ed.; Oxford University Press: Oxford, 1997, pp. 29-57.

36. T. Houthaeve, H. Gausepohl, M. Mann and K. Ashman, *FEBS Letters* **1995**, *376*, 91-94.

37. A. Shevchenko, O. N. Jensen, M. Wilm and M. Mann, *Proc. 45th ASMS Conf. on Mass Spectrometry and Allied Topics* **1996**, Portland, 331.

38. O. Vorm, P. Roepstorff and M. Mann, *Anal. Chem.* **1994**, *66*, 3281-3287.

39. O. Vorm and M. Mann, *J. Am. Soc. Mass Spectrom.* **1994**, *5*, 955-958.

40. P. A. Wigge, O. N. Jensen, S. Holmes, S. Soues, M. Mann and J. V. Kilmartin, *J. Cell Biol.* **1998**, *141*, 967-977.

41. O. N. Jensen, O. Vorm and M. Mann, *Electrophoresis* **1996**, *17*, 938-944.

42. B. Spengler, D. Kirsch, R. Kaufmann and E. Jaeger, *Rapid Commun. Mass Spectrom.* **1992**, *6*, 105-108.

43. J. K. Eng, A. L. McCormack and J. R. Yates, *J. Am. Soc. Mass Spectrom.* **1994**, *5*, 976-989.

44. M. Wilm, G. Neubauer and M. Mann, *Anal. Chem.* **1996**, *68*, 527-533.

45. H. Nakano, M. Shindo, S. Sakon, S. Nishinaka, M. Mihara, H. Yagita and K. Okumura, *Proc. Natl. Acad. Sci. USA* **1998**, *95*, 3537-3542.

46. M. Mann, *Trends Biol. Sci.* **1996**, *21*, 494-495.

47. G. Neubauer, A. King, J. Rappsilber, C. Calvio, M. Watson, P. Ajuh, J. Sleeman, A. Lamond and M. Mann, *Nature Genet.* **1998**, *20*, 46-50.

48. D. F. Hunt, J. R. Yates, J. Shabanowitz, S. Winston and C. R. Hauer, *Proc. Natl. Acad. Sci. USA* **1986**, *83*, 6233-6237.

49. K. M. McNagny, I. Petterson, F. Rossi, I. Flamme, A. Shevchenko, M. Mann and T. Graf, *J. Cell Biol.* **1997**, *138*, 1395-1407.

50. E. Bruyns, A. Marie-Cardine, H. Kirchgessner, K. Sagolla, A. Shevchenko, M. Mann, F. Autschbach, A. Bensussan, S. Meuer and B. Schraven, *J. Exp. Med.* **1998**, *188*, 561-575.

51. R.-H. Chen, A. Shevchenko, M. Mann and A. W. Murray, *J. Cell Biol.* **1998**, in press.

52. J. Lingner and T. R. Cech, *Proc. Natl. Acad. Sci. USA* **1996**, *93*, 10712-10717.

53. H. R. Morris, T. Paxton, A. Dell, J. Langhorn, M. Berg, R. S. Bordoli, J. Hoyes and R. H. Bateman, *Rapid Commun. Mass Spectrom.* **1996**, *10*, 889-896.

54. A. Shevchenko, I. Chernushevich, W. Ens, K. G. Standing, B. Thomson, M. Wilm and M. Mann, *Rapid Commun. Mass Spectrom.* **1997**, *11*, 1015-1024.

55. A. Shevchenko, I. Chernushevich and M. Mann, *Proc. 46th ASMS Conf. on Mass Spectrometry and Allied Topics* **1998**, Orlando, 237.

56. A. Goffeau, B. G. Barrell, H. Bussey, R. W. Davis, B. Dujon, H. Feldmann, F. Galibert, J D. Hoheisel, C. Jacq, M. Johnston, E. J. Louis, H. W. Mewes, Y. Murakami, P. Philippsen, H. Tettelin and S. B. Oliver, *Science* **1996**, *274*, 546-567.

57. R. W. King, J. M. Peters, S. Tugendreich, M. Rolfe, P. Hieter and M. W. Kirschner, *Cell* **1995**, *81*, 279-288.

58. W. Zachariae and K. Nasmyth, *Mol. Biol. Cell.* **1996**, *7*, 791-801.

59. W. Zachariae, T. H. Shin, M. Galova, B. Obermaier and K. Nasmyth, *Science* **1996**, *274*, 1201-1204.

60. A. D. Rudner and A. W. Murray, *Curr. Opin. Cell Biol.* **1996**, *8*, 773-780.

Questions and Answers

Steven Carr (SmithKline Beecham Pharmaceuticals)

How weak an interaction can you detect in these multiprotein complexes? Perhaps the more relevant question is how fast an off-rate can you deal with?

Answer. Here I used the term "weak interactions" only to point out that we are interested in the identification of components of the complex, purified in substoichiometric amounts compared with the core subunits. I did not intend any quantitative meaning when speaking about weak interactions.

Michael O. Glocker (University of Konstanz, Germany)

Is the TEV-protease commercially available, and what is the cleavage specificity of this protease?

Answer. The TEV protease is commercially available. The cleavage site recognized by the enzyme is unique, and preserves the components of a multiprotein complex from the degradation by the limited proteolysis.

Liang Tang (Affymax Research Institute)

How did you elute sample to the ESI tip after the washing of sample was done on a SPE column?

Answer. The capillary, containing 50-100 µl of POROS material is aligned in the special holder with the spraying needle. Then the holder is placed in a microcentrifuge, and the liquid is purged through the solvent layer directly into the spraying needle by gentle centrifuging.

Scott Patterson (Amgen, Inc.)

Can you generate spectra on the QqTOF by selecting a series of regions for fragmentation rather than careful examination of small peaks?

Answer. This approach — looking for daughter ions with m/z heavier than multiply charged precursor ions — has been tested at Sciex and reported at the last ASMS meeting in Orlando. It appeared to be by far less efficient than selecting of parent ions by careful inspection of "MS-only" spectra.

Investigation of Apoptosis-Involved Processes by Mass Spectrometric Identification of the Apoptosis-Associated Proteins in IgM-Induced Burkitt Lymphoma Cells

Eva-Christina Müller, Brigitte Wittmann-Liebold, and Albrecht Otto
Max-Delbrück-Centrum für Molekulare Medizin, Protein Chemistry, Robert-Rössle-Straße 10, D-13125 Berlin, Germany

In recent years there has been an explosion of interest in various processes involved in cell death, and new data have appeared. Although apoptosis, the programmed cell death plays a major role in development, homeostasis and immune response in multicellular organisms and controls lymphocyte growth and selection, this research area has not attracted great interest in the past for several reasons. First, the efficient disposal of apoptotic cells by phagocytosis is a very rapid process. The amount of apoptotic cells is a small fraction of the total. Second, death most often is a passive process for the entire organism, but the cells play an active roll in cell death. Apoptotic cell death can be distinguished from necrotic cell death by microscopic inspection. Necrotic cell death, the pathologic form of cell death, is caused by an acute cellular injury resulting in cellular swelling and loss of membrane integrity. In contrast, apoptotic cell death is characterized by controlled autodigestion of functional, important proteins of the cell connected with cellular shrinkage, membrane blebbing and chromatin condensation. The nuclear DNA of apoptotic cells is often fragmented. The membrane integrity of the cells is maintained, but changes within give a signal for neighboring phagocytotic cells to feed on the apoptotic cells. Apoptotic cell death is a process without induction of an inflammatory response, although this happens in necrotic cell death.

New name	Other name	SwissProt ID
Caspase-1	ICE	I1BC_Human
Caspase-2	ICH-1	ICE2_Human
Caspase-3	CPP32, Yama, apopain	ICE3_Human
Caspase-4	TX, ICH-2, ICE_{rel}-II	ICE4_Human
Caspase-5	TY, ICE_{rel}-III	ICE5_Human
Caspase-6	Mch2	ICE6_Human
Caspase-7	Mch3, ICE-LAP3, CMH-1	ICE7_Human
Caspase-8	Mach, Mch5, FLICE	ICE8_Human
Caspase-9	Mch6, ICE-LAP6	ICE9_Human
Caspase-10	Mch4	ICEA_Human

Table 1. Human caspases

Studies using *Caenorhabditis elegans* have provided clear evidence for cell death by an active process dependent on the activity of particular genes [1]. Two genes, *ced*-3 and *ced*-4, are required for the somatic cell death, which occurs during nematode development. Mutations of *ced*-3 and *ced*-4 abolish the apoptotic capability of cells that would otherwise be destined to die during *C. elegans* embryogenesis [2]. No homologs of the CED-4 protein have yet been identified, but numerous related proteins of CED-3 have been characterized. The first mammalian homolog of CED-3 identified was interleukin-1β-converting-enzyme (ICE) [3]. In 1996 Alnemri et al. [4] proposed using the name caspase as a root for all ICE family members. The selection of caspase was based on two catalytic properties of these enzymes. The "c" is thought to reflect a cysteine protease mechanism, and "aspase" refers to their ability to cleave after aspartic acid. To designate individual family members, the name caspase is followed by an arabic numeral, which is assigned based on its date of publication. Ten members of this family are known (Table 1). The greatest similarities show caspase-4 and caspase-5 with 74% identity, and caspase-3 and caspase-7 with 52%. Caspase-11 and caspase-12 of mouse are most similar to the human caspase-4 and caspase-5 (about 60%), but caspase-4 and caspase-5 are not known for mouse or rat.

Overexpression of ICE or CED-3 in Rat-1 fibroblasts induces apoptosis, suggesting that ICE is functionally, as well as structurally, related to CED-3 [5]. The nuclear lamins are cleaved early during apoptosis, and ICE-like proteases may play an important role in this process. Lamins are intermediate filament proteins that provide a structural framework for the inner nuclear membrane. Negative regulators of the cell death pathway have been found like the *C. elegans* *ced*-9 gene that antagonizes the function of *ced*-3 and *ced*-4 by blocking cell death. CED-9 is homologous to Bcl-2. The fact that *bcl*-2 can substitute functionally for *ced*-9, preventing cell death in the nematode emphasizes the highly conserved nature of the cell death machinery [6]. The Bcl-2 family is comprised of both inhibitors of cell death, such as Bcl-2 and Bcl-x_L, and promotors of cell death such as Bax and Bak [7, 8].

Recent investigations suggest that the cells which undergo apoptotic cell death might be involved in the pathogenesis of a variety of human diseases, including cancer, autoimmune diseases, and viral infections (Table 2). Further, a great number of diseases, such as AIDS (acquired immunodeficiency syndrome) and neurodegenerative disorders, characterized by cell loss, may result from accelerated rates of apoptosis [9]. Infection with the human immunodeficiency virus (HIV) leads to AIDS in about 10 years, characterized by cell loss and tissue atrophy in several organs, including the immune system, the bone marrow, and the brain. The major pathological feature of this disease is the progressive collapse of the two most complex regulatory networks of the human body, the immune system, leading to immune incompetence, and the central nervous system, leading to brain atrophy and dementia. AIDS induces the loss of $CD4^+$ T-cells in the immune system [10] and neuronal loss in the brain [11]. In neurodegenerative disorders cell death results in specific disorders of movement and central nervous system function. Oxidative stress, calcium toxicity,

Diseases associated with increased apoptosis	Diseases associated with inhibition of apoptosis
• AIDS	• Cancer Lymphomas Tumors with p53 mutations Hormone-dependent tumors
• Neurodegenerative disorders	• Autoimmune disorders immune-mediated glorulonephritis
• Alzheimer's disease Parkinson's disease Cerebrellar degeneration	• Viral infections Adenoviruses Herpesviruses
• Myelodysplastic syndromes Aplastic anemia	
• Ischemic injury Myocardial infarction Stroke	
• Toxin-induced liver disease Alcohol	

Table 2. Disorders of apoptosis

mitochondrial defects, and deficiency of survival factors have all been postulated to contribute to the pathogenesis of these disorders [12-14].

Cancer is a disease characterized by the accumulation of cells which can result from either increased proliferation or the failure of cells to undergo apoptosis in response to appropriate stimuli. Tumor cells must develop some degree of independence from the survival factors to survive in unusual sites. Several genes in the regulation of apoptosis in tumor cells have been found. The gene *bcl*-2 is present in most human follicular lymphomas [8]. Overexpression of specific genes prevent cells from initiating apoptosis [15]. Overexpression of *bcl*-2 or *bcl*-x_L has been found to confer resistance to cell death in response to chemotherapeutic agents such as cytosine arabinoside, vincristine, and cisplatin [16, 17]. Drug-induced Bcl-2 phosphorylation with concomitant apoptosis in malignant cells derived from a variety of hormone-dependent malignancies, leukemia, and lymphoma have been described by Haldar et. al. [18].

Burkitt lymphoma, a monoclonal B-cell tumor, is one of the most closely studied human diseases, from the level of its epidemiology right down to its molecular biology. It is an important model for studying oncogenesis and apoptosis. Most of our knowledge of the cell and molecular biology of Burkitt lymphoma has been acquired as a result of the success with which cell lines have been established *in vitro* from these tumors. The Burkitt lymphoma tumor cell line BL 60 has been used in our investigation of apoptosis. These cells are highly sensitive to IgM-mediated apoptosis. Plasma membrane changes have been shown to occur very early in the cell undergoing cell death. Martin et al. [19] have shown that in many murine and human cell types phosphatidylserine, which is normally located on the inner leaflet of the plasma membrane, is translocated to the outer leaflet. This offers the possibility to label apoptotic cells with AnnexinV-FITC, because it binds to phosphatidylserine. AnnexinV-FITC-positive cells can be enriched by magnetic cell sorting using anti-FITC magnetic beads.

The study of proteins and their modifications is imperative to understand the various cell processes, especially those involved in control of gene expression, regulation of cell-signaling pathways, and apoptosis. Separation of the complex protein mixture of entire cells by high resolution two-dimensional (2-D) electrophoresis [20, 21] is an established method for visualization of proteins contained in cell lysates and tissues [22-26]. Several mass spectrometric techniques have rapidly become increasingly important as a primary tool for identifying proteins from 2-D gels, which are fast and sensitive, and peptide mixtures can be directly measured. Peptide mass fingerprints of peptide mixtures may be used to search peptide-mass databases for protein identification [27-30]. However, modifications of some peptides, posttranslationally or induced by electrophoresis can complicate the unambiguous identification [31, 32]. Furthermore, the number of possible candidates as a result of the search in a database can be reduced only with a mass spectrometer of highest mass accuracy. Difficulties are encountered in determining the organism (rat, mouse, human) of a highly conserved protein by only evaluating peptide mass fingerprint results (unpublished results of a contest 1998, 5th German Workshop in Micromethods in Protein Chemistry, München). For instance, a mass difference of 16 can be caused by an oxidation of methionine or by an exchange of proline to leucine or isoleucine, respectively. This means additional sequence information in combination with mass fingerprinting is often necessary for a correct identification of the proteins.

The development of the nanospray source for triple-quadrupole mass spectrometry by Wilm and Mann [33] offered the possibility to obtain partial sequence information for sample amounts at femtomole concentrations. The extended time of measurement permits the sequencing of as little as 1 μl of the digestion mixture. The new hybrid quadrupole–orthogonal acceleration time-of-flight mass spectrometer (Q-Tof) equipped with a Z spray ion source is a powerful instrument with a higher accuracy and sensitivity than the present quadrupole devices. Routine accuracy of within 0.1 Da of the expected theoretical peptide mass values are obtained [34].

For the study of apoptosis and the identification of associated proteins of the Burkitt lymphoma cell line BL 60, we have employed high resolution 2-D gel electrophoresis, Edman sequencing of internal peptides after HPLC, MALDI-mass fingerprinting, and sequencing by MS/MS techniques with the Q-Tof instrument.

Preparation

The human Burkitt lymphoma cell line BL 60 was cultured as described by Rickers et al. [35]. The rate of apoptosis was measured by acridine orange or AnnexinV-FITC labeling. After 24 h of anti-IgM treatment the separation of apoptotic and non-apoptotic cells by AnnexinV-FITC labeling and subsequent magnetic separation with anti-FITC magnetic microbeads was performed.

The cell pellets were thawed and mixed with protease inhibitors according to Klose and Kobalz [36]. Additionally, the pellet was rapidly mixed with urea (9 M final concentration), then with DTT (70 mM final concentration); and finally with 2% carrier ampholytes for analytical gels or 6% carrier ampholytes for micropreparative gels. After 30 min of gentle stirring at room temperature, the samples were frozen at -70°C. The resulting protein concentration was determined by amino acid analysis as 15-30 mg/ml.

The proteins were separated by large gel (23 cm by 30 cm) high resolution 2-D electrophoresis [36]. The complete equipment for electrophoresis and the ready-to-use solutions were purchased from WITA GmbH (Teltow, Germany). The importance of the use of standardized solutions of high quality cannot be overestimated. The quality of the gels depends on preparation with high quality chemicals, a standard protocol, and experienced personel. About 60 μg of the total protein mixture was loaded on an analytical gel, about 800 μg on a preparative gel. The preparative gels were stained by Coomassie blue R-250 as described by Eckerskorn et al. [37], and the analytical gels by silver solution according to Blum et al. [38].

For micropreparative investigations up to ten Coomassie-stained spots of the same protein were combined. For identification by nanospray-ESI mass spectrometry only one Coomassie- or one silver-stained gel spot was necessary. The enzymatic in-gel digestion with trypsin, the extraction of the peptides from the gel, and desalting by reversed-phase material was performed as described by Otto et al. [39].

The separation of the peptides was performed using a microbore HPLC system with a column of 2.1 mm ID and 10 cm length purchased from Pharmacia-Biotech. A gradient with increasing concentration of acetonitrile in 0.1% v/v TFA and a flow rate of 100 μl/min at room temperature was used. The peptide fractions were concentrated in a Speed Vac concentrator and loaded onto a Biobrene-coated glass fiber filter of a Procise or 477 A pulsed-liquid phase sequencer for Edman sequencing.

Mass Spectrometry

MALDI-MS was performed with a VG TofSpec, equipped with a nitrogen laser (337 nm, pulse duration 4 ns) and a reflectron. A saturated solution of α-cyano-4-hydroxycinnamic acid in aqueous 50% acetonitrile and 0.1% TFA was used as a matrix. The sample (0.8 μl), the peptide mixture or one HPLC fraction, and the matrix (1.2 μl) were mixed on the target, air dried, and inserted in the mass spectrometer for mass measurement.

Tandem mass spectrometry (MS/MS) experiments were performed with a Q-Tof (Micromass, Manchester, UK) equipped with a nanoflow Z spray ion source. This was operated at a temperature of 30°C with a nitrogen drying gas flow of about 180 L/h. A potential of 1.4 kV was applied to the nanoflow borosilicate glass capillary. An aliquot of 1 μl was sufficient for one MS spectrum and five to six MS/MS experiments, employing a flow rate of 30 nl/min. A collision voltage of 28-35 V was applied for fragmentation of a peptide, the gas pressure in the collision cell was regulated to 6.0 x 10^{-5} mbar. The Q-Tof calibration with [Glu]-fibrinopeptide (Sigma) was performed daily.

DATABASE SEARCH

Some different approaches have been used to identify proteins. The search program MS-Fit by Clauser and Baker (http://falcon.ludwig.ucl.ac.uk/ucsfhtml/msfit.htm) for peptide masses offers the possibility to choose between different nucleotide and protein databases. Cysteine and methionine were considered as either normal or modified; serine, threonine, or tyrosine as unmodified or phosphorylated; glutamine at the *N*-terminus as pyroglutamic acid or without modification. For any database entry with a methionine the *N*-terminal peptide was calculated as either in its original form or in a shortened version with removal of the *N*-terminal methionine; the next amino acid was then calculated with or without acetylation. If MS/MS sequence information was available, we first used an approach which combines partial manual spectrum interpretation of about three amino acids (sequence tag) with the residual *N*-terminal and *C*-terminal masses of the interpreted region and the entire peptide mass [40]. This information was used to search in a non-redundant translated nucleotide database (http://www.protana.com/~pm/#PeptProtana). Modifications of methionine and cysteine were taken into account. The combination of the results from the tag-search and MS-Fit execution yielded an unambiguous identification if the protein was known in the available databases. The program FindMod (http://www.expasy.ch/sprot/findmod.html) was used as a helpful tool for unmatched peptide masses to find modifications and potential single amino acid substitutions.

Finally, manual interpretation of spectra (*de novo* sequencing) and searching the protein, nucleotide or combined databases like OWL [41] by FastA and TFastA algorithms [42] (http://vega.crbm.cnrs-mop.fr/bin/fasta-guess.cgi) was also performed. The Micromass software MassLynx 3.1 was successfully employed to interpret a MS/MS spectrum of samples without any modifications.

RESULTS

Figure 1 summarizes our strategy to find proteins changed during apoptosis. The proteins were separated by 2-D electrophoresis, and gels of untreated and treated cells were compared by subtractive analysis [43]. Interesting spots which changed their intensity or position were excised and subjected to tryptic in-gel digestion. If Edman sequencing and MALDI-mass fingerprinting were used for identification, protein spots from several preparative gels were pooled. This was not necessary for measurements with the nanospray-ESI technique, because one analytical or preparative gel was sufficient for identification of the proteins. Experimentally determined peptide masses or sequence tags were used in search programs with different databases via the Internet. Proteins were verified by using the sequences determined to search in the non-

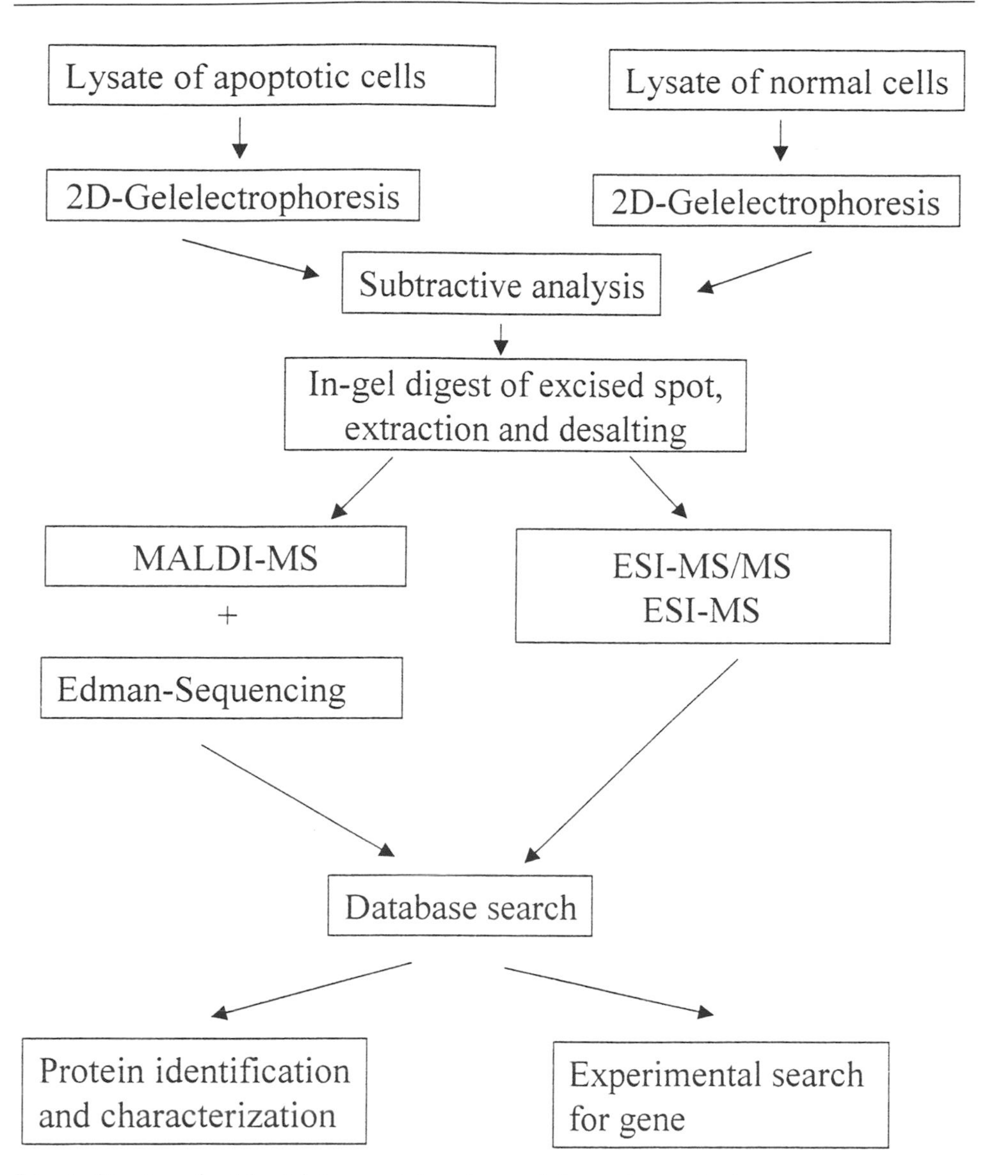

Fig. 1. Strategy for identifying proteins changed through apoptosis.

redundant protein sequence database OWL, using the search programs FastA or BLAST, or alternatively, searching the nucleotide databases with TFastA.

In Fig. 2, the most altered gel areas in two analytical gels containing i) nuclear proteins of the Burkitt lymphoma BL 60 cell line in the normal state (a), and ii) the IgM-induced apoptotic state (b) are compared. About 80 modified proteins have been found in our gels by this comparison [44]. As examples, the proteins D4-GDI (1), ribosomal protein P0 (2), neutral calponin (3), actin (4), and hnRNP C1/C2 (5) are labeled (Fig. 2a, b).

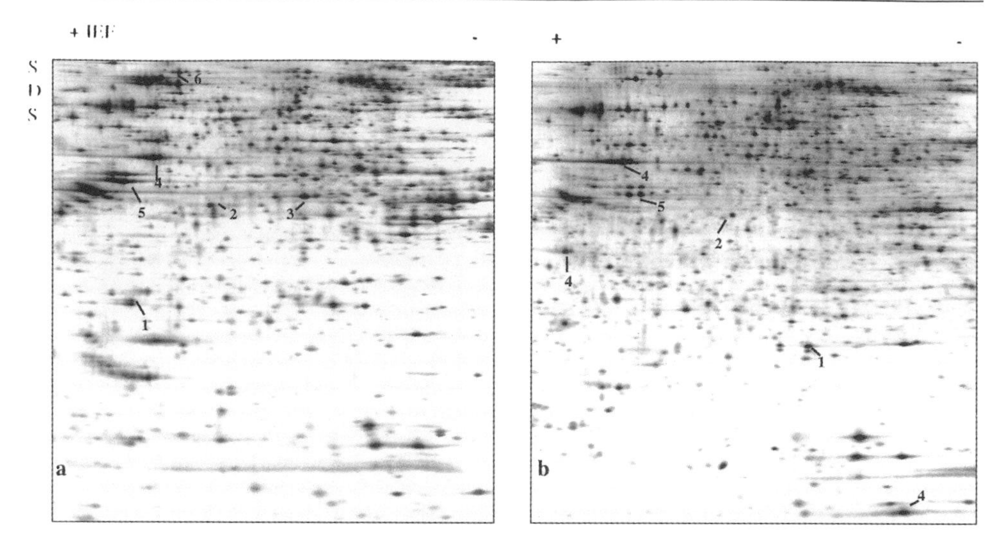

Fig. 2. Silver stained 2-DE analytical gels of the proteins derived from the nuclear fraction of Burkitt lymphoma BL 60 cells: a) normal cells, b) Anti-IgM-induced apoptosis cells. Examples of proteins altered after treatment: 1-D4-GDI, fragmented in b), 2 – ribosomal protein P0, pI shift in b), 3- neutral calponin (not found in the apoptotic gel), 4 – actin and actin fragments, 5 – hnRNP C1/C2, pI shift in b). HSP60 (6) was not changed after apoptosis. For more detailed description see Brockstedt et al. [44].

The chemical noise in an ESI-MS spectrum of a peptide mixture generated from a low concentration protein spot of one analytical gel was relatively high, and it turned out to be difficult to detect the relevant peaks. Triple quadrupole users may perform parent ion scans with an immonium ion of one or two amino acids, but that is not an efficient process in the Q-Tof. We have observed a collisional focusing with a clear improvement of the signal/noise ratio and a four- to ten-fold increased intensity for data acquisition with the collision gas argon switched on (Fig. 3).

The spectra of the ribosomal protein P0 spots (a double spot) (Fig. 4) was found at different places in the apoptotic-induced and normal gel. We expected to find two different peptides or variants thereof in the analysis. However, in both cases we found only the same set of peptides of this protein (marked with an asterisk) and autolysis products of trypsin (marked with +). Usually not all peptide masses of the protein can be seen in the MS spectrum, therefore it is likely that the changed peptide was not recovered. The MS/MS spectrum of one of these P0 peptides (Fig. 5) demonstrates the high sensitivity of the Q-Tof mass spectrometer. The complete set of the y"-ion series (the masses of the *C*-terminal peptide fragment ions) was recovered, as is often the case for tryptic peptides which carry a charged amino acid at the *C*-terminal end. The corresponding b_2 or a_2 ions (the *N*-terminal ion series) were in general easily detected. Furthermore, the immonium ions of the amino acids near the *N*-terminus are more intense than for internal amino acids. This knowledge is useful for *de novo* sequencing of unknown proteins.

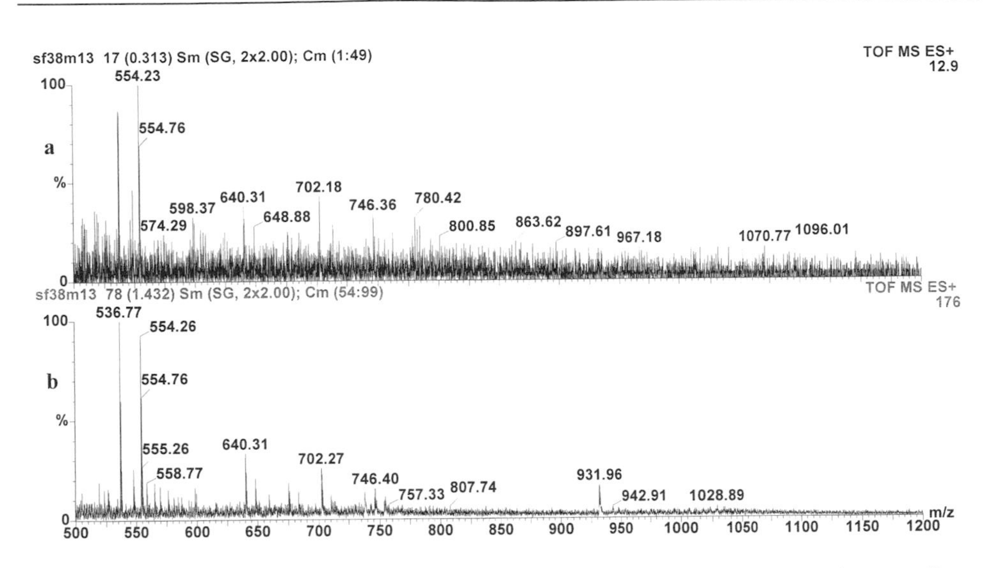

Fig. 3. Nanoflow ESI spectrum of spot 3, identified as neutral calponin, from a silver-stained gel, digested with trypsin, acquired as normal (a), and with the collision gas argon (b) switched on to raise the intensity. The base peak intensity was increased from 12.9 to 176. The signal/noise ratio was clearly improved.

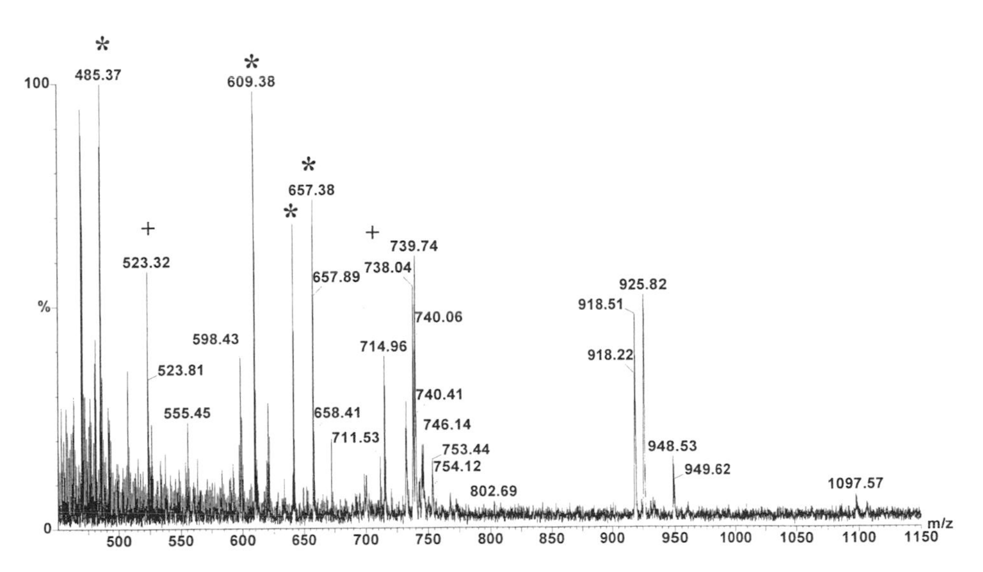

Fig. 4. Nanoflow ESI spectrum of spot 3 from a silver-stained gel, digested with trypsin. Fifty percent (1 µl) of the total digest were used for the MS-spectrum and six MS/MS spectra from the peptides thereof. Peptides marked with an asterisk have been identified as peptides of the ribosomal protein P0 (see Table 3), marked with + as trypsin, respectively.

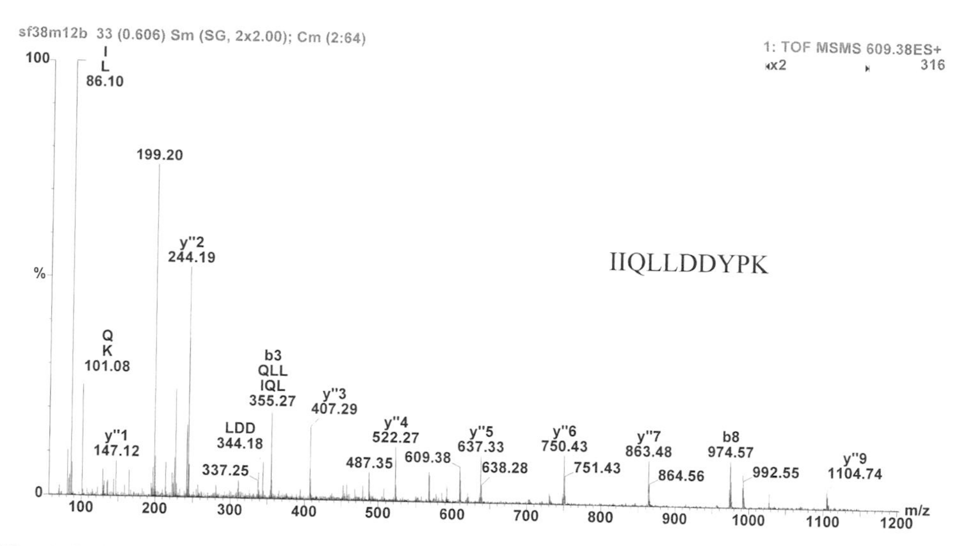

Fig. 5. MS/MS spectrum of the peptide $(M+2H)^{2+} = 609.4$ from the tryptic digest of spot 2. Relevant ions are labeled according to the nomenclature proposed by Roepstorff et al. [60], as modified by Biemann et al. [61].

Spot	Search result	Th. Mass	Delta Mass	Residues	Sequence
38M1/1	Heat shock	2560.25	0.27	97-121	LVQDVANNTNEEAGDGTTTATVLAR
	protein P60	1389.70	0.12	222-233	GYISPFINSK
Spot 6	P10809	1233.60	0.06	406-417	VGGTSDVEVNEK
in Fig.2a		960.51	0.05	421-429	VTDALNATR
		1215.66	0.08	482-493	NAGVEGSLIVEK
38M1/2	Ribosomal	1217.68	0.08	17- 26	IIQLLDDYPK
	protein P0	1280.59	0.11	27- 38	CamFIVGADNVGSK
Spot 2	RLA0_HUMAN	968.52	0.08	84- 92	GNVGFVFIK
in Fig.2a		1313.71	0.05	135-146	TSFFQALGITTK
38M1/3	Neutral calponin	1107.57	-0.05	9- 19	GPSYGLSAEVK
	EMBL:D83735	1279.59	0.03	56- 66	DGTILCamTLMNK
Spot 3		1072.60	0.04	135-145	GLQSGVDIGVK
in Fig.2a		1490.69	0.01	161-174	AGQCamVIGLQMGTNK
		1403.56	0.02	175-187	CamASQSGMTAYGTR
38M3/1	Ribosomal	1217.68	0.14	17- 26	IIQLLDDYPK
	protein P0				
Spot 2	RLA0_HUMAN				
in Fig.2b					

Table 3. Proteins identified by MS/MS measurements. Spots 1-3 were excised from a silver-stained non-apoptotic gel, the last one from an apoptotic gel, respectively.

Table 3 summarizes peptide mass data, sequences and database search results for spots of the silver-stained gels shown in Fig. 2. Three spots were analyzed from the untreated cells, the fourth one derived from the apoptotic gel. The mass accuracy of the measurements was at least 0.01%. In contrast to preparative gels, only modifications of cysteines with acrylamide (+71) were found; neither methionine oxidation (+16) nor oxidized cysteine with acrylamide

Protein	10 Coomassie-stained gels	1 Coomassie-stained gel	1 Silver-stained gel
	MALDI-MS + Edman sequencing	ESI-MS + ESI-MS/MS	ESI-MS + ESI-MS/MS
Actin	X	X	
Lamin B1	X	X	
Nuleolin	X		
D4-GDI	X	X	
hnRNP C1/C2	X		
Neutral calponin	X	X	X
LSP1	X		
HHR23P	X		
DUT-N	X	X	
hnRNP A1	X	X	
HP1 alpha	X		
Ribosomal protein P0	X	X	X

Table 4. Identification of apoptosis-associated proteins as done by the different methods available at the beginning of the work and two years later.

(+87) was seen. Both spots 38M1/1 and 38M3/1 were identified as ribosomal protein P0 (MS spectra not shown); the expected difference between the two proteins could not be found due to the small sample amount employed.

Proteins known to be apoptosis-associated and new candidates have been detected with Edman sequencing and MALDI-MS [35, 44] and 33 proteins excised from a Coomassie blue R-250-stained gel have been identified by the ESI technique [45]. The most recent list of the proteins identified as apoptosis-associated is shown in Table 4. Among those identified in correlation with induced apoptosis are proteins such as actin, lamin B, nucleolin, calponin, and D4-GDI (see below) which are either up- or downregulated, or which are identified as fragments due to caspase activities.

Actins are highly conserved proteins and are ubiquitously expressed in all eukaryotic cells. The beta and gamma actins co-exist in most cell types as components of the cytoskeleton and as mediators of internal cell motility. Mashima et al. [46] proved actin *in vitro* and *in vivo* to be a substrate of caspase-3 like protease in cell extracts. Caspase-3 can cleave actin into a 15 kDA and a 31 kDa fragment [46]. These results are also found in our 2-D gels (protein 4 on Fig. 2 a, b).

Lamin B1 is a component of the nuclear lamins in eukaryotic cells and known to be apoptosis-associated [47]. The spot intensity was found to increase in the apoptotic gel [44].

Nucleolin, a major RNA-binding protein of the nucleolus, was found fragmented in the apoptotic gel [44]. Pasternack et al. [48] have demonstrated that the serine protease granzyme A binds and cleaves nucleolin *in vitro*.

Spot 1 in Fig. 2 is the protein D4-GDI. It was initially identified by its homology to Rho-GDI [49]. Its expression is mainly restricted to cells of the hematopoietic system [50]. D4-GDI disappears after stimulation of apoptosis, and it appears in a much more basic region identified as a proteolytic cleavage product [35]. Sequence analysis revealed a caspase-3 consensus cleavage site

DExD at position D^{19}. These results showed that D4-GDI is a specific substrate of caspases during the process of anti-IgM-mediated apoptosis. Specific inhibition of caspase-3 by z-DEVD-fmk blocks cleavage of D4-GDI and completely abolishes apoptosis.

In our 2-D pattern of non-apoptotic cells the heterogeneous nuclear ribonucleoprotein C1/C2 (hnRNP C1/C2) is resolved in six protein spots (5 in Fig. 2a). These six spots are reduced to four spots in the protein pattern of apoptotic cells (5 in Fig 2b). However, the shift of the hnRNP C1/C2 proteins towards a neutral pH and lower mass range is the evidence for the fragmentation of these proteins in the apoptotic process. The hnRNP C1/C2 proteins are abundant nuclear proteins and thought to be involved in RNA splicing. The cleavage by ICE-like proteases during apoptosis in a Burkitt lymphoma cell line by a variety of stimuli, including ceramide, etoposide, and ionizing radiation was described [51].

Spot 3 in Fig. 2a was identified as neutral calponin (see also Table 3), which is a non-muscle isoform of calponin and may play a physiological role in cytoskeletal organization [52]. The intensity is significantly decreased in the 2-D pattern of the apoptotic cells. The assumed fragmentation of the human neutral calponin during the apoptotic process might be essential for the cytoskletal reorganization.

Lymphocyte specific protein (LSP1) has been suggested to interact with the cytoskeleton [53] and this could be the reason for its involvement in the apoptotic process and the decreased intensity. The intensity of the UV excision repair protein, protein RAD23 homolog (HHR23P), is decreased as well. It is involved in DNA excision repair *in vitro*, and may play a part in DNA recognition and/or in altering chromatin structure to allow access by damage-processing enzymes [54]. A role in the apoptotic process seems to be likely.

Two distinct forms of deoxyuridine triphosphate nucleotidohydrolase (dUTPase) have been isolated by Ladner et al. [55], the nuclear (DUT-N) and the mitochondrial (DUT-M) form. They differ in their *N*-terminal parts and were considered splicing variants. DUT-N is the more abundantly expressed form and was found both in non-apoptotic and with decreased intensity in apoptotic gels. The enzyme produces dUMP, the immediate precursor of thymidine and it decreases the intracellular concentration of dUTP so that uracil cannot be incorporated into DNA during replication or repair. An incorporation of uracil into DNA leads to DNA fragmentation and cell death.

The *N*-terminal fragment of heterogeneous nuclear ribonucleoprotein A1 (hnRNP A1) was only found in the pattern of apoptotic cells [44]. The hnRNP A1 is one of the major pre-mRNA/mRNA binding proteins in eukaryotic cells and an abundant protein in the nucleus. It is likely involved in mRNA transport from the nucleus to the cytoplasm [56]. One component of heterochromatin is the nuclear protein heterochromatin protein 1 homolog alpha (HP1 alpha), a DNA binding protein [57]. It was found only in the apoptotic pattern [44]. HP1 alpha can be associated with the transcriptional coactivators TIF1α and TF1β as well as the inner nuclear membrane protein LBR [58].

The 60 S ribosomal protein P0 plays an important role in polypeptide chain elongation during translation. Grabowski et al. [59] suggested an extra ribosomal function as a DNA repair protein and the P0 gene was shown to be overexpressed in colorectal tumors. It was identified in the normal and apoptotic nuclear cell fraction (2 in Fig. 2, Table 3) with a slightly increased pI in the apoptotic gel. Further investigations are needed to determine the modification in this case.

Summary

We have investigated the entire protein mixture of the total and the nuclear fraction of human Burkitt lymphoma BL60 cells and found significant changes on the high-resolved 2-D gels when comparing apoptotic to non-apoptotic cells. The changes concern the intensity as well as modifications like fragmentation, which effect a change in the position on the gel. The identification of the proteins is possible with nanospray-ESI Q-Tof mass spectrometry from only one Coomassie stained or one silver-stained gel. The high quality MS and MS/MS-spectra which were obtained in combination with the high accuracy of the measurements makes the assignment of apoptosis-induced protein variations possible with very little starting cell material. New candidates for apoptosis-associated proteins have been found. Some of them, like D4-GDI, could already be confirmed as apoptosis-associated, combining the results with data from cells that were obtained after caspase inhibition. After caspase-3 inhibition, the alteration of the protein D4-GDI was not found. Further investigations are necessary to find posttranslational modifications or reasons for differing positions of proteins on the gel. It is a challenge to solve these problems with a very small amount of sample material.

From the mass spectrometric data obtained it will be feasible to perform an entire proteome analysis of cells with and without apoptosis induction to generate a specific database. This data collection will be the basis for further studies on drug screening, early tumor markers, and other disease states.

References

1. R. E. Ellis, J. Yuan and H. R. Horvitz, *Annu. Rev. Cell Biol.* **1991**, *7*, 663-698.

2. J. Yuan and H. R. Horvitz, *Dev. Biol.* **1990**, *138*, 33-41.

3. J. Yuan, S. Shaham, S. Ledoux, H. M. Ellis and H. R. Horvitz, *Cell* **1993**, *75*, 641-652.

4. E. S. Alnemri, D. J. Livingston, D. W. Nicholson, G. Salvesen, N. A. Thornberry, W. W. Wong and J. Yuan, *Cell* **1996**, *87*, 171.

5. M. Miura, H. Zhu, R. Rotello, E. A. Hartwieg and J. Y. Yuan, *Cell* **1993,** *75*, 653-660.

6. M. O. Hengartner and H. R. Horvitz, *Cell* **1994**, *76*, 665-676.

7. G. Nunez and M. F. Clarke, *Trends Cell Biol.* **1994**, *4*, 399-403.

8. S. N. Farrow and R. Brown, *Curr. Opin. Gen. Dev.* **1996**, *6*, 45-49.

9. C. B. Thompson, *Science* **1995**, *267*, 1456-1462.

10. A. S. Fauci, *Science* **1988**, *239*, 617-622.

11. S. I. Savitz and D. M. Rosenbaum, *Neurosurgery* **1998**, *42*, 555-572.

12. I. Ziv, E. Melamed, N. Nardi, D. Luria, A. Achiron, D. Offen and A. Barzilai, *Neurosci. Lett.* **1994**, *170*, 136-140.

13. N. A. Simonian and J. T. Coyle, *Ann. Rev. Pharmacol. Toxicol.* **1996**, *36*, 83-106.

14. M. Leist, and P. Nicotera, *Rev. Physiol. Biochem. Pharmacol.* **1998**, *132*, 79-125.

15. T. Kobayashi, S. Ruan, K. Clodi, K. O. Kliche, H. Shiku, M. Andreeff and W. Zhang, *Oncogene* **1998**, *16*, 1587-1591.

16. J. C. Reed, *Semin. Hematol.* **1997**, *34*, 9-19.

17. G. P. Amarante-Mendes, A. J. McGahon, W. K. Nishioka, D. E. Afar, O. N. Witte and D. R. Green, *Oncogene* **1998**, *16*, 1383-1390.

18. S. Haldar, A. Basu and C. M. Croce, *Cancer Res.* **1998**, *58*,1609-1615.

19. S. J. Martin, C. P. M. Reutelingsperger, A. J. McGahon, A. Radar, R. C. van-Schie, D. M. LaFace and D. R. Green, *J. Exp. Med.* **1995**, *182*, 1545-1556.

20. J Klose, *Humangenetik* **1975**, *26*, 21-234.

21. H. P. O'Farell, *J. Biol. Chem.* **1975**, *250*, 4007-4021.

22. H. Rasmussen, J. VanDamme, M. Puype, J.E. Celis and J. Vandekerckhove, *Electrophoresis* **1992**, *13*, 960-962.

23. K. R. Clauser, S. C. Hall, , D. M. Smith, J. W. Webb, L. E. Andrews, H. M. Tran, , L. B. Epstein and A. L. Burlingame, *Proc. Natl. Acad. Sci. USA* **1995**, *92*, 5072-5076.

24. L. B. Epstein, D. M. Smith, N. M. Matsui, H. M. Tran, C. Sullivan, I. Raineri, A. L. Burlingame, K. R. Clauser, S. C. Hall and L. E. Andrews, *Electrophoresis* **1996**, *17*, 1655-1670.

25. D. Arnott, K. L. O'Connell, K. L. King and J. T. Stults, *Anal. Biochem.* **1998,** *258*, 1-18.

26. E.-C. Müller, B. Thiede, U. Zimny-Arndt, C. Scheler, J. Prehm, U. Müller-Werdan, B. Wittmann-Liebold, A. Otto and P. Jungblut, *Electrophoresis* **1996**, *17*, 1700-1712.

27. W. J. Henzel, T. M. Billeci, J. T. Stults, S. C. Wong, C. Grimley and C. Watanabe, *Proc. Natl. Acad. Sci. USA* **1993**, *90*, 5011-5015.

28. M. Mann, P. Hojrup and P. Roepstorff, *Biol. Mass Spectrom.* **1993**, *22*, 338-345.

29. D. J. Pappin, P. Hojrup and A. J. Bleasby, *Curr. Biol.* 1993, *3*, 327-332.

30. B. Thiede, A. Otto, U. Zimny-Arndt, E.-C. Müller and P. Jungblut, *Electrophoresis* **1996** *17*, 588-599.

31. S. D. Patterson and R. Aebersold, *Electrophoresis* **1995**, *16*, 1791-1799.

32. K. M. Swiderek, M. T. Davis and T. D. Lee, *Electrophoresis* **1998**, *19*, 989-997.

33. M. Wilm and M. Mann, *Anal. Chem.* **1966**, *68*, 1-8.

34. H. R. Morris, T. Paxton, A. Dell, J. Langhorne, M. Berg, R. S. Bordoli, J. Hoyes and R. H. Bateman, *Rapid. Comm. Mass Spectrom.* **1996**, *10*, 889-896.

35. Rickers, E. Brockstedt, M.Y. Mapara, A. Otto, B. Dörken and K. Bommert, *Eur. J Immunol.* **1998**, *28*, 296-304.

36. J. Klose and U. Kobalz, *Electrophoresis* **1995**, *16*, 1995, 1034-1059.

37. C. Eckerskorn, P. Jungblut, W. Mewes, J. Klose and F. Lottspeich, *Electrophoresis* **1988**, *9*, 830-838.

38. H. Blum, H. Beier and H. J. Gross, *Electrophoresis* **1987**, *8*, 93-99.

39. A. Otto, B. Thiede, E.-C. Müller, C. Scheler, B. Wittmann-Liebold and P. Jungblut, *Electrophoresis* **1996**, *17*, 1643-1650.

40. M. Mann and M. Wilm, *Anal. Chem.* **1994**, 66, 3281-3287.

41. A. J. Bleasby and J. C. Wotton, *Protein Eng.* **1990**, *3*, 153-159.

42. W. R. Pearson and D. J. Lipman, *Proc. Natl. Acad. Sci. USA* **1988**, *55*, 2444-2448.

43. R. Aebersold and J. Leavitt, *Electrophoresis* **1990**, *11*, 540-546.

44. E. Brockstedt, A. Rickers, S. Kostka, A. Laubersheimer, B. Dörken, B. Wittmann-Liebold, K. Bommert and A. Otto, *J. Biol. Chem.* **1998** (in press).

45. E.-C. Müller, M. Schümann, A. Rickers, K. Bommert, B. Wittmann-Liebold and A. Otto, *Electrophoresis* **1998** (accepted).

46. T. Mashima, M. Naito, K. Noguchi, D. K. Miller, D. W. Nicholson and T. Tsuruo, *Oncogene* **1997**, *14*, 1007–1012.

47. D. Wallach, M. Boldin, E. Varfolomeev, R. Beyaert, P. Vandenabeele and W. Fiers, *FEBS-Lett.* **1997**, *410*, 96-106.

48. M. S. Pasternack, K. J. Bleier and T. N. McInerney, *J. Biol. Chem.* **1991**, *266*, 14703-14708.

49. P. Scherle, T. Behrens and L. M. Staudt, *Proc. Natl. Acad. Sci. USA* **1993**, *90*, 7568-7572.

50. J. M. Lelias, C. N. Adra, G. M. Wulf, J. C. Guillemot, M. Khagad, D. Caput and B. Lim, *Proc. Natl. Acad. Sci. USA*, **1993**, *90*, 1479-1483.

51. N. Waterhouse, S. Kumar, Q. Song, P. Strike, L. Sparrow, G. Dreyfuss, E. S. Alnemri, G. Litwack, M. Lavin and D. Watters, *J. Biol. Chem.* **1996**, *271*, 29335-29341.

52. H. Masuda, K. Tanaka, M. Takagi, K. Ohgami, T. Sakamaki, N. Shibata and K. Takahashi, *J. Biochem. Tokyo* **1996**, *120*, 415-424.

53. J. Jongstra-Bilen, P. A. Janmey, J. H. Hartwig, S. Galea and J. Jongstra, *J. Cell. Biol.* **1992**, *118*, 1443-1453.

54. K. Sugasawa, C. Masutani, A. Uchida, T. Maekawa, P. J. van der Spek, D. Bootsma, J. H. Hoeijmakers and F. Hanaoka, *Mol. Cell. Biol.* **1996**, *16*, 4852-4861.

55. R. D. Ladner, D. E. McNulty, S. A. Carr and G. D. Roberts, *J. Biol. Chem.* **1996**, *271*, 7745-7751.

56. S. Piñol-Roma and G. Dreyfuss, *Mol-Cell-Biol.* **1993**, *13*, 5762-5770.

57. K. Sugimoto, T. Yamada, Y. Muro and M. Himeno, *J. Biochem. Tokyo* **1996**, *120*, 153-159.

58. Q. Ye, I. Callebaut, A. Peszhman, J. C. Courvalin and H. J. Worman, *J. Biol.Chem.* **1997**, *272*,14983-14989.

59. D. T. Grabowski, R. O. Pieper, B. W. Futscher, W. A. Deutsch, L. C. Erickson and M. R. Kelley, *Carcinogenesis* **1992**, *13*, 259-263.

60. P. Roepstorff, and J. Fohlmann, *Biomed. Mass Spectrom.* **1984**, *11*, 601.

61. K. Biemann, and H. A. Scoble, *Science*, **1987**, *237*, 992-998.

Questions and Answers

Marvin Vestal (PerSeptive Biosystems)

You showed that turning on the argon collision gas in your Q-TOF decreased the chemical noise and improved s/n for multiply charged peptides in the normal MS spectrum. Do you have an explanation for this effect? What was the collision energy when this effect was observed?

Answer. The collision energy for MS acquisition is normally set to 4V and it was also used for the acquisition with argon. I cannot explain what the reason could be for the higher intensity and the better s/n ratio.

Scott Patterson (Amgen, Inc.)

It has been published that IgM mediated apoptosis results in decreased amounts of Numatrin, and in other systems that it is a carpase substrate. Have you observed this?

Answer. We have not identified Numatrin up to now in our gels.

Scott Patterson (Amgen, Inc.)

Necrotic cells are also Annexin V positive. Are they bound by the magntic beads?

Answer. The picture I have shown was made after taxol treatment for 72 hours. In the case of anti-IgM mediated apoptosis we did not observe necrotic cells. It was checked with fluorescence microscopy.

IR-MALDI – Softer Ionization in MALDI-MS for Studies of Labile Macromolecules

R. Cramer and A. L. Burlingame
Ludwig Institute for Cancer Research, University College of London, London W1P 8BT, U.K.

In the last decade matrix-assisted laser desorption/ionization (MALDI) has become a powerful tool in biological mass spectrometry (MS). At present, electrospray ionization (ESI) and MALDI are the two major soft ionization techniques for the intact analysis of macromolecules by mass spectrometry. While MALDI-MS provides inherently fast data acquisition and spectra facile for data analysis, ESI is considered to be the softer of the two techniques and is therefore the method of choice for the analysis of labile substituents of posttranslationally modified peptides and proteins, as well as studies of non-covalent complexes.

Recently, it has been shown that the use of infrared (IR) lasers induces significantly less fragmentation in MALDI [1-4]. In contrast to UV-MALDI, peak tailing on the low mass side is virtually absent in spectra of high mass molecules, and even the reflector mode can be used for fragile macromolecules. Hence, biomolecules such as sialylated glycopeptides or phosphopeptides can be analyzed intact without losing labile substituents, while preserving all the benefits of the reflector mode operation [4].

Although there is evidence that IR-MALDI can provide valuable information complementary to UV-MALDI, it is currently less established than its sister technique. This may be caused by the later introduction of IR lasers in MALDI [5, 6] and their relatively high costs. Besides these more historic reasons, the macroscopic ablation (rather than desorption) characteristic of IR-MALDI makes it more dependent on sample morphology. As a consequence, sample consumption is relatively high, thus limiting the number of productive shots per sample spot location (~5-10). Hence, skillful sample scanning is required because of the variable shot-to-shot ion signal intensity. In addition, somewhat lower resolution is normally observed in the low mass range. However, with the emergence of new sample preparation techniques providing thin homogeneous films, more stable new laser designs and delayed extraction, shot-to-shot reproducibility and mass resolution comparable to UV-MALDI can be now easily achieved, i. e., more than 10,000 (or less than 10 ns time resolution). Based on these developments it may be anticipated that IR-MALDI will play a significant role in the characterization of macromolecules in the future.

In addition, use of the right matrix/wavelength combination provides the advantages inherent in the softer ionization characteristic of ESI MS. Hence IR-MALDI eliminates the limitations of conventional UV-MALDI that arise from the deposition of significant amounts of internal energy.

This chapter provides a brief overview of the current performance of IR-MALDI achievable using commercial instrumentation. The potential of IR-MALDI is presented as a summary of our recent research efforts exploring the advantages of IR-MALDI and its differences with UV-MALDI. These advantages and differences might prove to be immensely valuable, particularly in carrying out proteome projects that require concerted mass spectrometric effort to achieve high throughput analysis of 2D-gel separated proteins.

MALDI-MS

All MALDI-MS data was obtained with a Voyager Elite/Elite XL (PerSeptive Biosystems, Framingham, MA). Both instruments are equipped with delayed extraction and a reflector analyzer for improved mass resolution and accuracy. Data acquisition was achieved with a TDS 540C/744A digitizing oscilloscope (Tektronix, Inc., Beaverton, OR). No numerical data processing such as smoothing or baseline correction was performed on any of the spectra presented.

For UV-MALDI, the standard VSL-337ND nitrogen laser (Laser Science, Ltd., Newton, MA) was used as provided with the mass spectrometers. For IR-MALDI, a Speser 15Q Er:YAG laser (Spektrum GmbH, Berlin, Germany) was interfaced to the mass spectrometers and employed as reported earlier [4]. The laser beam was focused on the target to a spot size of approximately 1 x 10^{-3} cm^2. Laser irradiance was in the 10^6-10^7 W/cm^2 range. The spatial laser beam profile is specified as Gaussian by the manufacturer. The comparisons between IR- and UV-MALDI were conducted under identical or comparable conditions. It has to be noted that all spectra were recorded by a non-selective accumulation of at least 32 consecutive shots.

α-Cyano-4-hydroxycinnamic acid (CHCA) was purchased in solution from Hewlett-Packard GmbH (Böblingen, Germany) as a ready-to-use matrix. All other matrix solutions were made up in the laboratory with compounds purchased from Sigma (Poole, Dorset, U.K.) or Aldrich (Gillingham, Dorset, U.K.). Typically, saturated aqueous or 50:50 (v:v) ethanol/water solutions were prepared without any prior purification. In the case of succinic acid, this solution was diluted by a factor of 4 in the same solvent. Samples were prepared using the "dried droplet" method, which involves mixing 0.5 µl of the analyte solution with 0.5-1 µl of the matrix solution on the target and drying by means of a warm stream of air. For high resolution measurements, a thin film preparation technique was used alternatively. This was performed by making a saturated solution of succinic acid in acetone in which a ratio of 1-10:100 (v:v) of a 10-100 µM aqueous analyte solution was added. Aliquots of approximately 0.5 µl of the final solution were then spotted on the target, resulting in a quick spread and evaporation of the solvent.

Properties of IR-MALDI Employing Commercial Instrumentation

At present, most of the available data about the characteristics of IR-MALDI-MS has been reported by the inventors of MALDI (e. g., [1, 2, 5-10]). Only a few other groups have pursued research in IR-MALDI thus far (see [3, 11-13]).

	In-House Built Instruments (IMPB, Münster)	**Commercial Instruments (LICR, London & UCSF)**
Sample Consumption	100-1000 times more than in UV-MALDI	100-1000 times more than in UV-MALDI
Ion Signal Intensity	Single shot intensity is comparable to UV-MALDI	Single shot intensity is comparable to UV-MALDI, but no or almost no ion signal in special applications (glycerol as matrix or proteins from electroblotted membranes)
Analyte-to-Matrix Ratio on the Target	Particularly in low mass range optimal analyte concentration in IR-MALDI is 10-100 times higher than typical analyte concentration in UV-MALDI	Typically the same as in UV-MALDI
Mass Range	> 500 kDa in reflector mode	< 50 –150 kDa in reflector mode
Max. Mass Resolution	10000 for peptides (1-6 kDa) 1500 for proteins (cytochrome c)	10000-15000 for peptides (1-6 kDa) 2500-3000 for proteins (10-25 kDa)
Mass Accuracy	100-5000 ppm up to 30 kDa (prompt extraction, external calibration)	< 100 ppm for peptides < 50 ppm for proteins (delayed extraction, external calibration) < 30 ppm for peptides (delayed extraction, internal calibration)

Table 1. IR-MALDI-MS performance.

Hillenkamp *et al.* have reported data for the performance of IR-MALDI in the 3 μm wavelength range obtained with a mass spectrometer built in-house using erbium lasers [2]. Studies utilizing different laser sources suggest that in principal, with laser wavelengths around 5.5 – 6 μm (FEL laser) and at 10.6 μm (CO_2 laser) performance is comparable to the standard performance using the typical IR-MALDI laser wavelength of 2.94 μm (Er:YAG laser) [14, 15]. Other wavelengths investigated so far produced only inferior results [12, 14].

Experience using our instrumentation to obtain IR-MALDI results has been comparable to those reported by Hillenkamp's group with a few but important exceptions (see also Table 1). In our case high mass molecules in excess of 50 kDa can hardly be detected in the reflector mode operation. We have not been particularly successful in the use of liquid matrices such as glycerol or in the analysis of proteins electroblotted directly from membranes. We assume that these observations are due to different ion source and mass analyzer designs limiting the extraction and transmission of analyte ions under certain conditions.

However, these differences in design might also be one of the reasons for the somewhat better mass resolution that we have achieved. Our mass resolution in the reflector mode for protein standards in the 10 – 25 kDa mass range is a factor of 1.5 – 2 higher, e. g., for cytochrome c mass resolution is around 2500 (FWHM), and for myoglobin up to 3000. As a higher mass research example, the TATA binding protein (TBP) from *Pyrococcus woesei* is shown in Fig. 1. In the reflector mode a mass resolution of more than 2300 can be obtained readily. This

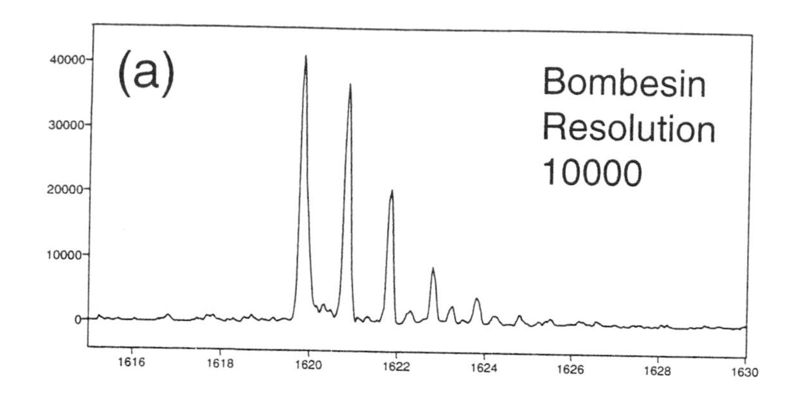

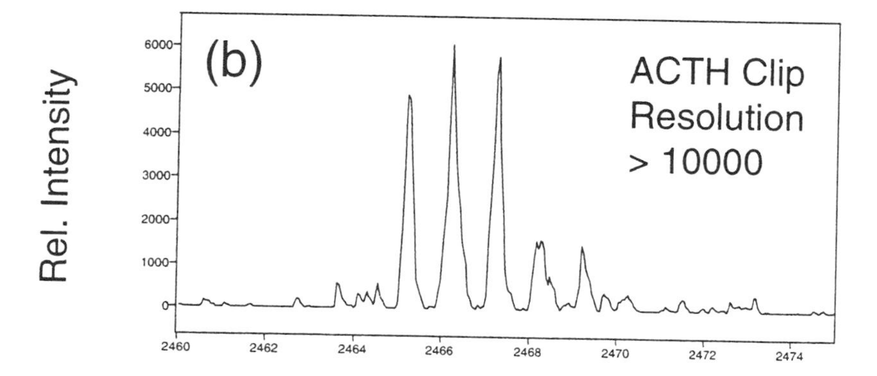

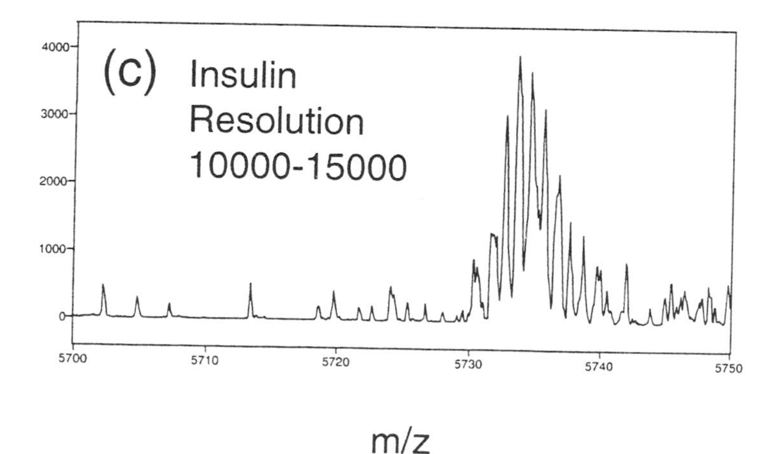

m/z

Fig. 1. MALDI mass spectra of TATA Binding Protein (Pyrococcus Woesei). Total amount of loaded analyte: ~ 5 pmol. (a) Reflector Mode, matrix: DHB, wavelength: 337 nm. (b) Reflector Mode, matrix: succinic acid, wavelength: 2.94 μm. (c) Linear Mode, matrix: DHB, wavelength: 337 nm. (d) Linear Mode, matrix: succinic acid, wavelength: 2.94 μm.

indicates that resolution is limited virtually only by the monoisotopic envelope. In an attempt to use UV-MALDI in this case, the signal intensity was not sufficient to achieve mass resolution at all. In contrast to the situation using UV-MALDI, the IR analog provides information that permits distinction between the two apparent species, TBP and TBP minus N-terminal methionine, even in the linear mode. Achieving this result with UV-MALDI is most likely thwarted by the higher metastable decay and/or higher adduct ion formation.

In our experience it is more difficult to obtain good resolution in the lower mass range (1 - 6 kDa). The investigation of different sample preparation techniques revealed that the use of thin sample films will facilitate the recording of spectra with a resolution comparable to that achievable using UV-MALDI. Hence, these results indicate that sample morphology has a strong influence on mass resolution in IR-MALDI. Obviously, such an influence can be explained by the fact that only a few laser shots are necessary to eradicate sample spots with crystals as large as 100 μm. This reflects the extreme spread of the locations of ion formation, and therefore much broader kinetic ion energy distributions in IR compared with UV-MALDI. Thus laser beam instabilities can be excluded as the main source for the impaired resolution often experienced. By using appropriate sample preparation techniques and/or careful selection of sample spots, mass resolution of 10,000 – 15,000 can be achieved, as shown in Fig. 2.

The high rate of sample ablation in IR-MALDI naturally influences not only mass resolution but also mass accuracy. Nonetheless, when employing α-cyano-4-hydroxycinnamic acid (CHCA) as matrix for measurements in the peptide mass range, the mass measurement accuracy is only somewhat (approximately a factor of 2) worse than in UV-MALDI. In this case the accuracy is typically better than 20 ppm (internal calibration). A slightly further deterioration can be observed with succinic acid as matrix. In the protein mass range the situation is quite different. The influence of high sample ablation appears to be negligible. In this case mass accuracy is comparable to or even surpasses that of UV-MALDI. For TBP, a mass accuracy of 30 ppm was achieved by external calibration using insulin, thioredoxin and myoglobin as calibrants (Fig. 1b). The typical value for proteins is 50 ppm with external calibration. This superior performance may be attributed to both the lower degree of fragmentation and adduct ion formation that both skew the centroid in an unpredictable manner.

IR-MALDI-MS of Posttranslationally Modified Peptides

While mass spectrometric technologies have become accepted as the methods of choice for the structural elucidation of posttranslational modifications of proteins, some of these present challenging tasks. Some biologically critical modifications such as phosphorylation and glycosylation represent naturally complex mixtures of modifications to a single amino acid sequence (protein) caused by differential site occupancy and stoichiometry, that may vary with the physiological status of the cell or tissue due to regulatory processes. Some of these structures are chemically labile, and thus prone to fragment easily when subjected to internal energy deposition during the ionization process. In this context electrospray ionization has proven to be extremely important,

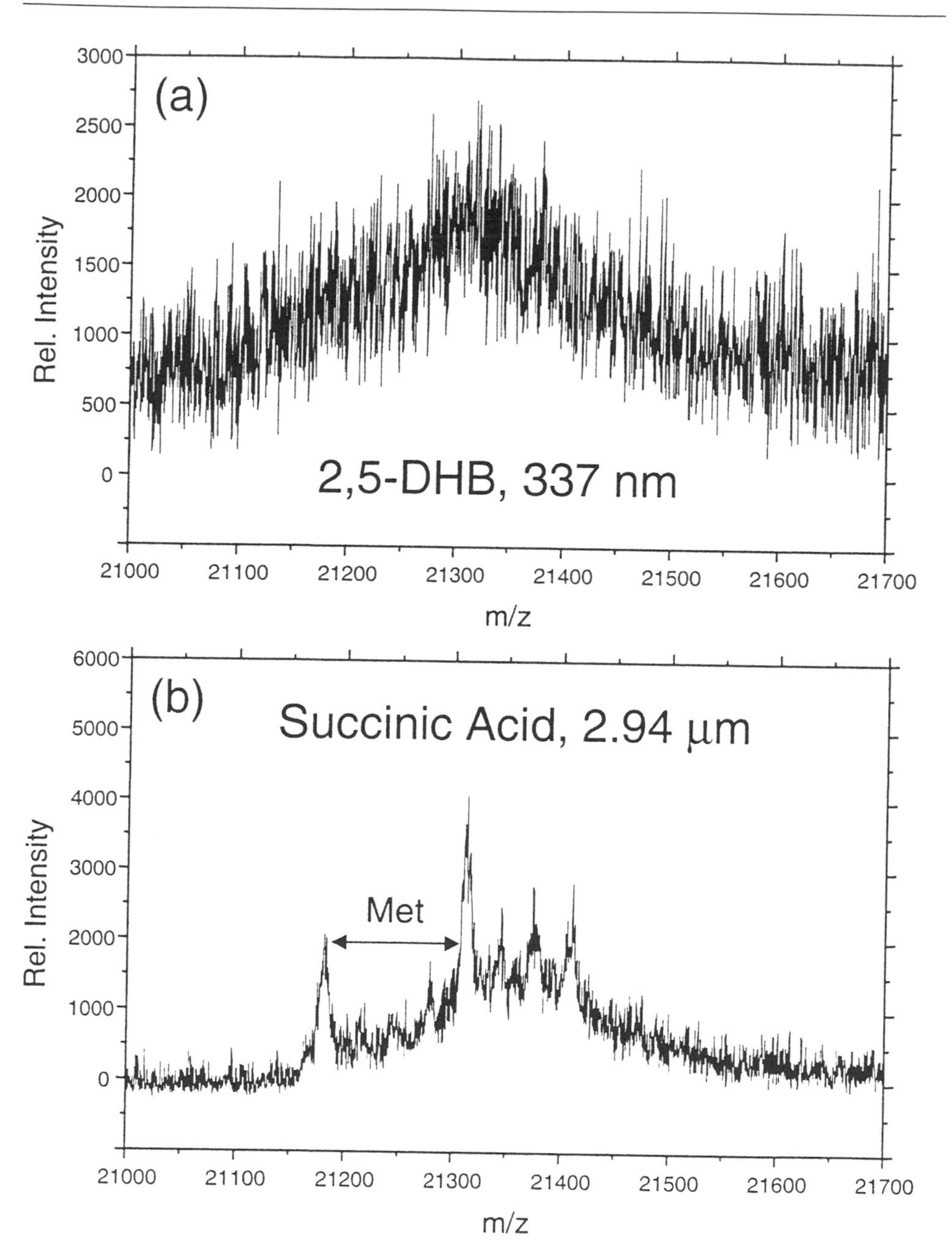

Fig. 2a and b. IR-MALDI mass spectra of bombesin (a), adrenocorticotropic hormone (ACTH) clip ^{18}R-F^{39}, and bovine insulin (c). Matrix: succinic acid, wavelength: 2.94 μm. Total amount of loaded analyte: < 1 pmol.

usually outperforming conventional MALDI whenever detection of intact native structures is required. As we have recently reported, IR-MALDI appears destined to close the gap between UV-MALDI and electrospray ionization by providing the necessary soft ionization conditions [4]. Figure 3 shows spectra of

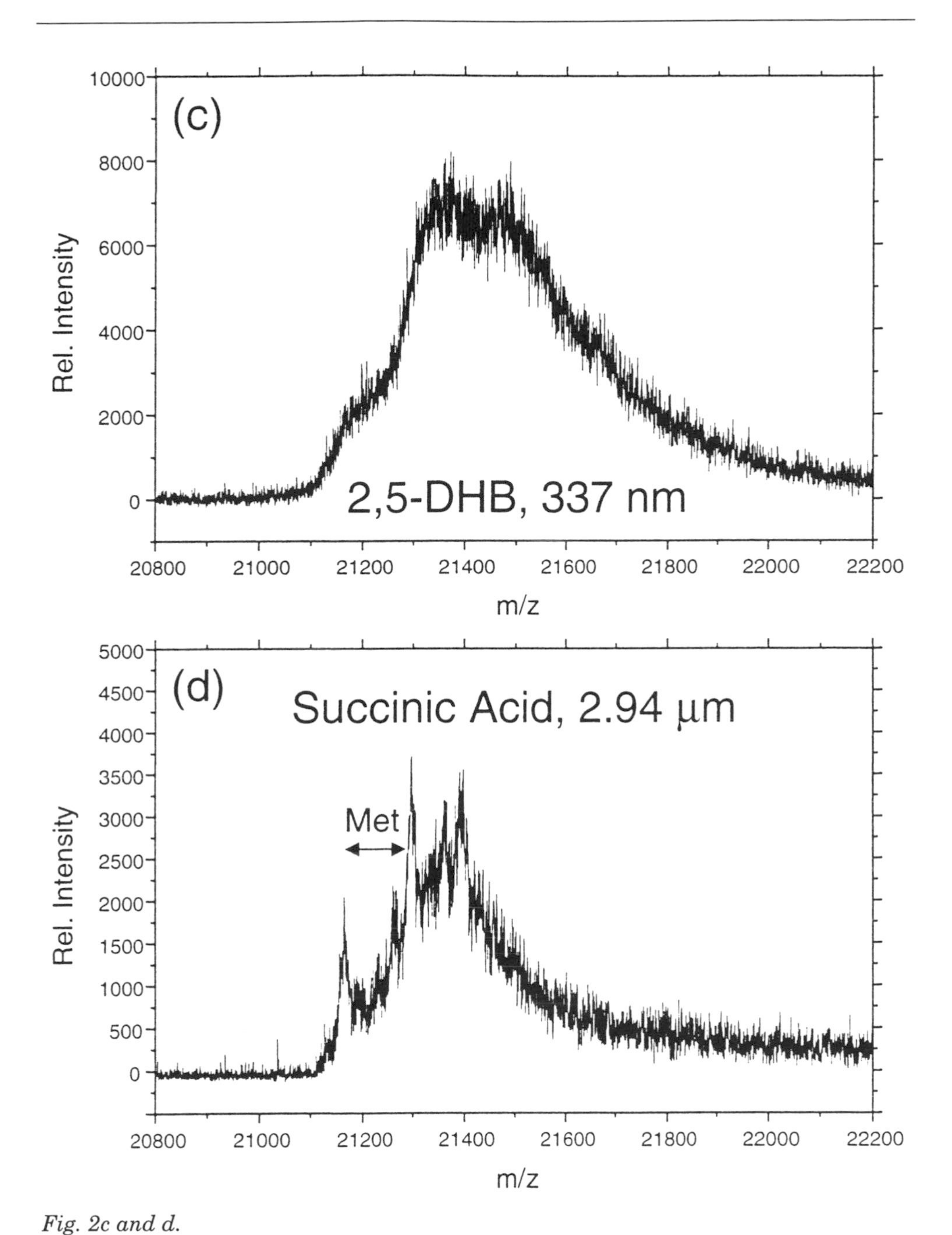

Fig. 2c and d.

an HPLC-separated fraction of a tryptic digest of bovine fetuin optimized for both IR- and UV-MALDI in the reflector mode. Although the latter shows extensive fragmentation, i. e., loss of sialic acid, only marginal degradation is observed in the IR-MALDI spectrum. In fact, a comparison to the linear mode spectrum (data not shown) reveals that the relative abundances of the ion signals are virtually identical.

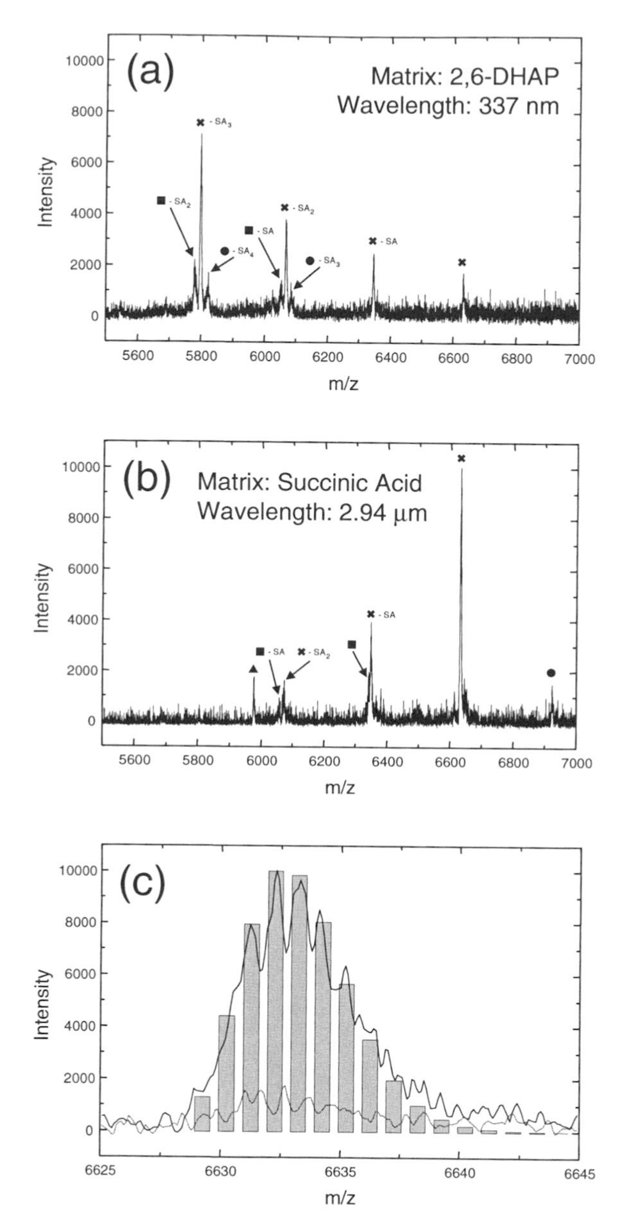

Fig. 3. MALDI reflector mass spectra of HPLC-separated tryptic glycopeptides from bovine fetuin. HPLC-separation and tryptic digestion are described in an earlier report [4]. All indicated losses of sialic acid (-SA_x) are a result of metastable decay. Total amount of loaded analyte: ~ 1 pmol. Peptide sequence: P = [54]RPTGEVYDIEIDTLETTBHVLDPTP LANBSVR[85] (B = S-β-4-pyridylethyl-cysteine). Glycopeptides[1]: [54]R-[85]R + $Hex_6HexNAc_5SA_4$ (●, MW_{av} = 6,923.2), [54]R-[85]R + $Hex_6HexNAc_5SA_3$ (✖, MW_{av} = 6,631.9), [54]R-[85]R + $Hex_6HexNAc_5SA_2$ (■, MW_{av} = 6,340.6), [54]R-[85]R + $Hex_5HexNAc_4SA_2$ (▲, MW_{av} = 5,975.3). (a) Matrix: 2,6-dihydroxyacetophenone (2,6-DHAP), wavelength: 337 nm. (b) Matrix: succinic acid, wavelength: 2.94 μm. (c) Enlarged section of mass spectra (a) (straight line) and (b) (dotted line) showing the glycopeptide [54]R-R[85] + $Hex_6HexNAc_5SA_3$ ion signal. Its theoretical isotope pattern (shifted by 0.4 u) is underlaid for comparison (columns) [4].

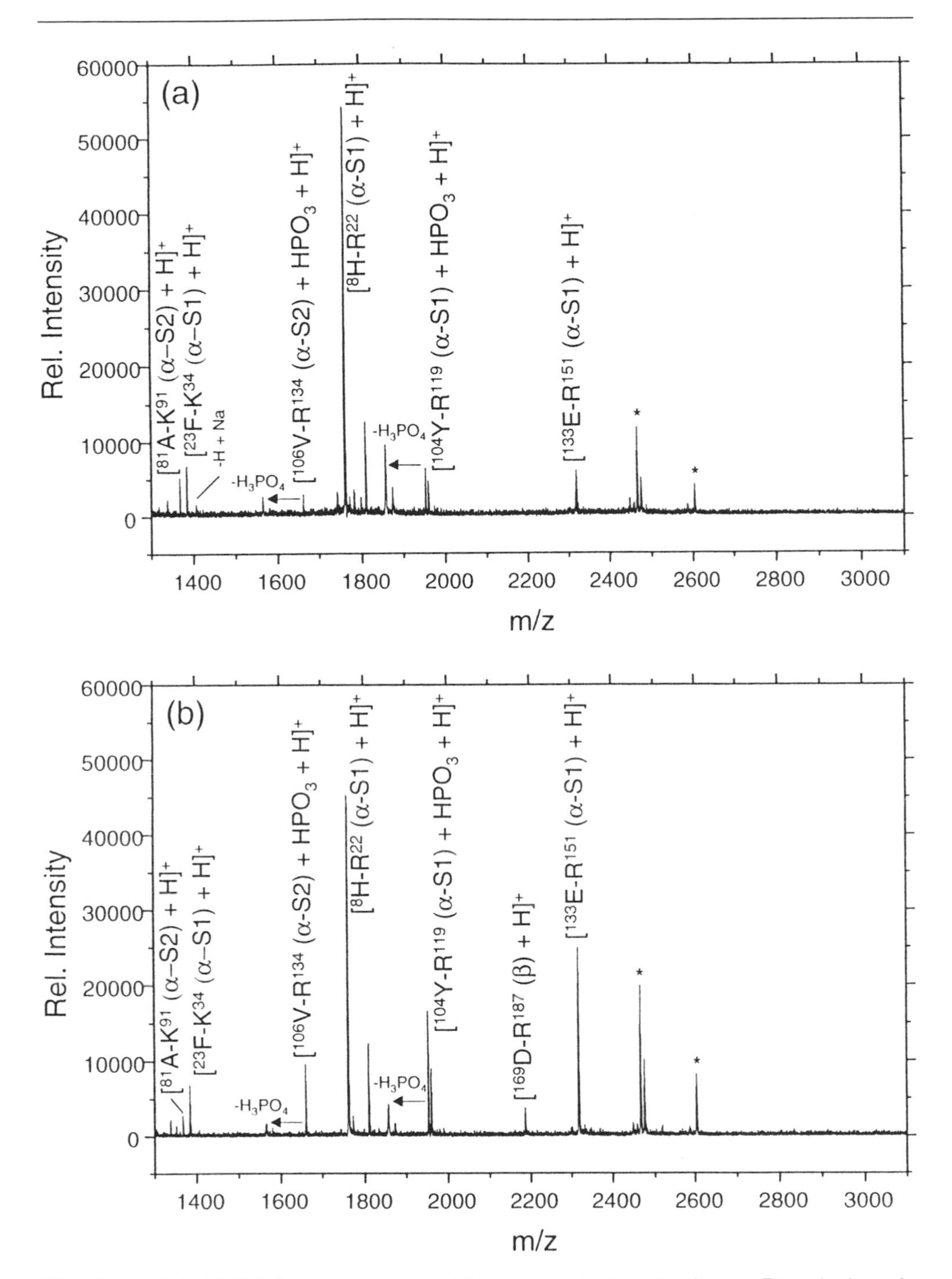

Fig. 4a and b. MALDI mass spectra of bovine casein tryptic digest. Descriptions in parenthesis refer to the different types of casein in the sample. Peaks marked with an asterisk indicate ion signals introduced by the added calibration solution. The tryptic digestion of bovine casein is described in an earlier report [4]. (a) Matrix: CHCA, wavelength: 337 nm. (b) Matrix: CHCA, wavelength: 2.94 µm. (c) Matrix: DHB, wavelength: 337 nm. (d) Matrix: succinic acid, wavelength: 2.94 µm.

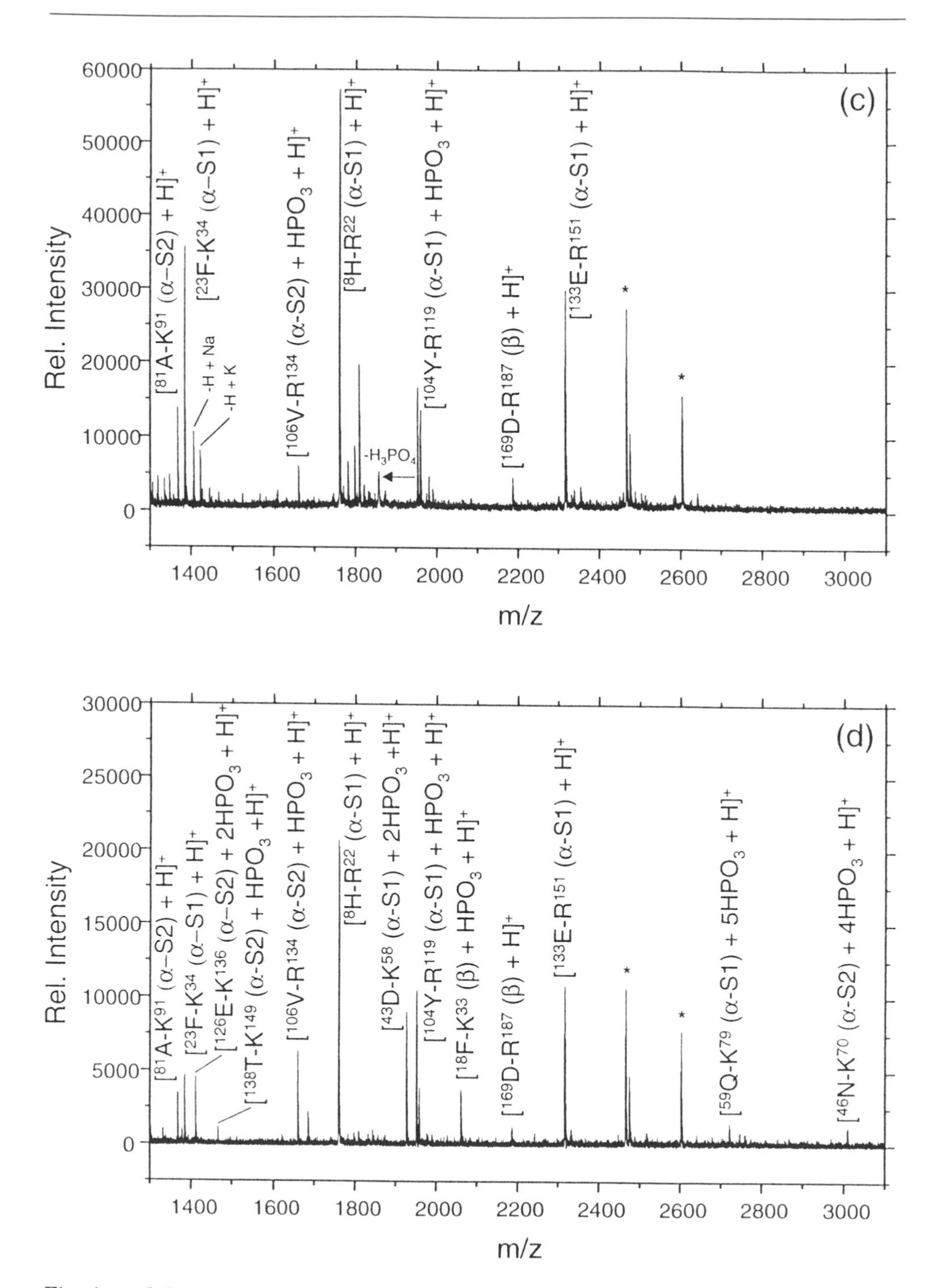

(c)
Rel. Intensity
m/z
[81A-K91 (α-S2) + H]+
[23F-K34 (α-S1) + H]+
-H + Na
-H + K
[106V-R134 (α-S2) + HPO3 + H]+
[8H-R22 (α-S1) + H]+
-H3PO4
[104Y-R119 (α-S1) + HPO3 + H]+
[169D-R187 (β) + H]+
[133E-R151 (α-S1) + H]+
(d)
[81A-K91 (α-S2) + H]+
[23F-K34 (α-S1) + H]+
[126E-K136 (α-S2) + 2HPO3 + H]+
[138T-K149 (α-S2) + HPO3 +H]+
[106V-R134 (α-S2) + HPO3 + H]+
[8H-R22 (α-S1) + H]+
[43D-K58 (α-S1) + 2HPO3 +H]+
[104Y-R119 (α-S1) + HPO3 + H]+
[18F-K33 (β) + HPO3 + H]+
[169D-R187 (β) + H]+
[133E-R151 (α-S1) + H]+
[59Q-K79 (α-S1) + 5HPO3 + H]+
[46N-K70 (α-S2) + 4HPO3 + H]+

Fig. 4c and d.

As has been reported earlier, the loss of labile substituents in the analysis of phospho- and glycopeptides by conventional MALDI is attributed to metastable decay. Although metastable decay has only little effect on ion detection in the linear mode, some degradation in spectral quality is unavoidable (compare Fig. 1c/d). However, extensive metastable decay makes it impossible to detect ions in the reflector mode, thus losing the benefits of high mass accuracy and resolution as well. Employing IR-MALDI these benefits are preserved as seen in Fig. 3c, which shows the ion signal comparison for the main glycoform of this particular HPLC-fraction.

The merits of IR-MALDI are also seen in the analysis of phosphopeptides as shown in Fig. 4. Three different matrices were used to analyze the peptide mixture of a bovine casein tryptic digest. With CHCA, a direct comparison between UV- and IR-MALDI using exactly the same sample can be attained (Fig. 4a/b). With 2,5-dihydroxybenzoic acid (DHB) and succinic acid, best conditions were created for UV- and IR-MALDI, respectively. In this comparison it can be seen that metastable decay depends to a greater extent on the correct choice of the matrix and not the wavelength. Although in the worst case (CHCA, 337 nm, Fig. 4a) only two phosphopeptides are detected with a corresponding metastable ion signal greater than the intact ion signal, employing succinic acid using the Er:YAG laser (Fig. 4d) eight phosphopeptides bearing up to five moles of phosphate are evident. Moreover, there are no significant metastable ion signals. In addition, it seems that for the ^{23}F-K^{34} (α-S1) peptide less alkali metal adduct ion formation occurs in IR-MALDI. Note that the DHB spectrum was recorded by accumulating shots from the sample rim, known to minimize such formation [16].

Summarizing the results obtained for glyco- and phosphopeptides with regard to the degree of metastable decay, the following scale can be drawn using the metastable-to-intact ion signal ratio: Succinic Acid (IR), Fumaric Acid (UV) <<<< DHB (IR) < 2,6- Dihydroxyacetophenone (UV) < DHB (UV) << CHCA (IR) <<< CHCA (UV) (the number of arrow characters indicates the degree of difference in metastable decay, starting with succinic acid exhibiting the least metastable decay).

The relative positions for DHB and 2,6-dihydroxyacetophenone are less determined and have to be permutated depending on sample preparation as well as analyte. Employing DHB in the infrared gives less ion signal than in all other cases, complicating an exact and reliable placement on this scale. As it can be seen, it is not just the wavelength but the combination of both wavelength and matrix determining the degree of metastable decay, i. e., fragmentation.

IR-MALDI-MS of SDS-PAGE In-Gel Digest Peptide Mixtures

Methodologies based on MALDI-MS have become indispensable for the rapid identification of protein at the subpicomole level. This development represents a critical component in mass mapping and database interrogation strategies aimed at rapid characterization of SDS-PAGE-separated proteins

Identification	Hit Rate for 10 Major Peaks			Hit Rate for Peaks with S/N > 3:1			
	DHB (UV)	Succinic Acid (IR)	CHCA (UV/IR)	DHB (UV)	Succinic Acid (IR)	CHCA (UV&IR)	All
Actin	6/10	10/10	-	13/28	13/19	-	15/34
Myosin*	9/10	9/10	7/10 (UV) 8/10 (IR)	31/44	39/46	37/48	65/85
Vimentin	5/10	9/10	-	13/37	23/55	-	25/78
Tropomyosin	2/10	10/10	7/10 (UV) 7/10 (IR)	11/39	18/27	12/32	20/71
Elongation Factor 1-β	3/10	0/10	1/10 (UV) 1/10 (IR)	5/39	4/27	4/32**	8/71

* Identified as non-muscle myosin heavy chain-A (*Rattus norvegicus*). No homologue for *Mus musculus* is present in database.
** 3/32 for UV-MALDI; 3/32 for IR-MALDI

Table 2. Characterization of SDS-PAGE separated proteins of immortalized dendritic mouse cells (cytoplasmic extract) by peptide mass mapping after in-gel digestion.

using unseparated proteolytic digest mixtures. Inherently MALDI offers the fastest spectral acquisition covering the whole peptide mass range in one spectrum, well suited to high throughput requirements. Unfortunately, it is well known that peptide signal suppression is a common feature in mixture analyses by MALDI [16-18].

As discussed above, analyses of labile posttranslationally modified proteins benefit directly from the employment of infrared lasers in MALDI. Therefore it was of interest to explore whether IR-MALDI could be employed to minimize suppression effects in peptide mixtures, particularly those obtained from in-gel digestion of protein bands. To this end SDS-PAGE-separated gel bands from an in-house project were investigated using both UV- and IR-MALDI employing a variety of matrices. The gel bands were cut out and subjected to in-gel digestion using a protocol reported elsewhere [19]. Using the mass values of the peptides detected in each case, database searches were carried out and the results obtained are summarized in Table 2. For these searches ion signals were required to match peptide masses with an accuracy of 50 ppm. Two different sets of detected ion signals from each spectrum were used for the four gel bands investigated. The first set of ion signals was selected by the peak height, whereas the second set was obtained by choosing those ion signals with a signal-to-noise ratio of at least 3:1. Known contaminants from keratin or the gel were excluded prior to selection. The identification of the proteins is based on the highest peptide mass match using the assigned peaks of all MALDI samples fulfilling the above signal-to-noise ratio. Interestingly, in almost all cases IR-MALDI generates a higher number of specific peptide ion signals than UV-MALDI. In

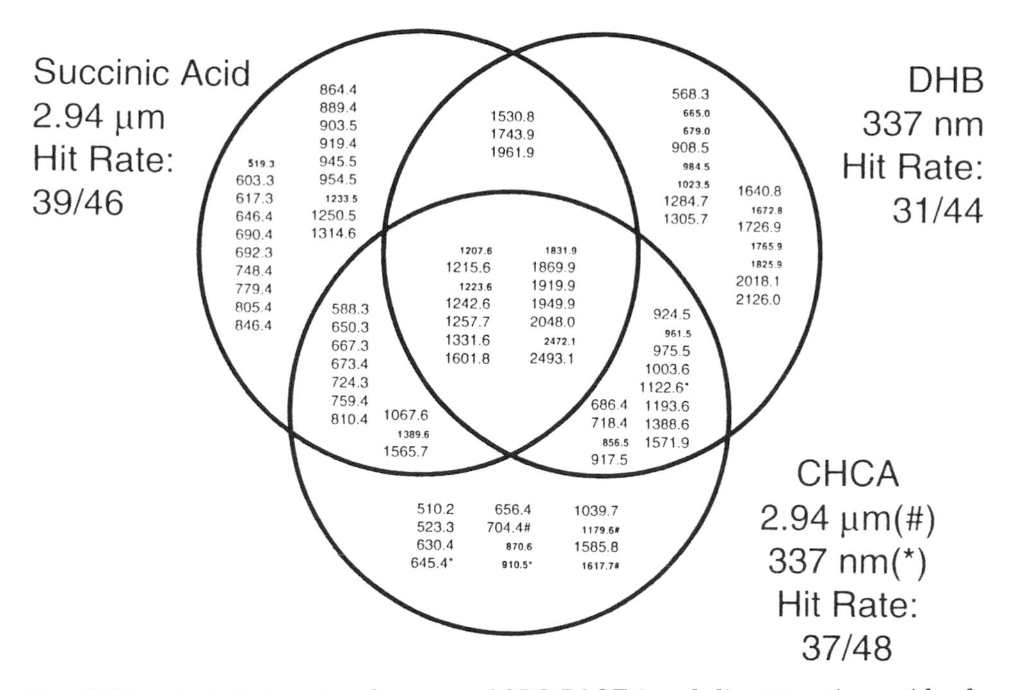

*Fig. 5. Monoisotopic ion signal masses of SDS-PAGE in-gel digest tryptic peptides for database search for protein identification classified according to their occurrence. Peptide selection criterion: S/N > 3:1. Protein identified as non-muscle myosin heavy chain-A. Same data as used for row 2 in Table 2. Large numbers indicate matched peptide masses for myosin; small numbers indicate no matches. For CHCA both UV- and IR-MALDI data were acquired and combined because of high redundancy. Masses marked with # or * appear only in IR- or UV-MALDI, respectively.*

particular, the ratio of specific to non-specific peaks is substantially higher in IR-MALDI. For vimentin a more stringent signal-to-noise criterion of 5:1 gives an even higher specific-to-non-specific peak ratio of 19/29, with a much higher absolute number of positive hits but lower number of selected peaks compared to UV-MALDI. The same can be seen for tropomyosin, with the signal-to-noise ratio of 3:1. The gel band containing tropomyosin actually consisted of at least one more protein. This was confirmed by analysis employing nanoelectrospray tandem mass spectrometry. For the second protein showing less peptide signals, UV-MALDI exhibits a higher hit rate on the basis of the peak height criterion. This is no surprise considering the 100% hit rate for the first protein in IR-MALDI. Taking the signal-to-noise criterion for ion signal selection, this imbalance then evens out.

In Fig. 5 the actual mass values of the peaks selected for database search are given with regard to their occurrence in the corresponding measurement for the myosin gel band. This plot shows that even without considering one or the other technique as superior in performance, a significant advantage can be gained by using more than one matrix/wavelength combination. For all bands analyzed the use of succinic acid with IR-MALDI provided additional information leading to a higher overall number of matched peptide mass values, i. e.,

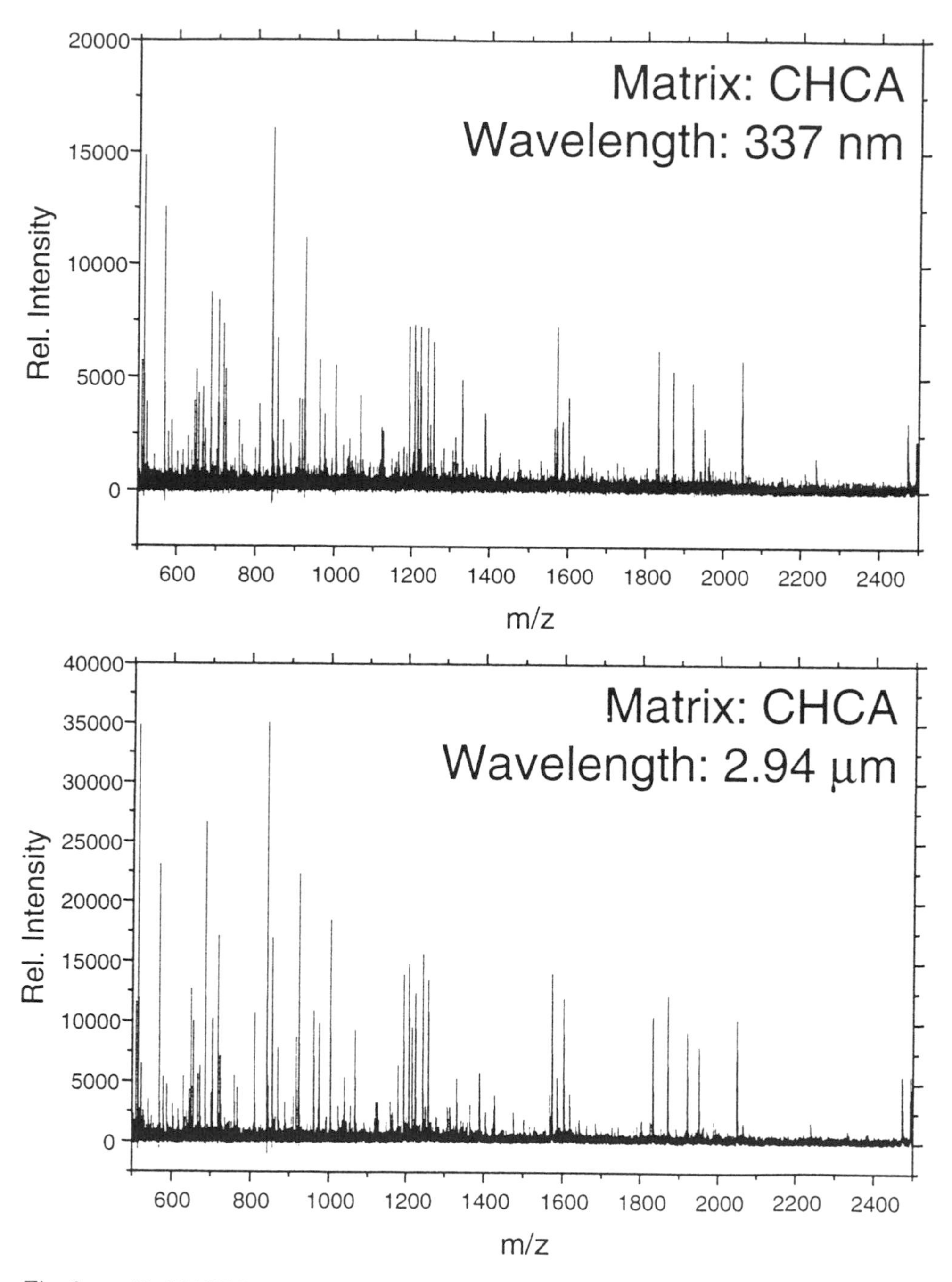

Fig. 6a and b. MALDI mass spectra of SDS-PAGE in-gel digest peptide mixture identified as tryptic non-muscle myosin heavy chain-A using ion signals from all spectra with a signal-to-noise ratio of more than 3:1 (compare with Fig. 5 and Table 2).

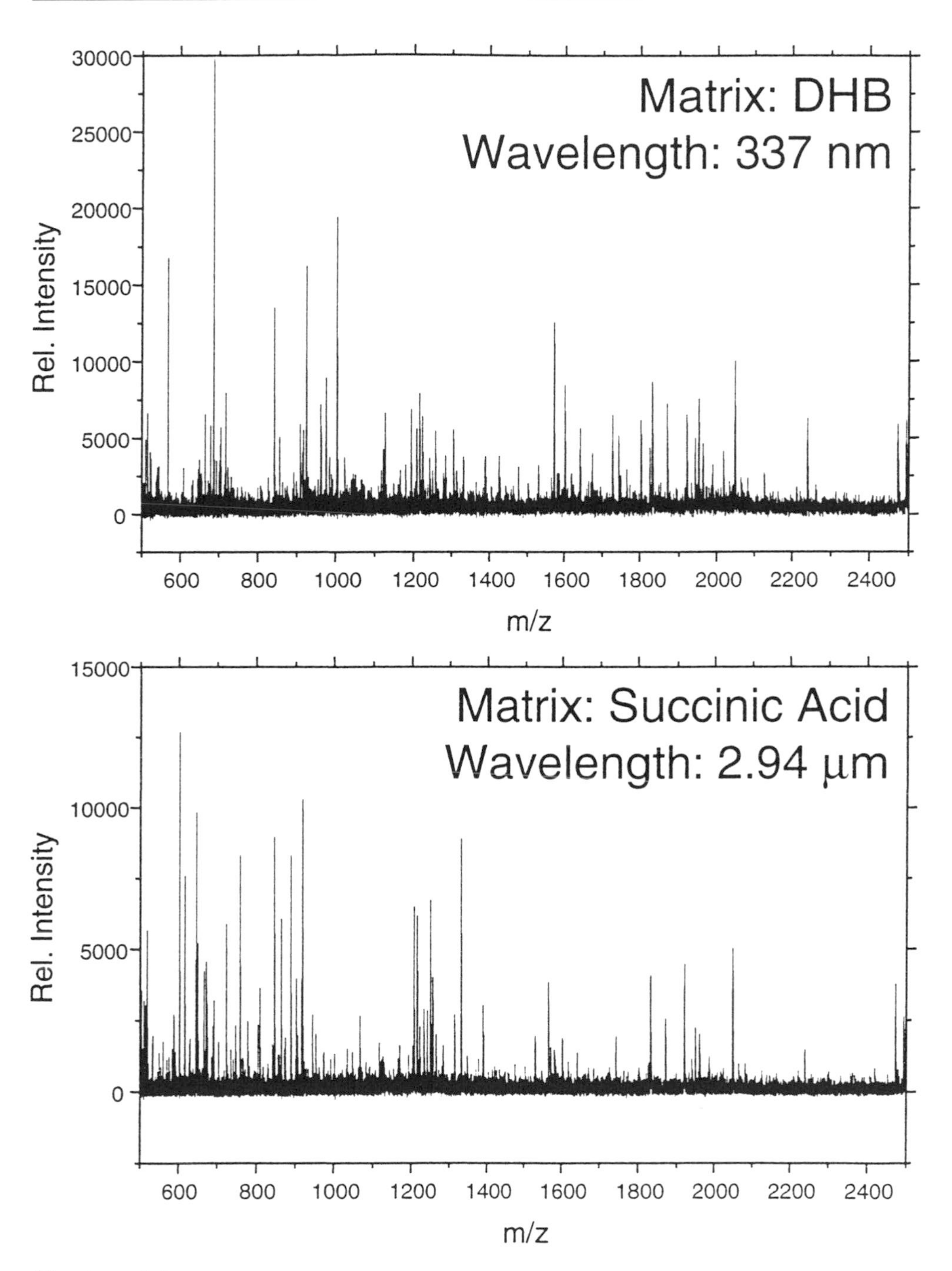

Matrix: DHB
Wavelength: 337 nm
Rel. Intensity
m/z
Matrix: Succinic Acid
Wavelength: 2.94 μm
Rel. Intensity
m/z

Fig. 6c and d.

higher coverage of the protein sequence and thus more confidence in protein identification.

Figure 6 shows the peptide mixture spectra of the myosin gel band. As observed in all measurements, the peptide ion signal patterns differ significantly comparing UV-MALDI using DHB as matrix to IR-MALDI with succinic acid as matrix. Almost no differences, however, can be found by direct comparison of both modes with CHCA as matrix. Again it seems that the matrix plays the more important role in the desorption/ionization process.

Conclusions

As presented in this chapter, IR-MALDI exhibits approximately the same performance as UV-MALDI with regard to mass resolution, mass accuracy and the analytically relevant sensitivity, i. e., total amount of loaded sample. There are two major differences between these two MALDI techniques. The first is the much higher overall sample consumption per laser shot in IR-MALDI, resulting in a significantly higher dependence on sample morphology and therefore the requirement of skillful sample spot scanning. On the other hand, IR-MALDI and the right choice of matrix induces significantly less metastable decay. This enables the intact detection of biomolecules with labile substituents such as sialylated glycopeptides and phosphopeptides even in reflector mode. Less metastable decay can also be assumed to be the main reason for the observed higher mass resolution in the high mass range.

In addition, IR-MALDI offers an inherently larger pool of potential matrices. As presented, the use of different matrices produces significantly more information for the analysis of tryptic peptide digest mixtures leading to more hits in a peptide mass mapping-based database search for protein identification. It seems that IR-MALDI with succinic acid as matrix generally provides more specific peptide ion signals and a higher hit rate in database searches than UV-MALDI using DHB or CHCA as matrices.

Acknowledgements

Financial support was provided by the Ludwig Institute for Cancer Reseach, London Branch. We wish to thank Ronan O'Brien for providing the TBP sample and Sorena Kiani-Alikhan for the cytoplasmic extract and his work on the in-gel digestion. Helpful discussions with Elaine Stimson are greatly appreciated.

References

1. F. Hillenkamp, M. Karas and S. Berkenkamp, *Proc. 43rd ASMS Conference on Mass Spectrometry and Allied Topics*, 21-26 May, Atlanta, GA, **1995**, 357.

2. S. Berkenkamp, C. Menzel, M. Karas and F. Hillenkamp, *Rapid Commun. Mass Spectrom.* **1997**, *11*, 1399-1406.

3. W. Zhang, S. Niu and B. T. Chait, *J. Am. Soc. Mass Spectrom.* **1998**, *9*, 879-884.

4. R. Cramer, W. J. Richter, E. Stimson and A. L. Burlingame, *Anal. Chem.*, **1998**, *70*, 4939-4944.

5. A. Overberg, M. Karas, U. Bahr, R. Kaufmann and F. Hillenkamp, *Rapid Commun. Mass Spectrom.* **1990**, *4*, 293-296.

6. A. Overberg, M. Karas and F. Hillenkamp, *Rapid Commun. Mass Spectrom.* **1991**, *5*, 128-131.

7. E. Nordhoff, F. Kirpekar, M. Karas, R. Cramer, S. Hahner, F. Hillenkamp, K. Kristiansen, P. Roepstorff and A. Lezius, *Nucleic Acids Res.* **1994**, *22*, 2460-2465.

8. R. Cramer, R. F. Haglund, Jr. and F. Hillenkamp, *Int. J. Mass Spectrom. Ion Proc.* **1997**, *169-170*, 51-68.

9. C. Eckerskorn, K. Strupat, D. Schleuder, D. Hochstrasser, J.-C. Sanchez, F. Lottspeich and F. Hillenkamp, *Anal. Chem.* **1997**, *69*, 2888-2892.

10. S. Berkenkamp, F. Kirpekar and F. Hillenkamp, *Science* **1998**, 281, 260-262.

11. M. Sadeghi, Z. Olumee, X. Tang, A. Vertes, Z.-X. Jiang, A. J. Henderson, H. S. Lee and C. R. Prasad, *Rapid Commun. Mass Spectrom.* **1997**, *11*, 393-397.

12. K. L. Caldwell, D. R. McGarity and K. K. Murray, *J. Mass Spectrom.* **1997**, *32*, 1374-1377.

13. S. Niu, W. Zhang and B. T. Chait, *J. Am. Soc. Mass Spectrom.* **1998**, *9*, 1-7.

14. R. Cramer, F. Hillenkamp and R. F. Haglund, Jr., *J. Am. Soc. Mass Spectrom.* **1996**, *7*, 1187-1193.

15. C. Menzel, S. Berkenkamp, F. Kirpekar, D. Schleuder, K. Strupat and F. Hillenkamp, *Proc. 46th ASMS Conference on Mass Spectrometry and Allied Topics*, Orlando, FL, **1998**, 797.

16. M. Kussmann, E. Nordhoff, H. Rahbek-Nielsen, S. Haebel, M. Rossel-Larsen, L. Jakobsen, J. Gobom, E. Mirgorodskaya, A. Kroll-Kristensen, L. Palm and P. Roepstorff, *J. Mass Spectrom.* **1997**, *32*, 593-601.

17. S. L. Cohen and B. T. Chait, *Anal. Chem.* **1996**, *68*, 31-37.

18. F. M. L. Amado, P. Domingues, M. G. Santa-Marques, A. J. Ferrer-Correia and K. B. Tomer, *Rapid Commun. Mass Spectrom.* **1997**, *11*, 1347-1352.

19. K. R. Clauser, S. C. Hall, D. M. Smith, J. W. Webb, L. E. Andrews, H. M. Tran, L. B. Epstein and A. L. Burlingame, *Proc. Natl. Acad. Sci. USA* **1995**, *92*, 5072-5076.

QUESTIONS AND ANSWERS

Joseph Loo (Parke-Davis Pharmaceuticals)

Who is the manufacturer of the IR laser you used?

Answer. The manufacturer of the infrared laser is Spektrum GmbH, Berlin, Germany.

Joseph Loo (Parke-Davis Pharmaceuticals)

How does the shot-to-shot reproducibility compare to the UV laser?

Answer. Because IR-MALDI is by far more dependent on sample morphology due to a penetration depth 100 times higher than in UV-MALDI, which then leads to much higher sample ablation, shot-to-shot reproducibility is worse than in UV-MALDI, if sample scanning on a heterogeneous sample is required. For succinic acid as matrix in combination with the standard "dried droplet" preparation technique, the shot-to-shot reproducibility is consequently worse. Nevertheless, with regard to the analyses of posttranslational modifications and peptide digest mixtures, a systematic sample scan without any particular spot selection leads to spectra with a signal-to-noise ratio only slightly lower than in the spectra presented. However, thin film preparation techniques or samples prepared with matrices such as CHCA, resulting in homogeneous samples, show almost the same good shot-to-shot reproducibility as in UV-MALDI, particularly compared to UV-MALDI with sample scanning. The growth of large crystals is another way to ensure good shot-to-shot reproducibility. In this case a sufficient number of single-shot spectra can be obtained from the same sample spot.

At the Ludwig Institute for Cancer Research the high performance of the infrared laser (<5% energy stability) is comparable to the nitrogen laser. Therefore, it can be assumed that energy fluctuation of the infrared laser has only marginal influence.

Joseph Loo (Parke-Davis Pharmaceuticals)

How many laser shots did you use to acquire the IR-MALDI data and how does this compare to your UV-MALDI data?

Answer. Spectra were recorded with a non-selective accumulation of either 32 or 64 consecutive laser shots. Within all UV-/IR-MALDI comparisons the same number of shots were accumulated.

Maria Person (UCSF)

Do you have any explanation for why different matrices favor different peptides? For example, is there a correlation with hydrophobicity or specific amino acid residues?

Answer. No. There have been numerous studies in UV-MALDI investigating peptide signal suppression. However, this phenomenon is still poorly understood. According to the presented results, peptide signal suppression is extremely dependent on the choice of matrix, not the wavelength. Therefore, it can be assumed that any understanding of this basis of peptide signal suppression in UV-MALDI is transferable to IR-MALDI. If there are answers in UV-MALDI, there will be answers in IR-MALDI. But the complexity of the problem with apparently several competing factors makes it difficult to explain or even predict peptide signal suppression.

Roman Zubarev (Cornell University)

Did you observe formation of "in-source decay" products, such as c and z· ions, in IR-MALDI?

Answer. We have not tried "in-source decay" yet. Surprisingly, it is possible to use the infrared laser for PSD. Prompt fragmentation can also be seen. Therefore, it is quite plausible that "in-source decay" product ions should be observable as well.

John Stults (Genentech, Inc.)

How does succinic acid compare with α-cyano matrix for salt and detergent tolerance?

Answer. We have not undertaken any systematic investigations on this subject. However, it seems that IR-MALDI with succinic acid is less sensitive to salt, detergent and buffer material than any major UV-MALDI matrix.

John Stults (Genentech, Inc.)

Has there been an improvement in IR lasers recently? They have a reputation for difficult and unreliable operation.

Answer. The laser system we use at the Ludwig Institute for Cancer Research imposes no difficulties in operation and has proved to be more reliable than the standard nitrogen laser. By now this laser system has been running for fifteen months and more than two million shots with one scheduled laser resonator upgrade, but without any repair in between. Laser energy and stability has not decreased significantly.

Identification of *in-vivo* Phosphorylation Sites with Mass Spectrometry

Xiaolong Zhang, Christopher J. Herring and Jun Qin
Laboratory of Biophysical Chemistry, National Heart, Lung and Blood Institute, NIH, Bethesda, MD 20892

Protein phosphorylation, principally on Ser, Thr and Tyr residues, is one of the most common cellular regulatory mechanisms [1-3]. Phosphorylation can modulate enzyme activity, alter affinity to other proteins, and transmit signals through kinase cascades that often are branched and interactive. To understand the molecular basis of these regulatory mechanisms, it is necessary to determine the sites that are phosphorylated *in vivo*. Mass spectrometry is becoming the method of choice for the identification of protein phosphorylation sites. This mass spectrometric approach offers at least three major advantages over the conventional biochemical and genetic approaches. First, it is accurate. There is no ambiguity once the phosphorylation site is identified through mass spectrometric sequencing of the phosphopeptide. Second, it is fast. The cycle of identifying phosphorylation sites is a few days. Third, it does not require ^{32}P labeling. Many improvements in this mass spectrometry-based methodology have been in the area of improving sensitivity, selectivity and sample handling [4-13].

With modern mass spectrometry using both MALDI and ESI as ionization methods, the inherent sensitivity of the mass spectrometer is high, typically in the low fmol to high attomol range for the detection of peptides. This sensitivity is comparable with immuno-methods, such as Western blotting. This would in principle allow us to identify phosphorylation sites with the sensitivity of a Western blot. However, this is not the case in reality. In contrast to immuno-methods, mass spectrometry has little selectivity. It is a highly general detection method. It will detect anything that can be ionized in the sample. It is difficult to identify the phosphopeptide ions in minute amount in a background of highly abundant ions of other peptides or contaminants. How to introduce and increase selectivity for the detection of phosphopeptides is therefore one of the key elements for the success of identifying phosphorylation sites with mass spectrometry. Most of the research effort in mass spectrometry for the identification of phosphorylation sites concentrates on this area [4-13]. Three major methods were tested and showed success to different degrees. They include: (1) the method of "parent-ion scan" [9] and the analogous stepped energy scanning [5]; (2) the method based on affinity chromatography for the enrichment of phosphopeptides [12]; and (3) the method of a phosphatase treatment [10, 13]. At present, these methods seem to be complementary rather than exclusive.

Sample handling may be the last and most difficult technical issue for the identification of phosphorylation sites. We define the process of preparing sample in a form that is suitable for MS analysis as sample handling. It includes at least: (1) how to handle minute amount of sample with minimal loss; (2) how to recover and detect all of the phosphopeptides; and (3) how to recover and detect most of the unphosphorylated peptides to evaluate the phosphorylation status of the entire sequence of the protein. This later issue is often referred to as sequence coverage. As regulatory proteins are often expressed only at low levels, any method for determining phosphorylation sites must include steps to purify phosphoproteins with small losses to achieve a high purification enrichment factor. This will yield partially purified proteins. Immuno-precipitation (IP) followed by one-dimensional SDS-PAGE seems to be the method of choice at present. The methods for the determination of phosphorylation sites must also include steps to recover, identify and prepare phosphopeptides in a form that is suitable for MS analysis with minimal loss. These steps are usually composed of: (1) *in-gel* digestion and extraction of the resulting peptides, (2) identification of the phosphopeptides through one of the three major methods mentioned above, and (3) sequencing of the phosphopeptides using either nanospray or on-line liquid chromatography electrospray tandem mass spectrometry (LC/ESI/MS/MS). Every step that starts from phosphoprotein purification and ends with phosphopeptide sequencing with mass spectrometry must be considered carefully to ensure a successful analysis. It is not enough just to emphasize one of the steps; all the steps must be approached collectively to ensure final success. Only such a comprehensive approach that takes every step of the process into account will make mass spectrometry the method of choice for the identification of phosphorylation sites. Once the phosphorylation sites are identified with mass spectrometry, they must be evaluated for their functions through genetics or biochemistry. This process and not the identification of phosphorylation sites is becoming the rate-limiting step now.

Site-specific phosphorylation stoichiometry is useful information, but in general this information cannot be obtained by mass spectrometry. Ionization efficiency and peptide extraction efficiency can be different for phosphopeptides and their unphosphorylated counterparts, therefore biological mass spectrometry is usually not quantitative. It may be possible to calibrate the response of phosphorylated and unphosphorylated peptides using synthetic peptides, but the peptide extraction efficiency during *in-gel* digestion for these two forms can be difficult to evaluate.

As the identification of phosphorylation sites must involve a comprehensive approach, it is important to treat the process as a whole. There always exists the compromise between resolution and sensitivity for both chromatography and mass spectrometry for all the steps involved in the analysis. A compromise must be made in every step for the best net gain of the entire process. For example, when dealing with limited amounts of sample, sensitivity has to be emphasized at the expense of the resolution. This compromise dictates what types of separation (chromatography) and mass analysis (mass spectrometry) to choose. One must also take the reliability of the analytical technique into consideration.

Only the most reliable and proven technique should be chosen when there exist parallel techniques. When these issues are correctly addressed, many *in vivo* phosphorylation sites can be identified, and more phosphorylation sites can be located to a stretch of amino acid residues using mass spectrometry. This allows us to study *in vivo* phosphorylation in the signal transduction pathways with reasonable effort. We will demonstrate this process using different examples emphasizing this comprehensive approach. We will also discuss the current limitations and technical difficulties in the identification of phosphorylation sites with mass spectrometry.

Experimental

In-gel Digestion of Gel-separated Proteins

Our previously published procedure for digestion of proteins in a gel slice [13] was modified as follows. After conventional SDS-PAGE, a Cu-stained gel piece containing 20 ng or more of protein was destained twice for 3 min each with the BioRad destain solution, soaked for 4 hours in the desalting solution (1:1 methanol:10% acetic acid solution) at room temperature and then in water for 30 min. Alternatively, a Coomassie blue stained gel piece was destained with the blue destain solution (25 mM NH_4HCO_3 in 1:1 methonal:water), then treated the same way as the Cu-stained gel piece. The cleaned, wet gel was finally washed with digestion buffer (50 mM NH_4HCO_3) and ground to a fine powder in 10 μl of digestion buffer with a fire sealed pipette tip. One μl of 200 ng/μl of trypsin in water was added, and the digestion tube was centrifuged briefly and digestion was carried out at 37°C for 90 min; smaller amounts of trypsin reduce the digestion efficiency significantly. To extract the peptides, three volumes of acetonitrile of the digestion solution was added, the tube vortexed for 2 min and the solution carefully removed with a gel-loading pipette tip. The above step was repeated for a second extraction with the addition of 10 μl of the digestion buffer, followed by 30 μl of acetonitrile. The extracted peptides were pooled and dried, and the dried peptides were redissolved in 10 μl of 50% acetonitrile in water for mass spectrometry. This procedure recovered > 60% of the radioactivity of a ^{35}S-labeled protein (the catalytic domain of Zap70) as soluble, extracted peptides (data not shown). We found that Cys containing peptides would be recovered without reduction and alkylation from a systematic study of optimizing our *in-gel* digestion procedure with ^{35}S-labeled protein containing 6 Cys residues. We generally found that prolonged sample handling at the level < 1 pmol of protein loaded in the gel tends to introduce contaminations, such as keratin. Therefore, the step of reduction and alkylation was usually omitted. Tryptic peptides extracted from the *in-gel* digestion are predominantly complete digestions and partial digestions with one miss (one internal R/K was not cut by trypsin). Partial digestions with up to four misses could also be observed (data not shown). We use this procedure for both protein identification/*de novo* sequencing and identification of posttranslational modifications. This procedure can be used for Lys-C, Asp-N and chymotrypsin digestions in the same fashion. We found that *in-gel* digestion with Glu-C usually yields too much autolysis product to be useful.

Phosphatase Treatment of Peptides

For phosphatase treatment of peptides, one-tenth (1 μl) of the dissolved peptide solution was mixed with 2 μl of 50 mM NH_4HCO_3 and 1 μl of 0.5 U/μl of calf intestine phosphatase (CIP, New England Biolabs, diluted from 10Uμl in storage buffer by 50 mM NH_4HCO_3). The resulting mixture was incubated for 30 min at 37°C (the CIP digestion buffer supplied by the manufacturer was not used), then dried on a SpeedVac and redissolved in 2 μl of 50% acetonitrile in water for mass spectrometry.

MALDI / TOF Mass Spectrometry

A 1.2-meter MALDI/TOF mass spectrometer with delayed extraction (Voyager-DE, PerSeptive Biosystems, Framingham, MA) was usually used. Data were collected with an external 2-GHz digital oscilloscope (Tektronix, Houston, TX). The working matrix solution was a two-fold dilution of a saturated solution of 2,5-dihydroxybenzoic acid in acetonitrile:water (1:1). Aliquots of 0.5 μl of the peptide mixture and 0.5 μl of the working matrix solution were mixed on the sample plate and dried in air prior to MS analysis. Two spectra were taken for each sample, one under conditions optimized for high mass resolution (to measure masses with high accuracy) and one under conditions optimized for sensitivity (to observe the weaker peaks and peaks with higher molecular weights). The high resolution spectrum is often acquired with another reflector instrument (the Voyager-DE-STR) to achieve a mass accuracy of better than 10 ppm. For the linear instrument, masses were calibrated internally with an added calibration standard (bradykinin fragment 1-5) and a peptide derived from trypsin autolysis. Masses of < 2000 Da were isotopically resolved so that even weak peaks could be measured with better than 0.2 Da mass accuracy. Most of the peaks in the spectrum of the tryptic digest could be easily assigned to unique peptides predicted from the protein sequence and the specificity of trypsin or to peptides formed by autolysis of trypsin. Peaks that could not be accounted for in this way were candidates for modified peptides or contaminants, and those whose observed masses were 80 Da (or multiples of 80 Da) higher than calculated for a predicted tryptic peptide were tentatively assigned as phosphopeptides. This assignment was confirmed by the absence of these peaks from the MALDI/TOF spectrum of the same peptide mixture after treatment with CIP, and the appearance of new peaks that are 80 Da (or multiples of 80 Da) lower in mass. In the event that (1) the stoichiometry of phosphorylation is low, (2) a peak that is 80 Da lower in mass exists prior to CIP treatment, or (3) a new peak appears without the disappearance of a peak of 80 Da higher in mass, the assignment of the phosphopeptide is considered to be only tentative, and further confirmation from tandem mass spectrometry is needed (see below). This step removed any ambiguity of peptide assignment due to mass degeneracy in which another unphosphorylated peptide happens to have the same mass as the phosphopeptide, or to other unknown modifications that might give rise to an unphosphorylated peptide with exactly the same mass as the phosphopeptide.

LC/ESI/MS/MS

An electrospray ion trap mass spectrometer (LCQ, Finnigan MAT, San Jose, CA) coupled on-line with a capillary HPLC (Magic 2002, Michrom BioResources, Auburn, CA) was used to sequence the phosphopeptide for the identification of phosphorylation sites. An 0.1 (or 0.2) x 20 mm-MAGICMS C18 column (5-µm particle diameter, 200 Å pore size) with mobile phases of A (methanol:water:acetic acid, 5:94:1) and B (methanol:water:acetic acid, 85:14:1) was used with a gradient of 2-98% of mobile phase B over 2.5 min followed by 98% B for 2 min at a flow rate of 50 µl/min. The flow was split with a Magic precolumn capillary splitter assembly (Michrom BioResources) and 1 µl/min directed to the 100 µm column. The steep gradient and narrow column resulted in narrow HPLC peaks (eluting in as short a time as 10 s) of small volume (0.2 µl), and thus high peptide concentration, which enhanced the ESI/MS response.

The ion trap mass spectrometer was operated in an unusual fashion, i. e., in a constant collision-induced dissociation (CID) mode for the entire LC run, with the mass spectrometer set to acquire CID spectra on a single m/z value. The charge state of the precursor ion was not measured directly, rather it was calculated by inspection of the sequence of the phosphopeptide identified in the MALDI/TOF measurements (usually it is doubly or triply charged for a tryptic peptide). We often found that peptides at < 2 fmol in a complex mixture are difficult to detect in the LC/MS run, but they can be detected in a LC/MS/MS experiment if the m/z value is known. The LCQ is rather insensitive to the choice of collision parameters for doubly and triply charged peptide ions, because there is no danger of further fragmenting fragment ions, in contrast to the triple-stage quadrupole mass spectrometer. Once the fragment ions are generated, they fall out of the resonance excitation frequency range and do not absorb energy anymore. This property of the LCQ mass spectrometer allows the easy acquisition of CID spectra from narrow LC peaks.

During a MS/MS measurement, a doubly or triply charged phosphopeptide is isolated in the ion trap and subjected to CID. Peptide bonds are cleaved to generate b or y fragment ions [14]. In the simplest application of this method, a fragment ion with mass 80 Da higher than the calculated mass of the expected fragment ion contains the phosphorylation site. And if the smallest phosphorylated fragment ion contains only one amino acid that could be phosphorylated (Ser, Thr or Tyr), then the phosphorylation site is identified. As illustrated in Fig. 2b, however, fragmentation of phosphopeptides in an ion trap is more complicated than this simple picture. Phosphopeptides that contain phosphoserine or phosphothreonine predominantly lose the elements of H_3PO_4 [15, 16], resulting in a loss of mass of 98 Da (49 Da for a doubly charged ion). Similarly, fragment ions containing the phosphate generated by CID also can lose the elements of H_3PO_4, creating ions with masses 98 Da less than the predicted mass; such ions are denoted as b_n^{Δ} or y_n^{Δ} (depending on whether they arise from b or y type ions). Thus, this 98-Da loss also serves as a signature for phosphopeptides. However, this loss of phosphate in an ion trap mass spectrometer is more complicated than stated here and is phosphoamino acid and charge state dependent [16].

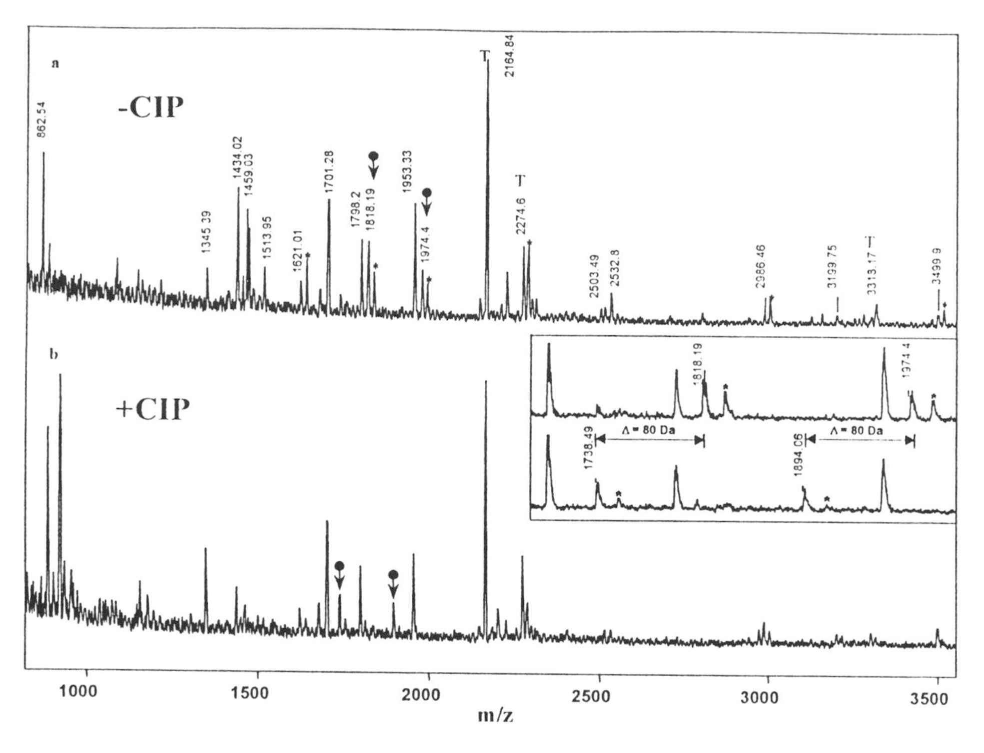

*Fig. 1. Identification of the phosphopeptides in the in-gel tryptic digest of 20 ng of the catalytic domain of MIHCK by MALDI/TOF spectrometry. (a) Before CIP treatment. Arrows indicate peptides that are modified. (b) After CIP treatment. Arrows indicate new peaks arising from dephosphorylation of the phosphopeptides. The insert shows detailed spectra of the region of phosphopeptides before (top) and after (bottom) CIP treatment. T: trypsin autolysis peaks; *: methionine-oxidized peptides of mass 16 Da higher.*

Results and Discussion

Documentation of the Procedure with a Known Phosphoprotein

We tested the procedure by analyzing the expressed catalytic domain of *Acanthamoeba* MIHCK, which has a single phosphorylated residue, Ser-627 [17]. An aliquot of 20 ng of protein (quantified by amino acid analysis) was subjected to SDS-PAGE on a 5-15% gradient gel and then visualized by Cu staining. After destaining, the gel slice containing the protein was digested with trypsin and the resulting peptides were extracted and analyzed by MALDI/TOF both before (Fig. 1a) and after (Fig. 1b) treatment with CIP. Most of the peaks in Fig. 1a were identified as peptides expected from trypsin digestion of MIHCK (accounting for 54% of the total protein sequence) or from trypsin autolysis (marked with T), except for two peaks at m/z values of 1818.19 and 1974.40 (arrows). The absence of these two peaks in Fig. 1b and the appearance of two new peaks at m/z values of 1738.49 and 1894.06 (arrows) unambiguously identified them as singly phosphorylated peptides (Fig. 1, insert). The

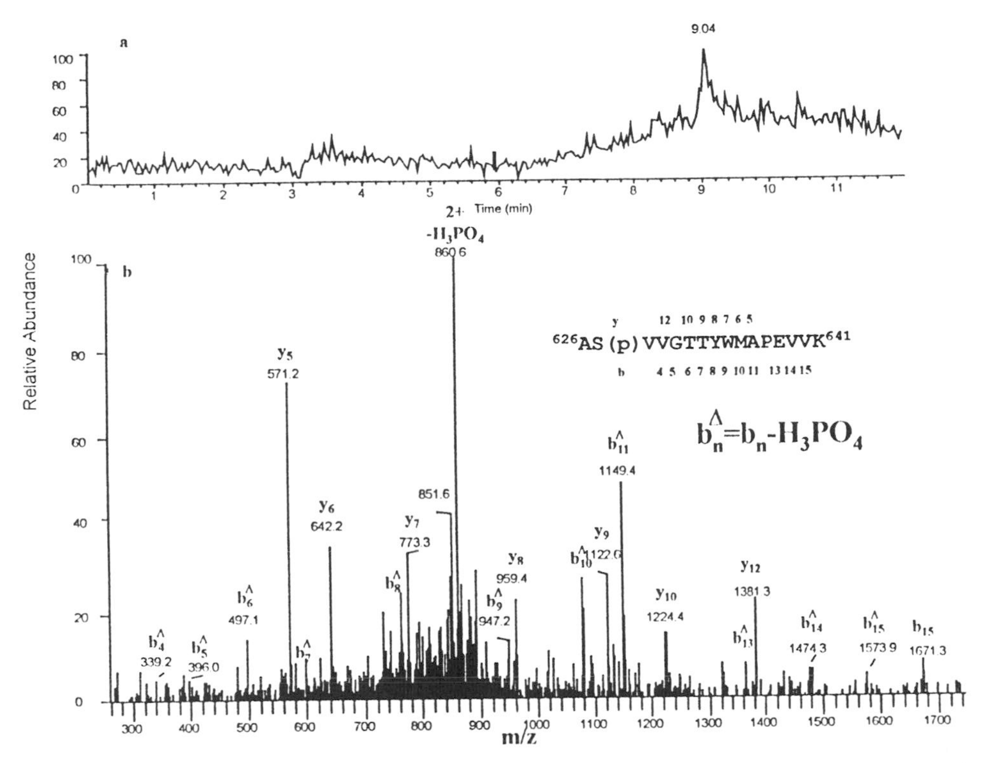

Fig. 2. Identification of the phosphorylated residue in the tryptic phosphopeptide from the catalytic domain of MIHCK by LC/ESI/MS/MS in an ion trap mass spectrometer. (a) Total fragment ion chromatogram of the doubly charged phosphopeptide (m/z 909.5) that was identified as a singly charged peptide, m/z 1818.19 in Fig.1. The ion trap mass spectrometer was operated in a constant CID mode with CID spectra obtained for the ion at m/z of 909.5 during the entire LC run. (b) The CID spectrum of the doubly charged phosphopeptide, which is the average of the 10 CID spectra that comprise the peak in Fig. 2a. Ions b_n and y_n are the N- and C-terminal fragments of length n as indicted in the sequence. Δ indicates the corresponding b_n or y_n ion minus the elements of H_3PO_4.

phosphopeptide with the higher m/z value corresponds to a partial tryptic fragment, so these two phosphopeptides contained the same phosphorylation site.

To determine the precise phosphorylation site within the phosphopeptide, the remainder of the original peptide mixture was dried, redissolved in 95% mobile phase A and 5% mobile phase B, and subjected to LC/ESI/MS/MS. MS/MS spectra were obtained during the entire LC run for the doubly charged phosphopeptide with m/z 909.5 produced by electrospray (Fig. 2). The total fragment ion chromatogram (i. e., the summation of all ion signals in every CID spectrum as a function of elution time) is shown in Fig. 2a. The averaged CID spectrum of the peak eluted at 9.04 min is shown in Fig. 2b (the LC pump has a delay time of 4 min to generate the gradient plus another 2 min to load the sample from a 5 μl sample loop at a flow rate ~ 3 μl/min for the 200 μm column used here). In Fig. 2b, the ion at m/z 860.6 is the doubly charged phosphopeptide with the sequence shown in the figure minus the elements of H_3PO_4. The ion at m/z 339.2, designated b_n^{Δ}, corresponds to the singly charged peptide of the sequence AS(**p**)VV minus the elements of H_3PO_4, thus establishing Ser-627 as the phosphorylation site. This assignment is confirmed by the observed y_{12} ion at m/z 1381.3, which corresponds to the sequence of the singly charged GTTYWMAPEVVK that is not phosphorylated. Thus neither of the two Thr residues nor the Tyr is phosphorylated, and this is further confirmed by the unambiguous assignments of all of the other observed b_n^{Δ} and y_n ions. These results obtained in a series of experiments using only 20 ng of protein in a gel slice are identical to those originally obtained by using hundreds of ng of protein [17]. This is the first example to our knowledge of a phosphorylation site identified by mass spectrometry using just 20 ng of protein loaded in a gel.

This procedure addresses the issues of sensitivity, selectivity and sample handling collectively. The key features are the optimized *in-gel* digestion, treatment with a phosphatase and the combined use of MALDI/TOF and LC/ESI/ion trap mass spectrometers. With our optimized *in-gel* digestion procedure, we could extract > 60% of the tryptic peptides with minimal contamination by peptides from trypsin autolysis and other sources (such as keratin) that could arise from prolonged sample handling. This procedure reflects the effort in sample handling. The use of a phosphatase to treat the entire mixture of extracted peptides in conjunction with MALDI/TOF measurements provides a simple, fast, sensitive, and robust method to identify phosphopeptide candidates for the subsequent tandem MS experiments to find the precise phosphorylation site. Digestion of the peptide mixture with a phosphatase allows the identification of phosphopeptides that could otherwise be easily overlooked due to mass degeneracy. Because the peptide mixture is treated with a phosphatase, the potential structure constraints that could prevent dephosphorylation by CIP in the protein are eliminated. We found that all three types of phosphoamino acids are removed quantitatively by CIP, with the exception that multiply phosphorylated peptides (> 3 phosphorylations) sometimes cannot be dephosphorylated completely (1 phosphate tends to remain under the experimental conditions).

The use of MALDI mass spectrometry for the identification of phosphopeptides is advantageous. MALDI is highly sensitive and tolerant of salt and biological buffers, thus eliminating the need for a desalting step and accompanying loss of sample, while still having a sufficiently high signal to noise ratio to allow the identification of low intensity ions. MALDI mass spectrometry requires only very small amounts of sample and there is no danger of losing the sample during the mass measurement. The use of a short, linear MALDI/TOF mass spectrometer (1.2 meter) with delayed extraction provides single-stage MS of the highest possible sensitivity, i. e., providing the potential for a high sequence coverage. We found that less abundant, higher molecular weight, and multiply-phosphorylated peptides often respond better in our short, liner DE instrument than our bigger reflector instrument (see Fig. 3), affording higher sensitivity and better sequence coverage. Fig. 3 shows a portion of the MALDI/TOF spectra of the same *in-gel* trypsin digestion of a phosphoprotein taken with

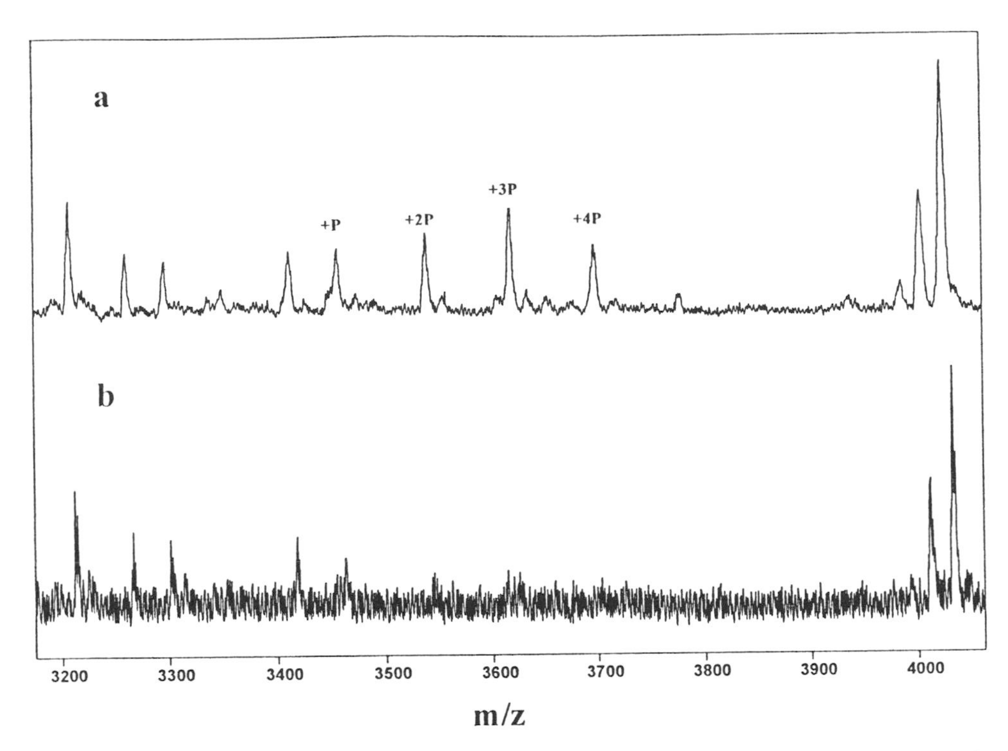

Fig. 3. The short linear MALDI / TOF mass spectrometer offers better sensitivity for the detection of multiply-phosphorylated peptides than the big reflector instrument. MALDI / TOF spectra of the in-gel trypsin digestion of a phosphoprotein taken with (a) the standing alone linear instrument and (b) the reflector instrument. The phosphopeptides were labeled with +P, +2P, +3P and +4P.

the linear instrument (Fig. 3a) and the reflector instrument (Fig. 3b). These spectra were taken from the same sample plate and with the same numbers of laser shot for averaging. It is apparent that the linear instrument yields a better S/N ratio as expected. It is not surprising that the multiply phosphorylated peptides are hardly detectable in the reflector instrument, but this raises the issue of what type of mass spectrometry is better suited for the identification of phosphopeptides. Phosphopeptides are known to be unstable in the reflector mode [8], and this situation is exacerbated for multiply phosphorylated peptides; higher molecular weight peptides also show lower sensitivity in the reflector mode than in the linear mode. We found that the linear MALDI/TOF instrument is much more useful for phosphopeptide mapping, although the reflector instrument offers much higher mass accuracy, facilitating the sequence assignment. We always use either an LC/MS/MS experiment or the reflector measurement as a second step, in which the region of a specific m/z value is measured more carefully for further confirmation of the assignment of phosphopeptides that were previously made in the linear measurement.

By contrast, electrospray mass spectrometry often requires a desalting step for *in-gel* digested samples that can lead to preferential loss of highly

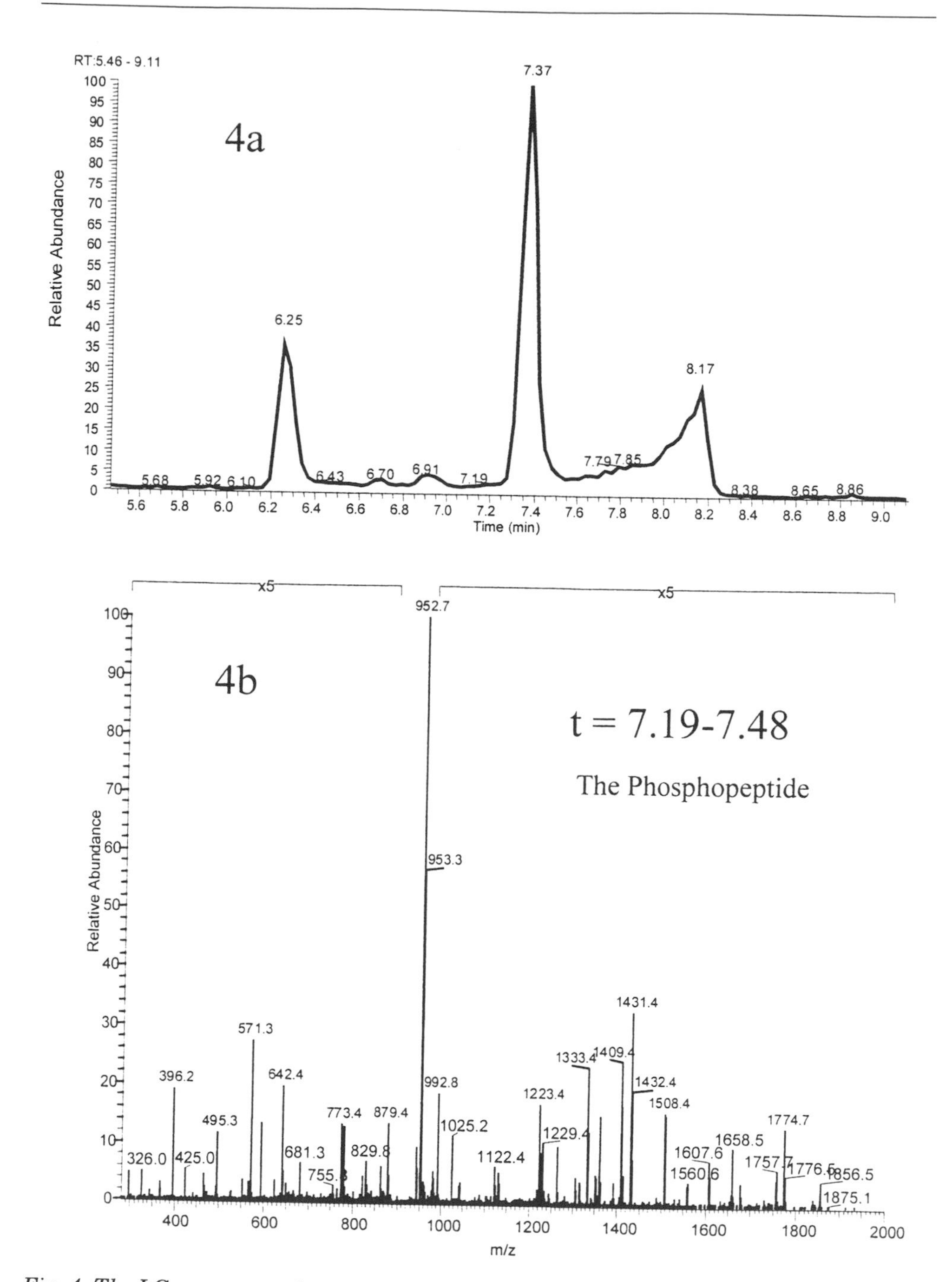

Fig. 4. The LC separates other peptides with the same m/z value as the phosphopeptide. (a) Total fragment ion chromatogram of the precursor ions (m/z 1001.5) from a loading of 200 fmol of trypsin digest of MIHCK. (b) The MS/MS spectrum of the phosphopeptide of interest eluting at 7.19 to 7.48 min.

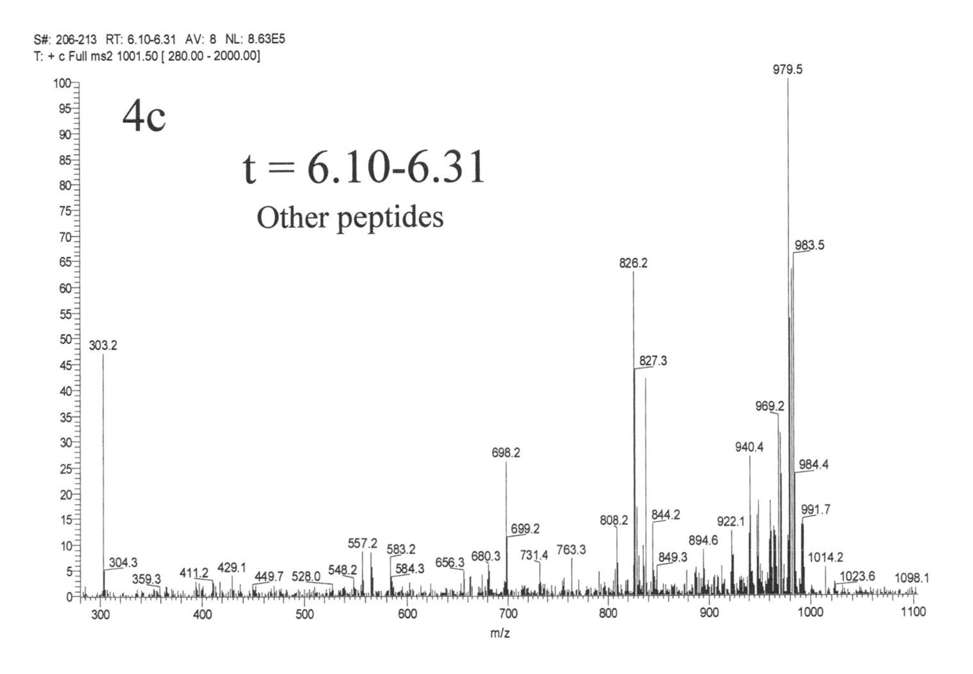

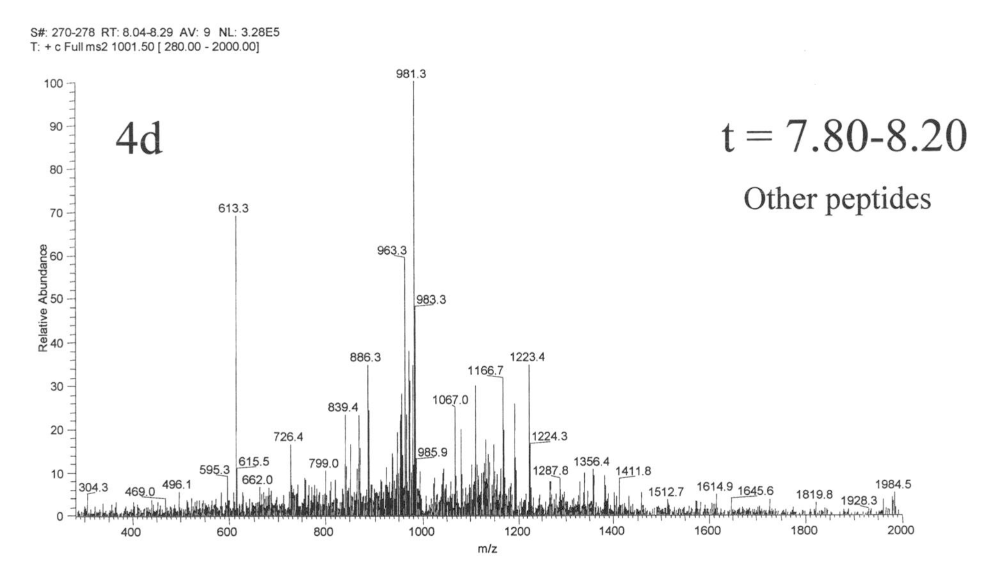

Fig. 4. The LC separates other peptides with the same m/z value as the phosphopeptide. (c) The MS/MS spectrum of a singly charged unphosphorylated peptide eluting at 6.10 to 6.31 min. (d) The MS/MS spectrum of other unphosphorylated peptides eluting at 7.80 to 8.20 min.

hydrophilic and hydrophobic peptides. There is also higher chemical background in electrospray than in MALDI, which often tends to mask the weaker peaks. We found that weaker peaks observed in MALDI/TOF are often missing in ESI/LC/MS using the LCQ mass spectrometer. However, electrospray combined with the ion trap mass spectrometer has a much higher sensitivity in the MS/MS mode than in the MS mode. Peptide ions that cannot be observed in the LC/MS run can be detected with good S/N ratio by a LC/MS/MS run if the m/z value is known. Therefore, using MALDI/TOF as the first stage MS to identify the phosphopeptides, followed by MS/MS in an ESI/ion trap provides a very sensitive means for the identification of phosphorylation sites. This approach allows sequencing of phosphopeptides that would not even be detected by LC/ESI/MS (see Figs. 4 and 5). Fig. 4 shows a LC/MS/MS run of ~ 200 fmol of the trypsin digest of MIHCK, monitoring only one m/z value of 1001.5 (corresponding to the 2+ charged phosphopeptide of m/z 1974.4 in Fig. 1). Fig. 5 shows a standard LC/MS run for the same sample. The phosphopeptide elutes at 7.19 to 7.48 min as indicated by the CID spectrum shown in Fig. 4b, but the corresponding mass spectrum from the LC/MS run at time of 7.09 to 7.55 min does not show a detectable peak at m/z 1001.5 (see Fig. 5c).

On-line LC/MS/MS is more advantageous than off-line techniques, such as nanospray. On-line LC can separate peptides of the same m/z value, thus allowing use of a wide mass isolation window (± 4 Da) for high sensitivity. The

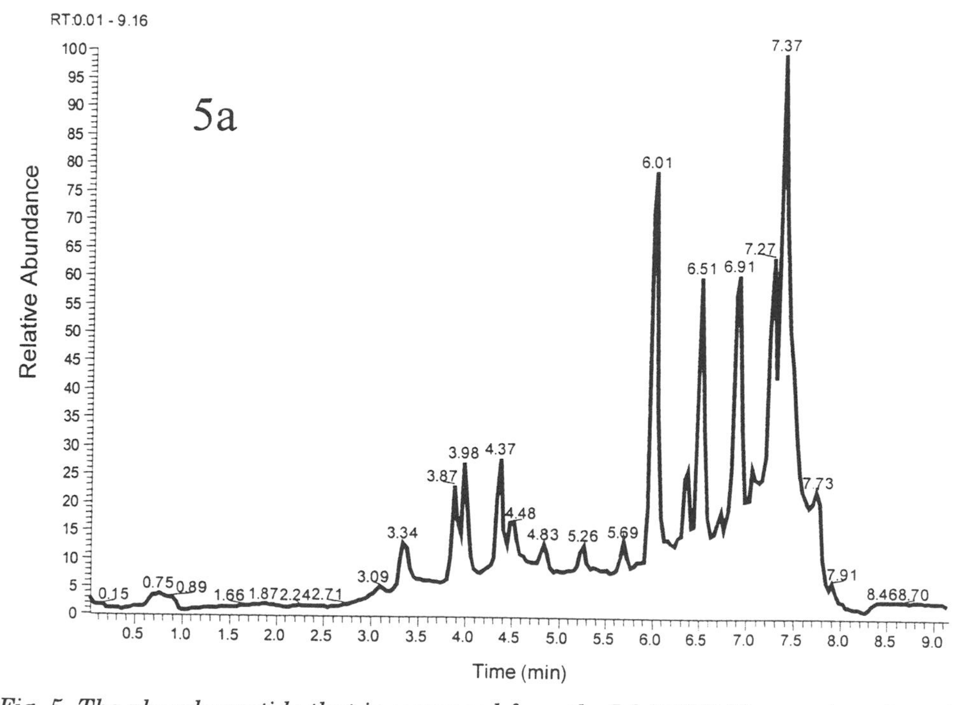

Fig. 5. The phosphopeptide that is sequenced from the LC/MS/MS run when the m/z value is known (see Fig. 4b) cannot be detected in the LC/MS run. (a) The HPLC chromatogram of 200 fmol of trypsin digest of MIHCK.

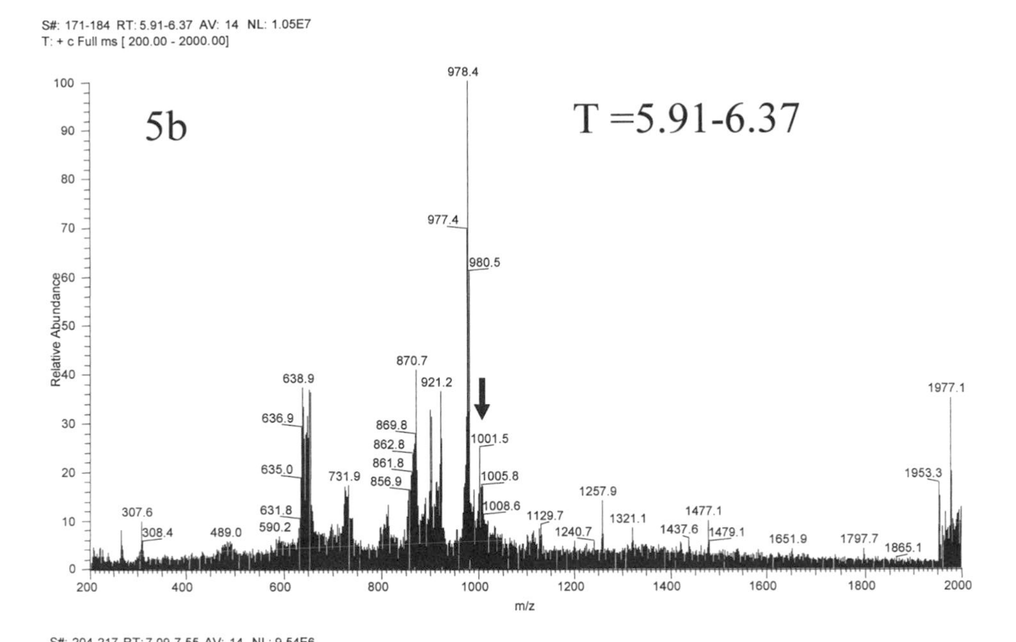

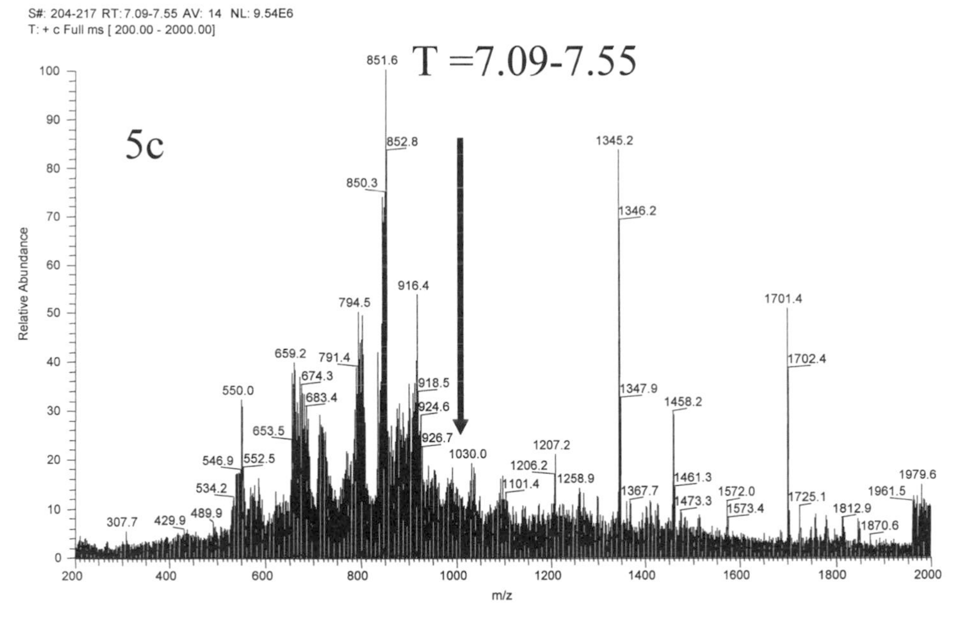

Fig. 5. The phosphopeptide that is sequenced from the LC/MS/MS run when the m/z value is known (see Fig. 4b) cannot be detected in the LC. (b) The MS spectrum of the peptides eluting at 5.91 to 6.37 min. The singly charged peptide ions with m/z of 1001.5 were detected as shown by the arrow. (c) The MS spectrum of peptides eluting at 7.09 to 7.55 min. The peptide ions with m/z of 1001.5 were not detected as shown by the arrow. The phosphopeptide of interest should elute here and its MS/MS spectrum was obtained when the m/z value was known and the LC/MS/MS run only monitored this value as shown in Fig. 4.

LC separates peptides with the same m/z values as the phosphopeptide of interest, thus simplifying the interpretation of the CID spectrum by eliminating unphosphorylated peptides of the same m/z (see Fig. 4c and 4d). The LC also concentrates peptides, and allows phosphopeptides that elute in the HPLC void volume and very hydrophobic phosphopeptides that elute only in a highly organic mobile phase (> 50% of mobile phase B) to be monitored, which would otherwise be lost for off-line techniques. The use of the simple methanol/water mobile phase without TFA and a steep gradient is a compromise between chromatography resolution and mass spectrometry sensitivity. The use of the 100 μm column is a compromise between reliability and sensitivity. In most cases, the amount of sample that is available allows one or two experiments, therefore the reliability (high success rate) is as important as any other criterion.

Our use of MALDI/TOF combined with a phosphatase treatment to identify phosphopeptides that otherwise cannot be detected in ESI/LC/MS is analogous to the elegant approach of parent-ion scanning to identify phosphopeptides developed by Carr's group [9]. The use of the phosphatase addresses the issue of selectivity of mass spectrometry. The two approaches are complimentary, yet are different from each other in one significant way. Our combined MALDI/TOF and LC/MS/MS strategy eliminates any off-line HPLC fractionation, thus minimizing sample handling and its associated sample losses, while the parent ion scan technique which uses nanospray, often requires desalting and HPLC fraction and fraction collection for complex mixtures. Although it has been shown that parent ion scan can be done with an unfractionated simple mixture, a very complex mixture that is derived from a large protein (often > 80 kDa) is most likely to be fractionated to separate peptides that have the same m/z values with the phosphopeptide. These peptides might interfere with the subsequent interpretation of the MS/MS spectrum. Parent-ion scan to identify phosphopeptides, of course, is more direct and less ambiguous than MALDI/TOF with a phosphatase treatment if collecting of HPLC fractions does not introduce sample losses to a degree that the phosphopeptides are below the detection limit. Our approach is also much more demanding in data analysis because the spectra before and after phosphatase treatment need to be compared. The peaks of interest can be the weak peaks, and they are not always obvious.

Potential pitfalls do exist in our approach as with any other. Phosphopeptides that cannot be ionized by MALDI cannot be identified. Small phosphopeptides are especially prone to this problem. Contrary to common belief, we do not observe significant "suppression effect" in MALDI/TOF when care is taken. The use of the DHB matrix without the addition of TFA or formic acid and a dilute peptide solution (< 100 fmol loading per peptide) helps to detect most of the peptides, including phosphopeptides, with a relatively uniform response. We sometimes find that phosphopeptides respond better in DHB than in 4HCCA, so we always use DHB as the matrix for MALDI experiments, particularly for *in-gel* digested samples. The reproducibility of MALDI sample preparation before and after CIP treatment is a potential problem, because some peaks can disappear and some peaks can appear after CIP treatment to give false positives. Fortunately, false positives are readily eliminated in the subsequent LC/MS/MS experiment.

Although large phosphopeptides are extracted with our *in-gel* digestion procedure and detected in the MALDI experiment, they are often lost, presumably on the LC column during sequencing to find the precise phosphorylation site. It usually requires more material to sequence large phosphopeptides with mass spectrometry; to make matters worse their extraction yield tends to be lower than small phosphopeptides. Therefore they are always more difficult to sequence. This problem is general for any MS technique. This inability to sequence large phosphopeptides prevents the identification of precise phosphorylation sites. Phosphorylation sites must be limited to only a few possibilities within a stretch of amino acid residues. We found that a sub-digestion with another enzyme is not usually successful in generating a smaller phosphopeptide without HPLC separation and fraction collection. Fraction collection and sub-digestion prove to be difficult when < 1 pmol of phosphoprotein is available. A better way to generate a smaller phosphopeptide is to use another enzyme to digest the full length protein in another gel piece. Asp-N is a good choice, and chymotrypsin works well for proteins < 30 kDa in spite of its broad specificity. Multiply phosphorylated peptides (> 2 phosphorylation sites) at low level (< 100 fmol) can be difficult to sequence in the ion trap as they are usually too labile, losing the phosphate group before the peptide bonds are broken. It is especially difficult when a series of S/T residues are closely spaced and they are multiply phosphorylated with sub-stoichiometry on most of the S/T residues. One might want to examine other types of mass spectrometry to exploit the stability of the phosphate group in different mass spectrometers. It is perhaps easier to use genetics to pinpoint the sites of phosphorylation, and of importance in the subsequent functional assay. The extremely hydrophilic phosphopeptides seem to be less of a problem, because they can be retained on the column when the proper stationary phase is chosen. We also found that they can be detected even when eluting in the void volume if the in-gel digested sample is relatively clean.

Another problem, which is universal for any technique, including conventional 2D peptide mapping with ^{32}P labeling, is the sequence coverage. If the phosphopeptides are not extracted from the gel piece, lost during sample handling, or cannot be ionized, they cannot be detected. One way to enhance sequence coverage is to generate different peptides using different enzymes, such as Asp-N. The combined sequence coverage of two digestions (usually trypsin or Lys-C and Asp-N) is often better than 85%. The remaining unobserved peptides usually do not contain any basic residues (Arg or Lys), are too small (< 5 residues), contain other modifications that are overlooked, or contain Cys residues that cannot be extracted from the gel piece. We get very mixed results for the extraction of Cys-containing peptides at the level < 1 pmol. Most of the time, we still fail to extract the Cys-containing peptides after reduction and alkylation, and the introduction of this reduction and alkylation step tends to introduce more contaminants due to the increased sample handling. Although it is difficult to detect peptides that do not contain basic residues for both MALDI and ESI, it is reasonable to speculate on an approach of derivatizing the peptides to introduce a positive charge. Much work remains to be done for this approach, especially for proteins at < 1 pmol level.

Sensitivity

We have shown that a single phosphorylation site was identified from 500 fmols of a phosphoprotein loaded in the gel. This is a favorable case because phosphorylation stoichiometry for this particular site is 100%. If the phosphorylation stoichiometry is low, as is almost always the case for *in vivo* phosphorylated proteins involved in regulation, the sensitivity will be expected to be lower. From our experience, our approach generally requires > 200 fmol of protein that is fully phosphorylated to identify a single phosphorylation site. If the phosphorylation stoichiometry is 10%, for example, in theory 2 pmol of total protein load in the gel may be required; if there are multiple sites, more sample may be needed as a function of the number of phosphopeptides detected. However, less phosphoprotein (200 fmol) may be sufficient for low phosphorylation stoichiometry, as the unphosphorylated protein can protect the phosphorylated protein from excessive loss during sample handing. It is also possible to increase the dynamic range in the MALDI/TOF measurement allowing the abundant ions to saturate so that the less abundant ions can be detected (see examples below). This can be

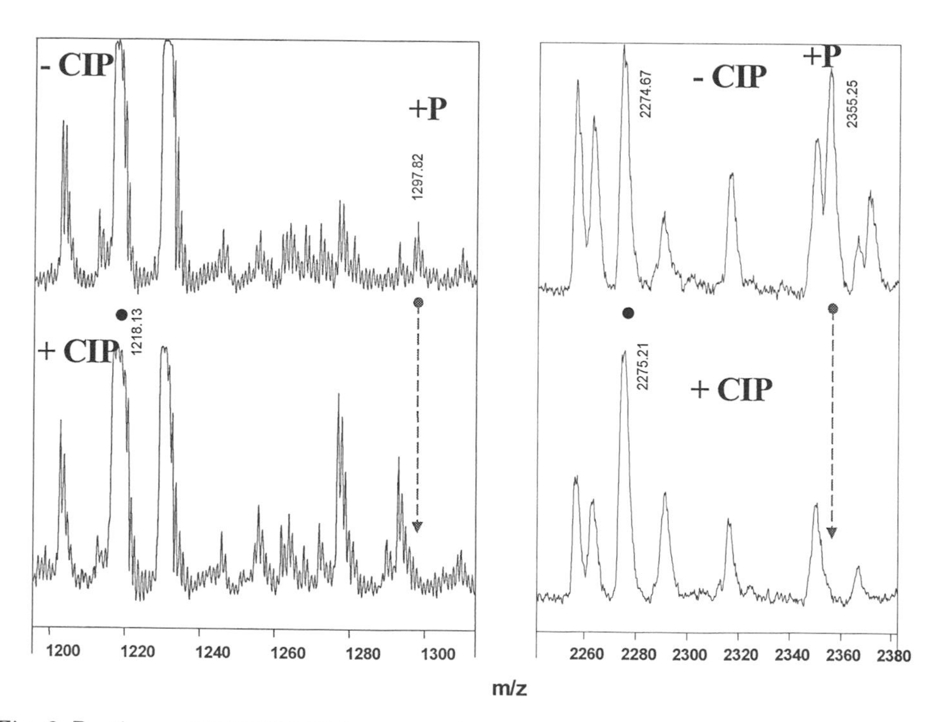

Fig. 6. Portions of MALDI/TOF spectra of an in-gel trypsin digestion of human Stat5A when activated by IL-2 that identify the phosphopeptides. Phosphopeptides are labeled with +P.

better accomplished in a linear instrument by increasing the laser power than in a reflector instrument where increasing laser power leads to phosphopeptide decomposition.

Phosphorylation Site Analysis of STAT5A: An Example of an Application

Signal transducers and activators of transcription (Stat) 1 and 3 require, in addition to an obligatory tyrosine phosphorylation, phosphorylation on serine 727 to boost their transactivating potential when activated by interleukins [18]. The presence or absence of this serine phosphorylation does not demonstrably affect DNA binding of either Stat1 or Stat3 [19]. Both the tyrosine and serine phosphorylation were identified by conventional 2D-peptide mapping [18, 19]. It has been difficult, however, to map the serine phosphorylation site of Stat5 by the same method.

In collaboration with Dr. Warren Leonard at NHLBI, we applied the above-described procedure to identify the tyrosine and the elusive serine phosphorylation site for human Stat5A activated by IL-2. The identification of this elusive serine phosphorylation site on Stat5A allows the hunt for the kinase responsible for this phosphorylation to begin. The Stat5A protein was immunoprecipitated with a monoclonal antibody from the whole cell lysate of 5L cultured cells 2 min. after the addition of IL-2 and resolved on SDS-PAGE. The Stat5A protein band was just visible when stained with Coomassie brilliant blue. This

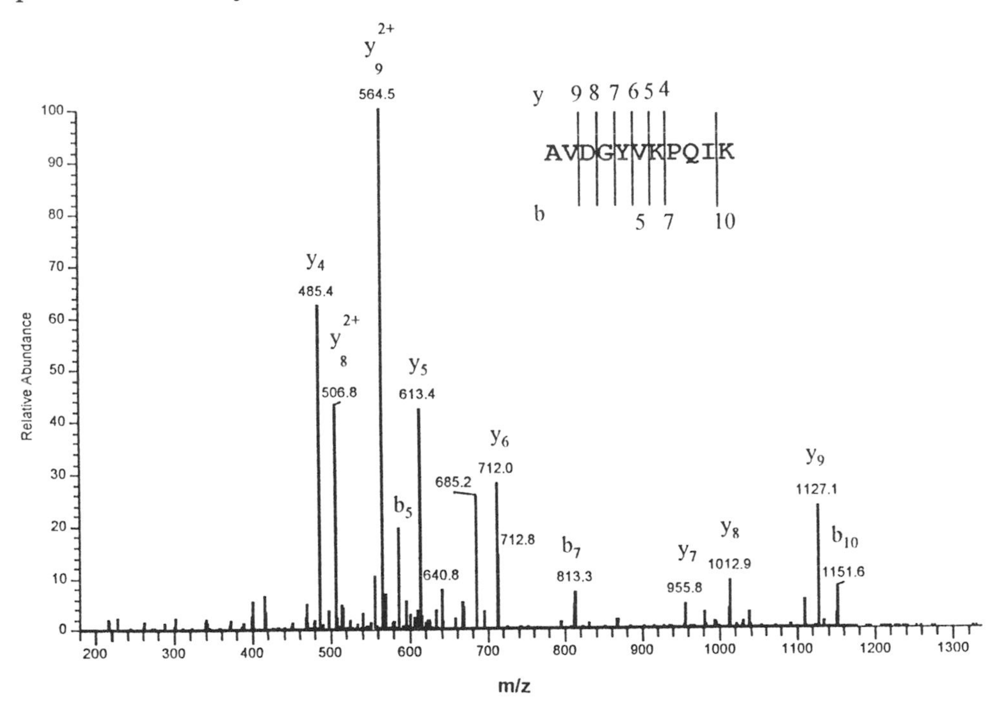

Fig. 7. The LC/MS/MS spectrum of the doubly charged small phosphopeptide as identified in Fig. 6. Y is the residue that is identified as phosphorylated.

band was cut out and subjected to mass spectrometric analysis. MALDI/TOF measurements of the trypsin digest before and after CIP treatment identified two phosphopeptides (Fig. 6). The LC/ESI/MS/MS spectrum is shown in Fig. 7 for the smaller phosphopeptide, which identified Y as the site of phosphorylation. This example illustrates the difficulty of the identification of phosphorylation sites of low stoichiometry. The tyrosine phosphorylation has such a low stoichiometry that the phosphopeptide is just barely visible in the linear MALDI/TOF measurements even when the corresponding unphosphorylated peptide is way over scale. This phosphotyrosine-containing peptide would not be detected in the reflector instrument. In contrast, the serine phosphorylation has a higher stoichiometry, rendering it easier to detect. This example also demonstrates the difficulty of data analysis for this procedure, as we have to compare two spectra to search for minute differences.

Conclusions and Perspectives

The ability to analyze phosphorylation and other posttranslational modifications is becoming more and more important in the post-genome era. It seems that mass spectrometry will replace the conventional 2D-phosphopeptide mapping as the method of choice for the identification of phosphorylation sites, although this technique is far from routine. The mass spectrometry based method is accurate, fast and does not require ^{32}P labeling. A comprehensive approach using mass spectrometry and proper separation that addresses sensitivity, selectivity, sample handling and reliability simultaneously often allows the identification of phosphopeptides using > 200 fmol of phosphoproteins when separated by SDS-PAGE. When the phosphopeptide is small (< 3 kDa), the precise phosphorylation site can be found in the same experiment. This capacity has already enabled us to identify many *in vivo* phosphorylation sites with reasonable effort. It is possible now to identify a phosphorylation site for proteins with abundance of > 1000 copy/cell when 10 billions of cells are available, assuming a phosphoprotein purification efficiency of > 10% and the phosphorylation stoichiometry of > 10%. This ability opens new possibilities in biological research when the phosphorylation is relevant. With the growing awareness and popularity of mass spectrometry among biologists, we believe that mass spectrometry will become an integral part of biological research.

References

1. P. Cohen, *Trends. Biochem. Sci.* **1992**, *17*, 408-413.

2. T. Hunter and M. Karin, *Cell* **1992**, *70*, 375-387.

3. J. Posada and J. A. Cooper, *Mol. Bio. Cell* **1992**, *3*, 583-592.

4. B. W. Gibson and P. Cohen, *Methods Enzymol.* **1990**, *193*, 480-501.

5. M. J. Huddleston, R. S. Annan, M. F. Bean and S. A. Carr, *J. Am. Soc. Mass Spectrom.* **1993**, *4*, 710-717.

6. J. Ding, W. Burkhart and D. B. Kassel, *Rapid Commun. Mass Spectrom.* **1994**, *8*, 94-98.

7. A. P. Hunter and D. E. Games, *Rapid Commun. Mass Spectrom.* **1994**, *8*, 559-570.

8. R. S. Annan and S. A. Carr, *Anal. Chem.* **1996**, *68*, 3413-3421.

9. S. A. Carr, M. J. Huddleston, and R. S. Annan, *Anal. Biochem.* **1996**, *239*, 180-192.

10. P. Liao, J. Leykam, P C. Andrews, D. A. Gage and J. Allison, *Anal. Biochem* **1994**, *219*, 9-20.

11. J. Qin and B. T. Chait, *Anal. Chem.*, **1997**, *69*, 4002-4009.

12. D. C. Neville, C. R. Rozanas, E. M. Price, D. B. Gruis, Verkman, A. S. and R. R. Townsend, *Protein Sci.*, **1997**, *6*, 2436-2445.

13. X. Zhang, C. J. Herring, P. R. Romano, J. Szczepanowska, H. Brzeska, A. G. Hinnebusch and J. Qin, *Anal. Chem.* **1998**, *70*, 2050-2059.

14. K. Biemann, *Biomed Environ. Mass Spectrom* **1988**, *16*, 99-111.

15. B. L. Gillece-Castro and J. T. Stults, *Proc. 43rd ASMS Conf. on Mass Spectrometry & Allied Topics*, 1995, p302.

16. J. P. DeGnore and J. Qin, *J. Am. Soc. Mass Spectrom.* **1998**, in press.

17. J. Szczepanowska, X. Zhang, C. J. Herring, J. Qin, E. D. Korn and H. Brzeska, *Proc. Natl. Acad. Sci. USA* **1997**, *94*, 8503-8508.

18. Z. Wen, Z. Zhong and J. E. Darnell, Jr., *Cell* **1995**, *82*, 241-250.

19. Z. Wen and J. E. Darnell, Jr., *Nucleic Acids Res.* **1997**, *25*, 2062-2067.

Questions and Answers

Christopher Yu (Chiron Corporation)

Is there a specific reference, which addresses the MALDI suppression effect and gives description on how to lower suppression by using DHB or lower concentration of samples?

Answer. I am not aware of it. There is an early reference in *Analytical Chemistry* that described this phenomenon. I think that the suppression effect in MALDI is not as serious as commonly believed. In fact, we have not observed it at all. We found that peptides tend to respond rather uniformly if DHB and a lower concentration of peptide (< 100 fmol) are used, and care is taken in sample preparation such that peptides do not precipitate before the crystallization of the matrices.

Roland Annan (SmithKline Beecham Pharmaceuticals)

Have you looked at the MALDI spectra of the phosphoprotein digestes in the negative ion mode?

Answer. We have. There is no difference between spectra in the postive ion and negative ion mode when DHB is used. They are actually mirror images of each other. Phosphopeptides do not necessarily respond better in the negative ion mode.

A. L. Burlingame (UCSF)

Since C-terminal Arg containing peptides are preferentially ionized/detected in MALDI, is there any concentration of Lys containing peptides with the "20%" missing coverage?

Answer. Statistically we do not see that Arg containing peptides are preferentially ionized/detected in MALDI when using DHB. Your conclusion might be based on using 4HCCA as the matrix. Arg and Lys containing peptides might respond differently when using 4HCCA. The only statistically significant conclusion that I can make is that peptides (1) without Arg or Lys, and (2) that are too short (< 6 amino acid residues) tend to be missing in the MALDI spectra. All the rest are always case-dependent.

Determination of Enzyme Mechanisms by Stopped-Flow ESI-MS

Frank B. Simpson and Dexter B. Northrop
Division of Pharmaceutical Sciences, School of Pharmacy, University of Wisconsin-Madison, Madison, WI 53706

Abstract

We suggest ESI-MS as the ideal method to characterize enzymatic catalysis. Observation of the low mass region of the spectrum will enable simultaneous monitoring of the time courses of every charged substrate and product of unique mass/charge (m/z) ratio, without reference to a chromophore. To date, most enzymes have been assayed by optical means, which has severely restricted the number and type of well-characterized enzymes. In the high mass region, because of the gentle conditions of electrospray ionization, both noncovalent and covalent interactions can be monitored. This attribute of ESI-MS will enable observation of mass changes due to interaction of substrates, products, and inhibitors with enzymes. By application of suitable rapid-reaction methods for sample introduction, reactions may be sampled continuously during presteady-state, steady-state, approach to equilibrium, and at equilibrium conditions. Subsequent analysis of mass spectra collected by each sampling method will enable direct determination of the binding constants, kinetic constants, reaction sequence, presence and identity of intermediates, and ultimately the free-energy reaction-coordinate diagrams of enzymes of interest.

Introduction

We suggest the use of ESI-MS to monitor directly the kinetics of enzyme-catalyzed reactions [1, 2].

The technique of ESI-MS has become well established during these last 10 or more years, and it has been shown to be an effective means to detect noncovalent interactions between ligands and their enzymes [3-5]. By this technique, an enzyme mixture may be injected through the ESI inlet, whereupon all of the liquid quickly is evaporated, leaving the charged, nonvolatile substrate, product, enzyme, buffer, etc., molecules to be transferred into the vacuum within the mass spectrometer's flight tube [4, 6]. Among the more conventional mass spectrometer designs, such as the magnetic sector or time-of-flight (TOF), an electrical potential applied to a series of accelerating plates accelerates the ionized molecules into the mass analyzer, enabling precise mass determination and quantitation of each nonvolatile species present in the injected solution. But the critical feature of ESI for the enzymologist is that gentle conditions of ionization and desolvation can be found such that noncovalent enzyme-ligand interactions remain intact [3-5].

An enzyme reaction is thought to be quenched upon transfer into the gas phase within the mass spectrometer. Most mass spectrometer designs, such as magnetic sector, quadrupole, and TOF instruments, maintain the electrospray ions in suspension in the vacuum within the flight tube for only micro to milliseconds, thus mass analysis is virtually instantaneous subsequent to sample introduction. In some other mass spectrometer designs, such as the ion trap or the Fourier transform mass spectrometer (FTMS), the vapor-phase ions and charged enzyme-ligand species produced by electrospray ionization are maintained within the mass spectrometer before and during mass analysis; it is not yet known whether a reaction can continue on an enzyme surface in the absence of solvent during this analysis period.

Many mass spectrometer designs today provide high resolution, high mass accuracy, and high mass range, suitable for the analysis of proteins. A 7 Tesla FTMS equipped with an ESI inlet has been used to obtain a value of MW 28,996.6 ± 0.1 Da for bovine carbonic anhydrase II [7], and this high performance suggests that intermolecular transfer of a single proton may be observed on this enzyme. Modern mass spectrometers in ESI mode are able to detect femtomole quantities of material [8]. Speakers at this proceeding have described attomole levels of detection.

A Primer of Enzyme Kinetics

A general reaction mechanism for an enzyme-catalyzed reaction is depicted in Scheme 1 [10].

$$E + S \underset{k_{-1}}{\overset{k_1}{\rightleftharpoons}} X1 \underset{k_{-2}}{\overset{k_2}{\rightleftharpoons}} \cdots \underset{k_{-n}}{\overset{k_n}{\rightleftharpoons}} Xn \underset{k_{-(n+1)}}{\overset{k_{(n+1)}}{\rightleftharpoons}} E + P$$

Scheme 1.

Intermediates X_1, ..., X_n are enzyme-bound. One goal of the enzyme kineticist is to describe the catalytic process by defining each of these reaction intermediates chemically, together with determining the numerical value of every rate constant which links these intermediates together. ESI-MS holds the potential to fulfill both of these tasks.

Two types of kinetic analyses often are employed in the attempt to determine these kinetic constants [11]. The first, steady-state kinetics, is useful for determining the order of binding and release of substrates and products, but the rate constants from k_2 and onwards (Scheme 1) are combined in a single parameter, k_{cat} or V_{max}. To extract rate constants for separate steps, one has to use presteady-state kinetics, which usually necessitates use of high, often stoichiometric concentrations of enzyme, and short incubation periods, often of less than one second. To achieve these short reaction times, fast, rapid-reaction, kinetic techniques are employed, such as continuous-flow, rapid-quench, push-push, stopped-flow spectrophotometry, temperature-jump, and others. We suggest that linking these fast, presteady-state kinetic methods to the ESI inlets of suitable mass spectrometers will revolutionize the study of enzyme-catalyzed reactions.

Steady-state and presteady-state catalysis can be contrasted by imagining a reaction sequence as a series of buckets linked together. A faucet fills these buckets with water at a constant rate, with each bucket representing a separate enzyme-bound reaction intermediate along the pathway from substrates to products. Early on the first bucket contains quite a bit of water, the second bucket only some water, and the third bucket contains almost no water. Although water is entering the system of buckets continuously from the faucet, no water has yet emerged from the outlet in this example. This is the presteady-state condition, in which the buckets still are filling, and outflow does not yet equal inflow. The steady-state begins when each of the buckets is filled to a constant level, and flow into the system equals flow out of the system. Observation of the system during steady-state allows us to determine the number of buckets (enzyme forms), and also to quantify the water level in each bucket (i. e., to quantify how much of each enzyme form is present). Observation of the buckets as they fill during the presteady-state, also enables determination of the number of enzyme forms, but additionally allows measurement of the rate of filling of each bucket, and their sequence of filling. We suggest that rapid-quench ESI-MS can sample and perform mass analysis of enzyme reaction mixtures at time points during presteady-state and steady-state catalysis, and also at equilibrium.

Anderson and Johnson [12] examined the enzyme EPSP synthase and determined all 12 kinetic constants along the pathway from shikimic acid-3-phosphate (S3P) plus phosphoenol pyruvate (PEP) to 5-enolpyruvoylshikimate 3-phosphate (EPSP), as displayed in Scheme 2.

$$\mathrm{E + S3P + PEP} \underset{4500\ \mathrm{sec^{-1}}}{\overset{650\ \mu\mathrm{M^{-1}s^{-1}}}{\rightleftharpoons}} \mathrm{E\text{-}S3P + PEP} \underset{280\ \mathrm{s^{-1}}}{\overset{15\ \mu\mathrm{M^{-1}s^{-1}}}{\rightleftharpoons}} \mathrm{E\text{-}S3P\text{-}PEP}$$

$$\underset{100\ \mathrm{s^{-1}}}{\overset{1200\ \mathrm{s^{-1}}}{\rightleftharpoons}} \mathrm{EI} \underset{240\ \mathrm{s^{-1}}}{\overset{320\ \mathrm{s^{-1}}}{\rightleftharpoons}} \mathrm{E\text{-}EPSP} \underset{0.07\ \mu\mathrm{M^{-1}s^{-1}}}{\overset{100\ \mathrm{s^{-1}}}{\rightleftharpoons}} \mathrm{E\text{-}EPSP} + \mathrm{P_i}$$

$$\underset{200\ \mu\mathrm{M^{-1}s^{-1}}}{\overset{200\ \mathrm{s^{-1}}}{\rightleftharpoons}} \mathrm{E + EPSP} + \mathrm{P_i}$$

Scheme 2.

Their studies combined information gathered from steady-state, presteady-state, and equilibrium studies of this enzyme. Their methods included equililibrium binding studies with fluorescent probes, stopped-flow spectrophotometry, internal equilibrium measurements with use of radiolabels, single-

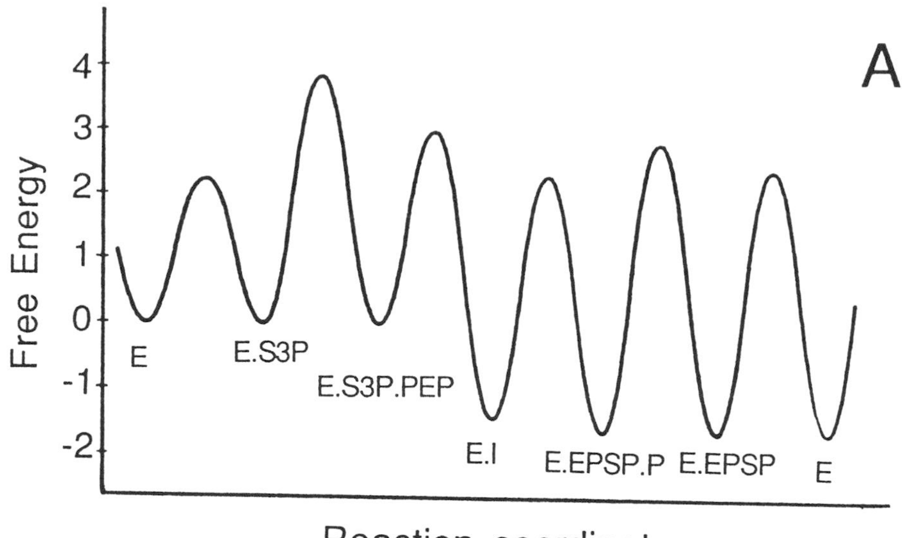

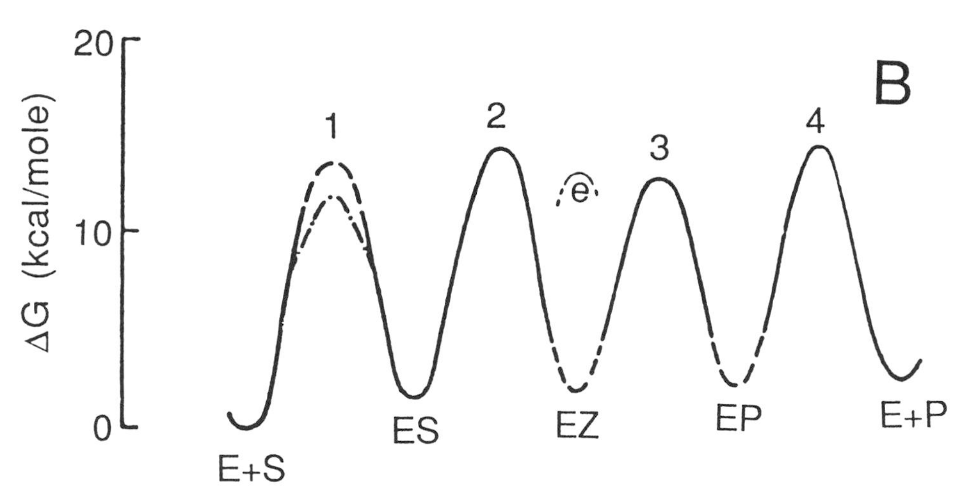

Fig. 1. A. Reaction free energy profile for EPSP synthase (reprinted with permission from [13]). B. The free energy profile for the reaction catalyzed by triose phosphate isomerase. The standard state is 40 μM [the known concentration of triose phosphates in vivo *(Williamson, 1965)]. Broken lines indicate limits on species that may be kinetically insignificant. The line —— for transition state 1 is the diffusion-controlled limit (assuming $k_1 = k_{-4}$). The barrier labeled "e" represents that for the exchange of the proton from the conjugate acid of the enzyme's catalytic base (in the enzyme-enediol intermediate) with a proton from the medium (reprinted with permission from [14]).*

turnover experiments with chemical quench and radiolabels, rapid-quench experiments to isolate sufficient tetrahedral intermediate for purification and subsequent identification by NMR, direct observation by NMR of the intermediate bound to the enzyme, and studies of the kinetics of rebinding of the

Table 1: Data for the Free-Energy Profile for Triose Phosphate Isomerase:

Value	ΔG (kcal mol^{-1})
$\frac{k_1 k_2 k_3 k_4}{k_{-1} k_{-2} k_{-3} k_{-4}}$ [a,b] = 0.0031	3.5
k_1/k_{-1} [a] = $1.3 \pm 0.4 \times 10^3\ M^{-1}$	1.8 [c]
$k_2/k_{-2} = 0.3 \pm 0.3$	>0.3
k_4/k_{-4} [a] > 1.1×10^{-5} M	<0.8 [c]
k_1 [a] > $10^7\ M^{-1}\ s^{-1}$	14.1 [c]
$k_2 = 1.7 \pm 0.5 \times 10^3\ s^{-1}$	13.3
$k_{-3}/k_4 = 21 \pm 6$	-1.8
k_{-4} [a] = $3.7 \times 10^8\ M^{-1}\ s^{-1}$	12.0
$\frac{k_5 k_{-3}}{k_3 k_4} = 14 \pm 3$	-1.6

[a] Values relating to the unhydrated substrate forms.
[b] Calculated from
$(k_1 k_2 k_3 k_4)/(k_{-1} k_{-2} k_{-3} k_{-4}) = [B_1(1 + K_s\text{-}1)]/[A_1(1 + K_p\text{-}1)]$.
[c] Calculated from a standard state of 40 μM.
(Data used with permission from Ref. [14].)

tetrahedral intermediate to the enzyme, to determine its dissociation constant. From these difficult, time-consuming studies, these workers constructed a free energy diagram for the catalytic mechanism of EPSP synthase [13] (Figure 1A). The free energy change and the apparent free energies of activation for each step of the reaction were calculated from the constants given for the forward reaction. The concentrations of substrates and products were set equal to their dissociation constants: [S3P] = 7 μM, [PEP] = 18 μM, [EPSP] = 1 μM, and $[P_i]$ = 1.43 mM, concentrations which are thought to approximate the physiological conditions. The relative apparent free energies of activation were calculated as $\Delta G^{\ddagger} = RT[\ln(k_B T/h) - \ln(k_{obsd})] - 10$ kcal/mol, where k_B is the Boltzmann constant. The value of 10 kcal/mol was chosen arbitrarily so that the relative magnitudes of the activation energies could be better illustrated in this graph. A free energy diagram also has been constructed by Albery and Knowles [14] for the enzyme triose phosphate isomerase (Figure 1B). These workers obtained the rate constants for this enzyme reaction (Table 1) by a series of 11 isotopic discrimination experiments combined with careful theoretical and mathematical analysis. To our knowledge such complete thermodynamic and kinetic discriptions of enzyme catalyzed reactions have been obtained in only three other cases, namely tetrahydrofolate reductase [15], T7 DNA polymerase [16], and human carbonic anhydrase isozymes I and II [17], thus bringing the total number of enzymes for which the complete catalytic mechanism has been characterized to about five. One reason that only about five enzymes have been described to this extent is because each enzyme necessitates use of different methods, and not all methods are universally applicable to all enzymes. For example, not all enzymes have a chromophoric substrate or product that can be monitored spectrophotometrically. Because of its potential to monitor by mass the time course of every reacting species, stopped-flow ESI-MS should enable identification of reaction intermediates, determination of kinetic constants, and the formulation of free energy diagrams, and may prove applicable for most enzyme systems.

ASPECTS OF ESI

In addition to the revolutionary fact that ESI enables transfer of proteins from the solution phase to the gas phase for mass spectral analysis [6], this technique leads to multiple protonation of proteins [6]. Protein molecules usually acquire not only a single proton, but a number of protons related to the number of accessible basic sites when ionized by electrospray. Thus, because mass spectrometers resolve ions according to the ratio of mass/charge (m/z), a family of peaks will appear in the mass spectrum for each enzyme molecule (Fig. 2). There are both advantages and complications to generating a family of multiply ionized enzyme species. One primary advantage is that one can monitor, for example, the +20 ion of an enzyme whose molecular mass is 100,000 Da. For a protein of this large size, the singly charged molecular ion of m/z = 100,000 will be poorly resolved and/or will be beyond the mass range of many mass spectrometers; however, the +20 species generated by ESI will appear in the mass spectrum at m/z = 5,000, which, though still beyond the m/z range of conventional quadrupole mass spectrometers (to circa m/z = 3000), is well within the m/z capabilities of many instruments currently available. One potential disadvantage of electrospray ionization is that quantitation of a particular protein molecule may be complicated by the presence of multiple peaks. However, the relative ratios among peaks should be comparable for the same molecule if conditions of electrospray ionization are maintained constant throughout an experiment; thus changes in the amplitude of a single m/z peak may be considered proportional to changes among the whole family of m/z ions of a given protein molecule. This empirically based principle of "reproducible proportionality" is utilized in the kinetic experiments of references [18] and [19], both of which will be discussed later in this chapter. Further, computer programs are

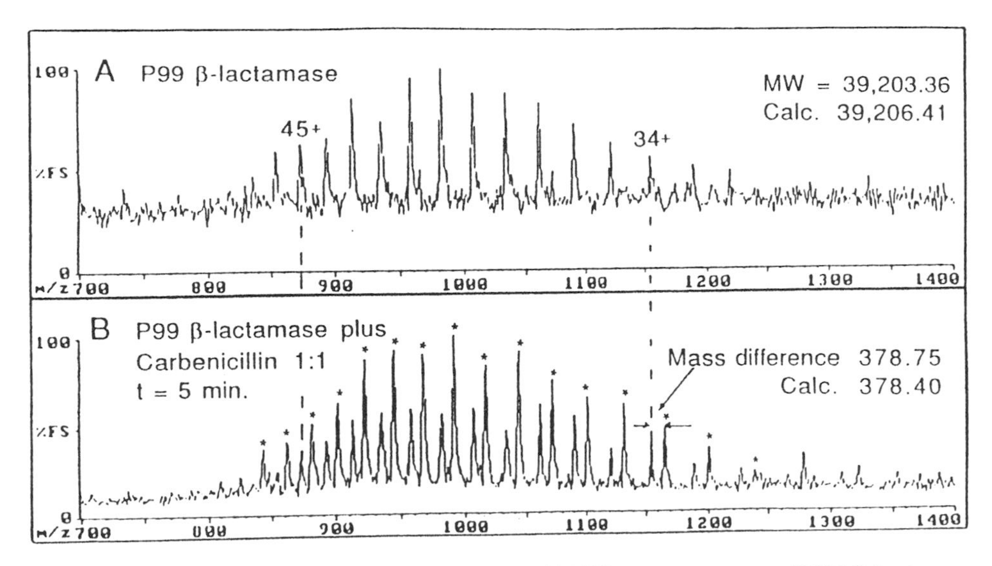

Fig. 2. a) ESI mass spectrum of P99 β-lactamase. b) ESI mass spectrum of P99 β-lactamase plus carbenicillin (indicates peaks assigned as acylated enzyme) (reprinted with permission from [20]).*

commonly available to deconvolute ESI spectra, and to present all of the members of an m/z family as an uncharged molecular species (we present the molecular species that would be generated by deconvolution in Figs. 4A and 4B, with reference to our illustrations with the enzyme hexokinase). A suitable, nonreactive, internal standard also may be included in enzyme mixtures to assist correction for variations in the distribution of ionization and ionization efficiency.

To demonstrate some of the properties of ESI when applied to kinetic analysis of enzyme reaction intermediates, we present two mass spectra obtained of the enzyme β-lactamase by Aplin, et al. 1990 [20](Figures 2A and 2B). In the upper spectrum (Figure 2A), the multiple ions generated during ESI analysis of native β-lactamase (MW 39,203.36) are centered around +40 protons, and the family of ion peaks spans from about +32 to +46. Aplin et al., incubated β-lactamase for 5 min, mixed 1:1 with the slow substrate carbenicillin (MW 378.40), which combines covalently with a serine residue of the enzyme during catalysis. Figure 2B shows the family of new, large satellite peaks (marked with *) displaced to higher mass, corresponding to the covalent, catalytic, carbenicillin- β-lactamase complex. The separation between the 34+ peaks for free enzyme and the satellite peaks for the enzyme-carbenicillin complex is about 11 m/z, which corresponds to the covalent adduct of MW 378.40 (378.40/34 = 11.13). This small m/z separation emphasizes the value of mass spectrometers with high mass resolution for detection of enzyme-bound reaction intermediates, particularly for observation of small ligands bound to highly ionized enzymes of high mass. Aplin *et al.* [20] reported that acylation of β -lactamase by carbenicillin was ca. 70% after 5 minutes incubation, and that acylation decreased to ca. 20% and then to zero after 10 min and 15 min incubation, respectively, thus showing that the transitory progress of reaction intermediates may be monitored with respect to time by monitoring the relative amplitudes of the acyl-generated satellite peaks.

Aplin et al. [20] first quenched their reactions with methanol-acetic acid-H_2O immediately prior to performing ESI mass analysis. Although chemical quench preserves many covalent interactions intact, it disrupts noncovalent enzyme-ligand interactions, thus preventing their direct observation. We will discuss evidence from the literature demonstrating that enzyme reactions also may be quenched by direct ESI introduction into the mass spectrometer, without need of prior chemical quench, thus permitting observation of both covalent and noncovalent enzyme-ligand interactions by mass spectrometry [18, 19, 21, 22].

As would be true of all covalent enzyme-ligand complexes, the covalent adduct of carbenicillin with β-lactamase is not cleaved by rigorous electrospray ionization at high cone voltages. However the same is not true for noncovalent enzyme-ligand interactions, whose stability may be destroyed by collisional activation during ESI. This property of rigorous electrospray ionization provides a distinguishing test for covalent versus noncovalent interactions [23, 24], and has been used to displace a transient, noncovalently bound, reaction intermediate from its enzyme to enable identification of the displaced ligand by mass spectrometry [22]. However, perhaps the most valuable characteristic of electrospray ionization for the enzymologist is that at low declustering voltages, noncovalent enzyme-ligand interactions often remain intact [3-5]. Thus ESI-MS at more gentle conditions of ionization can be useful to observe, and to quantify [7, 18, 25, 26] noncovalent, as well as covalent interactions of ligands with their enzymes.

We are interested primarily in employing the gentle conditions of electrospray ionization to monitor mass spectrometrically the kinetics of formation of both covalent and noncovalent interactions of enzymes with their substrates, reaction intermediates, and products during catalytic turnover. This hitherto little used application of ESI-MS potentially will extend the field of enzymology to enable extraction of the rate constants, plus construction of the free-energy diagrams for most enzyme-catalyzed reactions. To accomplish this experimentally, we suggest attachment of rapid-reaction mixing systems to the ESI inlets of suitable mass spectrometers, and use of these systems to monitor the kinetics of the formation and/or the disappearance of every reacting species. We suggest that the new method of rapid-reaction ESI-MS may be used, for example, to monitor for an enzyme-catalyzed reaction each of the satellite peaks corresponding to each enzyme-ligand species, similar to the time course of the appearance and disappearance of the satellite peaks observed by Aplin et al., [20] (Figure 2B) by ESI for the covalent adduct of carbenicillin with β-lactamase.

Theory and Application

We now review a simple application of ESI-MS to whet the imaginations of those experimentalists and theoreticians who may wish to pursue kinetic analysis of enzyme reaction mixtures. As our illustration, we will monitor by ESI the time course of phosphorylation of glucose to glucose-6-phosphate catalyzed by the enzyme hexokinase [1, 2] (Equation 1) [9], with the implication that the general principles presented for hexokinase will be applicable to any enzyme-catalyzed reaction.

$$\text{glucose} + \text{MgATP} \xrightleftharpoons{\text{hexokinase}} \text{glucose-6-P} + \text{MgADP}$$

Equation 1.

Hexokinase catalyzes the first reaction of the glycolytic pathway and is therefore familiar to most biologists. The substrates and products of this reaction are not colored, and no simple method is available to monitor them directly. To observe the progress of this reaction with a spectrophotometer, it must be coupled through a second reaction which produces a colored product, as also is the case with many enzyme-catalyzed reactions which lack a chromophoric substrate or product. For routine biochemical assay, production of glucose-6-phosphate is monitored spectrophotometrically at 340 nm by coupling to $NADP^+$/NADPH via the enzyme glucose-6-phosphate dehydrogenase [30]. Alternatively production of MgADP may be followed, again at 340 nm, by addition of excess phosphoenolpyruvate and NADH by coupling through two enzymes, pyruvate kinase and lactic dehydrogenase [31].

---

[1]We are aware of no published ESI mass spectrum for hexokinase at this time. At 102,000 Da, the ions of hexokinase may exceed the m/z range of typical quadrupole mass spectrometers, however, the ability to observe hexokinase will depend upon the extent of its ionization. Figures 4A and 4B depict deconvolution of hypothetical ESI mass spectra; in practice, each enzyme-ligand species will have multiply ionized forms. In addition, the binding of a charged ligand such as MgATP to an enzyme may affect the charge state of the complex, and it has been reported that ligands may mask identical 51,000 MW subunits, and appears to be able to bind one glucose and one MgATP per monomer, or 2 glucose and two MgATP per 102,000 MW holoenzyme dimer [39]. Thus addition of glucose to the hexokinase dimer will result in 2^2=4 possible combinations of interaction, namely free hexokinase (zero added glucose), glucose-subunit A (1 added glucose), glucose-subunit B (1 added glucose), and glucose-subunit A — glucose-subunit B (2 added glucose per dimer). Also, the dimer can dissociate to the monomer under some conditions [39]. Thus in using hexokinase as our illustration, we have oversimpligied the real situation and have considered binding of only one glucose and only one MgATP per one 102,000 MW holoenzyme.

[2] Molecules of neutral charge, such as glucose used in the example of hexokinase, do not perform well in ESI, and the technique is more effective for the monitoring of species like glucose-6-phosphate and MgATP, which are charged in solution at neutral pH. However, even if ESI proves ineffective for glucose, the ability to monitor directly only the one substrate MgATP, together with both products of the hexokinase reaction provides a vast improvement over any assay method presently available for this enzyme. The neutral molecule lactose was measured quantitatively by ion spray LC/MS [28], and neutral molecules may be derivatized to enable their observation by ESI [29]. We caution, however, that chemical modification of a molecule may affect its suitability as a substrate, due to the specificity of enzymes.

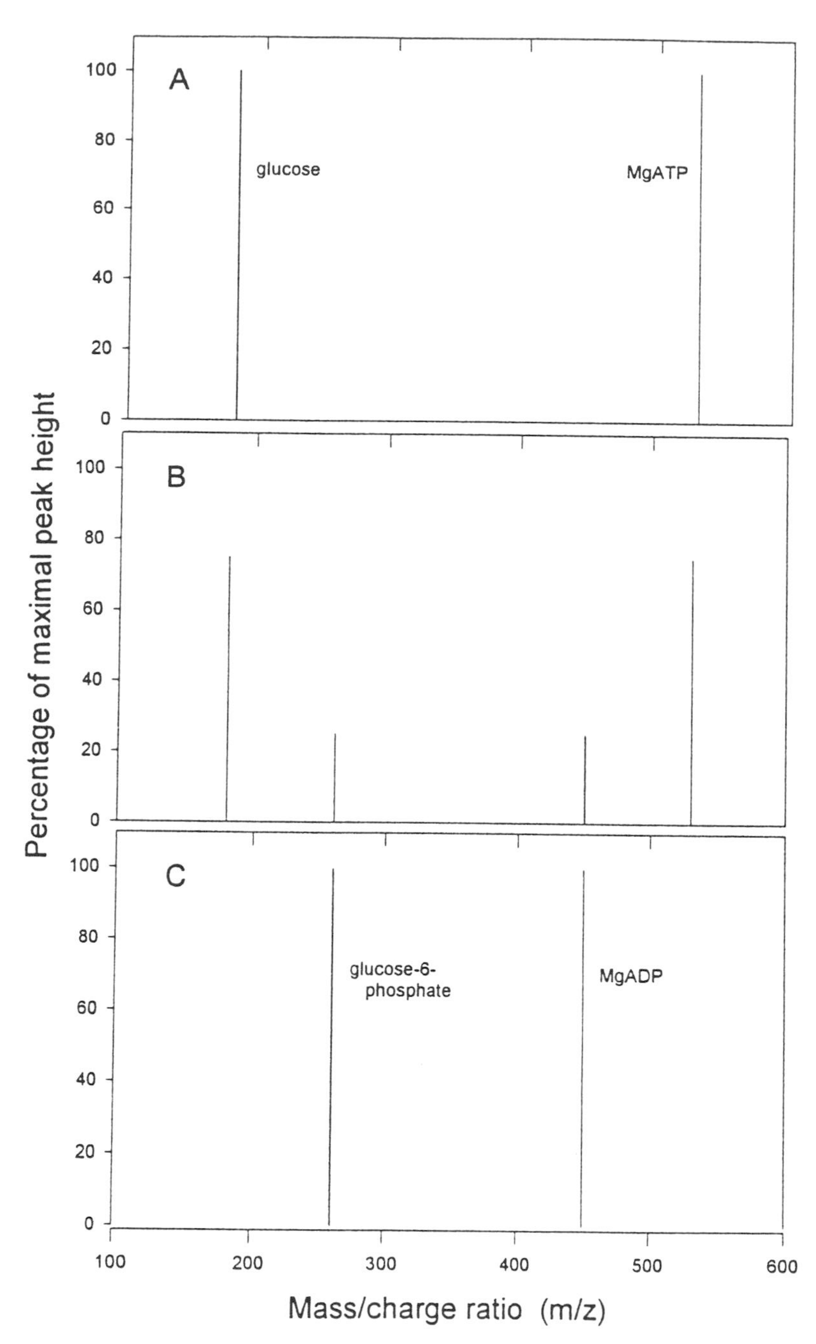

Fig. 3. Hypothetical representation of singly charged ions in the low mass region of the mass spectrum for the conversion of glucose to glucose-6-phosphate catalyzed by hexokinase. A) Glucose (m/z = 180) and MgATP (m/z = 530) at the start of the reaction. B) Glucose, glucose-6-phosphate (m/z = 260), MgADP (m/z = 450), and MgATP after conversion of 25% of the glucose to glucose-6-phosphate. C) Glucose, glucose-6-phosphate, MgADP, and MgATP at equilibrium (K_{eq} = 661) [32].

The Low M/Z Region of the ESI Spectrum

Let us now examine by ESI-MS, the progress of the hexokinase reaction as glucose and MgATP are converted to products. Were the enzymic reaction mixture injected into the mass spectrometer at early times during the reaction, by monitoring the lower mass range (m/z up to 600), we would anticipate observing large peaks for the substrates glucose (MW 180 Da) and MgATP (MW 530 Da), and negligible peaks for glucose-6-phosphate and MgADP (Figure 3A). As the reaction progressed, we would expect to observe the depletion of glucose and MgATP, and the appearance of the products, glucose-6-P (MW 260 Da) and MgADP (MW 450 Da), with time (Figure 3B). Upon completion of the reaction, because the equilibrium constant for the forward reaction at pH 7 is 661 [32], glucose-6-P and MgADP would be the predominant peaks (Figure 3C). Quantitation of the lower mass region therefore could enable us to monitor directly and simultaneously the respective concentrations of every substrate and of every product with time[2]. ESI-MS thereby may provide the first direct assay of the hexokinase reaction, and by analogy could provide an assay for any enzyme which catalyzes a mass change. Thus ESI-MS potentially may yield in a single method 1) a universal assay method for any enzyme-catalyzed reaction which involves at least one nonvolatile, charged substrate or product of unique m/z ratio, and 2) simultaneous monitoring of substrates and products. It also should be apparent that ESI-MS will provide a convenient means to monitor isotopically-labeled reagents during the progress of a reaction [22]. Additionally, in an historic step for enzymology, coupled assays would be rendered obsolete by ESI because a chromophore is not required.

The High M/Z Region of the ESI Spectrum

In addition to observing the complete kinetic patterns of substrate utilization and of product formation precisely by quantitating the low mass region of the ESI mass spectrum, we also can monitor the high mass region of the spectrum, even the mass spectrum of the reacting enzyme, itself; thus we can observe mass changes occurring on the enzyme during turnover[1]. (The high mass region of the mass spectrum is especially interesting because, as mentioned, under the gentle conditions of electrospray ionization, covalently and noncovalently bound ligands and enzymic reaction intermediates remain bound to the enzyme.) Thus, when glucose (MW 180 Da) binds to yeast hexokinase (MW 102,000 Da) to initiate turnover, the enzyme will increase in mass by approximately 180 Da, and this new complex should appear as a new peak in the deconvoluted mass spectrum at 102,180 Da (Fig. 4A). Similarly, when MgATP binds to free hexokinase, the enzyme will increase in mass by 530 Da to generate a peak at 102,530 Da. An additional peak then should appear at 102,710 Da when the second substrate (glucose or MgATP) binds, to yield the ternary hexokinase-glucose-MgATP complex. Then, if the mass resolution of the mass spectrometer is sufficient, mass changes occurring on the enzyme due to formation of reaction intermediates may be discernable. For the hexokinase reaction one might observe hydrolysis of MgATP due to addition of water (a mass change of 18 Da),

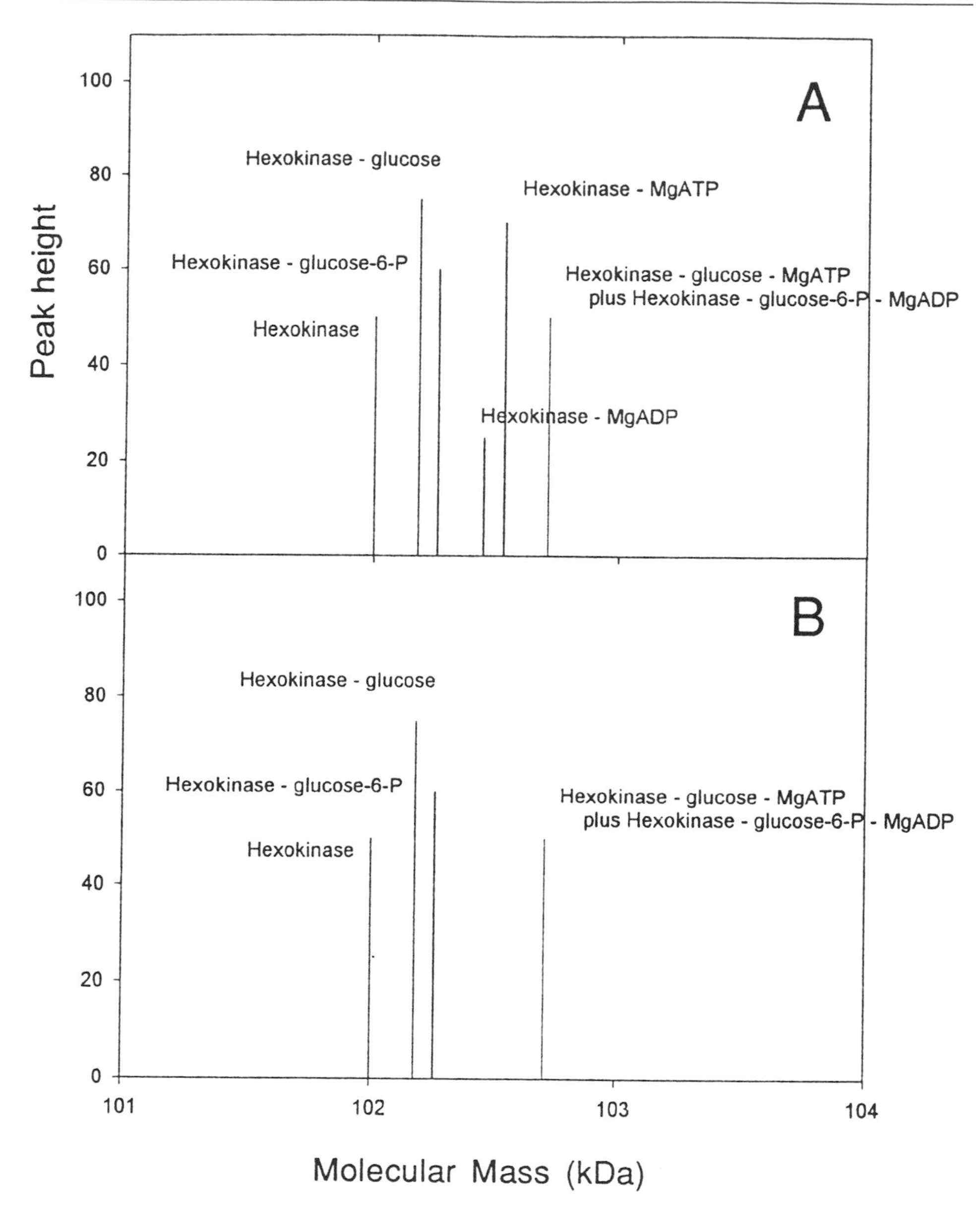

Fig. 4. a) Hypothetical deconvoluted ESI mass spectrum during the steady-state conversion of glucose to glucose-6-phosphate catalyzed by hexokinase. Random sequential binding and release of substrates and products is assumed, with no covalently bound intermediates [32]. Peaks represent free hexokinase (m/z = 102,000); hexokinase - glucose complex (m/z = 102,180); hexokinase - MgATP complex (m/z = 102,530); hexokinase - glucose - MgATP ternary complex (m/z = 102,710); hexokinase-glucose-6-phosphate-MgADP ternary complex (m/z = 102,710); hexokinase - glucose-6-phosphate complex (m/z = 102,260); hexokinase - MgADP complex (m/z = 102,450). Relative peak amplitudes will vary during the time course of the reaction as a function of the kinetic constants specific to a particular reaction step. b) Hypothetical deconvoluted ESI mass spectrum for which hexokinase is assumed to display ordered sequential binding and release of substrates and products, with glucose binding first and glucose-6-phosphate being released last, with no covalently bound intermediates [32].

or transfer of the phosphate to glucose, or formation of a covalent adduct with the enzyme, if it occurs. (Note that the hexokinase - glucose - MgATP complex, and the hexokinase - glucose-6-phosphate - MgATP complex have the same mass). Even the complexes of enzyme with the products MgADP (MW 450 Da) and/or glucose-6-phosphate (MW 260 Da) should be observable by monitoring changes in the mass of the enzyme as products are formed and subsequently are released. Thus during steady-state catalysis, as depicted in Fig. 4A, we predict peaks corresponding to every possible enzyme-substrate, enzyme-intermediate, and enzyme-product reaction form.

Random vs. Ordered Mechanisms in Steady-State Mass Spectra

Let us suppose that when we monitor by ESI a typical two substrate, two product enzyme reaction such as that of hexokinase, instead of seeing the mass spectrum of Fig. 4A, we instead obtain the spectum of Fig. 4B during steady-state catalysis. Note that in this new spectrum, although we have a peak at 102,180 corresponding to the hexokinase - glucose complex, the peak at 102,530 corresponding to hexokinase - MgATP is absent. Similarly, the peak corresponding to hexokinase - MgADP is missing from this second spectrum, also. We must now consider the possibility of an ordered sequential vs. a random sequential mechanism, as depicted in Fig. 5.

In a random sequential mechanism with two substrates, the substrates bind, one after the other, but either substrate may bind first, or second. Thus by a random mechanism (Fig. 5), both the hexokinase-glucose, and the hexokinase-MgATP complexes may form, and both complexes will be present during steady-state catalysis (Fig. 4A). However, in an ordered sequential mechanism (Fig. 5), glucose must bind first, or MgATP must bind first, followed subsequently by binding of the second substrate. Thus if glucose must bind first to the enzyme, and MgATP second, we should see a peak corresponding to hexokinase-glucose, and a peak for hexokinase-glucose-MgATP, but no hexokinase-MgATP complex would be present as depicted in Fig. 4B. Similarly, if MgATP must bind first, we would observe the peak for the hexokinase-MgATP complex, and no hexokinase-glucose complex would be present. The pathway for MgATP binding first is not supported by the spectrum of Fig. 4B. Thus by observation of the ESI spectrum of hexokinase during steady-state catalysis, we may be able to distinguish whether this enzyme has a random sequential, or an ordered sequential, mechanism, and, if ordered, what the order of binding is. The situation is similar for the release of products (Fig. 5). The mass spectrum of hexokinase of Fig. 4A suggests random release of products, whereas the spectrum of Fig. 4B suggests that MgADP is released first, and glucose-6-phosphate last.

Random vs. Ordered Mechanisms in Presteady-State Mass Spectra

Another method of determining whether an enzyme has an ordered or a random mechanism is to apply ESI-MS to investigate the presteady-state. As mentioned in the introductory section, presteady-state methods involve short reaction times and enable determination of both the rates and the sequences of formation of enzyme reaction intermediates. Thus we suggest adding a second

A Random or a Sequential Mechanism?

Steps of Substrate Binding:

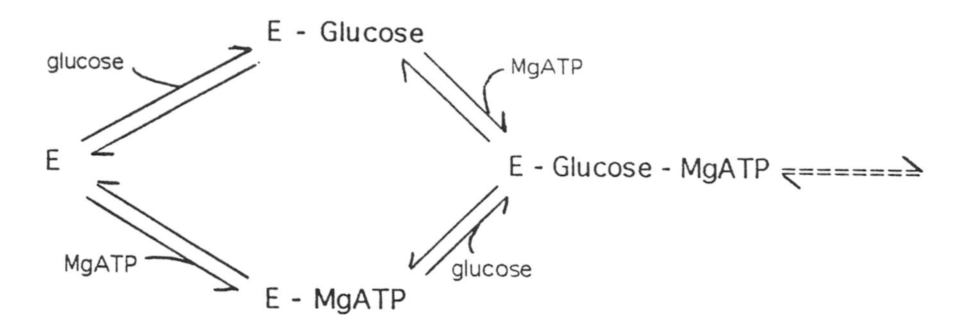

Chemistry:

E - glucose - MgADP ⇌ X_n? ⇌ E-glucose-6-P - MgADP

Steps of Release of Products:

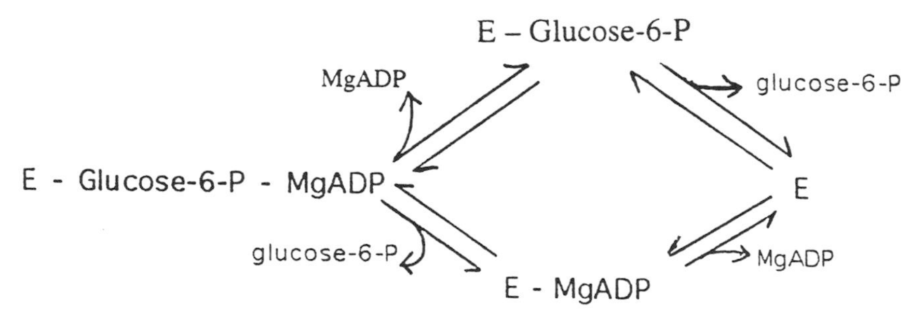

Fig. 5. Scheme for comparing an ordered sequential mechanism with a random sequential mechanism, with two substrates and two products.

syringe, a mixing chamber, and a variable-length delay line to present ESI delivery systems to adapt them for kinetic use, as in Fig. 6. Such introductory systems, which may be continuous-flow, or push-pause-push, will enable rapid quenching by ESI and immediate mass analysis of enzyme reaction mixtures, after variable periods of reaction. By following the time course of the hexokinase reaction in the high mass region, at intervals from the earliest times, in single turnover experiments, it should be possible to determine the order of binding of multiple substrates to the enzyme, analogous to the time courses of intermediate

ESIMS to Detect Transient Enzyme Intermediates

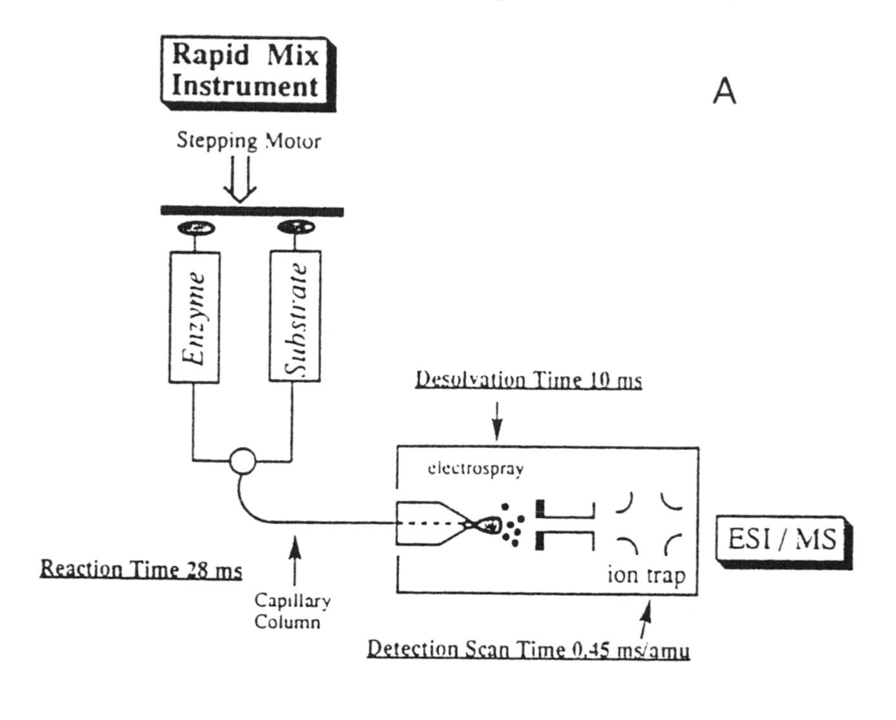

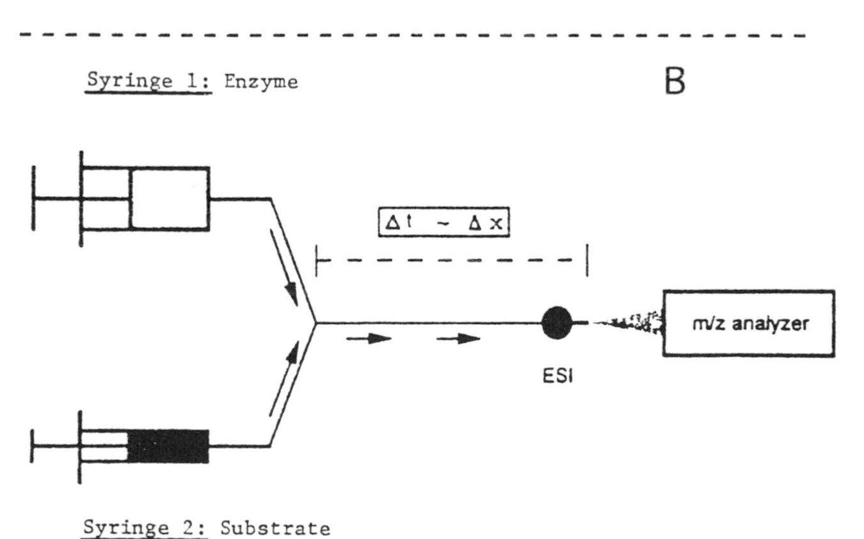

Fig. 6. Two continuous flow apparati for rapid mixing of reagents and incubation of reaction mixtures for precise time periods prior to quenching in an ESI mass spectrometer. A) After Paiva et al. [22]; only the 28 msec time point was examined with this system (reprinted with permission from Ref. [22]). B) After Douglas [19, 33, 34]; reaction periods were varied from 0.13 to 48 seconds by inserting 0.075 mm i.d. lines of different lengths between the point of mixing of reagents and the ESI inlet [19] (reprinted with permission from [33]).

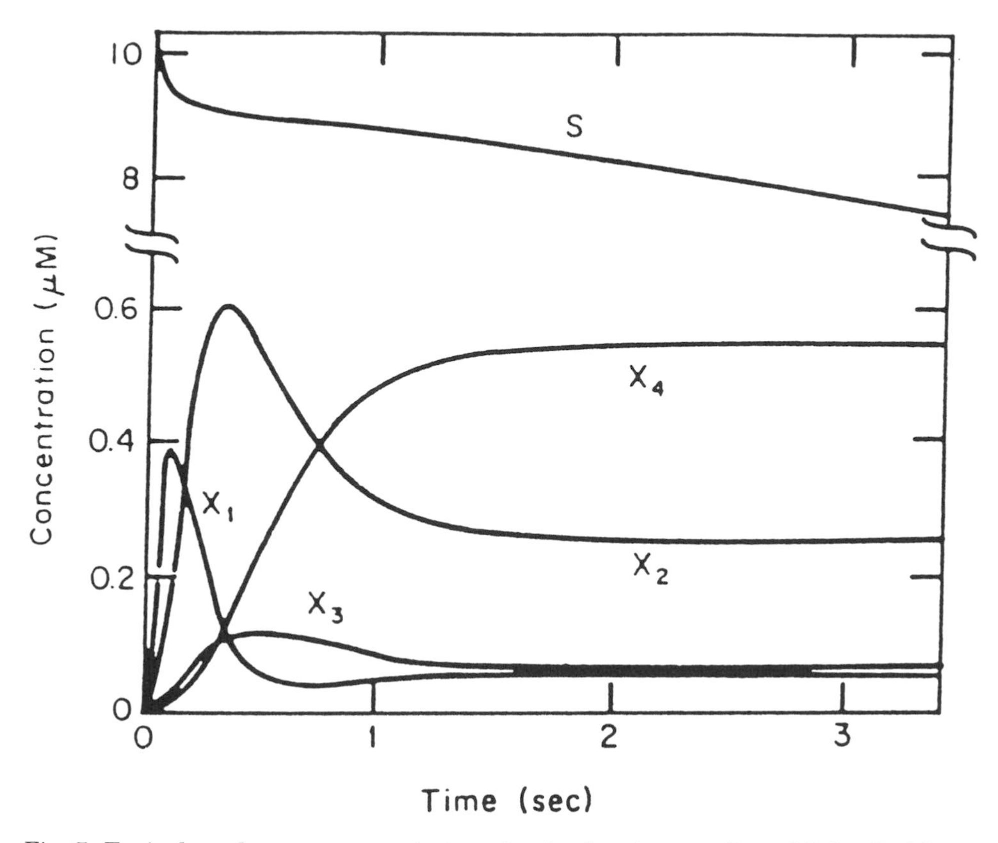

Fig. 7. Typical analog computer solutions for the four intermediate, Michaelis-Menten mechanism of Scheme 1, with $X_{n=4}$. The parameters used were S_o = 10 μM, E_o =1 μM, k_1 = 0.9 X 10^6 $M^{-1}sec^{-1}$, k_2 = 9 sec^{-1}, k_3 = 1 sec^{-1}, k_4 = 9 sec^{-1}, k_5 = 2 sec^{-1}, and k_i = 0 for all i (reprinted with permission from [10]).

formation simulated with the reaction sequence of Scheme 1, and presented in Fig. 7 [10]. Similarly, the sequential release of the products MgADP (MW 450 Da) and glucose-6-phosphate (MW 260 Da) might be observed directly in single turnover experiments by monitoring temporal changes in the mass of the enzyme, together with the times of appearance of peaks in the low mass region of the spectrum due to release from the enzyme of the products, themselves. A complete progress curve could be obtained for every reaction species of unique mass/charge (m/z) ratio, even enzyme-bound intermediates. If the results of such studies show that the X_1 curve of Fig. 7 corresponds to the time course for formation of the hexokinase-glucose complex, and if the X_2 curve corresponds to the time course for formation of the hexokinase-glucose-MgATP complex, then we will be able to diagram the hexokinase reaction as in Scheme 3, as an ordered sequential mechanism, with glucose binding first.

Although most researchers presently believe that hexokinase displays an ordered-sequential mechanism, with glucose binding first, this issue has been debated for at least 30 years [32], and it is not yet resolved, particularly because

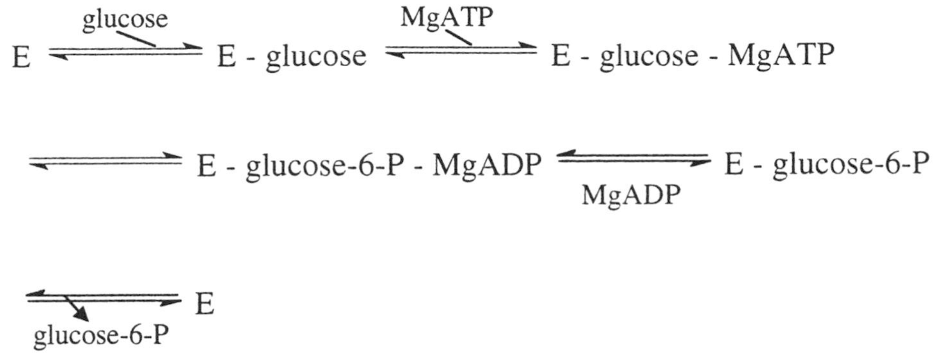

$$\mathrm{E} \underset{}{\overset{\text{glucose}}{\rightleftharpoons}} \mathrm{E\text{ - }glucose} \overset{\text{MgATP}}{\rightleftharpoons} \mathrm{E\text{ - }glucose\text{ - }MgATP}$$

$$\rightleftharpoons \mathrm{E\text{ - }glucose\text{-}6\text{-}P\text{ - }MgADP} \underset{\text{MgADP}}{\rightleftharpoons} \mathrm{E\text{ - }glucose\text{-}6\text{-}P}$$

$$\underset{\text{glucose-6-P}}{\rightleftharpoons} \mathrm{E}$$

Scheme 3.

MgATP is hydrolysed slowly by hexokinase in the absence of glucose. ESI-MS holds the potential for resolving this mechanistic issue, by either steady-state mass spectral analysis of enzyme intermediates, or by the presteady-state analysis just described (we suggest a third method to distinguish an ordered from a random mechanism by ESI-MS in Scheme 4). The ability to discern reaction mechanisms by ESI mass spectrometery may revolutionize our understanding of enzymes and catalysis.

$$\mathrm{E + S1} \underset{k_2}{\overset{k_1}{\rightleftharpoons}} \mathrm{ES1} \quad \mathrm{ES1 + S2} \underset{k_4}{\overset{k_3}{\rightleftharpoons}} \mathrm{ES1S2}$$

$$\mathrm{ES1S2} \underset{k_6}{\overset{k_5}{\rightleftharpoons}} \mathrm{EP1P2} \qquad \mathrm{EP1P2} \underset{k_8}{\overset{k_7}{\rightleftharpoons}} \mathrm{EP2 + P1}$$

$$\mathrm{EP2} \underset{k_{10}}{\overset{k_9}{\rightleftharpoons}} \mathrm{E + P2}$$

Scheme 4.

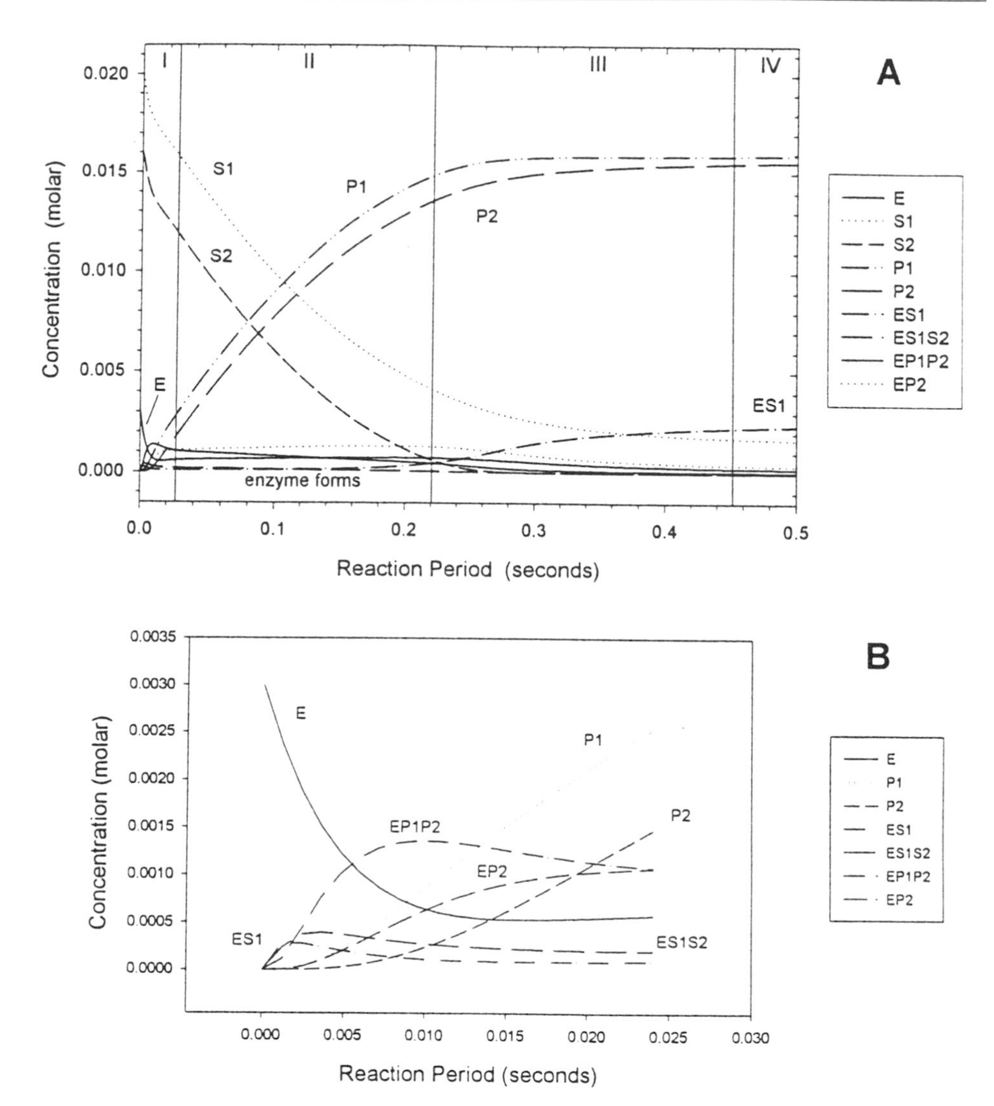

Fig. 8. The concentration of every substrate, product, and enzyme species as a function of time for a hypothetical ordered reaction analogous to the conversion of glucose to glucose-6-phosphate by hexokinase. A) A complete time course from time zero to near equilibrium. B) An expanded view of the presteady-state portion of the time course, but with substrates omitted. Curves were generated with the program MathCad for the mechanism displayed in Scheme 4. Rate constants were chosen as $k_1 = 10^4$, $k_2 = 1$, $k_3 = 10^5$, $k_4 = 10^2$, $k_5 = 10^3$, $k_6 = 10^2$, $k_7 = 10^2$, $k_8 = 10^3$, $k_9 = 10^2$, $k_{10} = 10^4$. Initial reactant concentrations were E = 0.003 M, S1 = 0.02 M, S2 = 0.016 M, P1 = 0 M, P2 = 0 M, i.e, S1 is present initially in slight excess to S2. The initial concentration of each enzyme intermediate was set to zero.

PRESTEADY-STATE VS. STEADY-STATE VS. POSTSTEADY-STATE VS. EQUILIBRIUM ANALYSIS

We next propose combining the ESI methods for hexokinase that we have discussed, and monitoring by rapid-reaction ESI-MS every reacting species of the enzyme-catalyzed reaction, simultaneously. In the low mass region of the spectrum, we would monitor every substrate and every product, and in the high mass region every enzyme intermediate species, from the earliest moments after rapidly mixing the reagents, to the end of the reaction. We would do this by coupling a rapid-reaction system of syringes and a mixer before the ESI inlet and by directly quenching the reaction mixture into the ESI mass spectrometer after various precisely determined periods of reaction, thereby immediately obtaining a complete mass spectrum for every time point during the enzyme-catalyzed reaction.

By this procedure, the progress curves of an enzyme reaction, such as that of hexokinase, can be obtained. We consider four periods of reaction: I. immediately after the reagents are mixed (single turnover, or presteady state), II. steady-state catalysis, III. poststeady-state (i. e., approach to equilibrium), or IV. after the reaction is completed (equilibrium) (Fig. 8A). Analysis of the presteady-state portion of the progress curves (Fig. 8B) enables determination of the number of enzyme intermediates, their order of formation, and by mathematical curve fitting, the kinetic constants of the rates at which they form. Analysis of reaction mixtures at later times, i. e., during steady-state or poststeady-state catalysis, or after equilibrium is attained (Fig. 8A), also enables determination of the number of enzyme-intermediate forms, and additionally allows comparison of the relative concentrations of these enzyme forms during these phases of catalysis. For this ordered mechanism, the addition of excess S1 leads to appreciable ES1 and negligible ES2 after the reaction reaches equilibrium, because binding of S1 preceeds binding of S2. By comparison, for this same ordered mechanism, addition of excess S2 would have led, instead, to negligible ES1, and again negligible ES2, at equilibrium. For a random mechanism an initial excess of S1 or S2 will result in an excess of either ES1 or ES2 at equilibrium, depending upon which substrate was in excess at the start of the reaction. Thus ordered and random mechanisms potentially may be distinguished by adding an excess of each substrate, SX or SY, and analyzing the equilibrium reaction mixture for evidence of excess ESX or ESY by ESI-MS.

A free energy diagram may be obtained at any point of the progress curve, (i) by determining the order of binding of each species (by steady-state analysis [see text], by the equilibrium analysis described above, or from temporal analysis of the presteady-state portion of the progress curve), (ii) by deriving the kinetic constants joining every enzyme form from mathematical analysis of the presteady-state kinetic curves for formation of each reactant species, and then (iii) by comparing the relative concentrations of each reaction species at that time point.

Determination of Dissociation Constants by ESI

The observation that ligands remain bound to their enzyme makes possible determination of dissociation constants of inhibitors and substrates with their enzymes. If the conditions of electrospray ionization can be tuned to eliminate disruption of the enzyme-ligand complex, then the relative concentrations of free ligand, free enzyme, and ligand-enzyme complex can be determined from a measurement of the calibrated peak heights of each of these species in the mass spectrum, and used to calculate K_d [26]. Ganem et al. [18] noted that the relative intensities of the mass spectral peaks obtained by ESI due to enzyme-substrate and enzyme-product complexes of hen egg white lysozyme correlated well with known association constants in solution. A study with ESI of the competitive interaction of carbonic anhydrase with a series of inhibitors with differing binding affinities, also enabled estimation of dissociation constants, and additionally afforded a powerful method of screening a library of potential inhibitors [7, 25]. When this ESI analysis was coupled with tandem mass spectrometry, the inhibitors were identified [7, 25].

Determination of K_d holds promise for the enzyme kineticist, who desires to compare K_d with the Michaelis constant, K_m, for enzyme-substrate interactions [1]. In most cases the K_m of a substrate for an enzyme can be determined from a steady-state analysis of initial reaction rates, v_o, as a function of the substrate concentration. But K_m and K_d are not identical; K_m differs from K_d by the rate constant k_{+2}, as in Scheme 5, below. K_d can be determined only with difficulty from steady-state measurements [1]. Now, in addition to measuring K_m, we also can determine K_d by ESI, and can compare these two terms.

$$E + S \underset{k_{-1}}{\overset{k_1}{\rightleftharpoons}} ES \underset{k_{-2}}{\overset{k_2}{\rightleftharpoons}} E + P$$

$$K_d = \frac{k_{-1}}{k_1} = \frac{[E]\,[S]}{[ES]}$$

$$K_m = \frac{k_{-1} + k_{+2}}{k_1}$$

Scheme 5.

Determination of Mechanisms of Inhibition

A competitive inhibitor will compete with a substrate for the same binding site on an enzyme. Noncompetitive inhibition implies that the substrate and the inhibitor bind to different sites on the enzyme. The observation that enzyme-ligand complexes remain intact during ESI suggests that competitve and noncompetitive inhibitors may be distinguished. In the competitive case, when E, I, and S are mixed, peaks for ES and EI should be observed, whereas mixing a noncompetitive inhibitor with E and S should lead to peaks for ES and EI, and additionally for EIS. Further, an increase in the concentration of a competitive inhibitor should result in a lowering of the concentration of ES, whereas an increase in the concentration of a noncompetitive inhibitor should not displace S from the ES complex. Ganem, Lee, and Henion [20] incubated hen egg white lysozyme with its substrate (N-acetyl-glucosamine)$_6$ (NAG_6) in the presence of the known competitive inhibitor tetra-N-acetylchitotetraose ∂-lactone (TACL). The authors noted that the peak for the enzyme-NAG_6 complex diminished in the presence of TACL, as would be expected since TACL is competitive versus NAG_6. However, we found it impossible to determine from their published mass spectra whether TACL might also be noncompetitive with NAG_6, i. e., to determine whether the ternary EIS, (enzyme-TACL-NAG_6)$^{8+}$ complex (m/z 2046.4), forms, because the large +7 charge state peak for native HEWL (m/z 2044.6) nearly coincides with the +8 peak expected for this hypothetical, noncompetitive complex, and thus precludes its observation.

Use of Harsher Conditions of Ionization by Electrospray

All of the previous ESI experiments have presumed gentle ionization, which preserves noncovalent enzyme-ligand interactions. From an experiment performed under a series of increasingly and/or decreasingly disruptive conditions of ionization, one may be able to estimate the relative binding constants of related reaction intermediates and/or inhibitors to an enzyme [27]. Further, once the enzyme intermediates are dissociated from the enzyme, the identity of these presently elusive reaction intermediates can be determined by the appearance of their new, characteristic peaks in the low mass region of the mass spectrum [22]. Subsequent, CID analysis of these new peaks, and/or of the parent enzyme-ligand complex, by tandem mass spectrometry, can lead to even more definitive identification of each intermediary reaction species [7, 22, 25]. The use of disruptive conditions of electrospray ionizatin additionally provides a distinguishing test for covalent vs. noncovalent interactions of ligands with their enzyme [23, 24].

Practical Aspects

The ESI inlets of most modern mass spectrometers designed for analysis of static solutions consist of a syringe pump which slowly and steadily pumps a single syringe containing the material of interest down a long, narrow-bore tube, and into the ionization region of the mass spectrometer. The simple expedient of adding a second syringe, a mixing chamber, and a variable-length delay line to present delivery systems will adapt them for kinetic use (Fig. 6). However, because slow pumping speeds are used unless comparatively slow reactions are monitored, or enzymic concentrations are adjusted to very low levels, most enzyme-catalyzed reactions will have attained equilibrium, or certainly already will have reached steady-state, by the time the reacting solution flows the necessary distance from the point of mixing of reactants to the inlet. Modern, slow-flowing ESI systems thus are easily adaptable for presteady-state investigations of slower reactions, or for study of rapidly reacting mixtures under steady-state and/or equilibrium conditions, though, until recently [1, 19, 22, 33, 34] few have considered adapting ESI inlets for these purposes [18, 21, 28]. On the other hand, the protocols necessary to introduce into the ESI mass spectrometer rapidly reacting enzyme samples for presteady-state kinetic analyses will necessitate higher flow rates, faster, perhaps pulsing, syringe pumps, splitters, much shorter (i.e. smaller volume) introductory lines, and/or mass spectrometers equipped with more powerful vacuum pumps. To our knowledge only one group has accomplished the linking of such a system to an ESI mass spectrometer [22]. We anticipate that this article may stimulate invention among experimentalists and manufacturers to this regard. For rapid reactions, mass spectrometers such as the TOF, with the ability to scan the mass range of the reaction mixture rapidly and repeatedly would be advantageous. For slower reactions or for reactions already at equilibrium, rapid mass analysis is less critical, and magnetic sector, quadrupole, ion trap, and FTMS mass spectrometers may be suitable.

Illustrations from the Literature

Below we review papers that describe the use of direct injection ESI-MS to quench enzyme catalyzed reactions, as these papers illustrate the feasibility of the techniques and principles that we have outlined. Additionally, we will include briefly some experiments in which reactions were quenched chemically prior to ESI-MS.

Kinetic Monitoring of the Substrates and Products of an Enzyme-Catalyzed Reaction in the Low Mass Region of the ESI Spectrum

The first report of use of ESI, actually ionspray, to monitor an enzyme catalyzed reaction was by Lee, Muck, Henion, and Covey in 1989 [21]. These workers simply and elegantly attached a syringe barrel to their ionspray (atmospheric pressure ionization for lactase) inlet, added the reaction mixtures to the syringe, and allowed the solution to drain continuously through their inlet and into a triple quadrupole mass spectrometer as the reaction progressed. Their "continuous-introduction ion-spray system" enabled continuous sampling of reaction mixtures and employed ESI both to quench the reaction mixtures and to introduce the reactants and products directly into the mass spectrometer for immediate mass analysis. These workers analyzed five reactions, two chemical, and three enzyme catalyzed. One of the enzymic reactions was the cleavage of o-nitrophenyl *B*-galactopyranoside by the enzyme lactase. The o-nitrophenolate anion produced by this reaction can be monitored optically, however Lee, et al., chose to monitor production of o-nitrophenolate anion continously at m/z 138. They determined the initial rates of reaction from the resulting progress curves, at five different initial concentrations of substrate, much as a steady-state kineticist would generate a series of progress curves with a spectrophotometer. From the initial rate data they constructed a simple Lineweaver-Burk plot and determined a K_m for o-nitrophenyl B-galactopyranoside of 4.2 mM, which was similar to literature values.

A second enzyme-catalyzed reaction [21], was cleavage of the peptide dynorphin by the serine protease chymotrypsin. They scanned the m/z range 300 to 1100 at 0.8 min/scan, scanning once per minute for 3 minutes. They observed the parent $(M+H)^+$ (m/z 981) and $(M+2H)^{++}$ (m/z 491) ions of dynorphin decrease in intensity, and the cleavage products of masses 443, 557, and 712 increase in intensity, as the enzymatic reaction progressed. They were able to monitor the relative concentrations of the substrate and all of the products, approximately simultaneously, throughout the progress of the reaction. These experiments provide clear evidence that both substrates and products can be monitored over time in the low mass region of the spectrum by direct quench ESI-MS, as we suggest for assay of hexokinase and most enzymatic reactions.

Kinetic Monitoring of Covalent Enzyme-Bound Reaction Intermediates in the High Mass Region of the ESI Mass Spectrum Following Chemical Quench

We already have described in our introduction the experiment of Aplin et al. [20], in which they observed the progress of formation and depletion of a covalent acyl enzyme intermediate formed by incubation of carbencillin with β-lactamase for 5, 10, and 15 minutes. These workers did not monitor the m/z of the substrate or the products; instead they monitored the high mass region for the enzyme peaks. As mentioned, the satellite peaks due to formation of the covalent carbenicillin- β-lactamase complex were observed after first quenching their samples, and then performing analysis by ESI (Fig. 2B). Brown et al., [35] also used chemical quench methods followed by ESI to construct a time course for covalent mass changes to TEM-2-β-lactamase upon incubation of this enzyme with the inhibitor tert-butylammonium clavulanate. Ashton et al., [36] have used quench methods after different periods of reaction, followed by ESI analysis of the quenched mixture, to determine that covalent acyl-enzyme intermediates are formed during reactions of the substrates indole acryloyl imidazole and cinnamoyl imidazole with chymotrypsin, subtilisin Carlsberg, and subtilisin BPN. Aplin et al. [37, 38] have used a Biologic QFM-5 quenched-flow apparatus to investigate interaction between a mutant β-lactamase, E166A, and the substrate penicillin V. After reaction periods of 5 msec, 200 msec and 1 sec, they quenched with formic acid. Using ESI with a triple quadrupole atmospheric pressure ionization mass spectrometer, they observed formation of a covalent acylenzyme intermediate which increased in concentration with time. It would be of interest to see the QFM-5 syringe system coupled directly to the ESI inlet, thus obviating the need for acid quench! We note that chemical quench methods permit subsequent mass analysis of covalent enzyme-ligand interactions by ESI, however chemical quench denatures noncovalent interactions, and prevents their observation.

Kinetic Monitoring of Noncovalent Enzyme-Bound Reaction Species in the High Mass Region of the ESI Mass Spectrum by Direct Introduction ESI Analysis

An especially fascinating study by Ganem, Li, and Henion [18] shows <u>noncovalently</u> bound enzyme-substrate and enzyme-product complexes of the enzyme, hen egg white lysozyme (HEWL, MW 14,305). These workers used a continuous-introduction ion spray MS system similar to that used earlier [21], except that the rate of sample introduction was constant at 2 μl/min. As before [21], reactions were not quenched chemically, but were quenched directly into the ESI mass spectrometer. The expectation was that the gentle conditions of electrospray ionization would maintain noncovalent interactions intact, and to this end they monitored mass changes of the enzyme, itself, in the high m/z region of the mass spectrum [18]. This was an important departure from their earlier work [21], which had investigated not the enzyme, but only depletion of substrates and/or formation of products in the low m/z region. As the +8 charge state was of maximal amplitude in the native ESI spectrum, only the +8 to +7 region was monitored to detect mass increases due to noncovalent interactions of the enzyme with substrates and products during catalysis, apparently in accord with the principle that any change in the amplitude of an 8+ species also would be proportional to the changes in amplitude of the remaining members of its m/z ion family.

To initiate the reaction these workers added as substrate, the hexasaccharide of N-acetylglucosamine, NAG_6, which is cleaved principally to yield NAG_4 and NAG_2. Figure 9 displays the +8 to +7 region of the ESI mass spectrum 1 min, 10 min, and 60 min after the enzyme and NAG_6 were mixed. Note the large peak for the enzyme-substrate complex, $(HEWL+NAG_6)^{+8}$, at m/z 1943.3 in spectrum A at 1 min, which decreases in intensity as substrate is consumed during the progress of the reaction, as one would expect. Note also that the $(HEWL+NAG_4)^{+8}$ enzyme-product complex is seen at m/z 1892.5, and that this peak increases in size as product is produced during the time course of the reaction, also as expected. The enzyme-NAG_2 complex was not observed because it binds approximately 100 times more weakly to the enzyme than does NAG_4. However even the $(HEWL+NAG_3)^{+8}$ complex is observed at the longer reaction times due to slow hydrolysis of NAG_4. The relative heights of the enzyme-substrate and enzyme-product peaks correlate qualitatively with the known binding constants for the various saccharide substrates to hen egg white lysozyme.

These steady-state experiments prove that noncovalent enzyme-substrate and enzyme-product complexes can be observed by direct introduction ESI-MS in the high mass region of the spectrum during the progress of an enzyme-catalyzed reaction, and illustrate our suggestion that similar experiments are feasible for other enzyme-catalyzed reactions with noncovalent intermediates. This work also suggests the superiority of direct introduction ESI-MS: chemical quench disrupts noncovalent enzyme-ligand complexes prior to mass analysis, whereas the gentle quench of ESI can permit their elucidation. (Both chemical quench and ESI quench should be equally effective for mass analysis of covalent, enzyme-ligand interactions, and it would be useful to perform an experiment to verify this.) These experiments also suggest that the peak intensities for noncovalent enzyme-ligand complexes correlate with the known binding constants and with the relative concentrations of these species in the reaction mixture.

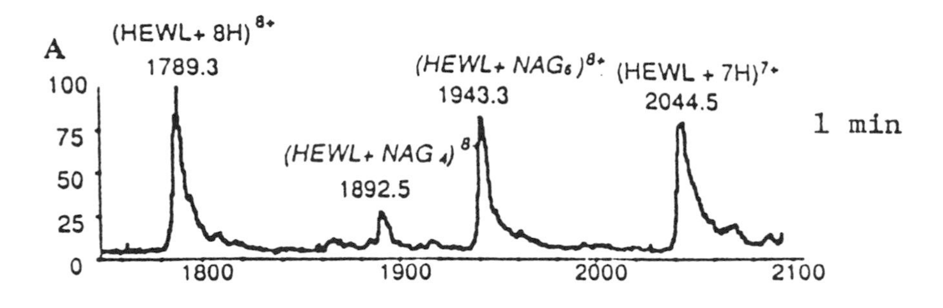

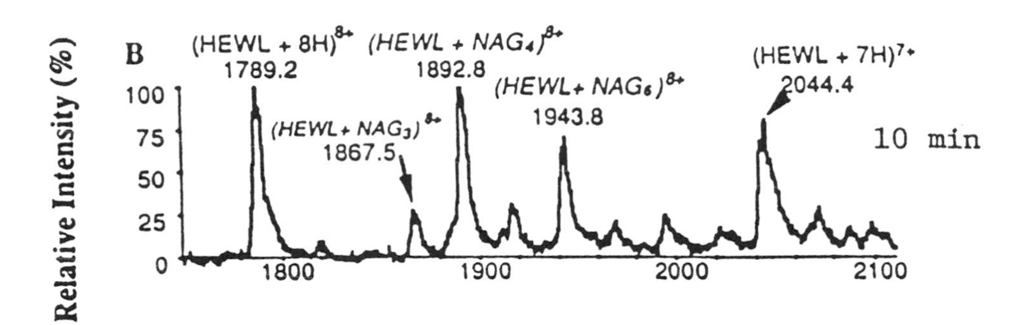

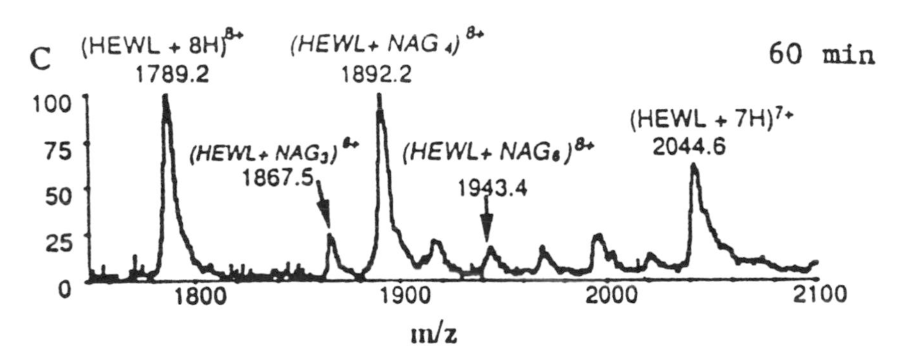

Fig. 9. An on-line, ion-spray MS time course study for the hydrolysis of NAG_6 by HEWL at room temperature over 60 min. A mixture of HEWL (3.17 X 10^{-5} M) and NAG_6 (4.41 X 10^{-4} M) in buffer (10 mM ammonium acetate, pH 5.0) was infused through the ion-spray interface at 2 µl / min. The partial mass spectra were obtained at 1 min (panel A), 10 min (panel B), and 60 min (panel C) (reprinted with permission from [18]).

Identification of a Transient Enzyme Reaction Intermediate in the Low Mass Region of the Mass Spectrum by Rapid-Quench ESI-MS

Karen Anderson's group from Yale University, sought to identify by ESI the noncovalently bound tetrahedral intermediate formed during conversion of shikimate-3-phosphate (S-3-P) plus phosphoenolpyruvate (PEP) to EPSP by EPSP synthase [Paiva et al., 22]. (The free-energy diagram and kinetic constants of EPSP synthase were described in the introductory section of this present article.) Their design employed a ram driven by a stepper motor to push two syringes (Fig. 6A), one containing the enzyme plus S3P, and the second PEP. This is a true rapid-reaction system, analogous to our suggestion. The reactant solutions were mixed rapidly through an Upchurch frit mixer, were passed at turbulent flow through an 11 cm, 0.01 cm i.d. fused capillary, and then were quenched by direct injection into an ESI ion trap mass spectrometer. Rather than introducing samples continuously, they pulsed the ram for about 0.3 sec to deliver 12 to 16 ul of reaction mixture into the mass spectrometer for each injection. The flow rate and the volume of the capillary were selected to provide a reaction period of 28 msec to achieve maximal presteady-state formation of the noncovalently bound intermediate on the enzyme.

Ganem et al. [18] selected declustering voltages so that the noncovalent enzyme-ligand complexes of HEWL would remain intact. However, Paiva et al. [22] deliberately chose a disruptive declustering voltage to dissociate the tetrahedral intermediate from EPSP synthase so as to enable identification of the tetrahedral intermediate in the low mass region of the ESI spectrum. Consequently, they monitored only the low mass region and not the m/z region of the enzyme. These workers report that they obtained better spectra with negative ion ESI than with positive ion mode.

In addition to the peaks observed for the substrates shikimic-3-phosphate and phosphoenol pyruvate, and the product EPSP, a peak was observed at m/z 421.0, corresponding to the tetrahedral intermediate that had formed on the enzyme during the reaction and was displaced from the enzyme during the disruptive ESI quench. A second experiment was performed to further characterize this peak. The substrate phosphoenolpyruvate was labled in the C-1 position with ^{13}C, and the rapid-quench experiment was repeated. The peaks for the substrate phosphoenol pyruvate, the product EPSP, and the tetrahedral intermediate (previously m/z 421.0, now 422.0) were displaced in mass by one additional amu (the PEP dimer increased in mass by 2 amu), while the mass of the unlabled substrate, shikimate-3-phosphate, remained unchanged. These results confirmed that the peak at m/z 422 is derived from the ^{13}C PEP via the enzyme catalyzed reaction. Further confirmation that the peak at m/z 421 was due to the tetrahedral intermediate was obtained with collision-induced dissociation (CID) which gave m/z 323, corresponding to the loss of a phosphate moiety by the tetrahedral intermediate. Incidentally, when these workers allowed the reaction to proceed to completion, after circa 1 second, the mass peak for the transient tetrahedral intermediate disappeared.

These fascinating experiments show that it is feasible to attach a rapid-reaction syringe system to an ESI mass spectrometer and to monitor successfully the formation of a transient enzyme reaction intermediate during presteady-state catalysis, 28 msec after mixing. Furthermore, it is not mandatory to use continuous injection ESI, but can be accomplished using pulsed injection. Additionally, ESI conditions may be used during the quench to disrupt noncovalent enzyme-ligand complexes, and to trap reaction intermediates for subsequent positive identification and characterization in the low mass region of the ESI spectrum by MS and, if desired, subsequent MS/MS.

Presteady-State Kinetic Analysis of Formation of a Transient, Covalently-Bound Enzyme Reaction Intermediate by ESI

D. J. Douglas and his research group at the University of British Columbia were the first of whom we are aware to attach a rapid-reaction, continuous-flow, syringe ram system to an ESI mass spectrometer, for direct injection, rapid-quench kinetic experiments of proteins [19, 33, 34]. Zechel et al. [19] were the first to monitor the presteady-state time course of formation of a transient enzyme bound intermediate, in this case one that attaches covalently to the enzyme during catalysis. Their system consisted of a rapid-reaction ram and two syringes joined into a "T" mixer (Fig. 6B). The flow rate was laminar and was maintained constant. The reaction period was varied by attaching lengths of 0.075 mm i.d. capillary of from 1 cm to 186 cm between the mixer and the ESI inlet to generate a range of time points from 0.13 to 47.8 seconds. Reaction samples were quenched by direct injection via ESI into a quadrupole mass spectrometer of their own construction.

Into one syringe they placed the enzyme, *Bacillus circulans* xylanase, in ammonium acetate buffer, pH 6.8, and into the second syringe the substrate, 2,5-dinitrophenyl B-xylobioside (abbreviated 2,5-DNPX$_2$). The reaction catalyzed by xylanase is depicted in Scheme 6.

$$E + 2,5\text{-}DNPX_2 \overset{K_d}{\rightleftharpoons} E\text{-}2,5\text{-}DNPX_2 \underset{2,5\text{-}DNP}{\overset{k_{+2}}{\rightleftharpoons}} E\text{-}X_2 \underset{H_2O}{\overset{k_{+3}}{\rightleftharpoons}} E + X_2$$

Scheme 6.

The first product, 2,5-DNP (2,5-dinitrophenol) is colored and they observed its production at 440 nm by stopped-flow spectrophotometry. In the wild-type enzyme the rate-limiting step is k_{+2}, which makes observation of the transient enzyme-bound intermediate, E-X$_2$ difficult, though these workers were able to detect a small amount of the E-X$_2$ intermediate bound to the wild type enzyme, 130 msec after mixing, with their rapid-reaction ESI mass spectrometer. However, Zechel et al. chose to investigate a mutant xylanase of *Bacillus circulan*, primarily because k_{cat} for the mutant enzyme is 480 times slower than that of the wild type enzyme, and also because the rate-limiting step for this mutant enzyme is not k_{+2}, but k_{+3}. Due to this different rate-limiting step, the mutant enzyme accumulated appreciable levels of the transient intermediate E-X$_2$, and more readily enabled measurement of the kinetics of accumlation of E-X$_2$ during the presteady-state.

The experiment consisted of four parts: 1) to monitor formation by the mutant enzyme of the E-X$_2$ intermediate during presteady-state catalysis by rapid-reaction ESI-MS, 2) to monitor presteady-state formation of the first product 2,5-DNP by stopped-flow spectrophotometry, 3) to develop computer fits

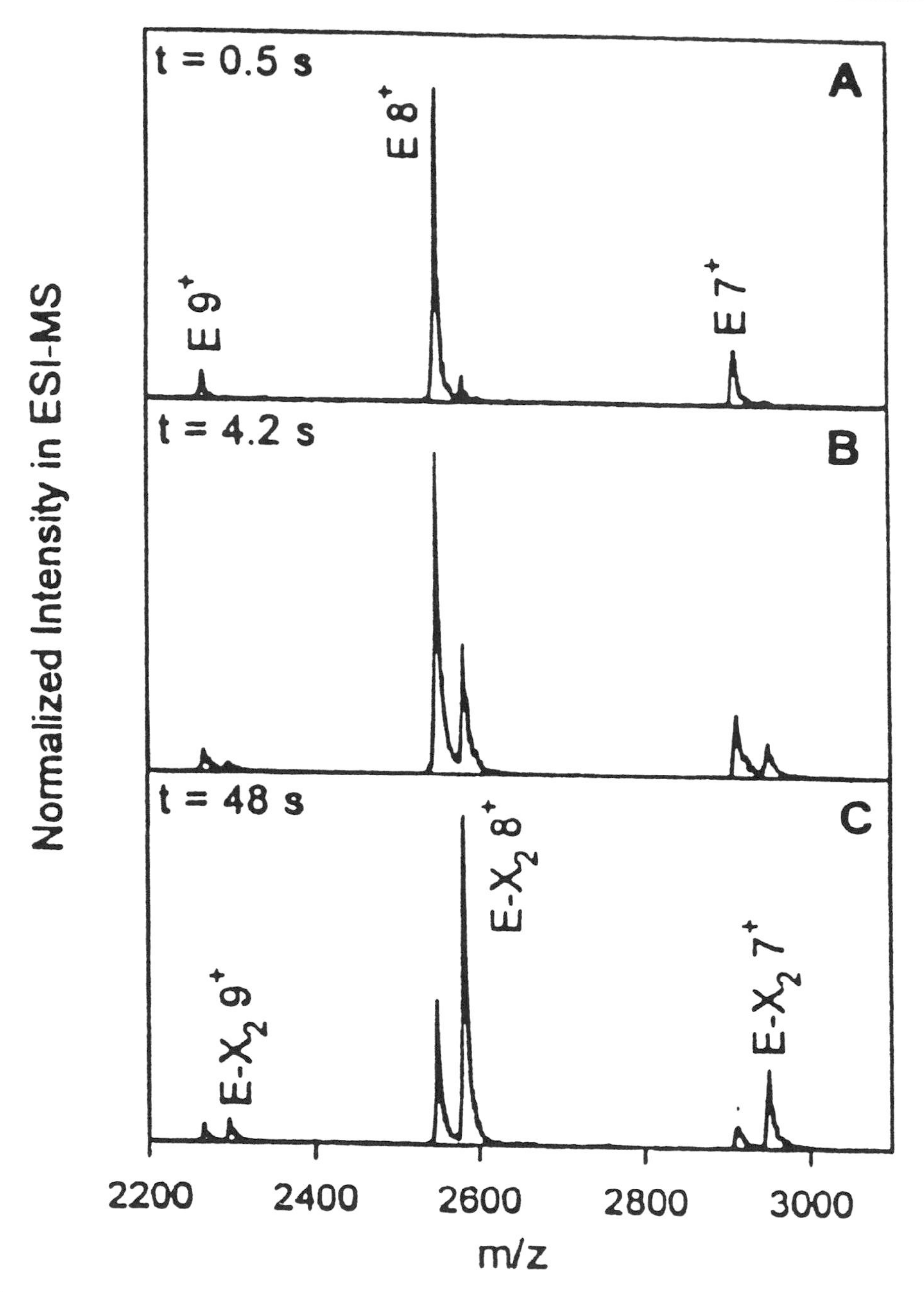

Fig. 10. Partial ESI mass spectra recorded 0.5 sec (A), 4.2 sec (B), and 48 sec (C) after mixing mutant xylanase enzyme, BCX Y80F (0.08 mg/ml, 4 μM), with substrate, 2,5-DNPX$_2$ (110μM). Peaks correspond to free enzyme (E) and the covalent xylobiosyl-enzyme intermediate (E-X$_2$) (reprinted with permission from [19.]).

to the presteady-state curves obtained by both methods so as to extract the rate constants, and 4) to compare the rate constants obtained by rapid-reaction ESI and by stopped-flow spectrophotometry. If rapid-reaction ESI-MS is a legitimate technique, the two methods will yield comparable presteady-state progress curves, and thus comparable rate constants.

Figure 10 displays three mass spectra in the region of the +9, +8, and +7 ions of the enzyme, obtained after incubation of the mutant xylanase BCX Y80F (MW 20,384) with the substrate 2,5-DNPX_2 for 0.5, 4.2 and 48 seconds. Note the small satellite peak displaced to higher mass in Fig. 10A, immediately to the right of the E^{8+} peak, due to formation of E-X_2^{8+} (MW 20,649). By 4.2 seconds (Fig. 10B) the satellite peak has risen appreciably, and by 48 seconds of reaction (Fig. 10C), the little satellite peak due to E-X_2^{8+} has increased to become the predominant species. Note the satellite peaks which are visible just to the right of the E^{7+} and the E^{9+} peaks as well, in all three panels.

Presteady-state time courses for formation of the E-X_2^{8+} intermediate were performed at several initial concentrations of 2,5-DNPX_2. Computer fits to the data points obtained by ESI-MS for formation of the transient E-X_2 intermediate enabled extraction of the rate constants, which were comparable to those obtained from similar curves for production of the colored product, 2,5-DNP by the stopped-flow spectrophotometric method.

This important paper was the first to use a rapid-reaction ESI mass spectrometer to investigate the presteady-state kinetics of formation of a transient enzyme reaction intermediate. Though the intermediate monitored in this experiment was covalently attached to the enzyme, these experiments, together with those of Ganem et al. [18] with noncovalent reaction species, suggest that ESI may enable presteady-state studies of both covalent and noncovalent enzyme reaction intermediates. Furthermore, rate constants can be obtained from presteady-state data obtained by ESI comparable with those obtained by stopped-flow spectrophotometry. Thus ESI appears to be an effective method for monitoring the presteady-state kinetics of this enzyme-catalyzed reaction.

PROSPECTUS

These few excellent contributions represent the published progress to date in applying ESI-MS to the kinetic investigation of enzymes and their reactions. We find these studies encouraging and supportive of the ideas that we have presented earlier in this article concerning the potential of ESI to address the fundamental questions of enzymatic catalysis [1]. Zechel et al. [19] have monitored successfully with rapid-quench ESI-MS the presteady-state kinetics of formation of a covalently bound enzyme reaction intermediate, but a similar presteady-state time course is still to be constructed experimentally for a noncovalent, enzyme-associated reaction intermediate. Paiva et al. [22] have applied rapid-quench ESI-MS successfully for the positive identification of the non-covalently bound, transient-state, enzyme reaction intermediate of EPSP synthase, thus opening the possibility of applying their method to the identification of the elusive reaction intermediates of other enzymes. Their study [22] also showed that rapid-quench ESI-MS is amenable to kinetic monitoring of isotopically-labeled substrates and products, which suggests to us that isotopic discrimination studies may be simplified by this technique [40].

We note that in each paper that we have described, either a substrate, or a reaction intermediate, or a product was monitored. But none has expanded their investigations to follow every substrate, and every enzyme reaction form, and every product as a function of time, or to monitor mass spectrometrically all of these species simultaneously from the same reaction mixture and/or set of mass spectra. Nor has any group attempted a complete presteady-state time course from the very beginning of the reaction, through the presteady-state, through the steady-state, until equilibrium is reached, even though complete kinetic analyses like these appear to be obtainable by methods recently available. To conduct studies of this nature with ESI on enzyme-catalyzed reactions probably would supply sufficient kinetic and thermodynamic information to define in a single experiment, by this single method, the catalytic free energy diagrams for most enzymes, just as was done by Anderson and Johnson for EPSP synthase [12, 13], and by Albery and Knowles for triose phosphate isomerase [14], with their more difficult, time-consuming methods. As we mentioned earlier, free energy descriptions of catalysis are available for only about five enzymes at this time. ESI holds the potential to assay and to reveal the catalytic mechanisms of most enzyme systems, i. e., to be nearly universally applicable. Theoretically, all that is needed is a mass change during the reaction, though other difficulties such as ionic strength considerations, etc., may thwart studies of some enzyme systems.

Figure 11 illustrates how free-energy diagrams might be derived from progress curves obtained with ESI-MS. The depths of the wells of the free energy diagram would be determined from the concentrations of each reactant form measured by ESI at equilibrium. The order of the wells would be determined from the order of appearance of each intermediate species during the presteady-state portions of the progress curves. The heights of the energy barriers between the wells can be calculated by extracting the kinetic constants from the presteady-state and the steady-state progress curves by computer fitting, and by then plugging these kinetic constants into the equations describing the Gibbs free energy.

Conclusions

We are confident that the technique of ESI-MS, when used to monitor directly enzyme-catalyzed reactions with respect to time, will contribute significantly to the field of enzymology. We also recognize that this technique may be adapted to study of virtually any nonvolatile, solution-phase, chemical reaction, be it enzyme-catalyzed, or otherwise. We anticipate that because of its obvious advantages for kinetic analyses of reacting enzyme mixtures, ESI-MS may supplant or enhance many presently available kinetic and thermodynamic techniques. We believe that ESI-MS holds the future of enzymology.

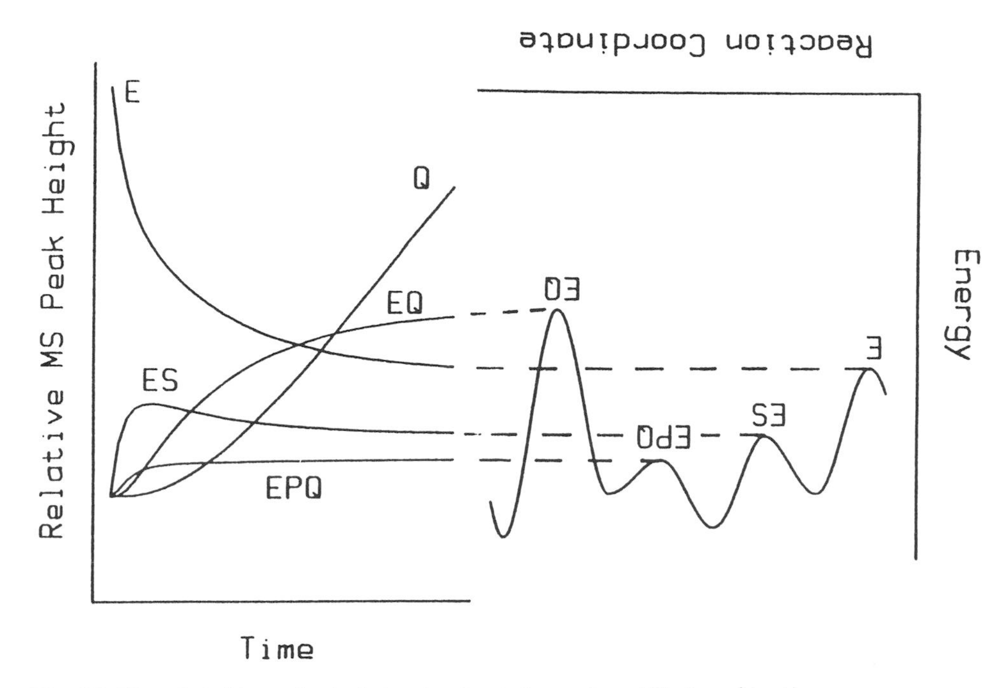

Fig. 11. Simulated transient-state, steady-state, and equilibrium kinetics progress curves from rapid-reaction ESI mass spectrometery, together with the resulting free-energy diagram for an hydrolytic reaction of the form shown in Scheme 7. Rate constants $k_1 = 2$, $k_{-1} = 4$, $k_2 = 3$, $k_{-2} = 5$, $k_3 = 6$, $k_{-3} = 1$, $k_4 = 0.7$. Relative concentrations of free Q were multiplied by 100. On the right is a simulated free-energy diagram plotted upside down to illustrate the inverse relationship between energy levels and concentrations. The equilibrium distribution defines the ground states of the free energy diagram, and the kinetics of their formation during transient-state conditions defines both the order of their appearance along the reaction coordinate and the energy levels of the transition states. Rotate 1/2 turn to view the energy diagram in the normal way (reprinted with permission from [1]).

$$E + S \underset{k_{-1}}{\overset{k_1}{\rightleftharpoons}} ES \underset{k_{-2}}{\overset{k_2}{\rightleftharpoons}} EPQ \underset{k_{-3}}{\overset{k_3}{\rightleftharpoons}} EQ \xrightarrow{k_4} E + Q$$

Scheme 7.

Acknowledgements

This project was supported by NIH grant GM46695. We wish to thank Professor Charles J. Sih for his clever suggestion that we inject our enzyme samples directly into a mass spectrometer, rather than past a membrane inlet, during our mass spectral analyses of the kinetics of CO_2 formation by carbonic anhydrase. We thank Mr. Gary G. Girdaukas for reading the manuscript.

References

1. D. B. Northrop and F. B. Simpson, *Bioorg. Med. Chem.* **1997,** *32*, 641-644.

2. R. M. Caprioli, in *Continuous-Flow Fast Atom Bombardment Mass Spectrometry*, R. M. Caprioli, Ed.; Wiley: Chichester, 1990; pp 63-92.

3. D. L. Smith and Z. Zhang, *Mass Spectom. Rev.* **1994,** *13*, 411-429.

4. R. D. Smith and K. Light-Wahl, *Biol. Mass Spectom.* **1993,** *22*, 493-501.

5. M. Przybylski and M. Glocker, *Angew. Chem. Int. Ed. Engl.* **1996,** *35*, 806-826.

6. J. B. Fenn, M. Mann, C. K. Meng, S. F. Wong and C. M. Whitehouse, *Science* **1989,** *246*, 64-71.

7. J. Gao, X. Cheung, R. Chen, G. B. Sigal, J. E. Bruce, B. L. Schwartz, S. A. Hofstadler, G. A. Anderson R. D. Smith and G. M. Whitesides, *J. Med. Chem.* **1996,** *39*, 1949-1955.

8. S. A. McLuckey, G. J. Van Berkel, G. L. Glish and J. C. Schwartz, in *Practical Aspects of Ion Trap Mass Spectrometry*, Vol. 2, R. E. March and J. F. J. Todd, Eds.; CRC Press: Boca Raton, 1995; pp. 89-141.

9. A. L. Lehninger, *Biochemistry*, Worth: New York, 1975.

10. G. G. Hammes and P. R. Schimmel, in *The Enzymes, Kinetics and Mechanism*, Vol. 2, 3rd Edition, P. D. Boyer Ed.; Academic Press: New York, 1970; pp. 67-114.

11. A. Cornish-Bowden, *Fundamentals of Enzyme Kinetics,* Revised Edition, Portland Press: London, 1995.

12. K. A. Anderson and K. A. Johnson, *Chem. Rev.* **1990**, *90*, 1131-1149.

13. K. A. Anderson, J. A. Sikorski and K. A. Johnson, *Biochemistry* **1988,** *27*, 7395-7406.

14. W. J. Albery and J. R. Knowles, *Biochemistry* **1976**, *15*, 5627-5631.

15. C. A. Fierke, K. A. Johnson and S. J. Benkovic, *Biochemsitry* **1987**, *26*, 4085-4092.

16. S. S. Patel, I. Wong and K. A. Johnson, *Biochemistry* **1991**, *30*, 511-525.

17. G. Behravan, P. Jonasson, B. H. Jonsson and S. Lindskog, *Eur. J. Biochem.* **1991**, *198*, 589-592.

18. B. Ganem, Y. Li and J. D. Henion, *J. Am Chem. Soc.* **1991**, *113*, 7818-7819.

19. D. L. Zechel, L. Konermann, S. G. Withers and D. J. Douglas, *Biochemistry* **1998**, *37*, 7664-7669.

20. R. T. Aplin, J. E. Baldwin, C. J. Schofield and S. G. Waley, *FEBS Lett.* **1990**, *277*, 212-214.

21. E. D. Lee, W. Muck, J. D. Henion and T. R. Covey, *J. Am Chem. Soc.* **1989**, *111*, 4600-4604.

22. A. A. Paiva, R. F. Tilton, Jr., G. P. Crooks, L. Q. Huang and K. S. Anderson, *Biochemistry* **1997**, *36*, 15472-15476.

23. M. Baca and S. B. H. Kent, *J. Am Chem. Soc.* **1992**, *114*, 3992-3993.

24. E. C. Huang, B. N. Pramanik, A. Tsarbopoulos, P Reichert, A. K. Ganguly, P. P. Trotta, T. T. Nagabhushan and T. R. Covey, *J. Am. Soc. Mass Spectrom.* **1993**, *4*, 624-630.

25. X. Cheng, R. Chen, J. E. Bruce, B. L. Schwartz, G. A. Anderson, S. A. Hofstadler, D. C. Gale, R. D. Smith, J. Gao, G. B. Sigal, M. Mammen and G. M. Whitesides, *J. Am Chem. Soc.* **1995**, *117*, 8859-8860.

26. M. J. Greig, H. Gaus, L. L. Cummins, H. Sasmor and R. H. Griffey, *J. Am Chem. Soc.* **1995**, *117*, 10765-10766.

27. J. A. E. Kraunsoe, R. T. Aplin, B. Green and G. Lowe, *FEBS Lett.* **1996**, *396*, 108-112.

28. F. Y. L. Hsieh, X. Tong, T. Wachs, B. Ganem and J. Henion, *Anal. Biochem.* **1995**, *229*, 20-25.

29. S. R. Wilson and Y. Wu, *J. Am. Soc. Mass Spectrom.* **1993**, *4*, 596-603.

30. D. L. DiPietro and S. Weinhouse, *J. Biol. Chem.* **1960**, *235*, 2542-2545.

31. A. Kornberg and W. E. Pricer, Jr., *J. Biol. Chem.* **1951**, *193*, 481-495.

32. C. Walsh, *Enzymatic Reaction Mechanisms*, W. H. Freeman: San Francisco, 1979.

33. L. Konermann, B. A. Collings and D. J. Douglas, *Biochemistry* **1977**, *36*, 5554-5559.

34. L. Konermann, F. I. Rosell, A. G. Mauk and D. J. Douglas, *Biochemistry* **1997**, *36*, 6448-6454.

35. R. P. A. Brown, R. T. Aplin and C. J. Schofield, *Biochemistry* **1996**, *35*, 12421-12432.

36. D. S. Ashton, C. R. Beddell, D. J. Cooper, B. N. Green, R. W. A. Oliver and K. J. Welham, *FEBS Lett.* **1991**, *292*, 201-204.

37. Y. Leung, C. V. Robinson, R. T. Aplin and S. G. Waley, *Biochem J.* **1994**, *299*, 671-678.

38. R. T. Aplin and C. V. Robinson, in *Mass Spectrometry in the Biological Sciences,* A. L. Burlingame and S. A. Carr, Eds.; Humana Press: Totowa, New Jersey, 1996; pp 69-84.

39. S. P. Colowick. in *The Enzymes*, P. D. Boyer, Ed.: Academic Press: New York, 1973; pp 1-48.

40. V. A. Weller and M. D. Distefano, *J. Am. Chem. Soc.* **1998**, *120*, 7975-7976.

Questions and Answers

Liang Tang (Affymax Research Institute)

How do you estimate the effects of solvation energy on the free energy of your reaction?

Answer. I am not a thermodynamacist, however equilibrium constants are affected by temperature. So solvation energy may translate to a change in temperature, which may affect reaction rates and equilibria. So the equilibrium of the equation $E + S \rightleftharpoons ES$ may be affected by the desolvation energy during the quench by ESI.

Hassan Fouda (Pfizer, Inc.)

Are you satisfied that in every case, gas phase complexation observable by ESI reflects what happens in solution?

Answer. If you are asking whether the structures of the protein-ligand complexes are identical in solution and in the gas phase, I would doubt it. If you are asking about complexation occurring between gaseous ions, I would think that they are sufficiently dispersed that such interactions would not occur.

Marvin Vestal (PerSeptive Biosystems)

The major concern is whether complexes observed in the gas phase are truly representative of those present in solution, because it is well known that clusters can be formed during the ESI process. How do you propose to exclude complexation in the spraying process?

Answer. Thank you for informing me of that fact, and for explaining it so well.

Peter Juhasz (PerSeptive Biosystems)

I have found these applications very fascinating. My comment is that the conclusions drawn from these kinetic measurements are valid only if the gas phase ion abundance of the enzyme-substrate complex is linearly proportional to its condensed phase concentration. So then the first step in the experiments should be the establishing of this linear proportionality.

Answer. I agree completely. Some form of internal calibration, or a standard curve is necessary. Although a standard curve is easily prepared for substrates and products, preparation of a standard curve for ES, EP, etc., types of complexes might require some ingenuity. If we are unable to establish some sort of predictable, reproducible calibration curve, then it will be difficult to quantify our spectra.

Peter Juhasz (PerSeptive Biosystems)

Are you aware of experimental solutions to address requirements of certain ionic strength or concentration levels of specific cations in enzyme reactions?

Answer. Many enzyme-catalyzed reactions have metal ion, cation, and ionic strength requirements. For example, the hexokinase reaction utilizes MgATP, as does nitrogenase. When we work with nitrogenase, we routinely use magnesium acetate, so perhaps the use of a volatile anion is one solution to satisfy the enzyme while maintaining low salt for effective ESI. I imagine that the Mg^{++} might remain complexed to the ATP during ESI. However, I fully expect that many enzymes will have cation requirements, and potentially many other requirements, that will prevent analysis of their reactions by ESI. I do not expect every enzyme to work. But some, perhaps many enzyme reactions, will yield to this method of analysis. If even one works, we will increase our present knowledge of catalysis by a sizable fraction.

Steve Barnes (University of Alabama Birmingham)

The chemical composition of the droplet changes from when it leaves the needle to the point where the ions (enzyme) are ejected from the droplet. Thus both the enzyme concentration and the substrate concentrations would be higher. How will this affect the kinetics? Will rearrangements occur in this time-scale?

Answer. You raise an excellent issue about ESI that I have not considered. Enzyme reaction rates most certainly are affected by substrate and product concentrations, as everyone knows. However you also mention the time scale. I do not know how long it takes for the ESI process to occur. But if it is on the order of 1 msec, then it still is a more rapid method of quench than freeze quench, or acid quench that we use and accept as valid now. Provided that the reaction period is long relative to the period of the quench, and provided that the reaction rates do not get too extreme during the quench, "concentration" period, results should be interpretable. I prefer methods to observe reactions that utilize no quench, as there always is the possibility that the quench is introducing another variable. As you and others correctly are pointing out, the extent of complexation observed between enzyme and substrate, enzyme and product, etc., may be affected by concentration changes, as well. Regardless, I think that the ESI method is worth pursuing with enzymes.

Steve Barnes (University of Alabama Birmingham)

There is the risk that the enzyme will be attracted differentially to the surface of the droplet. Can this be modeled?

Answer. This subject will be studied later.

Glycosylation of Proteins — A Major Challenge in Mass Spectrometry and Proteomics

Gerald W. Hart, Robert N. Cole, Lisa K. Kreppel, C. Shane Arnold, Frank I. Comer, Sai Iyer, Xiaogang Cheng, Jill Carroll and Glendon J. Parker
Department of Biological Chemistry, Johns Hopkins University School of Medicine, 725 N. Wolfe St., Baltimore, MD 21205-2185

Although we have known for many years that most cell surface and extracellular proteins are glycosylated, only recently have we come to appreciate that most proteins within the nucleus and cytoplasm are also dynamically modified by the addition and removal of saccharides [1]. Indeed, in eukaryotes most polypeptides are glycosylated. Extracellular protein-bound glycans are generally complex and large, whereas cytosolic and nuclear glycans are often modified by simple monosaccharides [2]. Each unique type of protein glycosylation presents special challenges to the structural analyses or identification of glycoproteins by mass spectrometry (MS)[3-9]. MS analyses of extracellular or cell-surface glycoproteins are complicated by the enormous structural diversity of the glycan side chains, by their large size, by the astonishing site-specific structural variability of glycans, and by the fact that many component monosaccharides have the same mass. Mass spectrometric analysis of O-GlcNAc-bearing nuclear and cytoplasmic glycoproteins is confounded by the highly-dynamic nature of the modification [10], causing sub-stoichiometric levels at single sites, by the inherent insensitivity of the method to glycopeptides as compared to unmodified peptides, and most importantly, by the lability of the saccharide linkage under most MS analytical conditions [11, 12]. Nonetheless, mass spectrometry of all types is the most powerful tool currently available to the glycoscientist interested in the structure/functions of posttranslationally modified proteins as they actually occur in biological systems.

THE GLYCOCALYX

Eukaryotic cell-surfaces are covered by an astonishingly complex array of protein- and lipid-bound carbohydrates. This glycocalyx dwarfs the lipid bilayer in size and represents a barrier through which small molecules, proteins or other cells must interact [13, 14]. Most of the glycans on glycoconjugates comprising the glycocalyx are highly negatively charged, terminating in sialic acids, sulfate esters or both. Even the simplest eukaryotic cells, such as the erythrocyte, have a complex glycocalyx. The glycocalyx has both the properties of gel filtration and ion-exchange resins. The complex glycans play a key role in cell-substratum interactions, cell-cell interactions and in modulating the activities of external regulatory proteins.

Extracellular Glycans Have Enormous Structural Diversity

Even though only a few different monosaccharides are commonly found attached to eukaryotic proteins, the structural diversity of the glycans dwarfs that for proteins. Extracellular glycoproteins display an almost infinite hierarchy of glycosylation [15, 16]. For example, collagens have only a few short monosaccharides or disaccharides attached to hydroxyproline residues. Mucins typically contain many clustered short saccharides often terminating in sialic acids or sulfate [17, 18]. N-linked glycoproteins typically have fewer glycan side-chains, but the chains are characterized by complex branched structures [19, 20]. Finally, at the extreme end of the spectrum are the proteoglycans, which can contain many highly negatively charged polysaccharides, termed glycosaminoglycans, as well as numerous O-linked and N-linked saccharide chains [21-23]. Proteoglycans are ubiquitous components of both the cell-surface and the extracellular spaces, and are likely the most structurally complex molecules in biology. Clearly, the large size and mind-boggling structural diversity of these molecules is a challenge to the ultimate goal of proteomics and to the mass spectrometric analysis of these molecules.

One basis for the structural diversity of complex glycans is their ability to form branched structures. For example, a simple disaccharide Gal-Man can have eight stereo isomers due to the chirality about the glycosidic bond and the four possible sites of attachment. In fact, an oligosaccharide of length ten, comprised of only eight different monosaccharides, can form over 100,000 more structures than a ten amino acid peptide potentially comprised of all twenty common amino acids. These minimal estimates do not take into account the over thirty-five different types of sialic acids [24, 25] and other modifications, such as sulfate esters [26].

Structural Characterization of Glycoforms is Biologically Relevant

The variation of complex saccharides at individual attachment sites not only confounds mass spectrometric analysis, but also may be the most biologically important aspect of protein glycosylation. Unlike nucleic acids and proteins, the complex glycan structures are not directly genetically encoded. Fidelity is controlled by the exquisite specificity of glycosyltransferases for both donor and acceptor substrates [27, 28]. Thus, even though a protein has the same polypeptide backbone in different cell-types, it is glycosylated differently in different cell-types, and differently in the same cell types under different growth conditions or developmental stage [15, 29, 30]. Recent studies suggest that the molecular species (called glycoforms) of glycoproteins display a highly reproducible pattern of carbohydrate modifications at the single site level, often leading to cell-specific molecular forms (glycotypes). An understanding of the site-specific glycosylation of proteins in different cell-types represents one of the major challenges to proteomics and to mass spectrometry of glycoproteins. However, the eventual elucidation of the biological functions of glycoproteins will depend upon such structural/functional analyses.

Glycoproteins are Classified by the Linkage of the Glycan to the Polypeptide

For the sake of convenience, glycobiologists classify glycoproteins by the nature of the protein-saccharide linkage. However, many glycoproteins contain multiple types of modifications. "Mucin-type" glycoproteins typically are linked via a GalNAc-serine linkage, most proteoglycans have a common core with a xylose-serine linkage, collagen type glycosylation involves galactose-hydroxylysine bonds, the most common form of nuclear and cytoplasmic glycosylation involves a N-acetylglucosamine linkage to serine or threonine, and N-glycans are attached by a glycosylamine linkage of N-acetylglucosamine to the amide nitrogen of asparagine. Each of these different types of glycans present a unique array of challenges to the mass spectrometrist, both in sample handling and in their characteristic physical properties.

Mucin-Type Glycans Play a Structural and Protective Role

GalNAc-serine linked glycans often occur in clusters in regions of proteins rich in serine/threonine and proline. Despite many years of effort from several laboratories, there is still no evident consensus sequence or obvious motif that dictates the addition of these glycans to the polypeptide. The highly clustered negative charges on these glycoproteins are thought to serve a structural role [31], creating a rigid "rod" that plays a role in keeping receptors above the rest of the glycocalyx and causing the glycoprotein to have properties of a lubricant and/or anti-desiccant [17, 32]. Until recently, the structural diversity and high levels of glycosylation of mucins have made these molecules intractable to conventional methods of protein analysis. The almost infinite structural diversity of the glycans on mucins produced by epithelial cells likely plays a key role in the prevention of infectious disease, by providing alternative high-affinity binding sites to viruses, bacteria, and parasites. Once bound to mucins, these infectious organisms are unable to enter cells and are washed away. Eventually, structural glycobiologists will need to sort through the morass of structures on mucins to better understand the roles of the molecules as a front line defense against the hazards of the environment surrounding our cells. The recent advances in the molecular biology of mucins and the glycosyltransferases involved in their synthesis is finally allowing the characterization of these important molecules to proceed at a reasonable rate.

Proteoglycans are the Most Complicated Molecules in Biology

A proteoglycan can be defined as any polypeptide that contains one or more glycosaminoglycan (GAG) side-chains. In fact, the structural complexity of this class of glycoproteins ranges from small polypeptides with only one GAG chain to the cartilage-type proteoglycans that contain nearly a hundred chondroitin sulfate and keratan sulfate, as well as N-linked and mucin-type O-linked glycan side-chains [23, 33, 34]. The large size of glycosaminoglycans, their repeating disaccharide structures, and their high negative charge due to the presence of numerous carboxyl and sulfate moieties, makes mass spectrometric analyses of these molecules very difficult. Proteoglycans are not only abundant in the extracellular spaces of virtually all eukaryotic cells, where they mediate collagen polymerization, cell adhesion, and development, but also they occur as membrane proteins, often attached to the membrane via GPI-anchors [35, 36]. These membrane-bound proteoglycans play a key role in modulating the activities of peptide growth hormones, such as fibroblast growth factor (FGF) and others. Detailed structural analyses of most proteoglycans is most likely out of the capabilities of current mass spectrometric methods. However, at the present rate of evolution in this area, it is likely that the method will be generally applicable to these most complex molecules in the future. In our efforts to develop strategies for proteomics [37, 38], especially approaches involving 2D-gel analyses, we should keep in mind that many of these large proteoglycans are not amenable to gel electrophoretic techniques, often not even entering acrylamide-based gels.

N-Glycans are Characterized by a Common Inner Core and Enormous Structural Variability in the Outer Structures

Mass spectrometry has played a key role in the structural elucidation of N-glycans on glycoproteins [3, 7, 39]. Interpretation of structural data is greatly aided by the fact that all N-linked glycans have a common $Man_3GlcNAc_2$ core and vary only in their outer monosaccharide constituents. In addition, the biosynthetic pathways for N-glycans are relatively well-characterized, placing known constraints on structural possibilities. The most obvious characteristic of N-glycans is their extensive branching produced by a number of different Golgi N-acetylglucosaminyltransferases attaching GlcNAc residues to the mannose core [20, 40]. Occasionally, one or more of these branches are elongated by large polylactosamine structures similar to keratan sulfates in saccharide structure [41]. Given the importance of N-glycans in the biology of virtually all cell-surface receptors, cell adhesion, and protein folding, it will be essential to elucidate the glycoforms of individual glycoproteins as an integral part to any program in proteomics. Mass spectrometry is especially powerful in the analysis of site-specific heterogeneity of these N-linked glycoproteins [4, 42].

Many Membrane Proteins are Anchored by Structurally Diverse Glycosyl Phosphatidylinositols (GPI-Anchors)

As late as the mid-1980's, it was realized that many membrane proteins are not attached to the membrane by hydrophobic stretches of amino acids, but rather are held in the membrane by glycosyl phosphatidylinositol anchors (GPI-anchors) [43, 44]. In fact, in prokaryotes GPI-anchor attachment is a more common mechanism of membrane anchoring than are hydrophobic domains of polypeptides [45]. Like N-glycans, GPI-anchors are pre-assembled and attached to polypeptides en bloc. Thus, all GPI-anchors from yeast to man have a common characteristic core structure composed of ethanolamine phosphate attached to an α-linked tri-mannosyl core, which in turn is attached to a non-acetylated glucosamine residue that is glycosidically attached to the inositol of phosphatidyl inositol. GPI-anchors are attached to polypeptides in the endoplasmic reticulum by a transpeptidase reaction, which cleaves a C-terminal GPI-signal peptide and forms an amide linkage between the phosphoethanolamine and the newly-exposed carboxy-terminal amino acid. Recent structural analyses of GPI-anchors from several sources have shown that these moieties have considerable structural variability in both their glycan and lipid components. Glycerol containing GPIs have considerable variation with the presence of both alkyl and acyl chains of various lengths and degrees of saturation. Yeast GPIs have been described that have ceramide as the lipid component [46]. The glycan portions of GPI-anchors have been found to be modified with additional mannosyl and ethanolamine phosphate moieties. GPI-anchors on prion proteins have been found to contain sialylated oligosaccharide substituents similar to those on more conventional glycans [47]. Clearly, as we learn more about these ubiquitous structures, more structural diversity will reveal itself. Mass spectrometry has played a central role in the structural elucidation of GPI-anchors. Because many proteins appear to be attached either by a transmembrane polypeptide or by a GPI-anchor, depending upon alternative mRNA splicing and the cell-type in which they are made, GPI-anchors greatly contribute to both the biological and structural diversity of these molecules. Table 1 summarizes the challenges presented by complex protein glycosylation to the mass spectrometrist interested in the characterization of molecular species of proteins that actually occur *in vivo*.

Table 1. Complex Glycosylation's Challenges to MS Analyses:

- **Enormous Structural Diversity.**
- **Large Size of Glycans on Glycoproteins.**
- **Site-Specific Heterogeneity - Glycoforms.**
- **Common Monosaccharides Have the Same Mass.**
- **Determination of Anomeric Configurations and Linkages is Difficult.**

Table 1.

Myriad Nuclear and Cytoplasmic Proteins are Dynamically Modified by O-Linked N-Acetylglucosamine (O-GlcNAc)

Until the mid-1980's, it was widely believed that only proteins localized in the lumen of cellular compartments or on the outside of cells were glycosylated [48]. The discovery of the abundant modification of nuclear and cytoplasmic proteins by N-acetylglucosamine monosaccharides O-glycosidically linked to the hydroxyl moieties of serine/threonine residues (termed O-GlcNAc) [49] has gradually changed this view. It is now clear that a multitude of nuclear and cytoskeletal proteins are dynamically modified by O-GlcNAc in virtually all eukaryotes, including plants and fungi. In fact, it appears that many, if not most of the cell's regulatory proteins, such as transcription factors, cytoskeletal regulatory proteins, and nuclear pore proteins are modified by O-GlcNAcylation. Table 2 lists some of the O-GlcNAcylated proteins identified to date. Studies during the last 14 years have supported the following overall view of O-GlcNAcylation: 1) O-GlcNAc is attached and removed by specific enzymes [50-52], much the same way as O-phosphate is attached by kinases and removed by phosphatases [53, 54]; 2) O-GlcNAc attachment sites are similar or identical to sites used by serine/threonine specific kinases; 3) there is a growing list of proteins where O-GlcNAc and O-phosphate have a reciprocal relationship; 4) like O-phosphorylation, O-GlcNAcylation is highly-responsive to the cell-cycle, cellular stimulants, and growth state of the cell. Pulse-chase studies indicate that O-GlcNAc turns over as much as fifty times in the life of a polypeptide. Recent data suggest that O-GlcNAc plays an important role in both the regulation of transcription and in the regulation of translation of proteins. Stoichiometry of O-GlcNAc on polypeptides ranges from over fifteen moles of saccharide per mole of protein on nuclear pore proteins and some transcription factors, to less than one mole per mole of protein on average for certain isolated proteins. Even for highly O-GlcNAcylated proteins, occupancy at single sites is generally substoichiometric, reflecting the highly-dynamic nature of the modification.

Table 2: Identified O-GlcNAcylated Proteins

Nuclear Proteins	Cytoskeletal Proteins	Other Proteins
Nuclear Pore Proteins	Cytokeratins 13, 8, 18	92kDa SER Protein
RNA Polymerase II	Neurofilaments H,M, L	p43/hnRNP
Many Transcription Factors	Human Erythrocyte Band 4.1	Adenovirus Fiber
c-myc Oncoprotein	Synapsin I	HCMV UL32
v-Erb-a Oncoprotein	Many Synaptic Vesicle Proteins	Rotavirus NS26
p53 Tumor Suppressor	MAP Proteins	Baculoviral gp41 tegument
SV40 T Antigen	Tau	p67 Translation Reg. Protein
Tyrosine Phosphatase	Talin	Malarial Proteins
Many Chromatin Protens	Vinculin	Trypanosome Proteins
Estrogen Receptors	Clathrin Assembly Protein AP3	GiardiaProteins
Fungal DNA-Binding Proteins	Aplasia	Entamoeba histolytica Proteins
	Neuron Proteins	Schistosome Proteins
	α-Crystallins	ß-Amyloid Precursor

Table 2.

Site-Mapping of O-GlcNAc has Suggested Reciprocity with O-Phosphate, and a Function For the Saccharide in Oncogenesis, Alzheimer's Disease, and in Synaptic Transmission

Much effort has been focused on identification and site-mapping of O-GlcNAcylated proteins. Site-mapping involves purification of the protein to homogeneity, and cleavage by proteases, generally trypsin. Glycopeptides are radiolabeled by the enzymatic attachment of tritiated-galactose to the GlcNAc moieties, using UDP-tritiated-galactose and bovine milk galactosyltransferase. Labeled glycopeptides are purified by successive rounds of reverse-phase HPLC and sites of O-GlcNAc attachment are determined by a combination of gas phase sequencing, manual Edman degradation, and mass spectrometry [11, 55]. Direct analysis of protein by standard colorimetric assays, and measurement of GlcNAc by pellicular ion-exchange (Dionex) chromatography with pulsed-amperometric detection is used to determine average stoichiometry [56]. Site-mapping is not only a necessary prelude to site-directed mutagenesis to evaluate function, but also in some cases immediately suggests an important function for the modification. For example, site mapping of the c-Myc oncogene protein localized a major O-GlcNAc site at threonine-58, which is also a major phosphorylation-site used by glycogen synthase kinase 3 (Fig. 1) [57]. Most importantly, threonine-58 is the mutational hotspot on c-Myc in human Burkitt's lymphomas [58] which causes the transcription factor to become an oncogenic protein. Site-directed mutagenesis of threonine-58 has confirmed the importance of this hydroxyl group in the oncogenic properties of the protein [59]. However, such analyses cannot separate or distinguish the respective roles of O-GlcNAcylation and O-phosphorylation of this key residue.

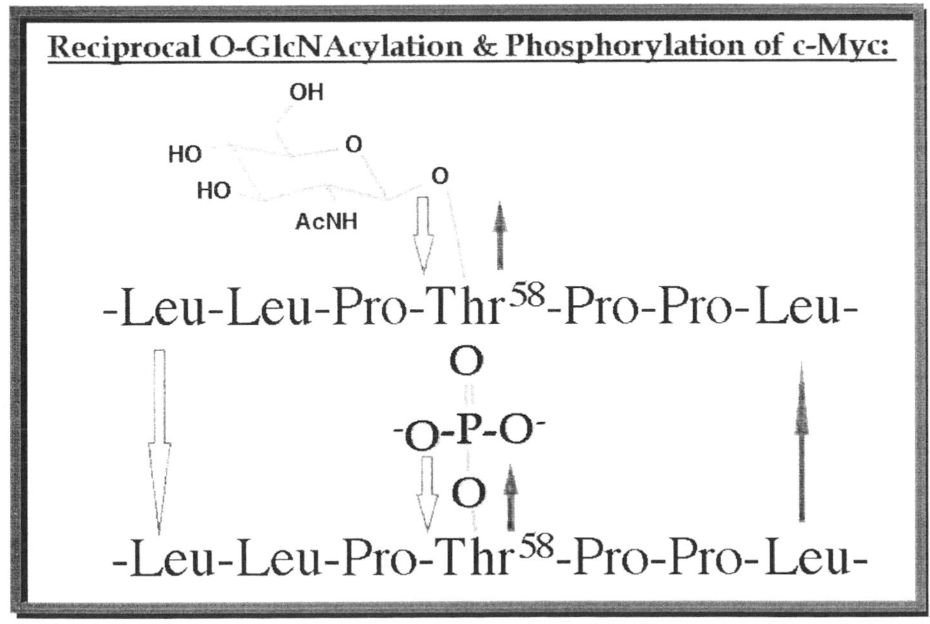

Fig. 1. Reciprocal O-GlcNAc and phosphate at Thr58 on the c-Myc oncogene.

During studies to map the sites of O-GlcNAc on neurofilament proteins [60, 61], the microtubule-associated protein, tau, was found to be extensively O-GlcNAcylated [62]. Tryptic maps suggest that at least twelve sites are modified by O-GlcNAc in tau from normal brains. Tau normally regulates tubulin assembly in the nerve axon. However, importantly, in Alzheimer's disease the tau protein becomes abnormally hyperphosphorylated and no longer binds to tubulin, but polymerizes with itself to form paired-helical filamentous tau (PHF-tau), which contributes to the death of the neuron [63, 64]. *In vitro* studies indicate that serine-262 phosphorylation accounts for up to seventy-percent of the propensity of tau to form PHF-tau [65]. Initial site-mapping studies indicate that serine-262 is also a major site of O-GlcNAcylation in normal tau. These studies suggest that hypo-O-GlcNAcylation of tau in aging neurons, perhaps due to lower glucose utilization, exposes kinase susceptible sites that ordinarily would not be accessible. Clearly, inhibitors that prevent O-GlcNAc removal have the potential to prevent hyperphosphorylation of tau. Recently, another major player in Alzheimer's disease, the β-amyloid precursor protein (APP), was also found to be modified by O-GlcNAc, suggesting that hypo-O-GlcNAcylation of APP might play a role in its abnormal degradation to the toxic β-amyloid peptide which is thought to initiate the cascade of events leading to neurodegeneration. Clearly, the site-mapping of O-GlcNAc residues on these proteins will be important to future functional studies.

Full MS of Equal Amounts of CTD, GlycoCTD, YAPAAPAK Peptides

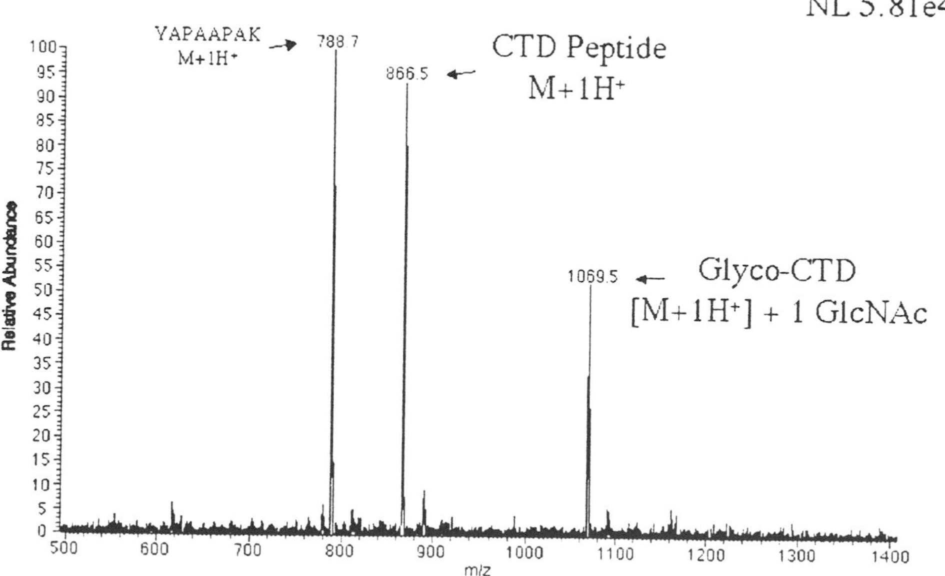

Fig. 2. ES-MS on LCQ of equal amounts of synthetic peptides and glycopeptides based upon the glycosylated sequences of RNA polymerase II. Note that the glycopeptide is detected with about two-fold less sensitivity than its unmodified counterpart.

Synapsin 1 is thought to regulate the association of secretory vesicles containing neurotransmitters to the cytoskeleton [66, 67]. Non-phosphorylated synapsin 1 mediates the binding of the vesicles to the actin cytoskeleton. Upon phosphorylation, the vesicles are released to the "releasable pool". Upon depolarization of the nerve cell, vesicles in the releasable pool rapidly fuse with the plasma membrane, releasing their neurotransmitters. Site mapping studies of O-GlcNAc on synapsin 1, which made heavy use of electrospray mass spectrometry, indicate that the seven major sites of O-GlcNAc on the protein bracket the CAM-kinase phosphorylation sites important to vesicular trafficking. These

Increased Orifice Potential Liberates the O-GlcNAc from the Glycopeptide

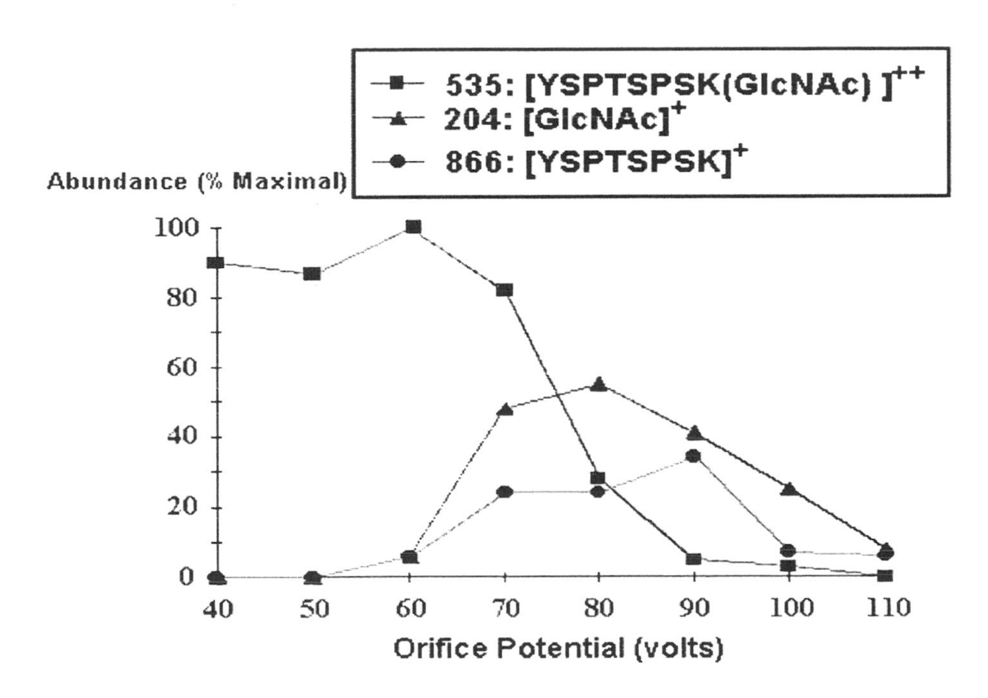

One nanomole of glycopeptide was assayed at various orifice potentials by direct infusion. The relative abundance of the indicated ions was determined from an average of 5 scans. Note that GlcNAc (204) is liberated beginning at 60 volts.

Fig. 3. Effect of orifice potential on loss of O-GlcNAc from glycopeptides.

findings again point to a possible role of the saccharide in the process. Studies with synthetic glyco- and phospho-peptides indicate that O-GlcNAc influences the phosphorylation of the enzyme by CAM-kinase *in vitro*.

O-GlcNAc Poses Special Challenges to Mass Spectrometric Analyses of Nuclear and Cytoplasmic Proteins

Mass spectrometry has already been the most important analytical tool in the analysis of O-GlcNAcylated proteins [10, 11, 68, 69]. During these studies several challenges have presented themselves, which helps explain why this ubiquitous protein modification has so often been missed by investigators not expecting to find it. Studies with chemically defined unmodified and identical peptides with a single O-GlcNAc attached, indicate that the sensitivity of MALDI-TOF using conventional peptide matrices, such as alpha-cyano-4-hydroxycinnamic acid, is reduced over five-fold by the addition of a single O-GlcNAc residue. This apparent loss in signal occurs over a wide range of laser-power and seems to be relatively independent of peptide composition. Furthermore, presence of unmodified peptides tends to suppress the signal of the glycosylated species. These problems become a serious matter considering the substoichiometric nature of the modification on natural proteins. More recent studies, as illustrated in Fig. 2, support a similar comparative lack of sensitivity for O-GlcNAc-modified peptides relative to the unmodified species when ES/MS-ion-trap instrumentation is employed.

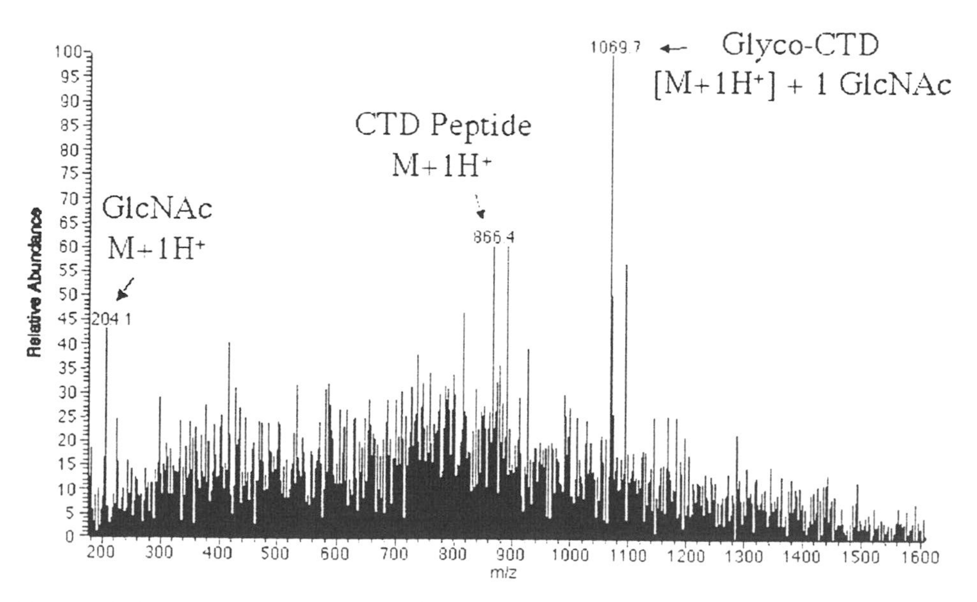

Fig. 4. LCQ MS analysis of a pure synthetic O-GlcNAc peptide based upon the glycosylation site of the C-terminal domain of RNA Polymerase II. Note the substantial loss of the O-GlcNAc residue.

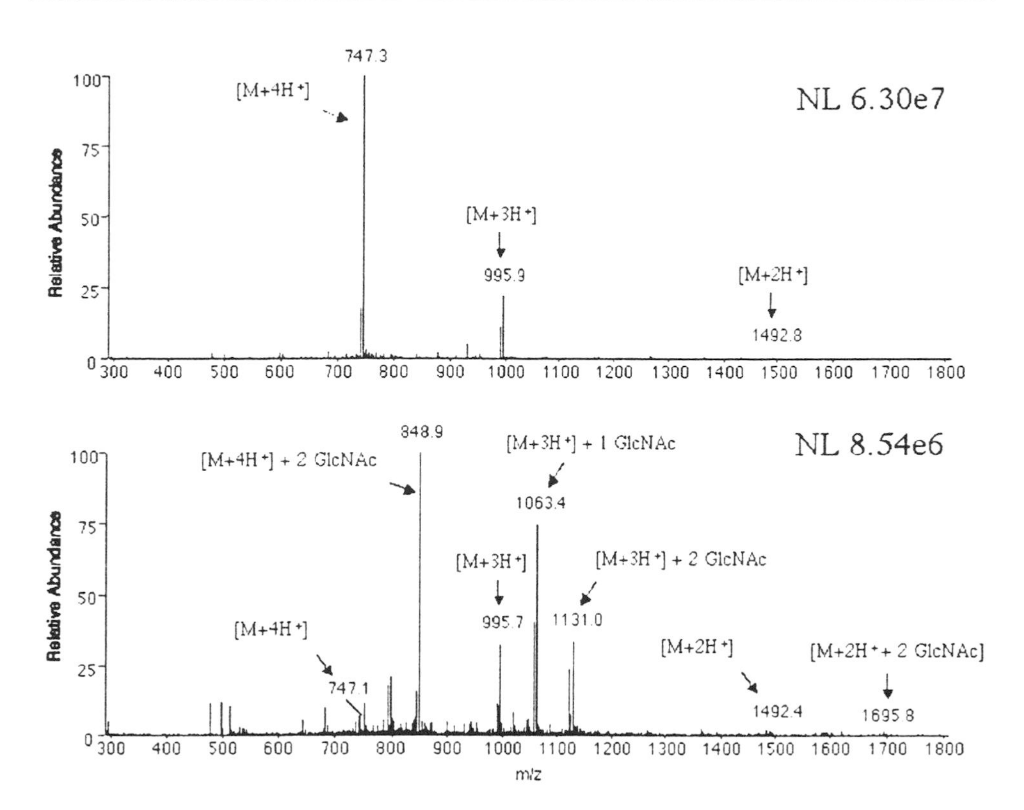

Fig. 5. Comparison of mass spectra from pure synthetic unmodified and di-GlcNAc forms of a 31-mer peptide based upon O-GlcNAc sites on Synapsin 1. Note the complex spectrum obtained from even a homogeneously O-GlcNAcylated peptide compared to the unmodified peptide. The loss of GlcNAc residues creates much complexity when multiple sites are substoichiometrically modified.

Studies using a triple-quadrapole ion-spray mass spectrometer indicate that the O-GlcNAc linkage is very labile, with substantial fragmentation at the glycosidic bond beginning at an orifice potential of only 60 volts (Fig. 3). This property is highly useful for precursor ion-scanning [70] to locate O-GlcNAc peptides in complex mixtures, but has the drawback of precluding direct ESMS-MS sequencing for site mapping. To get around this limitation, we have taken advantage of the high-sensitivity of the O-GlcNAc bond to alkali-induced β-elimination. Thus, controlled conditions for β-elimination allow release of the saccharide without peptide hydrolysis, and allow the localization of the attachment site by the loss of the water molecule [10]. LC-MS analysis of stoichiometrically modified synthetic glycopeptides based upon the glycosylation sites of RNA polymerase II [71] show that substantial amounts of unmodified peptide are generated even under the mildest conditions of electrospray ion-trap mass spectrometric analysis (Fig. 4). Under identical conditions, similar losses from stoichiometrically phosphorylated peptides do not occur. These issues become especially problematic on multiply glycosylated peptides. A simple case is illustrated for a stoichiometrically di-O-GlcNAcylated 31-mer synthetic peptide based upon synapsin 1. In Fig. 5, note how the spectra for the unmodified peptide is relatively simple, but that for the pure di-GlcNAcylated peptide is complex, containing various combinations of fragments in which one or more of the O-GlcNAc residues have been released. Note that this is a case where 100% of the peptide is modified at both sites. However, the analysis of a real-world sample will be complicated by different substoichiometiric levels in the starting peptide at each site. Many proteins have O-GlcNAc residues clustered together, so that the spectra for these various glycopeptides can become quite complex or even lost in the noise or overwhelmed by the much stronger signals from the unmodified peptides. Only very high-resolution HPLC methods are capable of separating the various glycoforms of O-GlcNAc modified peptides. Finally, in proteomic studies compared to analyses of protein phosphorylation, the methods for the detection of O-GlcNAc-modified proteins are on the order of two-hundred fold less sensitive than ^{32}P-detection. Hopefully, this situation will improve with the development of better anti-O-GlcNAc antibodies, and improvements in our methods of direct detection of the saccharide. Table 3 summarizes the challenges posed by O-GlcNAc to the mass spectrometrist studying intracellular regulatory proteins.

CONCLUSIONS

Mass spectrometry has evolved into the most commonly useful tool for the glycoscientist interested in the structure/functions of glycoproteins. However, the complexities of the extracellular molecules and their inherent physical properties defies simple notions of applying 2D-gels and mass spectrometry for

Table 3. O-GlcNAc: A Challenge in MS-Based "Proteomics"

- **Sub-stoichiometric at any single site.**
- **Abundance of O-GlcNAcases.**
- **Presence Greatly Reduces MS Sensitivity - Both MALDI and ESMS.**
- **O-GlcNAc Linkage is Very Labile in Electrospray MS - Good & Bad!**
- **Difficult to Detect on 2D Gels with Current Methods.**

Table 3.

complete proteomic analysis of a cell. Likewise, the dynamic properties of O-GlcNAcylation, the lability of its linkage, and the overall insensitivity of methods for its detection pose strong challenges to the molecular characterization of nuclear and cytoskeletal glycoproteins.

References

1. G. W. Hart, *Ann. Rev. Biochem.* **1997**, *66*, 315-335.

2. G. W. Hart, *Curr. Opin. Cell Biol.* **1992**, *4*, 1017-1023.

3. A. L. Burlingame, *Curr. Opin. Biotechnol.* **1996**, *7*, 4-10.

4. S. A. Carr, G. D. Roberts, A. Jurewicz and B. Frederick, *Biochimie* **1988**, *70*, 1445-1454.

5. R. A. Laine, *Methods Enzymol.* **1990**, *193*, 539-553.

6. D. J. Harvey, *Glycoconjugate J.* **1992**, *9*, 1-12.

7. A. Dell, A. J. Reason, K.-H. Khoo, M. Panico, R. A. McDowell and H. R. Morris, *Methods Enzymol.* **1994**, *230*, 108-132.

8. Y. M. Zhao, S. B. H. Kent and B. T. Chait, *Proc. Natl. Acad. Sci. USA* **1997**, *94*, 1629-1633.

9. C. E. Costello, *Biophys. Chem.* **1997**, *68*, 173-188.

10. G. W. Hart, L. K. Kreppel, F. I. Comer, C. S. Arnold, D. M. Snow, Z. Y. Ye, X. G. Cheng, D. DellaManna, D. S. Caine, B. J. Earles, *et al.*, *Glycobiology* **1996**, *6*, 711-716.

11. K. D. Greis and G. W. Hart, in *Methods in Molecular Biology, Vol. XX: Glycoanalysis Protocols*, E. F. Hounsell, Ed., Humana Press, Inc.: Totowa, NJ, 1997, pp. 0.

12. K. D. Greis, B. K. Hayes, F. I. Comer, M. Kirk, S. Barnes, T. L. Lowary and G. W. Hart, *Anal. Biochem.* **1996**, *234*, 38-49.

13. P. L. Debbage, *Acta Histochem.* **1996**, *98*, 9-28.

14. A. Frey, K. T. Giannasca, R. Weltzin, P. J. Giannasca, H. Reggio, W. I. Lencer and M. R. Neutra, *J. Exp. Med.* **1996**, *184*, 1045-1059.

15. P. M. Rudd and R. Dwek, *Crit. Rev. Biochem. Mol. Biol.* **1997**, *32*, 1-100.

16. A. Kobata, *Accounts Chem. Res.* **1993**, *26*, 319-324.

17. P. L. Devine and I. F. C. McKenzie, BioEssays **1992**, *14*, 619-625.

18. G. J. Strous and J. Dekker, *Crit. Rev. Biochem. Mol. Biol.* **1992**, *27*, 57-92.

19. J. C. Rouse, A. M. Strang, W. Yu and J. E. Vath, *Anal. Biochem.* **1998**, *256*, 33-46.

20. H. Schachter, S. Narasimhan, P. Gleeson and G. Vella, *Can. J. Biochem. Cell Biol.* **1983**, *61*, 1049-1066.

21. R. V. Iozzo and A. D. Murdoch, *FASEB J.* **1996**, *10*, 598-614.

22. M. Yanagishita, *Experientia* **1993**, *49*, 366-368.

23. T. E. Hardingham and A. J. Fosang, *FASEB J.* **1992**, *6*, 861-870.

24. A. Varki, *Glycobiology* **1992**, *2*, 25-40.

25. R. Schauer, *Adv. Carbohydr. Chem. Biochem.* **1982**, *40*, 131-234.

26. Z. F. Kanyo and D. W. Christianson, *J. Biol. Chem.* **1991**, *266*, 4264-4268.

27. J. U. Baenziger, *FASEB J.* **1994**, *8*, 1019-1025.

28. J. C. Paulson and K. J. Colley, *J. Biol. Chem.* **1989**, *264*, 17615-17618.

29. R. A. Dwek, *Biochem. Soc. Trans.* **1995**, *23*, 1-26.

30. G. Opdenakker, P. M. Rudd, C. P. Ponting and R. A. Dwek, *FASEB J.* **1993**, *7*, 1330-1337.

31. N. Jentoft, *Trends Biochem. Sci.* **1991**, *16*, 11-12.

32. W. Gevers, *S. Afr. Med. J.* **1987**, *72*, 39-42.

33. H. Kresse, H. Hausser and E. Schönherr, *Experientia* **1993**, *49*, 403-416.

34. G. David, *Adv. Exp. Med. Biol.* **1992**, *313*, 69-78.

35. J. T. Gallagher, J. E. Turnbull and M. Lyon, *Adv. Exp. Med. Biol.* **1992**, *313*, 49-57.

36. M. Yanagishita, *J. Biol. Chem.* **1992**, *267*, 9505-9511.

37. P. Roepstorff, *Curr. Opin. Biotechnol.* **1997**, *8*, 6-13.

38. M. R. Wilkins, J. C. Sanchez, K. L. Williams and D. F. Hochstrasser, Electrophoresis **1996**, *17*, 830-838.

39. S. A. Carr, M. J. Huddleston and M. F. Bean, *Protein Sci.* **1993**, *2*, 183-196.

40. H. Schachter, *Glycobiology* **1991**, *1*, 453-461.

41. C. H. Hokke, J. P. Kamerling, G. W. K. Van Dedem and J. F. G. Vliegenthart, *FEBS Lett.* **1991**, *286*, 18-24.

42. C.-T. Yuen, S. A. Carr and T. Feizi, *Eur. J. Biochem.* **1990**, *192*, 523-528.

43. M. A. J. Ferguson, *Biochem. Soc. Trans.* **1992**, *20*, 243-256.

44. D. M. Lublin, *Curr. Top. Microbiol. Immunol.* **1992**, *178*, 141-162.

45. M. G. Low, *Biochim. Biophys. Acta* **1989**, *988*, 427-454.

46. C. Fankhauser, S. W. Homans, J. E. Thomas-Oates, M. J. McConville, C. Desponds, A. Conzelmann and M. A. J. Ferguson, *J. Biol. Chem.* **1993**, *268*, 26365-26374.

47. T. Kinoshita, K. Ohishi and J. Takeda, *J. Biochem. (Tokyo)* **1997**, *122*, 251-257.

48. G. W. Hart, R. S. Haltiwanger, G. D. Holt and W. G. Kelly, *Annu. Rev. Biochem.* **1989**, *58*, 841-874.

49. C.-R. Torres and G. W. Hart, *J. Biol. Chem.* **1984**, *259*, 3308-3317.

50. R. S. Haltiwanger, M. A. Blomberg and G. W. Hart, *J. Biol. Chem.* **1992**, *267*, 9005-9013.

51. L. K. Kreppel, M. A. Blomberg and G. W. Hart, *J. Biol. Chem.* **1997**, *272*, 9308-9315.

52. D. L.-Y. Dong and G. W. Hart, *J. Biol. Chem.* **1994**, *269*, 19321-19330.

53. E. G. Krebs, *Biosci. Rep.* **1993**, *13*, 127-142.

54. E. H. Fischer, *Angew. Chem. (Engl)* **1993**, *32*, 1130-1137.

55. E. P. Roquemore, T.-Y. Chou and G. W. Hart, *Methods Enzymol.* **1994**, *230*, 443-460.

56. G. D. Holt, R. S. Haltiwanger, C. R. Torres and G. W. Hart, *J. Biol. Chem.* **1987**, *262*, 14847-14850.

57. T.-Y. Chou, G. W. Hart and C. V. Dang, *J. Biol. Chem.* **1995**, *270*, 18961-18965.

58. B. Smith-Sorensen, E. M. Hijmans, R. L. Beijersbergen and R. Bernards, *J. Biol. Chem.* **1996**, *271(10)*, 5513-5518.

59. L. M. Facchini and L. Z. Penn, *FASEB J.* **1998**, *12*, 633-651.

60. D. L.-Y. Dong, Z.-S. Xu, M. R. Chevrier, R. J. Cotter, D. W. Cleveland and G. W. Hart, *J. Biol. Chem.* **1993**, *268*, 16679-16687.

61. D. L.-Y. Dong, Z. S. Xu, G. W. Hart and D. W. Cleveland, *J. Biol. Chem.* **1996**, *271*, 20845-20852.

62. C. S. Arnold, G. V. W. Johnson, R. N. Cole, D. L.-Y. Dong, M. Lee and G. W. Hart, *J. Biol. Chem.* **1996**, *271*, 28741-28744.

63. M. L. Billingsley and R. L. Kincaid, *Biochem. J.* **1997**, *323*, 577-591.

64. V. M. Y. Lee, *Ann. NY Acad. Sci.* **1996**, *777*, 107-113.

65. G. Drewes, B. Trinczek, S. Illenberger, J. Biernat, G. Schmitt-Ulms, H. E. Meyer, E.-M. Mandelkow and E. Mandelkow, J. Biol. Chem. **1995**, *270*, 7679-7688.

66. P. Greengard, F. Valtorta, A. J. Czernik and F. Benfenati, *Science* **1993**, *259*, 780-785.

67. R. B. Kelly, *Curr. Biol.* **1993**, *3*, 59-61.

68. B. K. Hayes, K. D. Greis and G. W. Hart, *Anal. Biochem.* **1995**, *228*, 115-122.

69. E. P. Roquemore, M. R. Chevrier, R. J. Cotter and G. W. Hart, *Biochemistry* **1996**, *35*, 3578-3586.

70. M. J. Huddleston, M. F. Bean and S. A. Carr, *Anal. Chem.* **1993**, *65*, 877-884.

71. W. G. Kelly, M. E. Dahmus and G. W. Hart, *J. Biol. Chem.* **1993**, *268*, 10416-10424.

QUESTIONS AND ANSWERS

Dan Kassel (CombiChem)

How simplified is the glycopeptide analysis when separation tools, such as CI and SFC are incorporated to separate out the various heterogeneous forms of the glycan?

Answer. Physical separation of the glycopeptides greatly simplifies the interpretation of the data. The extraordinary resolving power of CZE has proven to be very useful. CZE can easily resolve peptides differing only by a single O-GlcNAc. Such a separation is much more difficult by RP-HPLC.

X. Christopher Yu (Chiron Corp.)

Are O-GlcNAc sites conserved across mammalian species?

Answer. O-GlcNAc attachment sites are often highly conserved. However, on occasion, as for Band 4.1 for example, they also occur at locations that are spliced-out in some cells but not others.

Elizabeth Komives (UCSD)

Phosphorylated proteins often cannot be expressed in a properly folded form without co-expression of the kinase, and even then often the protein does not fold correctly. Your results showing rapid glycosylation after lymphocyte stimulation suggest a role for glycosylation in protein folding during protein biosynthesis. Have you considered the possibility?

Answer. You propose an interesting idea that I have not considered. O-GlcNAc is added rapidly and co-, as well as posttranslationally. Therefore, it might play a role in folding of nascent chains.

Jasna Peter-Katalinic (University of Muenster)

Is there any evidence for competitive O-GlcNAc and O-GalNAc glycosylation on the same protein substrates?

Answer. No. However, apo-mucin is a good substrate for the O-GlcNAc transferase. O-GlcNAc is β-linked and O-GalNAc is α-linked and each occurs on a different side of the membrane-enclosed cellular compartments.

Site-specific Characterization of the N-linked Glycans of Murine PrP^Sc Using Advanced Methods of Electrospray Mass Spectrometry

Elaine Stimson[*†], *James Hope*[†], *Angela Chong*[†] *and Alma L. Burlingame*[*‡]
[*‡]Ludwig Institute for Cancer Research, 91 Riding House Street, London, UK; [†]BBSRC Institute for Animal Health, Compton, Berkshire RG20 7NN, UK; [‡]Dept. of Pharmaceutical Chemistry, UCSF, San Francisco, CA and Dept. of Biochemistry and Molecular Biology, UCL, London

Scrapie, a neurodegenerative disease of sheep, is one of a group of diseases of humans and animals termed prion diseases. The scrapie prion protein (PrP^{Sc}) accumulates in amyloid plaques in the brains of hamsters and mice artificially infected with scrapie. PrP^{Sc} is an abnormal isoform of a normal cellular protein (PrP^{C}), and is the only identified component of the infectious particles called prions [1]. PrP^{Sc} and PrP^{C} are both encoded by the same gene, and PrP^{Sc} is derived from PrP^{C} by posttranslational events [2, 3]. Many different strains of agent from natural cases of ovine scrapie have been characterized by their different phenotypes (relative incubation times and lesion profiles) in a panel of inbred mice [4]. The phenotypes can be replicated in a single strain of mouse producing a common amino acid sequence of the PrP polypeptide chain. For example, strains Me7, 22A and 301V all have different lesion and incubation periods in VM mice.

PrP^{Sc} is a glycoprotein giving three bands by sodium dodecylsulphate-polyacrylamide gel electrophoresis (SDS-PAGE) at 33-35 kDa. In different strains, these bands differ in intensity, indicating that the ratios of mono, di- and unglycosylated species of PrP^{Sc} vary in distinct strain models of murine scrapie [5, 6]. This link between glycosylation and strains of agent has prompted us to gain structural information about the N-linked carbohydrate structures on PrP^{Sc} to ascertain whether certain structures are unique to particular sites on strains. However, the site-specific characterization of the N-glycosylation of murine PrP^{Sc} is a challenging task for the following reasons:

- PrP^{Sc} is a typical membrane protein, and it is extremely difficult to isolate the protein in high yield and high purity from scrapie infected mouse brain. A single hamster brain contains less than 5 μg, i. e., 15 pmol of protein and one mouse brain contains only one third of that amount.

- The two N-linked consensus sites are only partially glycosylated, and the extent of glycosylation differs between strains.

- The glycans are extremely heterogeneous.

Strategies involving mass spectrometric methods have been extremely successful in the site-specific characterization of N-linked glycans (for review see [7]). It is ideally suited to analyze complex mixtures, and thus assess glycoform heterogeneity. Second, due to its high sensitivity, mass spectrometry offers a potential route for structural analysis of natural glycoproteins purified in limited amounts. Third, high accuracy mass measurement enables unambiguous assignment of the molecular weight of components, and hence glycan composition.

Mass spectrometric studies on the primary structure of PrP^{Sc} were first developed using the truncated glycoprotein isolated from hamster brain after treatment with proteinase-K [8]. This enabled the protein to be isolated in slightly higher yield than full-length PrP^{Sc} with less contamination by other proteins. The purified glycoprotein was reduced and carboxymethylated before digestion with endopeptidase Lys-C. The resulting mixture of peptides and glycopeptides was purified by high-pressure liquid chromatography (HPLC) prior to analysis by liquid secondary ionization (LSI) and electrospray ionization (ESI) mass spectrometry. ESI-MS using a triple quadrupole mass analyzer allowed the detection of a heterogeneous cluster of molecular ions corresponding to the glycosylated peptide containing Asn-181. Mass resolution was enhanced by Maximum Entropy deconvolution to reveal over 30 glycoforms attached to Asn-181. The identification of the glycosylated peptide spanning Asn-197 was not accomplished using the mass spectrometers available at that time. This glycopeptide was, however, present in quantities amenable to Edman sequencing, which suggests that the glycosylation is extremely heterogeneous, with glycopeptide species not sufficiently abundant to be detected by the mass analyzer.

Recently, a new type of mass spectrometer has been introduced that has facilitated a comprehensive site-specific characterization of the N-glycans on murine PrP^{Sc} [9]. In this study, both capillary liquid chromatography-electrospray mass spectrometry (LC/ESI-MS) and nano-electrospray tandem mass spectrometry were performed using orthogonal acceleration time-of-flight mass spectrometers (oaTOF-MS). This new ion optical configuration used in both conventional MS and tandem MS mode yields glycopeptide and peptide information of exceptional quality. The following chapter shows some examples of how these instruments may be used for the detailed characterization of extremely complex mixtures of glycopeptides.

Experimental

Purification of PrP^{Sc} Protein

Full-length PrP^{Sc} was purified from mouse brains by detergent lysis, differential centrifugation and size-exclusion chromatography according to the procedure described [10].

Alkylation and Tryptic Digestion

Purified PrP^{Sc} was reduced in dithiothreitol (Pierce) and alkylated with 4-vinyl pyridine (Sigma) in the presence of 6M Guanidine HCl pH 8.5. The alkylated protein was precipitated with methanol, redissolved in 4 M urea, 200 mM ammonium bicarbonate, pH 8.5 and digested with trypsin (Promega) (5% trypsin by weight overnight at 37°C).

Sequential Exoglycosidase Digestions

These were carried out on purified glycopeptides: Neuraminidase (from *Vibrio Cholerae*, EC 3.2.1.18, Boehringer Mannheim), broad specificity: 50 milliunits in 100 μl of 50 mM ammonium acetate buffer, pH 5.5 for 48 hours, with a fresh aliquot added after 24 hours; α-fucosidase (from *Xanthomonas manihotis*, New England Biolabs), specificity for α(1,3/4)-fucose: 0.2 units in 100 μl of provided incubation buffer for 24 hours; β-galactosidase (from *Diplococcus pneumonaie*, Boehringer Mannheim), specificity for β(1,4)Gal: 10 milliunits in 100 μl of provided incubation buffer for 24 hours with a fresh aliquot of enzyme added after 12 hours. All enzyme digestions were carried out at 37°C and terminated by centrifugal evaporation. Each digest mixture was re-suspended in 0.1% formic acid and an appropriate aliquot was analyzed by capillary HPLC/ ES-oaTOF-MS using the Mariner mass spectrometer.

HPLC/ESIMS

Microbore and capillary HPLC/ESI-MS experiments were performed with a PE Sciex 300 triple quadrupole and a Mariner ESI-oaTOF mass spectrometer, respectively. Glycopeptides were detected selectively by microbore LC-ESMS equipped with selected ion monitoring for sugar oxonium ions [11, 12]. This was performed using an ABI 140B dual syringe pump system set to deliver mobile phase at 40 μl/min. The tryptic digest mixture in urea was injected onto a Vydac microbore column (C18, 1mm x 150 mm, 5 μ.) and separation was achieved using an alcohol based mobile phase [13]. The column was equilibrated in solvent A containing 5% solvent B, and the gradient was started 10 minutes after injection then increased linearly to 60% B over 80 minutes. The column effluent was monitored at 214 nm by a UV detector (ABI 785A) equipped with a high sensitivity U-Z flow cell (LC Packings), after which it was split so that approximately 5 μl/min was directed into the Sciex mass spectrometer and 35 μl/ min was collected manually and saved for further characterization. The mass spectrometer was scanned in both the conventional and SIM mode. For the latter experiment the carbohydrate oxonium ions at m/z 204 ($HexNAc^+$), m/z 292 ($NeuAc^+$) and m/z 366 ($HexHexNAc^+$) (dwell time, 200 ms each) were monitored at a high orifice potential of 200 V. Full scans at m/z 383-2500 (0.125 AMU steps, scan time 5 s) were then acquired at a lower orifice potential of 75 V.

The collected HPLC fractions containing glycopeptides were pooled, dried down and re-analyzed by capillary HPLC coupled to a Mariner orthogonal acceleration electrospray-TOF mass spectrometer (PerSeptive Biosystems, Framingham, MA). Capillary HPLC was achieved using a 140C dual syringe pump flowing at 10-12 μl/min, which was stream split so that 1-2 μl/min flowed onto the column (180 μm x 50 mm, LC Packings). After sample injection, the column was eluted with a linear gradient of 0.5% per min using the alcohol based mobile phase described above, and the HPLC eluent was monitored at 210 nm

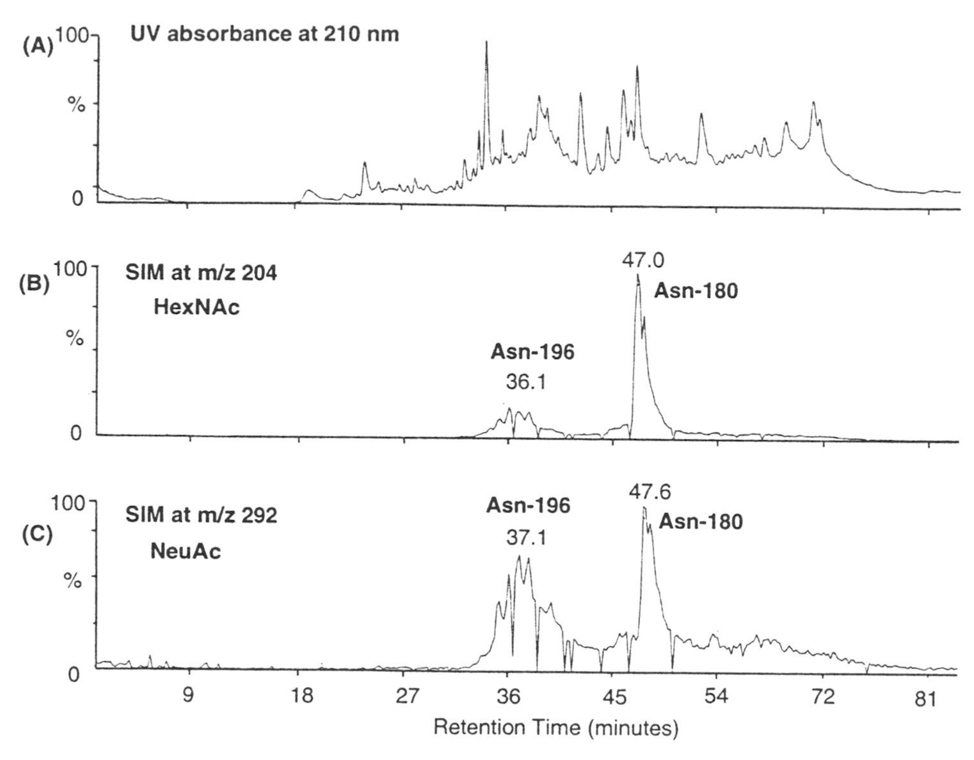

Fig. 1. (A) UV absorbance (210 nm) chromatogram obtained from microbore HPLC/ESIMS analysis of a 40 pmol injection of a tryptic digest of murine PrP^{Sc}. Chromatograms obtained from selected ion monitoring at m/z 204 (HexNAc$^+$) (B) and m/z 292 (NeuAc$^+$) (C). Voltage of the sampling cone was adjusted from 60 V to 180 V during each SIM experiment to induce dissociation.

using a UV detector (ABI 785A) which was fitted with a capillary U-Z flow cell (LC-Packings). The HPLC was interfaced to the electrospray source of the Mariner by a length of fused silica capillary tubing. A scan was acquired every 3 seconds as the result of 24,000 pulses, over a mass range of m/z 360-2000. Singly and doubly protonated ions of gramicidin S were used for a two point external calibration.

Nano-Electrospray Tandem Mass Spectrometry

Tandem mass spectrometry was carried out on HPLC purified glycopeptide mixtures using a high-resolution Sciex Qq-TOF mass spectrometer fitted with nano-electrospray source (Protana).

IDENTIFICATION OF N-LINKED GLYCOPEPTIDES IN A COMPLEX DIGEST MIXTURE

Glycopeptides separately containing glycosylation sites at Asn-180 and Asn-196 were isolated from tryptic digests of purified full-length PrP^{Sc} by microbore HPLC/ESIMS using a Sciex triple quadrupole mass spectrometer scanned in both the conventional (MS) and selected ion monitoring (SIM) mode.

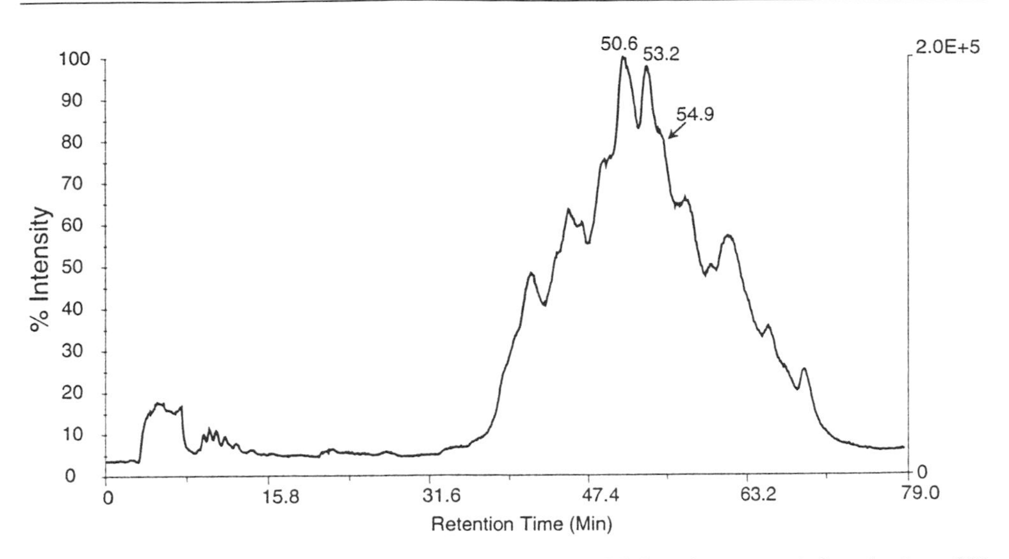

Fig. 2. TIC trace produced by re-analysis of pooled HPLC fractions containing the Asn-180 glycopeptides by capillary HPLC/ESI-oaTOF-MS.

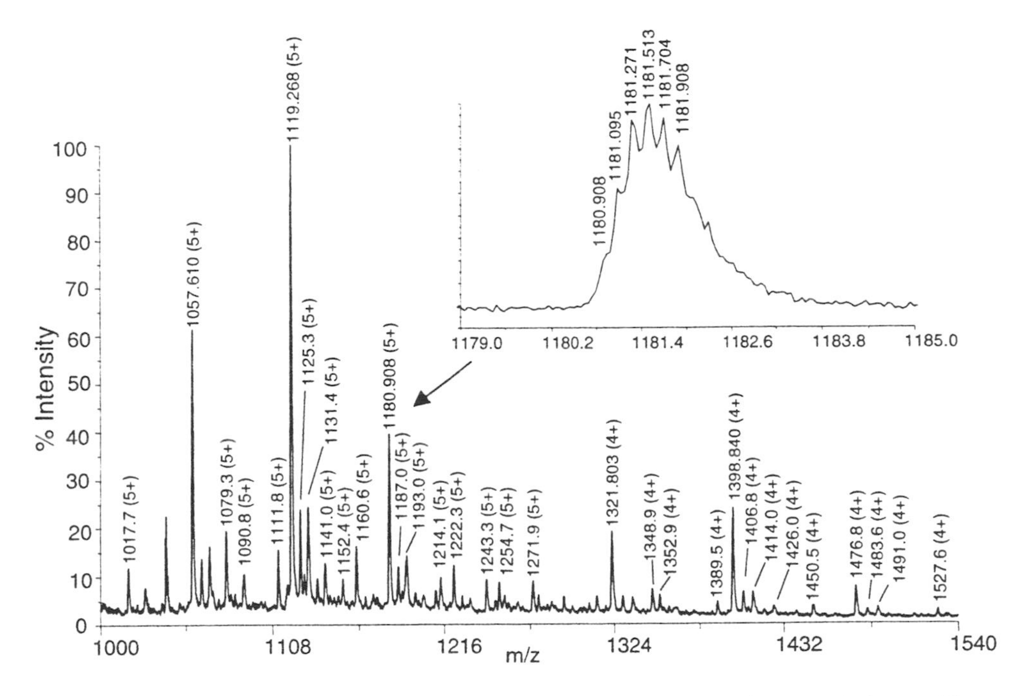

Fig. 3. ESI-oaTOF mass spectrum of the neutral tryptic glycopeptides containing residues Y_{156}-K_{184}, which contains a consensus sequence for N-linked glycosylation at Asn-180. The inset shows the resolution during this experiment, which immediately enables charge state assignment. The mass difference between the calculated and determined molecular mass for this glycopeptide was better than 20 ppm.

For the SIM experiment, the mass spectrometer was set to monitor the mass values representing specific carbohydrate fragment ions characteristic of generic glycopeptides [11, 12, 14]. These SIM data (Fig. 1) revealed the elution profiles of two different glycopeptide-containing fractions at 36 and 47 minutes. Summation of conventional (MS) scans across the glycopeptide-containing region at 47 minutes gave a collection of molecular ion signals that correspond to glycopeptides containing the amino acid sequence Y_{156}-K_{184} spanning Asn-180 (data not shown). As anticipated from previous mass spectrometric studies [8], characterization of the glycosylation at Asn-196 proved more difficult because molecular ion signals for glycopeptides were not observed in the former glycopeptide-containing region (37 min). However, the lack of glycopeptide signals in the conventional MS scans is not surprising, as the broad profiles of the SIM data suggest that extreme micro-heterogeneity exists at this site. Thus, to improve the detection of putative Asn-196 glycopeptides and to ensure the detection of all glycoforms at both N-linked sites, HPLC fractions spanning regions at 37 and 47 minutes were separately pooled. A small aliquot from both glycopeptide-containing samples (approximately 10-15 pmol total) was subjected to capillary LC/ESI-oaTOF-MS analysis.

The total ion current (TIC) chromatogram produced by re-analysis of the glycopeptides containing Asn-180 by capillary LC/ESI-oaTOF-MS is shown in Fig. 2. The alcohol based mobile phase employed for the capillary chromatography permits some glycoform separation, i. e., more hydrophilic species are retained longer [11]. Thus, three sets of mass spectral data were observed at 50.6, 53.2 and 54.9 minutes, which exhibited molecular ion signals corresponding to neutral (shown in Fig. 3), monosialyl and disialyl glycoforms, respectively, containing the amino acid sequence Y_{156}-K_{184}. The high-mass resolution afforded by the oaTOF mass spectrometer permits unambiguous charge state assignments for the most abundant glycopeptide signals (Fig. 3 inset and legend), and thus enables unambiguous molecular mass determination. However, poor ion statistics for neutral and sialylated glycopeptides of low abundance produced unresolved isotope peaks, and in these cases, the average masses for molecular ions were determined. By subtracting the monoisotopic or average mass of the tryptic peptide spanning residues 156-184 (MW_{mi} = 3635.726; MW_{avg} = 3637.1) from the measured monoisotopic or average mass of each glycopeptide, respectively, it was possible to deduce the composition of the carbohydrate moieties (summarized in Table 1). The structural class of each oligosaccharide was then assigned by comparison of the oligosaccharide compositions with the established carbohydrate structures previously observed on hamster PrP27-30 [15] and those typically found in mammalian glycoproteins. Over 30 N-linked complex glycans were detected, with the most abundant glycoforms consisting of neutral biantennary structures containing bisecting GlcNAc.

Re-analysis of the pooled HPLC fractions containing the Asn-196 glycopeptides by capillary HPLC/ESI-oaTOF-MS produced the TIC trace shown in Fig. 4. The dynamic range and high signal-to-noise ratio provided by this kind of instrument facilitated the detection of five sets of molecular ion signals for

m/z Monoisotopic	m/z Average	Intact Glycopeptide MW (Da)	Carbohydrate mass$_{avg}$ (Da)	Carbohydrate assignment	α-fucosidase digestion m/z	α-fucosidase digestion MW$_{avg}$ (Da)	β-galactosidase digestion m/z	β-galactosidase digestion MW$_{avg}$ (Da)
	1017.7 [5+] 1271.9 [4+]	5083.5_{avg}	1463.4_{avg}	Hex$_3$HexNAc$_4$Fuc	1017.7 [5+] 1271.9 [4+]	5083.5	1017.7 [5+] 1271.9 [4+]	5083.5
1057.610 [5+] 1321.803 [4+]	1058.3 [5+] 1322.7 [4+]	5283.150_{mi} 5286.7_{avg}	1665.324_{mi} 1666.6_{avg}	**Hex$_3$HexNAc$_5$Fuc**	1058.3 [5+] 1322.7 [4+]	5286.7	1058.3 [5+] 1322.7 [4+]	5286.7
	1079.3 [5+] 1348.9 [4+]	5391.7_{avg}	1771.6_{avg}	Hex$_4$HexNAc$_4$Fuc$_2$	1050.1 [5+] 1312.4 [4+]	5245.6	1017.7 [5+] 1271.9 [4+]	5083.5
	1352.9 [4+]	5407.7_{avg}	1787.7_{avg}	Hex$_5$HexNAc$_4$Fuc	1082.5 [5+] 1352.9 [4+]	5407.7	1017.7 [5+] 1271.9 [4+]	5083.5
	1090. 8 [5+]	5448.4_{avg}	1828.4_{avg}	Hex$_4$HexNAc$_5$Fuc	1090.8 [5+] 1363.1 [4+]	5448.4	1058.3 [5+] 1322.7 [4+]	5286.7
	1111.8 [5+] 1389.5 [4+]	5553.8_{avg}	1933.8_{avg}	Hex$_5$HexNAc$_4$Fuc$_2$	1082.5 [5+] 1352.9 [4+]	5407.7	1017.7 [5+] 1271.9 [4+]	5083.5
1119.268 [5+] 1398.840 [4+]	1120.0 [5+] 1399.7 [4+]	5591.340_{mi} 5594.9_{avg}	1973.614_{mi} 1974.8_{avg}	**Hex$_4$HexNAc$_5$Fuc$_2$**	1090.8 [5+] 1363.1 [4+]	5448.4	1058.3 [5+] 1322.7 [4+]	5286.7
	1131.4 [5+] 1414.0 [4+]	5652.0_{avg}	2032.0_{avg}	Hex$_4$HexNAc$_6$Fuc	1131.4 [5+] 1414.0 [4+]	5652.0	1099.0 [5+] 1373.5 [4+]	5489.8
	1426.0 [4+] 1141.0 [5+]	5700.7_{avg}	2080.7_{avg}	Hex$_5$HexNAc$_4$Fuc$_3$	1082.5 [5+] 1352.9 [4+]	5407.7	1017.7 [5+] 1271.9 [4+]	5083.5
	1152.4 [5+]	5757.0_{avg}	2137.0_{avg}	Hex$_5$HexNAc$_5$Fuc$_2$	1123.2 [5+] 1403.7 [4+]	5610.9	1058.3 [5+] 1322.7 [4+]	5286.7
	1160.6 [5+] 1450.5 [4+]	5798.1_{avg}	2178.1_{avg}	Hex$_4$HexNAc$_6$Fuc$_2$	1131.4 [5+] 1414.0 [4+]	5652.0	1099.0 [5+] 1373.5 [4+]	5489.8
1180.908 [5+]	1181.6 [5+] 1476.8 [4+]	5899.540_{mi} 5903.2_{avg}	2281.814_{mi} 2283.1_{avg}	**Hex$_5$HexNAc$_5$Fuc$_3$**	1123.2 [5+] 1403.7 [4+]	5610.9	1058.3 [5+] 1322.7 [4+]	5286.7
	1193.0 [5+] 1491.0 [4+]	5960.3_{avg}	2340.2_{avg}	Hex$_5$HexNAc$_6$Fuc$_2$	1163.8 [5+] 1454.5 [4+]	5814.1	1099.0 [5+] 1373.5 [4+]	5489.8
	1214.1 [5+]	6065.4_{avg}	2445.3_{avg}	Hex$_6$HexNAc$_5$Fuc$_3$	1156.6 [5+]	5778.1	1058.3 [5+] 1322.7 [4+]	5286.7
	1222.3 [5+] 1527.6 [4+]	6106.4_{avg}	2486.3_{avg}	Hex$_5$HexNAc$_6$Fuc$_3$	1163.8 [5+] 1454.5 [4+]	5814.1	1099.0 [5+] 1373.5 [4+]	5489.8
	1243.3 [5+]	6211.5_{avg}	2591.4_{avg}	Hex$_6$HexNAc$_5$Fuc$_4$	1156.6 [5+]	5778.1	1058.3 [5+] 1322.7 [4+]	5286.7
	1254.7 [5+]	6268.6_{avg}	2648.5_{avg}	Hex$_6$HexNAc$_6$Fuc$_3$	1196.3 [5+] 1495.1 [4+]	5976.3	1099.0 [5+] 1373.5 [4+]	5489.8

Table 1a. HPLC/electrospray mass spectrometry data obtained from sequential exoglycosidase digestion of N-linked glycopeptides (Y_{156}-K_{184}) containing Asn-180. Bold type indicates the major species identified based on the relative abundance of the molecular ions in the mass spectra.

Intact Glycopeptide m/z (Average)	MW$_{avg}$ (Da)	Carbohydrate mass (Da)	Carbohydrate assignment	α-sialidase digestion m/z	α-sialidase digestion MW$_{avg}$ (Da)	α-sialidase digestion Carbohydrate mass (Da)	α-fucosidase and β-galactosidase digestion m/z	α-fucosidase and β-galactosidase digestion MW$_{avg}$ (Da)
1149.0 [5+] 1436.0 [4+]	5740.1	2120.0	Hex$_4$HexNAc$_5$FucNeuAc	1090.8 [5+] 1363.1 [4+]	5448.4	1828.4	1058.3 [5+] 1322.7 [4+]	5286.7
1189.6 [5+] 1486.8 [4+]	5943.2	2323.2	Hex$_4$HexNAc$_6$FucNeuAc	1131.4 [5+] 1414.0 [4+]	5652.0	2032.0	1099.0 [5+] 1373.5 [4+]	5489.8
1210.7 [5+] 1513.1 [4+]	6048.3	2428.2	Hex$_5$HexNAc$_5$Fuc$_2$NeuAc	1152.4 [5+] 1440.3 [4+]	5757.0	2137.0	1058.3 [5+] 1322.7 [4+]	5286.7
1251.3 [5+] 1563.9 [4+]	6251.5	2631.5	Hex$_5$HexNAc$_6$Fuc$_2$NeuAc	1193.0 [5+] 1491.0 [4+]	5960.3	2340.2	1099.0 [5+] 1373.5 [4+]	5489.8
1272.3 [5+]	6356.5	2736.2	Hex$_6$HexNAc$_5$Fuc$_3$NeuAc	1214.1 [5+]	6065.4	2445.2	1058.3 [5+] 1322.7 [4+]	5286.7
1313.0 [5+]	6559.8	2939.8	Hex$_6$HexNAc$_6$Fuc$_3$NeuAc	1254.7 [5+]	6268.6	2648.5	1099.0 [5+] 1373.5 [4+]	5489.8
1228.3 [5+]	6136.4	2516.4	Hex$_5$HexNAc$_4$Fuc$_2$NeuAc$_2$	1111.8 [5+] 1389.5 [4+]	5553.8	1933.8	1017.7 [5+] 1271.9 [4+]	5083.5
1239.7 [5+]	6193.5	2573.5	Hex$_5$HexNAc$_5$FucNeuAc$_2$	1123.2 [5+] 1403.7 [4+]	5610.9	1990.9	1058.3 [5+] 1322.7 [4+]	5286.7
1268.9 [5+]	6339.6	2719.6	Hex$_5$HexNAc$_5$Fuc$_2$NeuAc$_2$	1152.4 [5+] 1440.3 [4+]	5757.0	2137.0	1058.3 [5+] 1322.7 [4+]	5286.7
1280.3 [5+]	6396.6	2776.5	Hex$_5$HexNAc$_6$FucNeuAc$_2$	1163.8 [5+] 1454.5 [4+]	5814.1	2194.0	1099.0 [5+] 1373.5 [4+]	5489.8
1301.3 [5+]	6501.6	2881.6	Hex$_6$HexNAc$_5$Fuc$_2$NeuAc$_2$	1184.8 [5+]	5919.1	2299.1	1058.3 [5+] 1322.7 [4+]	5286.7
1309.6 [5+]	6542.8	2922.6	Hex$_5$HexNAc$_6$Fuc$_2$NeuAc$_2$	1193.0 [5+] 1491.0 [4+]	5960.3	2340.2	1099.0 [5+] 1373.5 [4+]	5489.8
1321.0 [5+]	6599.8	2979.8	Hex$_5$HexNAc$_7$FucNeuAc$_2$	1204.5 [5+]	6017.3	2397.3	1139.6 [5+]	5693.0
1342.0 [5+]	6704.9	3084.8	Hex$_6$HexNAc$_6$Fuc$_2$NeuAc$_2$	1225.5 [5+]	6122.4	2502.4	1099.0 [5+] 1373.5 [4+]	5489.8
1350.2 [5+]	6746.0	3126.0	Hex$_5$HexNAc$_7$Fuc$_2$NeuAc$_2$	1233.7 [5+] 1541.9 [4+]	6163.5	2543.5	1139.6 [5+]	5693.0
1371.2 [5+]	6851.1	3231.0	Hex$_6$HexNAc$_6$Fuc$_3$NeuAc$_2$	1254.7 [5+]	6268.6	2648.5	1099.0 [5+] 1373.5 [4+]	5489.8
1382.6 [5+]	6908.2	3288.0	Hex$_6$HexNAc$_7$Fuc$_2$NeuAc$_2$	1266.1 [5+] 1582.4 [4+]	6325.7	2705.5	1139.6 [5+]	5693.0

Table 1b. HPLC-electrospray mass spectrometry data obtained from sequential exoglycosidase digestion of the sialylated glycopeptides (Y_{156}-K_{184}) containing Asn-180.

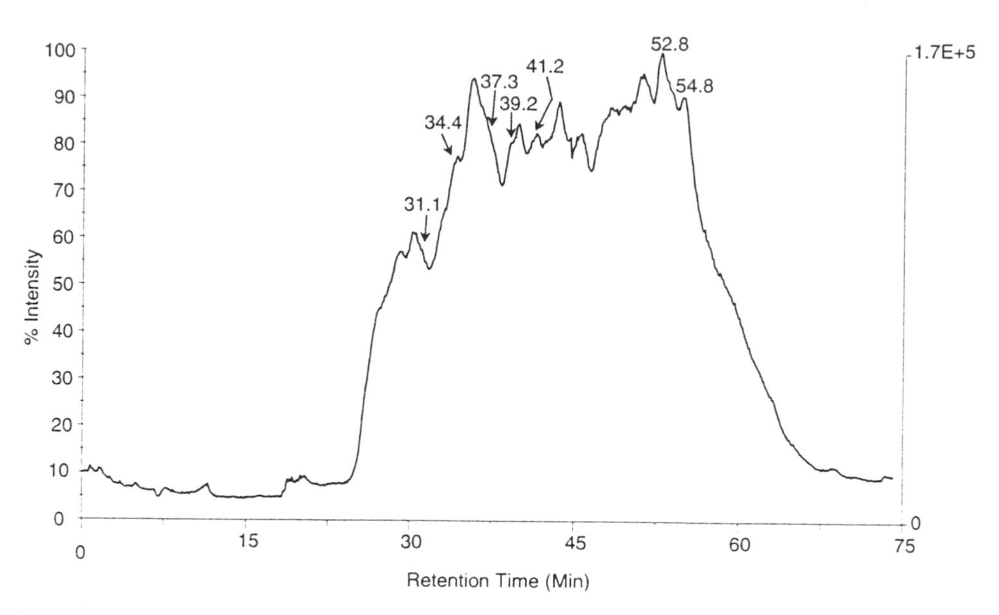

Fig. 4. TIC trace produced by re-analysis of pooled HPLC fractions containing the Asn-196 glycopeptides by capillary HPLC / ESI-oaTOF-MS.

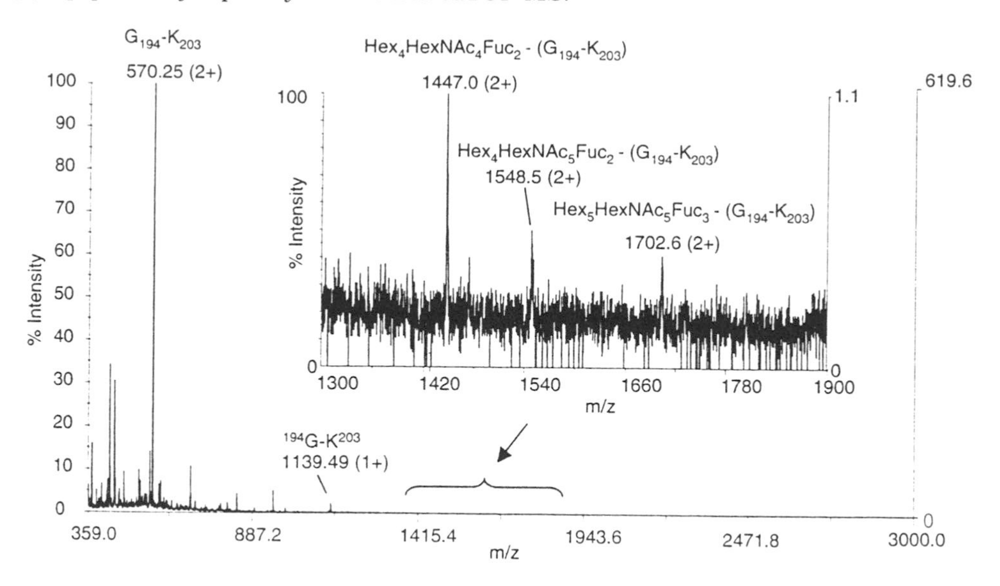

Fig. 5. ESI-oaTOF mass spectrum exhibiting signals for the neutral glycopeptides containing amino acid residues Gly_{194}-Lys_{203}.

glycopeptides at 31.1, 34.4, 37.3, 39.2 and 41.2 minutes. Summation of scans at 31.1 minutes produced the oaTOF spectrum shown in Fig. 5. The base peak at m/z 570.25 $[M + 2H]^{2+}$ corresponds to the unglycosylated peptide G_{194}-K_{203} spanning Asn-196, suggesting that a significant proportion of the protein is not glycosylated at this site. Further inspection of this spectrum (Fig. 5, inset) revealed extremely minor signals for the glycosylated counterparts of this

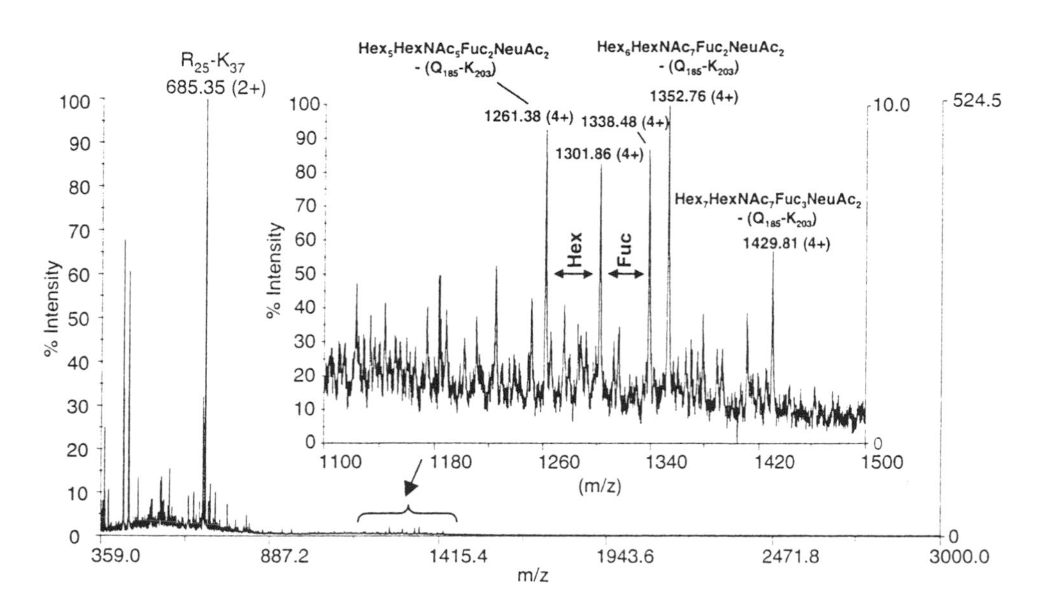

Fig. 6. ESI-oaTOF mass spectrum exhibiting signals for the disialylated glycopeptides containing amino acid residues Gln_{185}-Lys_{203}.

peptide (G_{194}-K_{203}), the peak height of the most abundant glycoform (at m/z 1447.0) is only 0.17% of that of the base peak (at m/z 570.25). The signals for these glycopeptides were inferior because the large, bulky glycosylation at Asn-196 prevented attack by trypsin at Lys-193. Consequently, only limited cleavage at this basic amino acid was observed which was confined to molecules that have no carbohydrate or only small neutral glycans. The major tryptic site in this glycosylated region occurs at Lys-184, liberating glycopeptides containing the amino acid sequence Gln_{185}-Lys_{203}. These glycopeptides are differentially sialylated: the neutral glycoforms eluted first (34.4 min) followed by the mono- (37.3 min), di- (39.2 min) and trisialylated (41.2 min) components. Figure 6 shows an ESI-oaTOF mass spectrum exhibiting small signals for disialylated glycopeptides containing the amino acid residues Gln_{185}-Lys_{203}. As seen in this spectrum, these disialyl glycoforms co-elute with the tryptic peptide R_{25}-K_{37} derived from PrP^{Sc} giving a major signal at m/z 685.35 $[M + 2H]^{2+}$. In addition to glycoform heterogeneity, further complexity was apparent due to *in vitro* conversion of a small proportion of these glycopeptides (Gln_{185}-Lys_{203}) to their $pyroGlu_{185}$-Lys_{203} counterparts. The $pyroGlu_{185}$-Lys_{203} glycoforms gave extremely minor molecular ion signals, and for simplicity are not discussed at this stage in the characterization. Instead, glycoforms containing the amino acid sequence Gln_{185}-Lys_{203} are discussed which were by far the most prevalent species. Nevertheless, average masses were assigned to these glycoforms (Gln_{185}-Lys_{203}) containing zero, one, two and three sialic acids because their overall low abundance produced unresolved isotope peaks. Therefore, subtraction of the calculated average mass of the peptide Gln_{185}-Lys_{203} (2137.3 Da) from the average molecular masses of the glycopeptides containing the amino acid sequence

Intact Glycopeptides				After α-sialidase (*V. cholerae*) digestion (Gln-185)			After α-fucosidase and β-galactosidase digestion (pyroGlu-185)	
m/z	Glyco-peptide MW_{avg}	Carbo-hydrate MW_{avg}	Proposed glycan composition	*m/z*	Glyco-peptide MW_{avg}	Carbo-hydrate MW_{avg}	*m/z*	Glyco-peptide MW_{avg}
1297.9 [3+]	3890.7	1771.6	$Hex_4HexNAc_4Fuc_2$	1297.9 [3+]	3890.6	1771.6	1189.5 [3+]	3565.4
1024.5 [4+] 1365.6 [3+]	4093.8	1974.8	$Hex_4HexNAc_5Fuc_2$	1024.5 [4+] 1365.6 [3+]	4093.8	1974.8	1257.2 [3+]	3768.6
1065.0 [4+]	4256.0	2137.0	$Hex_5HexNAc_5Fuc_2$	1065.0 [4+]	4256.0	2137.0	1257.2 [3+]	3768.6
1101.5 [4+] 1468.4 [3+]	4402.1	2283.1	$Hex_5HexNAc_5Fuc_3$	1101.5 [4+] 1468.4 [3+]	4402.1	2283.1	1257.2 [3+]	3768.6
1115.8 [4+]	4459.2	2340.2	$Hex_5HexNAc_6Fuc_2$	1115.8 [4+] 1487.4 [3+]	4459.2	2340.2	1324.9 [3+]	3971.7
1152.3 [4+]	4605.3	2486.3	$Hex_5HexNAc_6Fuc_3$	1152.3 [4+]	4605.2	2486.3	1324.9 [3+]	3971.7
1192.9 [4+]	4767.6	2648.5	$Hex_6HexNAc_6Fuc_3$	1192.9 [4+]	4767.6	2648.5	1324.9 [3+]	3971.7
1097.3 [4+]	4385.0	2266.0	$Hex_4HexNAc_5Fuc_2NeuAc$	1024.5 [4+] 1365.6 [3+]	4093.8	2137.0	1257.2 [3+]	3768.6
1137.8 [4+] 1516.8 [3+]	4547.2	2428.2	$Hex_5HexNAc_5Fuc_2NeuAc$	1065.0 [4+] 1419.7 [3+]	4256.0	2137.0	1257.2 [3+]	3768.6
1152.0 [4+]	4604.3	2485.3	$Hex_5HexNAc_6FucNeuAc$	1079.3 [4+] 1438.7 [3+]	4313.0	2194.0	1324.9 [3+]	3971.7
1224.9 [4+]	4895.5	2776.5	$Hex_5HexNAc_6FucNeuAc_2$	1079.3 [4+] 1438.7 [3+]	4313.0	2194.0	1324.9 [3+]	3971.7
1251.2 [4+]	5000.6	2881.6	$Hex_6HexNAc_5Fuc_2NeuAc_2$	1105.5 [4+] 1473.7 [3+]	4418.1	2299.1	1257.2 [3+]	3768.6
1324.0 [4+]	5291.9	3172.9	$Hex_6HexNAc_5Fuc_2NeuAc_3$	1105.5 [4+] 1473.7 [3+]	4418.1	2299.1	1257.2 [3+]	3768.6
1188.6 [4+]	4750.5	2631.5	$Hex_5HexNAc_6Fuc_2NeuAc$	1115.8 [4+]	4459.2	2340.2	1324.9 [3+]	3971.7
1261.4 [4+]	5041.6	2922.6	**$Hex_5HexNAc_6Fuc_2NeuAc_2$**	1115.8 [4+]	4459.2	2340.2	1324.9 [3+]	3971.7
1334.2 [4+]	5332.9	3213.9	$Hex_5HexNAc_6Fuc_2NeuAc_3$	1115.8 [4+]	4459.2	2340.2	1324.9 [3+]	3971.7
1214.8 [4+]	4855.2	2736.2	$Hex_6HexNAc_5Fuc_3NeuAc$	1142.0 [4+]	4564.2	2445.2	1257.2 [3+]	3768.6
1229.2 [4+]	4912.6	2793.6	$Hex_6HexNAc_6Fuc_2NeuAc$	1156.3 [4+]	4621.3	2502.3	1324.9 [3+]	3971.7
1301.9 [4+]	5203.8	3084.8	**$Hex_6HexNAc_6Fuc_2NeuAc_2$**	1156.3 [4+]	4621.3	2502.3	1324.9 [3+]	3971.7
1374.8 [4+]	5495.1	3376.1	$Hex_6HexNAc_6Fuc_2NeuAc_3$	1156.3 [4+]	4621.3	2502.3	1324.9 [3+]	3971.7
1265.7 [4+]	5058.8	2939.8	**$Hex_6HexNAc_6Fuc_3NeuAc$**	1192.9 [4+]	4767.5	2648.5	1324.9 [3+]	3971.7
1338.5 [4+]	5350.0	3231.0	**$Hex_6HexNAc_6Fuc_3NeuAc_2$**	1192.9 [4+]	4767.5	2648.5	1324.9 [3+]	3971.7
1279.9 [4+]	5115.8	2996.8	$Hex_6HexNAc_7Fuc_2NeuAc$	1207.1 [4+]	4824.5	2705.5	1392.6 [3+]	4174.9
1415.6 [4+]	5658.3	3539.3	$Hex_7HexNAc_6Fuc_2NeuAc_3$	1197.1 [4+]	4748.5	2629.5	1324.9 [3+]	3971.7
1352.8 [4+]	5407.0	3288.0	**$Hex_6HexNAc_7Fuc_2NeuAc_2$**	1207.1 [4+]	4824.5	2705.5	1392.6 [3+]	4174.9
1425.6 [4+]	5698.3	3579.3	$Hex_6HexNAc_7Fuc_2NeuAc_3$	1207.1 [4+]	4824.5	2705.5	1392.6 [3+]	4174.9
1452.1 [4+]	5804.4	3685.4	$Hex_7HexNAc_6Fuc_3NeuAc_3$	1233.7 [4+]	4930.6	2811.6	1324.9 [3+]	3971.7
1316.5 [4+]	5262.0	3143.0	**$Hex_6HexNAc_7Fuc_3NeuAc$**	1243.7 [4+]	4970.7	2851.7	1392.6 [3+]	4174.9
1357.0 [4+]	5424.1	3305.1	$Hex_7HexNAc_7Fuc_3NeuAc$	1284.2 [4+]	5132.8	3013.8	1392.6 [3+]	4174.9
1429.8 [4+]	5715.3	3596.3	**$Hex_7HexNAc_7Fuc_3NeuAc_2$**	1284.2 [4+]	5132.8	3013.8	1392.6 [3+]	4174.9
1466.1 [4+]	5860.5	3741.5	$Hex_7HexNAc_7Fuc_2NeuAc_3$	1247.7 [4+]	4986.7	2867.7	1392.6 [3+]	4174.9
1502.6 [4+]	6006.6	3887.6	$Hex_7HexNAc_7Fuc_3NeuAc_3$	1284.2 [4+]	5132.8	3013.8	1392.6 [3+]	4174.9

Table 2. HPLC/electrospray mass spectrometry data obtained from sequential exoglycosidase digestion of the glycopeptides (Q/pyroE185-K203) spanning Asn-196. Bold type indicates the major species identified based on the relative abundance of the molecular ions in the mass spectra.

Gln_{185}-Lys_{203} provided the saccharide compositions summarized in Table 2. These compositions suggest the presence of over thirty complex-type N-linked glycans at Asn-196, whose structures range from small neutral biantennary oligosaccharides to highly sialylated tetraantennary structures.

DETERMINATION OF NON-REDUCING STRUCTURE AND LINKAGE

To establish that the proposed glycan structures are correct for the compositions assigned and to gain linkage information, the remainder of the purified glycopeptide-containing fractions was incubated sequentially with specific exoglycosidases and analyzed by HPLC/ESI-oaTOF-MS after each step. The mass difference after each single exoglycosidase digestion and the monosaccharide specificity of each exoglycosidase was used to determine the sequences of the non-reducing termini. The glycopeptide mixtures were first digested with

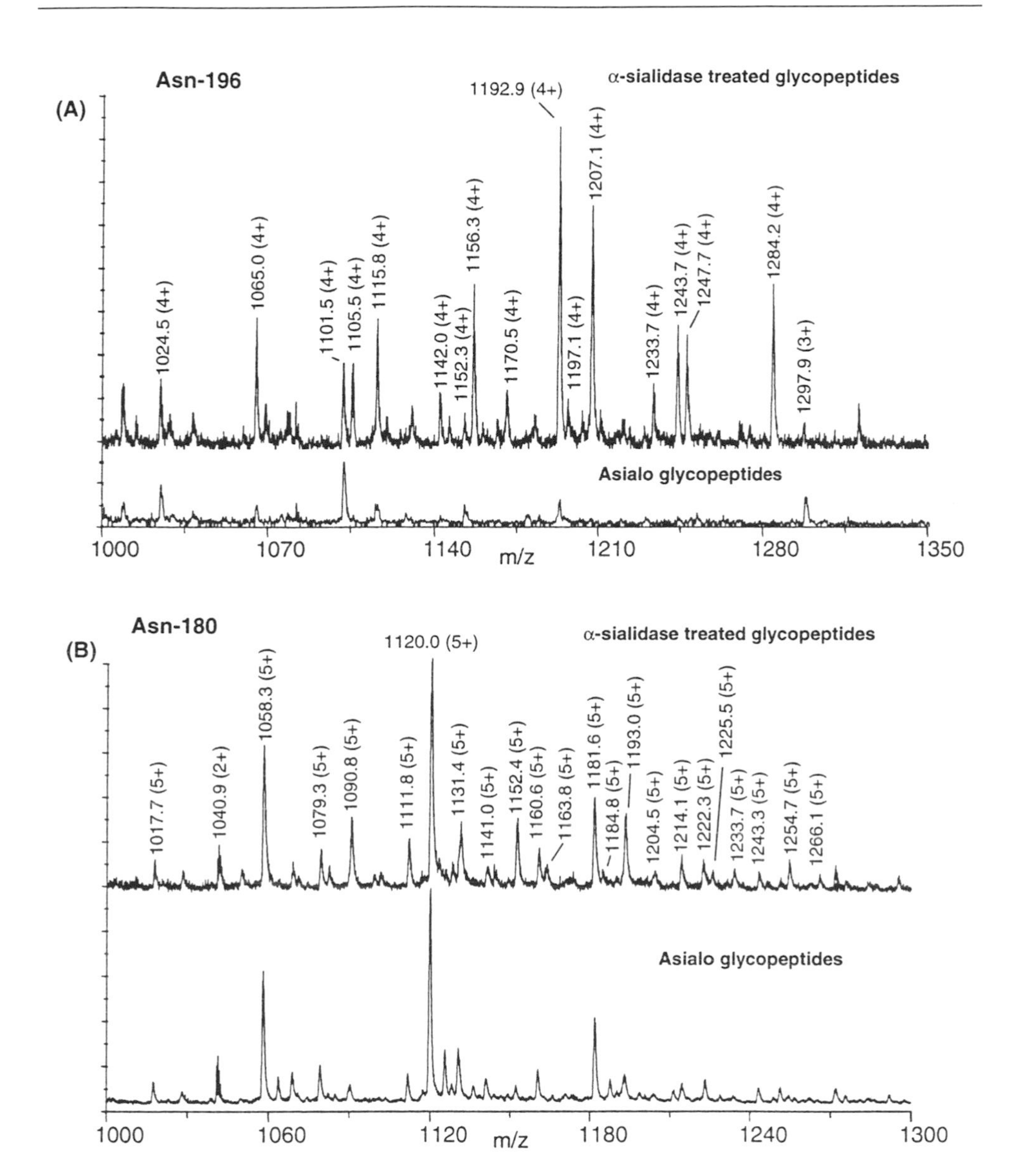

Fig. 7. Mass spectral comparisons of the asialo and α-sialidase treated glycopeptides at (A) Asn-196 and (B) Asn-180. The asialo and α-sialidase mass spectra were normalized by using the intensities of the assigned neutral glycopeptides in each spectrum as guides, i. e., no sialylated counterparts for $Hex_5HexNAc_5Fuc_3$ at m/z 1101.5 were previously detected, therefore the intensity of this signal in each spectrum was used as the guide.

the non-specific α-sialidase from *V. cholerae*. The resulting ESI-oaTOF mass spectra showed an absence of signals for the sialylated structures, concomitant with the appearance or increase in abundance of their non-sialylated counterparts (summarized in Tables 1 and 2) that now co-eluted with the neutral constituents. Comparison of the relative abundance of the molecular ions for the neutral and α-sialidase treated glycopeptides containing the amino acid se-

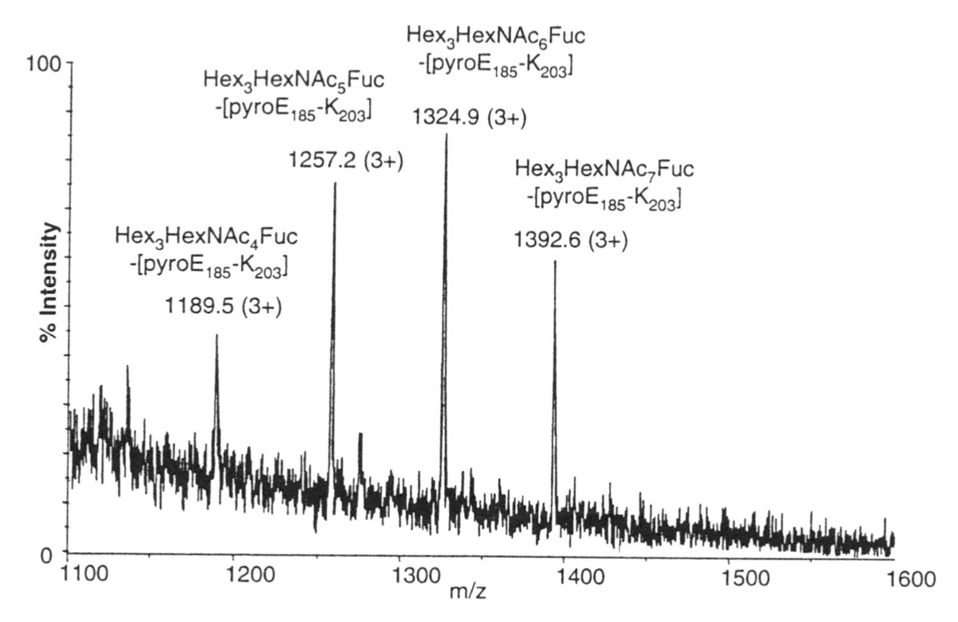

Fig. 8. ESI-oaTOF mass spectrum of the Asn-196 glycopeptides after digestion with α-sialidase, α-fucosidase and β-galactosidase.

quence Gln_{185}-Lys_{203} spanning Asn-196 (lower and upper spectrum of Fig. 7A, respectively) reveal that the major glycoforms at this site contain terminal sialic acid. For example, a significant increase in molecular ion abundance was observed for the asialo structures at m/z 1065.0 ($Hex_5HexNAc_5Fuc_2$), m/z 1115.8 ($Hex_5HexNAc_6Fuc_2$) and m/z 1192.9 ($Hex_6HexNAc_6Fuc_3$) after α-sialidase digestion, establishing that a large proportion of these glycans are capped with sialic acid. Furthermore, many sialylated structures do not have asialo counterparts as many new signals appear after neuraminidase digestion [e. g., m/z 1105.5 ($Hex_6HexNAc_5Fuc_2$), m/z 1156.3 ($Hex_6HexNAc_6Fuc_2$), m/z 1207.1 ($Hex_6HexNAc_7Fuc_2$) and m/z 1284.2 ($Hex_7HexNAc_7Fuc_3$)]. The relative abundance of most molecular ion signals for the asialo glycopeptides containing the amino acid sequence Y_{156}-K_{184} spanning Asn-180 and their α-sialidase digested counterparts (lower and upper spectrum of Fig. 7B, respectively) remained the same indicating that neutral glycoforms are the major components at this site. However, a significant difference in signal intensity occurred at m/z 1090.8 (5+) ($Hex_4HexNAc_5Fuc$), m/z 1152.4 (5+) ($Hex_5HexNAc_5Fuc_2$) and m/z 1193.0 (5+) ($Hex_5HexNAc_6Fuc_2$), indicating that these glycoforms are the major sialylated substituents at Asn-180.

Further characterization of the α-sialidase treated glycopeptides was achieved by exhaustive digestion with α(1-3,4) fucosidase (*X. manihotis*) followed by β-galactosidase (*D. pneumonaie*), with analysis by HPLC/ESI-oaTOF-MS after each step (summarized in Tables 1 and 2). These experiments apparently caused glycopeptides containing the amino acid sequence Gln_{185}-Lys_{203} to undergo cyclization via NH_3 loss, producing glycopeptides with an N-terminal pyroglutamate residue ($pyroGlu_{185}$-Lys_{203}) that were now the major species. This

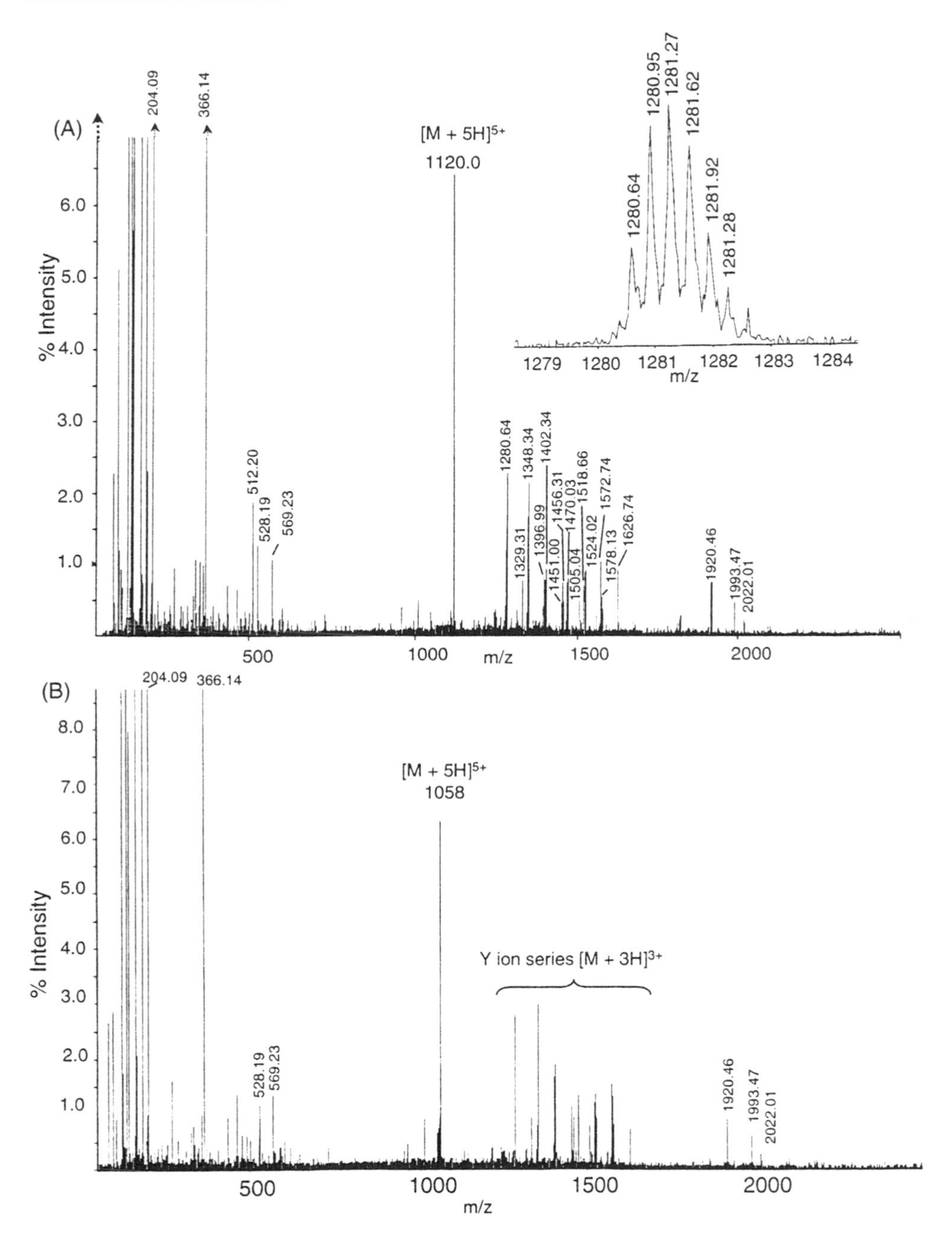

Fig. 9. CID spectra of the glycopeptide $Hex_4HexNAc_5Fuc_2$- (Y_{156}-K_{184}) at m/z 1120.0 $[M+5H]^{5+}$ (A) and $Hex_3HexNAc_5Fuc$- (Y_{156}-K_{184}) at m/z 1058 $[M + 5H]^{5+}$ (B). The singly charged ions at m/z 204.09 $(HexNAc)^+$, m/z 366.14 $(HexHexNAc)^+$, m/z 512.20 $(HexHexNAcFuc)^+$ and m/z 528.2 ($Hex_2HexNAc$) are non-reducing terminal fragments, whereas m/z 569.23 $(HexHexNAc_2)^+$ is an internal ion resulting from multiple cleavages. The absence of m/z 512.2 $(HexHexNAcFuc)^+$ in (B) is the only ion which distinguishes the two spectra.

(A)

$B_{2\alpha}$
Fuc
$Y_{4\alpha}$
Gal — GlcNAc — Man
$Y_{3\alpha}$
Fuc
$Y_{1\beta}$
GlcNAc — Man — GlcNAc — GlcNAc-Peptide
$Y_{3\gamma}$
Y_2
GlcNAc — Man
$Y_{3\beta}$
$Y_{1\alpha}$
$B_{1\beta}$
$Y_{4\beta}$

(B)

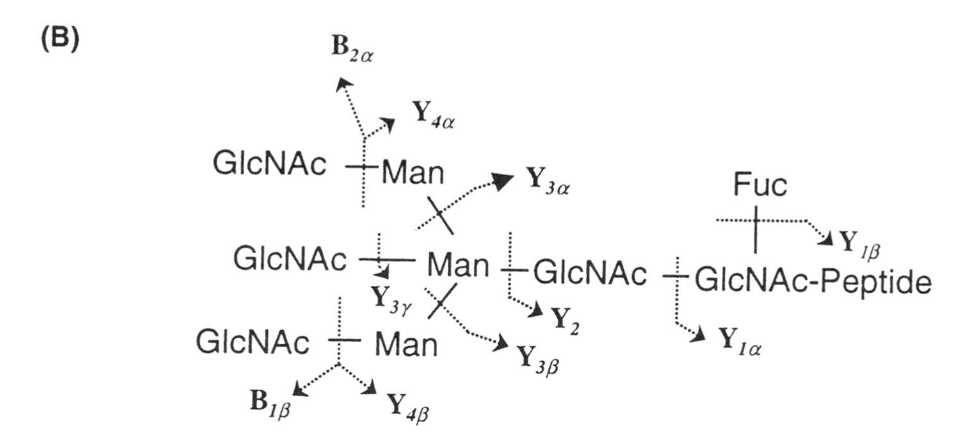

Fig. 10. Glycosidic cleavages found in the CID spectra of $Hex_4HexNAc_5Fuc_2$- (Y_{156}-K_{184}) (A) and $Hex_3HexNAc_5Fuc$- (Y_{156}-K_{184}) (B). The fragmentation scheme is based on that proposed by Domon and Costello [18].

conversion may have been due to extensive incubations in buffer solutions and/ or frequent re-suspensions and short periods of storage in 0.1% formic acid. Previous investigators have stated that *in vitro* conversion of N-terminal Gln to pyroGlu is most likely to occur in acid media [16]. Importantly, the saccharide moieties on glycopeptides spanning Asn-196 (pyroGlu$_{185}$-Lys$_{203}$) and Asn-180 (Y_{156}-K_{184}) were cut back to the following core structures: $Hex_3HexNAc_4Fuc$, $Hex_3HexNAc_5Fuc$, $Hex_3HexNAc_6Fuc$ and $Hex_3HexNAc_7Fuc$. The ESI-oaTOF spectrum exhibiting molecular ion signals for the fully digested glycopeptides containing the amino acid sequence pyroGlu$_{185}$-Lys$_{203}$ (spanning Asn-196) is shown in Fig. 8. All Asn-196 and Asn-180 glycopeptides have one fucose residue

Fragment ion Composition (Y series)	m/z	Observed MW_{mi} (Da)	Calculated MW_{mi} (Da)
Y_{156}-K_{184} - HexNAc	1280.64 $[M+3H]^{3+}$	3838.92	3838.81
	1920.46 $[M+2H]^{2+}$		
Y_{156}-K_{184} -HexNAcFuc	1329.31$[M+3H]^{3+}$	3984.93	3984.86
	1993.47 $[M+2H]^{2+}$		
Y_{156}-K_{184} -$HexNAc_2$	1348.34 $[M+3H]^{3+}$	4042.02	4041.88
	2022.01 $[M+2H]^{2+}$		
Y_{156}-K_{184} -$HexNAc_2Fuc$	1396.99 $[M+3H]^{3+}$	4187.97	4187.94
Y_{156}-K_{184} -$HexNAc_2Hex$	1402.34 $[M+3H]^{3+}$	4204.02	4203.94
Y_{156}-K_{184} -$HexNAc_2HexFuc$	1451.00 $[M+3H]^{3+}$	4350.00	4349.99
Y_{156}-K_{184} -$HexNAc_2Hex_2$	1456.31 $[M+3H]^{3+}$	4365.93	4365.99
Y_{156}-K_{184} -$HexNAc_3Hex$	1470.03 $[M+3H]^{3+}$	4407.09	4407.02
Y_{156}-K_{184} - $HexNAc_2Hex_2Fuc$	1505.04 $[M+3H]^{3+}$	4512.12	4512.05
Y_{156}-K_{184} - $HexNAc_3HexFuc$	1518.66 $[M+3H]^{3+}$	4552.98	4553.08
Y_{156}-K_{184} - $HexNAc_3Hex_2$	1524.02 $[M+3H]^{3+}$	4569.06	4569.07
Y_{156}-K_{184} - $HexNAc_2Hex_3Fuc$	1559.38 $[M+3H]^{3+}$	4675.14	7674.10
Y_{156}-K_{184} - $HexNAc_3Hex_2Fuc$	1572.74 $[M+3H]^{3+}$	4715.22	4715.13
Y_{156}-K_{184} - $HexNAc_3Hex_3$	1578.13 $[M+3H]^{3+}$	4731.39	4731.12
Y_{156}-K_{184} - $HexNAc_3Hex_3Fuc$	1626.74 $[M+3H]^{3+}$	4877.22	4877.18
Y_{156}-K_{184} - $HexNAc_3Hex_3Fuc_2$	1675.45 $[M+3H]^{3+}$	5023.35	5023.24

Table 3. Y series fragment ions observed in the CID spectrum of the N-linked glycopeptides (Y_{156}-K_{184})-$Hex_4HexNAc_5Fuc_2$ and (Y_{156}-K_{184})-$Hex_3HexNAc_5Fuc$ at m/z 1120.0 $[M+5H]^{5+}$ and m/z 1058.0 $[M+5H]^{5+}$, respectively.

that is resistant to α-fucosidase digestion that is consistent with α(1,6) fucose remaining on the chitobiose core. Also, the complete removal of terminal galactose by β-galactosidase (β1,4 specific) indicated the presence of type-2 antennae, and thus, the sialyl Lewis^x [NeuAcα2-3Galβ1-4(Fucα1-3)GlcNAc-] and Lewis^x [Galβ1-4(Fucα1-3)GlcNAc-] structural epitopes [17].

To strengthen the exoglycosidase data, selected glycopeptides containing Asn-180 were subjected to low-energy MS/MS using a high-resolution hybrid quadrupole-oaTOF (Qq-TOF) tandem mass spectrometer fitted with nano-electrospray ionization. The resolution provided by this type of instrument enables immediate assignment of the charge state of each fragment ion (see inset

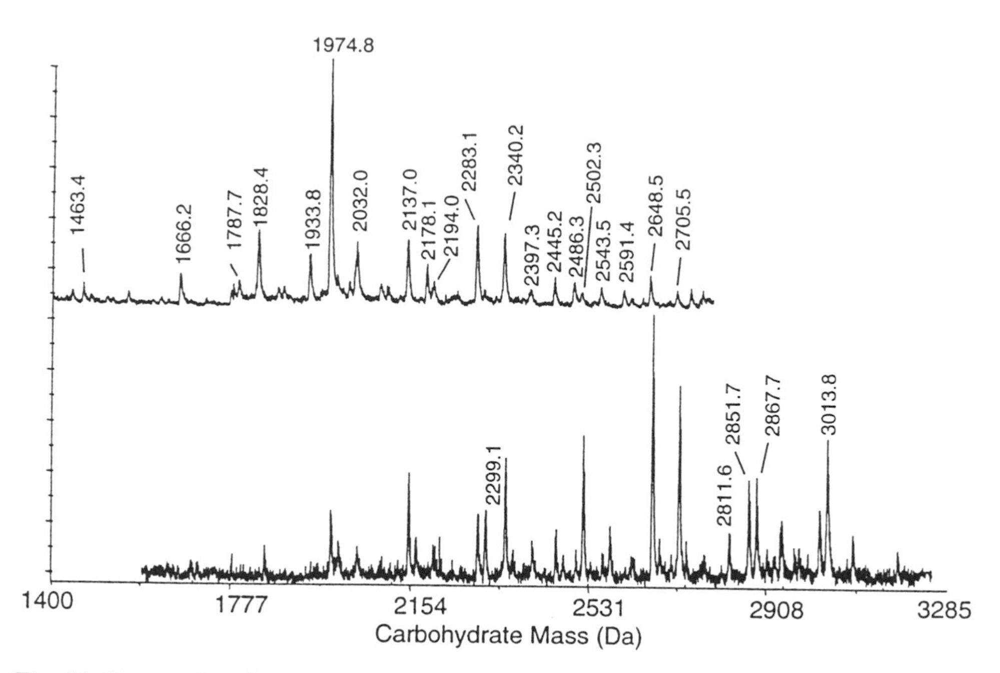

Fig. 11. Deconvoluted mass data comparing the α-sialidase treated oligosaccharides at Asn-180 (upper trace) and Asn-196 (lower trace). By plotting carbohydrate mass (Da) instead of glycopeptide mass (Da) along the x-axis, the glycans at both sites were directly compared. Intact carbohydrate mass was calculated by subtraction of the calculated mass of the unglycosylated peptides (Q_{185}-K_{203}; 2137.3 Da for Asn-196 and Y_{156}-L_{184}; 3638.1 Da for Asn-180) from glycopeptide mass and adding H_2O (18 Da).

in Fig. 9A) and its inherent mass accuracy gives high confidence in sequence assignments. The fragment ion spectrum of the most abundant glycoform at Asn-180 ($Hex_4HexNAc_5Fuc_2$-[Y_{156}-K_{184}]) is shown in Fig. 9A and the fragmentation scheme (based on Domon and Costello nomenclature [18]) is depicted in Fig. 10A. The MS/MS spectrum is dominated by oxonium ions (B ions) at low mass, which are derived from glycosidic cleavages at the non-reducing end of the carbohydrate chain. Importantly, the singly charged non-reducing B_2 fragment ion at m/z 512.20 (for $HexFucHexNAc^+$) supports the presence of the terminal Lewis^x epitope. Corroboration of this assignment is provided by the CID spectrum of the truncated biantennary glycoform, comprised of $Hex_3HexNAc_5Fuc$-[Y_{156}-K_{184}], that does not contain the Lewis^x structure and, as a result, the signal at m/z 512.20 is absent (Fig. 9B; the fragmentation scheme is shown in Fig. 10B). More sequence information is provided by a series of multiply charged Y ions which are all found in both MS/MS spectra (summarized in Table 3), whose formation is favored because charge is retained by the peptide attached to the reducing terminus. Importantly, the presence of the reducing terminal fragment ion Y_1 at 1329.31 $[M+3H]^{3+}$ in each product ion spectra shows that core fucose is present on both glycopeptides analyzed. Furthermore, product ions at m/z 1470.03 $[M + 3H]^{3+}$ ($HexNAc_3Hex$) and m/z 1518.66 $[M + 3H]^{3+}$ ($HexNAc_3HexFuc$) arise from multiple Y rearrangements ($Y_{3\alpha}$, $Y_{3\beta}$, $Y_{1\beta}$ and $Y_{3\beta}$, respectively) and confirm the presence of bisecting GlcNAc on β-mannose.

Site-specific Comparisons of Intact N-linked Glycans

The glycosylation at each of the two N-linked consensus sites was directly compared using transformed spectra with intact carbohydrate mass plotted on the x-axis instead of glycopeptide mass. Direct comparison of glycoforms with zero, one and two sialic acid residues at each of the N-linked sites reveals that many structures are common to both Asn-196 and Asn-180. The N-linked consensus site at Asn-196, however, has a significant amount of tri-sialylated tri- and tetraantennary glycans that were not detected at Asn-180. A comparison of the masses of the α-sialidase treated glycoforms from both sites (shown in Fig. 11) provides clear evidence of differing N-glycan populations. It can be clearly seen that most glycoforms at Asn-180 are complex bi- and triantennary structures falling within a mass range of approximately 1660-2340 Da. By contrast, the major glycan components at Asn-196 are larger, consisting of complex tri- and tetraantennary structures.

Conclusions

Capillary LC/ESIMS using an orthogonal acceleration time-of-flight mass spectrometer has been successfully employed to identify more than sixty N-linked glycans, site-specifically, on PrP^{Sc}. Sequential exoglycosidase digestions and low-energy MS/MS were used to determine the overall structure of each glycoform, although only the former method provides important linkage information (e. g., the $Lewis^{x}$ epitope). This paper has shown that the sensitivity and dynamic range of new oaTOF instrumentation provides the means to detect extremely minor glycopeptide species in very complex mixtures while only consuming low picomole amounts of protein sample. Using this sensitive strategy, additional studies are continuing to determine whether certain carbohydrate structures are unique to particular sites on other strains of murine scrapie.

Acknowledgements

Financial support was provided by the Ludwig Institute for Cancer Research, the BBSRC and a grant from the Biomedical Research Technology Program of the National Center for Research Resources (grant no. NIH NCRR BRTP RR01641).

References

1. S. B. Pruisner, *Science* **1986**, *21*, 136-144.

2. B. Cauhey and G. J. Raymond, *J. Biol. Chem.* **1991**, *266*, 18217-18223.

3. D. R. Borchelt, M. Scott, A. Taraboulos, N. Stahl and S. B. Pruisner, *J. Cell. Biol.* **1990**, *110*, 43-752.

4. M. Bruce, A. Chree, I. McConnell, J. Foster, G. Pearson and H. Fraser, *Philosophical Transactions of the Royal Society of London Series B-Biological Sciences.* **1994**, *343*, 405-411.

5. R. J. Kascsak, *J. Virol.* **1986**, *59*, 676-683.

6. R. A. Somerville and L. A. Ritchie, *J. Gen. Virol.* **1990**, *71*, 833-839.

7. A. L. Burlingame, *Curr. Opin. Biotechnol.* **1996**, *7*, 4-10.

8. N. Stahl, M. A. Baldwin, D. B. Teplow, L. Hood, B. W. Gibson, A. L. Burlingame and S. B. Prusiner, *Biochemistry* **1993**, *32*, 1991-2002.

9. E. Stimson, A. Chong, J. Hope, and A. L. Burlingame, **1998**, In press.

10. J. Hope, G. Multhaup, L. J. Reekie, R. H. Kimberlin and K. Beyreuther, *Eur. J. Biochem.* **1988**, *172*, 271-277.

11. S. A. Carr, M. J. Huddleson and M. F. Bean, *Protein Sci.* **1993**, *2*, 183-196.

12. M. J. Huddleson, M. F. Bean and S. A. Carr, *Anal Chem.* **1993**, *65*, 877-884.

13. K. F. Medzihradszky, D. A. Maltby, S. C. Hall, C. A. Settineri and A. L. Burlingame, *J. Am. Soc. Mass Spectrom.* **1994**, *5*, 350-358.

14. K. F. Duffin, J. K. Welpy, E. Huang and J. D. Henion, *Anal Chem.* **1992**, *64*, 1440-1448.

15. T. Endo, D. Groth, S. B. Prusiner and A. Kobata, *Biochemistry*, **1989**, *28*, 8380-8388.

16. S. A. Carr and K. Bieman, In *Methods in Enzymology*, Wold, F. and Moldave, K, Eds.; Academic: Orlando, Fl, **1984**; *106*, 29-58.

17. S.-I. Hakomori, *Histochem. J.* **1992**, *24*, 771-776.

18. B. Domon and C. E. Costello, *Glycoconjugate J.* **1988**, *5*, 397-509.

Questions and Answers

Michael Baldwin (UCSF)

Western blots show strain-specific differences in the amounts of the differently glycosylated forms of PrP^{Sc}. Are you able to carry out quantitative analysis?

Answer. No, I can not accurately calculate the amounts of unglycosylated versus glycosylated protein at each glycosylation site by the techniques described. However, I do observe differences in the ease of detection of the glycopeptides spanning Asn-196 from different strains, using similar quantities of protein. For instance, the glycosylation on PrP^{Sc} from strain Me7 SV is more readily detected than that on PrP^{Sc} from 22A SV, suggesting that the latter strain has less carbohydrate at Asn-196 than the former.

Michael Baldwin (UCSF)

2-D Western blots from different mouse brain regions show regional variation in glycosylation of PrP^{Sc}, which may be relevant to the variable distribution of the pathology arising from different strains. Do your methods have the sensitivity to analyze PrP glycoforms using dissected brain sections rather than homogenized whole brains?

Answer. That's a good question. I'm not sure if that would be possible because it is already very difficult to detect the glycosylation at Asn-196 on PrP^{Sc} purified from homogenized whole brains, and to analyze the glycosylation of PrP^{Sc} from dissected areas we would probably need more mouse brains. However, this type of characterization may also be easier because the glycosylation could be less heterogeneous on PrP^{Sc} purified from individual regions of the brain.

Jasna Peter-Katalinic (University of Münster)

Sialic acid variant expressed in mouse is Neu5Gc rather than Neu5Ac, as found in many gangliosides. Could you detect this variant in the PrP^{Sc} N-glycosylation as well?

Answer. I was aware that this variant may be present but could not detect it. The identification of Neu5Gc is difficult when only using LC/MS to attain intact masses for glycopeptides. This is because the mass increment between Neu5Ac (291 Da) and Neu5Gc (307 Da) is 16 Da and, the mass difference between a hexose (162 Da) and fucose (146 Da) residue is also 16 Da. Consequently, I discounted the presence of Neu5Gc using exoglycosidase digestions in conjunction with LC/MS.

Steven Carr (SmithKline Beecham Pharmaceuticals)

How quantitative are the relative peak heights of the sialylated versus neutral oligosaccharides relative to the actual glycoform heterogeneity at each specific site?

Answer. It is not feasible to compare the peak heights of neutral versus sialylated glycopeptides quantitatively. However, I could definitely tell from the relative peak heights of the neutral and α-sialidase digested glycopeptides containing amino acid residues Q_{185}-K_{203} (Asn-196) and Y_{156}-K_{184} (Asn-180) (shown in Figure 7A and B, respectively) that the glycans at Asn-196 are highly sialylated whereas those at Asn-180 are mostly neutral. From this experiment I can also estimate that the neutral structure $Hex_6HexNAc_6Fuc_3$ (at m/z 1192.9 $[4H^+]$) is approximately 10-12 fold less abundant than the sialylated counterparts of this glycoform at Asn-196. To be more accurate is difficult because there may be levels of sialylated glycopeptides still present after α-sialidase digestion, which are not detected by the mass spectrometer, also a portion of the sample is likely to be lost during the exoglycosidase digestion experiment. Furthermore, the neutral glycoforms of the protein were found to be a little more susceptible to digestion with trypsin at Lys-193, yielding neutral glycoforms containing both the amino acid residues G_{194}-K_{203} and Q_{185}-K_{203}, whereas the sialylated glycoforms mostly contained the latter amino acid residues (Q_{185}-K_{203}).

Lowell H. Ericsson (University of Washington)

Have you seen a difference between tryptic and Lys-C digestion when a larger steric blocking structure is close to a Lys site where you wish to obtain cleavage?

Answer. No, I haven't observed a difference between these two enzymes in their ability to cleave at the lysine residue (Lys-193) found close to the N-linked glycosylation at Asn-196 in mouse PrP^{Sc}. Previously, N. Stahl and M. Baldwin [8] also discovered that endo Lys-C cleavage of hamster PrP^{Sc} at Lys-194 was incomplete yielding the glycosylated peptide Q_{186}-K_{204}. They found that this glycopeptide could, however, be digested to completion after removal of the carbohydrate with peptide N-glycosidase F.

K.-A. Karlson (University of Gotenberg)

Is the glycosylation essential for biological activity?

Answer. In terms of conversion of PrP^{c} to PrP^{Sc}, it has been shown that asparagine-linked glycosylation is not necessary for the synthesis of PrP^{Sc} in cultured cells [1 , 2]. In fact PrP^{Sc} is produced more rapidly when scrapie-infected cells are treated with tunicamycin [19]. Previous studies to examine whether the unglycosylated PrP^{Sc} retains infectivity were not definitive, i. e., the infectivity introduced by *de novo* synthesis of unglycosylated PrP^{Sc} in the presence of tunicamycin could not be detected because of the presence of endogenous scrapie infectivity in cultured cells [19].

1. A. Taraboulos, M. Rogers, D. R. Borchelt, M. P. McKinley, M. Scott, D. Serban and, S. B. Pruisner. *Proc. Natl. Acad. Sci. USA.* **1990**, *87*, 8262-8266.
2. S. Lehmann and D. A. Harris. *J. Biol. Chem.* **1997**, *272*, 21479-21487.

Matrix-Assisted Laser Desorption/Ionization Mass Spectrometry of N-Linked Carbohydrates and Related Compounds

David J. Harvey, Bernhard Küster*, Susan F. Wheeler*, Ann P. Hunter*, Robert. H. Bateman† and Raymond A. Dwek**

*Oxford Glycobiology Institute, Department of Biochemistry, South Parks Road, Oxford, OX1 3QU, UK; †Micromass Ltd., Floats Road, Wythenshawe, Manchester, M23 9LZ, UK.

Matrix-assisted laser desorption/ionization (MALDI) mass spectrometry of N-linked carbohydrates (those linked to asparagine in glycoproteins) was first reported in 1991 by Mock *et al.* [1] and provides a simple and sensitive method for producing spectra from underivatized N-linked glycans. It has been estimated that MALDI is about 10 to 100 times more sensitive than fast-atom bombardment

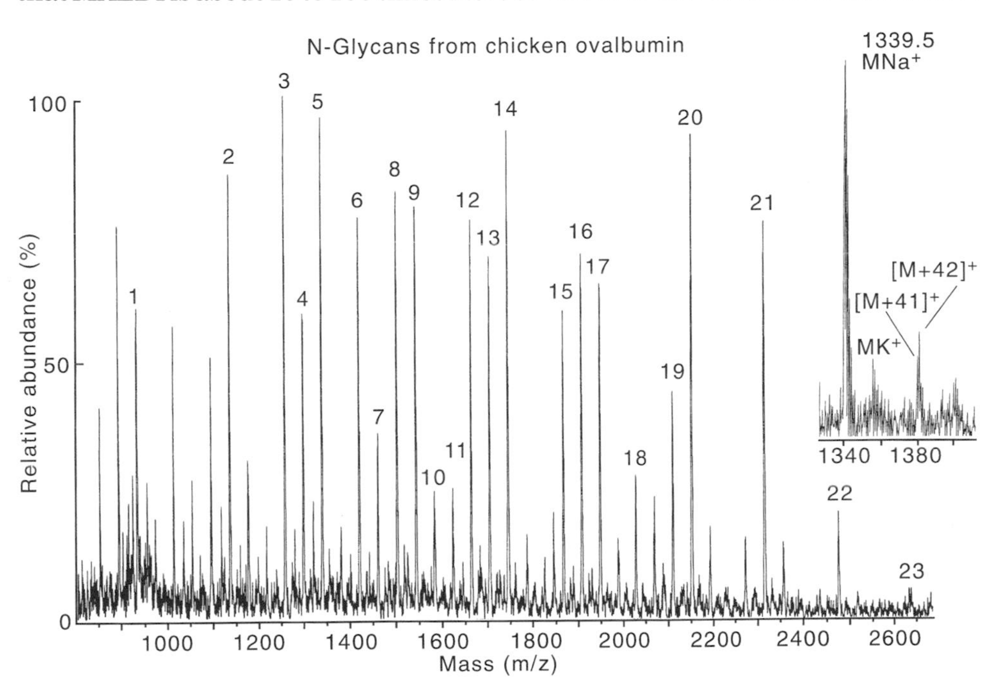

Fig. 1. Positive ion MALDI mass spectrum of N-linked glycans released from chicken ovalbumin (Sigma grade VI, containing other glycoproteins) by hydrazinolysis and recorded from 2,5-DHB with a delayed extraction, reflectron-TOF mass spectrometer. Structures of the constituent glycans are listed in Table 1. The inset shows an enlargement of one of the peaks with its accompanying MK+, [M + 41]+ and [M + 42]+ ions. The [M + 41]+ ion is produced by the N-acetyl-glycosylamine, whereas the [M + 42]+ ion is caused by a mono-O-acetyl derivative. Peaks in the m/z 1400 region correspond to acetyl-hydrazides and the potassium adducts of the [M + 41]+ and [M + 42]+ ions.

(FAB) for glycan analysis [2], and does not require derivatization by permethylation or peracetylation, as required by FAB, to obtain a signal. When recorded with linear time-of-flight (TOF) mass spectrometers, MALDI mass spectra of neutral N-linked glycans contain few, if any, fragment ions and so the technique provides a rapid method for glycan profiling. The spectrum of N-linked glycans released from chicken ovalbumin, shown in Fig. 1 illustrates the type of result that can be obtained. Structures of the glycans producing the peaks are shown in Table 1. The masses of the peaks can lead directly to the compositions of the constituent glycans in terms of their isobaric monosaccharide compositions, but determination of further structural details requires production of fragment ions or coupling to other methods such as exoglycosidase digestion as described below. Acidic N-linked glycans produce more fragmentation than neutral glycans and are generally more difficult to handle. Nevertheless, they provide excellent spectra under the right conditions. This paper summarizes work in this area and presents examples of data recorded in our laboratory.

INSTRUMENTATION

The first commercial instruments used for MALDI analysis were linear-TOF spectrometers in which the resolution was insufficient to define isotopes from N-linked glycans. We overcame this problem with a magnetic sector instrument (Micromass AutoSpec QFPD) fitted with an array detector to accommodate the signal from the pulsed laser ion-source [3]. The ion source was a conventional FAB source, and ionization was with a nitrogen laser mounted outside the vacuum system. Spectra had to be acquired in segments because of the limited mass range of the array detector, and sample consumption was high compared with that normally consumed by TOF instruments [4]. Resolutions in excess of 2000 (FWHM) were obtained. The instrument was used for glycan profiling from glycoproteins such as CD-59 [5] and hemaglutinin from the influenza-A virus [6]. Although convenient, an array detector is not essential for recording MALDI spectra on this instrument. We have shown that if the scan speed is slow enough so that about 6-10 laser shots are acquired across the mass peaks, the spectrum can be recorded with a conventional point detector. In Fig. 2a, each "peak" represents one laser shot. The ion peak profile can be extracted simply by smoothing with a factor appropriate to the peak width (Fig. 2b). This method enables the full resolution capability of the sector instrument to be utilized, and resolutions approaching 20,000 have been obtained.

Development of TOF instrumentation rapidly overtook these approaches, however, particularly with the introduction of reflectron-TOF instruments fitted with time-lag-focusing (delayed extraction) [7] which provide resolutions in the 10,000-20,000 range. Not only do these instruments improve resolution, but also allow fragmentation, both in-source and more importantly, post-source, to be observed. Such fragmentation is the result of relatively low energy processes, but high energy collision-induced decomposition (CID) can be observed with a magnetic sector instrument fitted with a tandem orthogonal-TOF analyzer as described below.

Structures and compositions of the B-linked glycans found in chicken ovalbumin

Peak No,[1]	Mass[2] Observed	Mass[2] Calculated	Composition[3]	Structure[4]
1	933.3	933.4	H_3N_2	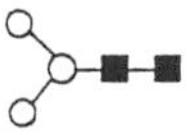
2	1136.4	1136.4	H_3N_3	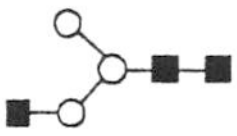
3	1257.4	1257.4	H_5N_2	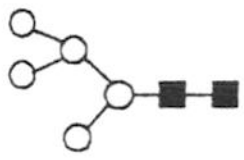
4	1298.4	1298.5	H_4N_3	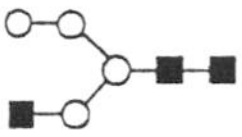
5	1339.5	1339.5	H_3N_4	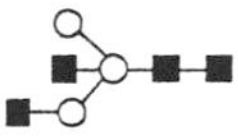
6	1419.5	1419.5	H_6N_2	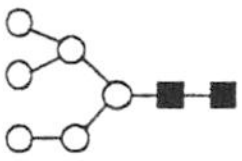
7	1460.5	1460.5	H_5N_3	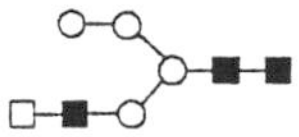
8	1501.6	1501.6	H_4N_4	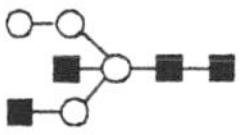
9	1542.6	1542.6	H_3N_5	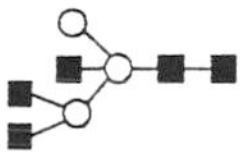
10	1581.6	1581.6	H_7N_2	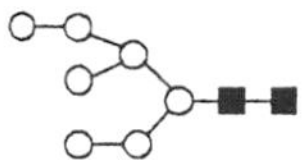
11	1622,5	1622.6	H_6N_3	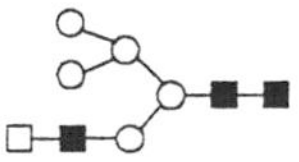
12	1663.5	1663.6	H_5N_4	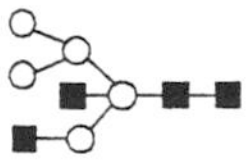
13	1704.5	1704.6	H_4N_5	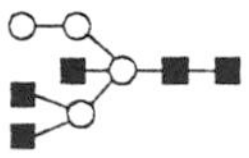

Table 1.

14	1745.6	1745.7	H_3N_6 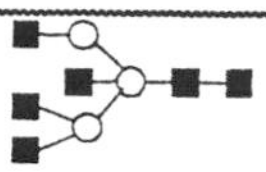
15	1866.6	1866.7	H_5N_5 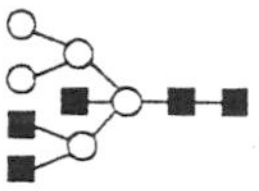
16	1907.7	1907.7	H_4N_6 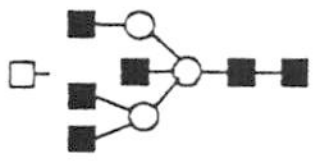
17	1948.7	1948.7	H_3N_7 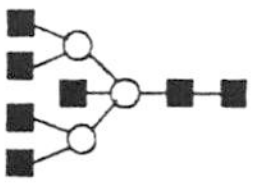
18	2028.7	2028.8	H_6N_5 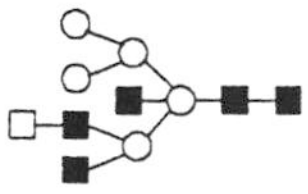
19	2110.7	2110.8	H_4N_7 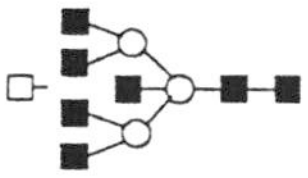
20	2151.7	2151.8	H_3N_8 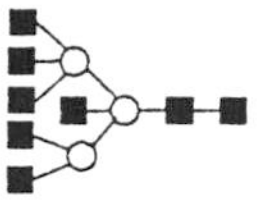
21	2313.8	2313,9	H_4N_8 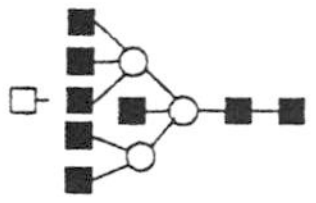
22	2475.8	2475.9	H_5N_8 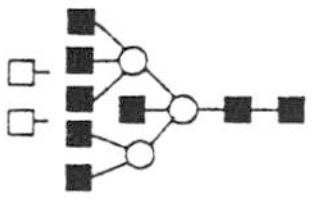

1) In Fig. 1
2) Monoisotopic mass of the MNa^+ ion
3) H = hexose (mannose or galactose), N = GlcNAc
4) Symbols for structural formulae: ■ = GlcNAc, □ = galactose, ○ = mannose. Where galactose residues are shown "floating", either the point of attachment has not been determined or the compound is a mixture of isomers.

Table 1 (cont.)

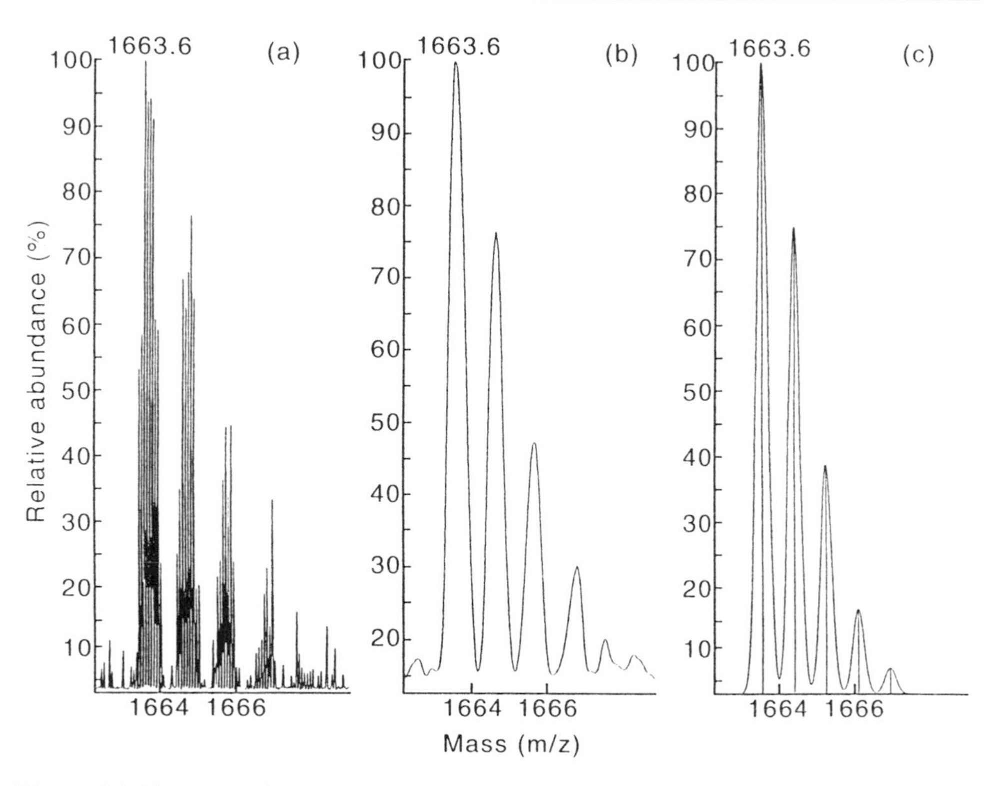

Fig. 2. (a) Slow scan (2000 sec/decade) acquisition of the molecular ion region of the biantennary N-linked glycan $(Gal)_2(Man)_2(GlcNAc)_2$ at a resolution of 2000 recorded from a mixture of α-cyano-4-hydroxycinnamic acid and 3-aminoquinoline [22] in positive ion mode. Each "peak" represents the signal from one laser shot. (b) The result of smoothing the data from Fig. 2a. (c) Calculated isotopic profile.

MATRICES

Neutral Glycans

The first matrix specifically designed for MALDI analysis of carbohydrates was 3-amino-4-hydroxybenzoic acid [1], but this compound was soon superseded by 2,5-dihydroxybenzoic acid (2,5-DHB) [8], the preferred matrix in use today. As with other matrices for carbohydrate analysis, it produces MNa^+ species as the major ion. A weaker MK^+ ion is also usually present. Other species, such as MLi^+, can be generated by the addition of the appropriate inorganic salt to the matrix [8]. Mohr *et al.* [9] have studied the relative affinities of different alkali metal ions for carbohydrates and have found that the affinity order is Cs > K > Na > Li > H. Cesium, although the most efficient at producing ions, is unable to ionize small carbohydrates [10]. Several investigators have added heavy metal salts to the matrix mixture to suppress the double peaks produced by the MNa^+ and MK^+ ions and also to inhibit fragmentation. It has been found that the ease of fragmentation varies inversely with the size of the adduct, with MH^+ ions being most prone to fragmentation [10, 11].

2,5-DHB typically crystallizes from mixtures of acetonitrile or ethanol and water as long, needle-shaped crystals that originate at the periphery of the target spot and project towards the centre. The central region usually contains an amorphous mixture of sugar, contaminants and salts and its diameter often reflects the amount on non-matrix material present. To produce a more even film of crystals [12], we re-dissolve the initially dried target spot in absolute ethanol and allow it to recrystallize. This technique not only produces a thin, even film of small crystals but also gives an increase in sensitivity of about an order of magnitude. This increased performance is probably the result of the more equal solubility of the carbohydrate and matrix in ethanol than in the original aqueous solvent and a better incorporation of the sample into the crystal lattice. Evidence for fractionation of the analyte in 2,5-DHB has been provided by Stahl *et al.* [13] who observed that, in a mixture of sugars and glycoproteins with 2,5-DHB, spectra of the sugars could be obtained from the amorphous central region of the target, whereas the glycoproteins gave spectra from the peripheral crystals.

Other isomers of DHB produce relatively poor signals from carbohydrates unless they contain an *ortho*-hydroxy group [12]. Krause *et al.* [14] have proposed that to be a matrix, the compound must be able to undergo an intramolecular hydrogen transfer between the phenolic group and the carboxyl group; only the *ortho*-substituted isomers are able to achieve this.

Several modifications to 2,5-DHB have been made with the aim of improving sensitivity or resolution and overcoming the problem of the large crystals. Thus, Karas *et al.* [15] have used small amounts of other substituted benzoic acids or related compounds as additives with the aim of causing disorder of the crystal lattice allowing "softer" desorption. The most effective additive was 10% of 2-hydroxy-5-methoxybenzoic acid, giving a mixture commonly referred to as "super DHB". An increase in sensitivity of 2-3 fold was reported for a standard dextran 1000 mixture, with a concomitant resolution increase being attributed to a reduction in metastable ion formation. Mohr *et al.* [9] have found that a mixture of 2,5-DHB and 1-hydroxyisoquinoline (HIQ) in the ratio of about 3:1 was very effective for carbohydrates. 1-Hydroxyisoquinoline itself is a poor matrix, but its presence in the 2,5-DHB causes much finer crystals to form than when DHB is used alone. This mixed matrix was found to be tolerant to the presence of a number of salts and additives and even to compounds such as sodium dodecylsulphate (SDS). Both of these binary matrices, super DHB and DHB/HIQ, subsequently have found considerable application in solving structural problems by MALDI analysis.

Metzger *et al.* [16] have used 3-aminoquinoline for ionization of plant inulins with masses of up to 6 KDa. A lower background and an increased resolution over that obtained with 2,5-DHB were reported. Papac *et al.* [17] report that this matrix sublimes too rapidly to be of great practical use. However, Stahl *et al.* [13] have reported that it is superior to 2,5-DHB for ionization of sialylated sugars, with a 5-10 fold increase in sensitivity being achieved. Mercaptobenzothiazoles have been examined by Xu *et al.* [18], and 5-chloro-2-

mercaptobenzithiazole (CMBT) found to be effective for carbohydrates. It was reported to be superior to DHB for analysis of high-mannose N-linked glycans in that it produced a better signal to noise ratio and somewhat better resolution.

Several β-carbolines such as harmane and harmaline have been found to be useful matrices for both cyclic (cyclodextrins in the 1000 Da region) and acyclic (MW 342-828) carbohydrates [19]. Unusually, they were reported to produce MH^+ rather than MNa^+ ions from cyclodextrins in the positive ion mode, and furthermore to give $[M-H]^-$ ions in the negative ion spectra. No negative ion signals are normally observed from neutral sugars with matrices such as 2,5-DHB. Hao *et al.* [20] have also observed negative ion spectra from cyclodextrins, but in their case the compounds were much larger (7500Da). Interestingly, the ions were produced from 2,5-DHB, an observation which led the authors to believe that the MALDI process for polysaccharides must differ somewhat from that for smaller sugars. In our hands, these β-carbolines give mainly MNa^+ ions from N-linked glycans in the positive ion mode, accompanied by a prominent ion caused by loss of water.

Osazones have also been reported to give spectra superior to those produced by 2,5-DHB [21]. In particular, resolution was improved and the signal to noise ratio was lower, leading to better detection limits (50 fmoles) for neutral sugars. Osazones synthesized from D- or L-arabinose with phenylhydrazine were found to be the best. Those from hexoses were too insoluble in water to be used with aqueous sugar solutions.

A major problem with all of these matrices is the inhomogeneous nature of the surface, which causes variations in signal strength depending on the position illuminated by the laser. Liquid matrices would overcome this problem but few single compounds have proved to be satisfactory. Nevertheless, several binary mixtures appear to be effective. Kolli and Orlando [22], for example, have used a mixture of 3-aminoquinoline and α-cyano-4-hydroxycinnamic acid to provide long-lasting signals from *N*-linked glycans. The matrix has the appearance of a viscous gum rather than a liquid but enables a signal to be recorded from a single spot for several minutes. This matrix was used in our experiments to record MALDI mass spectra with a magnetic sector instrument, noted above, and in fact was developed for this purpose, although the investigators recorded their spectra in a different way. A mixture of potassium hexacyanoferrate and glycerol, developed by Zöllner *et al.* [23], also appears to be effective.

α-Cyano-4-hydroxycinnamic acid itself will ionize N-linked glycans but with somewhat less efficiency than 2,5-DHB and with the production of many more fragment ions, particularly from sialylated glycans. Another "hot" matrix is 2-(*p*-hydroxyphenylazo)benzoic acid (HABA). Although efficient at ionizing glycans, it produces much fragmentation and also matrix-related ions up to about m/z 1000 [9], thus limiting its use for low mass glycans. Although α-cyano-4-hydroxycinnamic acid, 2,5-DHB and HABA are also effective for MALDI analysis of peptides and small proteins, the most useful protein matrix, sinapinic acid, is unsatisfactory for glycans [12].

Acidic Glycans

Acidic glycans, such as the sialylated *N*-linked carbohydrates, generally give poor MALDI spectra when ionized with matrices such as 2,5-DHB. Both positive (MNa^+) and negative ions ($[M - H]^-$) are produced and multiple peaks are present as the result of sodium and potassium salt formation. Fragmentation by loss of CO_2 and sialic acid can also be a problem. Papac *et al.* [17] have found that 6-aza-2-thiothymine is a reasonably satisfactory matrix for sialylated glycans; biantennary *N*-linked glycans were reported to give a detection limit of 50 fmole (signal to noise ratio = 6:1). However, fragmentation by loss of CO_2 was seen in linear mode from the $[M - H]^-$ ion, and by loss of sialic acid when spectra were recorded in reflectron mode.

Another matrix that has proved to be useful for sialylated glycans is a mixture of 2,4,6-trihydroxyacetophenone (THAP) and ammonium citrate (1 mM THAP in 1 ml of a 1:1 mixture of acetonitrile:20 mM ammonium citrate) [17]. This mixture gave a single peak from sialylated *N*-linked glycans at the 10 fmole level in linear mode with no evidence of fragmentation. However, in reflector mode loss of sialic acid was seen. Several other substituted acetophenones, such as 2,5-dihydroxyacetophenone, have been shown to be efficient matrices for neutral glycans. As with the substituted benzoic acids, only those compounds possessing an *ortho*-hydroxy group were effective. 2,5-Dihydroxyacetophenone, however, was found to be ineffective for sulphated carbohydrates [24]. Pitt and Gorman [25] have confirmed the utility of 2,6-dihydroxyacetophenone mixed with di-ammonium hydrogen citrate (10%, 1M), as a useful matrix for glycoproteins.

Carbohydrates containing uronic acids appear to be stable under MALDI conditions [26], with oligomers containing as many as 12 galacturonic residues giving strong signals in positive ion mode with 2,5-DHB as the matrix [27].

Sulphated glycans are particularly difficult to study by MALDI, partly because some sulphate linkages are labile. Signals are produced with most of the normal carbohydrate matrices but a number of other compounds have been found to be particularly useful for sulphated carbohydrates. Thus, Dai *et al.* [28] have found that the neutral matrix 7-amino-4-methyl coumarin gives good signals from monosulphated disaccharides and that this compound, mixed with 6-aza-2-thiothymine, can be used with tri- and tetra-saccharides, even when sialic acids are present. In our experience, the lability of the sulphate group is very dependent on its position of substitution; for example, sulphates attached to the 3-position of galactose moieties are labile whereas those linked at C-4 are stable.

Release of N-Linked Glycans From Glycoproteins

Although glycan profiles can be obtained from small glycoproteins by direct MALDI analysis, most glycoproteins are too large or heavily glycosylated for the glycoforms to be resolved. In addition, no information is available as to the site of attachment of the glycans. Above about 30 kDa, MALDI analysis can do little but detect the presence of glycosylation by the difference between the measured mass and that predicted by the amino-acid sequence of the protein or the mass as measured after digestion with endoglycosidase. Consequently, glycan analysis is most often performed either on glycopeptides obtained by digestion with a suitable enzyme such as trypsin, on the asparagine derivative of the glycan following complete digestion of the protein with pronase, or on the glycans themselves after they have been removed from the glycoprotein. The former method has the advantage that, by suitable choice of the enzyme, glycosylated sites can be isolated to individual glycopeptides and their glycans analyzed individually. However, analysis is more difficult than with the free glycan because it is essential that the mass of the peptide be accurately known so that the glycan mass can be found by difference. Many digests result in incomplete digestion and peptides may well carry additional posttranslational modifications. We recently encountered a situation in which a glycan mixture appeared to consist of a series of unfucosylated glycans together with their fucosylated analogues. However, accurate mass measurements indicated a mass difference of 1 Da between the two series of glycans, and they were later found to differ in the peptide portion, with one containing a di-peptide as the result of incomplete digestion. This example emphasizes the importance of making accurate mass measurements on all samples to minimize misinterpretation of data.

Release of N-linked glycans from glycoproteins is usually accomplished either chemically or enzymatically. Both N- [29, 30] and O-linked [30] glycans can be removed with hydrazine, a reagent which cleaves peptide bonds including that between the glycan and the asparagine. O-Linked glycans are specifically released at 60°C whereas 95°C is needed to release the N-linked sugars. Although all types of glycans can be released in this way, the method has several major disadvantages. Because all peptide bonds are destroyed, all information relating to the protein is lost. In addition, the acyl groups are cleaved from the *N*-acetylamino sugars and sialic acids, to be replaced chemically with acetyl groups, thus losing information as to their identity. Also, the reducing terminus of some of the glycans contains a hydrazide or amino group which, after acetylation, produces a mass increment equivalent to that between a hexose and an *N*-acetylaminohexose. Bendiak and Cumming [31] have examined the hydrazinolysis/reacetylation reaction in detail and have calculated that as much as 25% of the total glycans are converted into products bearing nitrogen-containing groups at the reducing terminus, and that these compounds can never be converted into the parent sugar. Peaks due to such compounds can be seen in Fig. 1 at 41 and 56 mass units higher than those due to the native glycans. Peaks appearing 42 mass units higher are the products of *O*-acetylation, present as by-products of the re-*N*-acetylation step. Finally, if the hydrazinolysis conditions are too vigorous, the *N*-acetylamino group can be removed from the reducing terminus or the reducing-terminal GlcNAc residue can be removed.

Several enzymes are available for releasing *N*-glycans. The most popular is peptide *N*-glycosidase F (PNGase-F) [32], which cleaves the intact glycan as the glycosylamine leaving aspartic acid in place of the asparagine at the N-linked site of the protein. The mass difference of one unit is readily detectable by mass spectrometry and provides information on the extent of occupancy at that site. The released glycosylamine readily hydrolyzes to the glycan except if the reaction is performed in ammonium-containing buffers [33] when a considerable amount of residual glycosylamine has been detected by MALDI. PNGase-F releases most glycans intact except those containing fucose α1-3 linked to the reducing-terminal GlcNAc [34]. In these situations, PNGase-A is effective. Endoglycosidase-H is more specific in releasing only high-mannose and hybrid glycans. By cleaving between the GlcNAc residues of the chitobiose core, it leaves the reducing-terminal GlcNAc residue, together with any attached groups, attached to the protein. Both hydrazine and enzymatic methods of glycan release preserve the reducing terminus of the glycan, which can subsequently be labelled with a fluorescent or other tag to facilitate detection by, for example, chromatographic systems.

We have developed a method whereby glycoproteins are separated or isolated on SDS-PAGE gels and the glycans are cleaved from the glycoprotein with PNGase-F within the gel [35] (Fig. 3). The glycoprotein is reduced and alkylated either before or after gel-separation to allow maximum release of the sugars, and incubated overnight with PNGase-F. The glycans are then extracted with water and cleaned by passage through a mixed bed resin column of AG-3 (removal of anions), AG-50 (removal of cations) and C_{18} (removal of organic compounds) packed into a gel-loader pipette tip. Samples are then examined directly by MALDI, or after fluorescent labelling, by HPLC. The AG-3 resin is essential for the production of MALDI signals but, unfortunately, it removes most acidic glycans. Consequently, if sialic or uronic acid-containing compounds are present in the sample, these must be methylated using the method developed by Powell and Harvey [36] prior to clean-up. The glycans can then be examined directly by MALDI or after fluorescent labelling, by HPLC. Sufficient material is available from 100 pmoles of glycoprotein for sequencing by exoglycosidase digestion as described below.

After glycan removal, the protein remains intact inside the gel where it can be cleaved with trypsin to produce peptides whose masses, measured by MALDI, can be used to identify the protein by database matching.

The most sensitive method for glycan release reported to date is that of Papac *et al.* [37]. MALDI spectra of sialylated N-linked glycans were obtained from as little as 100 ng of recombinant tissue plasminogen activator. The method relied on absorption of up to 50 μg of the glycoprotein onto the polyvinylidene difluoride (PVDF, Immobilon P) membrane of a 96-well MultiScreen IP plate (pore size 0.45 μm), where it was reduced and alkylated before the glycans were released with PNGase-F. Use of the multiScreen plate enabled the solvents and reagents to be removed rapidly by application of gentle suction and enabled the glycoproteins to be concentrated from relatively large volumes. Any free binding surface was blocked by treatment with polyvinylpyrrolidone (PVP) prior to incubation with PNGase-F (25 U/ml). Tris-acetate buffer (10 mM, pH 8.5) rather than the more usual sodium phosphate was used for the incubation to avoid

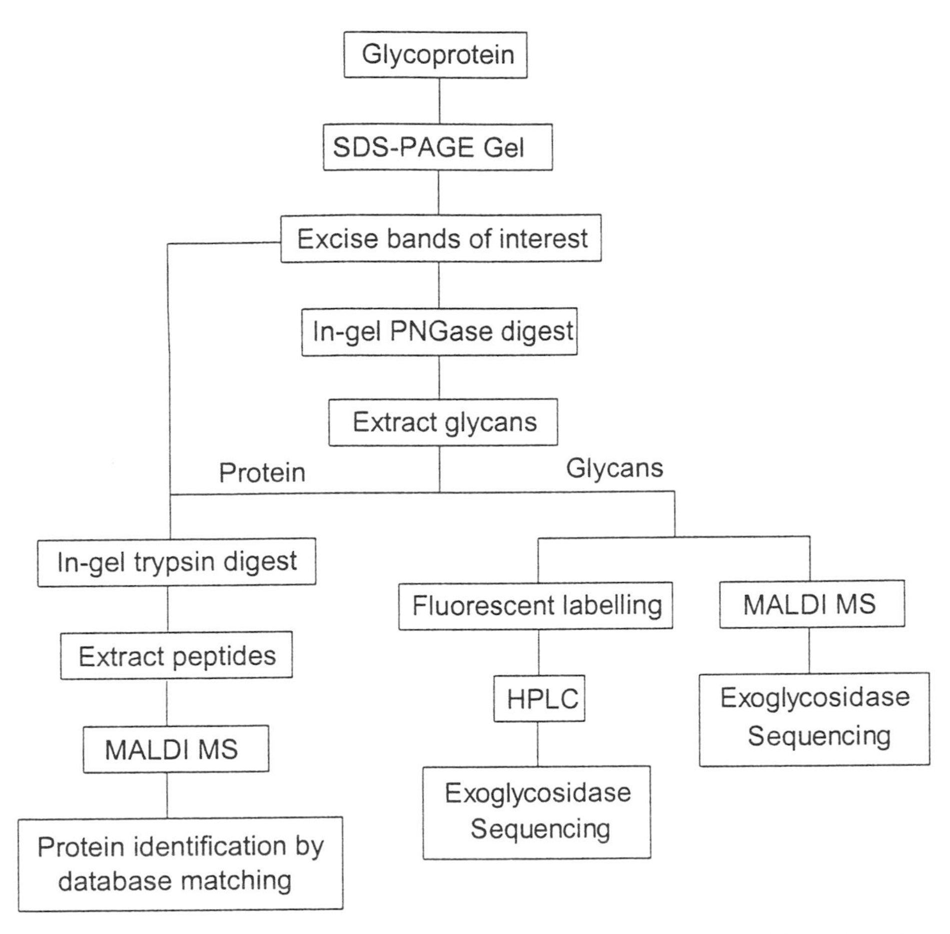

Fig. 3. Scheme for enzymatic N-linked glycan release from within an SDS-PAGE gel and subsequent analysis by MALDI mass spectrometry.

lactone formation from sialic acid residues [17]. Glycan release was complete in three hours. Following release, the glycans were incubated at room temperature for 3 hours with 150 mM acetic acid to ensure that any of the glycosylamine released by the enzyme was converted into the free sugar. Although this treatment did not cause any desialation, subsequent vacuum drying caused a 15% loss of sialic acid. Samples were then desalted with AG50W-X8 resin; this process was found to cause about 3% loss of sialic acid. MALDI analysis was performed with THAP/ammonium citrate (negative ion) or with "super DHB".

SAMPLE PREPARATION FOR MALDI ANALYSIS

Sample preparation is a very important aspect of obtaining good MALDI results from glycans extracted from biological mixtures. Many such mixtures contain only minute amounts of glycans but often relatively large amounts of other compounds such as buffers, salts, residual proteins, antioxidants and compounds of unknown origin. Many of these compounds will produce signals in the mass spectrum whereas others will suppress the ionization process. It is important, therefore, that as much of this extraneous material is removed before MALDI analysis or better, that it is not introduced in the first place. Simple but obvious precautions need to be taken, such as using apparatus of a size comparable to the amount of material being analyzed, and distilling solvents (including water) before use. It cannot be emphasized too strongly that decisions on how the sample is to be prepared be made at the start of a project to avoid presenting the mass spectrometry laboratory with an unnecessarily contaminated sample. During sample preparation, it is important to ensure that isolation and purification techniques do not cause fractionation or decomposition of the glycans with loss of quantitative information. Sialic acids, for example are often lost from glycoproteins if the pH becomes too low or the sample temperature is too high.

MALDI analysis of proteins and glycoproteins is less affected by the presence of contaminants than most other forms of mass spectrometry. The maximum tolerated amounts of common buffers, etc., for protein analysis have been investigated by Mock *et al.* [38]. Carbohydrates appear to be somewhat more susceptible to the presence of these contaminants, although the inclusion of small amounts of sodium, or other alkali metal salts, is essential for ionization. Thus, if present, these contaminating substances should be removed.

Carbohydrates may be desalted satisfactorily by drop dialysis on a membrane with a reasonably low molecular weight cut-off [39, 40]. We usually use either a 500 or a 100 Da cut-off membrane with dialysis times varying from 10 min to several hours. Börnsen *et al.* [41] have used a similar technique with a Nafion-117 membrane pretreated with strong acid. This membrane has the additional advantage of adsorbing peptides and proteins. We have used a combination of ion-exchange followed by treatment with a Nafion membrane for obtaining glycan profiles from mixtures released by automated hydrazinolysis. Well resolved glycan peaks were obtained after treatment whereas, before dialysis, only a broad hump was present in the spectrum (Fig. 4).

Other investigators have used a variety of techniques for sample preparation. For example, Rouse and Vath [42] have deposited the sample on the MALDI target, dried it with a stream of air and then added the matrix, super DHB together with beads of various resins to absorb the contaminants. After drying the sample, the resin beads were loosened and removed with a forced air stream.

Derivatization

In addition to the introduction of a chromophore or fluorophore, reducing-terminal derivatization, usually by reductive amination, has frequently been investigated as a method to increase the sensitivity of detection of glycans by introducing either a group with a high proton affinity or a constitutive charge.

Reductive Amination. Takao *et al.* [43] have prepared 4-aminobenzoic acid 2-(diethylamino)ethyl esters of the *N*-linked glycan, Man-8, maltoheptaose

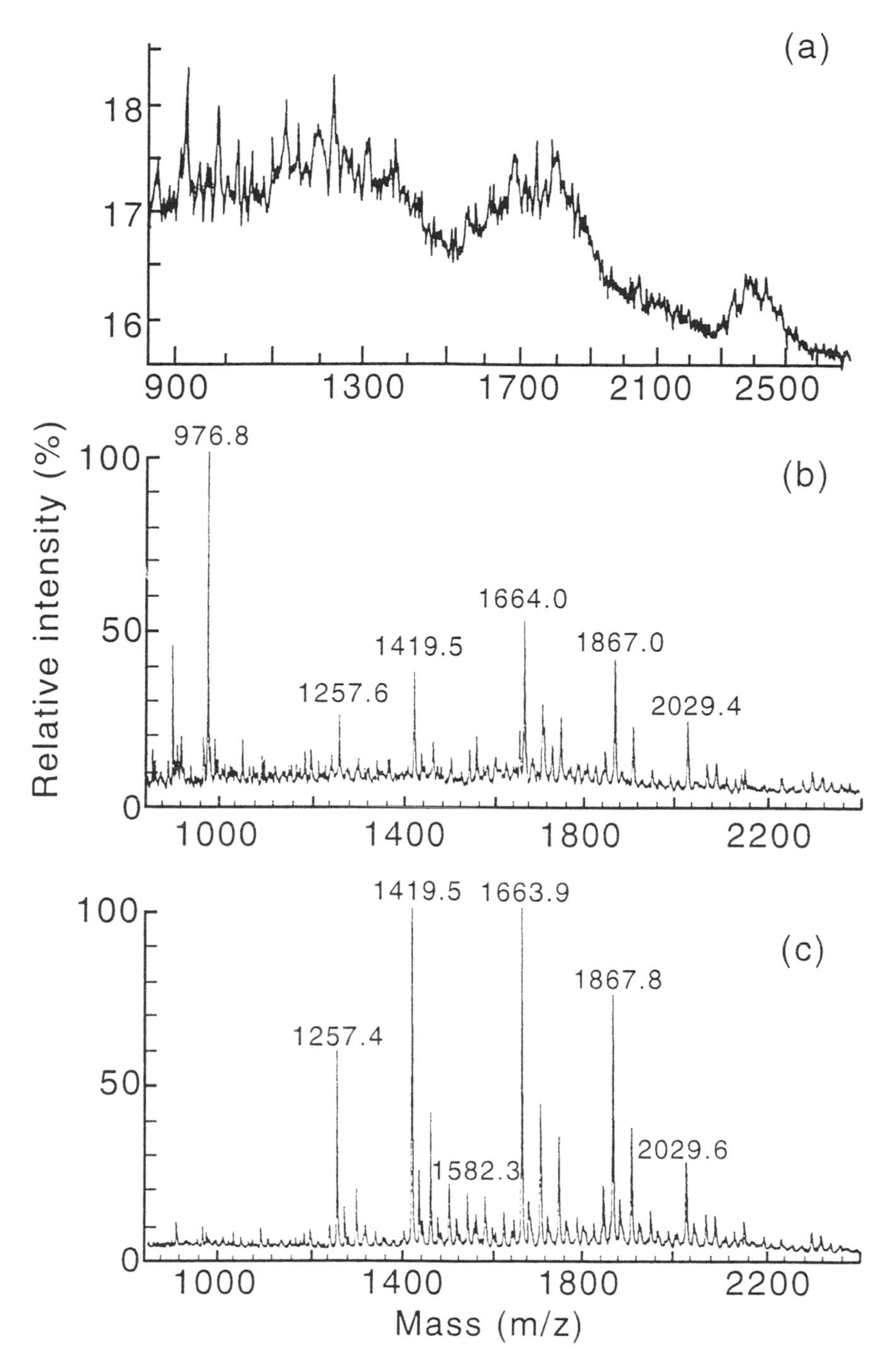

Fig. 4. a) MALDI mass spectrum (2,5-DHB) of N-linked glycans from ovalbumin (sample shown in Fig. 1 with contaminating glycoproteins removed) released by automated hydrazinolysis and examined without further purification. b) The same sample after desalting by mixed ion exchange chromatography. c) After further treatment on a Nafion-117 membrane.

and dextran by reductive amination and report sensitivity increases of 1000 fold over that of the free glycan following formation of a MH^+ rather than a MNa^+ ion. However, this increase in sensitivity does not appear to have been matched by other derivatives. Okamoto *et al.* [44] have compared the signals from 2-aminopyridine (2-AP), 4-aminobenzoic acid ethyl ester (ABEE) and trimethyl-(*p*-aminophenyl)amino (TMAPA) derivatives of maltopentaose and found a 100-fold, a 30-fold and a 10-fold increase in sensitivity, respectively, when spectra were recorded from DHB. These values are close to those found by other investigators for similar derivatives.

A well used reducing-terminal derivative for chromatography is that formed from 2-aminopyridine [45], but the derivative has not found extensive use for MALDI mass spectrometry. The 2-aminobenzamide (2-AB) derivatives are a later development and are popular for conferring fluorescence to the glycans for HPLC detection [46]. Unlike the 2-AP derivatives, they do not produce any appreciable MH^+ ions and do not appear to offer any advantage for MALDI detection, such as increasing sensitivity. Indeed, they are disadvantageous for many fragmentation studies as several important cross-ring fragmentations of the reducing-terminal sugar are suppressed. The related 2-aminobenzoic acid (2-AA) provides greater fluorescence than 2-AB, shows no specificity in its labelling of a wide range of structurally different glycans [47], and allows both neutral and acidic glycans to be observed as negative ions in the same spectrum. Another derivative that has been developed for HPLC separations and gives good MALDI spectra is that formed with 2-aminoacridone (AMAC) [48].

Oximes and Related Carbonyl Derivatives. Zhao *et al.*[49] used substituted-oxime formation to add a basic peptide residue in its aminooxyacetyl form to the reducing terminus of several neutral N-linked glycans, and reported sensitivity increases of between 50 and 1000 fold. Although this technique appears to work well with pure sugars, samples derived from biological sources invariably contain other carbonyl-containing compounds that appear to react preferentially with the reagent. We have investigated substituted hydrazones in an attempt to avoid the reduction step of reductive amination with its subsequent problems of reagent removal [50]. The reaction with Girard's T reagent was found to be particularly beneficial in producing reasonable increases in sensitivity because of constituent cationic charge carried by this reagent. However, the increase of 10-fold was lower than expected from results on the use of charged derivatives for glycan detection in FAB spectra, but were comparable to that found for the charged TMAPA derivative described above.

Other Derivatives. Reaction of reducing sugars with 1-phenyl-3-methyl-5-pyrazolone (PMP) under basic conditions leads to the formation of a di-substituted PMP derivative [51] which provides good chromatographic and detection properties. Pitt and Gorman [52] have found that this derivative is also useful for MALDI analysis of maltooligosaccharides and N-linked glycans. 2,5-DHB and α-cyano-4-hydroxycinnamic acid yielded MNa^+ ions, but 2,6-dihydroxyacetophenone in the presence of diammonium hydrogen citrate produced predominently MH^+ ions. The derivatives were claimed to be particularly useful for obtaining signals in the presence of contaminating material.

As an alternative approach to improving the MALDI spectra of sialic-acid-containing carbohydrates, we have converted the carboxylic acid group of the sialic acid into its methyl ester to produce a neutral sugar [36]. This was accomplished by first converting the acid into its sodium salt with an AG-50 ion-exchange resin in its sodium form, and then reacting the salt with methyl iodide in dry DMSO for two hours. The free acid can also be methylated but the reaction takes 48 hours to complete. Formation of the methyl ester improves the signal in four ways. First, it converts all of the ion current into the positive mode, thus avoiding the splitting of the signal between positive and negative ionization. Second, it enables the sialic acid-substituted glycans to be measured in the same spectrum as the neutral glycans with equivalent ionization efficiency, and thus quantitative relationship. Third, it prevents salt formation with the result that only one peak is produced. Fourth, it stabilizes the sialic acid and prevents loss by fragmentation. Permethylation, although conferring the same properties as methyl ester formation and somewhat increasing the sensitivity, is less useful for glycan profiling because of the formation of abundant in-source fragment ions, some of which are isobaric with neutral sugars.

GLYCAN PROFILING USING MALDI COUPLED WITH EXOGLYCOSIDASE DIGESTIONS

Use of exoglycosidase digestions to remove non-reducing terminal residues and monitoring of the products is one of the most important techniques in N-linked glycan analysis. Structural information is derived from the specificity of the enzyme and the number of residues removed under specific conditions. Monitoring of the products can be performed chromatographically or, more recently, by MALDI analysis. Although resolution of isomers is not achieved by MALDI, the general resolution of glycans (defined as the number of different species detected) is generally higher than can be achieved by HPLC. MALDI analysis of the products of exoglycosidase digestions was first demonstrated by Sutton *et al.* [53] for the glycoprotein, tissue inhibitor of metaloproteinase (TIMP), following cleavage with trypsin and examination of the tryptic peptides from α-cyano-4-hydroxycinnamic acid. We showed later in the same year that the technique could be used with mixtures of hydrazine-released glycans from glycoproteins such as immunoglobulin G [4] using 2,5-DHB as the matrix. In both of these approaches incubations were performed with single enzymes applied sequentially. An alternative approach is the array method whereby enzymes are applied as mixtures. Several parallel incubations are performed with a set of enzymes, with each incubation containing some enzymes of the set missing. Incubation proceeds to the point of the missing enzyme. Enzymes suitable for glycosidase sequencing are summarized in Table 2.

To handle the small amounts of glycans typically encountered in the examination of glycoproteins, several investigators have attempted to miniaturize this method. Our first approach involved performing enzyme digests directly on the MALDI target [54]. By working in low volumes, concentrations of the substrates approach the Michaelis constant (K_m) of the enzyme enabling reac-

Enzymes used for glycan sequencing

Enzyme	EC Number	Source	Specificity
α-D-Sialidase	3.2.1.18	*Arthrobacter ureafaciens*	NeuNAc/GCα2→6 > 3, 8
		Newcastle disease virus	NeuNAc/GCα2→3, 8
		Clostridium perfringes	NeuNAc/GCα2→3, 6, 8
		Vibrio cholerae	NeuNAc/GCα2→3, 6, 8
β-D-Galactosidase	3.2.1.23	Bovine testis	Galβ1→3 > 4 > 6
		Streptococcus pneumoniae	Galβ1→4
		Jack bean (*Canavalia ensiformis*)	Galβ1→6 > 4 >> 3
		Escherichia coli	Galβ1→4Glc
α-D galactosidase	3.2.1.22	Green coffee bean	Galα1→3, 4, 6
β-N-Acetyl-D-hexosaminidase	3.2.1.30	Jack bean (0.01 U/mL)	Glc(Gal)Nacβ1→2
		(10 U/mL)	Glc(Gal)Nacβ1→2, 3, 4, 6
α-N-Acetyl-D-hexosaminidase	3.2.1.49	Chicken liver	GalNAcα1→
α-D-Mannosidase	3.2.1.24	Jack bean	Manα1→2. 3. 6
		Aspergillus phoenicis I	Manα1→2
β-D-Mannosidase	3.2.1.25	*Helix pomatia*	Manβ1→4
		Achatina fulica	Manβ1→4
α-L-Fucosidase	3.2.1.51	*Charonia lampas*	Fucα1→2, 3, 4, 6
		Bovine epididimis	Fucα1→6 > 2, 3, 4
		Almond emulsin II	Fucα1→2
		Almond emulsin III	Fucα1→3, 4
β-D-Xylosidase		*Charonia lampas*	Xylβ1→2
Endo-β-D-galactosidase	3.2.1.103	*Escherichia freundii*	*R*Galb1→4GlcNAc

Table 2.

tions to proceed rapidly. In some cases, as little as 10 minutes was sufficient for complete digestion. As no reaction products were formed which could alter the pH of the incubation mixture, there was no need for strongly buffered systems, and thus 20 mM sodium acetate was used. Following incubation, the sample solution (1-3 µl) was transferred to a drop dialysis membrane (500 Da cut-off) for 10 min to remove buffer salts and then mixed with 2,5-DHB for MALDI analysis. The matrix was then removed by further drop-dialysis and the sample re-incubated with the next exoglycosidase. The procedure was repeated until sequencing was complete. Starting with 100 pmoles of glycan, it was possible to conduct three successive enzyme digestions before the sample became too weak to give a MALDI signal. It was estimated that about 25 pmoles of material was lost in each round of exoglycosidase digestion and measurement.

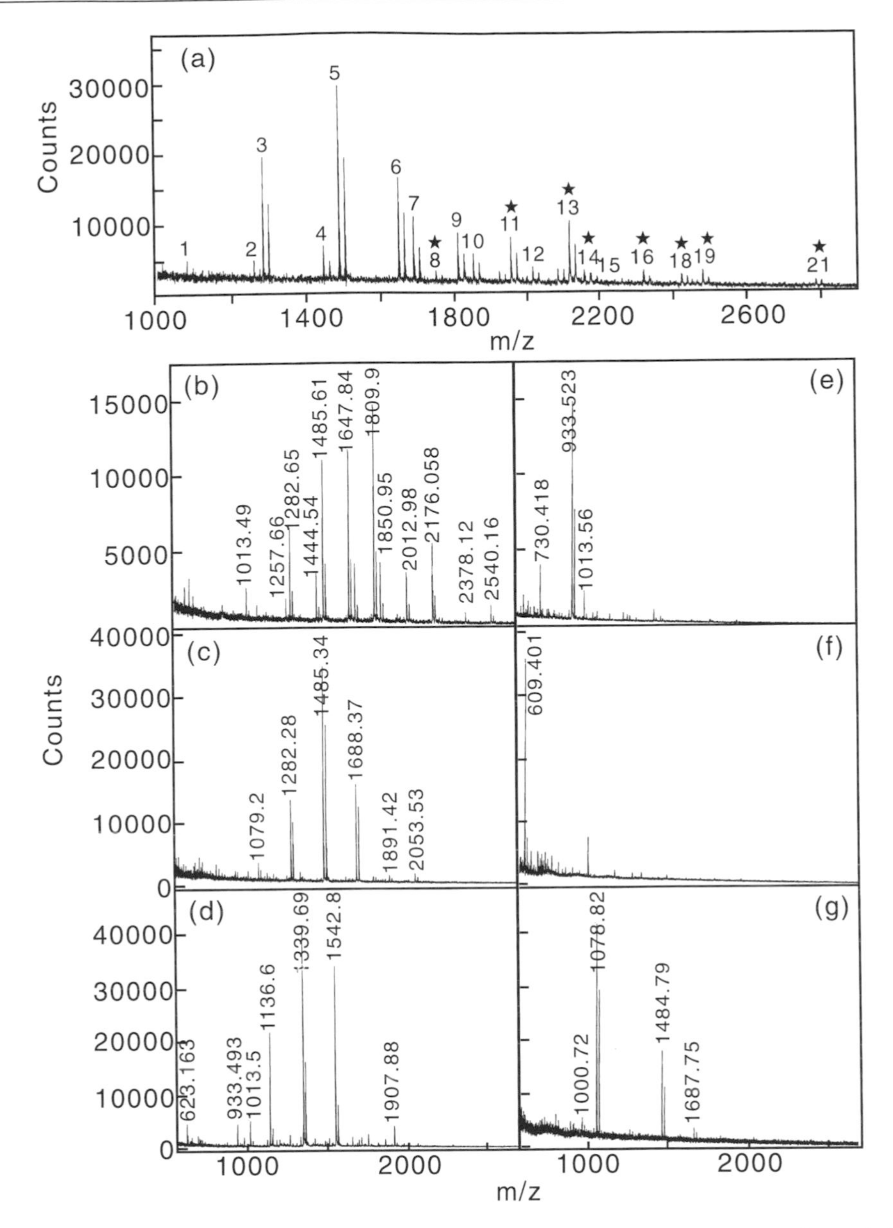

Fig. 5. (a) Positive ion MALDI mass spectrum (2,5-DHB) of glycans released from recombinant TIMP-1. Sialic acids are present as their methyl esters and peaks produced by these compounds are marked with asterisks. (b) MALDI profile obtained by incubation of unmethylated glycans with Arthrobacter ureafaciens *sialidase. (c) MALDI profile obtained by incubation of unmethylated glycans with sialidase plus* S. pneumoniae *β-galactosidase. (d) MALDI profile obtained by incubation of unmethylated glycans with sialidase, β-galactosidase and bovine epididimis fucosidase. (e) MALDI profile obtained by incubation of unmethylated glycans with sialidase, β-galactosidase, fucosidase and Jack bean N-acetylhexosaminidase. (f) MALDI profile obtained by incubation of unmethylated glycans with sialidase, β-galactosidase, fucosidase, N-acetylhexosaminidase and Jack bean α-mannosidase, g) MALDI profile obtained by incubation of unmethylated glycans with sialidase, β-galactosidase, N-acetylhexosaminidase.*

Mechref and Novotny [55] have developed a method which contains no extraction steps and in which all reactions are performed on the MALDI target with enzyme arrays and 10 mM phosphate buffer. PNGase-F was used to release the glycans on the target and array digestions were performed with *A. ureafaciens* neuraminidase, *D. pneumoniae* β-galactosidase, and *D. pneumoniae* β-D-N-acetylglucosaminidase. Incubations were conducted for three hours, after which time the glycans were examined by MALDI from arabinosazone.

We have used our in-gel release method, described above, followed by sequential or array digestion with exoglycosidases to examine a large number of glycoproteins including ovalbumin, gp-120 from the human immunodefficiency virus [35] and α1-acid glycoprotein from several species [56]. We have now examined N-linked glycans from recombinant TIMP-1 by this method and the results are shown below. TIMP is a heavily cross-linked 24-34 kDa glycopeptide with two N-linked glycosylation sites. Glycans are mainly bi- and tri-antennary compounds, some of which carry sialic acid [53]. The glycoprotein was run out on a 17.5% gel using a Mini PROTEIN II cell (Bio-Rad) at 200 V constant voltage in 25 mM Tris buffer, 192 mM glycine, 0.1% SDS, pH 8.5. The gel was stained with Coomassie blue, the glycoprotein band was removed and destained. After being reduced and alkylated, the glycans were released by incubation with PNGase-F overnight. Glycans were extracted with three portions of water (200 μL) with sonication for 30 mins, followed by a further extraction with acetonitrile (200 μL). The combined extracts were filtered through a Millipore syringe filter type FH, (0.45 μm). Because the glycans contained sialic acid, they were converted into their methyl esters [36] prior to MALDI analysis.

Figure 5a shows the MALDI profile of the intact glycans; sialylated sugars are indicated with an asterisk. Structural information was obtained by array digestion using the following enzymes and mixtures: 1) *Arthrobacter ureafaciens* sialidase, 2) sialidase plus *S. pneumoniae* β-galactosidase, 3) sialidase, β-galactosidase and bovine epididimis fucosidase, 4) sialidase, β-galactosidase, fucosidase and Jack bean *N*-acetylhexosaminidase, 5) sialidase, β-galactosidase, fucosidase, *N*-acetylhexosaminidase and Jack bean α-mannosidase and 6) sialidase, β-galactosidase and *S. pneumoniae* *N*-acetylhexosaminidase. The MALDI spectra obtained at each stage are shown in Fig.5 (b-g) and peaks are identified in Table 3. Array 6 was used to identify triantennary glycans with a branched 6-arm.

Quantitative Aspects

We have found that the signal strength of glycans produced by MALDI appears to reflect the amount of material on the target closely, providing that the correct matrix is chosen [12]. We have also found that 3-amino-4-hydroxybenzoic acid is a poor quantitative matrix for N-linked glycans as it appears to show a saturation effect at relatively low analyte concentrations. 2,5-DHB, on the other hand, does not show such an effect and produces a linear response from sample over several decades of concentration (Fig. 6).

Masses, compositions and exoglycosidase sequencing data for the N-linked oligosaccharides released from recombinant TIMP-1.

Peak[1]	Mass[2] Measured	Calc.	Composition[3]	Structure[4]	Array1[5]	Array2[6]	Array3[7]	Array4[8]	Array5[9]
1	1079.4	1079.4	H_3N_2F	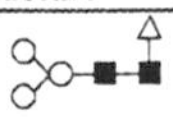	H_3N_2	H_3N_2	H_3N_2	H_3N_2	H_3N_2
2	1257.4	1257.4	H_5N_2	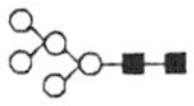	H_5N_2	H_5N_2	H_5N_2	H_5N_2	H_3N_2
3	1282.4	1282.4	H_3N_3F	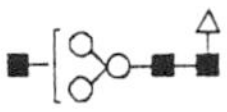	H_3N_3F	H_3N_3F	H_3N_3	H_3N_2	H_3N_2
4	1444.5	1444.5	H_4N_3F	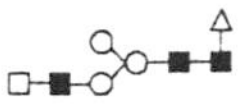	H_4N_3F	H_4N_3F	H_4N_3	H_3N_2	H_3N_2
5	1485.5	1485.5	H_3N_4F	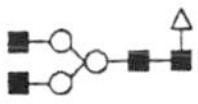	H_3N_4F	H_3N_4F	H_3N_4	H_3N_2	H_3N_2
6	1647.6	1647.6	H_4N_4F		H_4N_4F	H_3N_4F	H_3N_4	H_3N_2	H_3N_2
7	1688.6	1688.6	H_3N_5F	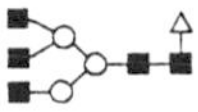	H_3N_5F	H_3N_5F	H_3N_5	H_3N_2	H_3N_2
8	1749.7	1749.6	H_3N_3FS	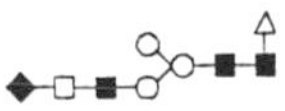	H_4N_3F	H_3N_3F	H_3N_2	H_3N_2	H_3N_2
9	1809.6	1809.7	H_5N_4F	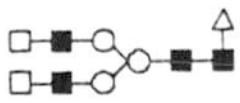	H_5N_4F	H_3N_3F	H_3N_3	H_3N_2	H_3N_2
10	1850.5	1850.7	H_4N_5F	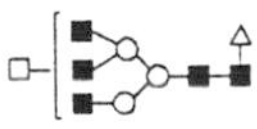	H_4N_5F	H_3N_5F	H_3N_5	H_3N_2	H_3N_2
11	1952.7	1952.7	H_4N_4FS	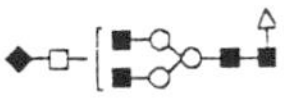	H_4N_4F	H_4N_3F	H_4N_3	H_3N_2	H_3N_2
12	2012.8	2012.7	H_5N_5F	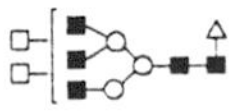	H_5N_5F	H_3N_5F	H_3N_5	H_3N_2	H_3N_2
13	2114.8	2114.8	H_5N_4FS		H_5N_4F	H_3N_4F	H_3N_4	H_3N_2	H_3N_2

Table 3.

Peak[1]	Mass[2] Measured	Calc.	Composition[3]	Structure[4] 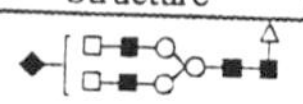	Array1[5]	Array2[6]	Array3[7]	Array4[8]	Array5[9]
14	2155.8	2155.8	H_4N_5FS	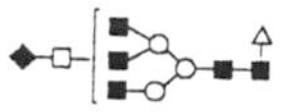	H_4N_5F	H_3N_5F	H_3N_5	H_3N_2	H_3N_2
15	2174.8	2174.8	H_6N_5F	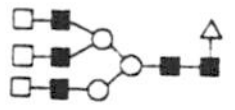	H_6N_5F	H_3N_5F	H_3N_5	H_3N_2	H_3N_2
16	2317.8	2317.8	H_5N_5FS	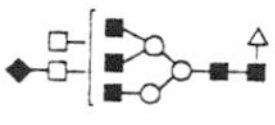	H_5N_5F	H_3N_5F	H_3N_5	H_3N_2	H_3N_2
17	2378.0	2377.9	-	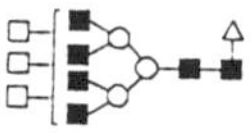	H_6N_6F	H_3N_6F	H_3N_6	H_3N_2	H_3N_2
18	2419.9	2419.8	$H_5N_4FS_2$	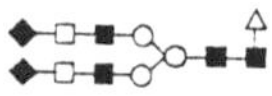	H_3N_4F	H_3N_4F	H_3N_4	H_3N_2	H_3N_2
19	2479.9	2479.9	H_6N_5FS	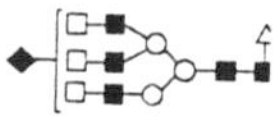	H_6N_5F	H_3N_5F	H_3N_5	H_3N_2	H_3N_2
20	2540.0	2540.0	-	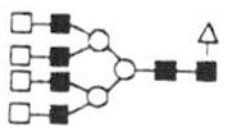	H_7N_6F	H_3N_6F	H_3N_6	H_3N_2	H_3N_2
21	2785.9	2785.0	$H_6N_5FS_2$	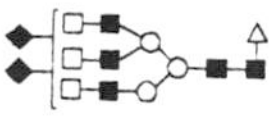	H_6N_5F	H_3N_5F	H_3N_6	H_3N_2	H_3N_2

1) From Fig. 5
2) Monoisotopic mass of the MNa^+ ion.
3) Composition [H = hexose (mannose or galactose), N = HexNAc (GlcNAc), F = deoxyhexose (fucose)]
4) Synbols: ■ = GlcNAc, □ = galactose, ○ = mannose, Δ = fucose, ◆ = Esterified sialic acid
5) *Arthrobacter ureafaciens* sialidase
6) Sialidase plus *S. pneumoniae* β-galactosidase
7) Sialidase, β-galactosidase and bovine epididimis fucosidase
8) Sialidase, β-galactosidase, fucosidase and Jack bean *N*-acetylhexosaminidase
9) Sialidase, β-galactosidase, fucosidase, *N*-acetylhexosaminidase and Jack bean α-mannosidase

Table 3 (cont.)

It is essential for quantitative work to compensate for the inhomogeniety of the target surface by averaging signals from several laser shots fired at multiple spots. We use five shots fired at each of 16 spots for each measurement for quantification on a Thermo Bioanalysis LaserMat instrument, and Bartsch

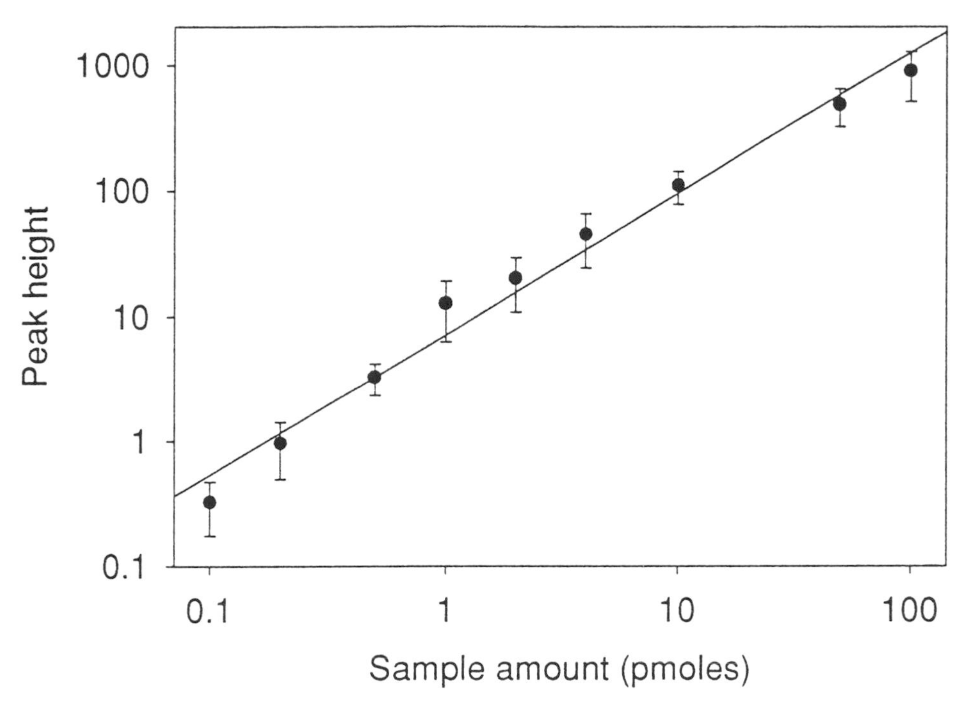

Fig. 6. Peak height of the MNa⁺ ion from different amounts of the biantennary N-linked glycan $(Gal)_2(Man)_3(GlcNAc)_4$ loaded onto the target. Each point is the accumulated signal from 5 shots fired at 16 laser spots ± standard deviation.

et al. [57], in a study with cyclodextrins, have used ten shots fired at ten spots. Such protocols usually generate standard deviations in the region of 6-10% from sugars and other compounds.

There appears to be little mass-dependent variation in ion abundance up to a mass of about 10,000 [8, 58], and for inulins in this mass range MALDI and HPAE chromatogram have been found to give similar profiles [58]. However, for larger polymer mixtures, there appears to be a fall in signal intensity with increasing mass. Garozzo *et al.* [59] have compared the MALDI responses given by β(1-2)-cyclic glycans with molecular weights from 2-4 kDa on three commercial instruments and found no significant difference. Furthermore, the profiles did not show any significant difference from that recorded from the same compounds by HPLC. Similarly, the profile of high mannose glycans released from ribonuclease B and examined by MALDI showed no significant differences from those obtained by capillary electrophoresis, gel filtration (P-4) and high performance anion exchange chromatography [60].

Several investigators [8, 61] have reported the fact that there appears to be no significant effect of structure on the efficiency of ionization of these glycans.

This is probably because the ionization process is similar for all compounds, unlike the situation with peptides where ionization depends on the proton affinity of the constituent amino acids. A study of the signal strength produced by equimolar amounts of 25 *N*-linked glycans [61] has shown no significant effect of structure on ion intensity, with the exception that the signal strength of glycans with masses below about 1 kDa fell progressively with decreasing molecular weight when recorded on a linear TOF instrument. This effect was attributed to transient saturation of the detector by the matrix ions.

FRAGMENTATION

Although few, if any, fragment ions are seen in spectra recorded with linear, continuous extraction, TOF mass spectrometers, a considerable amount of fragmentation does take place in the flight tube and can be recorded as post-source decay (PSD) fragments on TOF instruments fitted with reflectrons. Alternatively, fragmentation may be induced by collision. Under some circumstances, as described below, fragment ions may be formed in the ion source, in which case they are focused and observed along with the parent ions. The types of ions formed are very similar to those seen with other ionization techniques and consist predominantly of the products of glycosidic cleavages. Under higher energy conditions, a larger number of cross-ring fragments appear. These latter ions are particularly useful as they provide information on chain branching and linkage. The accepted nomenclature for describing these ions is that proposed by Domon and Costello [62]. Cleavages retaining the charge on the reducing terminus are termed A, B, or C depending on whether the cleavage is across a sugar ring, on the non-reducing side of the linking oxygen or on the reducing side, respectively. Cleavages producing the corresponding ions that retain the charge on the non-reducing terminus are termed X, Y and Z. Superscript numbers are used to denote the bonds broken in formation of the A and X cross-ring cleavage ions and subscripts are used for all ions to indicate the sugar residues involved.

In-Source (ISD) Fragmentation

We have observed ISD fragment ions on both magnetic sector or TOF instruments. These ions are the result of fragmentations that occur in time periods that are short enough to produce fragments before the ions leave the ion source. The ions are relatively weak when observed on magnetic sector instruments [63] and tend to be obscured by matrix ions below about *m/z* 600. However, relatively prominent ions resulting from losses of antennae linked to the 3-position are often produced, and there is usually a prominent B-cleavage due to loss of the reducing-terminal GlcNAc residue, with its attached fucose if present. Two prominent cross-ring fragments ($^{O,2}A$ and $^{3,5}A$) are also produced by cleavage of the reducing-terminal GlcNAc residue. On TOF instruments fitted with delayed extraction, the relative abundance of ISD fragments increases with the delay, reflecting the longer in-source residence time of the ions [64].

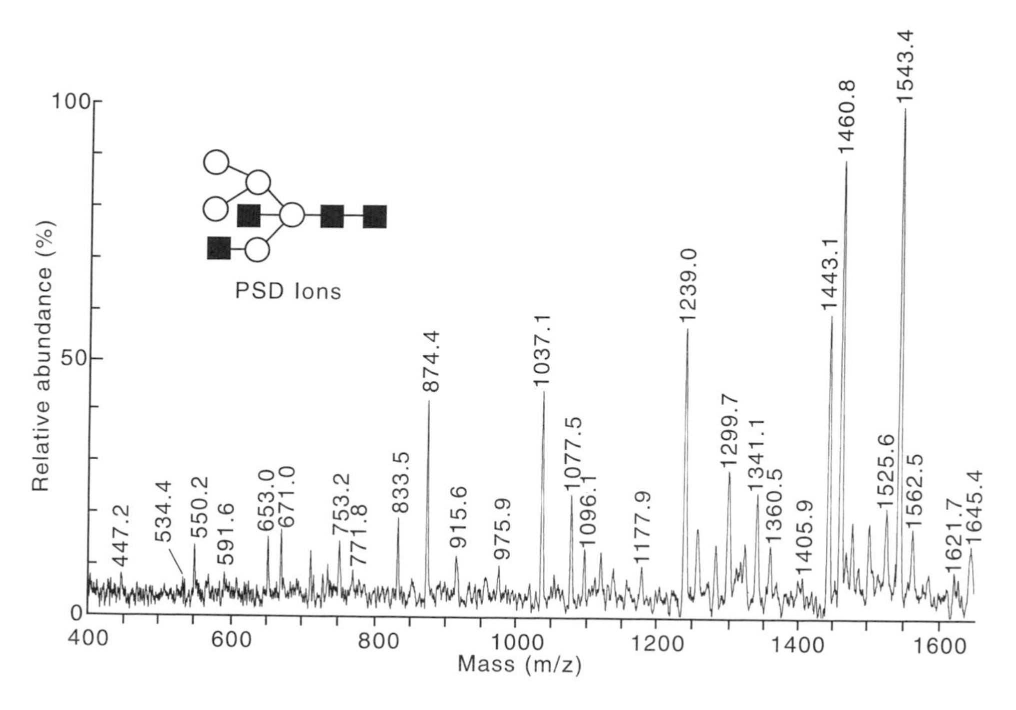

Fig. 7. PSD spectrum of the hybrid N-linked glycan $(Gal)_2(Man)_3(GlcNAc)_4$ released from chicken ovalbumin. PSD ions have been normalised on the largest fragment. Symbols for the structural formulae are as shown in Table. 1.

Post-Source Decay (PSD)

PSD spectra of N-linked glycans were first reported by Huberty *et al.* [2], who noted ready loss of sialic acid residues from biantennary glycans in a reflectron instrument. Spengler *et al.* [65], who made the first detailed study of the PSD spectra of N-linked glycans, found the spectra to be dominated by B and Y ions with only very weak ions produced by cross-ring cleavages. Internal fragments arose from several sites and confused spectral interpretation. Isomer differentiation could be made, however, by observing relative abundance differences between ions. Isomeric pentasaccharides, for example, differing only in the α1-3- or α1-6-linkage between two mannose units gave spectra in which the most significant difference was in the relative abundances of the C and Z ions formed by cleavage of the mannose-mannose bond. In the spectrum of the α1-6-linked isomer, the C-ion was the more abundant, whereas in the spectrum of the α1-3-isomer the position was reversed. Rouse *et al.* [66] have found that in the PSD spectra of N-linked glycans, ions formed by elimination of the 3-antenna tend to be more abundant than those formed by loss of the 6-arm. We have made similar observations and also found that an internal fragment produced by loss of the 3-antenna and the chitobiose core, termed *ion a* , is particularly useful in determining antenna composition. Figure 7 shows the PSD spectrum of a hybrid glycan from chicken ovalbumin (Peak 12, Fig. 1) and this ion can be seen at m/z 874.4. The presence of terminal mannose and GlcNAc residues in this ion is verified by losses of 162 mass units (hexose) and 203 mass units (GlcNAc) to give the ions at *m/z* 712.5, 550.2 and 671.0. A related ion containing an extra GlcNAc residue is present at *m/z* 1077.6 and appears to be formed by a loss of the 3-antenna together with the reducing-terminal GlcNAc residue. Evidence for its structure is provided by the loss of the mannose residues, as above, to give *m/z* 915.6 and 753.2. Other major ions are formed as follows: *m/z* 1543.4, cross-ring ^{0,2}A ion from the reducing terminal GlcNAc; *m/z* 1460.8, Y-type loss of GlcNAc, *m/z* 1443.1, B-type loss of the reducing terminal GlcNAc; *m/z* 1239.0, internal fragment resulting from a Y-type loss of a GlcNAc residue from *m/z* 1443.1; *m/z* 1037.1, loss of two GlcNAc residues from *m/z* 1443.1. Thus most of the major ions can be seen to be relatively undiagnostic internal fragment ions.

Our studies have also shown that PSD ions can often be observed in normal reflectron-TOF spectra as unfocused peaks appearing a few mass units higher than the focused (in-source) fragment ion (Fig. 8). These ion pairs are particularly prominent in the spectra of sialic acids as the result of the favourable loss of the sialic acid moiety. The relative abundance of these "metastable" ions decreases as the ion extraction delay is increased because there is more time for the fragmentation to occur in-, rather than post-source. Consequently, we have found that it is advantageous to record PSD spectra using short delays. The mass separation between the focused and unfocused ions increases with the delay time and varies depending on the instrument used to record the spectrum. Consequently, the measured mass of the metastable ion cannot be used directly to link a fragment with its parent as is the case with metastable ions recorded on magnetic sector instruments. Nevertheless, a formula, (Formula 1) containing a term "r" which reflects the relative path lengths of the instrument and delay time, has been developed and can be used to predict the metastable mass [67].

$$M_x = M_a \left[\frac{1 + \left(\frac{M_b}{M_a} \right) . r}{(1 + r)} \right]^2$$

Formula 1

Formula 1 may be rearranged to formulae to predict the parent (Formula 2) or fragment (Formula 3) ion masses.

$$M_a = \left(\frac{M_c}{2} \right) .(1 + r)^2 - r.M_b + (1 + r). \sqrt{\left(\frac{M_c}{2} \right)^2 .(1 + r)^2 - r.M_b.M_c}$$

Formula 2

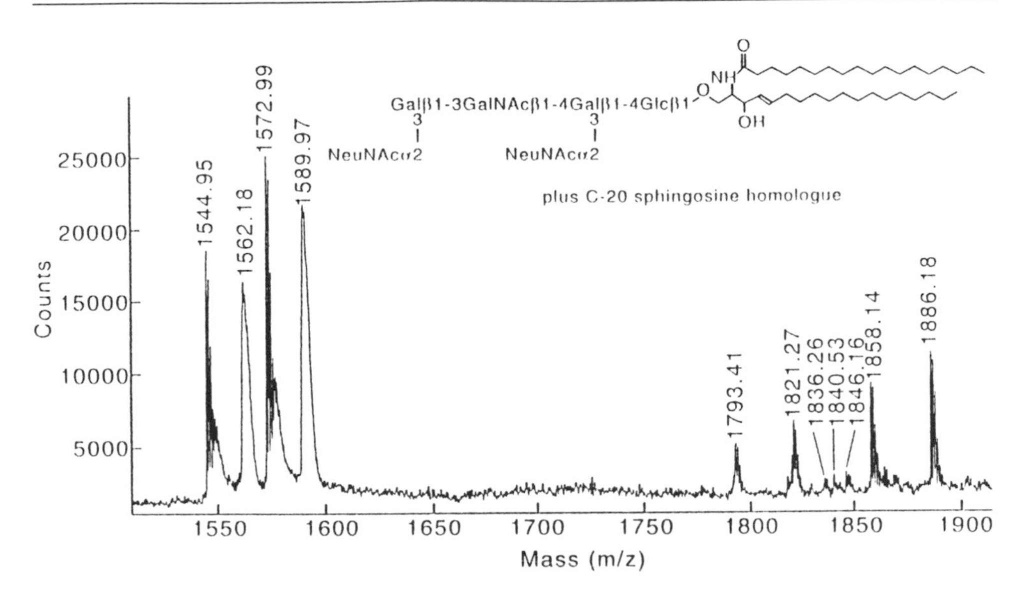

Fig. 8. Positive ion delayed-extraction reflectron-TOF MALDI mass spectrum of the ganglioside GD_{1a} showing parent, fragment and metastable ions.

$$M_b = \frac{1}{r}\sqrt{M_c . M_a . (1 + r)} - M_a$$

Formula 3

Where M_x = Metastable ion mass, M_a = Parent ion mass, M_b = Fragment ion mass, r = instrumental constant.

Use of the formulae is illustrated in the following example: the negative ion reflectron spectrum of the two homologous gangliosides of type GD_{1a} illustrated in Fig. 8, shows molecular ions at *m/z* 1858.1 and 1886.2 ($[M - H]^-$) due to the monosodium salts. Fragment ions caused by loss of one of the sialic acids plus sodium are present at *m/z* 1544.9 and 1573.0, and these are accompanied by metastable ions at *m/z* 1562.2 and 1590.0. Applying Formula 3 to the parent and metastable ion pair at *m/z* 1886.2 and 1590.0, with a value for r of 0.972, obtained from another spectrum, gives a mass for the fragment ion of 1572.9 (observed 1573.0). Formula 2 applied to the metastable and fragment ion pair at *m/z* 1562.2 and 1544.9, with the same value for r, gives a predicted parent ion mass of 1858.4 (observed, 1858.1). The precision for parent ion prediction is less than that for prediction of the fragment ion because of imprecision in the mass of the unfocused metastable ion and its mass separation from the parent. In Fig. 8, this mass approximates to the monoisotopic mass, but when this is not the case, better results are obtained if the spectrum is smoothed to give average masses for all ions. Formula 2 is particularly useful for spectra which do not

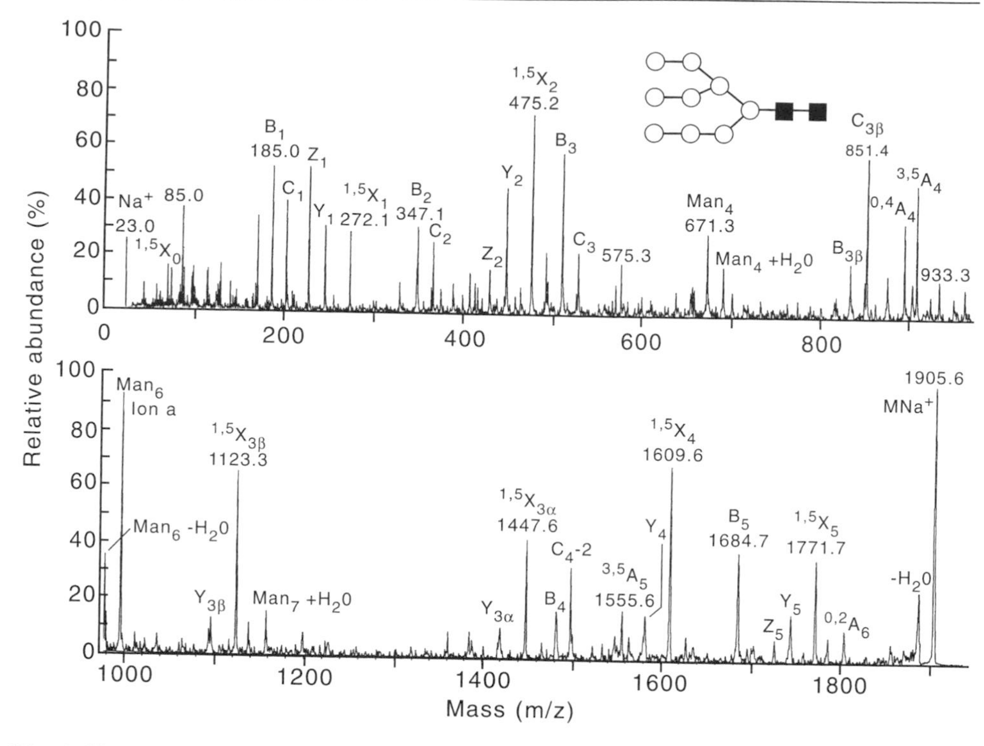

Fig. 9. High-energy CID spectrum of the high-mannose glycan (Man)$_9$(GlcNAc)$_2$ recorded on a magnetic sector mass spectrometer fitted with an orthogonal-TOF analyser. Symbols for the structural formulae are as shown in Table 1. Reproduced with permission from [68].

contain molecular ions, as the compound mass can be predicted from the masses of the fragment ions (focused and unfocused). An example is given in [67].

High-Energy Collisional-Induced Decomposition (CID)

The spectra described above are mainly the result of relatively low energy processes and carry limited linkage information about the glycan. More cross-ring cleavages can be induced by use of high-energy collisions using a collision cell fitted either to a TOF or a magnetic sector instrument. In our case, we have used a Micromass AutoSpec magnetic sector mass spectrometer fitted with a LSI nitrogen laser operated at 10 Hz, and an orthogonal TOF analyzer [68]. Samples (10-100 pmoles), in DHB were prepared as above. The molecular ions (MNa$^+$) were selected with MS1 and transmitted to the collision cell, which used xenon, and was operated to give a centre-of-mass collision energy of 800 eV. The push-out pulse for the orthogonal-TOF was synchronized with the laser pulse to give a delay equivalent to the flight-time from the ion source. Ions were accumulated by the orthogonal-TOF detector until a satisfactory signal to noise ratio was obtained. This process usually took about 20-30 min.

Spectra were characterized by a large number of abundant glycosidic and cross-ring fragment ions across the entire mass range (Fig. 9). Cross-ring fragments, particularly ^{1,5}X fragments, were particularly abundant and much more prominent than in the ISD or PSD spectra. The ^{1,5}X fragments could be used

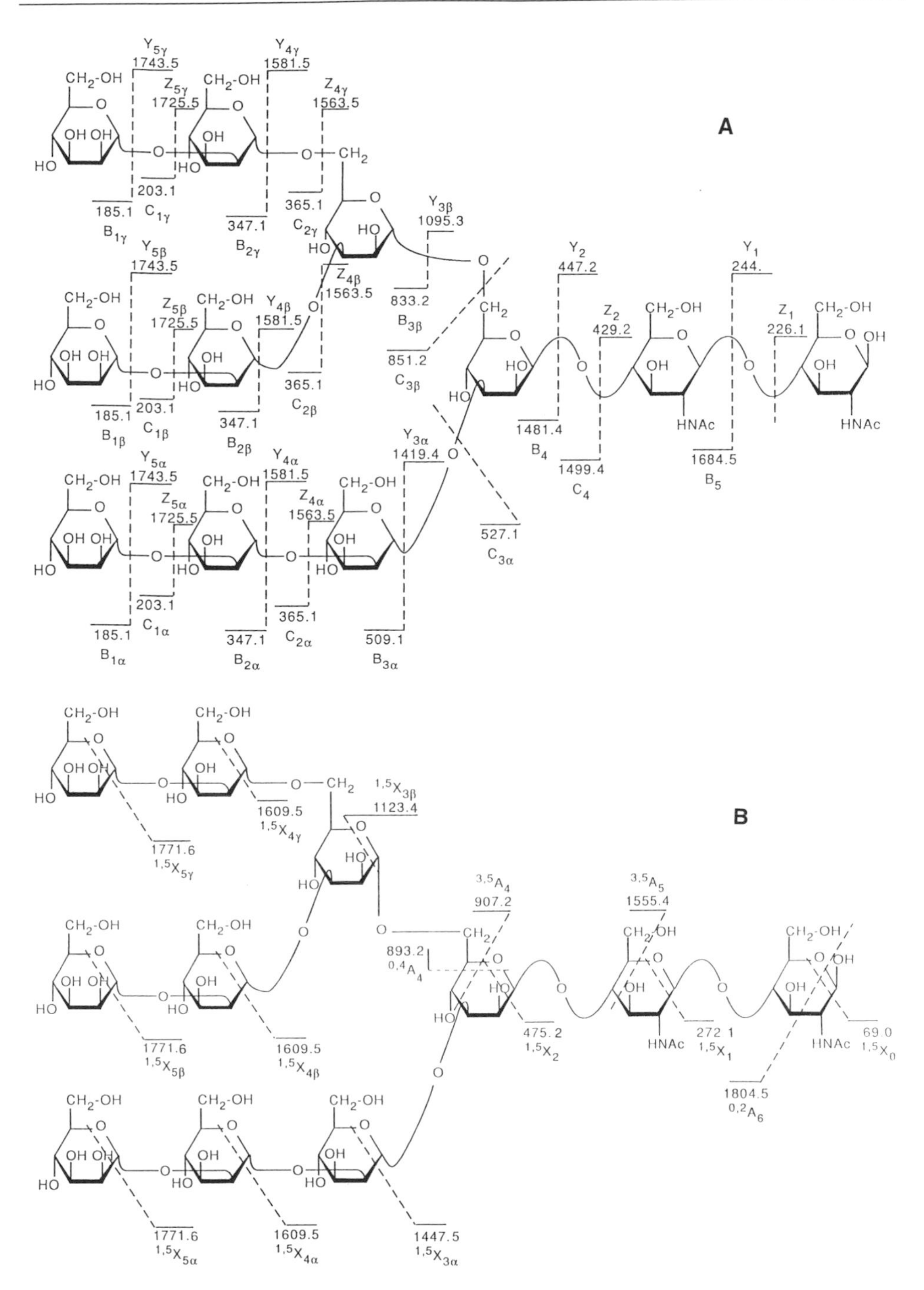

Fig. 10. Scheme showing the formation of the fragment ions in the spectrum shown in Fig. 9, top — glycosidic cleavages, bottom — cross-ring cleavages. Reproduced with permission from [68].

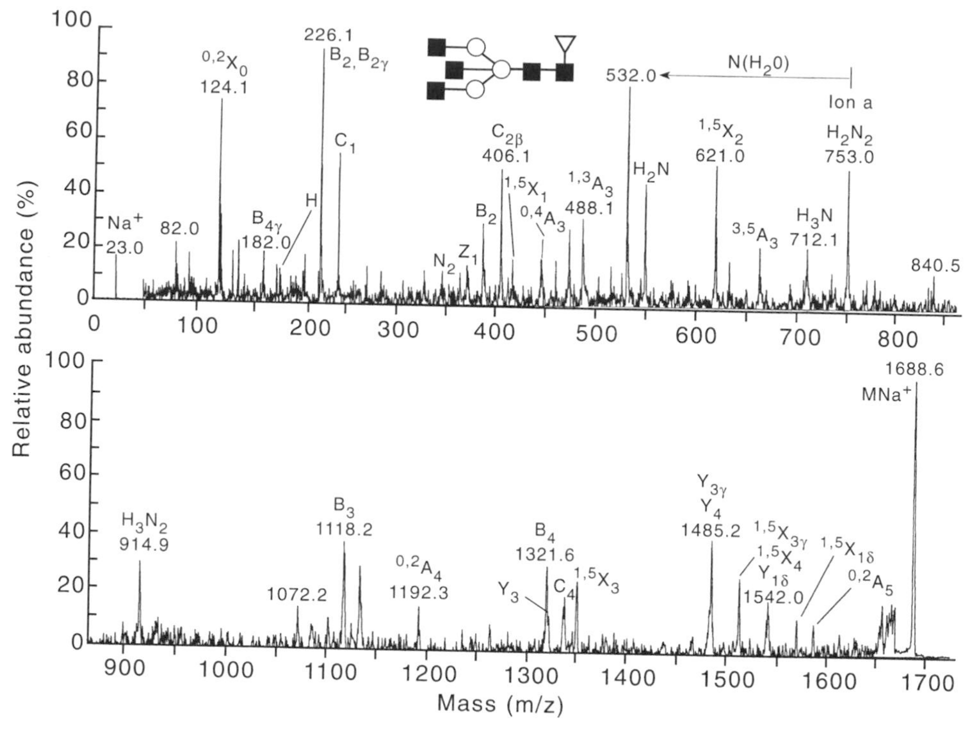

Fig. 11. High-energy CID spectrum of the complex glycan $(Man)_3(GlcNAc)_5(Fuc)_1$ recorded on a magnetic sector mass spectrometer fitted with an orthogonal-TOF analyser. Symbols for the structural formulae are as shown in Table. 1. Plus Δ = fucose. Reproduced with permission from [68].

to define the chain branching of high mannose glycans such as $(Man)_9(GlcNAc)_2$, whose high-energy CID spectrum is shown in Fig 9. Thus, $^{1,5}X$ fragments at each mannose residue resulted in losses of 1, 2, 3, 5 or 9 mannose units to give the ions at *m/z* 1771.7, 1609.6, 1447.6, 1123.3 and 475.2, respectively. Another prominent cross-ring fragment was the ion at m/z 907.2 representing a $^{3,5}A$ cleavage of the core mannose. This ion is particularly significant as it allows the distribution of the mannose residues between the two antennae to be determined. Ion *a* at m/z 995.4 was the most abundant ion in the spectrum and lost water to give the ion at m/z 977.4. The cleavages producing the most abundant ions in the spectrum are summarized in Fig. 10.

Spectra of the complex glycans contained similar structural features (Fig. 11). The presence of GlcNAc residues in the antennae produced abundant B-cleavages, particularly when they were substituted with galactose. Thus, the ion at m/z 338 was usually the most abundant in the spectra of neutral bi-, tri- and tetra-antennary glycans. Bisecting GlcNAc residues gave rise to a prominent loss of an intact GlcNAc molecule (221 mass units rather than a GlcNAc residue of 203 mass units) (Fig. 11). The presence of reducing terminal derivatives have a significant affect on the fragmentation [69]. Asparagine terminated glycans tend to produce spectra with more cross-ring cleavage ions than the free glycan, but 2-AB-labelled glycans give spectra with fewer diagnostic ions than the free sugars.

Conclusions

The above work shows that MALDI is now a valuable technique for the analysis of N-linked glycans and related compounds. It is generally more sensitive than FAB mass spectrometry and has the advantage that glycan profiles can be obtained from underivatized compounds. However, sensitivity is not as high as that obtained for peptides, probably as the result of the general absence of a proton acceptor site and the consequent formation of MNa^+ rather than MH^+ ions. Thus, it is unlikely that major increases in sensitivity will be found by the use of different matrices. Although the work cited above shows that some improvement can be obtained, the newer matrices have not provided any dramatic increases over those used in early studies. A more promising approach would appear to be the provision of a proton acceptor or constitutive charge in the form of a reducing-terminal derivative, as shown by a number of reported studies. Most of the reported derivatives are produced by reductive amination with the consequent necessity to remove salts and other reagents. The ideal derivative would be one that could be synthesized from volatile reagents, preferably without the reduction step, and which would provide good HPLC (fluorescence) as well as mass spectrometric detection.

Sample preparation is critical to obtaining good results with this technique but there is still much scope for improvement. As only a very small fraction of the sample currently applied to the MALDI target is consumed, it should be possible to examine much smaller amounts of sample than is normally attempted. To do this, better sample preparation techniques employing miniature apparatus and low-volume separation methods will have to be devised. Many current problems arise from the presence of compounds other than the target molecules in the sample. Avoiding the presence of these compounds becomes more difficult as the sample size is reduced, but it is essential that work continues in this area so that meaningful data can be obtained from smaller and more biologically relevant samples.

The resolution provided by modern MALDI instruments is vastly better than that achieved by the early TOF mass spectrometers. Even so, much greater resolution is still desirable for examination of glycoforms of intact glycoproteins. Site analysis, in terms of glycan profile and occupancy, can usually be achieved for individual sites by use of current instrumentation following protease digestion and examination of the resulting fragments. However, such experiments do not reveal how the specific glycoforms at one site affect glycosylation at other sites. This information can best be obtained through examination of the intact molecules with instruments providing high resolution. One obvious problem with the use of high resolution at high mass is the absence of the monoisotopic mass ion and ions containing low numbers of ^{12}C, etc. Nevertheless, if the protein sequence is known, the isotopic profile can be calculated, and thus meaningful mass data should be obtainable. In many experiments, only the mass difference between peaks and a knowledge of the protein sequence (and possibly other posttranslational modifications) would be needed to define glycosylation profiles. Thus the argument that it is not possible to measure the absolute mass directly becomes irrelevant.

Although the mass of a glycan can reveal its isobaric monosaccharide composition, it provides no information on other structural features, although an informed guess can be made with a knowledge of the compound's origin (N-linked, etc.). Isolated mass measurements can sometimes be misleading, however, as coincidences with the masses of other compounds can occur. In general, the more accurately the mass measurement is made, the less is the chance of misinterpreted data. Other checks on composition can be made such as observing the isotopic profile; because sugars contain a high percentage of oxygen, and thus a lower percentage of carbon than most other naturally occurring organic compounds, the ^{12}C-isotopic peak remains the most abundant at higher masses than with other compounds.

More detailed information requires the acquisition of fragment ion spectra or the intervention of other techniques. Exoglycosidase digestion is convenient and widely used but suffers from the unavailability of certain desirable enzymes and its dependence on a ready source of the currently available enzymes. Fragmentation provides much information on sequence and linkage, but spectra are often difficult to interpret because of the presence of "internal fragments" formed by losses from different parts of the molecule. The use of high energy collisions appears to produce more abundant linkage-revealing, cross-ring cleavages than the lower energy collisions provided by in- and post-source decay fragmentation, but no fragmentation method is yet able to provide information on the specific constituent monosaccharides. This information must still be obtained by hydrolysis and chromatography or GC/MS. Because of the complexity of N-linked carbohydrates, it is doubtful if any single-stage fragmentation method will be able to provide this information, even with specific forms of derivatization. However, multi-stage fragmentation methods such as those reported for ions produced by electrospray on ion-trap [70] or ICR instruments may provide at least some of this information. Implementation of such MS^n experiments following MALDI ionization will probably be the next step forward in the determination of glycan structure by mass spectrometry.

ACKNOWLEDGMENTS

We are grateful to Prof. E. M. Southern (Biochemistry, Oxford) and Amersham International for making the PerSeptive Voyager Elite mass spectrometer available to us. This work was supported, in part, by grants from the Biotechnology and Biological Science Research Council and from Deutscher Akademischer Austauschdienst grant D/94/14920 (to BK)

REFERENCES

1. K. K. Mock, M. Davy and J. S. Cottrell, *Biochem. Biophys. Res. Commun.* **1991**, *177*, 644-651.

2. M. C. Huberty, J. E. Vath, W. Yu and S. A. Martin, S. A. *Anal. Chem.* **1993**, *65*, 2791-2800.

3. R. S. Bordoli, K. Howes, R. G. Vickers, R. H. Bateman and D. J. Harvey, *Rapid Commun. Mass Spectrom.* **1994**, *8*, 585-589.

4. D. J. Harvey, P. M. Rudd, R. H. Bateman, R. S. Bordoli, K. Howes, J. B. Hoyes and R. G. Vickers, *Org. Mass Spectrom.* **1994**, *29*, 753-765.

5. P. M. Rudd, B. P. Morgan, M. R. Wormald, D. J. Harvey, C. W. Van den Berg, S. J. Davis, M. A, J. Ferguson and R. A. Dwek, *J. Biol. Chem.* **1997**, *272*, 2729-7244.

6. S. Y. Mir-Shekari, D. A. Ashford, D. J. Harvey, R. A. Dwek and I. T. Schulze, *J. Biol. Chem.,* **1997** *272*, 4027-4036.

7. M. L. Vestal, P. Juhasz and S. A. Martin, *Rapid Commun. Mass Spectrom.* **1995**, *9*, 1044-1050.

8. B. Stahl, M. Steup, M. Karas and F. Hillenkamp, *Anal. Chem.* **1991**, *63*, 1463-1466.

9. M. D. Mohr, K. O. Börnsen and H. M. Widmer, *Rapid Commun. Mass Spectrom.* **1995**, *9*, 809-814.

10. M. T. Cancilla, S. G. Penn, J. A. Carroll and C. B. Lebrilla, *J. Am. Chem. Soc.* **1996**, *118*, 6736-6745.

11. L. C. Ngoka, J.-F. Gal and C. B. Lebrilla, C. B. *Anal. Chem.* **1994**, *66*, 692-698.

12. D. J. Harvey, *Rapid Commun. Mass Spectrom.* **1993**, *7*, 614-619.

13. B. Stahl, T. Klabunde, H. Witzel, B. Krebs, M. Steup, M. Karas and F. Hillenkamp, *Eur. J. Biochem.* **1994**, *220*, 321-330.

14. J. Krause, M. Stoeckli and U. P. Schlunegger, *Rapid Commun. Mass Spectrom.* **1996**, *10*, 1927-1933.

15. M. Karas, H. Ehring, E. Nordhoff, B. Stahl, K. Strupat and F. Hillenkamp, *Org. Mass Spectrom.* **1993**, *28*, 1476-1481.

16. J. O. Metzger, R. Woisch, W. Tuszynski and R. Angermann, *Fresenius J. Anal. Chem.* **1994**, *349*, 473-474.

17. D. I. Papac, A. Wong and A. J. S. Jones, *Anal. Chem.* **1996**, *68*, 3215-3223.

18. N. Xu, Z.-H. Huang, J. T. Watsom and D. A. Gage, *J. Am. Soc. Mass Spectrom.* **1997**, *8*, 116-124.

19. H. Nonami, K. Tanaka, Y. Fukuyama and R. Erra-Balsells, *Rapid Commun. Mass Spectrom.* **1998**, *12*, 285-296.

20. C. Hao, X. Ma, S. Fang, Z. Liu, S. Liu, F. Song and J. Liu, *Rapid Commun. Mass Spectrom.* **1998**, *12*, 345-348.

21. P. Chen, A. G. Baker and M. V. Novotny, *Anal. Biochem.* **1997**, *244*, 144-151.

22. V. S. K. Kolli and R. Orlando, *Rapid Commun. Mass Spectrom.* **1996**, *10*, 923-926.

23. P. Zöllner, E. R. Schmid and G. Allmaier, *Rapid Commun. Mass Spectrom.* **1996**, *10*, 1278-1282.

24. H. Nonami, S. Fukui and R. Erra-Balsells, *J. Mass Spectrom.* **1997**, *32*, 287-296.

25. J. J. Pitt and J. J. Gormon, *Rapid Commun. Mass Spectrom.* **1996**, *10*, 1786-1788.

26. S. D. Simpson, D. A. Ashford, D. J. Harvey and D. J. Bowles, *Glycobiology* **1998**, *8*, 579-583.

27. P. J. H. Daas, P. W. Arisz, H. A. Schols, G. A. De Ruiter and A. G. J. Voragen, *Anal. Biochem.* **1998**, *257*, 195-202.

28. Y. Dai, R. M. Whittal, C. A. Bridges, Y. Isogai, O. Hindsgaul and L. Li, *Carbohydrate Res.* **1997**, *304*, 1-9.

29. S. Takasaki, T. Misuochi and A. Kobata, *Methods Enzymol.* **1982**, *83*, 263-268.

30. T. Patel, J. Bruce, A. Merry, C. Bigge, M. Wormald, A. Jaques and Parekh, R. *Biochemistry* **1993**, *32*, 679-693.

31. B. Bendiac and D. A. Cumming, *Carbohydrate Res.* **1985**, *144*, 1-12.

32. A. L Tarentino, C. M. Gómez and T. H. Plummer, Jr., *Biochemistry* **1995**, *24*, 4665-5671.

33. B. Küster and D. J. Harvey, *Glycobiology* **1997**, *7*, vii-ix.

34. V. Tretter, F. Altmann and L. März, *Eur. J. Biochem.* **1991**, *199*, 647-652.

35. B. Küster, S. F. Wheeler, A. P. Hunter, R. A. Dwek and D. J. Harvey, *Anal. Biochem.* **1997**, *250*, 82-101.

36. A. K. Powell and D. J. Harvey, *Rapid Commun. Mass Spectrom.* **1996**, *10*, 1027-1032.

37. D. I. Papac, J. B. Briggs, E. T. Chin and A. J. S. Jones, *Glycobiology* **1998**, *8*, 445-454.

38. K. K. Mock, C. W. Sutton and J. S. Cottrell, *Rapid Commun. Mass Spectrom.* **1992**, *6*, 233-238.

39. R. Marusyk and A. Sergeant, *Anal. Biochem.* **1980**, *105*, 403-404.

40. H, Görisch, *Anal. Biochem.* **1988**, *173*, 393-398.

41. K. O. Börnsen, M. D. Mohr and H. M. Widmer, *Rapid Commun. Mass Spectrom.* **1995**, *9*, 1031-1034.

42. J. C. Rouse and J. E. Vath, *Anal. Biochem.* **1996**, *238*, 82-92.

43. T. Takao, Y, Tambara, A. Nakamura, K.-I. Yoshino, H. Fukuda, M. Fukuda and Y. Shimonishi, *Rapid Commun. Mass Spectrom.* **1996**, *10*, 637-640.

44. M. Okamoto, K. Takahashi, T. Doi and Y. Takimoto, *Anal. Chem.* **1997**, *69*, 2919-2926.

45. S. Hase, T. Ibuki and T. Ikenaka, *J. Biochem. (Tokyo)* **1984**, *95*, 197-203.

46. J. C. Bigge, T. P. Patel, J. A. Bruce, P. N. Goulding, S. M. Charles and R. B. Parekh, *Anal. Biochem.* **1995**, *230*, 229-238.

47. K. R. Anumula and S. T. Dhume, *Glycobiology* **1998**, *8*, 685-694.

48. G. Okafo, L. Burrow, S. A. Carr and G. D. Roberts, *Anal. Chem.* **1996**, *68*, 4424-2230.

49. Y. Zhao, S. B. H. Kent and B. T. Chait, *Proc. Natl. Acad. Sci. USA* **1997**, *94*, 1629-1633.

50. T. J. P. Naven and D. J. Harvey, *Rapid Commun. Mass Spectrom.* **1996**, *10*, 829-834.

51. S. Honda, E. Akao, S. Suzuki, M. Okuda, K. Kakehi and J. Nakamura, *Anal. Biochem.* **1989,** *180*, 351-357.

52. J. J. Pitt and J. J. Gorman, *Anal. Biochem.* **1997,** *248*, 63-75.

53. C. W. Sutton, J. A. O'Neill and J. S. Cottrell, *Anal. Biochem.* **1994,** *218,* 34-46.

54. B. Küster, T. J. P. Naven and D. J. Harvey, *J. Mass Spectrom.* **1996a,** *31*, 1131-1140.

55. Y. Mechref and M. V. Novotny, *Anal. Chem.* **1998,** *70*, 455-463.

56. B. Küster, A. P. Hunter, S. F. Wheeler, R. A. Dwek and D. J. Harvey, *Electrophoresis*, in press.

57. H. Bartsch, W. A. König, M. Stra_ner and U. Hintze, *Carbohydrate Res.* **1996,** *286*, 41-53.

58. B. Stahl, A. Linos, M. Karas, F. Hillenkamp and M. Steup, *Anal. Biochem.* **1997,** *246*, 195-204.

59. D. Garozzo, E. Spina, L. Sturiale, G. Montaudo and R. Rizzo, *Rapid Commun. Mass Spectrom.* **1994,** *8*, 358-360.

60. P. M. Rudd,, I. G. Scragg, E. Coghill and R. A. Dwek, *Glycoconjugate J.* **1992,** *9*, 86-91.

61. T. J. P. Naven and D. J. Harvey, *Rapid Commun. Mass Spectrom.* **1996b,** *10*, 1361-1366.

62. B. Domon and C. E. Costello, *Glycoconjugate J.* **1988,** *5*, 397-509.

63. D. J. Harvey, T. J. P. Naven, B. Küster, R. H. Bateman, M. R. Green and G. Critchley, *Rapid. Commun. Mass Spectrom.* **1995,** *9*, 1556-1561.

64. T. J. P. Naven, D. J. Harvey, J. Brown and G. Critchley, *Rapid Commun. Mass Spectrom.* **1997,** *11*, 1681-1686.

65. B. Spengler, D. Kirsch, R. Kaufmann and J. Lemoine, *J. Mass Spectrom.* **1995,** *30*, 782-787.

66. J. C. Rouse, A.-M. Strang, W. Yu and J. E. Vath, *Anal. Biochem.* **1998,** *256*, 33-46.

67. D. J. Harvey, A. P. Hunter, R. H. Bateman, J. Brown and G. Critchley, *Int, J. Mass Spectrom. Ion Processes* **1998**, in press.

68. D. J. Harvey, R. H, Bateman, and M. R. Green, *J. Mass Spectrom.* **1997**, *32*, 167-187.

69. B. Küster, T. J. P. Naven and D. J. Harvey, *Rapid Commun. Mass Spectrom.* **1996**, *10*, 1645-1651.

70. A. S. Weiskopf, P. Vouros and D. J. Harvey, *Rapid Commun. Mass Spectrom.* **1997**, *11*, 1493-1504.

QUESTIONS AND ANSWERS

Brad Gibson (UCSF)

As you showed in your talk, the importance of de-salting highly anionic oligosaccharides is very important for obtaining decent signals under MALDI conditions. Do you know whether IR-lasers might generate less sodiated molecular ion species as suggested by Dr. R. Cramer's talk yesterday?

Answer. No, I am not aware of any work suggesting that less sodium salt is formed under these conditions. It would probably be more of a reflection of the composition of the sample.

Karl-Anders Karlsson (Gothenburg University)

The methyl ester you make of the sialic acid is rather labile; does that cause any problems, especially in the clean-up step 2?

Answer. No, these methyl esters appear to be stable under these conditions.

A. L. Burlingame (UCSF)

How much sample was required to obtain the AutoSpec CID spectra?

Answer. The spectra were obtained from about 100 pmoles of glycan applied to the target. The spectra were recorded at the Micromass factory and there was not time for an accurate determination of the detection limit. However, in one experiment 12 pmoles of a biantennary glycan was used and the spectra were found to have signal to noise ratio about the same as for the more concentrated samples. However, I should say that much more material is consumed in their experiments than with spectra recorded on time-of-flight instruments. On the AutoSpec, about 50% of the sample is normally consumed, whereas when spectra are recorded with time-of-flight instruments only a very small fraction of the glycan is used.

Electrophoretic and Mass Spectrometric Strategies for the Identification of Lipopolysaccharides and Immunodeterminants in Pathogenic Strains of *Haemophilus influenzae*; Application to Clinical Isolates

P. Thibault, J. Li*, A. Martin*, J. C. Richards*, D. W. Hood*[†] *and E. R. Moxon*[†]

*Institute for Biological Sciences, 100 Sussex Dr., Ottawa, Ontario, Canada, K1A 0R6; [†]Institute for Molecular Medicine, John Radcliffe Hospital, Headington, Oxford 0X3 3DU, UK

Abstract

The application of capillary electrophoresis coupled to electrospray mass spectrometry (CE-ES-MS) for the analysis of complex lipopolysaccharides (LPS) is presented. Electrophoretic conditions conducive to both negative and positive ion detection were developed, and facilitated the separation of closely related glycoforms and isoforms from O-deacylated LPS of different strains of *Haemophilus influenzae*. To aid the identification of specific functionalities and immunodeterminants of LPS such as pyrophosphoethanolamine, phosphocholine and N-acetyl neuraminic acid, a mixed scan function was used to promote the in-source formation of selected fragment ions under high orifice voltage conditions, while enabling the detection of multiply-charged ions using low orifice voltage. By using such scanning functions, trace levels of O-deacylated LPS containing a single phosphocholine group were detected at an estimated level of 5% of the overall LPS population. The sensitivity and specificity of the mixed scan function also facilitated the identification of trace levels of sialylated LPS from isolates originally obtained from otitis media. The detection of positive ions from anionic O-deacylated LPS was made possible in CE-ES-MS experiments using ammonium acetate buffers, thereby enabling the structural characterization of oligosaccharide branching by on-line tandem mass spectrometry using a quadrupole/time-of-flight instrument. The combination of high resolution with high sensitivity mass spectrometric detection provided an efficient analytical tool for probing the subtle structural changes occurring in the diverse population of LPS from *H. influenzae*.

Introduction

The Gram-negative bacterium *Haemophilus influenzae* comprises both capsular and non-typable (acapsular) strains of significant virulence and pathogenicity to humans. Capsular strains (mainly serotype b strains) are associated with invasive infections of the blood stream including septicemia, meningitis and pneumonia [1, 2] whereas non-typable strains of *H. influenzae* are primary pathogens in otitis media and respiratory tract infections [2, 3]. While the exact mechanisms of colonization and invasion are still poorly understood, a number of reports have documented evidence suggesting that lipopolysaccharides (LPS) are a major virulence determinant of this organism [4-6], and can associate with mucus and damaged epithelium of the human nasopharyngeal tissue [7].

As for other Gram-negative pathogens, the LPS of *H. influenzae* is an outer component of the bacterial cell wall, and consists of two parts of different properties: a hydrophilic carbohydrate component containing acidic (3-deoxy-D-manno-2-octulosonic acid, KDO) and neutral residues (glucose, galactose, heptose) bonded to a hydrophobic lipid A component composed of a glucosamine disaccharide to which are attached O- and N-linked fatty acids [8]. However, the LPS structure of *H. influenzae* contrasts with that of other Gram-negative bacteria by the lack of O-repeating antigens characteristic of many enteric pathogens. A structural model has been previously proposed for *H. influenzae* LPS in which a conserved heptose-containing inner core tri-saccharide is attached to a single KDO 4-phosphate residue [9]. Figure 1 shows the conserved inner core region based on this earlier model [9] and additional data obtained from this laboratory. Chain elongation can take place at each heptose (Hep) within this triad, thus conferring significant molecular diversity to the outer core oligosaccharide. Structural variability can also be observed in the extent to which functional groups such as phosphates (P), pyrophosphates (PP), phosphoethanolamine (PE) and phosphocholine (PC) are appended. In some cases, the presence of specific residues or functional groups has been related to the virulence and invasiveness of particular strains. For example, the PC substitution was found to be phase variable, and its occurrence was paralleled with the persistence of *H. influenzae* in the human respiratory tract [10].

Structural studies conducted on *H. influenzae* serotype b strains have indicated that LPS from this pathogen displays a heterogeneous population of complex glycans showing both intra- and interstrain variability [11-13]. This structural variability is partly accounted for by molecular mechanisms generat-

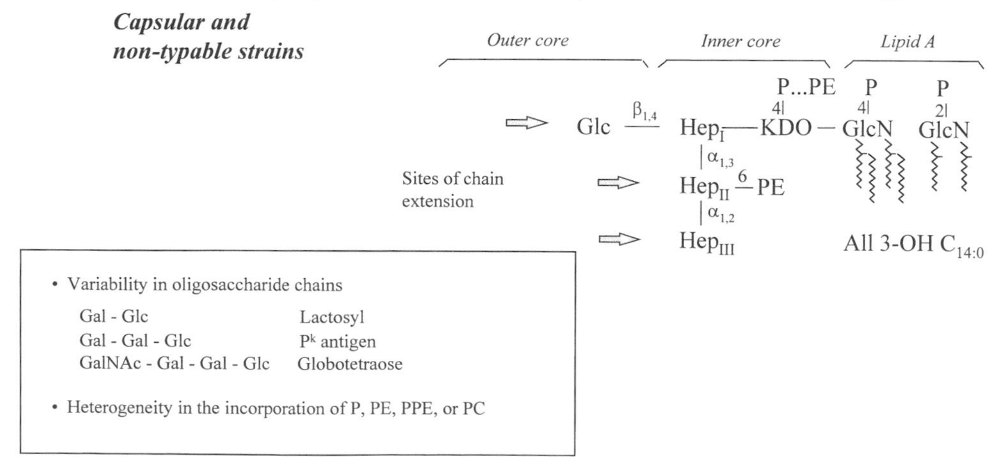

Fig. 1. Structure of the conserved core oligosaccharide from H. influenzae, *adapted from [11].*

ing phase variation of terminal epitopes identified in the genes *lic 1A*, and *lic 2A* [14-17]. The *lic 2A* gene contains multiple copies of the tetramer 5'-CAAT-3', and the number of these repeats can vary as a result of slipped-strand mispairing. Loss or gain of a single repeat moves upstream initiation codons in and out of frame with the remainder of the gene, thus creating a translational switch that affects the expression of the terminal α-Gal (1-4)-β-Gal disaccharide epitope [13-15]. The implications of these structural features in relation to function are that the antigenic variation of the terminal saccharides of *H. influenzae* LPS provide an important mechanism whereby *H. influenzae* can adapt to changing environmental conditions encountered during the pathogenic cycle of infection.

The availability of the complete 1.83 Mb pair sequence of the *H. influenzae* strain Rd genome [18] has had a significant impact on the further understanding of LPS biosynthesis. Construction of mutant strains and characterization of expressed LPS products using specific monoclonal antibodies, band fingerprinting on gel electrophoresis and electrospray mass spectrometry (ES-MS) enabled the identification of key LPS genes. By searching the *H. influenzae* database, with sequences of known LPS biosynthetic genes from other organisms, more than 20 candidate genes were identified in strain Rd [19].

To probe the subtle changes in LPS structure as a result of phase variation or as a consequence of site-directed mutation, it becomes important to develop sensitive analytical tools which provide the required level of specificity and selectivity. To this end ES-MS has played a pivotal role in the characterization of these complex carbohydrates. Applications of ES-MS have been demonstrated for the structural characterization of O-deacylated LPS from *Haemophilus*, *Neisseria*, and *Salmonella* strains [9, 20]. Most mass spectrometric investigations have been performed directly on samples purified by either gel-filtration or ion exchange chromatography. However, the lack of on-line separation techniques often precludes the determination of sample heterogeneity associated with substituents attached to the oligosaccharide backbone (i. e., phosphates, phosphoethanolamines, acetyl groups, and substituted carbohydrate chains). Recent investigations from this laboratory have focused on the development of electrophoretic separation techniques compatible with ES-MS, and their application to the analysis of LPS derived from *Moraxella catarrhalis* [21], *Pseudomonas aeruginosa* serotype O6 [22], and for the characterization of O-polysaccharide chains from *Yersinia ruckerii* [23]. The coupling of capillary electrophoresis to electrospray mass spectrometry (CE-ES-MS) has provided unparalleled resolution and identification of glycoform populations and substituted phosphorylated functionalities present in LPS. In an effort to enhance the sensitivity of this technique for trace level analysis of O-deacylated LPS, recent studies have presented adsorption-preconcentration approaches for on-line enrichment of glycolipids prior to CE-ES-MS [24]. In combination with enzymatic releasing methods using DNase and proteinase K, this technique provided excellent sensitivity and enabled the identification of LPS surface antigens from as little as five bacterial colonies of *H. influenzae* strain Eagan [24].

The present study outlines the analytical potentials of CE-ES-MS to separate closely related glycoform families in O-deacylated LPS from different strains of *H. influenzae* based on their characteristic band patterns observed in the contour profile of m/z *vs.* time. This type of data representation not only provided structural relationship between families of glycolipids of extended chain length, but also facilitated the visualization of LPS isoforms differing by the location of phosphorylated groups or hexose residues. The structural characterization of these isoforms was facilitated using on-line tandem mass spectrometry on a hybrid quadrupole/time-of-flight mass spectrometer (Q-TOF). Specific scanning functions were also developed to identify important immunodeterminants such as PC and N-acetyl neuraminic acid (Neu5Ac) present at trace levels in extracts of O-deacylated LPS.

EXPERIMENTAL

Chemicals and Materials

Fused-silica capillaries were purchased from Polymicro Technologies (Phoenix, AZ). The enzymes Proteinase K, deoxyribonuclease I (DNase) and ribonuclease (RNase) were obtained from Sigma Chemical Co. (St. Louis, MO). Methanol, and isopropyl alcohol were from Em Science (Gibbstown, NJ). The hydrazine reagent, morpholine and other buffers were obtained from Aldrich Chemical Company Inc. (Milwaukee, WI).

Bacterial Strains and Growth Conditions

H. influenzae RM 7004 strain is an encapsulated bacterium and was originally isolated from the cerebrospinal fluid of children with meningitis. A mutant strain referred to as RM 7004 *lic 2B* was produced by site-directed mutagenesis, and was shown to be deficient in the chain extension of the outer core [25] as detected from immunoblotting experiments using LPS-specific monoclonal antibodies [15]. *H. influenzae* 375 strain was originally isolated as part of a Finnish study of otitis media during a pneumococcal vaccine trial. These strains were cultured in Brain Heart Infusion (BHI) broth (Oxoid) supplemented with 10 μg/mL haemin (Sigma) and 2 μg/mL NAD (Sigma). Solid BHI medium was prepared by the addition of Bacto agar (1% w/v) and Levinthal base [26]. The *H. influenzae* strain 319 PC^- is an encapsulated strain Eagan, and was obtained from Dr. J. N. Weiser (U. of Pennsylvania), and the RM 118 isolate is an acapsular strain of *H. influenzae* Rd^-, a derivative of a type d strain. Culture of these strains was achieved at 37°C on chocolate agar plates (Quelab, Montreal, Canada) according to conditions described previously [11].

Preparation and Extraction of LPS

Small scale extraction of LPS was conducted by washing the colonies from the plates and dispersing them in 1.5 mL tubes each containing 500 μL of deionized water. The cells were freeze-dried overnight , dissolved in 360 μL of deionized water, and a 40 μL aliquot of a 25 μg/mL solution of proteinase K was added to each vial. The suspended cell solutions were incubated at 37°C for 90 min and the digestion was stopped by raising the temperature to 65°C for 10 min. The solutions were allowed to cool at room temperature and were subsequently freeze-dried. The cells were further digested by incubating them at 37°C for 4 hours in a 200 µL solution of 20 mM ammonium acetate buffer pH 7.5 containing DNase (10 µg/mL) and RNase (5µg/mL). The released LPS were dialysed against water and freeze-dried overnight.

Preparation of O-deacylated LPS

The freeze-dried and digested cells containing the free LPS were dissolved in 200 µL of hydrazine and incubated at 37°C for 50 min with constant stirring to release O-linked fatty acids [27]. The reaction mixtures were cooled (0°C), the hydrazine destroyed by addition of cold acetone (600 µL), and the final product was obtained by centrifugation. The pellets were washed with 2 x 600 µL of acetone, dissolved in 100 µL of water and 400 µL acetone, centrifuged and then lyophilized from water.

CE-ES-MS and CE-ES-MS-MS analyses

A Crystal model 310 CE instrument (ATI Unicam Boston, MA) was coupled to the mass spectrometer via a co-axial sheath-flow interface [21, 24, 28]. Mass spectral analyses were conducted using either a API 300 triple quadrupole mass spectrometer (Perkin Elmer/Sciex, Concord, Canada) for analyses involving orifice stepping functions (see below) or a Q-TOF (Micromass, Manchester, U.K.) hybrid quadrupole/time-of-flight instrument [29] for high resolution and on-line tandem mass spectrometric experiments. A sheath solution (70:30, isopropanol:methanol) was delivered at a flow rate of 1.5 µL/min to a low dead volume tee (250 µm i. d., Chromatographic Specialties Inc., Brockville, Canada). All aqueous solutions were filtered through a 0.45 µm filter (Millipore, Bedford, MA) before use. An electrospray stainless steel needle (27 gauge) was butted against the low dead volume tee and enabled the delivery of the sheath solution to the end of the capillary column. Separations were obtained on 90 cm length bare fused silica using 30 mM aqueous morpholine/acetate pH 9, containing 5% methanol for negative ion detection and 30 mM aqueous ammonium acetate pH 8.5, containing 5% methanol for positive ion detection. A voltage of 25 kV was typically applied at the injection end of the capillary. The outlet of the fused-silica capillary (185 µm o. d. x 50 µm i. d.) was tapered to 75 µm o. d. (20 µm i. d.).

To probe specific functionalities and residues such as PPE, PC and Neu5Ac, a mixed scan function was developed in which the high/low orifice voltage stepping was incorporated as one of the mass spectral acquisition parameters. For negative ion detection this was achieved by acquiring the fragment ions m/z 220 (PPE) and 290 (Neu5Ac) under selected ion monitoring (SIM) at an orifice voltage of 120 V (150 msec duration), together with the full mass scan (m/z 600-1500) at a lower orifice voltage of 50 V (3.5 s duration). Similarly, positive ion detection of PC and Neu5Ac was obtained by recording the fragment ions m/z 328 (Hex-PC) and m/z 292 (Neu5Ac) under SIM mode (orifice voltage: 120 V, 150 msec duration), together with the full mass scan (orifice voltage: 50 V, 3.5 s duration). A Macintosh Power PC 900 computer was used for instrument control, data acquisition, and data processing.

All combined CE-MS-MS analyses were conducted using the Micromass Q-TOF, and separation conditions were identical to those stated above. Precursor ions identified in a preliminary survey scan were selected by the first quadrupole while a pusher electrode was pulsed (≈16 kHz frequency) to transfer fragment ions formed in the R.F.-only hexapole cell to the time-of-flight analyzer. The detector is a dual-stage microchannel plate, and acquisition is made through a time-to-digital convertor (TDC) operating at 1 GHz (Precision Instruments, Knoxville, TN). Mass spectral resolution was typically 4,000-5,000. A scan duration of 1 s and 2 s was set for conventional and MS-MS mass spectral acquisition, respectively. Collisional activation was performed using argon collision gas at an energy (laboratory frame of reference) of typically 75 eV for negative ion mode and 55 eV for positive ion mode. Data were acquired and processed in the Mass Lynx Windows NT based data system. The calculation of oligosaccharide molecular masses was facilitated using the computer program "Gretta's carbos" developed by W. Hines (UCSF, San Francisco, CA)

RESULTS AND DISCUSSION

Profiling the Microheterogeneity of LPS Glycoforms Using CE-ES-MS

Direct mass spectrometric analysis of intact LPS is often impractical, in part due to the hydrophobic nature of the sample, its heterogeneity, and the difficulty in finding buffers compatible with the operation of ES-MS. The lack of solubility of the LPS in aqueous buffers also prevents its analysis by combined CE-ES-MS. Traditionally, the analysis of O-deacylated glycolipids arising from mild hydrazinolysis of intact LPS has been one of the preferred chemical methods for obtaining hydrophilic products that are amenable to direct flow injection. Mild hydrazinolysis can be used on small amounts of material to quantitatively remove all O-linked acyl moieties, although structural information on ester linked substituents (i. e., O-acetyl groups) is lost as a result of this reaction. Alternatively, the characterization of the fully-acylated lipid A and the core oligosaccharide can be achieved by analyzing the products released from mild acid hydrolysis of the intact LPS. This latter procedure yields a hydrophilic core oligosaccharide often containing ionic functionalities such as phosphate groups and/or acidic KDO residues imparting a net negative charge to the molecule, and thus favoring its analysis by combined CE-ES-MS.

Earlier investigations from this laboratory [24] have focused on the development of electrophoretic conditions conducive to the analysis of O-deacylated LPS from *H. influenzae* strain Eagan by CE-ES-MS. Separations conducted using aqueous morpholine buffer generally gave enhanced resolution of the different LPS glycoforms compared to ammonium acetate or ammonium formate buffers of similar concentrations. The morpholine buffer also provided lower electrophoretic currents, and low intensity of adduct ions in negative ion

ES-MS. However, significant peak broadening was noted on individual ion electropherograms, possibly suggesting the adsorption of the analytes on the capillary surface. Evidence of this was also noted in experiments involving the injection of serial dilution of the same LPS preparation, where poor linearity and analyte response were observed at concentrations below 0.1 mg/mL. To overcome this problem, an organic modifier was added to the separation buffer. An aqueous solution of 30 mM morpholine containing 5% (v:v) methanol provided the best analytical performance with limits of detection in the order of 10 μg/mL (on-column injection of 300 pg) of total O-deacylated LPS. Further improvement in sensitivity can be achieved for O-deacylated LPS using on-line adsorption preconcentrator, and a 40-fold enhancement of concentration detection limits compared to that observed in CE-ES-MS was reported recently [24].

The difficulty in obtaining proper resolution of each glycoform present in the diverse population of O-deacylated LPS products also arises from the instrumental limitations of the scanning mass spectrometer. The combination of a high-resolution separatory technique with a scanning mass spectrometer often requires some compromises in terms of mass range, step increments, and dwell times to attain adequate detection limits. Obviously, wide mass scan acquisition in CE-ES-MS experiments (as typical for profiling the glycoform distribution of LPS) imposes considerable demands on the mass spectrometer detection capabilities, and duty cycles of typically 3-4 s/scan are often used on quadrupole

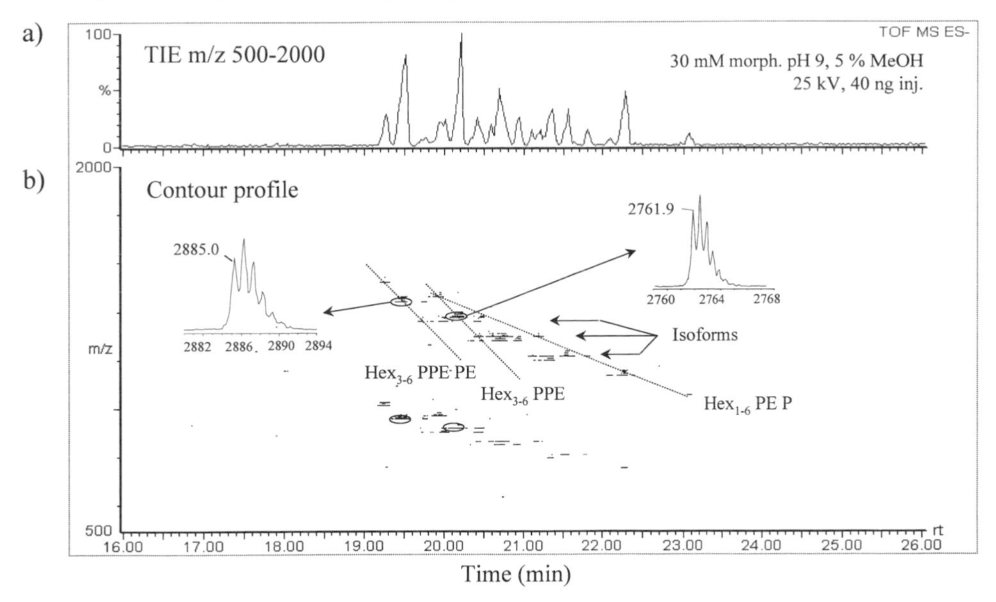

Fig. 2. CE-ES-MS (- ion mode) analysis of O-deacylated LPS from H. influenzae *strain 7004 lic 2B. (a) Total ion electropherogram (m/z 500-2000), (b) Contour profile of m/z vs. time. Dotted lines indicate family of closely related glycoforms generally composed of sequential addition of Hex residues whereas isoforms are isomeric glycolipids corresponding to substitution of functional group. Separation conditions: Bare fused silica (90 cm x 50 μm i.d.), 30 mM morpholine, pH 9.0, 5% methanol, + 25 kV, 30 ng of total glycolipid injected.*

instruments. However, such limitations are significantly reduced with the availability of TOF analyzers (and to some extent ion trap mass spectrometers), which not only offer faster acquisition time (<1 s/scan) and resolution ($M/\Delta M$ in excess of 5,000) but also enable the collection of wide mass range with excellent sensitivity.

An example of CE-ES-MS separation conducted on a Q-TOF instrument is shown in Fig. 2 for the analysis of O-deacylated LPS from a mutant strain of the clinical isolate RM 7004, where site-directed mutagenesis was targeted toward the expression of the *lic 2B* gene. The total ion electropherogram (TIE) corresponding to the full scan analysis (m/z 500-2000) is presented on the top panel of Fig. 2, along with the contour intensity plots as a function of m/z *vs.* time in the bottom panel. The contour profile shows a series of doubly- and triply-deprotonated molecules from which the molecular mass of the different analytes can be calculated. One of the most prominent glycolipids, observed at 20.2 min, displayed $[M-3H]^{3-}$ and $[M-2H]^{2-}$ ions at m/z 919.6 and m/z 1379.9, respectively. The reconstructed molecular mass profile for this peak is shown as an inset in Fig. 2b, with an observed molecular mass (M_{obs}) of 2761.8 Da for the monoisotopic ^{12}C component. Based on this mass measurement and on previous structural assignments [11], this glycolipid was assigned to a structure composed of PPE Hex_5 Hep_3 KDO $GlcN_2$ with two N-linked 3-OH myristic acid ($C_{14}H_{27}O_2$) and two phosphate groups ($M_{calc.}$: 2761.9 Da). It is noteworthy that an alternate structural candidate of the same molecular mass would be a O-deacylated LPS with P and PE groups rather than a single PPE functionality (see later).

The contour representation provides a valuable analytical tool to the analyst because it offers the possibility of identifying closely related families of glycolipids based on the appearance of diagonal lines. This phenomenon arises from the fact that progressive extension of Hex residues on the core structure results in a regular increase of molecular mass and a concurrent decrease in electrophoretic mobility. For example, the O-deacylated LPS, having a molecular mass of 2761.8 Da, is a member of a glycoform family extending from Hex_3 to Hex_6, all of which comprise a single PPE group presumably on the KDO residue. Similarly, a second set of glycoform having the same glycan distribution but with an additional PE group is observed in Fig. 2b as a parallel diagonal line shifted to earlier migration time. The lower mobility observed for this set of glycoforms is consistent with an increase in molecular mass, and the parallel slope substantiates that no change in the net charge of the O-deacylated LPS took place as a result of the PE addition. Interestingly, another glycan series of higher mobility is also indicated in Fig. 2b. The molecular mass of this set of glycoforms is identical to the Hex_{3-6} PPE family except that this glycolipid series is placed on a diagonal of shallower slope. This change in migration pattern suggests an increase in the effective charge of the corresponding O-deacylated LPS, possibly arising from the incorporation of an additional ionizable group. Based on earlier structural assignments on *H. influenzae* strain Eagan [11], this observation could be explained by the presence of a phospho-KDO, and the remaining PE group would be located on the second Hep residue. It is not clear at present if this

Time (min)	$M_{calc.}$ (Da)[1]	$M_{obs.}$ (Da)[2]	Assignment[3]			
			P	PE	PPE	Hex
19.3	3047.0	3047.0	-	1	1	6
19.3	3086.1	3086.3	1	1	-	7
19.5	2885.0	2885.0	-	1	1	5
19.5	2924.0	2924.0	-	-	1	6
19.7	2722.9	2723.0	-	1	1	4
19.8	2885.0	2885.0	1	2	-	5
19.9	2924.0	2924.0	1	1	-	6
20.0, 20.4*	2722.9	2722.8	1	2	-	4
20.2	2761.9	2761.9	-	-	1	5
20.3, 20.4*	2560.9	2560.9	1	2	-	3
20.4, 20.6*	2599.9	2599.5	-	-	1	4
20.5	2761.9	2762.0	1	1	-	5
20.7, 20.9*	2560.9	2560.8	-	1	1	3
20.7, 21.3*	2599.9	2599.8	1	1	-	4
21.2, 21.4*	2437.8	2437.8	-	-	1	3
21.4	2398.8	2399.6	-	1	1	2
21.6,21.8*	2437.8	2437.8	1	1	-	3
22.1	2275.8	2275.6	-	-	1	2
22.3, 22.4*	2275.8	2275.8	1	1	-	2
23.1	2113.7	2113.6	1	1	-	1

[1] ^{12}C monoisotopic component.
[2] Based on the atomic masses: ^{12}C: 12.0000, ^{1}H: 1.0078, ^{16}O: 15.9949, ^{14}N: 14.0031
[3] Glycans and functional groups appended to a core oligosaccharide comprising Hep_3, KDO, and a Lipid A composed of $GlcN_2$, P_2, and 2 N-linked 3-OH myristic acid ($C_{14}H_{27}O_2$). Hex, GlcN, Hep and KDO designate hexose (Glc, Gal), glucosamine, heptose, and 3-deoxy-D-manno-2-octulosonic acid, respectively.
* Glycolipids of identical compositions are presumed to come from the occurrence of isoforms differing by the location of Hex residues on the Hep triad.

Table 1. Assignment of the different O-deacylated LPS from H. influenzae *RM 7004 lic 2B strain (Fig. 2).*

heterogeneity arose from a side reaction of the mild hydrazinolysis treatment and/or from the natural distribution of the LPS. For convenience, the assignment of the major O-deacylated LPS observed in Fig. 2 is summarized in Table 1. Mass accuracy observed on the different glycolipids was typically within ± 0.1 Da of the predicted molecular mass.

In view of the complex LPS biosynthetic pathways, the occurrence of glycolipids of identical molecular mass (isoforms) is not entirely unexpected. Indeed, a number of glycolipid isoforms were observed in the O-deacylated LPS from *H. influenzae* RM 7004 *lic 2B* as indicated in the contour profile of Fig. 2b by the band patterns of identical m/z values. Furthermore, the reconstructed ion electropherograms of selected molecular species such as that of 2723.0, 2560.9, and 2398.6 Da displayed two doublets (data not shown). To investigate the nature of these isoforms, further structural characterization was achieved using on-line tandem mass spectrometry. The total ion electropherogram of the product ions of $[M-3H]^{3-}$ at m/z 811.6 (Hex_3 PPE or Hex_3 P PE) is presented in Fig. 3b, together with the extracted MS-MS spectra of the different peaks. It is noteworthy that as a result of the higher electroosmotic flow, the analysis depicted in Fig. 3 shows poorer isoform resolution compared to the corresponding reconstructed ion profile of m/z 811.6 in Fig. 2 (data not shown). The MS-MS spectra shown in Fig. 3b-d are marked by similar fragmentation patterns, with

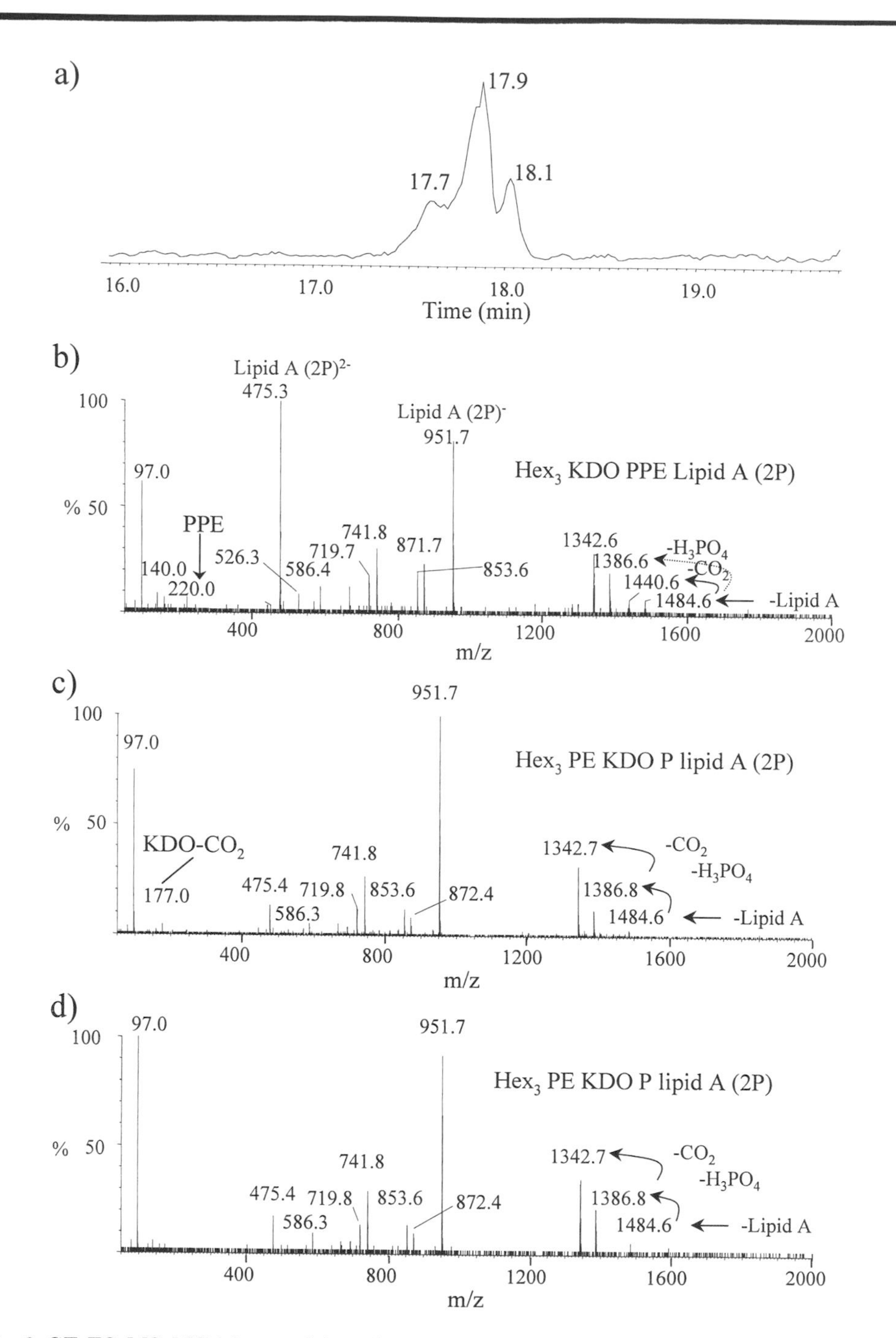

Fig. 3. CE-ES-MS-MS (- ion mode) analysis of O-deacylated LPS from H. influenzae *strain 7004 lic 2B product ion of m/z 811.6. (a) Total ion electropherogram showing isomeric components, MS-MS spectra of peaks identified at 17.7 min (b), 17.9 min (c) and 18.1 min (d). Separation conditions as for Fig. 2 except Ar collision gas, $E_{lab.}$: 90 eV (laboratory frame of reference).*

variation in the abundance of specific fragment ions. Typically, MS-MS spectra of O-deacylated LOS yielded limited fragmentation with very little information on the oligosaccharide branching. In all cases the major fragment ion observed corresponded to the cleavage of the glycosidic bond between the KDO and the glucosamine of the lipid A. The dissociation process is possibly due to a charge separation yielding singly-and doubly-charged fragment ions. The more intense of the two complementary charge-separation fragments corresponded to the lipid A fragment ion. This cleavage is similar to that observed for other underivatized oligosaccharides, where the fragmentation gives rise to both B- and Y-type ions [30]. Interestingly, the doubly-charged Lipid A fragment ion at m/z 475.3 is more abundant in Fig. 3b than in panel c or d, whereas the reverse situation is observed for the singly-charged fragment at m/z 951.7. Although the composition of the Lipid A appears identical in all cases, this observation may reflect the different chemical environment and stabilization effect of neighboring groups such as PE or PPE. The dissociation of the core oligosaccharide fragment ion at m/z 1484.8 gave rise to consecutive losses of ethanolamine, phosphate and CO_2, the order of which depended on the nature of the substituent attached to the KDO. For example, the product ion spectrum of the early migrating peak (Fig. 3b) afforded fragment ions at m/z 1440.6, 1386.6 and 1342.6 arising from losses of ethanolamine and/or H_3PO_4 moieties, which suggests the presence of PPE on the KDO residue. This proposal was further supported by the observation of a fragment ion at m/z 220 corresponding to PPE. In contrast, the MS-MS spectra of Figures 3c and d only showed the consecutive loss of H_3PO_4 followed by ethanolamine or CO_2, thereby suggesting the occurrence of a phosphorylated KDO residue. A fragment ion at m/z 177 was assigned to the loss of CO_2 from the dephosphorylated KDO.

The similarity of the MS-MS spectra of Figs. 3c and d suggests the presence of closely related isoforms differing by the location of Hex residues. Indeed, isomeric glycolipids would also be anticipated from chain extensions at each of the Hep residues, thus conferring significant molecular diversity to the outer core oligosaccharide. In the present case, the distribution of O-deacylated LPS was expected to be altered through mutation of the *lic 2B* gene, thereby affecting the incorporation of additional Gal disaccharide on the second Hep residue. Heterogeneity in the Hex_3 isoforms, as reflected by the peak doublet at 17.9 and 18.1 min in Fig. 3a, could be accounted for by a substituted core comprised of a Hex residue at each of the Hep triad and a disaccharide on Hep I, plus a single Hex at Hep II. Although no molecular modeling studies are available at present to corroborate this proposal, the different frictional drag expected from these isoforms would presumably account for their difference in migration times.

Probing Specific Functionalities of O-deacylated LOS Using Diagnostic Fragment Ions

The characterization of the diverse glycolipid population from extracts of O-deacylated LPS is a challenging task and requires analytical procedures offering sensitivity, high separation efficiencies and good mass resolution for proper structural assignment. This is particularly true not only for the identification of isoforms as evidenced in the previous section, but also for the monitoring of trace level glycolipids present as a small subset of a wider bacterial extract. Current methods for probing these structural changes involve gel electrophoresis and immunoblotting experiments with a number of specific monoclonal antibodies to profile the LPS epitopes [10, 15-17, 19]. These methods offer excellent sensitivity for the identification of structural variants at the single colony level, but rely on the availability of multiple monoclonal antibodies for adequate structural coverage.

In an effort to develop analytical tools providing both structural specificity and sensitivity, we investigated the potentials of CE-ES-MS for probing characteristic functional groups and residues present on the LPS of different *H. influenzae* strains. Particular emphasis was placed on the identification of immunodeterminants such as PC, a substituent that was linked to the persistence of phenotypes on the mucosal surface, and to its possible evasion from innate immunity mediated by C-reactive protein [10]. The presence of PC on the LPS structure of *H. influenzae* was found as a unique feature of this pathogen [31, 32], and its expression appears to be phase variable [31]. Recent studies have shown that PC is acquired from the environment and linked to a hexose residue on the outer core region of the LPS [31-33].

Tandem spectral analyses on core oligosaccharides of *H. somnus* enabled the identification of PC via the observation of characteristic fragment ions at m/z 184 and 328 corresponding to the hydrated PC and the Hex-PC cations, respectively [34]. The presence of an intense fragment ion at m/z 328 is

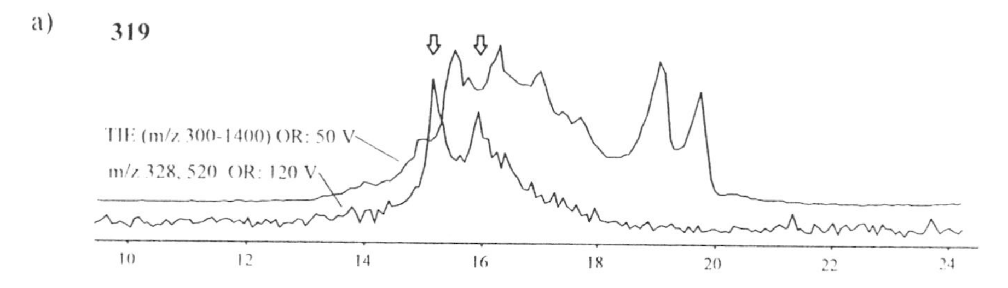

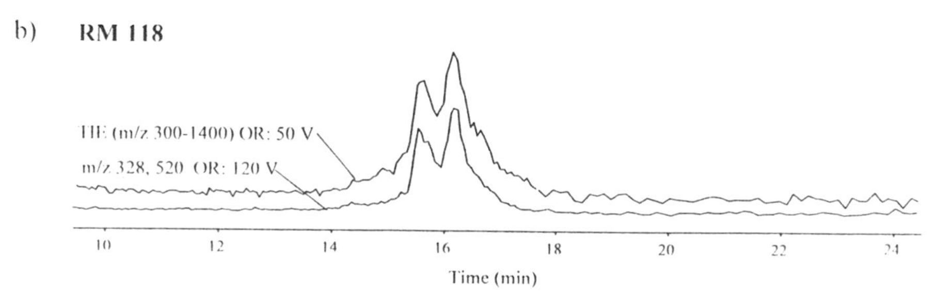

Fig. 4. CE-ES-MS (+ ion mode) analysis of O-deacylated LPS from H. influenzae *strain 319 (a) and RM 118 (b). Conditions: Mixed scan SIM m/z 328 (orifice voltage 120 V), full scan (m/z 300-1400, orifice voltage 50 V), other conditions as for Fig. 2 except that a 30 mM morpholine pH 9 (no methanol) and a separation voltage of +30 kV were used.*

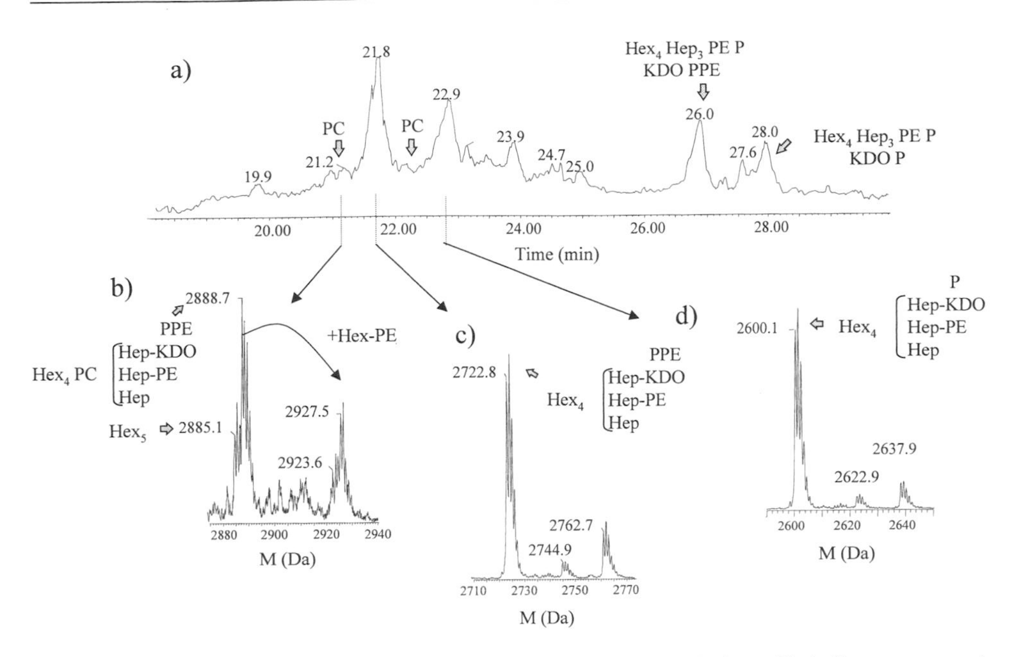

Fig. 5. CE-ES-MS (- ion mode) analysis of O-deacylated LPS from H. influenzae *strain 319. (a) TIE (m/z 500-2000), reconstructed molecular mass profiles for peak identified at 21.2 (b), 21.8 (c) and 22.9 min (d). Conditions as for Fig. 2.*

characteristic of glycolipids containing PC, and this feature was used advantageously in CE-ES-MS experiments to probe its occurrence amongst the diverse population of LPS products. To alternatively detect characteristic fragment ions together with multiply-charged species, a mixed scan function was developed, in which the high/low orifice voltage stepping was incorporated as one of the mass spectral acquisition parameters. An example of this is shown in Fig. 4 for the analysis of O-deacylated LPS from *H. influenzae* strains RM118 and 319 previously identified from monoclonal antibody experiments as PC positive and PC negative phenotypes, respectively. In both cases the electropherogram corresponding to the fragment ion m/z 328 formed at an orifice voltage of 120 V is presented, together with the TIE of the full scan acquisition at an orifice voltage of 50 V. Under the present electrophoretic conditions, the strain RM 118 gave a relatively narrow distribution of glycolipids, most of them showing the expression of the PC epitopes (Fig. 4b). In contrast, the strain 319 exhibited a broader population of O-deacylated LPS with a minor series of glycolipids presenting the PC group at 15.2 and 16.0 min in Fig. 4a.

To improve the peak resolution and mass assignment of these different components, this analysis was repeated on the Q-TOF instrument using an electrolyte containing 30 mM morpholine pH 9 and 5% methanol to prevent analyte adsorption. The corresponding CE-ES-MS analysis (negative ion mode) is shown in Fig. 5 for the analysis of O-deacylated LPS from strain 319. The position of peaks previously identified as PC-substituted glycolipid (Fig. 4a) is indicated in Fig. 5a at the expected time. The most abundant peak observed at 21.8 min in Fig. 5a corresponds to a glycolipid having a molecular mass of 2722.9

Da, and was tentatively assigned to a Hex_4 PPE PE glycoform ($M_{calc.}$: 2722.9 Da) appended to the conserved core oligosaccharide composed of Hep_3 and KDO (Fig. 1). The reconstructed molecular mass profile for this peak (Fig. 5c) also showed a co-migrating peak of $M_{obs.}$: 2762.7 Da corresponding to the Hex_5 P PE glycoform. Consistent with previous assignments, the P and PE groups were assumed to be located on the KDO and Hep II residues, respectively. Heterogeneity of substituent located on the KDO was noted by the appearance of a lower molecular mass glycolipid ($M_{obs.}$: 2600.1 Da) migrating at 22.9 min and assigned to the KDO-P (Fig. 5d).

As observed in Fig. 4a from the reconstruction ion profile for m/z 328, the *H. influenzae* strain 319 yielded two minor peaks showing the expression of the PC epitope. Trace level glycolipids containing this functionality were observed at 21.2 and 22.3 min in Fig. 5a. Based on the relative intensity of the multiply-charged ions, these components were estimated to be present at a level of less than 5% of the entire glycolipid population from this sample. The reconstructed molecular mass profile of co-migrating components of the first peak is presented in Fig. 5b, and glycolipids having a mass of 2888.7 and 2927.5 Da were assigned to the Hex_4 PPE PE PC ($M_{calc.}$: 2888.9 Da) and the Hex_5 P PE PC ($M_{calc.}$: 2928.0 Da) glycoforms, respectively. It is noteworthy that closely related components approximately 4 Da lower were observed for each of these two glycolipids. These satellite peaks corresponded to the substitution of a PC by an Hex giving rise to Hex_5 PPE PE and Hex_6 PPE glycoforms [ΔM=166 (PC)-162 (Hex)]. The second PC-containing glycolipid of higher mobility (22.3 min) has a molecular mass of 123 Da lower than the Hex_4 PPE PE PC glycoform, and was again related to the heterogeneity of KDO substituent. Adequate separation of these different glycolipids requires mass resolution in excess of 2,500, a situation which is rarely

Time (min)	$M_{calc.}$ (Da)[1]	$M_{obs.}$ (Da)[2]	Assignment[3]				
			P	PE	PPE	PC	Hex
21.2	2888.9	2888.7	-	1	1	1	4
21.2	2885.0	2885.1	-	1	1	-	5
21.2	2928.0	2927.5	1	1	-	1	5
21.2	2925.0	2923.6	1	1	-	-	6
21.8	2722.9	2722.8	-	1	1	-	4
21.8	2762.9	2762.9	1	1	-	-	5
22.3	2765.9	2765.7	1	1	-	1	4
22.9	2599.9	2600.1	-	-	1	-	4
22.9	2638.9	2638.1	1	-	-	-	5
23.2	2599.9	2599.5	1	1	-	-	4
26.0	2802.9	2803.0	1	1	1	-	4
27.6, 28.0*	2679.9	2680.0	2	1	-	-	4
28.0	2718.9	2718.9	2	-	-	-	5

[1] ^{12}C monoisotopic component.
[2] Based on the atomic masses: ^{12}C: 12.0000, ^{1}H: 1.0078, ^{16}O: 15.9949, ^{14}N: 14.0031
[3] Glycans and functional groups appended to a core oligosaccharide comprising Hep_3, KDO, and a Lipid A composed of $GlcN_2$, P_2, and 2 N-linked 3-OH myristic acid ($C_{14}H_{27}O_2$). Hex, GlcN, Hep and KDO designate hexose (Glc, Gal), glucosamine, heptose, and 3-deoxy-D-manno-2-octulosonic acid, respectively.
* Glycolipids of identical compositions are presumed to come from the occurrence of isoforms differing by the location of Hex residues on the Hep triad.

Table 2. Assignment of the different O-deacylated LPS from H. influenzae *strain 319 (Fig. 5).*

met on quadrupole instrument without significantly compromising the sensitivity. The proposed compositions for the major components observed in Fig. 5 are summarized in Table 2.

Separations conducted using morpholine as an electrolyte provided adequate resolution of isoforms and glycoforms compared to other buffers commonly used in CE-ES-MS experiments. However, analyses conducted with such an electrolyte are not entirely compatible with positive ion detection of multiply-protonated molecular species. Although fragment ions formed in the sampling cone region of the mass spectrometer such as that of Hex-PC at m/z 328 are readily identifiable in the corresponding electropherogram (Fig. 4), the glycoform assignments proved to be more difficult, due to the presence of intense adduct ions corresponding to extensive addition of morpholine (1-6 molecules) to the multiply-protonated species (data not shown). The abundance of these cluster ions could not be attenuated to any reasonable extent in spite of attempts to raise the cone/skimmer voltage. For these reasons alternate buffers were selected for positive ion detection of O-deacylated LPS.

An example of separation conducted with 30 mM ammonium acetate pH 8.5 containing 5% methanol is shown in Fig. 6 for the analysis of O-deacylated LPS from strain RM 118. Interestingly, the use of this buffer in flow injection experiments did not provide successful ionization of O-deacylated LPS. This observation may be explained by the tendency of these analytes to form aggregates and salt adducts in solution, a situation which can be minimized during the electrophoretic separation. As mentioned earlier, the RM 118 strain comprises a high proportion of PC-bound glycolipids with a narrow distribution

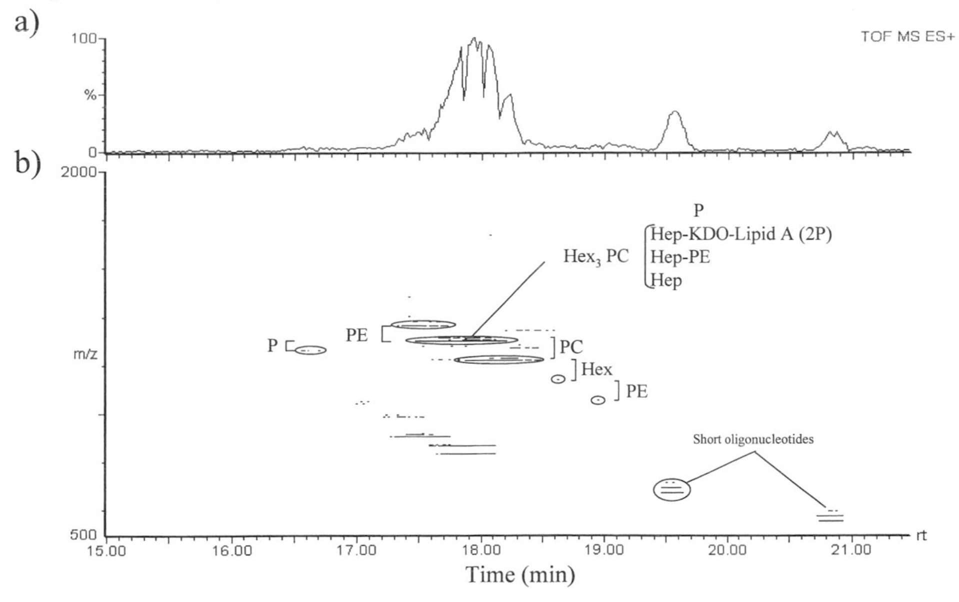

Fig. 6. CE-ES-MS (+ ion mode) analysis of O-deacylated LPS from H. influenzae *strain RM 118. (a) TIE (m/z 500-2000), (b) Contour profile of m/z vs. time. Separation conditions: Bare fused silica (90 cm x 50 µm i.d.), 30 mM ammonium acetate, pH 8.5, 5% methanol, + 30 kV, 30 ng of total glycolipid injected.*

of glycoforms. This is evidenced in Fig. 6a from the TIE profile of the corresponding analysis which shows a broad peak centered at 17.8 min. Additional singly-charged peaks of higher mobilities are observed at 19.6 and 20.8 min and were associated with short oligonucleotide contaminants often encountered as part of the sample preparation procedures. The most abundant glycolipid observed at 17.8 min gave rise to $[M+3H]^{3+}$ and $[M+2H]^{2+}$ ions at m/z 869.0 and m/z 1303.0, respectively. The molecular mass associated with this glycolipid (M_{obs}: 2604.0 Da) was in close agreement with that expected from the Hex_3 P PE PC glycoform (M_{calc}: 2603.9 Da). Further heterogeneity in the abstraction of a single P and the addition of a PE group from the Hex_3 P PE PC glycoform were also observed in the contour profile by the presence of peaks at 16.6 and 17.5 min, respectively. It is noteworthy that additional components devoid of a PC epitope were identified on the tailing edge of the peak at 17.8 min. These components were assigned to the consecutive losses of PC (M_{obs}: 2437.8 Da), Hex (M_{obs}: 2275.8) and PE (M_{obs}: 2152.6 Da) from the 2604.0 Da glycolipid.

The capability of detecting positive ions from species that are normally anionic in solution opens up new avenues in the structural characterization of O-deacylated LPS. Indeed, the fragmentation of cationic oligosaccharides typically proceeds by cleavage at the glycosidic bonds, thus enabling branching information to be obtained. This is illustrated in Fig. 7 for the on-line CE-ES-MS-MS analysis of the Hex_3 P PE PC glycoform previously identified in Fig. 6. The product ion spectrum of the doubly-protonated ion m/z 1302 (Fig. 7a) shows a number of fragment ions associated with the consecutive cleavage of glycosidic bonds and was used in the present study to confirm the sequences of carbohydrates forming the core structure. Such a fragmentation process is in distinct contrast with the product ion spectra obtained for the multiply deprotonated precursors of the O-deacylated LOS (Fig. 3) which exhibited very limited cleavage of the glycosidic bond in a similar collision energy regime.

The product ion spectrum of Fig. 7a shows an abundant doubly-charged fragment ion at m/z 1253.9 arising from the expulsion of a neutral H_3PO_4 moiety, which subsequently loses the tri-saccharide Hep-Hex_2 and an additional H_3PO_4 group to yield a doubly-charged fragment ion at m/z 995.3. Cleavages of the glycosidic bond generally proceeded via the formation of complementary fragment pairs of B and Y ions. The most prominent series was observed for the dissociation of the KDO-GlcN bond, giving rise to m/z 1651.4 and 953.7, respectively. Further loss of H_3PO_4 and an anhydro-KDO from m/z 1651.4 yielded fragment ions at m/z 1553.2 and 1351.3. These observations supported an earlier proposal suggesting the presence of a monophosphorylated KDO residue. The lipid A fragment ion at m/z 953.7 gave rise to product ions at m/z 855.4, 468.2 and 388.2, corresponding to the consecutive loss of H_3PO_4, a N-acylated glucosamine (GlcN-FA) and a second H_3PO_4 group, respectively. The location of the PC epitope was assigned to the Hex residue attached to Hep I, based on the presence of a fragment ion at m/z 722 arising from the addition of an anhydro KDO to the PC-Hex-Hep at m/z 520. This proposal was also supported by the observation of fragment ions at m/z 520.2 and 835.2 correspond-

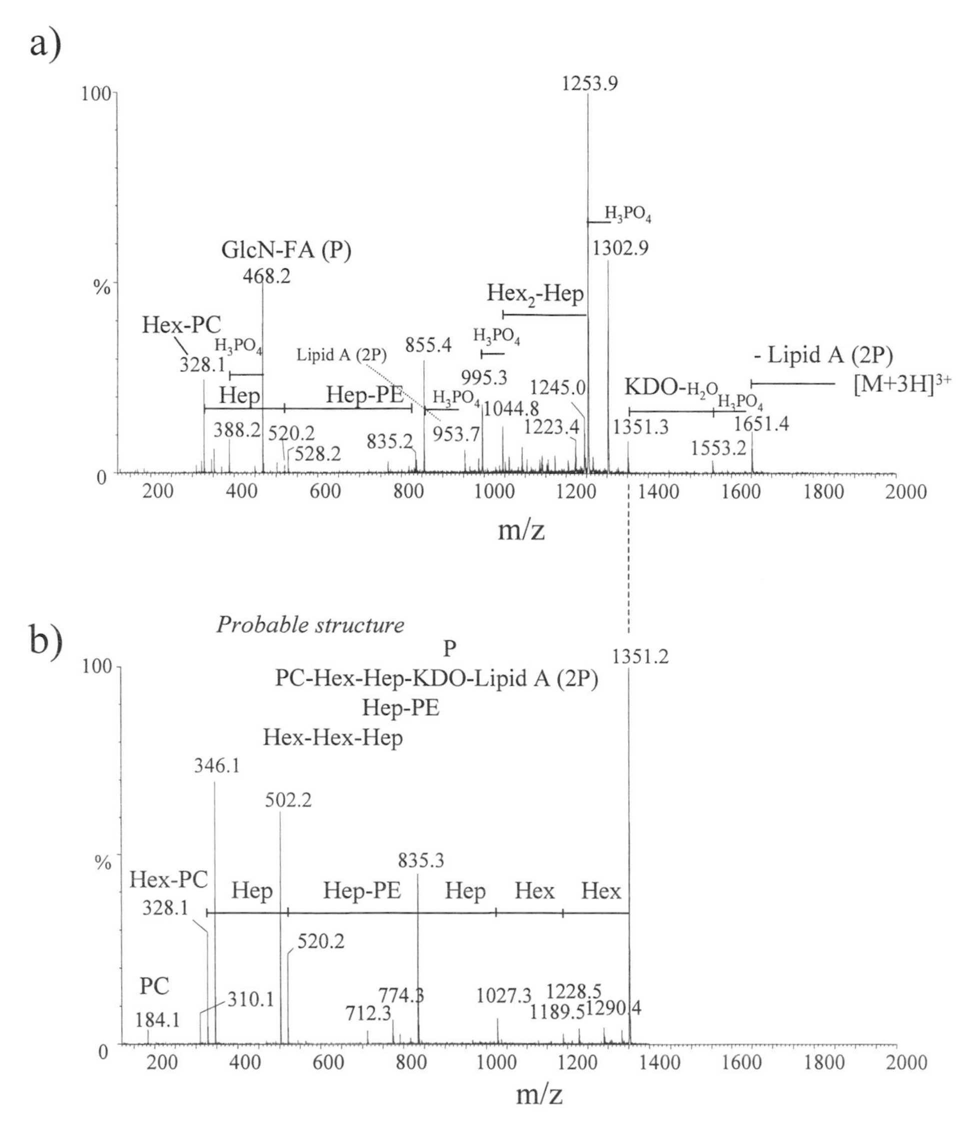

Fig. 7. CE-ES-MS-MS (+ ion mode) analysis of O-deacylated LPS from H. influenzae *strain RM 118. (a) Product ion of [M+2H]$^{2+}$ m/z 1302, (b) MS-MS spectrum of fragment ion m/z 1351 promoted using an orifice voltage of 120 V. Separation conditions as for Fig. 6 except Ar collision gas, E_{lab}: 60 eV (laboratory frame of reference).*

ing to the consecutive addition of Hep and Hep-PE residues from the Hex-PC at m/z 328.

Further evidence of this structural assignment was obtained in a separate set of MS-MS experiments where the singly-charged m/z 1351 formed at a high cone voltage was selected as a precursor (Fig. 7b). This ion corresponds to the core oligosaccharide, and interpretation was simplified by the lack of fragmentation associated with the lipid A. The MS-MS spectrum of m/z 1351 was

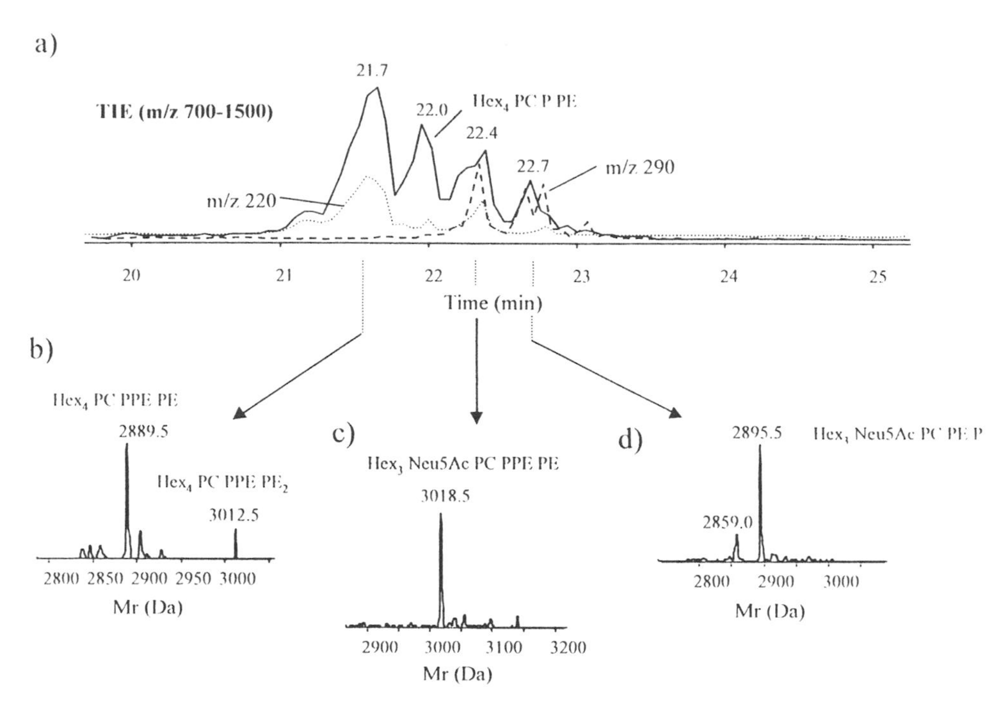

Fig. 8. CE-ES-MS (- ion mode) analysis of O-deacylated LPS from H. influenzae *strain 375. (a) TIE (m/z 700-1500), SIM for m/z 220 (dotted line), and m/z 290 (dashed line). Reconstructed molecular mass profiles for peak identified at 21.7 (b), 22.4 (c) and 22.7 min (d). Conditions as for Fig. 2 except that approximately 20 ng of total glycolipid injected.*

dominated by simple cleavage of each glycosidic bond yielding a direct sequence assignment. The linear sequence consisting of PC-Hex-Hep-Hep (PE)- Hep-Hex-Hex was confirmed by the observation of fragment ions at m/z 328, 520, 835, 1027, and 1189, 1351, respectively. These results were consistent with recent structural studies (Risberg *et al.*, submitted) indicating that a lactose unit β-Gal-(1-4)- β-Glc was attached to the Hep III residue. It is interesting to note that the location of the PC substituent found in the present RM 118 isolate differs from that observed in other Rd⁻ derived mutant labeled RM 118-28, where this immunodeterminant was assigned to a Gal residue linked to the Hep III [32]. The presence of PC epitope on different residues suggests a variability in acceptor specificities [33] and/or the involvement of more than a single transferase.

In addition to phase variation, some strains of *H. influenzae* have been found to mimic blood group antigens such as P^k epitope (α-Gal(1-4)- β-Gal (1-4)-Glc) and paragloboside (β-Gal(1-4)- β-GalNAc (1-3)- β-Gal (1-4)-Glc) [35]. Such epitopes are also good acceptors for sialyl transferases, although the occurrence of sialylated LPS of *H. influenzae* has only been reported for the A2 strain where the Neu5Ac residue was presumed to be attached to the terminal N-acetyl lactosamine [36]. The biological importance of sialylated LOS on the pathogenicity and virulence of certain bacterial strains of both gonococci and meningococci was reviewed recently [37]. LPS sialylation was found to affect the pathogenicity of meningococci, though to a lesser extent to that observed in gonococci. Most serotype groups of meningococci (B, C, W and Y) have the ability to synthesize the substrate cytidine monophosphate Neu5Ac which can be incorporated into capsular polysaccharides, and to terminal galactose residues of LPS [38]. It is not clear at present if the same pathogenesis mechanisms are operative for sialylated LPS from *H. influenzae*, because the complex glycolipid heterogeneity prevented the evaluation of defined carbohydrate epitopes in bacterial virulence and pathogenic processes such as cell adhesion, invasion and dissemination.

To facilitate the identification of the Neu5Ac residue from complex glycolipid extracts, a method employing a mixed scan function with a high/low orifice voltage stepping was developed for CE-ES-MS experiments, and successfully applied to the detection of sialylated LPS from *N. meningitidis* immunotype L1 [39]. Fragment ions at m/z 220 and 290 formed under high orifice voltage conditions were used as diagnostic ions for the presence of PPE and Neu5Ac moieties, respectively. This method was evaluated for the identification of Neu5Ac in O-deacylated LPS from a bacterial isolate obtained from a child otitis media. The corresponding CE-ES-MS analysis (negative ion mode) is shown in Fig. 8 for a sample of approximately 20 µg of LPS obtained from the equivalent of 10^9 bacterial cells (≈ 100 colonies). For convenience, the reconstructed ion electropherogram for the fragment ions m/z 220 (PPE) and m/z 290 (Neu5Ac) are superimposed with the TIE (Fig. 8a). The most prominent peak observed at 21.7 min in Fig. 8a contains a PPE group, but did not yield a detectable signal in the m/z 290 corresponding to the Neu5Ac residue. The molecular mass of this O-deacylated LPS product was 2889.5 Da (Fig. 8b), and was consistent with a glycoform composed of Hex_4 PC PPE PE (isotope averaged mass, Mr: 2890.4 Da). The presence of a PC group on this glycoform was confirmed in separate CE-ES-MS experiments conducted in positive ion mode using the diagnostic fragment ion m/z 328, characteristic of Hex-PC (data not shown). The peak migrating at 22.0 min in Fig. 8a has a molecular mass precisely 123 Da lower than the first component, and was assigned to the glycoform Hex_4 PC P PE based on its mass, its migration position and the lack of a significant signal in the m/z 220 trace. Two sialylated components were observed at 22.4 and 22.7 min in Fig. 8a, and gave Mr values of 3018.5 Da (Fig. 8c) and 2895.5 Da (Fig. 8d), respectively. Heterogeneity in the substitution of a PPE for a P group was again at the basis of the molecular difference between these two glycolipids. Both species gave detectable ion signals for the presence of a Hex-PC moiety, as identified in the positive ion CE-ES-MS of the corresponding analysis (data not shown). Based on their molecular mass and the presence of detectable functionalities such as Neu5Ac, PC, PPE and/or P, these components were assigned to the Hex_3 Neu5Ac PC PPE PE (Mr: 3019.5 Da) and Hex_3 Neu5Ac PC P PE (Mr: 2896.5 Da) glycoforms.

It is noteworthy that both Neu5Ac and PC immunodeterminants are present on these glycoforms. Also interesting is the fact that in contrast to earlier studies on the A2 strain, sialylation does not occur on the terminal N-acetyl lactosamine [36], but on a different epitope. This might reflect a different acceptor specificity or the involvement of a different transferase. Assuming that the PC is linked to one Hex residue (presumably Glc), the remaining glycolipid epitopes would either be a Hex_3 or a Hex_2 Neu5Ac saccharide. On a biosynthetic basis, one could presume an ambivalence in the chain elongation beyond the Hex_2 glycan, whereby further extension can proceed via the addition of a Hex (major pathway) or a Neu5Ac (minor pathway) residue. Further studies are presently under way to fully characterize the structure of these short LPS and their relationships to pathogenicity.

CONCLUSIONS

The Gram-negative bacteria *H. influenzae* displays a complex population of LPS components, varying in the composition and linkages of glycoforms appended to a conserved inner core region composed of a heptose trisaccharide, to which is attached a 3-deoxy-D-manno-octulosonic acid (KDO) residue The biosynthesis of these glycolipids is very complex and involves an intricate number of glycosyltransferases. The expression of some of these are subject to molecular mechanisms generating phase variation of terminal epitopes. Through this molecular switch, *H. influenzae* has the ability to express a number a closely related structural variants, which enable this pathogen to adapt to changes in environmental conditions within or between individual hosts.

Structural characterization of this diverse population of glycolipids is not only important for the further understanding of the pathogenesis processes and biosynthetic mechanisms leading to their expression, but also for the development of antibodies specifically targeted toward these immunodeterminant structures. To this end, the combination of capillary electrophoresis coupled to electrospray mass spectrometry has been a valuable analytical tool for probing subtle structural changes taking place on the LPS molecules. The extension of hexose residues to the LPS structure gives rise to a decrease in electrophoretic mobility and an increase in molecular masses, which are easily visualized by characteristic diagonal lines of negative slopes in contour profiles of m/z vs. time. Similarly, the occurrence of substituted LPS differing by the location or presence of functional groups such as P, PE, PPE or PC are reflected by changes of electrophoretic mobilities and masses. The availability of high resolution mass spectrometers such as the Q-TOF instrument has facilitated the sensitive detection of closely related glycoforms, and enabled the assignment of potential glycan compositions with a higher precision than that typically available on quadrupole instruments.

Specific scanning functions were also developed and integrated to CE-ES-MS analyses to identify the presence of characteristic functionalities such as PPE, PE, PC and Neu5Ac on the structures of LPS. This was possible using a mixed scan function, where a high/low orifice voltage stepping was incorporated as one of the mass spectral acquisition parameters. By using such scanning functions, trace levels of O-deacylated LPS comprising a single PC group were detected at an estimated level of 5% of the overall LPS population of *H. influenzae* RM 319 strain Eagan. Electrophoretic conditions were also developed for the detection of positive ions from O-deacylated LPS species, and enabled the characterization of oligosaccharide branching through the formation of stable fragment ions corresponding to cleavage at each glycosidic bond. Application of this was demonstrated using on-line tandem mass spectrometry on the Q-TOF instrument for the analysis of O-deacylated LPS from *H. influenzae* RM 118 strain Rd$^-$, and helped to locate the PC substituent on the core oligosaccharide structure. The sensitivity and specificity of the mixed scan function also facilitated the identification of trace level sialylated LPS derived from a clinical isolate of otitis media. Such analytical advances now permit the detection of glycolipids obtained from selected bacterial isolates, and opens up new avenues in immunological investigations designed to probe the molecular basis of bacterial virulence.

ACKNOWLEDGMENTS

The authors gratefully acknowledge Drs. Andrew Cox, Malcolm Perry and Hussein Masoud for valuable discussions, Douglas Griffith for technical assistance in preparing the bacterial cultures, and Mary Deadman for the extraction of some LPS used in this study. We are also most grateful to Micromass for access to the Q-TOF instrument and for their generous support and assistance during this study.

REFERENCES

1. D. C. Turk, in *Haemophilus influenzae, Epidemiology, Immunology and Prevention of Disease*, S. H. Sell and P. F. Wright, Eds.; Elsevier: New York, 1981; pp 3-9.

2. R. D. Moxon, in *Principles and Practice of Infectious Diseases* (Fourth Edition), G. L. Mandrell, J. E. Bennett and R. Dolin Eds.; Churchill Livingstone Inc: New York, 1995; pp 2039-2045.

3. T. F. Murphy and M. A. Apicella, *Rev. Infect. Dis.,* **1987**, *9*, 1-15.

4. A. Kimura and E. J. Hansen, *Infect. Immun.* **1986**, *51*, 69-79.

5. A. Zwahlen, L. G. Rubin, and E. R. Moxon, *Molec. Pharmacol.* **1986**, *1*, 465-473.

6. A. Kimura, C. C. Patrick, E. E. Miller, L. D. Cope, G. H. J. McCracken and E. J. Hansen, *Infect. Immun.* **1987**, *55*, 1979-1986.

7. A. D. Jackson, D. Maskell, E. R. Moxon and R. Wilson, *Microb. Pathogen.* **1996**, *21*, 463-470.

8. S. G. Wilkinson, *Rev. Infect. Dis.* **1983**, *5*, S941-S949.

9. B. W. Gibson, N. J. Phillips, W. Melaugh and J. J. Engstrom, in *Biochemical and Biotechnological Applications of Electrospray Ionization Mass Spectrometry*, P. Snyder Ed., ACS symposium series vol. 619, 1996, pp. 166-184.

10. J. N. Weiser, N. Pan, K. L. McGowan, D. Musher, A. Martin and J. Richards., *J. Exp. Med.* **1998**, *187*, 631-640.

11. H. Masoud, E. R. Moxon, A. Martin, D. Krajcarski and J. C. Richards, *Biochemistry* **1997**, *36*, 2091-2103.

12. J. N. Weiser, *Microb. Pathogen.* **1992**, *13*, 335-339.

13. J. N. Weiser, A. Williams and E. R. Moxon, *Infect. Immun.* **1990**, *58*, 3455-3457.

14. J. N. Weiser, J. Love and E. R. Moxon, *Cell* **1989**, *59*, 657-665.

15. N. J. High, M. E. Deadman and E. R. Moxon, *Mol. Microbiol.* **1993**, *9*, 1275-1282.

16. D. J. Maskell, M. J. Szabo, P. D. Butler, A. E. Williams and E. R. Moxon, *Mol. Microbiol.* **1992**, *5*, 1013-1022.

17. E. R. Moxon, P. B. Rainey, M. A. Nowak and R. E. Lenski, *Curr. Biol.* **1994**, *4*, 24-33.

18. R. D. Fleishman et al., *Science* **1995**, *269*, 496-512.

19. D. W. Hood, M. E. Deadman, T. Allen, H. Masoud, A. Martin, J. R. Brisson, R. Fleishmann, J. C. Venter, J. C. Richards and E. R. Moxon, *Mol. Microbiol.* **1996**, *22*, 951-965.

20. B. W. Gibson, W. Melaugh, N. J. Phillips, M. A. Apicella, A. A. Campagnari and J. M. Griffiss, *J. Bacteriol.* **1993**, *175*, 2702-2712.

21. J. F. Kelly, H. Masoud, M. B. Perry, J. C. Richards and P. Thibault, *Anal. Biochem.* **1996**, *233*, 15-30.

22. S. Auriola, P. Thibault, I. Sadovskaya, E. Altman, H. Masoud and J. C. Richards, in *Biochemical and Biotechnological Applications of Electrospray Ionization Mass Spectrometry*, P. Snyder Ed., ACS symposium series vol. 619, 1996, pp.149-165.

23. K. P. Bateman, J. H. Banoub and P. Thibault, *Electrophoresis* **1996**, *17*, 1818-1828.

24. J. Li, P. Thibault, A. Martin, J. C. Richards, W. W. Wakarchuk and W. van der Wilp, *J. Chromatogr. A.,* in press.

25. N. High, M. E. Deadman, D. W. Hood and E. R. Moxon, *FEMS Microbiol. Lett.* **1996**, *145*, 325-331.

26. H. E. Alexander, in *Bacterial and Mycotic Infections of Man*, R. J. Dubos, and J. G. Hirsch, Eds.; Pitman Medical: London, 1965; pp 724-741.

27. H. Masoud, E. Altman, J. C. Richards and J. S. Lam, *Biochemistry* **1994**, *33*, 10568-10578.

28. R. D. Smith, C. J. Barinaga and H. R. Udseth, *Anal. Chem.* **1988**, *60*, 1948-1952.

29. H. R. Morris, T. Paxton, A. Dell, J. Langhorne, M. Berg, R. S. Bordoli, J. Hoyes and R. H. Bateman, *Rapid Commun. Mass Spectrom.* **1996**, *10*, 889-997.

30. B. Domon and C.E. Costello, *Glycoconjugate J.* **1988**, *5*, 397-409.

31. J. N. Weiser, Z. M. Shchepetov and S. T. H. Chong, *Infect. Immun.* **1997**, *65*, 943-950.

32. A. Risberg, E. K. H. Schweda and P.-E. Jansson, *Eur. J. Biochem.* **1997**, *243*, 701-707.

33. E. K. H. Schweda, H. Masoud, A. Martin, A. Risberg, D. W. Hood, E. R. Moxon, J. N. Weiser and J. C. Richards, *Glycoconj. J.* **1997**, *14*, S23.

34. A. D. Cox, M. D. Howard, J. R. Brisson, M. van der Zwan, P. Thibault, M. B. Perry and T. J. Inzana, *Eur. J. Biochem.* **1998**, *253*, 507-516.

35. R. E. Mandrell, R. McLaughlin, Y. Abu Kwaik, A. Leese, R. Yamasaki, B. Gibson, S. M. Spinola and M. A. Apicella, *Infect. Immun.* **1992**, *60*, 1322-1328.

36. B. W. Gibson, W. Melaugh, N. J. Phillips, M. A. Apicella, A. A. Campagnari and J. M. Griffiss, *J. Bacteriol.* **1993**, *175*, 2702-2712.

37. H. Smith, N. J. Parsons and J. A. Cole, *Microbiol. Pathogenesis* **1995**, *19*, 365-370.

38. R. E. Mandrell, J. M. Griffiss, H. Smith and J. A. Cole, *Microbiol. Pathogenesis* **1993**, *14*, 315-320.

39. W. W. Wakarchuk, M. Gilbert, A. Martin. Y. Wu, J. R. Brisson, P. Thibault and J. C. Richards, *Eur. J. Biochem.* **1998**, *254*, 626-633.

QUESTIONS AND ANSWERS

A. L. Burlingame (UCSF)

What has the *H. influenzae* genome information made possible in understanding structural diversity and phase variation in these bacterial cells?

Answer. The availability of the genome from *H. influenzae* Rd- strain made it possible to construct mutant strains obtained through site-directed mutagenesis of specific genes. The functions of these genes with respect to lipopolysaccharide biosynthesis was uncovered by characterizing the structures of LPS for each of the respective strains using techniques such as mass spectrometry, and NMR spectroscopy.

Jasna Peter-Katalinic (University of Münster)

Phosphocholine moiety was found in lipoteichoic acids of Gram-positive bacteria like Streptococcus. Would the virulence mechanisms related to PC be the same for Gram-positive and Gram-negative ?

Answer. The presence of phosphocholine was found to be phase variable in a number of non-typable strains of *H. influenzae*. Recently, the group of Weiser has reported that the occurrence of this immunodeterminant was also related to the persistence of PC+ phenotypes in human respiratory tract. Unfortunately, no such data is presently available for Gram-positive bacteria.

Brad Gibson (UCSF)

Would capillary electrochromatography offer any advantage over CZE in terms of sample loadings of LOS ?

Answer. Sample introduction in CEC is achieved using electrokinetic injection and would not offer any practical advantage in terms of enhancement of sensitivity and detection limits compared to adsorption or isotachophoretic preconcentration as demonstrated in this presentation.

Brad Gibson (UCSF)

What do you consider to be the limiting factor towards the analysis of LOS glycoforms directly from *in vivo* source (without external growth)?

Answer. First, extraction and clean-up procedures become very important in minimizing sample losses when dealing with trace level analyses. In such situations, we favor enzymatic means of releasing LPS from the membrane surface. Mild hydrazinolysis can also be used to produce soluble O-deacylated LPS in quantitative yields. Another important aspect is how many cells can we collect from clinical isolates, and the number of LPS molecules attached to them. Our current detection limits for full scan analysis are approximately 50-100-fmols. If such amounts can be isolated successfully from the targeted cells, then it would be conceivable to analyze these samples using adsorption preconcentration in combination with CE-ES-MS.

Mycobacterial Lipoglycans: Structure and Roles in Mycobacterial Immunity

J. Nigou, B. Monsarrat, M. Gilleron, T. Brando, R. Albigot and G. Puzo
IPBS, CNRS, 205 Rte de Narbonne, 31077 Toulouse Cedex, France

Tuberculosis is an ancient disease, and remains today a major human health problem. *Mycobacterium tuberculosis* infects one third of the world population and is alone responsible for three million deaths annually. Moreover, tuberculosis has also re-emerged as an important health problem in many developed countries [1].

Tuberculosis is typically a lung disease. The mycobacteria are inhaled and when they arrive in the lung they are phagocytosed by the alveolar macrophages. The non-virulent mycobacteria are promptly killed, whereas the virulent ones survive and multiply within the macrophages. This survival is in part explained by the fact that phagosomes containing the virulent mycobacteria do not fuse with the lysosomes. Thus mycobacteria attachment and phagocytosis are important steps in the infectious process.

Specific receptor-ligand interactions mediate this internalization. Moreover, it has been postulated that some families of receptors play an important role in the early survival of intracellular pathogens by blocking the stimulation of the release of oxygen metabolites and cytokines (TNF-α, IL-12...). Schlesinger's studies [2] have demonstrated the importance of the complement receptors (CR1, CR3 and CR4) in the phagocytosis of attenuated and pathogenic *M. tuberculosis* stains by human alveolar macrophages. In contrast to CR, the same group has determined that the macrophage mannose receptor (MR) mediates adherence of the Erdman and H37Rv but not the attenuated H37Ra [3]. The lipoarabinomannan (LAM) from *M. tuberculosis* was found to be the ligand that interacts with the MR [4]. Independently, we have found [5] that lipoarabinomannan from *Mycobacterium bovis* BCG also binds to the MR. In contrast, the level of binding drastically decreased when LAM from *M. smegmatis* was used.

LAM from pathogenic mycobacteria are also involved in protective immunity. Protective immunity against tuberculosis is dominated by the macrophages and αβ T cells (TH1 cells and $CD8^+$ cells) and, in this case, antibodies play little role if any. This process requires the activation of specific T cells. For more than a decade, the central dogma has been that αβ T cells recognize mycobacterial peptides in the context of MHC class I and II molecules. However, recent studies [6] have begun to reveal that double negative and $CD8^+$ αβ T cells recognize non-peptidic antigens presented by the CD1 molecules. Two classes of mycobacterial antigens presented by CD1 molecules have been identified: free mycolic acids and glycosylated mycolates [7] and lipoglycans, particularly the LAM from *M. tuberculosis* and *M. leprae* [8].

Understanding the structural requirements and recognition mechanisms of the LAM binding to the mannose receptor and to CD1 requires knowledge of the LAM chemical structure. In this report, we describe the use of mass spectrometry to determine the structure of the functional domains of the lipoarabinomannans from *Mycobacterium bovis* BCG.

Material and Methods

LAM Preparation

All the lipoarabinomannan preparations were isolated according to procedures described in detail elsewhere [13]. Acyl-glycerol was prepared from the LAM by treatment with anhydrous acetic acid/acetic anhydride at 110°C for 12 h.

Sample Preparation

Monosaccharides and glucose oligomers (maltose, maltotriose, maltotetraose) were purchased from Sigma, trisodium APTS from Interchim (Paris). The dried carbohydrates (approx. 5 nmol) were labeled by reductive amination by adding 0.5 μl 0.2M APTS in 15% acetic acid and the same volume of $NaBH_3CN$ 1M in THF. The mixture was incubated for 90 min at 55°C and then diluted with deionized water to stop the reaction.

Capillary Electrophoresis

Capillary electrophoresis separations were performed on a P/ACE capillary zone electrophoresis system (Beckman Instruments, Inc.) with the cathode on the injection side and the anode on the detection side. The samples were injected by applying 0.5 psi pressure for 5 s. The separations were monitored on a 470 mm x 50 μm (I.D) uncoated fused-silica capillary column (Sigma, Division Supelco, St. Quentin, France) with a Beckman laser-induced fluorescence (LIF) detection system using a 4 mW argon-ion laser with an excitation wavelength of 488 nm and an emission filter of 520 nm. The temperature of the capillary in the P/ACE instrument was maintained at 25°C. The LIF detection was determined at a capillary length of 40 cm. The electropherograms were acquired and stored on a Dell/Pentium computer using the system Gold software package (Beckman Instruments, Inc.). The capillary was flushed with HCl 0.1N. Separations were carried out using different buffers, prepared from aqueous solutions of acetic acid and formic acid at pH < 4 adjusted with an aqueous solution of TEA.

Mass Spectrometry

GC/MS analysis. GC was performed on a Girdel series 30 equipped with an OV1 capillary column (ID 0.22 mm x 25 m). GC/MS analyses were performed on a Hewlett-Packard 5889 X mass spectrometer (electron energy, 70eV) working on both electron impact (EI) and chemical ionization modes using NH_3 as reagent gas (CI/NH_3), coupled with a Hewlett-Packard 5890 Gas chromatograph series II fitted with a similar OV1 column (0.30 mm x 12 m). Acetolysis products were analyzed on a 0.35 m length column using a temperature separation program from 160 to 300°C at a speed of 8°C/min. The injector and interface temperature were 290°C.

CE-Electrospray analysis. All these experiments were carried out using the Beckman CE described above via a floated Finnigan Mat ESI source to a Finnigan Mat triple quadrupole instrument (TSQ-700). The CE capillary replaced the sample capillary of the ESI source and the electrospray process was performed from the tip of this capillary. The CE capillary length was extended beyond the outlet of the cartridge for connection to the ESI source. Thus, the most convenient length to use was 80 cm. In all the experiments, up to 1 cm of polyimide was removed from the spray tip of the capillary. This tip was extended beyond the sheath liquid needle by 0.5 mm and this distance was empirically determined from the ESI spray current using a micrometer head adaptor kit for CE/MS (Finnigan Mat). The outlet end of the capillary was maintained at 4 kV which was the voltage used during the ESI. The voltage allowing electrophoretic separation of the analytes was the difference between the CE (20 kV) and the ESI (4 kV). The sheath liquid (isopropanol/water, 75/25 v/v) was supplied at a flow rate of 3 μL/min and a sheath gas (nitrogen, 30 psi) were coaxially delivered to the CE capillary. In all cases, the sheath liquid should be degassed daily by sonication. During the CE/ESI-MS experiments, the CE current and ESI current spray should be monitored and used diagnostically. Electropherograms of the APTS-carbohydrate derivatives were obtained using the mixture of acetic or formic acid adjusted to low pH with triethylamine.

MALDI-TOF analysis. All MALDI-TOF experiments were recorded on a Voyager-DE STR instrument (PerSeptive Biosystems, Framingham, MA). Spectra were mass assigned by using a two point external calibration. The LAM samples were run in recrystallized 2,5-dihydroxybenzoic acid (DHB).

Results

Lipoarabinomannan: Current Structural Model

The LAM is ubiquitously found in the envelope of mycobacteria. Figure 1 represents the current structural model of the LAM, namely mannosylated LAM (ManLAM) isolated from the virulent and pathogenic mycobacteria species [9, 10]. The ManLAM is composed of three domains assigned to a polysaccharidic core, a phosphatidyl-*myo*-inositol anchor, and finally the capping motifs. The polysaccharidic core is formed by two homopolysaccharides of D-mannan and D-arabinan. The D-mannan is a linear segment of 6-O-linked α-D-Man*p* highly branched with single α-D-Man*p* units linked to the C2. The arabinan chain is composed of 5-O-linked α-D-Ara*f* with branched chains at various sites which emerge from the 3,5-di-O-linked α-D-Ara*f* units.

Their non-reducing terminals are capped by mannose residues leading to the so-called ManLAM. These mannose caps are exclusively composed of mono-, di- and trimannosyl units. The latter oligomannosides are α→2 linked.

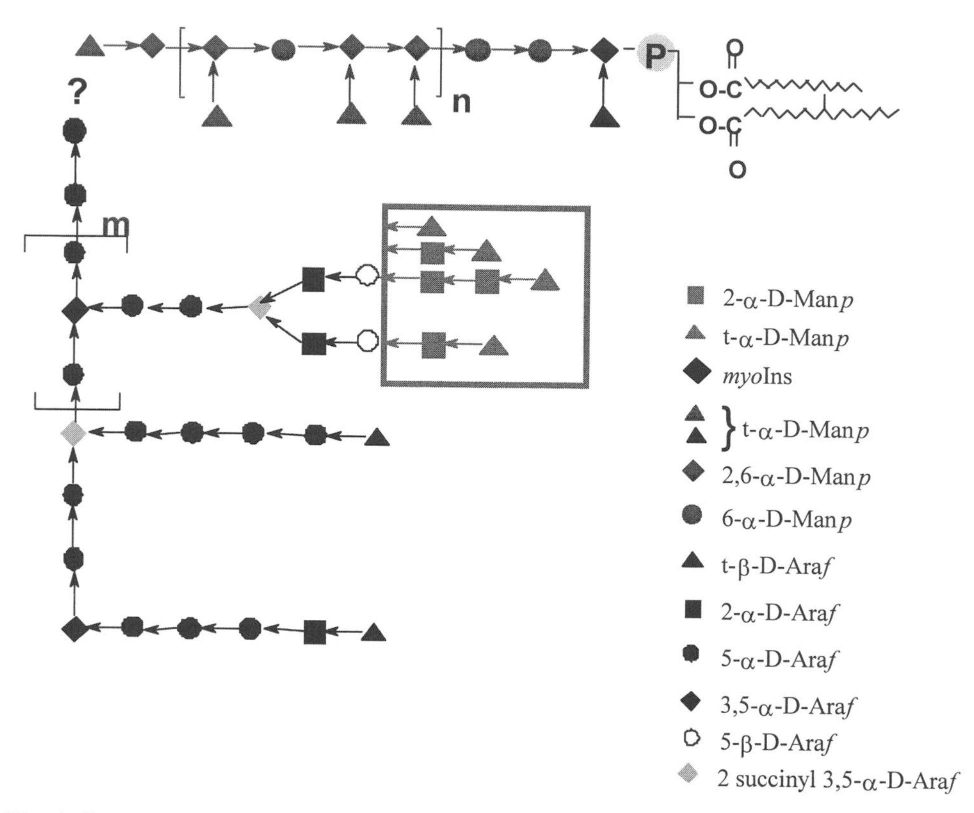

Fig. 1. Current structural model of ManLAM from virulent and pathogenic mycobacteria.

Finally, the last domain is based on glycerol, esterified mainly by palmitic (C16) and tuberculostearic (C19) acids and by a phospho-*myo*-inositol which is glycosylated at the C2 by one single α-D-Man*p* unit and at the C6 by the mannan. More recently the presence of extra fatty acids located on the *myo*-inositol and on the Man*p* unit was determined (unpublished data). This structural model corresponds to the ManLAM from *M. tuberculosis*, *M. leprae* and *M. bovis* BCG, and to date the only significant structural differences described between these ManLAM are restricted to the frequency of capping.

In contrast, it was found that the LAM from non-virulent mycobacteria differs from the previous one by the cap structure, which was assigned to phospho-*myo*-inositol [11, 12].

Parietal and Cellular M. bovis *BCG ManLAM*

Recently, a new extraction approach for the ManLAM from *M. bovis* BCG was developed, giving rise to two types of ManLAM, called parietal and cellular [13]. The parietal one arises from ethanol/ water extraction of delipidated mycobacteria. Afterwards, the resulting cells are disrupted and then extracted again by ethanol/water to give the fraction containing the cellular ManLAM. An important step in this work was the development and successful application of

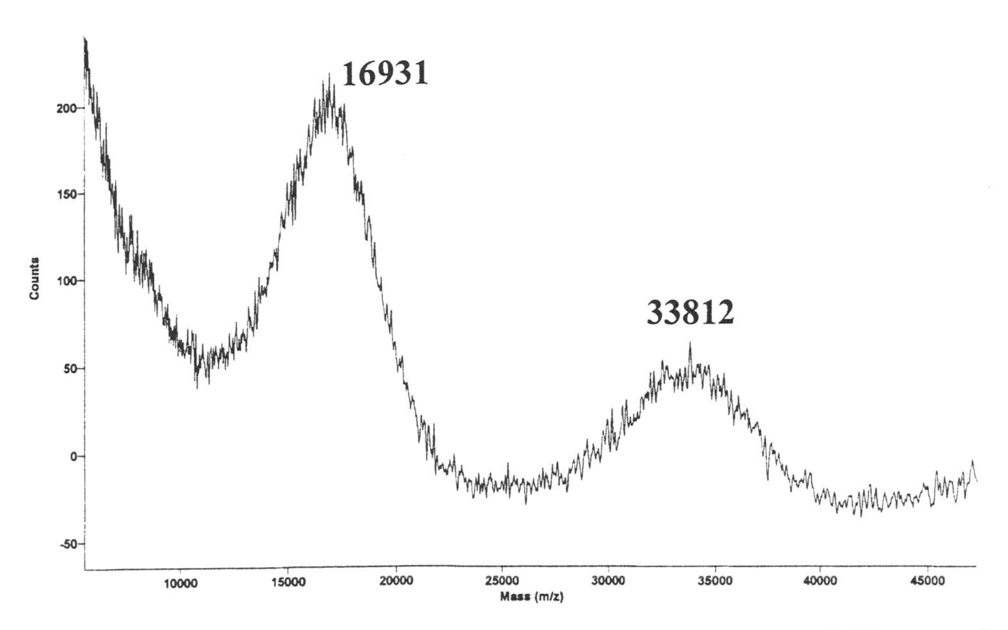

Fig. 2. Positive-ion MALDI-TOF spectrum of the BCG cellular ManLAM. DHB was used as matrix, linear detection.

the use of Triton X-114 to yield ManLAM and lipomannan (LM) devoid of the arabinomannan polysaccharide.

Despite this fractionation, parietal and cellular ManLAM show the same behavior in SDS-PAGE analysis, characterized by broad bands revealing that ManLAM is composed of many glyco- and acylforms [13]. Therefore to determine precisely their molecular weights and heterogeneity, these various ManLAM's were analyzed by MALDI-TOF.

MALDI-TOF Analysis of the Parietal and Cellular ManLAM's

Figure 2 shows the positive-ion MALDI-TOF spectrum of BCG cellular ManLAM obtained using DHB as matrix and run with delayed extraction time and linear detection. As seen in the Figure 2, the spectrum shows two broad peak shapes. These peaks were centered at approximately 17 kDa and 34 kDa and tentatively assigned to the singly charged molecular ions and dimeric molecular ions. This assignment was supported by the negative-ion MALDI-TOF spectrum (Figure 3), dominated by one peak at 16.5 kDa. From these data a molecular mass of 17 kDa was assigned for the most abundant glyco- and acylforms of the cellular ManLAM. Likewise, negative-ion MALDI-TOF mass spectrum of the parietal ManLAM shows a large peak centered at 16 kDa.

In agreement with the SDS-PAGE analysis, the peak width at half-height of approximately 5 kDa, reflects in part the molecular heterogeneity of the ManLAM analyzed. This heterogeneity arises from molecular species which differ mainly by the number of monosaccharide units on both arabinan and mannan, and also from the degree of acylation. Prompt fragments of higher

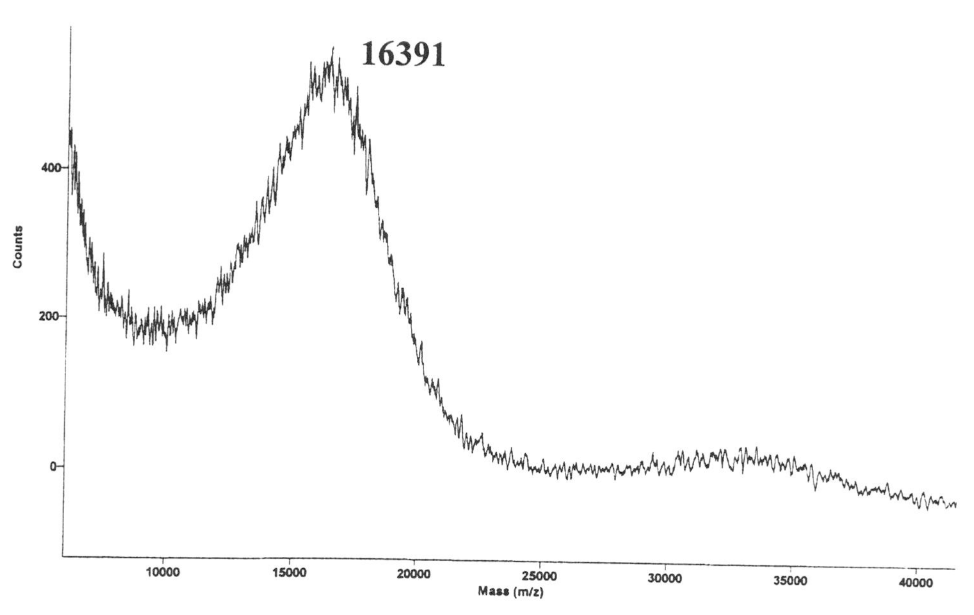

Fig. 3. Negative-ion MALDI-TOF spectrum of the BCG cellular ManLAM. DHB was used as matrix, linear detection.

masses and salt or matrix adducts could also contribute to the width of the peak. However, in negative- and positive-ion modes the mass spectra show the same broad peak shapes, which suggests that fragment ions, cations and matrix adducts of the peak may be considered as minor contributors to the peak shape.

In linear mode, the mass resolution is poor, probably due to metastable decay preventing the characterization of the different glyco- and acyl-forms. However, under reflectron conditions the resulting spectra in both positive- and negative-ion modes were devoid of significant peaks arising from the analyzed molecules. Despite our efforts to improve the sample preparation by desalting using both cation exchange and dialysis [14], the data were unfortunately not improved.

By routine analysis, it was found that the cellular ManLAM contains approximately two palmitic acids and one tuberculostearic acid, but these fatty acids were missing in the parietal form. Ara/Man ratios of 1.4 and 1.2 were determined for the cellular and parietal ManLAM's, respectively. From these data, we estimate the arabinan domain to be around 74 residues and the mannan with the caps to be 43 units, leading to a total of 117 monosaccharide units for the major glyco- and acylform of the cellular ManLAM. Likewise, we found 43 Man residues and 63 Ara units for the parietal ManLAM.

Acyl-Glycerol Structures of the Parietal and Cellular ManLAM

As mentioned above, ManLAM activates αβ T cells. This process requires the binding of the ManLAM to the CD1 molecules. There is now structural evidence that the lipid part of the ManLAM is involved in this binding. Indeed,

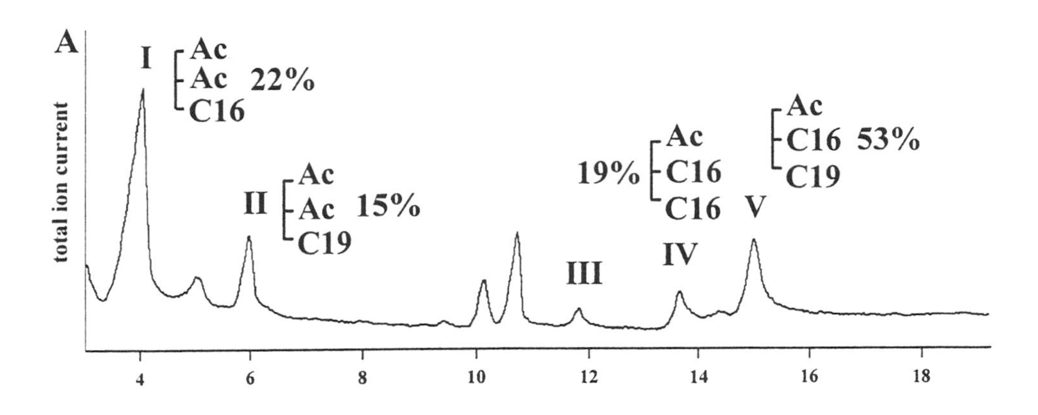

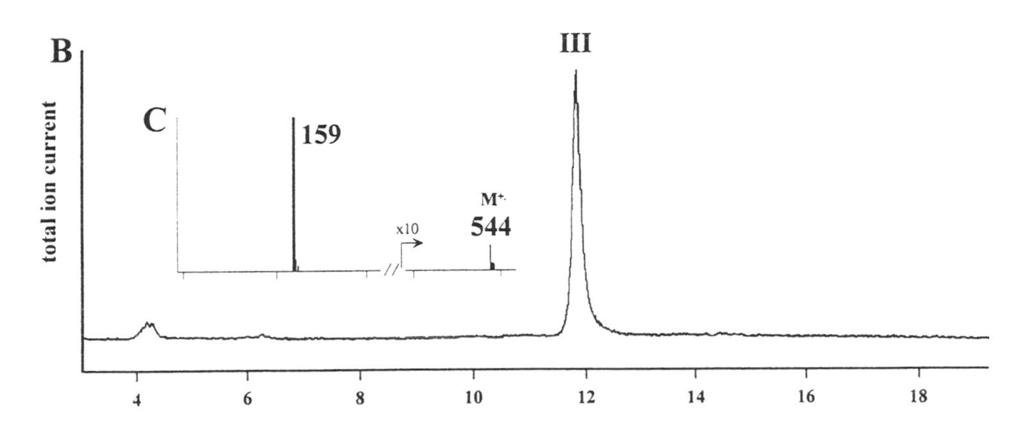

Fig. 4. Acyl-glycerol structures from parietal and cellular ManLAM. (A) EI total ion chromatogram of the cellular acyl-glycerol; Ac, Acetyl residue; (B) EI total ion chromatogram of the parietal acyl-glycerol; (C) mass spectrum of the parietal acyl-glycerol.

the crystal structure of the mouse CD1 molecule reveals an antigen-binding groove [15], exclusively composed of hydrophobic amino-acids, which is suitable to bind the LAM through the fatty acyl chains. Finally, this assumption was supported by the inhibition of LAM binding to CD1 by phosphatidyl-*myo*-inositol mannosides [16].

The parietal and cellular ManLAM's were subjected to acetolysis. The acetolysis reaction preserves the acyl groups but cleaves preferentially the acyl-glycerol-phosphoester linkage as well as the 6-O-linked Man*p* residues. The partially O-acetylated acyl-glycerols were extracted with cyclohexane and the mixture was analyzed by GC/MS using EI and CI modes. Figure 4A shows the EI-total ion chromatogram of the cellular acyl-glycerols. Compounds I and II were assigned to monoacyl-glycerol carrying one C16 [($M^{+\cdot}$ – CH_3CO_2H), m/z

354], ($M+NH_4^+$, m/z 432) and one C19 [($M^{+\cdot} - CH_3CO_2H$), m/z 396], ($M+NH_4^+$, m/z 474) respectively. CI spectra showing adducted molecular ions ($M+NH_4^+$) at m/z 628 and m/z 670 indicate that compounds IV and V correspond to diacylated glycerol with two C16 and with (1)C16/(1)C19, respectively.

In contrast, the EI-total ion chromatogram (Figure 4B) of parietal ManLAM reveals the presence of only one compound with a molecular weight of 544 mu. In addition, from the fragment ions at m/z 159 (Figure 4C) arising from the loss of the carboxylate radical we have determined a molecular weight of 386 for the fatty acid esterifying the glycerol. This fatty acid was released by alkaline hydrolysis, derivatized to the methylester, trimethylsilylated and analyzed by GC/MS using CI/NH_3 and EI ionization modes. A molecular weight of 300 Da was deduced for this fatty acid, from the CI mass spectrum showing $(M+H)^+$ ions at m/z 387 and $(M+NH_4)^+$ ions at m/z 404. In addition, from the EI fragment ions at m/z 187 and m/z 301 the structure 12-hydroxy-stearic acid was proposed. However, its molecular weight is 86 mass units lower than calculated from the molecular weight of the parietal acyl-glycerol. This mass difference is probably due to the loss of an acyl residue during the alkaline hydrolysis but which is preserved by acetolysis. We propose that a methoxypropanoic acid esterifies the 12 –hydroxy-stearic acid leading to the structure 12-O-methoxypropanoyl-12-hydroxy-stearic acid.

In summary, acyl-glycerol analysis reflects the heterogeneity of the cellular ManLAM due to the fatty acyl chains carried on their glycerol moiety. The diacylated glycerol represents the dominant acylforms (72%) among which the glycerol carrying the expected C16 and C19 substituents are the most abundant. In contrast, only one acylform was identified for the parietal ManLAM.

More recently, NMR analysis of the cellular ManLAM revealed extra fatty acids located on the *myo*-inositol and on the Man*p* residues which glycosylate the *myo*-inositol, indicating the presence of tri- and tetraacylated forms of the cellular ManLAM (unpublished data). To identify these fatty acids, the tri- and tetraacylated phosphatidyl-*myo*-inositol mannosides called PIM_2 were analyzed. The carbohydrate moiety of PIM_2 is composed of two Man*p* units located on the C2 and C6 of the *myo*-inositol unit, and it was proposed that LAM's are multiglycosylated forms of the acylated forms of the PIM_2 [9].

Characterization of the Acyl Chains of the Tri- and Tetraacylated Forms of PIM_2

Tri- and tetraacylated PIM_2 were purified from BCG cells and analyzed by ESI-MS in negative mode. The negative-ion ESI spectrum of the triacylated (Fig. 5) is dominated by one peak at m/z 1414.5 assigned to the deprotonated molecular ion. It is deduced that the triacylated form carries exclusively two C16 and one C19. The MS/MS spectrum of the precursor ions m/z 1414.5 shows fragment ions arising from the loss of diacylated anhydroglycerol, which indicates that the glycerol mainly bears C16 and C19. The remaining C16 was located by NMR on the *myo*-inositol residue (unpublished data).

The ESI mass spectrum (Fig. 6) of the tetraacylated forms of PIM_2 is more complex. Three peaks at m/z 1652.6, 1680.6 and 1694.4 assigned to

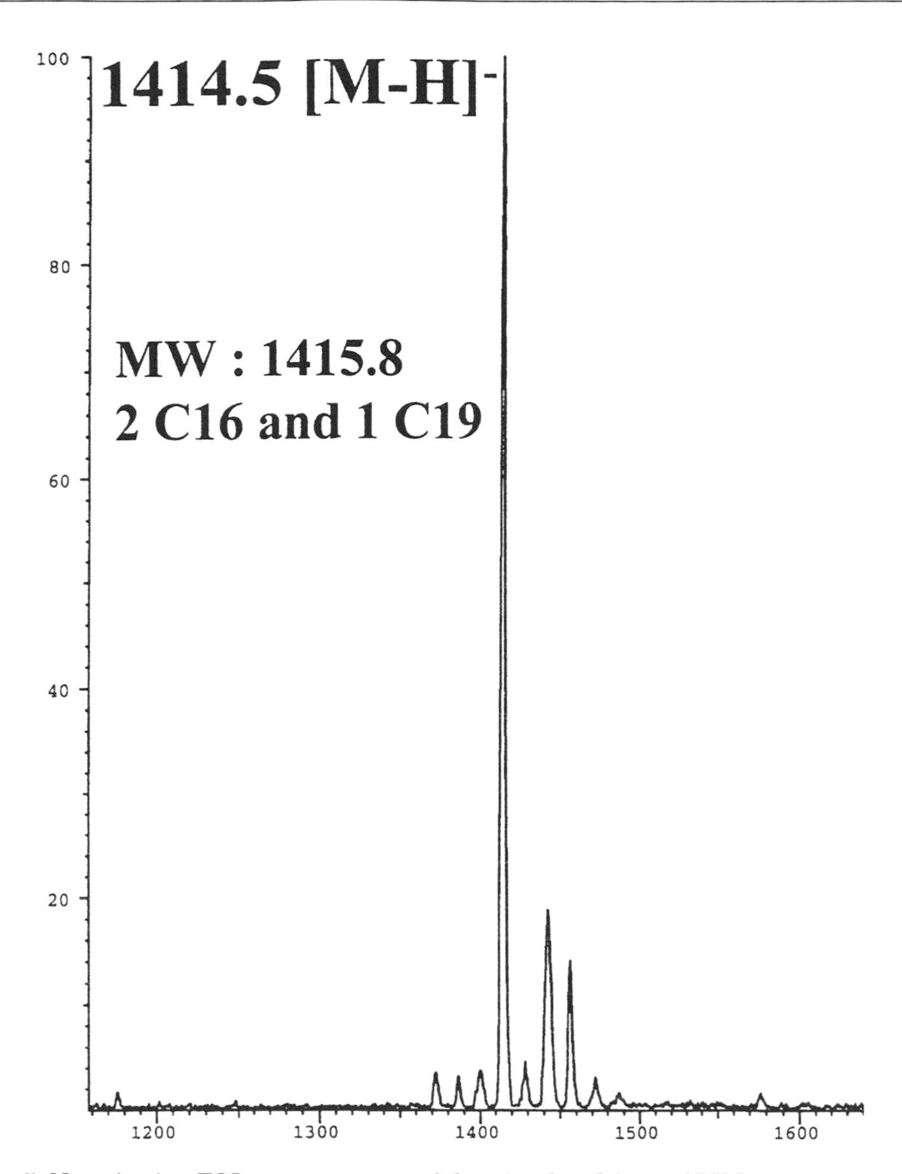

Fig. 5. Negative-ion ESI mass spectrum of the triacylated form of PIM_2.

deprotonated molecular ions reveal the presence of at least three acylforms. The deprotonated molecular ions at m/z 1652.6, 1680.6 and 1694.4 bear (3)C16/(1)C19, (2)C16/(1)C18/(1)C19 and (2)C16/(2)C19, respectively. By MS/MS, it was found that glycerol is mainly esterified by C16 and C19. In addition, from the triacylated PIM_2 structure it was tentatively advanced that the fourth acyl chain C16, C18 or C19 is located on the Man*p* unit.

These tri- and tetraacylated forms of PIM_2 may correspond to the tri- and tetraacylated anchor structures of the cellular ManLAM. However, in the future these ManLAM acylforms must be fractionated for precise structural determination.

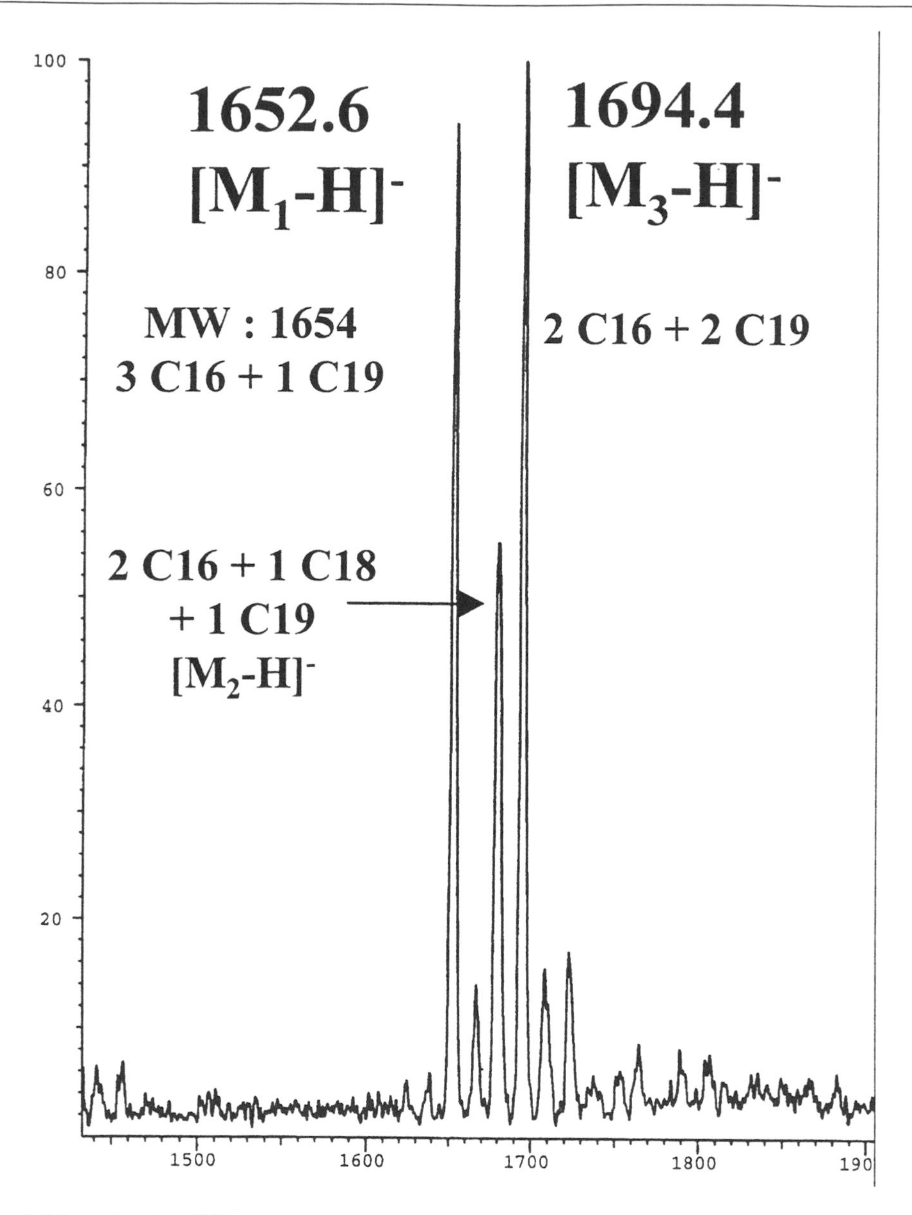

Fig. 6. Negative-ion ESI mass spectrum of the tetraacylated forms of PIM_2.

Characterization of the Mannooligosaccharide Caps by Capillary Electrophoresis/Electrospray Mass Spectrometry (CE/ESI-MS)

ManLAM is exclusively composed of neutral monosaccharides which do not significantly absorb in the UV light. The monosaccharides are then tagged via a reductive amination reaction with the fluorescent dye 1-aminopyrene-3,6,8-trisulfonate (APTS) (17). In addition, the three negative charges carried by the sulfonic groups account for the electrophoretic mobility of the sugar-APTS derivative.

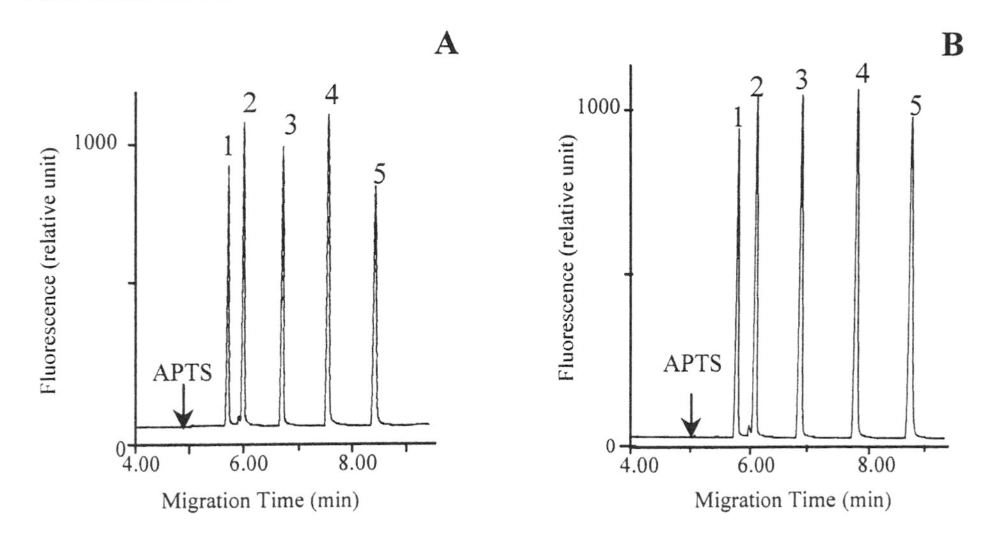

*Fig. 7. LIF electropherogram of APTS sugar derivatives. (A) Triethylammonium acetate buffer, pH 3.5 (1% v/v acetic acid, 30 mM TEA). Peak assignments: **1** (Arabinose-APTS, 5.82 min); **2** (Mannose-APTS, 6.04 min); **3** (Maltose-APTS, 6.74 min); **4** (Maltotriose-APTS 7.56 min); **5** (Maltotetraose-APTS, 8.32 min); (B) Triethylammonium formate buffer, pH 2.7 (1% v/v formic acid, 30 mM TEA).*

Triethylammonium phosphate buffer is currently used for the CE separation of APTS sugar derivatives. However, such a buffer prevents electrospray ionization of the carbohydrate-APTS derivatives. Therefore more volatile acids such as formic and acetic acids were investigated, and the pH was adjusted using TEA. The electropherogram in Figure 7A illustrates the separation of the APTS-derivatized arabinose, mannose, maltose, maltotriose and maltotetraose on uncoated fused-silica capillaries, at pH 3.5, using triethylammonium acetate buffer. The samples were introduced at the cathode and detected at the anode by Laser-Induced-Fluorescence (LIF). As depicted in the Figure 7A, these carbohydrate-APTS are well resolved under such conditions and, as expected, the separation occurs according to their mass because the net charge is the same for all the derivatives. The separation remains similar using triethylammonium formate (Figure 7B). This buffer, which is more volatile, was selected for CE/ESI analysis. A sheath liquid of isopropanol/water (60/40, v/v) at a flow rate of approximately 3µl/min was required to improve the formation of highly charged droplets. In addition, the position of the capillary towards the ESI needle was critical; it was located between 0.1 to 0.5 mm beyond the ESI needle [18].

The total ion electropherogram (Figure 8A) reveals higher migration times due to the length of the capillary (80 cm instead of 40 cm), but also a drastic decrease in the CE chromatography resolution. This point is reflected by the fact that Ara- and Man-APTS derivatives appear as one peak. Nevertheless, maltose-, maltotriose- and maltotetraose-APTS derivatives are still separated.

All the negative-ion ESI mass spectra are dominated by singly and doubly deprotonated molecular ions (see Table 1). However, in the case of the

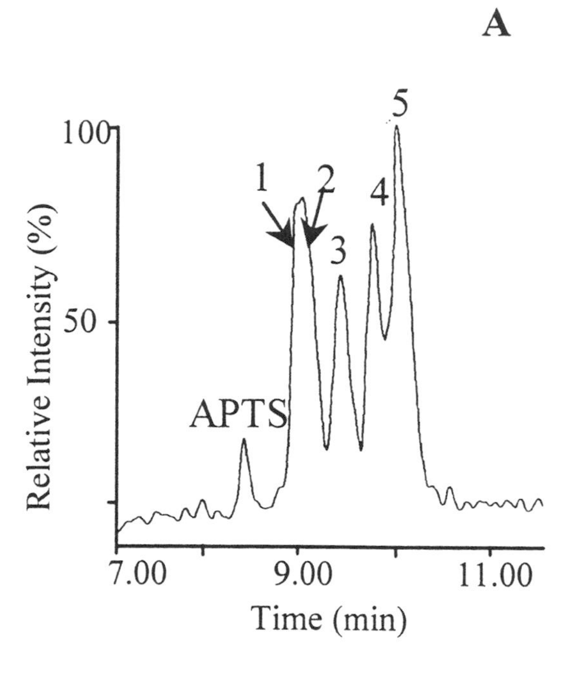

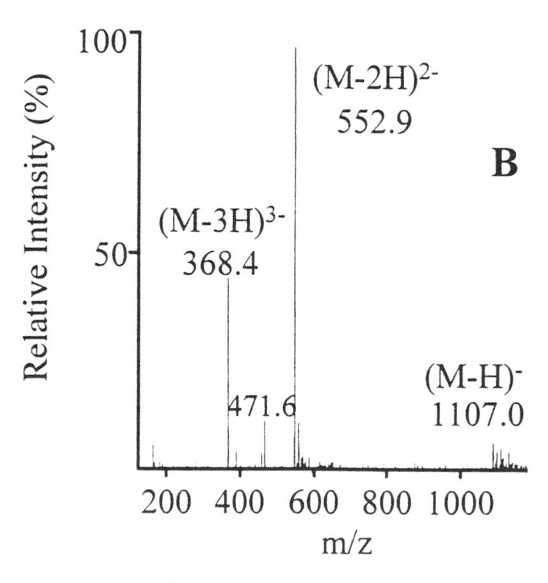

*Fig. 8. CE/ESI-MS analysis of APTS-derivatives from standards. (A) TIC Electropherogram profile of APTS-labeled standards. CE conditions: triethylammonium formate buffer, MS conditions : ESI 4 KV, sheath liquid (isopropanol/water 75/25 v/v; 3 µl/min), sheath gas N_2. Peak assignments: **1** Arabinose-APTS; **2** Mannose-APTS; **3** Maltose-APTS; **4** Maltotriose-APTS; **5** Maltotetraose-APTS; (B) Maltotetraose-APTS negative-ion mass spectrum of compound 5.*

Peak	Migration time (Minutes)			Structure	MW calculated
		$(M-H)^-$	$(M-2H)^{2-}$		
1	8.90	590.1	294.5	APTS-Ara	591.01
2	9.20	620.1	309.6	APTS-Man	621.02
3	9.40	782.3	390.5	APTS-Glc-Glc	783.08
4	9.81	944.4	471.6	APTS-Glc-Glc-Glc	945.13
5	10.05	1106	552.9	APTS-Glc-Glc-Glc-Glc	1107.18

Table 1. Characterization of APTS-carbohydrate standards by CE/ESI-MS analysis. Molecular weights correspond to the calculated exact molecular mass.

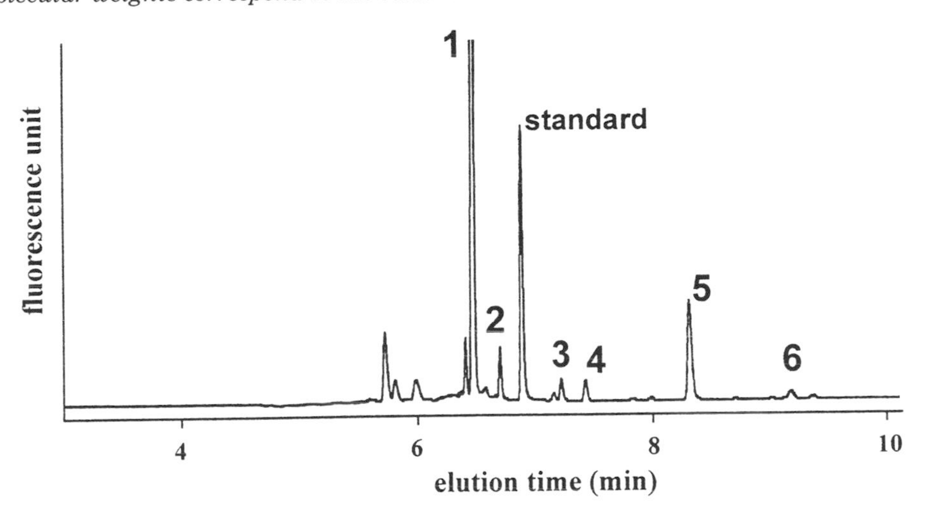

Fig. 9. LIF electropherogram of the APTS sugar derivatives arising from mild hydrolysis of the cellular ManLAM. Hydrolysis condition (0.1N HCl, 30 min, 110°C).

maltotetraose-APTS derivative the ESI negative-ion spectrum (Figure 8B) shows, beside the latter ions, triply charged deprotonated molecular ions and fragment ions at m/z 471.6 arising from the doubly charged deprotonated molecular ions by the loss of anhydro-Glc. The ionization and capillary electrophoresis parameters noted above were selected for the analysis of the mannooligosaccharide caps from the parietal and cellular ManLAM.

The mannooligosaccharide caps were released from 2 µg of cellular ManLAM by mild acid hydrolysis (0.1N HCl, 30 min, 110°C). Under such conditions, the arabinan domain of the ManLAM is preferentially cleaved, giving rise to Ara, mannan linked to the anchor, and mannooligosaccharide caps linked at their reducing ends to Ara. Then the mixture was tagged with APTS and analyzed by CE using the triethylammonium formate buffer. Figure 9 depicts the LIF electropherogram profile revealing the presence of six types of carbohydrate-APTS derivatives. Using a standard, the most abundant APTS derivative (peak 1) was unambiguously assigned to Ara-APTS. Likewise, peak 2, in lower amounts, corresponds to Man-APTS. Finally, the other carbohydrate-APTS derivatives, such as compounds 3 and 4, display electrophoretic mobilities close to those of a disaccharide-APTS standard, and compounds 5 and 6 have similar migration times to tri- and tetrasaccharide-APTS standards, respectively.

To identify these oligosaccharide-APTS derivatives, they were analyzed by CE/ESI-MS. Again, a drastic decrease of the chromatographic resolution is observed (Figure 10A). However, the negative-ion ESI spectra unambiguously allow the identification of the carbohydrate-APTS derivatives. This was reflected by the two mass spectra extracted from the unresolved peak at 13.3 min showing doubly charged deprotonated molecular ions at 294.6 and 309.6, typifying Ara-APTS and Man-APTS, respectively (Table 2). Likewise (Table 2), from the mass values of the doubly charged deprotonated molecular ions, compounds 3 and 4 were assigned to Ara-Ara-APTS and Man-Ara-APTS, respectively. Finally, the ESI spectrum of compound 5 (Figure 10B) is dominated by the peak, at m/z 456.6, assigned to doubly charged deprotonated molecular ions identifying a trisaccharide-APTS derivative composed of two Man and one Ara residue. The presence of a less intense peak is observed at m/z 375.6 arising from the loss of anhydro-Man, indicating that a Man residue is located at the non-reducing end. From these results the following sequence was proposed for compound 5: Man-Man-Ara-APTS.

This analytical approach was also applied to the parietal ManLAM, leading to the same structure of the mannooligosaccharide caps in similar relative abundances.

Conclusion

From a structural point of view, now much is known about the caps and anchor structures of the ManLAM which contribute as virulence and/or protective epitopes. By the combination of acetolysis and conventional EI and CI mass spectrometry, we were able to identify the acyl-glycerol forms of the parietal and cellular ManLAM. In addition, from the ESI-MS analysis of the tri- and tetraacylated PIM_2 we were able to characterize the extra fatty acids which esterified the *myo*-inositol and the Man*p* units of the anchor.

CE/ESI-MS provides a powerful analytical approach for the separation and the characterization of the mannoligosaccharide caps. This approach was successfully applied to a complex mixture, thus no purification step was required after ManLAM hydrolysis. CE/ESI-MS was conducted at a nanomolar range of ManLAM, and this amount could be decreased by a factor of one hundred when LIF was used.

The ManLAM structural study using MALDI-TOF, CE/ESI-MS and conventional mass spectrometry tools has revealed the glyco- and acylforms of the parietal and cellular ManLAM. The next step will be their fractionation and the determination of their precise structure.

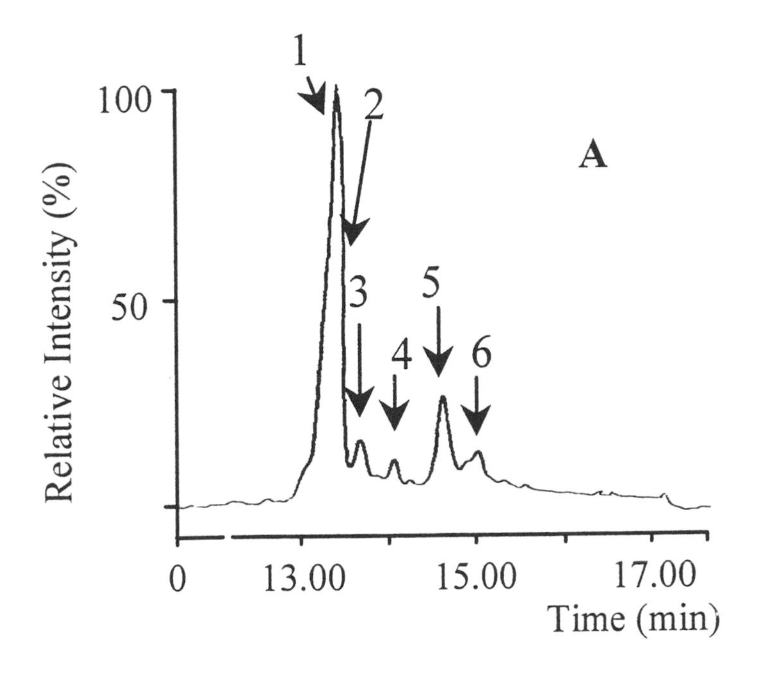

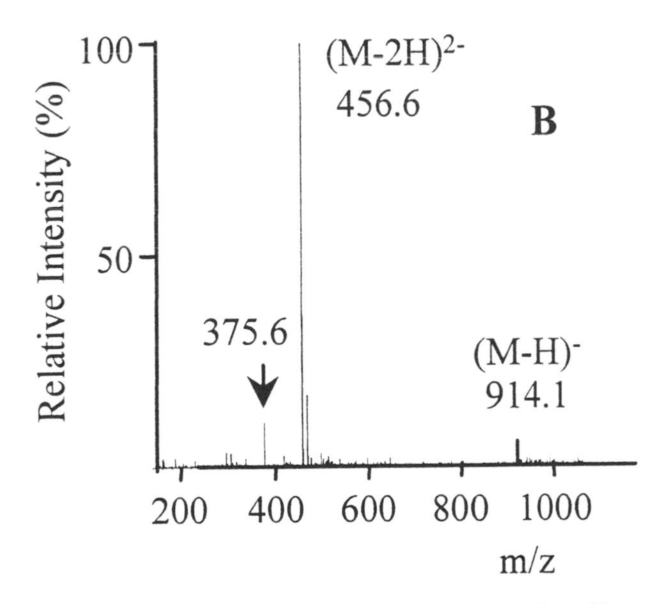

Fig. 10. Analysis of the APTS-mannooligosaccharide caps from the cellular ManLAMs of M. bovis *BCG. (A) TIC profile from CE / ESI-MS: CE and MS conditions are the same as those described in the legend of Figure 8; (B) Extracted ion negative mass spectrum of peak 5.*

Peak	Migration time (Minutes)			Structure	MW calculated
		$(M-H)^-$	$(M-2H)^{2-}$		
1	13.25	590.1	294.6	APTS-Ara	591.01
2	13.40	620.3	309.6	APTS-Man	621.02
3	13.60	722.6	360.3	APTS-Ara-Ara	723.06
4	14.05	_	375.6	APTS-Ara-Man	753.07
5	14.60	914.1	456.6	APTS-Ara-Man-Man	915.12
6	15.01	1046.8	522.9	APTS-Ara-Ara-Man-Man	1047.15

Table 2. Identification of the APTS-labeled mannooligosaccharide caps from cellular ManLAM of M. bovis *BCG by CE/ESI-MS analysis.*

To date, the structural models proposed for the LAM have been carried out on relatively large scale preparations isolated from mycobacteria grown on liquid culture media. In the future, we will need to improve the sensitivity to the subpicomole range to perform LAM analysis from animal and human sources without external culture. In this context, the fractionation step could be overcome by direct analysis of the LAM from electrophoresis gels by MALDI-TOF and CE/ESI-MS.

ACKNOWLEDGEMENTS

The authors would like to thank Mrs. Elisabeth Ryan from PerSeptive Biosystems for recording the MALDI-TOF mass spectra of the LAM. This work was supported by grants from the Mission Scientifique et Technique du Ministère de l'Education Nationale, de l'Enseignement Superieur et de la Recherche, action Microbiologie.

REFERENCES

1. B. R. Bloom and G. J. Murray, *Science* **1992**, *257*, 1055-1064.
2. L. S. Schlesinger, *Curr. Topics Microbiol. Immunol.* **1996**, *215*, 71-96.
3. L. S. Schlesinger, *J. Immunol.* **1993**, *150*, 2920-30.
4. L. S. Schlesinger, T. M. Kaufman, S. Iyer, S. R. Hull and L. K. Marchiando, *J. Immunol.* **1996**, *157*, 4568-75.
5. A. Venisse, J. J. Fournie and G. Puzo, *Eur. J. Biochem.* **1995**, *231*, 440-7.

6. E. M. Beckman, S. A. Porcelli, C. T. Morita, S. M. Behar, S. T. Furlong and M. B. Brenner, *Nature* **1994**, *372*, 691-4.

7. D. B. Moody, B. B. Reinhold, M. R. Guy, E. M. Beckman, D. E. Frederique, S. T. Furlong, S. Ye, V. N. Reinhold, P. A. Sieling, R. L. Modlin, G. S. Besra and S. A. Porcelli, *Science* **1997**, *278*, 283-6.

8. P. A. Sieling, D. Chatterjee, S. A. Porcelli, T. I. Prigozy, R. J. Mazzaccaro, T. Soriano, B. R. Bloom, M. B. Brenner, M. Kronenberg, P. J. Brennan, *et al.*, *Science* **1995**, *269*, 227-30.

9. D. Chatterjee and K. H. Khoo, *Glycobiology* **1998**, *8*, 113-20.

10. G. Nigou, A. Vercellone and G. Puzo, Frontiers in Biosciences **1998**

11. M. Gilleron, N. Himoudi, O. Adam, P. Constant, A. Venisse, M. Riviere and G. Puzo, *J. Biol. Chem.* **1997**, *272*, 117-24.

12. K. H. Khoo, A. Dell, H. R. Morris, P. J. Brennan and D. Chatterjee, *J. Biol. Chem.* **1995**, *270*, 12380-9.

13. J. Nigou, M. Gilleron, B. Cahuzac, J. D. Bounery, M. Herold, M. Thurnher and G. Puzo, *J. Biol. Chem.* **1997**, *272*, 23094-103.

14. B. W. Gibson, J. J. Engstrom, C. M. John, W. Hines and A. M. Falick, *J. Am. Soc. Mass Spectrom.* **1997**, *8*, 645-658.

15. Z. Zeng, A. R. Casta, B. W. Segelke, E. A. Stura, P. A. Peterson and I. A. Wilson, *Science* **1997**, *277*, 339-45.

16. W. A. Ernst, J. Maher, S. Cho, K. R. Niazi, D. Chatterjee, D. B. Moody, G. S. Besra, Y. Watanabe, P. E. Jensen, S. A. Porcelli, M. Kronenberg and R. L. Modlin, *Immunity* **1998**, *8*, 331-340.

17. R. A. Evangelista, M. S. Liu and F. T. A. Chen, *Anal. Chem.* **1995**, *67*, 2239-2245.

18. K. L. Johnson and A. L. Tomlinson, *Rapid Commun. Mass Spectrum.* **1996**, *10*, 1159-1160.

Questions and Answers

Pierre Thibault (National Research Council)

Is there any selective binding of the different LAM structures having heterogenous Man population?

Answer. It was established by Sieling *et al.* (*Science* **1995**, *269*, 227) that the double negative αβ T cell line, namely "LDN4", obtained from the skin lesion of a leprosy patient responds selectively to the ManLAM from *M. leprae* but not to the ManLAM from *M. tuberculosis*. To date, the discrimination of these ManLAM by T cell clones was tentatively explained by the difference of their degree of capping. Indeed, it was found by the Brennan group that the mannose capping is lower in the ManLAM of *M. leprae* compared with *M. tuberculosis*. These ManLAM are presented to the T cell receptors by the CD1 molecules which are expressed in the membrane of antigen presenting cells. We now have evidences that the ManLAM binding to CD1 occurs via the acyl chain moiety of the ManLAM, and it can be speculated that the different acylforms of the ManLAM revealed by this work will bind the CD1 with different affinities. Thus discrimination could also occur during the binding of the ManLAM to CD1.

Pierre Thibault (National Research Council)

In vaccine formulation would you recommended a multivalent vaccine that incorporates both LAM structures and antigenic peptides?

Answer. There is now a consensus that protective immunity against *M. tuberculosis* requires cooperative interactions between macrophages and $CD4^+$ and $CD8^+$ αβ T cells. The role of these DN αβ T cells *in vivo* in protective immunity is not yet clearly established. If, in the future, it is found that these T cells play a key role, then CD4 peptides added to the protective epitope of the ManLAM could be tested.

K. A. Karlsson (Gothenburg University)

So the model includes taking up of the glycoconjugate and processing it for a transport out in the antigen-binding site for a presentation to the T-cell; does this mean that the fatty acids you have identified are essential for a hydrophobic anchoring in the site, and also potentially important in a vaccine?

Answer. We now have evidence from the crystalline structure of the CD1 binding site and Surface Plasmon Resonance experiments that the fatty acyl residues are essential for the ManLAM binding to the CD1 groove. However, to date these experiments were conducted, as shown by this work, on a mixture containing mono-, di-, tri- and tetraacylated forms of the ManLAM. In the future, we plan to separate these acylforms to study their interaction with CD1. Polysaccharides called Arabino-Mannans from the mycobacteria envelope which are devoid of the anchor do not express T cell activity, highlighting the central role of the fatty acid residues in this process. However, the anchor alone is not sufficient to activate these T cells.

The Impact of Drug Metabolism in Contemporary Drug Discovery: New Opportunities and Challenges for Mass Spectrometry

Thomas A. Baillie and Paul G. Pearson
Department of Drug Metabolism, Merck Research Laboratories,
West Point, PA 19486

Abstract

Current developments in contemporary drug discovery techniques offer the potential to discover and develop new chemical entities to treat a spectrum of diseases or medical conditions for which either no current therapies exist, or for which existing therapies are inadequate. In particular, advances in synthetic medicinal chemistry (i. e., combinatorial chemistry and rapid analog synthesis), the availability of new molecular targets derived from genomics-based discovery, and high-throughput screening technology have the potential to identify new chemical entities (NCEs) with potent and selective pharmacological properties. To realize the full potential of these techniques, it is absolutely essential that a contemporary pharmacokinetics and drug metabolism program be implemented as a fully-integrated component of the drug-discovery process.

Introduction

At the present time, the cost to discover and develop an NCE ranges from \$250-500 million and may take up to 15 years from the initial identification of a disease-related molecular target. The actual magnitude of cost and time reflects the complexity of the drug development process in terms of the intended routes of administration, duration of administration and target patient population. In addition, execution of a successful drug discovery program frequently requires the evaluation of multiple lead compounds in Phase I and II clinical trials prior to selection of an NCE with optimal safety, pharmacokinetic and pharmacodynamic properties. During this process, compounds with undesirable properties may be eliminated at various stages of development (i. e., preclinical safety assessment, Phase I and Phase II clinical evaluation) for a variety of reasons. A joint benchmark study conducted in 1991 by the Food and Drug Administration and Pharmaceutical Manufacturers Association (now PhRMA) identified several factors which lead to termination of drug development programs (Fig. 1). At the time of that study, inappropriate pharmacokinetics was identified as the leading reason (40%) for discontinuation. Other prominent causes included preclinical toxicity (11%) and adverse events in man (10%); in some cases, these adverse events were ascribed to the formation of chemically-reactive drug metabolites. It is readily apparent, therefore, that a detailed early metabolic and pharmacokinetic characterization of lead compounds has the potential to eliminate poor compounds as early as possible in the discovery

PMA/FDA Preclinical Meeting May 21, 1991

Reason for Failure	Incidence
Pharmacokinetics	40 %
Lack of Efficacy	30 %
Animal Toxicity	11 %
Adverse Effects in Man	10 %
Commercial Reasons	5 %
Miscellaneous	4 %

Fig. 1. FDA/PMA (now PhRMA) survey-1991: Factors which lead to termination of drug development programs.

phase. In addition, this philosophy, when embraced in a prospective manner, can result in the discovery of NCEs with superior metabolic and pharmacokinetic characteristics.

A comprehensive metabolic characterization of each lead compound requires that a broad range of studies be performed to address critical issues in drug metabolism [1] (Fig. 2). Therefore, it is important that relatively high throughput assays be employed for this purpose so that large numbers of lead compounds can be evaluated for suitability as drug candidates [2, 3]. The overall goal of this approach is to provide a rational selection of lead compounds at the drug metabolism/discovery interface (Fig. 3). Specific preclinical metabolism objectives include optimization of pharmacokinetic properties (notably terminal half-life and oral bioavailability) and obtaining information to predict the potential for undesirable drug-drug interactions [1]. The availability of this information at an early stage facilitates lead selection and dictates the shortest path to clinical proof of concept testing (Fig. 3). In most cases, these studies require highly sensitive and specific analytical methods for the quantitation of lead compounds and for structural characterization of drug metabolites. Tandem mass spectrometry, when used in conjunction with atmospheric pressure ionization techniques and liquid chromatography, has become an essential tool for quantitation and structural elucidation of drug candidates and metabolites, respectively. Indeed, LC-MS/MS has become established as an indispensable tool for quantitative applications in pharmaceutical research, including the very sensitive, highly specific analysis of mixtures [4, 5]. Furthermore, rapid structural elucidation of drug metabolites by mass spectrometry provides timely information on biotransformation pathways that may be responsible for extensive metabolic clearance and inappropriate pharmacokinetic properties. In

Critical Issues for Drug Metabolism

- Bioavailability
  - Poor absorption versus pre-systemic elimination
  - Metabolic basis for clearance
- Pharmacokinetics consistent with intended use
- Drug-drug interactions
  - Enzyme induction/inhibition
- Active/toxic metabolites
  - Reactive intermediate formation
- Early evaluation of large numbers of lead compounds

Fig. 2. Critical drug metabolism and pharmacokinetic issues to be addressed in the selection of lead compounds during the drug discovery process.

Drug Metabolism/Discovery Interface

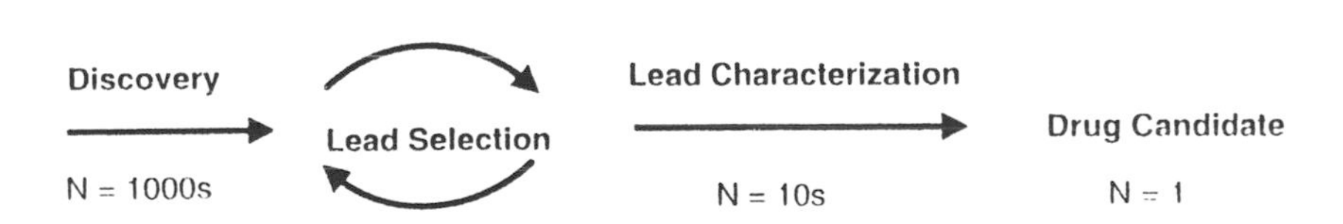

- Rational selection of lead compounds
  - Eliminate poor compounds ASAP
  - Optimize PK, bioavailability
  - Minimize drug interactions
- Facilitate the shortest path to the clinic
  - Proof of concept, reduce cycle time for backup compounds

Fig. 3. The role of drug metabolism at the interface with drug discovery in the selection, characterization and optimization of lead compounds

addition, the characterization of glutathione conjugates provides insights into the structure of short-lived, reactive metabolites that may serve as mediators of drug-induced toxicities [6]. Therefore, the availability of information on the metabolism of lead compounds at an early stage in the discovery process often can be exploited in directing structural modification of a lead compound to

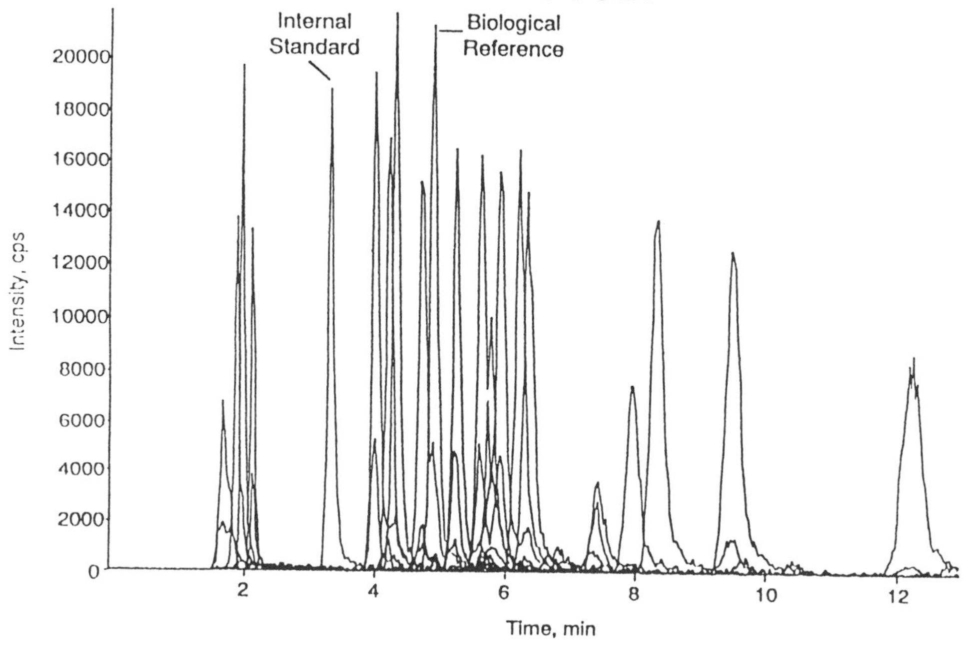

Fig. 4. Selected reaction monitoring chromatogram of an extract of plasma from a dog simultaneously dosed orally with twenty drug candidates.

enhance metabolic stability or circumvent formation of chemically-reactive drug metabolites. In this chapter, examples are presented of the utility of LC-MS/MS in the drug discovery process with regard to issues of a quantitative, qualitative or mechanistic nature.

QUANTITATIVE APPLICATIONS

Early and rapid pharmacokinetic characterization of lead compounds in animal models serves as a screening tool for the rank-ordering and selection of lead compounds. The sensitivity and selectivity of LC-MS/MS techniques, in conjunction with protocols based on cassette dosing regimens permits the simultaneous pharmacokinetic characterization of a large number of compounds [7-9]. In conventional pharmacokinetic studies, single compounds are administered to individual animals in a crossover study design, and quantitative analysis of drug in plasma then is conducted to derive the pharmacokinetic parameters (i. e., clearance, volume of distribution, oral bioavailability). While this approach yields detailed pharmacokinetic information, it is "low-throughput" and not amenable to analysis of combinatorial libraries. In contrast, the cassette-dosing approach involves the administration of multiple compounds (e. g., n=20) to single animals in parallel study designs to evaluate multiple

Single *Vs.* Multiple Component PK Screening

Program (Clearance)	Route of Admin.	Key PK Parameters	Changes from Single to Multiple Component Dosing
Program 1 (low)	I.V.	$t_{1/2}$	Generally: 20-30% increase in $t_{1/2}$
Program 2 (moderate)	P.O./I.V.	% F, $t_{1/2}$, Cl_p	Generally: 20-50% increase $t_{1/2}$ and reduction of Cl_p Occasionally: 2 to 3-fold increase in % F
Program 3 (moderate)	P.O.	C_{max}, AUC, $t_{1/2}$	Generally: 20-60% increase in C_{max} and AUC; 30% increase in $t_{1/2}$ Exceptionally: 5-fold increase in AUC and $t_{1/2}$
Program 4 (moderate-high)	P.O./I.V.	% F, $t_{1/2}$	Generally: 2-fold increase in % F and $t_{1/2}$ Occasionally: 8 to 15-fold increase in % F
Program 5 (high)	P.O.	C_{max}, AUC, $t_{1/2}$	Generally: 10 to 15-fold increase in C_{max} and AUC; 2 to 3-fold increase in $t_{1/2}$

Table 1. A comparison of pharmacokinetic parameters derived from single and cassette dosing studies for Merck Investigational compounds. Observed changes in key pharmacokinetic parameters on cassette dosing are classified according to the plasma clearance characteristics of the compounds from Programs 1-5.

administration routes. The cassette dosing approach takes advantage of the sensitivity and selectivity of LC-MS/MS methodology to provide quantitative information on plasma concentrations of individual components to permit the derivation of pharmacokinetic parameters (Fig. 4). The simultaneous pharmacokinetic characterization of multiple compounds serves to eliminate inter-individual variability, saves time, reduces the number of animals and is readily applicable to the quantitative pharmacokinetic characterization of small combinatorial libraries.

The limitations of this approach are based on the potential for metabolism based drug-drug interactions, which may occur when mixtures of uncharacterized compounds are administered simultaneously, leading to erroneous estimates of pharmacokinetic parameters. For example, oral bioavailability may be overestimated when one component of the mixture may inhibit competitively, or irreversibly, a metabolic clearance process for a second component of the mixture, resulting in increased systemic exposure of the second component. A typical example of this situation is where one component of the mixture inhibits a cytochrome P450 enzyme which plays an important role in the metabolism of a second component. As a consequence, the elimination half-life of the second compound is prolonged and its pharmacokinetics, in general, appear to be superior to those obtained following single compound administration ("false positive").

A second limitation of the cassette dosing approach is encountered when one or more components of the mixture display non-linear pharmacokinetics. In this case, metabolic clearance may be overestimated ("false negative"), as cassette studies typically employ doses that are one-tenth (for a mixture of ten components) of those administered in single compound dosing studies. In general, the compounds that are prone to "false-negative" or "false-positive" pharmacokinetic results are those with high plasma clearance values, and these effects are more pronounced following oral administration (Table 1). To overcome these limitations, an ideal cassette dosing study will contain a biological internal standard in the form of a structurally-representative compound, which has well characterized pharmacokinetic properties derived from single-administration studies employing a range of doses and administration routes. Therefore, a comparison of the pharmacokinetic profile of this biological internal standard across a large number of cassette-dosing experiments serves to report on potential metabolic-based drug-drug interaction that may be occurring. Periodically, it may be necessary to change biological internal standards, as the structural analog series evolves chemically to yield lead compounds of diverse structure and improved pharmacokinetic properties. Furthermore, the cassette-dosing approach places high demands on mass spectral sensitivity and resolution, together with chromatographic resolution to separate components of analytical interest from metabolites formed in the biological experiment (Fig. 4). Alternative approaches, including post-dose pooling of plasma samples [10] and continous blood withdrawal [11], have also been explored to increase throughput in pharmacokinetic studies.

Qualitative Applications

Early structural characterization of metabolites during the discovery process provides valuable information on metabolic "soft-sites", which may be chemically "blocked" by informed chemical intervention to enhance metabolic stability and improve pharmacokinetic properties. In addition, identification of pharmacologically-active metabolites facilitates adequate patent protection and yields "structural opportunities" for new lead compounds. LC-MS/MS is a powerful analytical tool for the structural characterization of drug metabolites isolated from biological matrices.

Classical approaches to the recognition and identification of metabolites at early stages in drug discovery, prior to the availability of radiolabeled compound, require tedious visual inspection of individual mass spectra acquired during lengthy LC-MS/MS analyses, assignment of fragmentation pathways and confirmatory experiments to support the proposed structure (e. g., high resolution mass spectrometry). In contrast, the application of newer automated spectral analysis techniques, including Correlation Analysis [12], eliminates the need to manually inspect chromatograms and mass spectra, and automates the recognition of metabolites derived from the compound of interest. Correlation Analysis employs pattern recognition techniques to automate the comparison of the product ion spectrum of a parent drug with product ion spectra of putative drug metabolites recorded from biological matrices [12]. An initial LC-MS analysis of a pre-dose biological sample (e. g., urine or control hepatic microsomal

MICROSOMAL INCUBATION OF A MIXTURE OF 5 MERCK COMPOUNDS

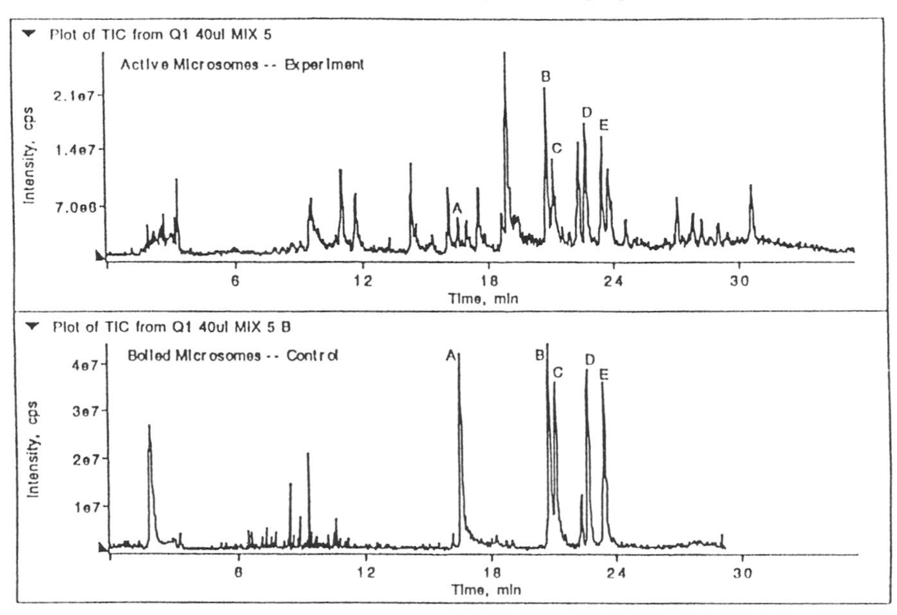

Fig. 5. Representative total ion chromatograms of metabolites of five Merck Investigational compounds as a mixture of human liver microsomes (top panel) and heat deactivated microsomes (lower panel).

incubations) is used to establish a control database of molecular ions and retention times of endogenous molecular entities. A second LC-MS analysis of an appropriate urine collection following drug administration, or microsomal incubation of the investigational compound(s), facilitates the recognition of "drug treatment-related" components. A third experiment invokes an LC-MS/MS analysis to record product ion mass spectra of "treatment-related components". The product ion data set thus generated subsequently, is subjected to a correlation analysis against the product ion spectrum of the parent, which affords a similarity index based upon automated pattern recognition techniques. A numerical correlation value is established, which serves to rank the "treatment-related" components with respect to their structural similarity to the parent drug.

Correlation analysis has been developed at Merck for this purpose and employed for automated characterization of metabolites of lead compounds from a drug discovery program [13]. In this manner, five Merck investigational compounds (A-E) were incubated with either metabolically-competent liver microsomes, or a heat inactivated microsomal preparation (Fig. 5). Using the approach previously described, the MH^+ ions of the drug-related metabolites were recognized in the microsomal incubation and their product ion mass spectra were subsequently recorded. A correlation analysis was conducted to

ASSIGNMENT OF SUBSTRATE-METABOLITE RELATIONSHIPS CORRELATION ANALYSIS

UNKOWN COMPONENT MOLECULAR ION	CORRELATION VALUE				
	A 551	B 553	C 535	D 589	E 492
264	0.121	0.225	**0.658**	0.192	0.465
282	0.096	**0.539**	0.232	0.172	0.377
282b	**0.220**	-0.017	-0.021	-0.034	0.182
290	-0.012	**0.293**	0.212	0.272	0.026
318	0.015	0.212	0.458	**0.666**	0.580
324	0.123	0.012	0.531	0.014	**0.845**
383	**0.445**	0.062	0.141	0.05	0.137
404	0.071	0.156	0.272	**0.429**	0.293
506	0.010	0.214	0.278	0.256	**0.938**
507a	0.464	0.197	0.275	0.142	**0.577**
507b	0.082	1.090	**1.244**	1.112	0.171
508	0.327	0.404	0.390	0.191	**0.864**
509	0.208	0.509	**0.861**	0.471	0.207
522	0.167	0.173	0.203	0.125	**0.816**
524	0.315	0.297	0.420	0.296	**0.893**
533	0.114	1.147	**1.300**	1.033	0.315
569	0.397	**1.230**	0.744	0.545	0.349

Table 2. Assignment of substrate-metabolite relationships through correlation analysis of the product ion mass spectra of drug-related components vs. each of the parent drugs (A-E). The highest correlation values within a substrate-metabolite series are respresented in bold type.

assign substrate-metabolite relationships within the individual parent drug-metabolite series (Table 2). In this investigation, seventeen drug-related metabolites were detected in the microsomal incubation and their metabolic origin was assigned to one of the five progenitor metabolic substrates. In this manner, correlation analysis facilitated the recognition of drug-related components in a complex biological system [12]

Mechanistic Studies

The timely execution of mechanistic drug metabolism studies in the drug discovery process can yield pivotal information to direct synthetic medicinal chemistry efforts. Mechanistic studies typically are designed to address specific critical issues for drug metabolism (see Fig. 2). The information thus derived can serve to identify potential liabilities associated with a specific chemical class of lead compound. For example, an undesirable feature of a lead compound is the potential for the formation of chemically-reactive drug metabolites that may serve as mediators of drug-induced toxicities. These toxicities may be manifested in definitive preclinical toxicology studies, or alternatively may become apparent only at late stages of clinical development in the form of idiosyncratic drug-induced toxicities. Thus, the idiosyncratic reactions of a number of NSAID drugs

HIV Protease Inhibitors

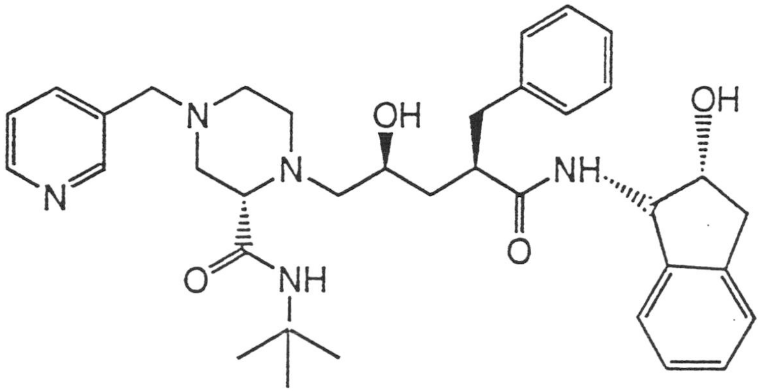

L-753,524 (MK-0639; Indinavir)

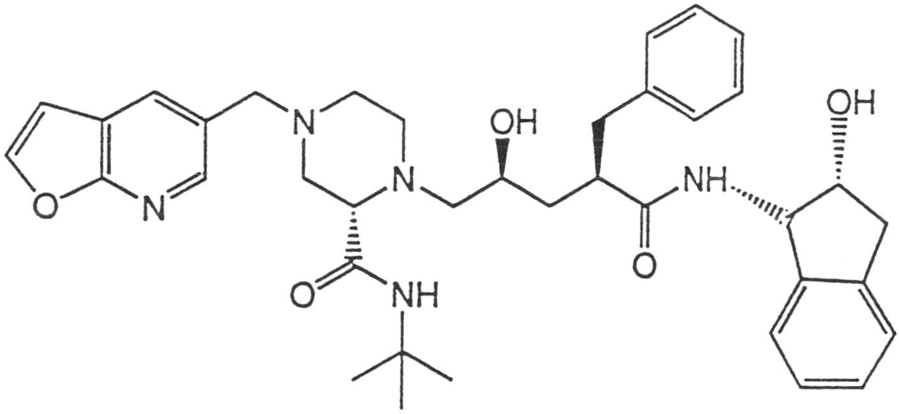

L-754,394

Fig. 6. Structures of HIV-1 protease inhibitors, Indinavir (L-753,524) and L-754,394450 (reproduced with permission from [18]).

has been associated retrospectively with the formation of reactive acyl glucuronide metabolites; and it becomes important, therefore, to evaluate for chemical reactivity, acyl glucuronide metabolites of candidate drugs at an early stage of development. Clearly, timely studies to characterize chemically-reactive drug metabolites can facilitate informed decisions on the selection of lead compounds for clinical development.

Merck has made significant contributions to the treatment of Acquired Immunodeficiency Syndrome by the discovery and development of the HIV protease inhibitor, Indinavir (Crixivan ™, Fig. 6). Continuing discovery efforts have sought to develop an analog of Indinavir with a longer plasma half life to circumvent the three times daily dosing regimen required to sustain Indinavir plasma concentration at effective anti-viral levels. During these efforts, a lead compound was identified (L-754,394, Fig. 6) which exhibited an increased plasma half-life in animals. However, upon further evaluation this apparent attribute was shown to be related to a non-linearity in the pharmacokinetics of L-754,394, which gave rise to substantial accumulation of the drug during multiple-dose administration studies in animal models [14]. These pharmacokinetic characteristics, when viewed together, are consistent with the irreversible inhibition of a metabolic clearance mechanism. Inspection of the chemical structure of L-754,394, reveals the presence of a furan ring, a structural entity that may undergo metabolic activation to a chemically-reactive intermediate [15, 16]. Mechanistic studies conducted *in vitro* with human liver microsomal preparations, indicated that incubation of L-754,394 resulted in a time-dependent loss of activity of the cytochrome P450 isoform, CYP3A [17]. This observation is consistent with mechanism-based inactivation of CYP3A by a chemically-reactive drug metabolite. Additional *in vitro* studies were conducted in which microsomal incubations were fortified with glutathione or methoxylamine, and the trapped electrophilic species was characterized by mass spectrometry as the corresponding glutathione conjugate or oxime derivative, respectively [18] (Fig. 7). The result of these experiments indicated that the furan ring of L-754,394 underwent metabolic activation to a presumed electrophilic epoxide intermediate that caused mechanism based inactivation of cytochrome P450, a long apparent half-life of the parent drug and the observed non-linearities in pharmacokinetics [14, 18]. The structural information thus obtained was critical in guiding synthetic chemistry efforts and resulted in elimination of L-754,394 from further consideration as a drug candidate.

Future Role of Mass Spectrometry in Drug Metabolism

The availability of sophisticated triple quadrupole and ion trap mass spectrometers equipped with rugged atmospheric pressure interfaces has provided powerful MS/MS tools for qualitative and quantitative analysis of drugs and metabolites. Continuing improvements in high throughput screening and automated synthetic techniques, including combinatorial chemistry, will yield a continually increasing number of potential drug candidates. To keep pace with these developments, it will be necessary for Drug Metabolism groups to increase

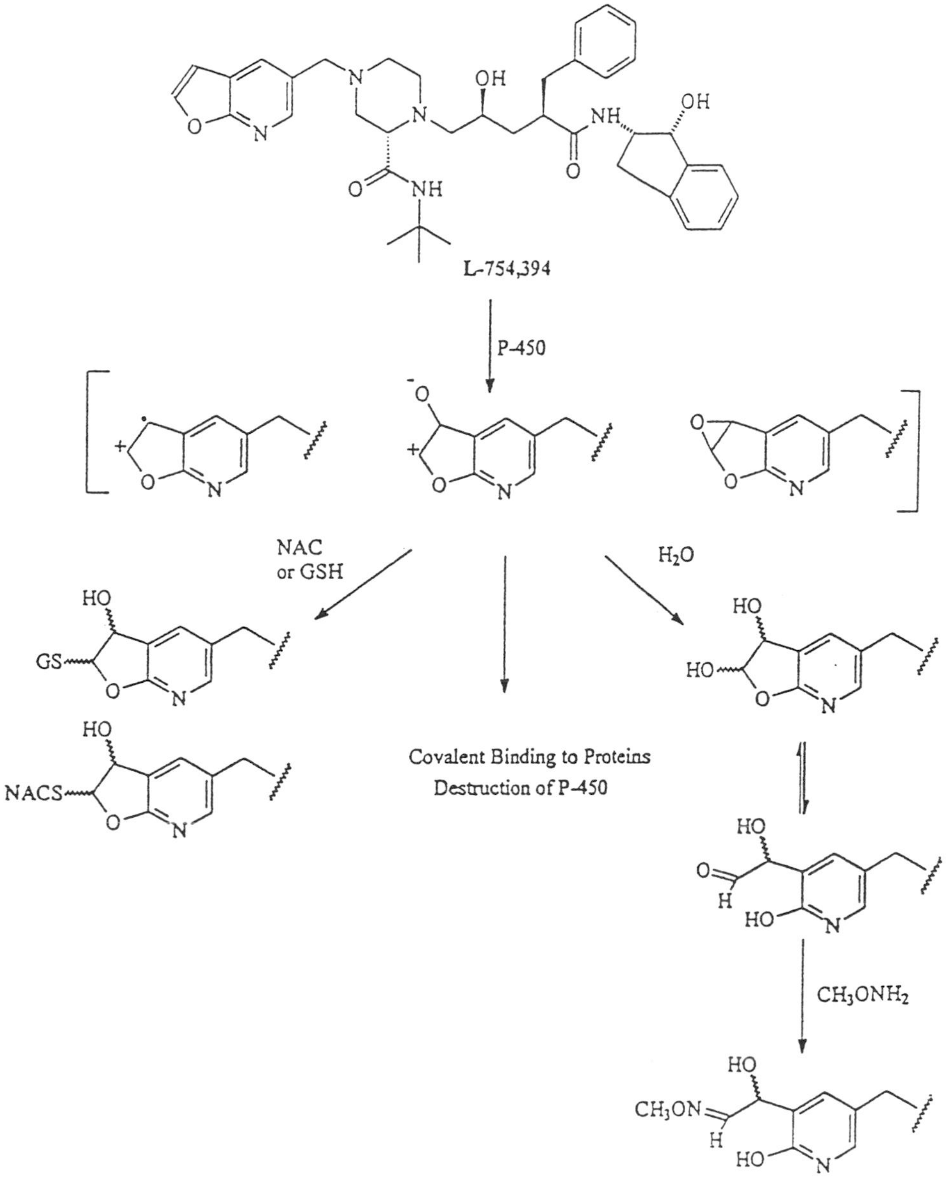

Fig. 7. Proposed pathway for metabolic activation of L-754,394 by cytochrome P-450 (reproduced with permission from [18]).

their capacity to generate qualitative and quantitative information, such that drug metabolism becomes more fully integrated into the discovery process.

Advances in instrument design have led to the availability of commercial time-of-flight (TOF) mass spectrometers with rugged interfaces. The attributes of TOF instruments (enhanced mass resolution, decreased cycle time and enhanced sensitivity) are ideally suited to high throughput drug metabolism

Future Role of Mass Spectrometry in Drug Metabolism

- New applications of existing instruments (TOF, QTOF)
  - Enhanced resolution, cycle time, sensitivity - HTS applications
- New techniques
  - LC-NMR-MS
  - Accelerator mass spec - trace quantities of labeled compounds
  - Chemical Reaction Interface MS (CRIMS) - stable isotopes
- Data Handling
  - Integrated technologies for MS
  - Databases and automated data processing
  - Expert systems - expanded application of automated techniques

Fig. 8. The future role of drug metabolism and mass spectrometry in contemporary drug discovery.

applications including the qualitative and quantitative analysis of mixtures. These advances, when incorporated into a tandem hybrid configuration such as a QTOF instrument, provide a powerful tool for the structural elucidation of drug metabolites. Furthermore, the availability of high-resolution product ion mass spectra serves to confirm the structure of proposed product ions that report on biotransformations.

Qualitative structural information also will be available from the combined use of LC-MS techniques with NMR spectroscopy to provide the ability to conduct LC-NMR-MS studies for metabolite profiling [19]. The complementary information obtained from NMR and MS will yield a definitive set of metabolism data to guide synthetic medicinal chemistry efforts. This application of hyphenated LC-NMR-MS techniques is especially powerful in the hyphenated format in which MS detection of a metabolite may be used to initiate a stopped-flow NMR spectral acquisition for full structural characterization.

A significant future challenge for mass spectrometry is to develop and utilize methodologies that will replace radiotracer studies for the definitive quantitative elucidation of metabolic clearance and elimination mechanisms of drugs. Successful implementation of MS technologies in this area will provide the opportunity to study the detailed metabolic fate of drugs in specialized human populations (i. e., pediatric, geriatric, hepatic and renal dysfunction, and various disease states). Accelerator mass spectrometry (AMS) permits the quantitative analysis of trace levels of carbon-14 and tritium in biological matrices, and holds significant future promise [20]. Chemical reaction interface mass spectrometry (CRIMS) can provide information complementary to the

AMS technique, and uses stable-isotope labeled drugs (e. g., carbon-13 and nitrogen-15) to provide chromatographic profile information following conversion of all labeled metabolites to $^{13}CO_2$ or ^{15}NO in a microwave plasma interface [21]. In this manner, the combination of gas chromatography or liquid chromatography with CRIMS can serve as a universal detection technique to generate quantitative chromatographic profiles of drug metabolites [22]. Since the information provided by both CRIMS and AMS is totally devoid of structural information, these techniques are regarded as complementary to the current array of MS instrumentation available to the Drug Metabolism scientist.

The future success of mass spectrometry in contemporary drug discovery efforts seems assured due to the need to fully integrate pharmacokinetic evaluation into the selection of lead compounds. Therefore, the comprehensive application of MS techniques at the interface between drug metabolism and drug discovery will continue to grow and will lead to increased efficiency in selection of drug candidates with optimal pharmacokinetics, metabolic and safety characteristics. It is anticipated that this proactive drug metabolism philosophy will serve to expedite the discovery and development of new therapeutic agents with superior pharmacokinetic properties.

References

1. J. H. Lin and A. Y. H. Lu. *Pharmacol. Rev*. **1997**, *49*, 403-449.

2. A. D. Rodrigues, *Pharmacol Res.* **1997**, *14*, 1504-1510.

3. M. H. Tarbit and J. Berman, *Current Opin. Chem. Biol.* **1998**, *2*, 411-416.

4. D. M. Dulik, W. H. Schaefer, J. Bordas-Nagy, R. C. Simpson, D. M. Murphy and G. R. Rhodes, in *Mass Spectrometry in the Biological Sciences*, A. L. Burlingame and S. A. Carr, Eds.; Humana Press: Totowa, NJ, 1996; pp 425-449

5. J. D. Gilbert, T. V. Olah and D. A. McLoughlin, in *Biochemical and Biotechnological Applications of ESI-MS*, A. P. Snyder, Ed.; ACS Symposia Series 619, 1996, pp 330-350.

6. T. A. Baillie and M. R. Davis, *Biol. Mass Spectrom.* **1993**, *22*, 319-325.

7. J. Berman, K. Halm, K. Adkison and J. Shaffer, *Med. Chem.* **1997**, *40*, 827-829.

8. M. C. Allen, T. S. Shah and W. W. Day, *Pharm. Res.* **1998**, *15 (1)*, 93-97.

9. D. A. McLoughlin, T. V. Olah and J. D. Gilbert, *J. Pharm. Biomed Anal.* **1997**, *15*, 1893-1901

10. C. E. C. A. Hop, Z. Wang, Q. Chen and G. Kwei, *J. Pharm. Sci.* **1998**, *87*, 901-903

11. W. G. Humphreys, M. T. Obermeier and R. A. Morrison, *Pharm. Res.* **1998**, *15 (8)*, 1257-1261

12. K. G. Owens, *Appl. Spectrosc. Revs.* **1992**, *27*, 1-49

13. C. L. Fernandez-Metzler, K. G. Owens, T. A. Baillie, and R. C. King, *Drug Metab. Disp.* ***1999***, *27*, 32-40.

14. J.H. Lin, I.-W. Chen, K. J. Vastag, J. A. Nishime, B. D. Dorsey, S. R. Michelson and S. L. McDaniel, *J. Pharmacol. Exp. Ther.* **1995**, *274*, 264-269.

15. L. T. Burka and M. R. Boyd, in *Bioactivation of Foreign Compounds*, M. W. Anders, Ed.; Academic Press: New York, 1985; pp 243-257.

16. V. Ravindranath, L. T. Burka and M. R. Boyd, *Science* **1984**, *224*, 884-886.

17. M. Chiba, J. A. Nishime and J. H. Lin, *J. Pharmacol. Exp. Ther.* **1995**, *275*, 1527-1534.

18. Y. Sahali-Sahly, S. K. Balani, J. H. Lin and T. A. Baillie, *Chem. Res. Toxicol.* **1996**, *9*, 1007-1012.

19. J.P. Shockcor, S.E. Unger, I.D. Wilson, P.J.D. Foxall, J.K. Nicholson and J. D. Lindon. *Anal. Chem.* **1996**, *68*, 4431-4435.

20. B. Kaye, R. C. Gardner, R. J. Mauthe, S. P. H. T. Freeman and K. W. Turteltaub. *J. Pharm. Biomed. Anal.* **1997**, *16*, 541-543.

21. F. P. Abramson, *Mass Spectrom. Rev.* **1994**, *13*, 341-356.

22. F. P. Abramson, Y. Teffera, J. Kusmierz, R. C. Steenwyk and P. G. Pearson, *Drug Metab. Dispos.* **1996**, *24*, 697-701.

QUESTIONS AND ANSWERS

R. R. Zhu (BASF Bioresearch Corporation)

Is the correlation analysis based on MS/MS spectra of each unknown compared to MS/MS of the parent drug?

Answer. Yes, that is correct. The overall approach is based on the premise that the MS/MS spectra of unknown metabolites will retain elements of the MS/MS characteristics of the parent drug, and therefore will be detected by the pattern recognition algorithm in the Correlation Analysis software.

Arthur Moseley (Glaxo Wellcome)

I was impressed with your correlation analysis of MS/MS spectra of complex metabolite mixtures for correlating potential metabolites with their parent drug. One of the favorable aspects of peptide MS/MS interpretation is that peptides fragment at logical positions along the peptide backbone. Small drug molecules do not fragment in easily predicted locations. What can you share with us in your efforts to automate the interpretation of the MS/MS spectra of these drug metabolites?

Answer. The Correlation Analysis approach functions to detect potential metabolites of a compound-of-interest, but not to identify them. Spectral interpretation is conducted largely by visual inspection of the MS/MS data by a trained user. While efforts to automate at least a portion of the interpretative work are ongoing, the assembly of a large in-house database of MS/MS spectra is viewed as a prerequisite for developing suitable spectral interpretation algorithms. It may be some time before we are in a position to implement fully automated procedures for metabolite detection and identification.

Dan Kassel (CombiChem, Inc.)

Can you compare reliability of the correlation method with predictive software? Which plots "typical" expected metabolites?

Answer. Although predictive software routines can be very helpful, they often fail in situations where a molecule undergoes either an unexpected route of metabolism or is cleaved at an internal site (e. g., by a heteratom dealkylation reaction) to yield fragments of low molecular weight. In our limited experience with the Correlation Analysis software, such products are recognized as long as they retain structural elements of the parent drug which direct collision-induced fragmentation.

Brad Gibson (UCSF)

Once you identified to bioactivation of the furan ring attached to pyridine moiety as the metabolite responsible for the inhibition of P-450 in your last drug molecule, could this then be used as a predictive factor for similar undesired metabolism in other drugs, or was there something about the remainder of this particular drug that allowed for this reaction to occur?

Answer. In the case of L-754,394, metabolism of the furan ring system led to the generation of a reactive intermediate which covalently modified proteins, including the enzyme system (CYP3A4) which catalyzed the oxidation reaction itself. Because bioactivation of simple substitute furans is well known from the literature, it might be argued that the incorporation of this ring system into drug candidates is unwise. However, with our present understanding of drug metabolizing enzymes, it is not always possible to accurately predict sites of metabolic attack based on the structure of the drug substance alone. A case in point was encountered with an analog of L-754,394 in which the fused pyridine and furan rings were interchanged; no metabolism occurred on the furan moiety of this compound, and no bioactivation was observed.

P. Juhasz (PerSeptive Biosystems)

When carrying out correlation analysis on a mixture of substrates, what do you do if two or more substrates have the same "most likely metabolite"?

Answer. A shortcoming of the Correlation Analysis approach is that a common metabolite of two or more foreign compounds will be recognized as being drug-related, but no information is provided on the precursor(s) of the metabolite. Thus, the used needs to be alert to such possibilities, and to conduct appropriate follow-up studies on the individual potential precursors, where necessary.

A. L. Burlingame (UCSF)

What effort goes on in bio-assay "receptor" array diversity in combinatorial screening?

Answer. All drug candidates are subject to an extensive battery of *in vitro* biological activity screens, usually long before they are selected for pharmacokinetic and metabolic evaluation.

Steven Carr (SmithKline Beecham Pharmaceuticals)

Are there computational methods that can reliably be used as predictors of success or failure of the bioavailability or pharmacokinetics of lead molecules?

Answer. Many attempts have been made to achieve this (elusive) goal, and several "rules-of-thumb" have been proposed to guide the selection of compounds with desirable pharmacokinetic and metabolic characteristics. I think that it is fair to say that, at the present time, we do not have any computational methods that adequately address this important issue, and therefore we are compelled to conduct the appropriate preclinical *in vitro* and *in vivo* studies in an effort to cautiously predict the behavior of a drug candidate in man.

Electrospray Mass Spectrometric Analysis of Lipid Mediators Derived from Arachidonic Containing Membrane Phospholipids

Robert C. Murphy and Tatsuji Nakamura
Department of Pediatrics, Division of Basic Sciences, National Jewish Medical and Research Center, 1400 Jackson Street, Denver, CO 80206

Arachidonic acid (5,8,11,14(Z,Z,Z,Z)- eicosatetraenoic acid) is one of the most common polyunsaturated fatty acids present within virtually all mammalian cells. Arachidonic acid plays an important role in biological chemistry because of its unique physical properties such as influencing membrane fluidity as well as serving as a source of a diverse family of oxidized metabolites known to play important physiological and pathophysiological roles within tissues. Even though a very small percentage of the total arachidonic acid within cells is ever involved in the oxidative reactions that lead to the formation of compounds such as prostaglandins [1] and leukotrienes [2], a considerable amount of effort has been placed in designing agents that either inhibit production of these molecules or block their actions. Prostaglandins and leukotrienes exert their biological activity through G-protein linked membrane receptors on the surface of cells, and these specific receptors are linked to numerous signal transduction processes, which lead to a myriad of biological responses mediated by these lipid products. The enzymes that form these lipid mediators utilize arachidonic acid only as a free carboxylic acid, and do not act on arachidonic acid while it is esterified to membrane phospholipids. This is rather curious because the majority of all arachidonate within cells exists in the plasma and organelle membranes in various phospholipid classes esterified to the sn-2 position [3]. The other curious feature of the enzymes responsible for the initial oxidation of arachidonic acid involves the formation of free radical intermediates which can readily react with molecular oxygen. The enzymes cyclooxygenase, 5-lipoxygenase, 15-lipoxygenase, and 12-lipoxygenase utilize iron in various bound forms in the active site to initiate these free radical reactions. Within the active site, the chemical reactivity of these radical species is controlled, resulting in a stereospecific reaction mechanism.

Arachidonic acid can also participate in free radical reactions not controlled by enzymatic reactions. Within biological tissues there are a host of processes which can generate free radical species ranging from quite reactive radicals such as the hydroxyl radical [4], to radical species such as nitric oxide (NO), which itself is not particularly reactive; however, NO can be activated by other species such as superoxide anion to form the highly reactive peroxynitrite radical [5]. The general process of lipid peroxidation typically involves formation of a carbon-centered radical at one of many potential positions along the acyl chains of phospholipids. However, it is the formation of the resonance stabilized *bis*allylic radical which imparts a reasonable half-life to this radical species, and enables it to encounter dissolved oxygen and undergo formation of the hydroperoxy radical product. The energetics of formation of the initial radical center in the arachidonate structure is outlined in Scheme 1 showing the various carbon-hydrogen bond dissociation energies. The lowest bond energy is the *bis*allylic position (75-80 kcal/mol) for which arachidonic acid has three [6]. It has been suggested that the oxidizability of any polyunsaturated fatty acid is largely a function of number of *bis*allylic hydrogen atoms [6].

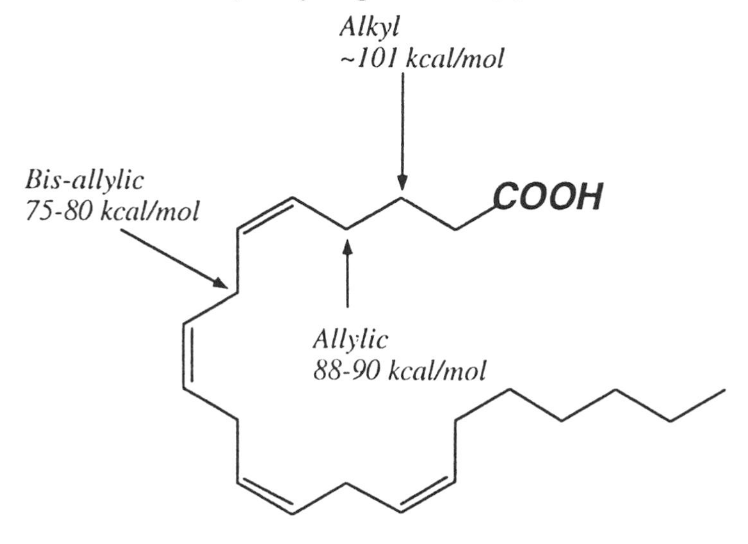

Scheme 1: Arachidonic Acid

An important feature of the free radical reactions initiated within tissues and cellular membranes is that the arachidonic acid and other polyunsaturated fatty acids present within cells exist as glycerophospholipids and radical reactions therefore result in the formation of oxidized phospholipids. However, there is abundant enzymatic capability to hydrolyze such oxidized phospholipids [7, 8], yielding oxidized arachidonic acid as a free carboxylic acid. Therefore, a fundamental difference exists between the formation of oxidized arachidonic acid by enzymatic and free radical processes, namely that free radical processes involve esterified arachidonic acid whereas enzymatic oxidation of arachidonic acid utilizes free arachidonate as substrate. Another important difference between these two modes of oxidation of arachidonic acid is that enzymatic oxidation of free arachidonic acid yields typically a single product, whereas free radical oxidation of esterified arachidonic acid yields a large number of closely related molecules, including stereo and regioisomers.

Recent studies have revealed that free radical initiated events can lead to quite complex structures derived from esterified arachidonic acid, including those which mimic the structure of cyclooxygenase products (prostaglandins), termed the isoprostanes [9]. The formation of compounds isomeric to the

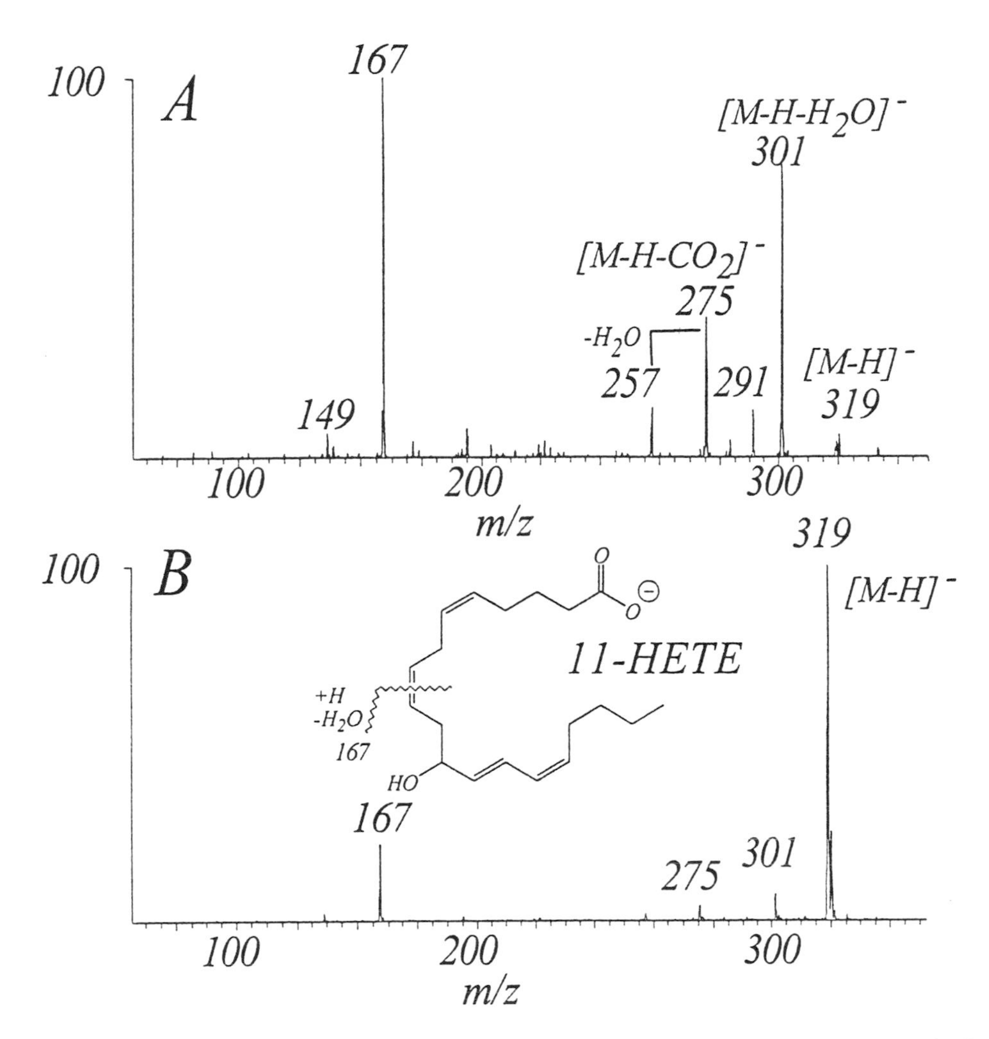

Fig. 1. Collisional activation of the carboxylate anion (negative ions) of 11-HETE (m/z 319) in an ion trap mass spectrometer. (A) Collisional activation of m/z 319 ± 1.0 daltons. (B) Collisional activation of m/z 319 ± 20 daltons.

leukotrienes has also been observed [10]. Less complex chemical species such as the hydroxy, hydroperoxy, and oxo derivatives of arachidonic acid have been observed [11, 12], and often represent the most abundant stable products of arachidonate peroxidation.

To investigate the formation of these free radical products it has been necessary to employ powerful means to deduce chemical structures. Thus mass spectrometry has played a central role in advancing our understanding of the free radical oxidation of arachidonic acid. Many initial studies were carried out using gas chromatography/mass spectrometry with electron ionization and chemical ionization (negative ions) techniques [13, 14]. More recently, electrospray mass spectrometry was found to be quite useful in the analysis of such compounds by generating abundant carboxylate anions (negative ions) which could be studied by collisional activation in the tandem quadrupole mass spectrometer [15-18]. We describe here the collision induced decomposition of carboxylate anions derived from various oxidized arachidonate species using an ion trap mass spectrometer as an alternative means to generate useful structural information. A second part of investigations reported here summarizes recent results in the detection of oxidized arachidonate species derived from phospholipids isolated from biological membranes.

Collision Induced Decomposition of Monohydroxyeicosatetraenoic Acids

The monohydroxyeicosatetraenoic acids (HETEs) are reduced products of the reaction of molecular oxygen with the stabilized *bis*allylic arachidonoyl radicals with six regioisomers possible, 5-,8-,9-,11-,12-, and 15-HETE. The collision induced decomposition of the carboxylate anions from these six species (m/z 319) generated by electrospray ionization has been previously studied in a tandem quadrupole mass spectrometer [15, 16]. The collisional activation of m/z 319 within the ion trap was found to be qualitatively similar to the tandem quadrupole instrument except for the relative abundance of the ions, specifically corresponding to loss of water $[M\text{-}H_2O]^-$ at m/z 301. This is exemplified by the collisional activation of m/z 319 from 11-HETE (Fig. 1A). The structurally significant ion at m/z 167 observed in this mass spectrum, corresponded to cleavage adjacent to the hydroxyl moiety (likely as the alkoxy anion) by a charge driven vinyllic fragmentation mechanism with loss of a neutral, unsaturated aldehyde (Scheme 2).

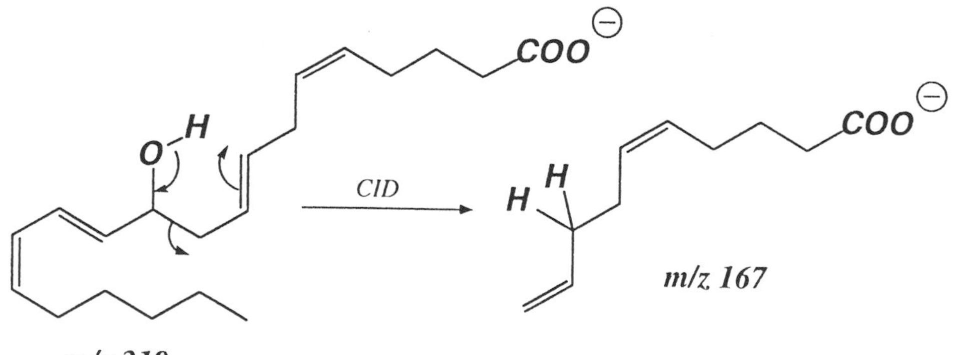

COO⊖
O
H
CID
COO⊖
H
H
m/z 167
m/z 319

Scheme 2:

A CID mass spectrum could be generated by the ion trap with reasonable similarity to that obtained by a triple quadrupole mass filter for 11-HETE when both m/z 319 and m/z 301 were simultaneously activated by broad band excitation (Fig. 1B). The ion trap CID mass spectra of m/z 319 $[M\text{-}H]^-$ derived

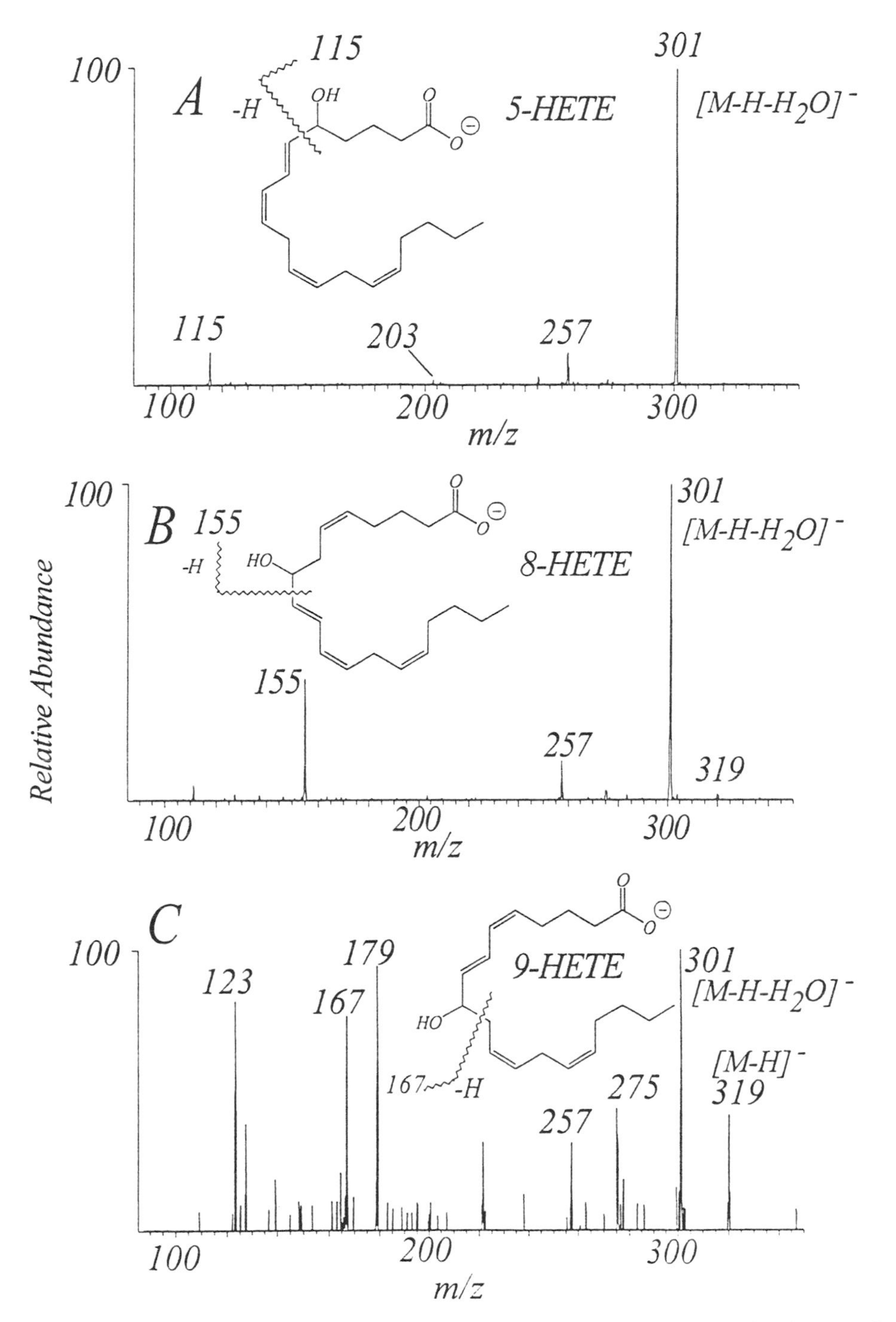

Fig. 2. Collision induced dissociation in an ion trap mass spectrometer of m/z 319, the carboxylate anion derived from (A) 5-HETE, (B) 8-HETE, (C) 9-HETE.

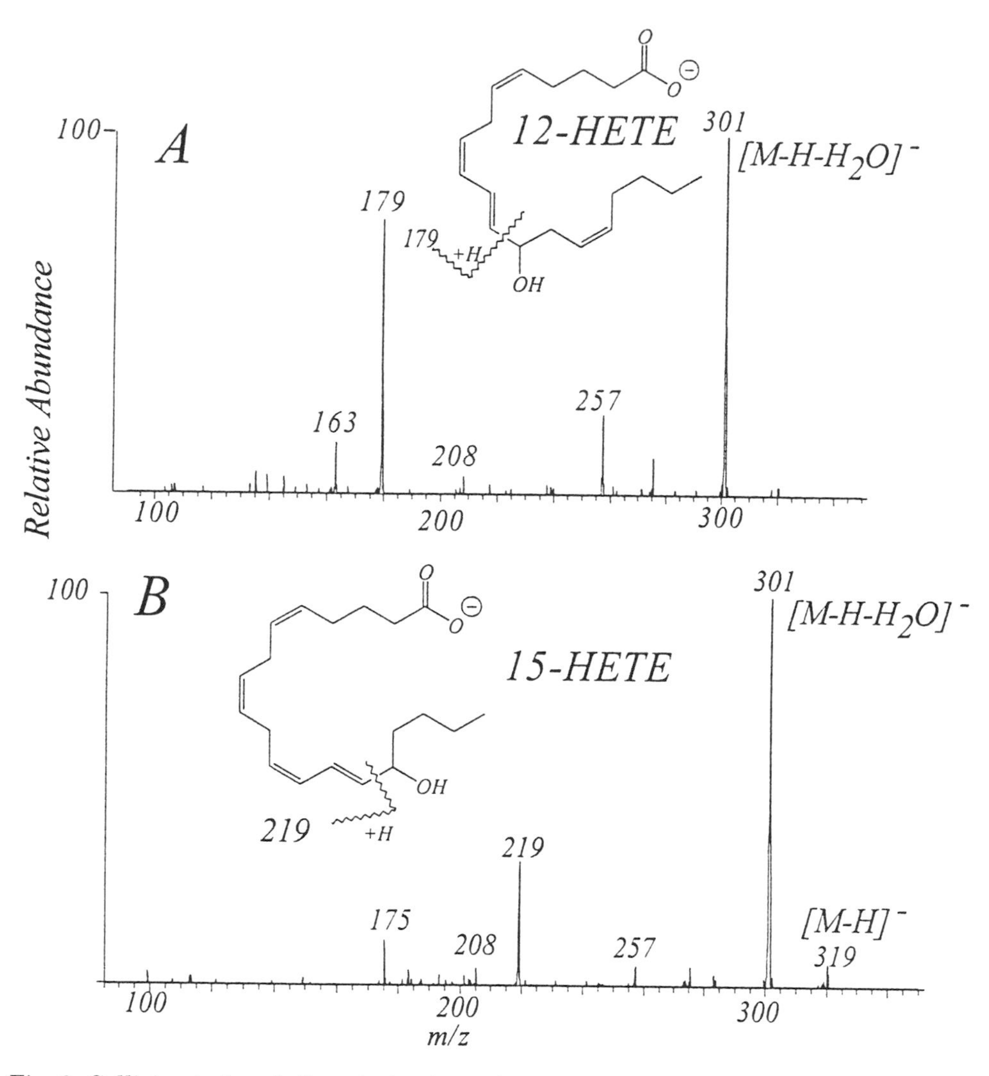

Fig. 3. Collision induced dissociation in an ion trap mass spectrometer of m/z 319, the carboxylate anion derived from (A) 12-HETE and (B) 15-HETE.

from the five additional monohydroxyeicosatetraenoic acid regioisomers are shown in Figs. 2 and 3, with a suggested origin for each of the structurally significant ions. These sites of fragmentation have been previously studied with ions generated by fast atom bombardment ionization [19].

Epoxyeicosatrienoic Acids

The epoxyeicosatrienoic acids (EET) result from the addition of an oxygen atom to one of the double bonds of arachidonic acid, without migration of the remaining three original double bonds. Four regioisomers can result from such reactions, which include 5,6-, 8,9-, 11,12-, and 14,15-EET [20]. The carboxy-

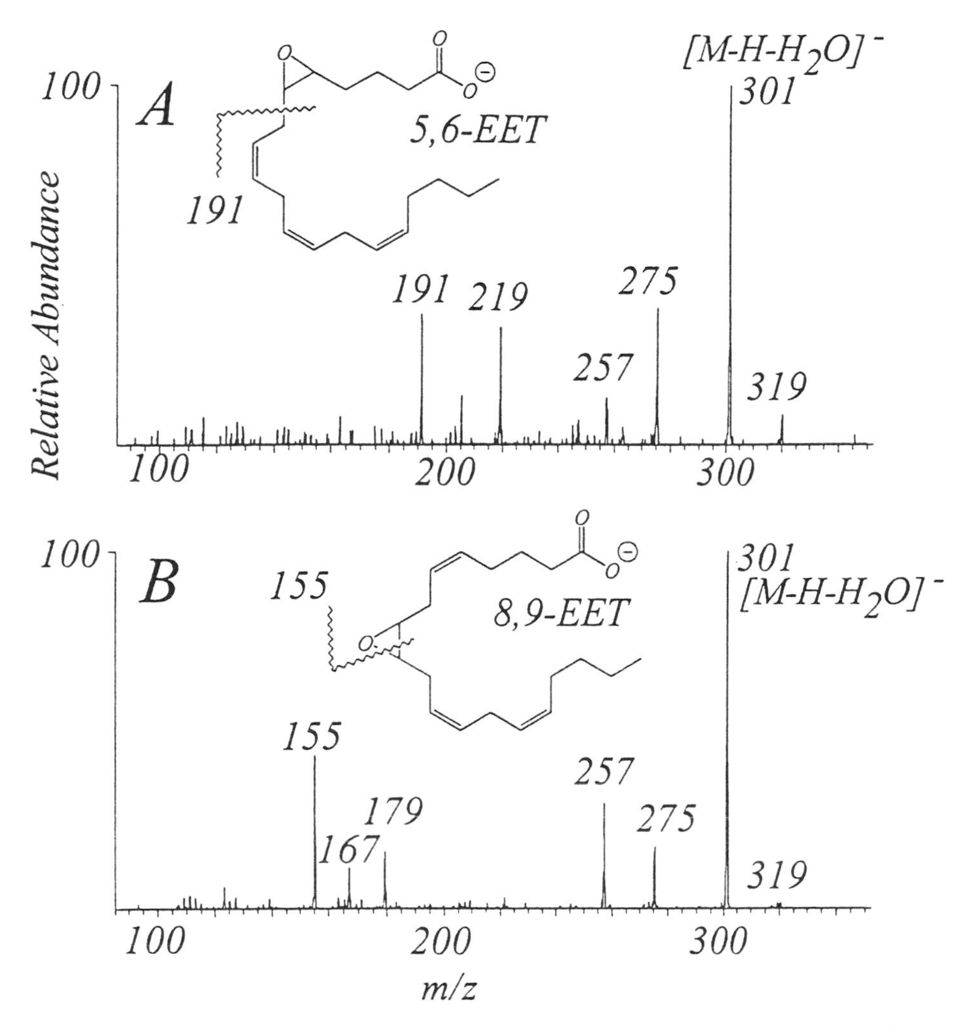

Fig. 4. Collision induced dissociation in an ion trap mass spectrometer of carboxylate anions from epoxyeicosatrienoic acid regioisomers at m/z 319 for (A) 5,6-ETE and (B) 8,9-ETE.

late anions from each EET regioisomer are observed at m/z 319, and thus are isobaric to the HETE products of arachidonic acid oxidation. In the ion trap mass spectrometer, specific as well as non-specific product ions result following collisional activation of m/z 319. The non-specific ions include loss of water (m/z 301) and loss of water and carbon dioxide (m/z 257). The specific ions for each of the regioisomers involve carbon-carbon bond cleavage driven by the position of the epoxide moiety and formation of stabilized anions as previously observed in the tandem quadrupole mass spectrometer [17]. As was the case for the HETE regioisomers, the relative abundance of these specific ions to each other was similar whether the experiment was carried out in the ion trap or the tandem quadrupole mass spectrometer; however, the relative abundance of the loss of

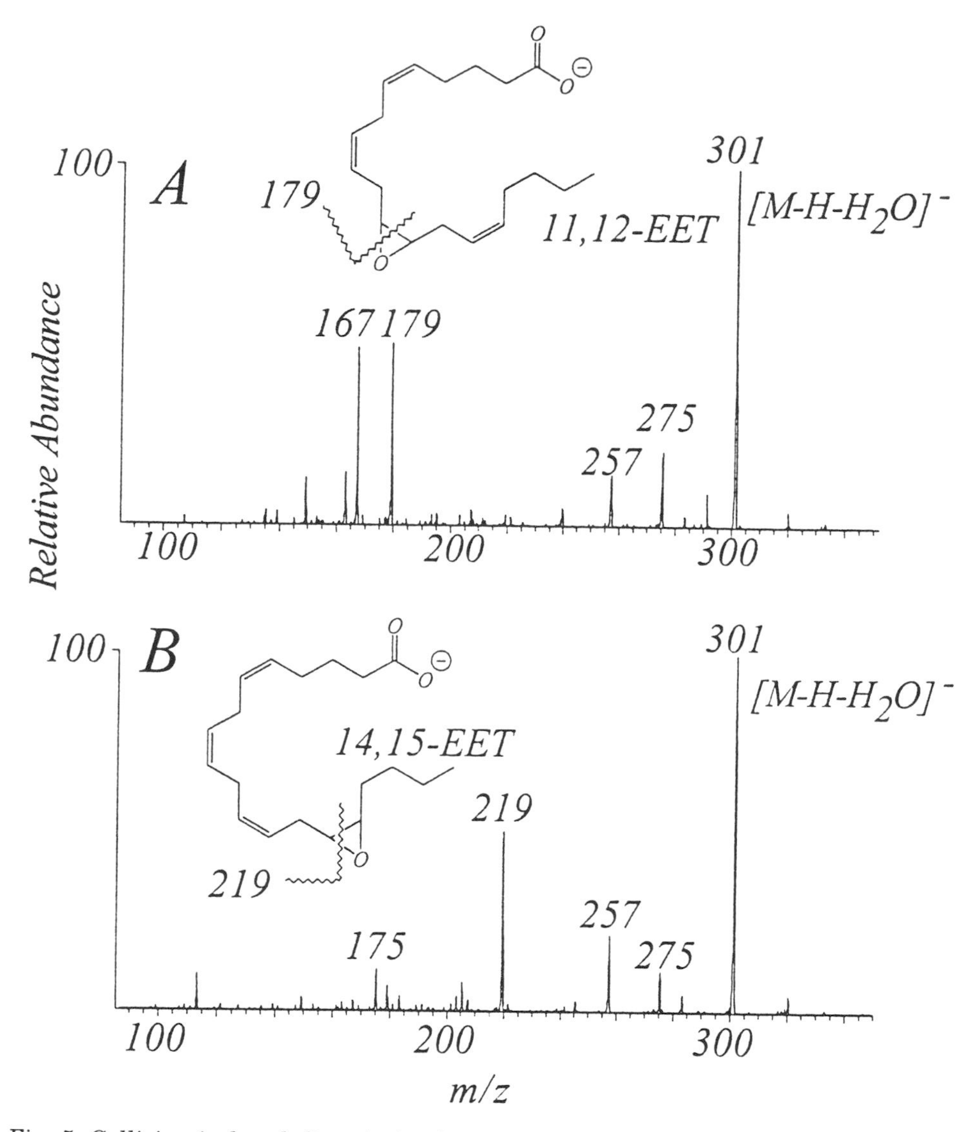

Fig. 5. Collision induced dissociation in an ion trap mass spectrometer of carboxylate anions from epoxyeicosatrienoic acid (EET) regioisomers at m/z 319 for (A) 11,12-ETE and (B) 14,15-ETE.

water was substantially higher (Figs. 4 and 5). It is likely that initial excitation of the carboxylate anion rapidly results in the loss of water, which does not undergo subsequent excitation and decomposition reactions. In the tandem quadrupole instrument, this would not be the case for a very rapid reaction product ion such as this dehydrated ion, which would be further activated through collisions in the second quadrupole region resulting in products of the dehydrated ion. The origin of each characteristic decomposition ion for the EET regioisomers has been previously described in some detail [17].

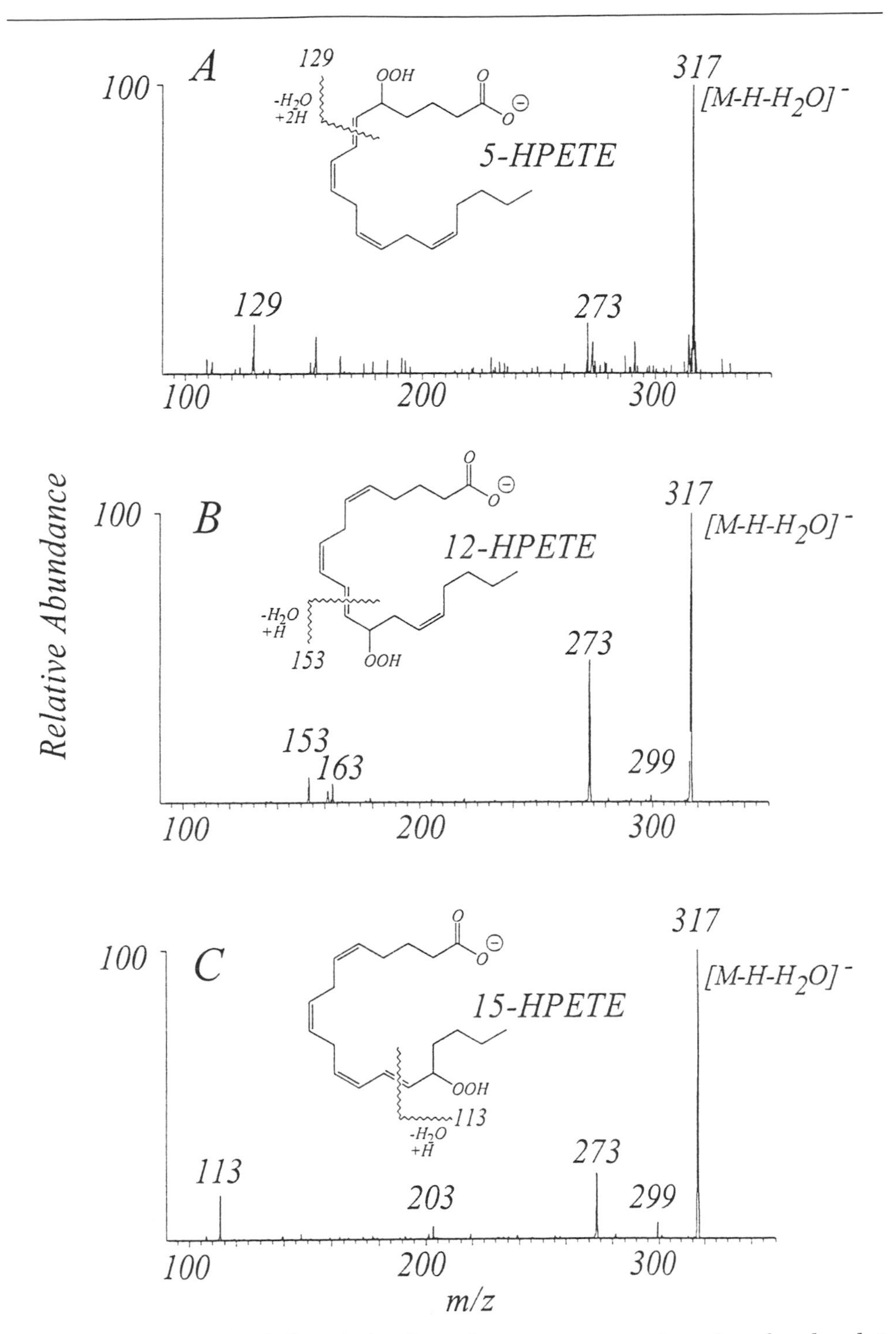

Fig. 6. Collision induced dissociation in an ion trap mass spectrometer of carboxylate anions from hydroperoxyeicosatrienoic acid (HpETE) at m/z 335 from (A) 5-HpETE, (B) 12-HpETE, and (C) 15-HpETE.

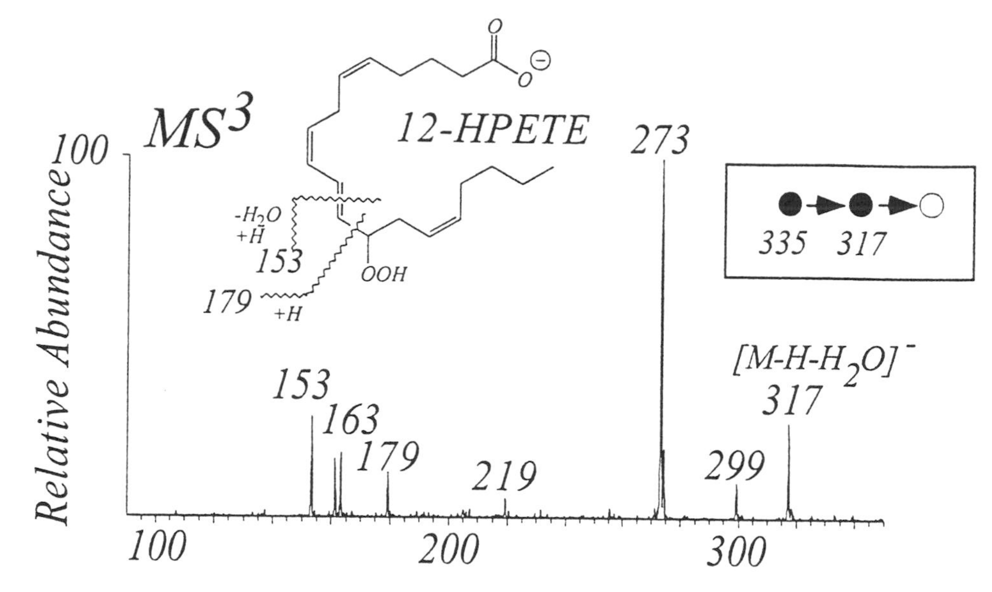

Fig. 7. Collisional activation and MS³ in an ion trap mass spectrometer of m/z 335 317 product ions for 12-hydroperoxyeicosatetraenoic acid.

HYDROPEROXYEICOSATETRAENOIC ACIDS

The hydroperoxyeicosatetraenoic acids (HpETE) are initial products of the addition of molecular oxygen to the *bis*allylic arachidonoyl radical species. Until recently, such compounds were found to be difficult if not impossible to analyze by gas phase mass spectrometric methods, such as electron ionization or electron capture ionization. Electrospray ionization is sufficiently gentle, not requiring volatilization during the ionization process, that molecular anions $[M-H]^-$ can be derived from this reactive, lipid hydroperoxy species [18]. Collision induced decomposition of $[M-H]^-$ within a tandem quadrupole mass spectrometer was previously reported to be characterized as proceeding through the $[M-H_2O]^-$ anion, likely the carboxylate anion of the corresponding oxo eicosatetraenoic acid (oxoETE) carboxylate anion [18]. Not surprisingly, the collisional activation of the carboxylate anion derived from hydroperoxy regioisomers of arachidonic acid results in abundant ions corresponding to the loss of water (m/z 317) in much more abundant yield than that observed in the tandem quadrupole mass spectrometer (Figs. 6 and 7). Excitation of both m/z 335 and m/z 317 results in a product ion spectrum similar to that observed in the tandem quadrupole mass spectrometer (Fig. 7).

Specific ions for the 5-, 12-, 15-hydroperoxyeicosatetraenoic acid were observed at m/z 129, 153, and 113 respectively, in the ion trap mass spectrometer. These same ions were previously reported as characteristic ions generated in the triple quadrupole instrument [18]. The characteristic ion for 12-HpETE observed at m/z 153 in the tandem quadrupole instrument was also observed as a reasonably abundant ion in the ion trap mass spectrometer. The previously suggested formation of this ion involved cleavage of C_{10}-C_{11} bond [18], with charge retention on the aliphatic portion of the molecule. The mechanism of formation of this ion likely involved an initial loss of water with formation of the enolized 12-oxo-ETE carboxylate anion following double bond migration. A conjugated species could then undergo cyclization followed by a charge driven allylic fragmentation to yield a highly stabilized anion shown in Scheme 3.

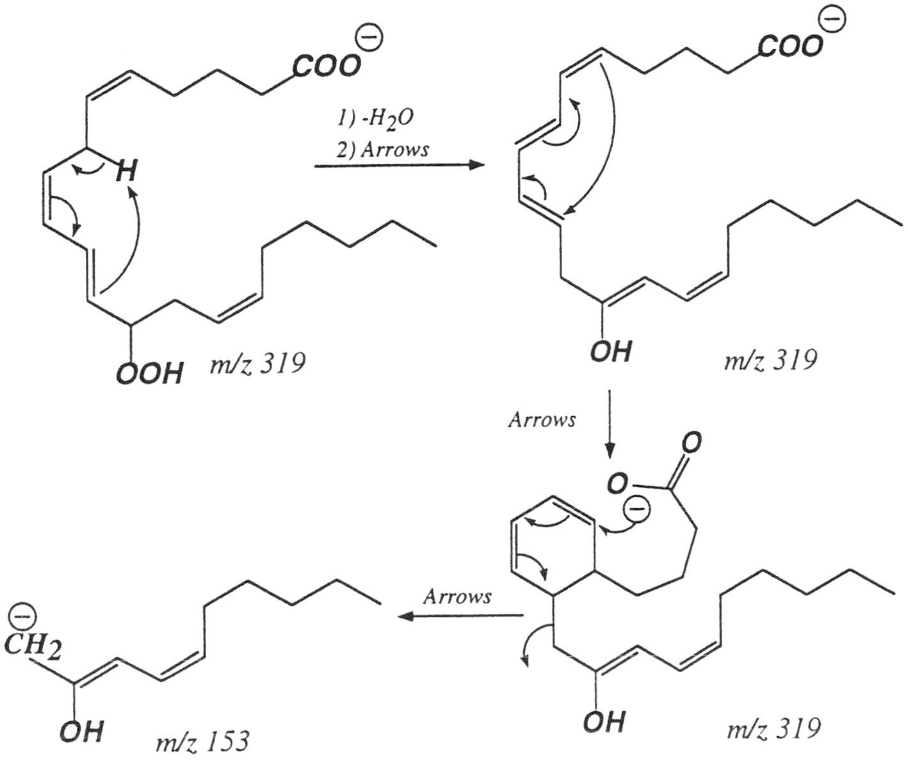

Scheme 3:

FORMATION OF OXIDIZED ARACHIDONATE CONTAINING PHOSPHOLIPIDS FROM NATURALLY OCCURRING BIOLOGICAL MEMBRANES

As a model of lipid peroxidation taking place within cells that contain robust antioxidant mechanisms, red blood cells (RBCs, 2 x 10^8/mL in phosphate buffer saline pH 7.4) were retreated with t-butylhydroperoxide (tBuOOH) (10 mM) at 37° for 90 min. After incubation, the cells were centrifuged and the pellet washed three times. Phospholipids were extracted from the cell pellets following the method of Rose and Oklander [21]. The phospholipids were then separated by normal phase HPLC into individual glycerophospholipid classes, including glycerophosphoethanolamine (GPE), glycerophosphoserine (GPS), and glycerophosphocholine (GPC) lipids as the major phospholipids present in the RBC. The normal phase HPLC fractions were then hydrolyzed in 1 M sodium hydroxide at room temperature for 90 min, fractions acidified to pH 3, and 10 ng of [$^{18}O_2$]12-HETE added as internal standard. The liberated free carboxylic acids

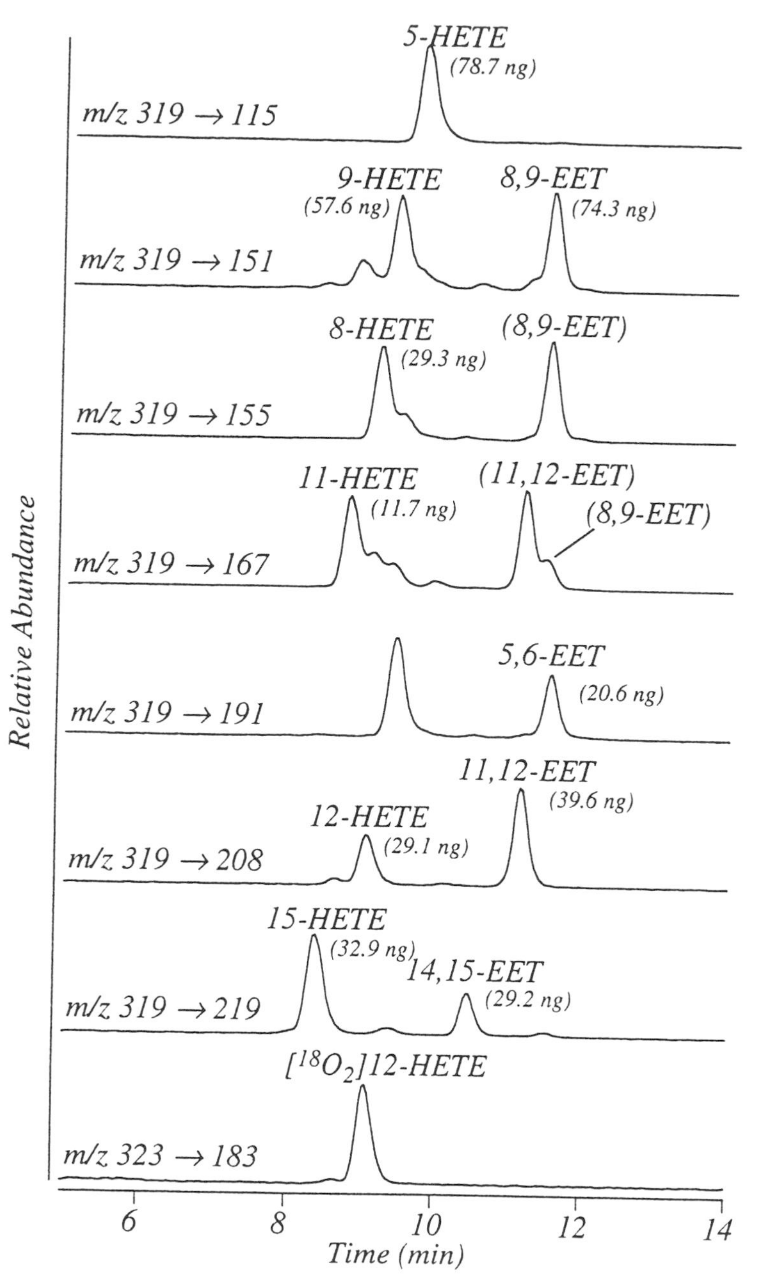

Fig. 8. Reverse phase LC/MS/MS analysis (negative ion transitions) of oxygenated arachidonate derived from glycerophosphoserine lipids isolated from red blood cells treated with tBuOOH. Phospholipids were extracted and GPS separated by normal phase HPLC then saponified into the free carboxylic acids prior to negative ion analysis. Specific ion transitions monitored can be observed in Figs. 1-4. The quantitation indicated in parentheses is based on the ratio of each ion transition to that of the [$^{18}O_2$]12-HETE internal standard and calibration curve.

were extracted with hexane/isopropanol and separated by reverse phase HPLC. A 5 μ C18 (250 x 1.0 mm I.D.) column was eluted with a gradient from 20% 6.5 mM ammonium acetate (pH 5.7) programmed to 100% methanol/acetonitrile (35/65) over a 15 min period at a flow rate of 50 μL/min.

The reverse phase LC/MS/MS analysis of the fatty acids derived from the GPS fraction is shown in Fig. 8. The oxidized arachidonate compounds were initially identified by monitoring the abundant carboxylate anion at m/z 319 and specific decomposition ions corresponding to HETEs and EETs. This data shows the elution of all the expected six HETE isomers, as well as the elution of the oxygen-18 labeled internal standard which was used for quantitative analysis purposes (m/z 323 183). The HETE isomers eluted from this column between 8 and 10 min, whereas more lipophillic components containing similar ion transitions were found to elute between 10.5 and 12 min. Inspection of the collision induced mass spectra of m/z 319 revealed that these later eluting compounds were in fact the EET isomers. This observation provided the first characterization of the EET formed within intact phospholipids in a lipid peroxidation model system [11]. When comparing the quantitative data for each of the monooxygenated arachidonate species to that obtained for control samples not treated with tBuOOH, the formation of oxidized arachidonic acid esterified to glycerophospholipids in treated RBCs was found to significantly increase for both the esterified EETs in the GPE, GPS, and GPC classes (49-, 34-, and 59-fold respectively), as well as for esterified HETE in GPE, GPS, and GPC (3-, 4-, and 11-fold, respectively) compared to untreated RBCs [11].

Interestingly, there was no direct evidence for the formation or release of hydroperoxyeicosatetraenoic acids in this lipid peroxidation model system. This was likely due to the abundance of glutathione within the red blood cell and multiple antioxidant mechanisms which would readily lead to the reduction of initially formed HpETEs into the observed HETEs. These antioxidant mechanisms include glutathione peroxidases as well as lipid hydroperoxide reductases [22, 23].

A physiologically relevant model of lipid peroxidation involved the treatment of mouse peritoneal macrophages with zymosan or yeast cell wall. The macrophage, being a phagocytotic cell, engulfs the yeast particles and during this process initiates the formation of reactive oxygen species including superoxide anion and nitric oxide [24]. In addition to the generation of free radical species, there is substantial evidence supporting the activation of phospholipases which release arachidonic acid that can serve as substrate for lipoxygenases, including the leukocyte-type 12-lipoxygenase, cyclooxygenase, and 5-lipoxygenase [25, 26]. Therefore, the murine peritoneal macrophage serves as an interesting model to investigate not only the enzymatic formation of oxidized metabolites of arachidonic acid, but also the possible formation of nonenzymatic metabolites of arachidonic acid during phagocytosis.

Following incubation of zymosan particles (50 particle/macrophage) with murine peritoneal macrophages (1 x 10^6 cells/mL) suspended in KRPD for 60 min at 37° C, reactions were quenched with two volumes of cold methanol

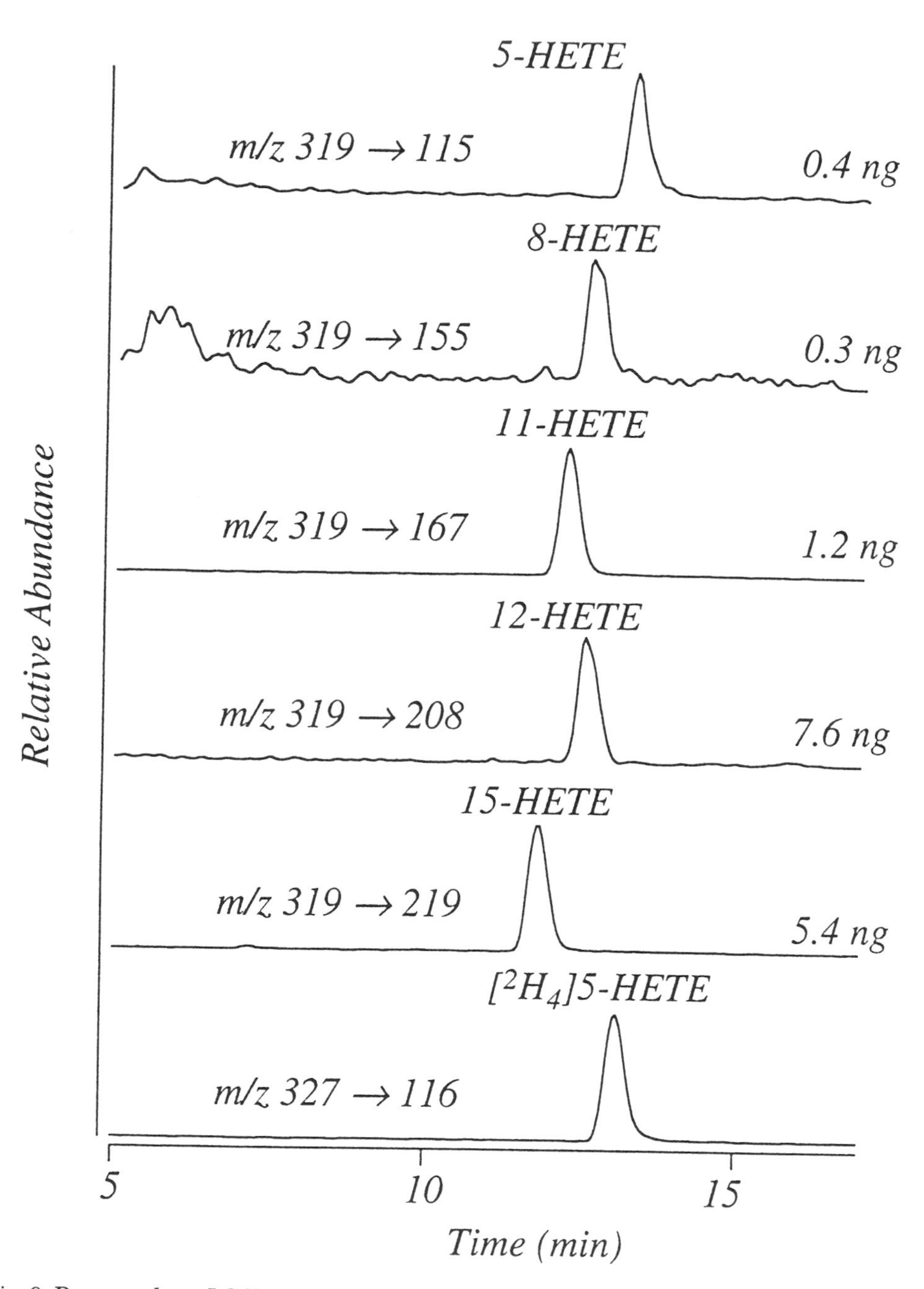

Fig. 9. Reverse phase LC/MS/MS analysis of oxygenated arachidonate isolated following exposure of murine peritoneal macrophages to zymosan particles. Free carboxylic acids were extracted then separated under reverse phase conditions during the LC/MS/MS analysis. Specific ion transitions monitored can be observed in Figs. 1-4. The quantitation indicated is based on the ratio of each ion transition to that of the [$^2H_4$]5-HETE internal standard and calibration curve.

containing 10 ng of [$^2H_4$]5-HETE as internal standard and eicosanoids were extracted by C18 solid-phase extraction columns described previously [27]. The extracted eicosanoids were then subjected to reverse phase LC/MS/MS analysis using conditions described above.

With this model system, it was possible to observe an abundant formation of several monohydroxyeicosatetraenoic acids species, including 15-, 12-, and 11-HETE (Fig. 9). These products likely were a result of enzymatic oxidization of free arachidonic acid by 15- and 12-lipoxygenase, yielding the 15-oxygenated species, as well as cyclooxgyenase, yielding the 11- and 12-oxygenated species.

The other HETE components formed during zymosan phagocytosis, 5- and 8-HETE regioisomers, were in substantially lower abundance than the 11- and 15-HETE regioisomers. These compounds would be candidates for lipid peroxidation products, but only 8-HETE could be formed by non-enzymatic processes, because there are no enzymatic processes yet observed which generate this HETE regioisomer in the mammalian cells and in particular in the murine peritoneal macrophage. The quantity of these components produced were 5-30 times less than the enzymatically derived oxidized arachidonoyl containing phospholipids, but nonetheless may represent directly oxidized glycerophospholipids formed during phagocytosis and initiation of the macrophage respiratory burst. It is also important to note that the specificity in the measurement of these molecules is based not only on the ion transitions observed, but also the unique HPLC retention times for each eicosanoid. In those experiments with the murine macrophage, there was no evidence for the formation of EET phospholipids during phagocytosis.

Conclusion

Electrospray mass spectrometry has proven to be extraordinarily powerful as a method to obtain both qualitative and quantitative information about lipid peroxidation taking place within biological membranes. The complexities of free radical oxidation of arachidonic acid render such studies a formidable task, in that numerous isobaric species are formed which require several stages of analytical separation, including HPLC, as well as specific ion transitions monitored by mass spectrometry. The ion trap mass spectrometer was found to generate virtually identical ions to those observed in the tandem quadrupole instrument, albeit in different relative ion abundances. In part, this was likely due to the unique events taking place within the ion trap compared to the collision cell of the tandem mass spectrometer where a single ion in the former instrument is uniquely activated to yield corresponding product ions. In the tandem quadrupole instrument, if an initial decomposition reaction takes place with sufficient rapidity, collisions of secondary ions can result which provide additional product ions. The ion trap mass spectrometer does offer unique specificity in being able to carry out MS^n that should enable one to follow specific ion transitions not possible in the tandem quadrupole mass spectrometric experiment.

ACKNOWLEDGEMENTS

This work was supported, in part, by a grant from the National Institutes of Health (HL34303) and support from Banyu Tsukuba Research Institute (Tsukuba, Japan).

REFERENCES

1. W. L. Smith, *Am. J. Physiol.* **1992**, *263*, F181-F191.

2. A. W. Ford-Hutchinson, M. Gresser and R. N. Young, *Ann. Rev. Biochem.* **1994**, *63*, 383-417.

3. R. F. Irvine, *Biochem. J.* **1992**, *204*, 3-16.

4. N. A. Porter, S. E. Caldwell and K. A. Mills, *Lipids* **1995**, *30*, 277-290.

5. W. A. Pryor and G. L. Squadrito, *Am. J. Physiol.* **1995**, *268*, L699-L722.

6. B. A. Wagner, G. R. Buettner and C. P. Burns, *Biochemistry* **1994**, *33*, 4449-4453.

7. M. G. Salgo, F. P. Corongiu and A. Sevanian, *Arch. Biochem. Biophys.* **1993**, *304*, 123-132.

8. M. G. Salgo, F. P. Corongiu and A. Sevanian, *Biochim. Biophys. Acta* **1992**, *1127*, 131-140.

9. J. D. Morrow and L. J. I. Roberts, *Biochem. Pharmacol.* **1996**, *51*, 1-9.

10 K. A. Harrison and R. C. Murphy, *J. Biol. Chem.* **1995**, *270*, 17273-17278.

11. T. Nakamura, D. L. Bratton and R. C. Murphy, *J. Mass Spectrom.* **1997**, *32*, 888-896.

12 L. M. Hall and R. C. Murphy, *J. Am. Soc. Mass Spectrom.* **1998**, *9*, 527-532.

13. R. C. Murphy, *The Handbook of Lipid Research*, F. Snyder, Ed.; Plenum Press: New York, 1993.

14. I. A. Blair, *Methods Enzymol.* **1990**, *187*, 13-23.

15. J. L. Kerwin and J. J. Torvik, *Anal. Chem.* **1997**, *237*, 56-64.

16. P. Wheelan, J. A. Zirrolli and R. C. Murphy, *Biol. Mass Spectrom.* **1993**, *22*, 465-473.

17. K. Bernstrom, K. Kayganich, and R. C. Murphy, *Anal. Biochem.* **1991**, *198*, 203-211.

18. D. K. MacMillan and R. C. Murphy, *J. Am. Soc. Mass Spectrom.* **1995**, *6*, 1190-1201.

19. P. Wheelan, J. A. Zirrolli and R. C. Murphy, *J. Am. Soc. Mass Spectrom.* **1996**, *7*, 129-139.

20. D. C. Zeldin, J. D. Plitman, J. Kobayashi, R. F. Miller, J. R. Snapper, J. R. Falck, J. L. Szarek, R. M. Philpot and J. H. Capdevila, *J. Clin. Invest.* **1995**, *95*, 2150-2160.

21. H. G. Rose and M. Oklander, *J. Lipid Res.* **1965**, *6*, 428.

22. S. S. Singhal, M. Saxena, H. Ahmad, S. Awasthi, A. K. Haque and Y. C. Awasthi, *Arch. Biochem. Biophys.* **1992**, *299*, 232-241.

23. F. Ursini, M. Maiorino and A. Roveri, *Biomed. Environ. Sci.* **1997**, *10*, 327-332.

24. G. M. Rosen, S. Pou, C. L. Ramos, M. S. Cohen and B. E. Britigan, *FASEB J.* **1995**, *9*, 200-209.

25. M. Peters-Golden, *Am. J. Respir. Crit. Care Med.* **1998**, *157*, S227-S231.

26. W. Pruzanski, E. Stefanski, P. Vadas, B. P. Kennedy and H. van den Bosch, *Biochim. Biophys. Acta* **1998**, *1403*, 47-56.

27. P. Wheelan and R. C. Murphy, *Anal. Biochem.* **1997**, *244*, 110-115.

Questions and Answers

Bradford W. Gibson (UCSF)

Is there any evidence for these lipid arachidonic acid radical oxygenated species reacting with membrane proteins or other macromolecules?

Answer. This is a largely unexplored area of investigation. However, investigations by Solomon in Cleveland and Ian Blair in Philadelphia have shown that various oxidized products of polyunsaturated fatty acids can covalently bind to macromolecules. We think you have asked an important question which needs to be further addressed.

Jasna Peter-Katalinic (University of Münster)

Do you use in general acyl ketene elimination fragment ion to assign the sn-2 position in glycerophospholipids in negative ion mode ESI/MS/MS?

Answer. The acyl ketene elimination of the sn-2 fatty acyl group is by far the preferred loss compared to the acyl ketene elimination from the sn-1 position. This was shown very nicely by the Michigan State group (Douglas Gage) using various glycerophospholipids. At the present time there are no known exceptions to this general rule, therefore we and others use this elimination to unambiguously assign the sn-2 fatty acyl substituent.

Mass Spectrometry After the Human Genome Project

Charles R. Cantor
Sequenom, Inc., San Diego, CA 92121

Almost all of the other speakers in this symposium are using mass spectrometry as a powerful tool to expedite the analysis of problems in biochemistry or cell biology. In contrast, my major focus is genetics. Past contributions of mass spectrometry to genetics have been almost nil. This is about to change dramatically, because recent improvements in the technology now permit the routine use of mass spectrometry to analyze the variability of DNA sequences. This variability is responsible for all genetic phenomena. Because the path that I have followed in turning to mass spectrometry is so different from most of the other speakers, I cannot resist beginning this presentation with some of the history.

The Past

I last presented a lecture at this symposium more than five years ago. At that time, my exposure to mass spectrometry had been very slight. Direct mass spectrometry of nucleic acid fragments large enough to be analytically useful seemed a distant future dream. Contemporary thinking focused on the use of mass spectrometry as a detector of isotopic mass labels after the target nucleic acid was incinerated. The great attraction of mass spectrometry even then was the potential availability of a plethora of different, distinguishable mass labels. Because four-color fluorescent nucleic acid sequencing was already in vogue and demonstrated to be a powerful technique, one imagined that mass spectrometry, in forty colors, might be developed to have ten times the throughput. This approach is still being pursued by others in several different variations. However, my view of mass spectrometry, and my involvement with it have both changed radically.

In the fall of 1992, I had the pleasure and the privilege to spend three months at Cornell University as the Baker lecturer. One of my hosts was Fred McLafferty, and I spent considerable time with him and his student, Daniel Little, who was just then obtaining absolutely beautiful Fourier Transform Ion Cyclotron Resonance Electrospray mass spectra on transfer RNA and similar sized (50 to 100 base pair) DNA fragments [1]. I was intrigued by both the beauty of the data and the potential of the methodology, but I was daunted by both the complexity of the data and the formidable appearance of the instrumentation required to produce it.

At about the same time, I participated in a small workshop on new techniques for nucleic acid research, on the outskirts of Oslo. There I heard Lloyd Smith and others report tantalizing but frustrating glimpses of Matrix-Assisted Laser Desorption Ionization Time-of-Flight (MALDI-TOF) mass spectra of short single-stranded oligonucleotides. The results were tantalizing because the spectra were so simple. The frustration was in the apparent unreliability of the method. Some compounds gave good clean spectra; others gave poor data or no data at all. Data quality was strongly dependent on which of the four bases (A, T, G, C) predominated in the compound under study, and I remember thinking at the time that there seemed to be some correlation with pKa's for protonation, but no one present seemed to have a clear vision of how to improve the situation.

My major scientific preoccupation during this period was nucleic acid arrays. Edwin Southern at Oxford [2] Radoje Drmanac and Radomir Crkvenjakov in Belgrade [3], Andrei Mirzabekov in Moscow [4], and William Bains had independently conceived the notion of DNA sequencing by hybridization. In this approach short oligonucleotides (8 bases was the current vogue) would be used to probe a target for the presence of the complementary sequence. By compiling and assembling a list of 8-letter word content, the text of the entire DNA target might be reconstructed. Certain ambiguities would prevent a perfect de novo sequencing effort, but for anything less than that the method had great potential appeal. The problem was making it work well enough to compete with already accepted gel electrophoretic methods. This involved two technical challenges. One was synthesis of arrays of oligonucleotides: a complete set of octanucleotides would entail more than 65,000 different compounds, and to make such an array useable in a cost effective manner it would have to be miniaturized. The second challenge was the hybridization step itself. Because the oligonucleotides would form short helices, base pairing at their ends would not be perfect, and considerable ambiguity in sequence recognition was noted in pilot experiments. An additional problem was the much greater thermodynamic stability of G-C pairs than A-T pairs. This made it difficult to find a single set of environmental conditions where G-C rich and A-T rich sequences could be probed reliably.

To circumvent the problems encountered in short oligonucleotide hybridization, I had conceived a scheme where enzymes could be used to proofread the edges of the double strand. DNA ligase would check that one edge was correctly based paired, while DNA polymerase would check the fidelity of the other edge [5]. I named this scheme positional sequencing by hybridization, and while I lectured at Cornell, coworkers at Berkeley performed some pilot steps to prove that the particular enzyme reactions needed were practical. Armed with this information, and anxious to reveal the scheme to others, I arranged for a patent attorney, James Remenick, to visit me in my Cornell office, and in a few intense days of effort (mostly his) we ground out a lengthy patent. Doing his job well, Remenick pressed me for details and extrapolations so the claims could be crafted as broadly as possible. I dutifully added mass spectroscopy to the list of detection methods that might prove suitable for nucleic acid arrays, because based on what I had heard in Oslo and seen at Ithaca, this seemed totally practical, powerful and attractive, even though no one had actually ever done it in the laboratory. This turned out to be a fateful inclusion that would impact on the next phases of my career in ways far broader than anything I might have imagined at the time.

The Baker lectures finished, I moved to Boston University in December of 1992, to be greeted by one of the fiercest Nor'Easters of all time. Five months after that I received a Sunday afternoon visit from Hubert Koester, a professor at the University of Hamburg. We had been introduced by mutual friends, Bob Molinari and Nola Masterson. They thought we might have complementary technology. Their insight was absolutely correct, and the meeting between Koester and myself was a key step in the formation of Sequenom, Inc., a company dedicated to high throughput mass spectroscopic analysis of DNA and other macromolecular samples. Robert Cotter of Johns Hopkins soon joined us as the third member of the initial scientific team.

Koester, by clever and thoughtful chemistry, had managed to find procedures that yielded reliable MALDI-TOF spectra of oligonucleotides up to sixty bases in length. This did not yet open the way to efficient de novo DNA sequencing by conventional approaches, with mass spectroscopy replacing gel electrophoresis as the separation technique. In de novo sequencing, read length is a premium because it greatly facilitates the assembly of individual sequence reads into final completed sequence. However, as a detection scheme for hybridization arrays, sixty base reads were far more than adequate. Thus evolved the notion of DNA MassArray™ technology, in which an array of DNA samples produced by any means would be analyzed by scanning the array by MALDI-TOF mass spectrometry. Five years after its conception, this notion is now reality.

The Present

One of the stated goals of the human genome project is to provide a complete DNA sequence of the approximately 3 billion base pairs of a single (haploid) human genome. This project is well underway, and neglecting a few regions in the genome with sequences too repetitive for any current methods to untangle, we can expect a successful conclusion in or around 2005. The second major goal is to find all the genes. This task is largely complete already, mostly through the private sector cloning and sequencing efforts of Incyte Pharmaceuticals, Inc. and Human Genome Sciences, Inc. All of these efforts rely on gel electrophoresis, which has been continually optimized and improved. Through incremental advances, gel-based technology has reached the point where the goals of the genome project will be achieved at an acceptable cost. However, the moment one looks beyond the project, the needs for nucleic acid sequence analysis and other related measurements will require throughput and accuracy which gel-based methods seem incapable of achieving. This has promoted a renewed interest in alternative technologies, and mass array methods loom as one of the most promising, perhaps the single most promising solution to the ever accelerating demand for faster, cheaper data acquisition.

The task that Sequenom had set for itself was to develop an automated platform to scan DNA arrays by mass spectrometry. This entailed three separate technical challenges. Array surfaces needed to be constructed that were suitable platforms for MALDI-TOF MS. Methods of immobilization of DNA sequences on these arrays at high density had to be created. Enzymatic manipulation of targets hybridized to these arrays had to be perfected so that the hybridization could be proofread. Finally, an efficient way of automatically scanning the arrays by MALDI needed to be found. All of these goals have now been accomplished. In the course of the work, several of the techniques developed turned out to be strongly synergistic.

For efficient DNA manipulation, we needed relatively small arrays with sample sizes and spacings much smaller than common in conventional MALDI. Small samples reduce the sample preparation cost, but we discovered that they also dramatically improved the reliability of the MALDI process. With conventional samples, matrix crystals are large and one typically has to search among the sample for sweet spots that yield good spectra. We had miniaturized the samples to the point where they were about the same size as the spot illuminated by the laser. Thus the sample was uniformly illuminated, and once conditions were found that produced matrices that yielded good spectra, all of the spectra were good. The actual data quality also improved. For example, with 200 micron samples mass accuracy is about an order of magnitude better than with conventional samples, both in terms of the standard error of the mean and the magnitude of the largest outliers seen.

As a support for the arrays, we focused on single crystal silicon. This has proven to have two advantages. All of the power of silicon micro-fabrication can be brought to bear to make arrays with nanoliter wells, textured surfaces, coated surfaces, and so on. Once efficient chemistry was developed to functionalize the surfaces, it was found that DNA probe densities on these surfaces exceeded those on more conventional array surfaces by up to two orders of magnitude. Moreover, up to half of the immobilized DNA would function in hybridization or other biochemical processes; this performance was far better than other arrays that have thus far been described. Silicon as a surface also shows a remarkably low background of non-specific DNA binding compared with other surfaces commonly used. All of these properties have led to very efficient performance of DNA arrays on silicon surfaces.

A third key step in the technology was fortuitous. Early experiments with Kai Tang, Cotter and Koester showed that MALDI conditions resulted in efficient DNA strand separation [6]. Thus if one DNA strand were covalently attached to a surface, and a second hybridized to it non-covalently, in MALDI-TOF MS the noncovalent strand could be detected efficiently while the covalent strand was never seen. This finding persisted even if biotin-streptavidin attachment was used instead of covalent attachment for the probe strand.

As the technology matured, it became practical to design and construct MALDI-TOF instruments with high resolution array scanning capability. Borrowing from the semiconductor industry, arrays are termed chips, and Sequenom called its particular form of arrays SpectroChips™. Cotter built the first chip-scanning instrument that was used for all of the experiments needed to validate the technology and develop pilot results for a variety of applications. More recently, commercial instrumentation has been developed and will be available as a product to the general public by the end of 1998. Typical chips will hold 96 or 384 samples, but far denser and larger arrays are usable. Up to ten chips can be loaded simultaneously and scanned automatically with no manual intervention. While many specialized chip designs are under development for particular nucleic acid applications, what will be available shortly are chips preloaded with matrix on which any desired samples can be deposited. While all of the original Sequenom technology was developed on DNA samples, it has not escaped our attention that mass arrays will be useful for RNA, protein, or other samples. The technology that has been developed is general and should see a wide range of applications. However, the remainder of this chapter will focus on nucleic acid applications.

With the continued development of Sequenom's high throughput MS platform, it became increasingly attractive for the scientists who had been instrumental in developing the methodology to play a more active role in deciding how it should be used. Koester left his Hamburg position to join the company full time as CEO in 1997, and a month ago I moved to San Diego to become Sequenom's CSO.

The Future

Sequenom's mass array technology can scan about 10,000 samples per day automatically and produce a high quality MALDI-TOF spectrum from each of them automatically. This represents a capacity that has not before been practical in mass spectrometry. The issues that now need to be resolved are what sorts of measurements in the post-genome project era need such a large number of mass spectra, and how can the tool be optimized to provide these measurements most effectively. Although I will focus on MALDI-TOF, it is clear that other forms of gentle ionization and mass measurement are also destined to play important roles if their throughputs can be enlarged to meet the demand.

Several general considerations can be offered about throughput. With high capacity now available, the cost of an instrument and the labor to run it become totally insignificant. The major costs that remain are in sample preparation and data analysis. Improvements in both of these areas are critical if the full power of automated MS is to be used advantageously. Reducing the size of samples will cut the costs of sample preparation considerably. In addition, automated sample preparation modules will reduce or all but eliminate manpower costs. At Sequenom, we are establishing automated sample processing lines whose capacity matches the demands of the automated MALDI-TOF instruments. Data analysis of MALDI-TOF spectra from nucleic acid samples should be relatively easy to automate and streamline. The mass of each desired peak is calculable in advance from the composition of the four bases. Peaks corresponding to doubly charged ions or compounds that have lost a purine (typically a G, a common artifact) can also be calculated in advance. Thus we expect that given the very high quality of the data routinely obtained in automated mode (resolutions of 500 to 1000, signal to noise of 20:1 or better), automated analysis at a rate sufficient to keep up with the sample stream will not pose formidable problems.

The problem that remains unsolved is how the user will look at the results of such an enormous daily load of mass data. This will not be possible manually. It will only be possible in an automated mode which will depend intimately on the particular experimental design used to create the samples, and the results desired from that design. In general, in biology it is a bad idea to have a technique and go looking for suitable applications. However MS is a general platform technology, a replacement for macromolecular gel electrophoresis that has pervaded almost all of molecular biology. Hence, one can systematically review existing experimental protocols and designs and ask which are amenable to MS detection and what would the advantages be. In nucleic acids, four areas currently have high throughput needs. For each I will discuss the relative merits of implementing automated MS to serve these needs.

Detection of Known Mutations or Polymorphisms

A single human genome sequence is just a reference point. Most of the interesting human-genetic phenomena will only be revealed by studying patterns of DNA sequence differences. The human species is amazingly diverse. When one individual is compared against a reference, on average there is a DNA sequence difference every thousand bases, or three million per haploid complement. However when the entire human population is compared, the frequency of sites that vary in at least part of the population is much higher. A recent study shows variants every 200 bases, and this involved only three diverse ethnic groups [7]. The true level of human DNA polymorphism will be much higher.

Polymorphisms are measured to correlate particular genetic variants with their observable physical consequence, called the phenotype. In the most straightforward form of analysis, the DNA sequence variation itself is the cause of the phenotype. For example, a single nucleotide change converts normal hemoglobin A into the variant hemoglobin S responsible for sickle cell disease. DNA diagnostics consists in the measurement of such variations. What makes the problem challenging is that some diseases may arise from mutations at many different places in a gene. Thus the magnitude of the task of mutation detection can approach that of full gene sequencing. Today most genetic diseases where the gene and alleles responsible are known are rare. Diagnostic DNA sequencing or allele-specific testing is an important tool both in the research laboratory and in the clinic. But current demand is not enormous and commercial opportunities to exploit this demand have to be evaluated cautiously.

Most inherited human diseases are the result of the concerted action of several genes. Finding these genes and developing diagnostic assays is a far more difficult task than what is required for single gene diseases. It is here where mass spectrometry seems destined to play a key role both in the original process of gene discovery, as well as the eventual development of robust clinical diagnostic tests. In traditional human gene discovery, a disease is traced in one or more families. In the absence of any specific leads, known DNA variations are chosen at random and their pattern of inheritance is compared with that of the disease. A similar pattern is evidence for what is called linkage. A minimum of thirty individuals is usually needed to provide the first convincing statistical evidence for linkage. Much larger numbers of individuals need to be examined to narrow down the location of a disease gene once preliminary evidence for its existence has been obtained.

For many diseases sufficient family data is simply not available. Then a large population-based study must be done directly. The aim is to find DNA variations that seem to be associated with the disease. This is more difficult than a direct family linkage study, but it is still possible in many cases. A reader new to this field may think that it is necessary to find the variant that is directly causal for the disease. This is the desired end goal, but it is not a necessary intermediate step. Some of the variations in our genome structure are relatively young, as are many variations that cause specific diseases. Thus it is possible that a statistical correlation can exist between an innocuous variation and a disease-causing one if both are new enough to the population that their distribution has not had time to randomize (equilibrate). Such a correlation will occur only when the two variants are near each other in the genome, because the process of equilibration becomes much more efficient as the distance along the DNA increases. However, to spot such a lack of equilibration requires the analysis of large numbers of samples. It does not, however, require any prior knowledge about the biology of the disease beyond that necessary to perform a correct diagnosis from the symptoms.

In most linkage and association studies, one treads on the brink of disaster. The statistical power in most data sets is usually barely adequate to provide convincing evidence that the neighborhood of a gene has been reached. A few errors in allele typing can destroy the entire study. Here the power of mass spectrometry is not just its throughput but also its accuracy. Unlike conventional gel-based assays, MS analysis of DNA is usually decisively convincing; typical errors, such as artifacts in DNA amplification, and sample mix-up reveal themselves readily.

Two types of DNA polymorphisms are commonly used to track variations in population studies. The first are simple sequence repeats. These are stretches like $(CAG)_n$ that can occur either in coding or non-coding regions. Most simple repeats are genetically unstable and have a high mutation rate that typically adds or deletes one or more copies of the repeat. As a result these sequences are much more divergent in the population than most other sequences. Because a high degree of variation is informative in population studies, simple repeats were early favorites for human genetic mapping. However many repeated sequences behave non-ideally in polymerase chain reaction (PCR) amplification, the method almost universally used to purify specific genomic sequences prior to any further study. Their complex behavior appears to result from stable secondary structure in the single-stranded sequences and from the ability of most polymerases to stutter and add or subtract bits of the sequence when confronted with repeats. The result is a complex spectrum of PCR amplification products which confuses subsequent analysis. For this reason current trends lie away from simple repeats for most applications. An exception is forensics, where high variability is so critical that some of the complications can be tolerated.

The second common type of DNA variation useful for genetic or population studies is a single nucleotide polymorphism (SNP). These usually involve just two alleles at a given position in the DNA sequence, say A or C. The disadvantage of such simple polymorphisms is that individuals are frequently homozygous at such locations — both copies of the genome each of us carry may have the same allele, A and A. In this case it is not possible to tell in a linkage study which copy of the genome was passed to an offspring, and the marker is called uninformative. However, SNPs are so easy and reliable to measure that it more than compensates for their lack of information content. A major demand in current planning for post human genome project studies of human variability is the need to measure large numbers of SNPs in large numbers of individuals. For several reasons, MALDI-TOF mass spectrometry appears to be an ideal technology for the high throughput, accurate measurement of SNPs.

An example of the need for enormous numbers of SNP measurements is the segmentation of patients on clinical trials. In a typical critical phase in the development of a new pharmaceutical, several thousand people will be treated either with a new drug or with a placebo. In some the drug will appear to be efficacious; in others there will be no effect; in some others there may be undesirable side effects. Some of this difference in response is due to the genetic variability of the population in the drug study. If one had a way to analyze this variation and find specific loci which correlated with efficacy or side reactions, it would provide a justification for separating the trial population and maintaining therapy only for those individuals for whom it was efficacious. The underlying reason is that the DNA variants that correlated with desirable drug responsiveness could be used on the general population to predict in advance who should be given the drug and who should not be given it. This is a winning scenario for all the parties concerned in improved medical care. It reduces the cost of the clinical trial; it allows drugs that are efficacious for only part of the population to become available for that cohort, and it provides a diagnostic test that can be used to segment the population prior to attempting therapy. However, with present technology the cost of measuring the needed SNPs in the clinical trial population is prohibitive.

If nothing at all is known about the genetic basis of a variation, such as differential drug responsiveness, one must scan the entire genome to find a predictive allele. For the population sizes used in typical clinical trials it is estimated that such a scan could involve up to 100,000 different SNPs. Thus a total of 100 to 200 million individual SNP measurements are required. With current technology each measurement costs several dollars. To make the basic strategy realistic, costs must be reduced by one to two orders of magnitude. The small sample sizes and automated sample production and handling techniques developed by Sequenom should make these cost reductions achievable over the next few years. This will also enable similar studies in agricultural genetics. Plants are more difficult targets than humans for genetic studies because the genomes are larger, and in a typical species of interest the variation is likely to be greater. For example, the gross physical appearance of tomatoes, including size, shape, color and texture is far more diverse than humans.

Mass spectrometry can be used to measure SNPs in several different ways. The most comprehensive is DNA sequencing, which would reveal any SNP whether or not it was previously known. However, DNA sequencing even by mass spectrometry is too expensive for most SNP analysis. An analog of sequencing is far more cost effective when the SNP of interest is known in advance. This is a form of mini-sequencing; in the format used by Sequenom it is called the PROBE (BiomassPROBE™), where the acronym stands for primer oligo base extension. This reaction is illustrated in Fig. 1. A primer is chosen that sits just upstream of the polymorphism to be measured. DNA polymerase is allowed to extend the primer using the information coded for by the target template strand. Three normal nucleoside triphosphate substrates are provided to the polymerase, but the fourth is replaced by a di-deoxy derivative which serves as a terminator and blocks chain extension once it is incorporated. In one allele the termination occurs after a single nucleotide is added; with the other allele at least one additional nucleotide is added before termination occurs. In this manner the polymorphism is presented as a pair of peaks in the mass spectrum that differ by at least one nucleotide, nominally 300 Da, and in most cases by two or three times this amount. With the current resolution of MALDI-TOF, such mass shifts are very easily detected as shown by the example in Fig. 1.

There are two reasons why the PROBE method is far preferable to DNA sequencing for measuring polymorphisms. Each DNA target strand yields just one mass peak. Thus depending on whether the allele is homozygous or heterozygous, at most two peaks will be seen in the spectrum. This dramatically improves sensitivity when compared to sequencing, because for a typical 100 base target, a sequencing reaction will result in around 25 different peaks. The small number of peaks produced by the PROBE reaction also means that many different SNPs can be measured in the same mass spectrum. All that is required is to adjust the sizes of the primers used for each different SNP so that the resulting extension products all have very distinct masses [8]. What is typically done is to perform several multiplex PCR amplifications, each of which can sample a few loci, and then pool the amplification products prior to mass spectrometry. In this way up to fifteen SNPs can be analyzed in a single spectrum. When combined with the fast chip scanning capability developed by Sequenom, this will allow a single instrument to measure 150,000 SNPs in one day. Relatively simple enhancements should allow the throughput to be increased by an order of magnitude. At these rates the mass spectrometry itself is no longer likely to be the rate limiting step in the overall process.

Discovery of Unknown Mutations or Polymorphisms

DNA variants are usually discovered today by conventional gel-based sequencing. This is an expensive and tedious process. To find a variation that occurs on average in 10% of a population, one must examine the same region of DNA in about 6 to 12 people (12 to 24 genome equivalents, because each of us has two genome equivalents). With perfect data one should be able to pool samples

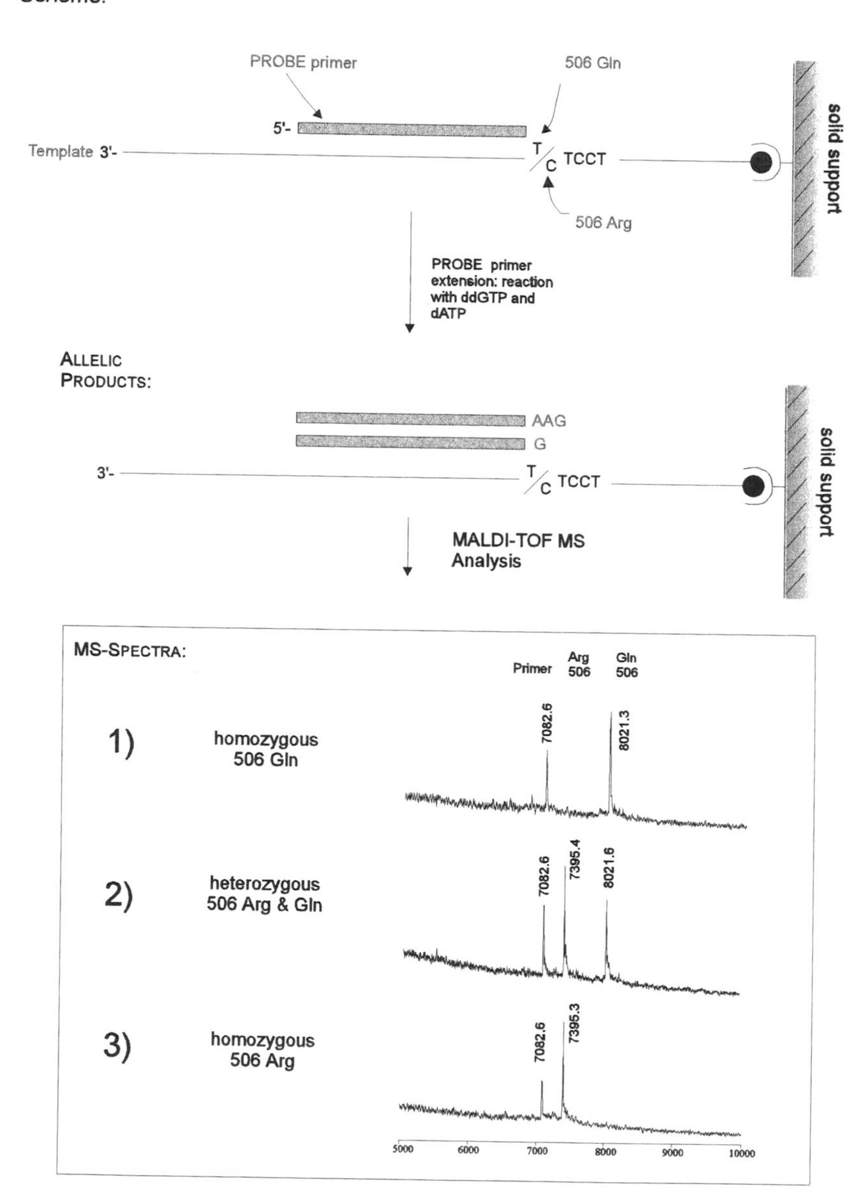

Fig. 1. An example of the PROBE technique used to analyze a single nucleotide polymorphism (SNP) in the Factor-V gene. The schematic reaction scheme is shown at the top. The actual mass spectra that result are shown at the bottom. Figure kindly provided by Andreas Braun and Brigitte Darnhofer, Sequenom, Inc.

from the set of individuals, amplify the corresponding region in all of them at once, and run a single set of sequencing reactions. As a reference, a single individual also would be examined. Differences between the two samples would signify polymorphisms. However, most gel-based sequencing data is not capable of revealing a polymorphism at the 10% level. Thus smaller pools, or even no sample pooling at all must be used to perform the analysis. The advantage of finding variants by sequencing is that all variants are found. The disadvantage is cost. Depending on the extent of pooling used, 2000 to 10,000 bases must be sequenced to find a polymorphism (assuming a frequency of 1:1000). At current sequencing costs of about $0.05 to $0.10 per base, this would require $100 to $1000 for each polymorphism. Current demands are for hundreds of thousands of new polymorphisms; hence some reduction in cost is urgently needed.

The cost of mass spectrometric DNA sequencing today is not much cheaper than conventional sequencing. The MS approach is much younger, and quite a few avenues for significant cost reductions are likely to prove successful. But even with current costs, MS will be more effective because the considerable increase in data accuracy will allow much more aggressive sample pooling strategies to be implemented.

A number of alternative approaches to finding polymorphisms are under intensive exploration. Most of these involve the formation of heteroduplexes. If a mixture of DNA strands from the same region in different individuals is allowed to re-anneal, the two complementary DNA strands will re-assort randomly. If any sequence differences exist in the population, these will result in heteroduplexes — DNA molecules that lack perfect base pairing at the site of the sequence variation. Various enzymatic or separation procedures can distinguish between perfect and imperfect duplexes. However, all of these methods ultimately require a DNA sizing step to prove that a heteroduplex has been detected. Mass spectrometry could be used in place of gel electrophoresis or chromatography for many of these procedures. However, there appear to be even more powerful ways to use mass spectrometry to find variations. I will describe the simplest approach. DNA from the target region of a pool of individuals can be amplified and cut enzymatically into a discrete set of fragments in the size range where highly accurate MALDI-TOF molecular weights can be obtained. Any fragments containing DNA variations will appear as two bands instead of one, split by the mass difference caused by the variation. In the worst case, substitution of an A by a T (or vice versa) the mass change is 9 Da. In all other possible cases it is significantly larger. Thus the mass spectrum reveals that a variant has been found, identifies by mass the particular fragment on which it lies, and indicates the chemical nature of the variation. If the location of the variation is needed, sequencing will then be required, but this need be done only on the single fragment that showed a variation, not on the whole target. Thus a single mass spectrum can reduce the sequencing burden by an order of magnitude. Further reduction in sequencing can be achieved by using intelligent pooling strategies, so that with several mass spectra of fragment mixtures one will know which individual DNA sample to sequence.

Analysis of patterns of gene expression: transcript profiling.

In just two years, an entire industry has developed based on several different high throughput strategies to measure the relative extent of gene expression in cells in well-defined physiological states. These developments have been spurred by the increasing numbers of known human genes, as well as the complete genomic DNA sequences now available for many bacteria, a yeast (*S. cerevisiae*), and soon for the nematode *C. elegans*. Studies to date have concentrated on the analysis of mRNA levels. This involves converting the mRNA to more stable cDNA and then comparing populations of cDNA from cells in particular physiological states. These comparisons were first done by gel separations called differential display or by randomly sampling cloned cDNAs and sequencing them. More efficient comparisons can now be done by array hybridization methods or by a very elegant mini-sequencing approach [9] called serial analysis of gene expression (SAGE). Both approaches provide fairly accurate representations of gene expression levels. However, it is by no means clear that mRNA levels correlate well with corresponding protein levels, and the latter are really needed for almost all biological investigations.

The high throughput of mass spectrometry makes it an attractive alternative to hybridization or gel-based sequencing for gene expression profiling. However, one potential limitation in contemporary mass spectrometry may limit its usefulness here. Most techniques for expression profiling require a quantitative comparison between two populations of cDNA species. Ideally, a factor of two or better quantitation is needed, and this must be maintained over more than a thousand fold dynamic range in concentration. While automated MALDI-TOF in our hands shows the necessary quantitation for peaks closely spaced in mass, its dynamic range does not yet approach that needed to implement simple one step expression profiling protocols.

The best MS approaches for transcript profiling are likely to be some kind of sampled sequencing. Less than 20 bases of DNA sequence will serve as a unique identifier for almost any site in the human genome. Twelve to fourteen bases are sufficient to uniquely characterize any cDNA. The power of MS to generate highly accurate short stretches of DNA sequence is considerable, and this should allow very effective analysis of sampled cDNAs. The SAGE method is a variety of sampled sequencing that is particularly effective. In SAGE, enzymatic protocols are used to cleave and purify a unique constant length short fragment from each cDNA. Additional steps are used to ligate and amplify pairs of cDNAs from two different samples to be compared. The frequency of occurrence of specific pairs allows their relative abundance to be computed. Because the pairs are formed before PCR amplification, any distortion of the original population of cDNAs by that amplification is avoided. The resulting biplex probes are still less than 30 bases, and thus easily sequenced by MS. It should be possible to modify the original SAGE protocols to optimize them for MS analysis.

De Novo DNA Sequencing

The power of the first few full genome DNA sequences available was greater than most expectations. Interesting biological phenomena have been discovered by mere inspection and comparison of the sequences. This has fueled a tremendous increase in the demand for additional full genome sequences, far beyond the task originally envisioned in the Human Genome Project. The short read lengths available in current MALDI-TOF MS of DNA are not ideal for de novo sequencing. However, the great accuracy of MALDI-TOF data will compensate in part for the short read lengths. Franz Hillenkamp and colleagues have recently shown that the use of IR lasers and compatible matrices may extend the size range of high resolution DNA MS considerably. As this technology develops it is likely to make MS fully competitive with gel-based sequencing techniques. Hillenkamp reports extremely high sensitivity with his IR procedures [10]. This will allow very tiny samples to be prepared for sequencing, which should keep the costs of sample preparation at or below that needed even for the newly emerging generation of capillary electrophoretic sequencing instruments.

CONCLUSIONS

Mass spectrometry is almost ideally suited for the analysis of DNA. Nature must have had the mass spectrometrist in mind when she invented DNA, because the four natural bases have very distinct masses. The past few years have brought tremendous technical advances in DNA MALDI-TOF MS. The stage is now set for the large scale implementation of MS methods for a variety of molecular genetic analyses. However, it is important to realize that almost all of the MS methods ready for use with DNA are simply MS versions of traditional gel electrophoretic tools. There is likely to be a whole new generation of MS methods conceived specially for DNA analysis, once the general applicability of this powerful technique becomes better appreciated by the biological community.

REFERENCES

1. D. P. Little, T. W. Thannhauser and F. W. McLafferty, *Proc. Natl. Acad. Sci USA* **1995**, *82*, 2318-22.

2. E. M. Southern, U. Maskos and J. K. Elder, *Genomics* **1992**, *13*, 1008-17.

3. R. Drmanac, S. Dramanac, Z. Strezoska, T. Paunesku I. Labat, M. Zeremski, *et al.*, *Science* **1998**, *260*, 1649-52.

4. K. R. Kharapko, Y. P. Lysov, A. A. Khorlyn, V. V. Shick, V. L. Florentiev and A. D. Mirzabekov, *FEBS Letts.* **1989**, *256*, 118-122

5. N. E. Broude, T. Sano, C. L. Smith and C. R. Cantor, *Proc. Natl. Acad. Sci. USA* **1994**, *91*, 3072-76.

6. K. Tang, D. Fu, S. Koetter, R. J. Cotter, C. R. Cantor and H. Koester, *Nucleic Acids Res.* **1995**, *23*, 3126-31.

7. D. A. Nickerson, S. L. Taylor, K. M. Weiss, A. G. Clark, R. G. Hutchinson, J. Stengard, *et al. Nature Genetics* **1998**, *19*, 233-40.

8. A. Braun, D. P. Little and H. Koester, *Clin. Chem.* **1997**, *43*, 1151-7.

9. V. E. Velculescu, L. Zhang, B. Vogelstein and K. W. Kinzler, *Science* **1995**, *270*, 484-7.

Questions and Answers

James McCloskey (University of Utah)

Does miniaturization of the mass spectrometer also play a role in these kinds of developments?

Answer. No. Although the sample size has been miniaturized, the mass spectrometer used is of standard size. Robert Cotter, one of our advisors, is in the process of perfecting miniaturized TOF instruments, and so eventually it may be possible to combine our small samples with his small instruments.

Marvin Vestal (PerSeptive Biosystems)

Do you think mass spectrometry will play a role in large scale genomic sequencing?

Answer. Whether or not MS plays a role in large scale DNA sequencing depends on what kind of sequencing one is talking about. For comparative or diagnostic sequencing such as needed for horizontal clinical studies, MS is here and now. For de novo, discovery sequencing, it will depend on the trade-offs between accuracy and read length. Current de novo sequencing relies on long read lengths to assist the assembly of finished sequence. It is conceivable that the superior accuracy of MS sequencing will allow comparable overall efficiency and throughput with shorter read lengths. However, this remains to be proven.

Steven Carr (SmithKline Beecham Pharmaceuticals)

Where will the next factor of 10x-100x increase in sample throughput come from?

Answer. Large increases in throughput will come from several different sources. Sample multiplexing can easily raise throughputs five fold for most kinds of experiments. Switching to 384 well micro-liter plate formats will give another four fold increase in throughput. Current software on the automated instrument is not yet very efficient and a factor of 2 to 4 additional throughput will be gained once it is optimized.

Steven Carr (SmithKline Beecham Pharmaceuticals)

How important is the ability to quantitate, and how do the MS methods compare to purely genetic approaches like in-site hybridization?

Answer. Quantitation is very important in one particular type of genetic experiment, monitoring the levels of expression of many genes simultaneously. MS actually shows excellent quantitation when carried out in our miniaturized sample format. However, dynamic range is still somewhat limiting. Today we have at most a hundred fold dynamic range; for broadly applicable gene expression monitoring, a ten thousand fold dynamic range would be desirable.

Problems and Prospects in the Characterization of Posttranscriptional Modifications in Large RNAs

Pamela F. Crain, Duane E. Ruffner, Yeunghaw Ho, Fenghe Qiu, Jef Rozenski and James A. McCloskey*
Departments of Medicinal Chemistry, and Pharmaceutics & Pharmaceutical Chemistry*, University of Utah, Salt Lake City, UT 84112

The central importance of RNA in biology derives principally from the key functions of four RNAs at and in the ribosome, at the precise sites of decoding the genetic message and of peptide bond formation: transfer RNA (tRNA), messenger RNA (mRNA), small subunit ribosomal RNA (16S–18S rRNA), and large subunit ribosomal RNA (23S–28S rRNA). Rapidly accumulating evidence for *direct* participation of the rRNAs in translation [1, 2] has led to a remarkable reversal of fortune: protein synthesis is largely under control of specific domains of rRNA rather than of ribosomal proteins as originally thought. This has left rRNA relatively understudied with regard to the three-dimensional structure of the ribosome, and the precise roles of single rRNA nucleotides in the complex mechanism of translation.

The functions of most RNAs, including the two large rRNAs, are fine-tuned by a large number of posttranscriptional modifications, 29 of which are presently known in rRNA in all three evolutionary domains [3]. However, detailed knowledge of the influence of base and ribose modification on RNA structure-function is derived mainly from extensive studies of tRNA (e. g., [4]). In contrast, relatively little is known about the role of rRNA modifications (see Table 1), and in fact only two primary rRNA structures are believed known at the fully processed RNA (rather than rDNA) level: 16S rRNA (11 modification sites) and 23S rRNA (23 modifications) from *E. coli* [3]. The present report provides a summary of current activities in our laboratory to further develop mass spectrometry-based methods (summarized in Fig. 1) for application to the characterization of modifications in large RNAs, such as those which occur in the ribosome.

role	reference
Subunit assembly and processing	[31-37]
Fidelity of decoding	[38]
Catalysis of peptidyl transfer[a]	[39,40]
Maintenance of loop structure[b]	[7]
Thermal stabilization[c]	[21]

[a] Involving pseudouridine
[b] Involving dihydrouridine
[c] Involving ribose methylation

Table 1. Established and proposed roles of posttranscriptional modification in ribosomal RNAs.

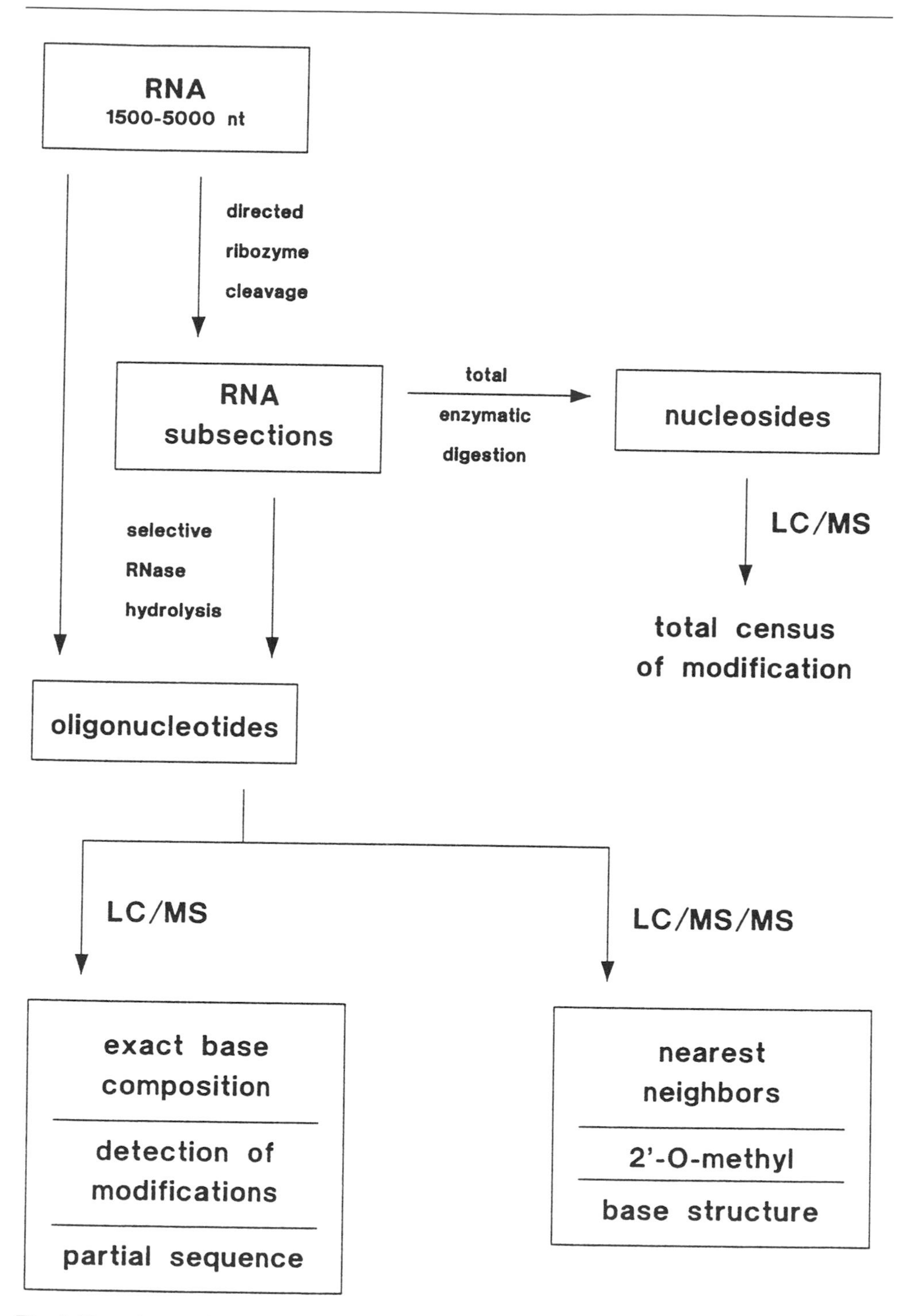

Fig. 1. Experimental strategy for characterization of posttranscriptional modification sites in large RNAs. These procedures represent an extension of earlier methods [16] developed before the advent of routine LC/MS of oligonucleotides.

The Problem

Completion of the correct modification maps for the *E. coli* rRNAs required nearly 30 years [5]. Why have there been such limitations in the determination and availability of sequence/modification data from the large[1] ribosomal RNAs[1]? There appear to be three principal reasons. (i) In early work, rRNA gene sequence data was not available, so that correct placement of small sequenced segments of RNA (e. g., fragments from RNase T1 digestion) was difficult. (ii) Prior to the development of mass spectrometry-based techniques, inadequate methods were used for identification of modified residues. For discussion of this problem relevant to rRNA see [6]. For instance chromatographic mobility has been commonly used for identification, but fares poorly when unexpected or structurally unknown nucleosides are encountered. (iii) Absence of *a priori* information as to the types and range of nucleoside structures likely to be encountered sometimes resulted in structural misassignments. On the other hand information concerning sequence locations of modifications has in general been more reliable (although not completely error-free) because such information can be established from "stops" due to modification in the reverse transcription of RNA to DNA. In any event, the overall problem of correct structure assignment and placement of modifications in large RNAs is exacerbated by molecular size of the RNA[1] and the complexity of oligonucleotide mixtures which result from nuclease cleavage. Using RNase T1 (cleaves after Gp), for example, digestion of 16S rRNA from *E. coli* yields 141 unique sequences from a total of 488 oligonucleotides, and the problem becomes significantly greater in the case of the eukaryotic 18S and 28S rRNAs.

Aside from the sheer complexity of the mixture resulting from selective RNase cleavage, compositional redundancies and even sequence redundancies become a problem. In *E. coli* 23S rRNA (2904 nt), for example, the gene sequence predicts that in the 6-mer pool of RNase T1 hydrolysis products there are 15 sets of oligonucleotides that are compositional isomers and thus may have slightly different HPLC retention times but identical molecular masses (if unmodified). Further, these sets of isomers range in complexity from single occurrences (six isomers) to 11 different (A_2,C_2,U)Gp isomers. Obviously this problem is of less concern if all isomers are unmodified, although keeping a tally of which unmodified fragments have been accounted for remains difficult. We have found significant relief from this problem using a practical means of initial sectioning of the rRNA molecule as described below.

Reduction in rRNA Sequence Complexity by Ribozyme Cleavage

To a considerable extent the problems described above can be alleviated by cleavage of RNA into subsections, using the corresponding gene sequence as a guide to mapping of desired cut sites. For use in conjunction with mass spectrometry the cleavage method must produce high yields, sharp cut sites and ease of recovery of the resulting fragments, preferably with low salt content. We

For example: 16S rRNA contains approximately 1500 nucleotides (nt), M_r 490 x 10^3; 18S, 2000 nt, M_r 650 x 10^3; 23S, 2900 nt, M_r 950 x 10^3; 28S, 5000 nt, M_r 1.6 x 10^6.

previously used a nuclease protection approach [7], in which the region of interest is protected by hybridization of a complementary DNA oligonucleotide, and then the remainder of the RNA is digested away by single-strand specific nucleases. This approach is suitable if only a specific region is of interest, but of necessity, the remainder of the RNA is lost. RNase H (digests the RNA portion that is hybridized to the DNA) can also be used to cleave RNA, but the source of the enzyme introduces an aspect of potential variability that is undesirable [8]. Accordingly, we have developed a method based on targeted rRNA cleavage using self-catalytic hammerhead ribozymes.

Ribozymes are catalytic RNAs that have the ability to cleave RNA molecules after the H at NUH sites, where H is A, C or U, but not G [9]. Methylation at O-2′ of residue H blocks the cleavage reaction. Ribozymes have an invariant catalytic domain, flanked by variable sequences that can be chosen to be complementary to the sequences flanking the NUH site in the target RNA. The ribozyme can then be synthesized using standard phosphoramidite chemistry. An example of the approach is given by recent work in our laboratory on mapping modification sites in 16S rRNA of the thermophilic archaeon *Sulfolobus solfataricus* P2 (1495 nt). We identified two suitable GUA sites [10] where cleavage would break the RNA into three pieces as illustrated in Fig. 2. Both

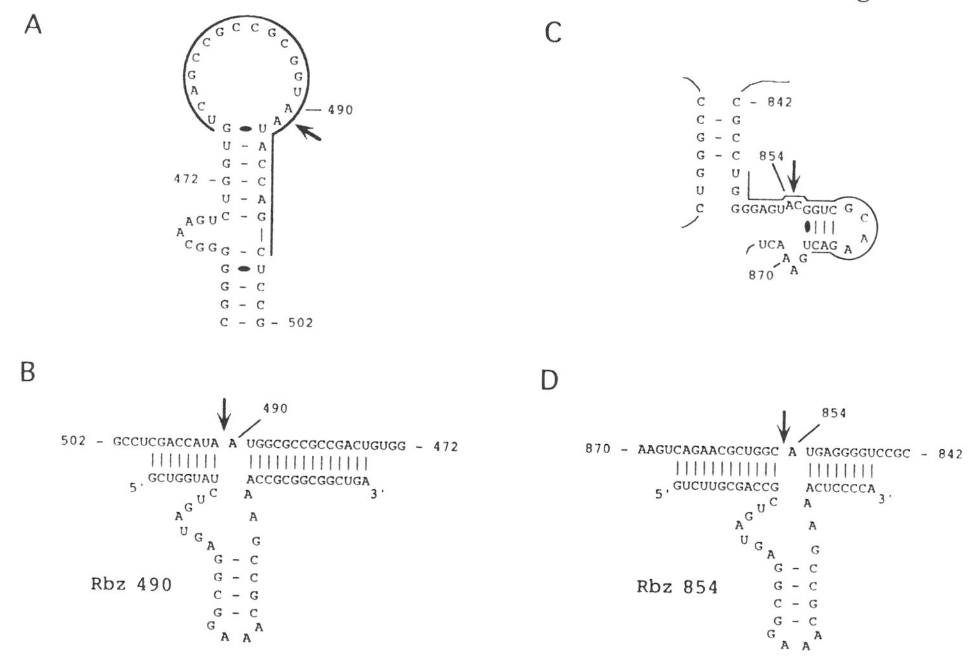

Fig. 2. Subdivision of RNA for mass spectrometry by directed ribozyme cleavage, shown for Sulfolobus solfataricus *16S rRNA. Cleavage sites are denoted by arrows. (A) Secondary rRNA structure in the region of residues 458-502, showing the ribozyme hybridization sites (dark line). (B) Ribosomal RNA (upper strand) hybridized to ribozyme (lower strand and structure) prior to cleavage. (C) Secondary rRNA structure in the region 842-873 showing ribozyme hybridization sites (dark line). (D) Hybridized rRNA (upper strand) and ribozyme (lower strand and structure) prior to cleavage.*

ribozymes were hybridized under high temperature conditions (see Methods section) owing to the thermal stability of the *Sulfolobus* rRNA; addition of Mg^{++} then initiated cleavage. As is apparent from the analytical electropherogram of the products shown in Fig. 3, three sharp, well-separated bands are produced, from which RNA can be extracted following preparative PAGE, for further mapping by selective RNase hydrolysis as outlined in Fig. 1.

The Key to LC/MS: The HPLC Solvent System

In many respects the problems associated with characterization of complex oligonucleotide mixtures derived from large RNAs are those of classical mixture analysis, and as is often the case, a key solution lies in combined chromatography-mass spectrometry. Although LC/MS has earlier been successfully applied to oligonucleotides (e. g., [11]), we view a major advance in this area to have been the recent introduction of the 1,1,1,3,3,3-hexafluoro-2-propanol

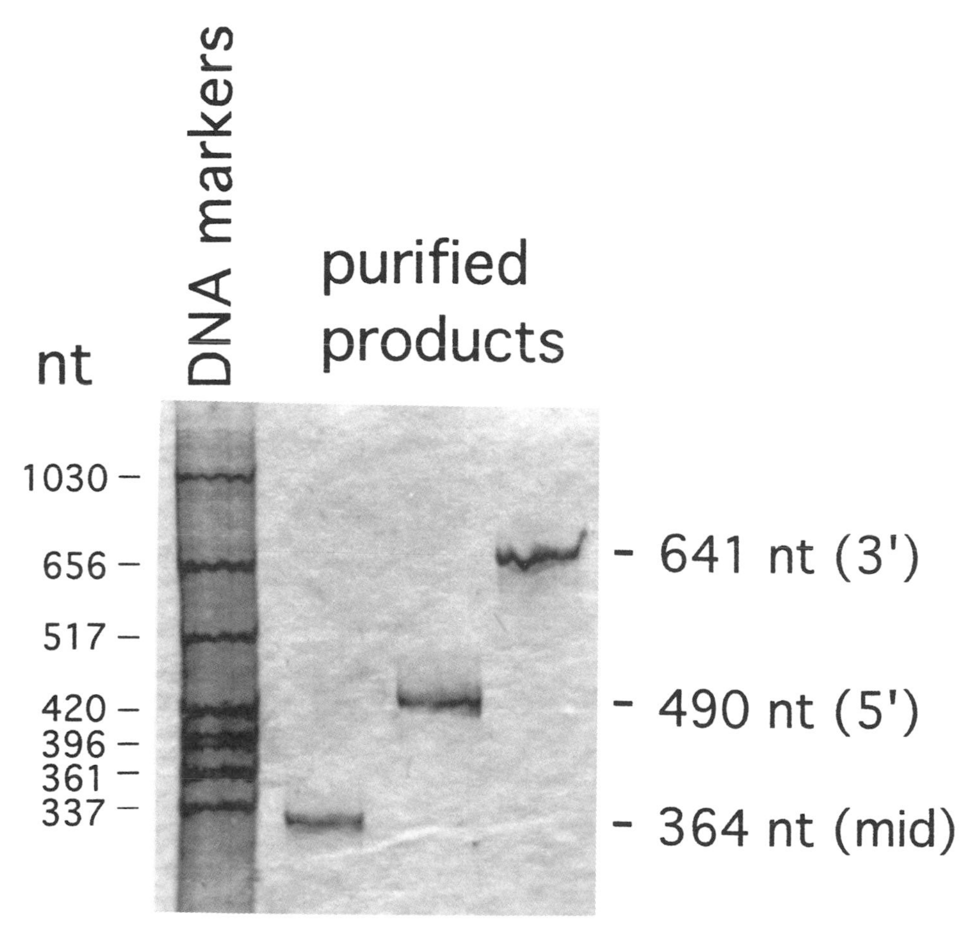

Fig. 3. Analytical electropherogram of the three ribozyme cleavage products resulting from the two cleavage reactions shown in Fig. 2.

(HFIP)/triethylamine-based HPLC buffer system for use with electrospray ionization LC/MS of DNA [12]. For applications to natural RNA [13, 14] this solvent system offers low picomole-level sensitivity and excellent chromatographic dispersion both between and within n-mer oligoribonucleotides.

Importantly, the HFIP system permits extension of previous protocols involving mass-to-base composition measurements [15, 16] to mixtures of greater complexity, with reduction in overall time and sample quantities. In turn, the expanded capabilities of LC/MS have permitted development of new routines for integration into existing protocols, as summarized in Fig. 1 and discussed in later sections.

A test of the overall utility of the HFIP-based LC/MS system using a 300 mm x 1 mm C18 reversed phase column and a quadrupole mass analyzer is provided by analysis of the RNase T1 digestion products of *E. coli* 16S rRNA, which contains 1542 nt, 11 of which are modified [3]. RNase T1 cleaves each of the 496 Gp residues to yield 488 oligonucleotides (dimers through one 16-mer), which are represented by 141 unique oligonucleotide sequences, all of which have Gp as the 3′ terminus. The entire digest can be injected directly onto the LC column, without any clean-up to remove enzymes and buffers or to recover specific oligonucleotides [14], resulting in the chromatogram shown in Fig. 4. Comparison of UV and total ion current detection patterns shows excellent correspondence, indicating retention of good chromatographic fidelity in the LC-electrospray ionization interface, of importance in distinguishing by mass spectrometry the numerous partially separated components. The complexity of

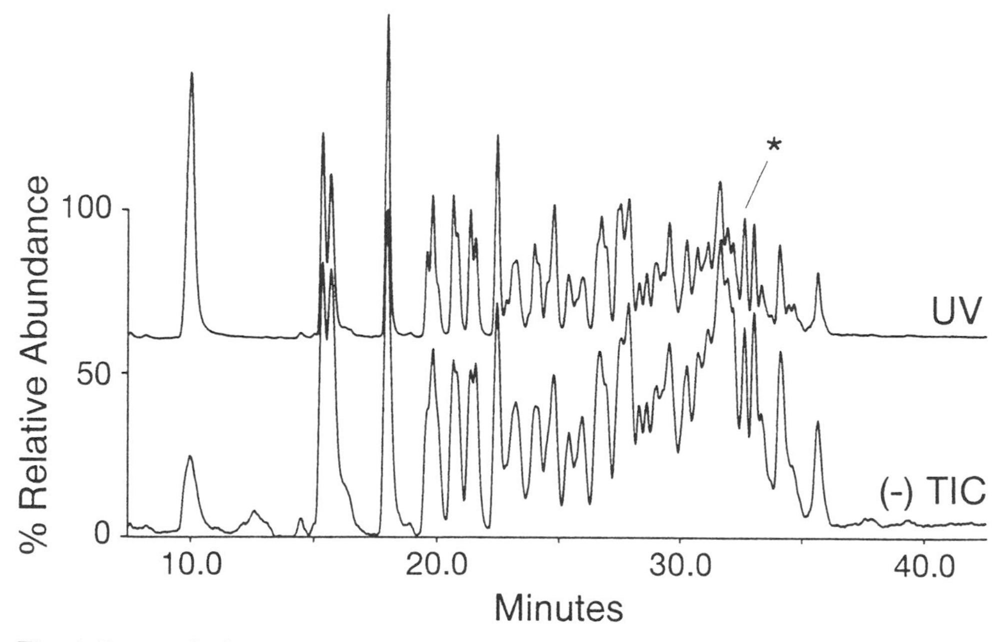

Fig. 4. Reversed phase HPLC/electrospray MS analysis of the products of RNase T1 hydrolysis of E. coli *16S rRNA. Top trace, UV detection at 260 nm; bottom, total negative ion current.*

the mixture is represented by the mass spectral data (Fig. 5) acquired for the peak eluting at 32.5 min (asterisk in Fig. 4). The sharp, symmetrical LC peak contains, in fact, five different oligonucleotides, whose masses could be determined with errors of less than 0.2 Da using a linear quadrupole mass analyzer. Base compositions [15] and implied sequence locations are then assigned in a separate step [16].

Although the primary data from LC/MS analysis of RNase digests consists of molecular mass, the on-line chromatographic fractionation of components serves another major advantage: time alignment of molecular mass and UV detection channels with ion signals related to information-bearing dissociation products formed in the ionization interface. Otherwise, the correlation between numerous low mass fragment ions and individual components in the mixture cannot be defined, unlike measurements carried out in the tandem (MS/MS) mode, which provides a degree of such information through precursor-product ion relationships. Experimentally, one data acquisition cycle is typically completed every 2–3 seconds and includes separate scan functions for molecular

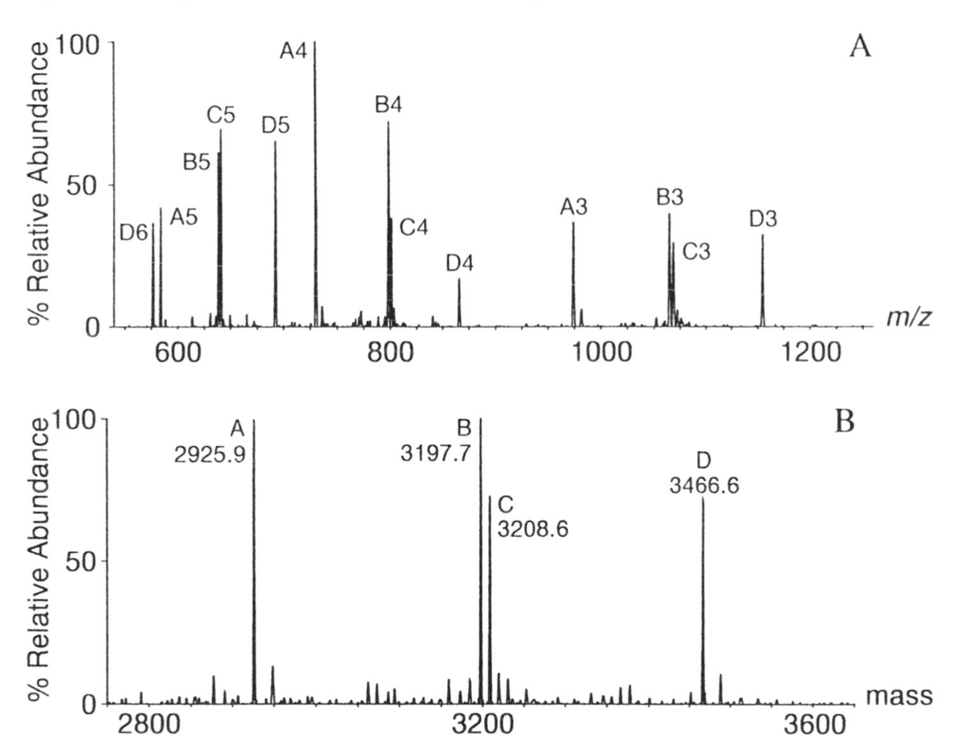

Fig. 5. Electrospray ionization mass spectrum of the 32.5 min. eluant denoted by the asterisk in Fig. 4. (A) Mass spectrum, with components A–D and charge states computer-assigned. (B) Mass profiles and M_r values reconstructed (Maximum Entropy) from data in panel A. Oligonucleotide compositions were determined directly from the M_r values [15]; locations inferred by correlation with the corresponding gene sequence [16]: A, 1145–AACUCAAAG–1153 and 1245–CAUACAAAG–1253 (2925.8 calc.); B, 1406–UCACACCAUG–1415 + methyl (3197.9 calc.); C, 149–AUAACUACUG–158 (3208.9 calc.); D, 1443–CUUAACCUUCG–1453 (3467.0 calc.).

mass, low m/z regions of interest, and MS/MS settings or scans. A summary of reactions and ions based on interface region dissociations used for this purpose is presented in the following section. Some measurements are carried out in the one-dimensional mode (single mass analyzer), offering a significant sensitivity gain, while others are made using two mass analyzers (MS/MS) at lower sensitivity but with greater selectivity. In either case, time alignment of data channels permits a range of useful measurements, and the design of new ones, applied directly to the oligonucleotide mixture for characterization of selected structural features without the necessity of intermediate isolation steps.

Oligonucleotide Dissociation Reactions in the Electrospray Interface

Electrospray is generally considered to be a low energy form of ionization [17], however, dissociation reactions in the ionization region can be induced by increasing the voltage potential between surfaces in the nozzle and skimmer, thus providing collision-induced dissociations in this relatively high pressure region. The products of such reactions are somewhat similar to those produced in a gas collision cell (for examples involving oligonucleotides, see [18]), but of course without the well-defined precursor-product relationship inherent in traditional tandem mass spectrometry. In the case of these so-called nozzle-skimmer (N/S) fragmentation reactions, precursor-product relationships can be inferred through the time domain available from the LC/MS experiment. A listing of such reactions recently developed or refined in our laboratory for applications utilizing LC/MS for oligonucleotide mixtures is given in Table 2.

Base Anion Release

Observation of the base anion released on oligonucleotide dissociation [16, 19] provides a means for detection of modified bases, in the uncluttered region of the spectrum below about m/z 250 [20]. An example is given in Fig. 6, showing detection of N^4-acetylcytosine (ac^4Cyt) anion in an oligonucleotide resulting from RNase T1 digestion of the 3′-most 641 nucleotides of *Sulfolobus solfataricus* 16S rRNA, produced from ribozyme cleavage of the intact molecule as shown in Fig. 2B. Reconstruction and verification of correct time alignment for high mass ions representing molecular mass charge states, showed M_r 1057.1 for the ac^4C-containing oligonucleotide. This mass corresponds to the composition ac^4Cac^4CGp, resulting from posttranscriptional acetylation of ~CCG~, a sequence which occurs 15 times in the intact molecule and six times in the ribozyme-generated fragment, according to the gene sequence. Note that the acetyl substituent mass (42 Da) is indistinguishable from three methyl groups (3 x 14 Da), but the total nucleoside census (see Fig. 1) had indicated the presence of N^4-acetylcytidine (but not trimethylcytidine) in the RNA [21]. Under most circumstances a correlation must exist between the mass of the modified base and the molecular mass, such that an allowable base composition $A_wC_xU_yG_z$ (usually with z = 1), must result [15].

Table 2

acquisition mode	oligonucleotide fragments formed in ionization region	MS-1	product transmitted and detected by MS-2	measurement	comments
MS	base anion	scans the base *m/z* region	not used	base identity	1
	base side chain fragment	scans the low mass region	not used	presence of specific modification	2
	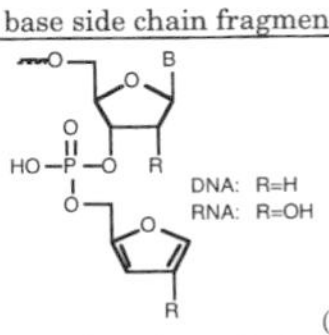 DNA: R=H RNA: R=OH (a – base ion series)	scans *m/z* 340–1200	not used	sequence 5. → 3.	3
	~pNpNp-3. (w ion series)	scans *m/z* 300–1200	not used	sequence 3. → 5.	4
MS/MS	N^1pN>p, NpN^2>p	selects each of four dinucleotide masses for N^1, N^2 = A,C,G,U	N>p	nearest neighbors of N	5
	pNm	selects each of four pNm masses for N = A,C,G,U	base anion	presence of specific 2.-O-methylribonucleotide	6
	pNm	selects each of four pNm masses for N = A,C,G,U	PO	presence of specific 2.-O-methylribonucleotide	7
	pNm	selects each of four pNm masses representing bases A,C,G,U	methylribose phosphate – H_2O (*m/z* 225)	presence of specific 2.-O-methylribonucleotide	8
	methylribose phosphate – H_2O (*m/z* 225)	selects *m/z* 225	PO	generic presence of 2.-O-methylribose	9

Comments
1. Note that isomers cannot be distinguished by base mass alone, e.g., 1- vs. 2-methyladenine.
2. Relatively little work in this area has been carried out to date.
3. For nomenclature and review of sequence-related ions see [23,41].
4. The w series ions [24] are formed but the extent to which they can be used to deduce an unknown sequence is not known.
5. See [25]. The notation >p does not distinguish 2.,3.-cyclic phosphate from various anhydro isomers.
6. See [28]. Because the base mass is designated as the product ion, this reaction pathway is useful for detection of Nm residues modified also in the base.

7,8. See [28].

9. U-rich oligoribonucleotides may give a weak signal using this reaction pathway because *m/z* 225 also arises from U-related minor fragment ions [28].

Table 2. Examples of electrospray ionization interface reactions for use with LC/MS of oligonucleotides.

Determination of Partial Sequence

The formation of sequence-related ions in the electrospray interface, as earlier demonstrated [18, 22] (by direct infusion of a single compound) indicates the potential for sequence determination without use of a collision cell. However, it has not yet been established whether the quantitative properties of the N/S mass spectra are sufficiently similar to those produced in a conventional collision cell to permit sequencing [23] if the structure is not known in advance. We find that the conventional rules [23] do apply for identification of the 3′ terminus residue (ion w_1 in the most widely used nomenclature [24]). In the case of analysis of nuclease digestion products, determination of this ion type in the hydrolytic mixture provides a rapid means for verification that correct cleavage has taken place (e. g., producing ~Gp-3′ from RNase T1 hydrolysis). When enzymes of poor specificity are used (e. g., RNase U2, which cleaves primarily at A, but also at G), recognition of the terminus is important in the derivation of base composition from molecular mass, which is made difficult if the number of any one type of base is not known (such as G = 1 from RNase T1 hydrolysis).

Figure 7 shows a simple example of the utility of partial sequencing using LC/MS for sequence assignments from a partially resolved mixture of five

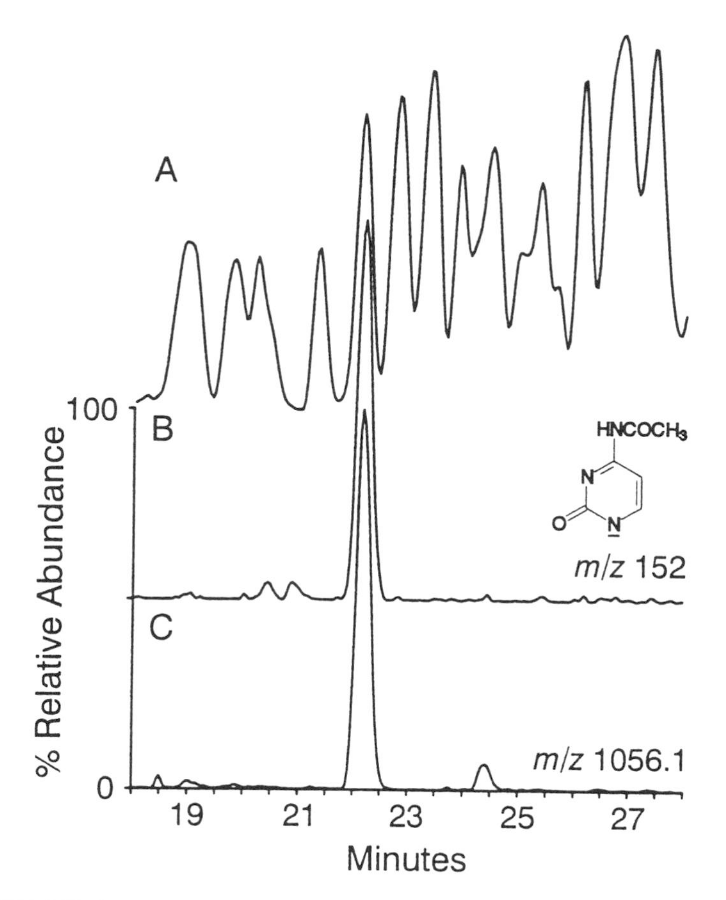

Fig. 6. LC/ES-MS detection of N^4-acetylcytidine in an oligonucleotide of M_r 1057.1 in an RNase T1 digest of nucleotides 5′-491–1495-3′ from ribozyme-cleaved S. solfataricus *16S rRNA. The approximate lengths represented in the partial chromatogram shown encompass dimers through heptamers. (A) Total negative ion current. (B) Mass channel 152 representing acetylcytosine base formed in the ionization region. (C) Mass channel 1056.1, corresponding to the $(M - H)^-$ ion of an oligonucleotide of composition ac^4Cac^4CGp.*

isomers containing dC, dT, dA and dG (Fig. 7, bottom panel). Reconstructed ion chromatograms of N/S-generated fragment ions representing the four 3′-terminal nucleotides (dpA, dpC, dpG, dpT) provide an indication of the elution order of the five components. Two peaks appear in the *m/z* 330 channel, reflecting two dA 3′- termini, which from these data alone cannot be distinguished. However by recording the 3′-terminus dimer channels (ion w_2) the elution order is shown to be d(CGTA), d(TCGA) (data not shown). This approach is presently being extended to longer sequences to determine the level of reliability available for determination of partial sequences. When used in conjunction with gene sequence data, the derivation of only two or three terminal residues would provide considerable constraints for the differentiation of isomeric nuclease cleavage fragments.

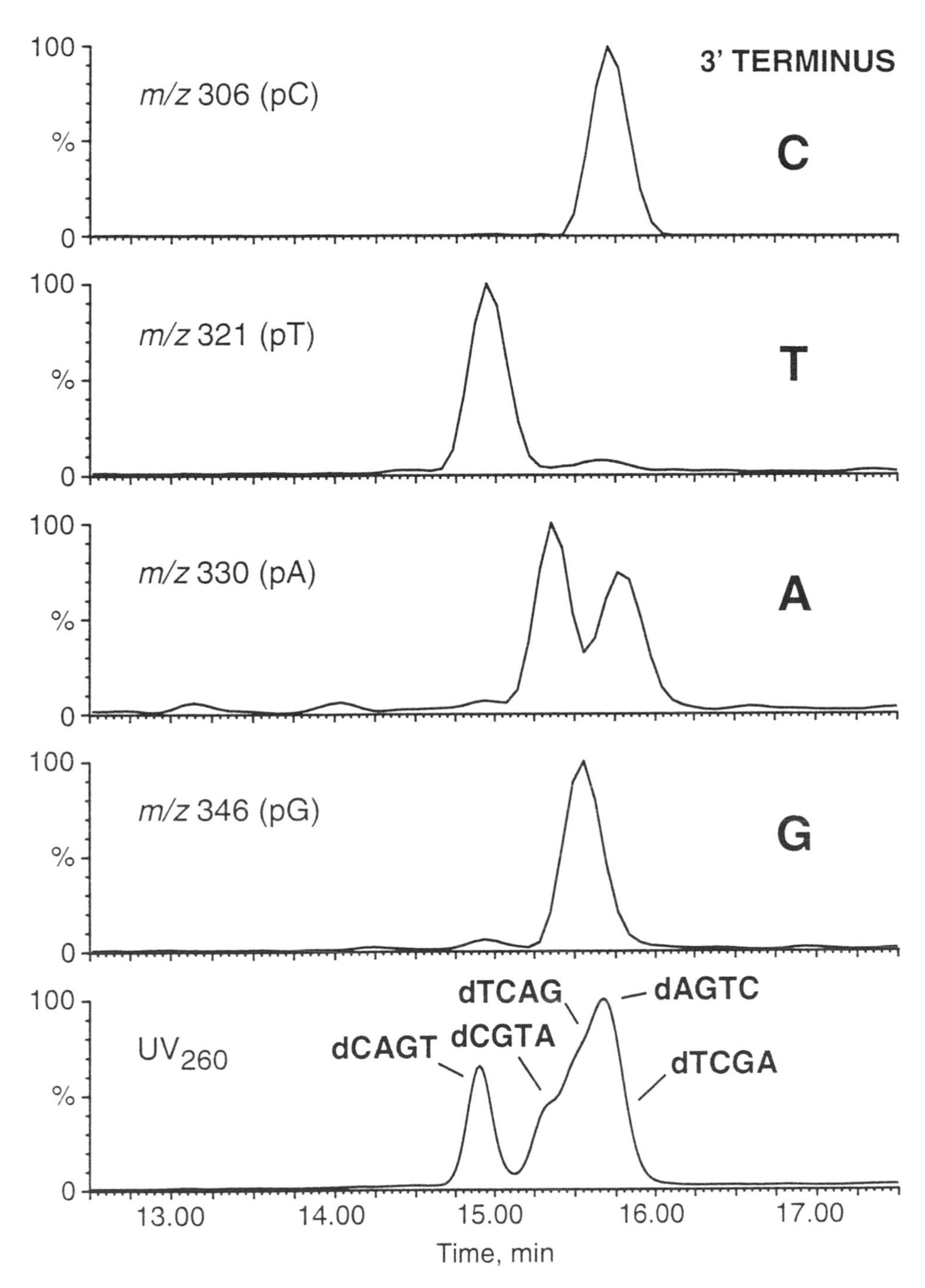

Fig. 7. Assignment of components in a mixture of five isomeric oligodeoxytetranucleotides by detection of the 3′-terminal nucleotide fragment ions formed in the electrospray ionization region. Top panels: mass channels corresponding to each of the four possible nucleotide residues. Bottom: UV detection at 260 nm of the partially resolved mixture.

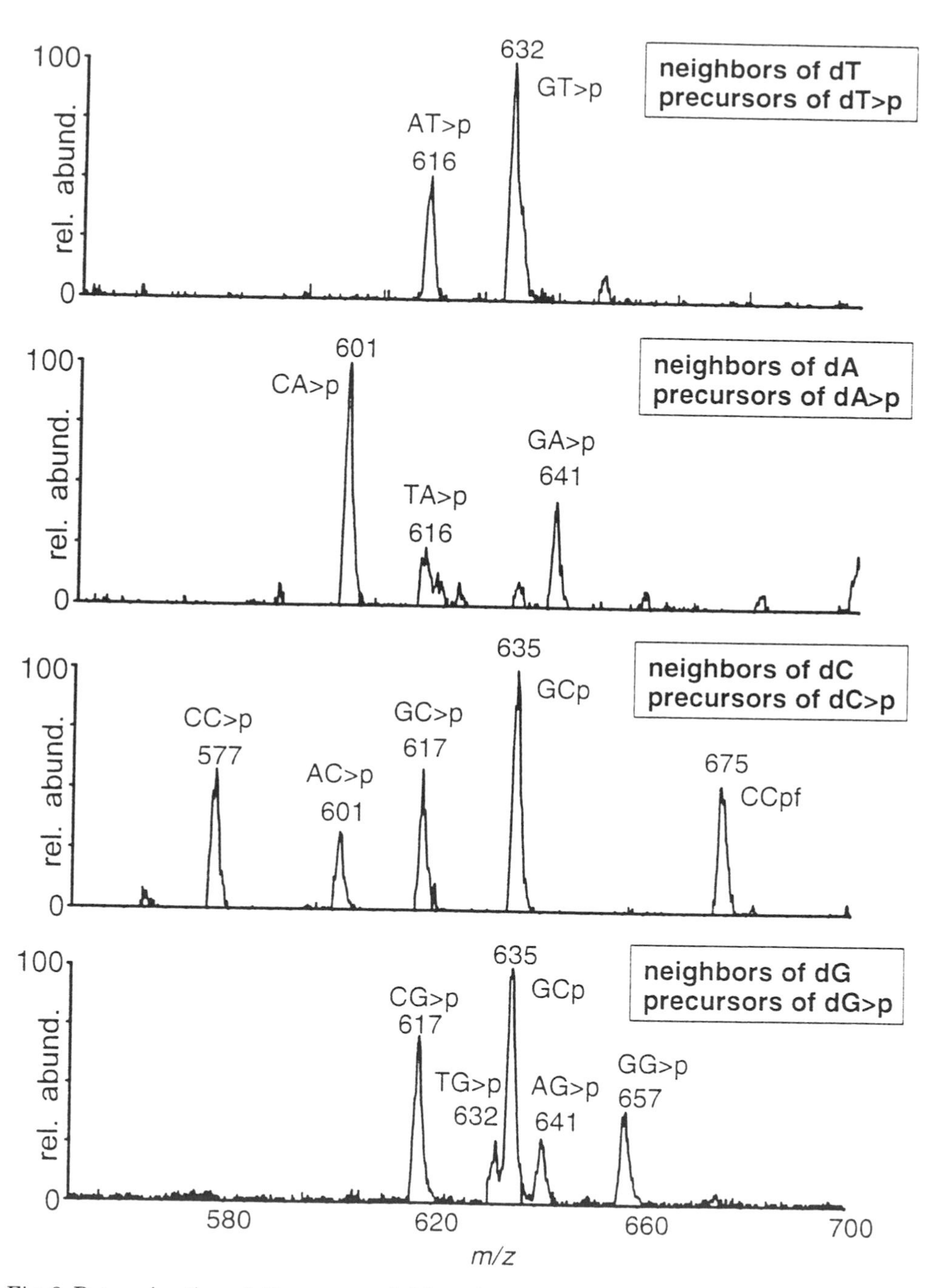

Fig. 8. Determination of all nearest neighbors for each of the four possible residues (dT, dA, dC, dG) in d(CCAGGTACGC). Sample introduced by simple infusion. Nomenclature: >p, substituent equivalent in mass to cyclic phosphate or dehydrophosphate; f, furanyl moiety resulting from base loss (see [41] and structure in Table 2).

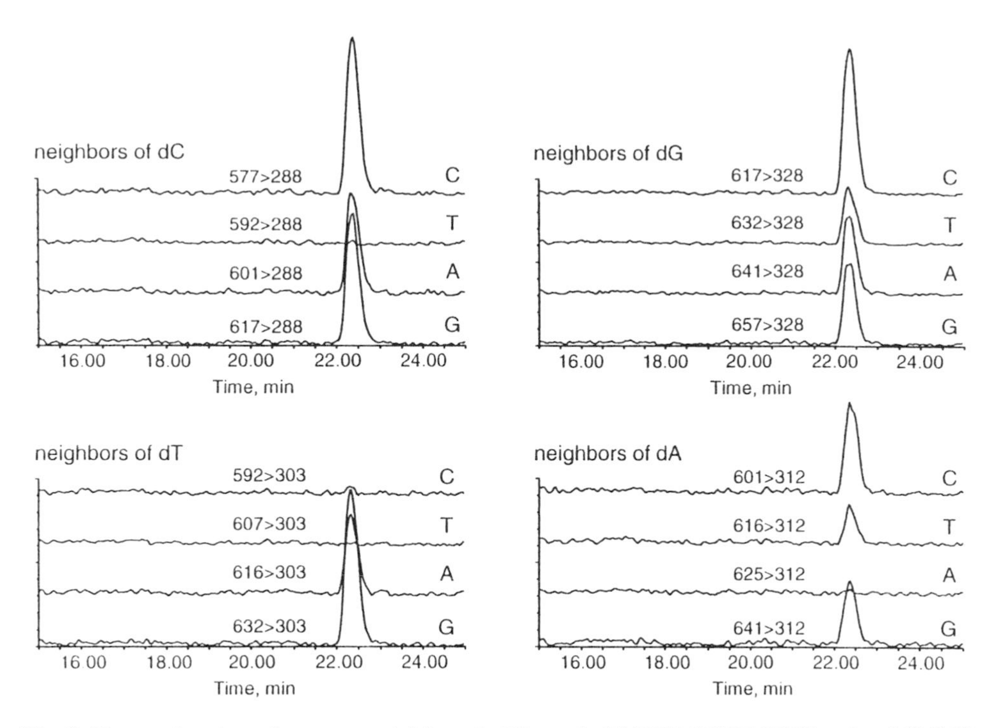

Fig. 9. Determination of nearest neighbors in 25 pmol of d(CCAGGTACGC) using LC/MS, based on selected ion monitoring of dimer precursor ions for each of the four possible residues (dT, dA, dC, dG) (cf. Fig. 8).

Nearest Neighbor Determination

Systematic exploration of the reaction pathways of the numerous [20] small nucleotides generated by dissociation of the oligomer chain revealed that ions equivalent in mass to mononucleotide cyclic phosphates (designated N>p, without implying the 2′,3′-cyclic phosphate structure) are derived from dissociation of both possible dimers N^1pN>p and NpN^2>p which are formed in the ionization region (see Table 2). Thus a method was conceived using tandem mass analyzers to establish the identities, through masses of the dimer parents, of the adjacent residues [25]. The results of this measurement are illustrated by the four mass spectra in Fig. 8, representing each of the four nucleotides in the 10mer d(CCAGGTACGC). In this experiment MS-2 is set to the mass of N>p^- and MS-1 is scanned to sample the population of low mass ions formed in the electrospray interface region that contain the dinucleotide precursor ions. The ion(s) detected establish the identities of adjacent nucleotides but not their sequence, i.e., in the top panel of Fig. 8 the *m/z* 632 signal does not distinguish d(GT>p) from d(TG>p). It may be noted that other ions (some of which may contain sequence information [18]) are also precursors to the N>p ion of interest, however, the only mass channels of direct interest are those representing the four possible nucleotide neighbors in the generic precursors N^1pN>p or N^2pN>p. Thus the measurement is amenable to LC/MS by monitoring the channels of interest as demonstrated

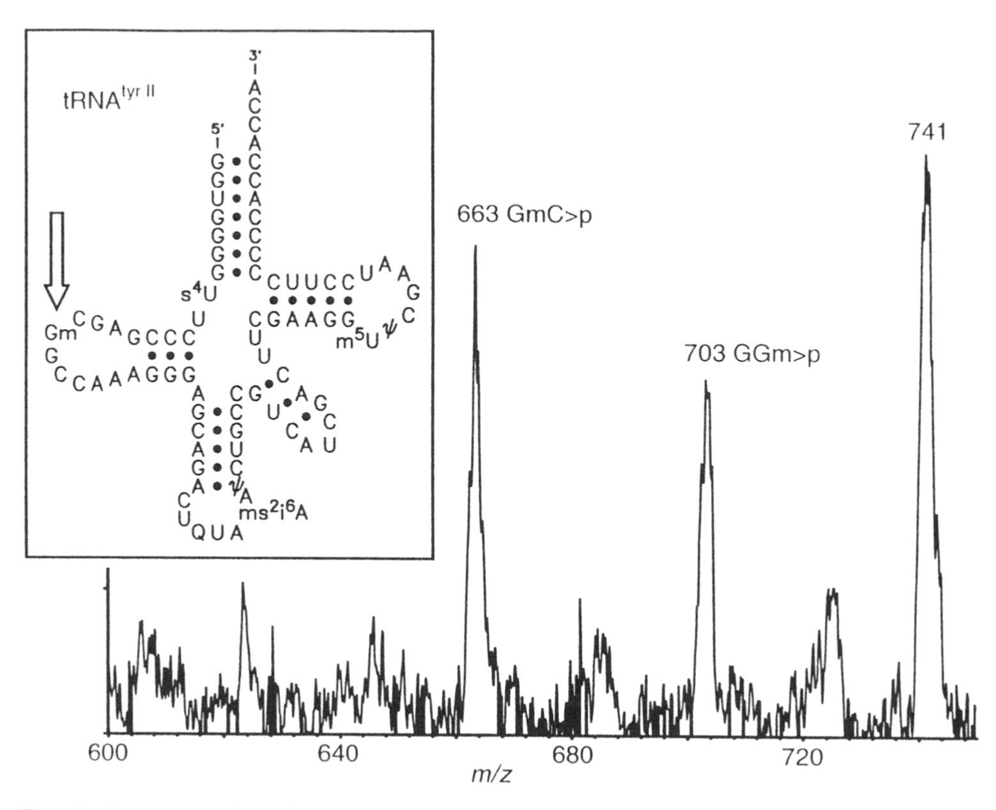

Fig. 10. Determination of nearest neighbors of the single 2′-O-methylguanosine (Gm, see arrow) in intact E. coli *tRNA*^{Tyr}*. Sample introduced by simple infusion.*

in Fig. 9. For example, (top left panel) a signal is obtained at the appropriate elution time corresponding to the reaction d(CpC>p) → d(C>p) (m/z 577 → 288) showing C to be a neighbor of C, i. e., presence of adjacent Cs. No signal is observed in the d(TpC>p) or d(CpT>p) reaction channel (m/z 592 → 288), showing T not to be a neighbor of C, in accord with the sequence. Analogous correct correlations are observed for each of the four types of residues in the 10-mer. The possible presence of modified neighbors can be tested for by adjustment of the mass channels sampled by MS-1. While the nearest neighbor measurement will often not provide a definitive answer for a given problem, it provides valuable constraints on sequence that can be applied directly to complex mixtures of oligonucleotides.

Interestingly, we find that the production of small nucleotide fragments in the high pressure, multi-collision ionization region appear to be less influenced by chain length than are the single-cut backbone cleavages typically used to derive sequence information. As a result, the nearest neighbor reaction channels can be detected in molecules up to at least tRNA (85-mer) lengths. As an illustration, Fig. 10 shows a region of the electrospray mass spectrum of intact *E. coli* tRNA^{Tyr} for detection of nucleotides adjacent to the single 2′-*O*-methylguanosine (Gm) residue at position 17 [26]. In this case MS-1 scans the mass range suitable for transmission of dimers formed in the ionization region, while MS-2 is set for the mass of monomethylated G>p, *m/z* 358. Signals are observed showing G (*m/z* 703) and C (*m/z* 663) neighbors, but not A (*m/z* 687) or U (*m/z* 664) although interferences in the latter channels and signals from other chain cleavage products (such as *m/z* 741) are notable. If the modified nucleotide residue has unusually high mass, the dimer precursor ions will tend to occur in a higher and more interference-prone mass region of the spectrum. Additionally, the presence of both C and U as neighbors, which differ in mass by 1 Da, will produce a relatively broad peak in the resulting mass spectrum, necessitating use of higher resolving power in MS-1.

Detection of 2´-O-Methyl Modifications

Ribose methylation is one of the most common forms of RNA modification, particularly in eukaryotic rRNA [27]. Implied in the correct recognition of ribose methylation is the ability to distinguish base methylation, also a relatively common motif in many types of RNA. This measurement can be accomplished using any of several dissociation pathways starting with the ribose-methylated mononucleotide ion Nmp (not distinguished for the present purpose from pNm) formed by oligonucleotide dissociation in the ionization region [28] (see Table 2). Channels which may be selected for detection of ribose methylation include dissociation of Nmp to either the base or phosphate anions or to the methylribose phosphate ion *m/z* 225 [29]. Alternatively, the transition *m/z* 225 → 79 does not specify the base mass in either the precursor or product ion and so is useful in generic detection of Nm-containing oligonucleotides.

The manner in which the measurement is made is demonstrated in Fig. 11, using a RNase T1 digest of *E. coli* $tRNA^{fMet}$ for detection of the single residue of 2´-*O*-methylcytidine (Cm). The base compositions of the hydrolysis fragments can be confirmed or in most cases deduced from molecular mass measurements [13, 15, 16], made using a single mass analyzer. Taken in conjunction with the reported tRNA sequence [30] specific assignments to HPLC peaks are readily made (Fig. 11A). Within each data acquisition cycle, the reaction Cmp → cytosine (*m/z* 336 → 110) was monitored using tandem mass analyzers. As shown in Fig. 11B, one clear signal results, showing the 33.0 minute eluant to contain the 2´-*O*-methylcytidine residue. This conclusion is supported by the molecular mass of the 33.0 minute component, 3503.6 (3503.1 calc. for the sequence shown). In this relatively simple example the identification of the Cm containing HPLC peak is straightforward. However, in very complex mixtures in which 5–10 oligonucleotides contribute to a 20–30 second elution profile, the time alignment of data channels is more critical and more difficult.

Methods

Materials

Oligonucleotides and ribozymes were synthesized by the University of Utah DNA/Peptide Core Facility. 16S rRNA was isolated as described [21] from

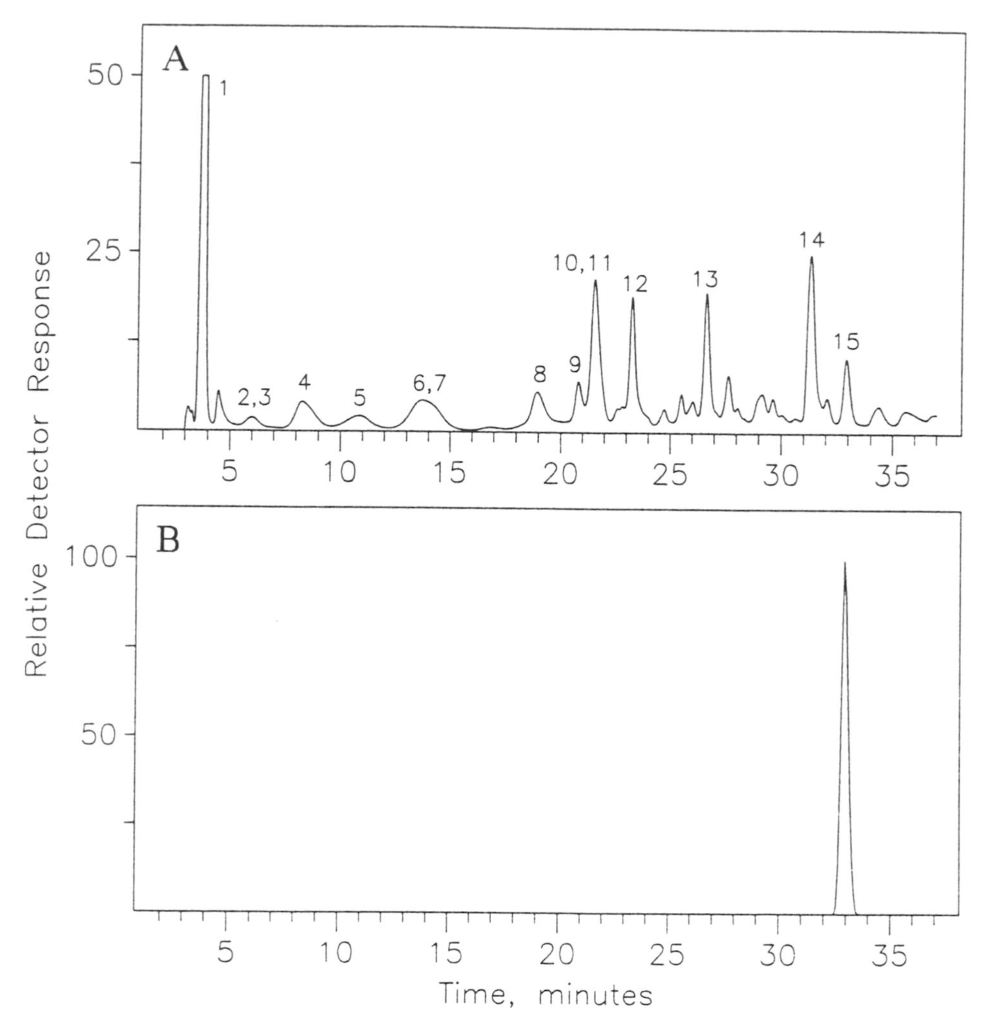

Fig. 11. Detection of the single 2′-O-methylcytosine (Cm) residue in an RNase T1 digest of E. coli *tRNA^{fMet}. (A) UV detection, 260 nm. 1, Gp; 2, CGp; 3, s^4UGp; 4, AGp; 5, UCGp; 6, DAGp; 7, CAGp; 8, m^7GUCGp; 9, AAGp; 10,11, CCUGp and CUCGp; 12, CAACCA-OH-3.; 13, CCCCCGp; 14, m^5UΨCAAAUCCGp; 15, CmUCAUAACCCGp. Unassigned peaks are judged to be from other isoaccepting tRNAs present in the sample. (B) Selected reaction channel for Cmp → cytosine, showing HPLC peak 15 to contain Cm.*

S. solfataricus (grown in our laboratory) and frozen *E. coli* cells (Grain Processing, Muscatine, IO; this product is no longer available). Tyrosine tRNAs were purchased from Sigma Chemical Co. (St. Louis, MO). 1,1,1,3,3,3-Hexafluoro-2-propanol and triethylamine were obtained from J. T. Baker and Aldrich Chemical Co., respectively. RNase T1 was purchased from Ambion (Austin, TX), and used as described to digest natural RNAs [16].

LC/ES-MS of Oligonucleotides

Oligonucleotide mixtures were analyzed on a Fisons Quattro II triple quadrupole instrument (Micromass; Beverly MA), to which a Hewlett-Packard

1090 liquid chromatograph with diode array detector is interfaced. rRNA digests and synthetic oligonucleotide mixtures (nearest-neighbor studies) were fractionated on a 300 x 1 mm LC-18S column (Supelco, Inc.; Bellefonte, PA) at a flow rate of 60 µl/min using a previously described HFIP/TEA solvent system [12], adapted for application to complex mixtures [14]. Digests were not cleaned up prior to analysis. tRNA digests and oligonucleotide mixtures used for 2′-O-methylation studies were analyzed on a 150 x 1 mm PhaseSep column (ChromTech; Apple Valley MN) using simple linear gradients and HFIP solvent system at a flow rate of 40 µl/min. The solvent was conducted into the mass spectrometer without splitting.

The type of analysis dictates the combination of mass spectrometer scan functions used. In any case, one acquisition channel is used to acquire full scans in continuum mode over a mass range ~500–1500 in three sec, typically with cone voltage offset about 50 volts. For base release, nearest neighbor and 2′-O-methyl detection, the cone voltage offset was 100–150 V. MS/MS techniques used Ar as the collision gas, and collision energies 15–30 eV (E_{lab}).

Parameter Optimization for Nearest Neighbor Studies

Electrospray mass spectra obtained in the negative mode by infusion were acquired on a Sciex API III+ triple quadrupole mass spectrometer (Concord, Ontario, Canada) with electrospray ionization, equipped with a home-made microspray device. The needle voltage was set to –2000V to –2200V, orifice sampling voltage to –100V, and collision energy to 10–20eV. Ar was used as collision gas at a thickness of ~2.5 x 10^{15} atoms/cm^2. Resolution of the first quadrupole (MS-1) was set to resolve the isotope peaks from monoisotopic peaks at m/z 800 at half width. The dwell time for parent ion spectra was set to 3 msec and the mass interval to 0.1 m/z units. Depending on signal intensities 10 to 50 spectra were accumulated in multi channel analysis (MCA) mode. MS-2 was set to the mass of the cyclic phosphate (N>p) of the nucleotide of interest.

Ribozyme Cleavage of S. solfataricus *16S rRNA and Purification of Products*

Ribozymes were synthesized by the University of Utah DNA/Peptide Facility and purified by electrophoresis in denaturing 20% polyacrylamide gels, from which RNA was recovered by electroelution and concentrated by EtOH precipitation. The ribozymes were dissolved in diethylpyrocarbonate (DEPC)-treated water and stored at –20°C.

Preparative cleavage of 16S rRNA was performed as follows. A 150 µl reaction was prepared containing the following reactant concentrations: 8 µM rRNA (~160 µg), 2.2 µM each ribozymes 490 (rbz490) and 854 (rbz854) (Fig. 2), 50 mM Tris-HCl (pH 7.5), 20% DMSO (optimum concentrations for ribozymes 490 and 854 are 30% and 10%, respectively), and 0.1% SDS. The reaction mixture was overlaid with mineral oil, placed at 90°C for 30 sec and incubated at 50°C for 2.5 hr. The cleavage reaction was started by addition of $MgCl_2$ to a final concentration of 10 mM. The reaction was incubated at 37°C for 2.5 hr. Under these conditions, cleavage by rbz854 was complete but that by rbz490 was incomplete. The reaction was concentrated by EtOH precipitation. The pellet was suspended in a final volume of 50 µl of 50 mM Tris-HCl (7.5), 30% DMSO, 0.1% SDS and an additional 36 nmol of rbz490 was added. The reaction mixture was overlaid with mineral oil, placed at 90°C for 30 sec and incubated at 50°C for 3 hr and 20 min. $MgCl_2$ (1 M) was added to 10 mM and the reaction incubated at 37°C for 4 hr. An aliquot was removed and analyzed on a denaturing 5% polyacrylamide gel. Cleavage by rbz490 was judged to be greater than 80% complete under these conditions. The reaction was concentrated by EtOH precipitation.

The cleavage products were preparatively separated by electrophoresis in a denaturing 5% polyacrylamide gel, and the three rRNA products were located by UV shadowing. Three bands containing each of the fragments were excised and the RNAs were recovered by electroelution. The recovered RNAs were concentrated by EtOH precipitation and suspended in 300 µl of DEPC-treated water.

For analysis of the final products, 0.15 µl of each purified RNA was loaded onto a denaturing (7% urea) 5% (19:1) polyacrylamide gel, along with DNA markers. Electrophoresis was performed in 1 X TBE buffer. The gel was visualized by silver staining (Fig. 3).

Acknowledgments

We thank E. Bruenger and K. R. Noon for culture of *Sulfolobus* cells and for isolation of ribosomal RNAs in our laboratory. W. F. Doolittle and C. Sensen, and M. Schenk, all of Dalhousie University, Nova Scotia, are thanked for providing the unpublished sequence of the *S. solfataricus* P2 16S rRNA gene, and a slant of the strain, respectively. S. C. Pomerantz assisted with preparation of figures. The University of Utah DNA and Peptide Core Facility is supported by NIH Grant CA42014. The support of NIH Grant GM21584 is gratefully acknowledged.

References

1. R. Green and H. F. Noller, *Annu. Rev. Biochem.* **1997**, *66*, 679–716.

2. I. Itta, Y. Kamada, H. Noda, T. Ueda and K. Watanabe, *Science* **1998**, *281*, 666–669.

3. J. Rozenski, P. F. Crain and J. A. McCloskey, *Nucleic Acids Res.* **1999**, in press.

4. G. Björk, J. U. Ericson, C. E. D. Gustafsson, T. G. Hagervall, Y. H. Jönsson and P. M. Wikstrom, *Annu. Rev. Biochem.* **1987**, *56*, 263–287.

5. P. Fellner, *Eur. J. Biochem.* **1969**, *11*, 12–27.

6. P. A. Limbach, P. F. Crain and J. A. McCloskey, *Nucleic Acids Res.* **1994**, *22*, 2183–2196.

7. J. A. Kowalak, E. Bruenger and J. A. McCloskey, *J. Biol. Chem.* **1995**, *270*, 17758–17764.

8. J. Lapham, Y. T. Yu, M. D. Shu, J. A. Steitz and D. M. Crothers, *RNA* **1997**, *3*, 950–951.

9. K. R. Birikh, P. A. Heaton and F. Ekstein, *Eur. J. Biochem.* **1997**, *245*, 1–16.

10. J. Haseloff and W. L. Gerlach, *Nature* **1988**, *334*, 585–591.

11. K. Bleicher and E. Bayer, *Chromatographia* **1994**, *39*, 405–408.

12. A. Apffel, J. A. Chakel, S. Fischer, K. Lichtenwalter and W. S. Hancock, *Anal. Chem.* **1997**, *69*, 1320–1325.

13. P. F. Crain, in *Modification and Editing of RNA*, H. Grosjean and R. Benne, Eds.; ASM Press, Washington, DC, 1998; pp 47–57.

14. B. Felden, K. Hanawa, J. F. Atkins, H. Himeno, A. Muto, R. F. Gesteland, J. A. McCloskey and P. F. Crain, *EMBO J.* **1998**, *17*, 3188–3196.

15. S. C. Pomerantz, J. A. Kowalak and J. A. McCloskey, *J. Am. Soc. Mass Spectrom.* **1993**, *4*, 204–209.

16. J. A. Kowalak, S. C. Pomerantz, P. F. Crain and J. A. McCloskey, *Nucleic Acids Res.* **1993**, *21*, 4577–4585.

17. J. B. Fenn, M. Mann, C. K. Meng, S. F. Wong and C. M. Whitehouse, *Science* **1989**, *246*, 64–71.

18. R. Lotz, M. Gerster and E. Bayer, *Rapid Commun. Mass Spectrom.* **1998**, *12*, 389–397.

19. S. A. McLuckey and S. Habibi-Goudarzi, *J. Am. Soc. Mass Spectrom.* **1994**, *5*, 740–747.

20. P. F. Crain, J. M. Gregson, J. A. McCloskey, C. C. Nelson, J. M. Peltier, D. R. Phillips, S. C. Pomerantz and D. M. Reddy, in *Mass Spectrometry in the Biological Sciences*, A. L. Burlingame and S. A. Carr, Eds.; Humana Press, Totowa, NJ, 1996; pp 497–517.

21. K. R. Noon, E. Bruenger and J. A. McCloskey, *J. Bacteriol.* **1998**, *180*, 2883–2888.

22. D. P. Little, R. A. Chorush, J. P. Speir, M. W. Senko, N. L. Kelleher and F. W. McLafferty, *J. Am. Chem. Soc.* **1994**, *116*, 4893–4897.

23. J. Ni, S. C. Pomerantz, J. Rozenski, Y. Zhang and J. A. McCloskey, *Anal. Chem.* **1996**, *68*, 1989–1999.

24. S. A. McLuckey, G. J. Van Berkel and G. L. Glish, *J. Am. Soc. Mass Spectrom.* **1992**, *3*, 60–70.

25. J. Rozenski and J. A. McCloskey, *Anal. Chem.* **1999**, in press.

26. M. Sprinzl, C. Horn, M. Brown, A. Ioudovitch and S. Steinberg, *Nucleic Acids Res.* **1998**, *26*, 148–153.

27. B. E. H. Maden, *Progr. Nucleic Acids Res. Mol. Biol.* **1990**, *39*, 241–303.

28. F. Qiu and J. A. McCloskey, *Nucleic Acids Res.* **1999**, submitted.

29. D. R. Phillips and J. A. McCloskey, *Int. J. Mass Spectrom. Ion Processes* **1993**, *128*, 61–82.

30. S. K. Dube and K. A. Marcker, *Eur. J. Biochem.* **1969**, *8*, 256–262.

31. M. H. Vaughan, R. Soeiro, J. R. Warner and J. E. Darnell, *Proc. Natl. Acad. Sci. USA* **1967**, *58*, 1527–1534.

32. P. R. Cunningham, C. J. Weitzmann, D. Nègre, J. G. Sinning, V. Frick, K. Nurse and J. Ofengand, in *The Ribosome: Structure, Function, and Evolution*, W.E. Hill, A. E. Dahlberg, R. A. Garrett, P. B. Moore, D. Schlessinger and J. R. Warner, Eds.; ASM Press, Washington, DC, 1990; pp 243–252.

33. P. R. Cunningham, R. B. Richard, C. J. Weitzmann, K. Nurse and J. Ofengand, *Biochimie* **1991**, *73*, 789–796.

34. M. Rydén-Aulin, Z. Shaoping, P. Kylsten and L. A. Isaksson, *Mol. Microbiol.* **1993**, *7*, 983–992.

35. K. Sirum-Connolly and T. L. Mason, *Science* **1993**, *262*, 1886–1889.

36. K. Sirum-Connolly, J. M. Peltier, P. F. Crain, J. A. McCloskey and T. L. Mason, *Biochimie* **1995**, *77*, 30–39.

37. R. Green and H. F. Noller, *RNA* **1996**, *2*, 1011–1021.

38. M. O'Connor, C. L. Thomas, R. A. Zimmerman and A. E. Dahlberg, *Nucleic Acids Res.* **1997**, *25*, 1185–1193.

39. B. G. Lane, J. Ofengand and M. W. Gray, *FEBS Lett.* **1992**, *302*, 1–4.

40. J. Ofengand and A. Bakin, *J. Mol. Biol.* **1997**, *266*, 246–268.

41. S. A. McLuckey and S. Habibi-Goudarzi, *J. Am. Chem. Soc.* **1993**, *115*, 12085–12095.

QUESTIONS AND ANSWERS

Pat Limbach (Louisiana State University)

Can you explain the appearance of the non-integral amounts of modification in rRNA?

Answer. There are two reasons for fractional values. First is the inherent inaccuracy of using HPLC peak areas for accurate molar ratio measurements. For example, it will be difficult to distinguish 9.4 nucleoside residues in the entire rRNA molecule from 10.0. Second, some of the sites are modified substoichiometrically in response to cell culture temperature, e. g., 75°C vs. 83°C.

Pat Limbach (Louisiana State University)

Will a parent ion scan work for modified pseudouridines?

Answer. I presume so.

Robert C. Murphy (National Jewish Hospital)

Can you carry out your nearest neighbor analysis of nucleotide bases using elevated skimmer voltage in an ion trap?

Answer. Probably yes, but it would be necessary to check it experimentally to determine whether sufficient levels of energy are available for the initial dissociation reactions.

Appendix

The Meaning and Usage of the Terms Monoisotopic Mass, Average Mass, Mass Resolution, and Mass Accuracy for Measurements of Biomolecules

S. A. Carr, A. L. Burlingame and M. A. Baldwin

Much confusion surrounds the meaning of the terms *monoisotopic mass, average mass, resolution* and *mass accuracy* in mass spectrometry, and how they interrelate. The following brief primer is intended to clarify these terms and to illustrate the effect they have on the data obtained and its interpretation. We have highlighted the effects of these parameters on measurements of masses spanning the range of 1000 to 25,000 Da because of important differences that are observed. We refer the interested reader to a number of excellent articles that have dealt with aspects of these issues [1-6].

MONOISOTOPIC MASS VERSUS AVERAGE MASS

Most elements have a variety of naturally occurring isotopes, each with a unique mass and natural abundance. The monoisotopic mass of an element refers specifically to the lightest stable isotope of that element. For example, there are two principle isotopes of carbon, ^{12}C and ^{13}C, with masses of 12.000000 and 13.003355 and natural abundances of 98.9 and 1.1%, respectively; thus the value defined for the monoisotopic mass of carbon is 12.00000 on this mass scale. Similarly, there are two naturally occurring isotopes for nitrogen, ^{14}N, with a mass of 14.003074 (monoisotopic mass) and a relative abundance of 99.6% and ^{15}N, with a mass of 15.000109 and a relative abundance of ca. 0.4%. The *monoisotopic mass* of a molecule is thus obtained by summing the monoisotopic masses (including the decimal component, referred to as the mass defect) of each element present. Of course the *measured* molecular species (a measurement made on a large, statistical ensemble of molecules) will consist not only of species having just the lightest isotopes of the elements present, but also some percentage of species having one or more atoms of one (or more) of the heavier isotopes. The contribution of these heavier isotope peaks in the molecular ion cluster depends on the abundance-weighted sum of each element present. The theoretical appearance of these isotope clusters may be precisely calculated by solving a polynomial expression [6], and there are now many commercially available programs that will do this automatically. On the other hand the chemical average mass of an element is simply the sum of the abundance-weighted masses of all of its stable isotopes (e. g., 98.9% ^{12}C and 1.1% ^{13}C, to give the isotope weighted average mass of 12.011 for carbon). The *average mass* of a molecule is then the sum of the chemical average masses of the elements present. The *peak top mass* is the mass of the peak maximum of the isotope cluster. The relationship

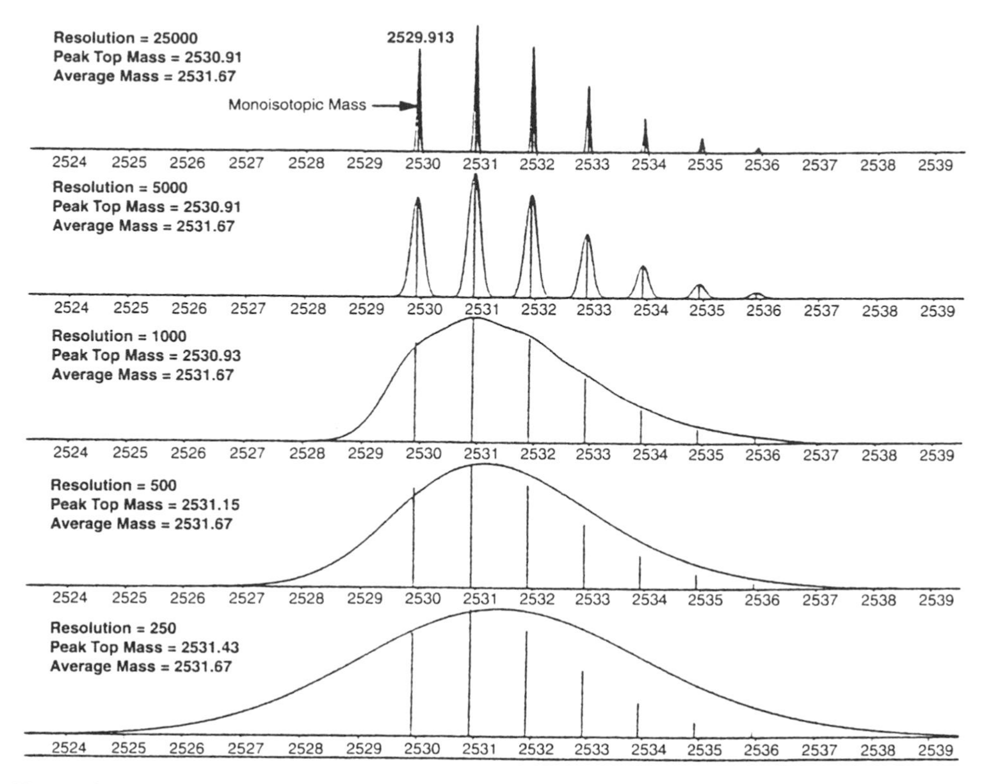

Fig. 1. Oxidized b-chain, insulin. Formula $C_{97}H_{151}N_{25}O_{46}S_4$. The molecular ion cluster for the oxidized b-chain of insulin is shown at various resolutions. The asymmetry of the cluster becomes less apparent as the resolution is decreased and the peak top mass and the average mass become almost identical.

between the monoisotopic mass, average mass, and peak top mass (sometimes referred to as the maximum mass) is illustrated in Figures 1 and 2 for a peptide of mass ca. 2500 and a protein with a mass of ca. 25,000, respectively. One important consequence of the contribution of the heavier isotope peaks is that for peptides with masses greater than ca. 2000, the peak corresponding to the monoisotopic mass is no longer the most abundant in the isotopic cluster (Fig. 1). With increasing molecular weight, the peak top mass continues to shift upward relative to the monoisotopic mass. Above masses of ca. 8000, the monoisotopic mass has an insignificant contribution to the isotopic envelope (Fig. 2). Whether the monoisotopic mass or the average mass should be used when measuring and reporting molecular weights will depend on the mass of the substance and the resolving power of the mass spectrometer.

RESOLUTION

The ability to separate and mass measure signals for peptides, proteins, etc. of similar, but not identical molecular mass is affected by the resolving power

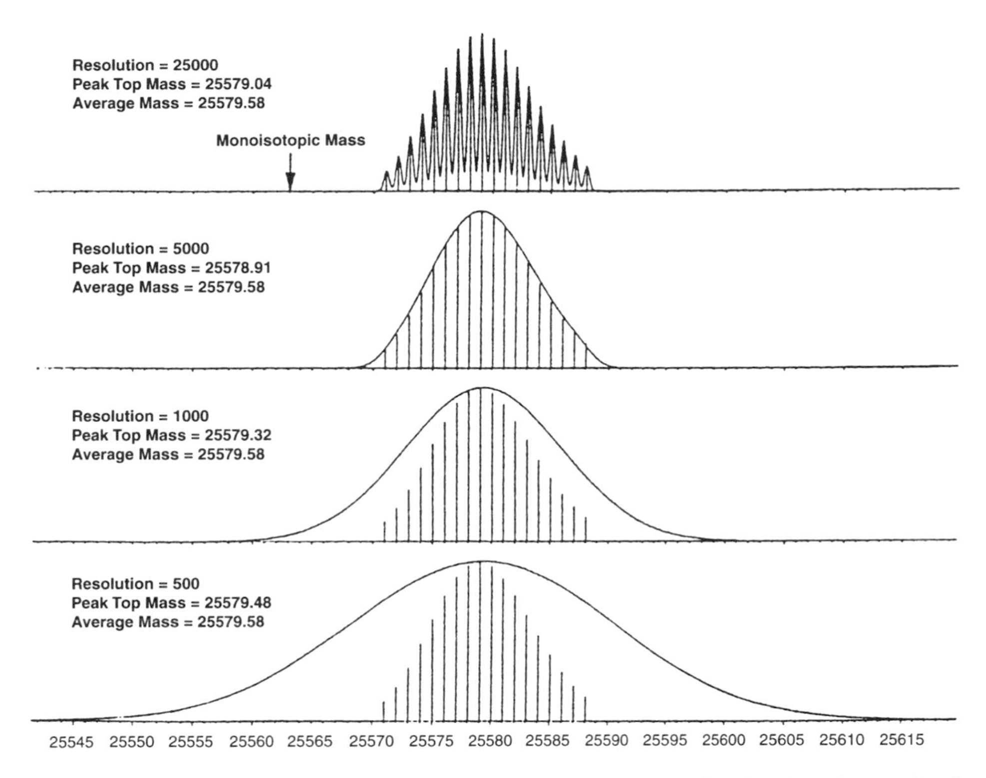

Fig. 2. Protein HIV-p24. Formula $C_{1129}H_{1802}N_{316}O_{335}S_{13}$. The molecular ion cluster for the protein HIV-p24 is shown at various resolutions. The position of the monoisotopic mass is indicated by the arrow.

of the mass analyzer. Because peaks in a mass spectrum have width and shape, it is necessary to evaluate the extent of overlap between adjacent peaks when determining the resolution. Mass resolution is often expressed as the ratio m/Δm where m and m + Δm are the masses (in atomic mass units or Da) of two adjacent peaks of approximately equal intensity in the mass spectrum (Fig. 3). The height of the "valley" between the two adjacent peaks, expressed as a percentage of the height of peak "m", is a measure of the degree of overlap. There are presently two definitions in widespread use, and it is essential to know which is being used when resolution figures are quoted. Historically, the first of these is the so-called 10% valley definition in which the two adjacent peaks each contribute 5% to the valley between them (Fig. 3). In practice, it is often necessary to determine the resolution of the mass analyzer using a single peak (either one peak in a mass-resolved isotopic peak cluster, or one peak corresponding to the envelope of unresolved isotopic peaks, see below). This is accomplished by measuring the peak width (in Da) at the 5% level, and dividing this number into the mass (in Da) of the peak. This definition is most frequently used for molecules with masses below 5000 Da where it is possible on certain analyzers to obtain

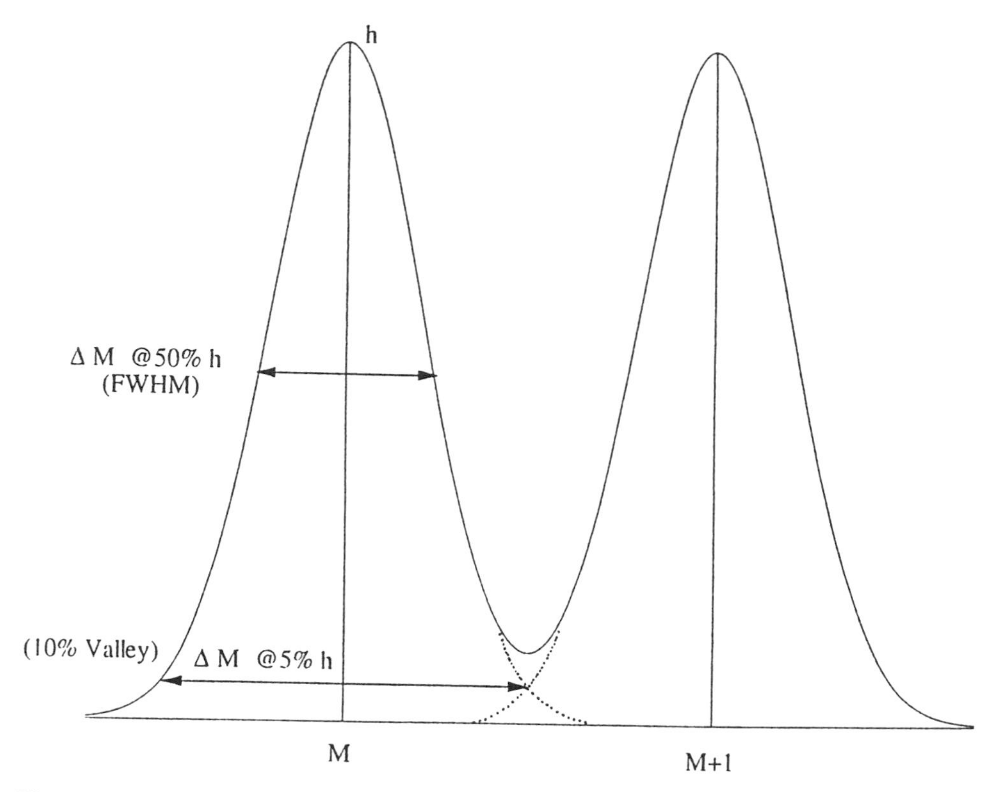

Fig. 3.

resolution of the isotope peaks. For unresolved isotopic peak envelopes the "full-width, half-maximum" (FWHM) definition has come into popular use. The resolution of a peak using this definition is the mass of the peak (in Da) divided by the width (in Da) of the isotopic envelope measured at the half-height of the peak (see Fig. 3). A useful rule of thumb is that the value for the resolution determined using the FWHM definition is approximately twice that obtained using the 10% valley definition (equivalent for singly charged ions to the full-width at 5% peak height for symmetrical peaks). For example, a resolution of 1000 using the 10% valley definition is approximately equivalent to resolution of 2000 using the FWHM definition: clearly, while the resolution value in the latter case is larger, the measured degree of separation of the peaks in Da is identical. It should also be noted that at a resolution of 1000 (FWHM) a peak at mass 1001 will essentially be *unresolved* from the peak at mass 1000.

However, for proteins, resolving the isotopes in the protonated molecular ion envelope is generally not possible (except for FTMS of electrospray-produced molecular ions), nor is it generally useful in practice to be able to resolve the isotopes of large molecules like proteins. Furthermore, increasing the resolving power of a mass spectrometer usually reduces the sensitivity of detection. Thus, in sample limited situations, the resolving power of the mass spectrometer may have to be reduced below the level necessary for resolution of the isotopic peaks to allow detection of the protonated molecular ion.

Mass Accuracy

The accuracy of the mass measurement is often stated as a percentage of the measured mass (e. g., molecular mass = 1000 +/- 0.01%), or as parts-per-million (e. g., molecular mass = 1000 +/- 100 ppm). As the mass being considered increases, the absolute mass error corresponding to the percent or ppm error will also increase proportionally (e. g., 0.01% or 100 ppm = 0.1 Da at m/z 1000, or 0.5 Da at m/z 5000, or 5 Da at m/z 50,000). Mass accuracy is dramatically affected by how the data is treated by the data system. Very significant error can be introduced in the mass range of 1000 to ca. 5000 if unresolved isotope clusters must be measured. In this mass range, the unresolved clusters will be asymmetric in shape, reflecting the asymmetry of the intensity distribution of the underlying isotopic masses in the cluster (see Fig. 1). If the mass being reported is the average mass, one must be sure to measure the centroid of the distribution and not the peak top mass, as these will not generally be the same in this mass range. Errors in measuring the centroid can occur when the ion intensities are weak (due to the inaccurate definition of the peak profile due to poor ion statistics) or due to other distortions of the peak profile of the cluster. Such distortions can be caused by the presence of unresolved adducts with the protonated molecular ion (with, e. g., Na^+, K^+, photochemical adducts of MALDI matrices, etc.), or by other proteins of similar but not identical mass that are not resolved. In this case, it will be impossible to obtain an accurate mass assignment unless the instrument resolving power can be increased or the contributing adduct ions can be removed by cleanup of the sample. As the mass of the protein increases, the peak profile (for a pure protein) becomes more symmetrical and the average mass and the peak top mass become almost identical. On the other hand, it becomes more and more difficult to resolve the substance from other substances of very similar molecular masses, and the possibility of adduct formation increases, with a concomitant effect on mass measurement accuracy.

How Much Resolution Is Required?

A resolution of at least several thousand (FWHM) is desirable for good mass measurement in the mass range of 1000 to ca. 5000 Da. Why do we care about having sufficient resolution to measure monoisotopic masses? Why not centroid the unresolved peak profile and obtain the chemical average mass? While in theory this should work, in practice it is often difficult with real data to precisely determine the centroid of an unresolved isotope cluster. In order to obtain an accurate centroid it is necessary to define the peak down to the baseline. Unfortunately, this is not routinely achieved as the bottom parts of the clusters are often distorted or poorly defined due to background sample (or matrix in MALDI) components, noise, and poor ion statistics. These factors can easily contribute errors of 1 Da or more in the mass range of 1000 to 5,000 Da.

In order to use monoisotopic masses you must have adequate resolution to separate adjacent isotope peaks sufficiently that their masses may be individually measured. Resolution sufficient to resolve single mass unit differences is also important for 1) deconvolution of mixtures of very. similar mass (e.g., Asn vs. Asp); 2) being able to distinguish charge-state directly from isotope peak spacing in ESMS (see below), etc.

Thus, in the mass range up to ca. 5000 Da it is recommended that as high a resolution as possible be used in order to facilitate mass assignment. If unresolved clusters must be measured, the centroid of the isotope cluster should be used to calculate the chemical average mass (while bearing in mind the warnings given above). For example, the resolution obtainable with quadrupole and time-of-flight mass analyzers is insufficient to resolve the individual isotope peaks of the elements for molecules with masses in excess of ca. 4000 Da. In this case, an unresolved peak envelope is obtained for a protein, the centroid of which is the abundance-weighted sum (chemical average mass) of the isotopes of the elements present. FTMS is unique in being able to provide higher resolution for large molecules (see chapter by Alan Marshall, this volume). Having ultra-high resolution does not help with this problem however, as the monoisotopic peak is of such low intensity that it cannot be identified at these masses. In this higher mass range it is necessary to use the chemical average mass.

However, the above statements do not mean that you need only very low resolution at masses above 10,000 Da. The lower the resolution, the lower your ability to detect variants of your protein that differ slightly in mass. Higher resolution also allows the identification of the charge state of an isotope cluster based on the spacing of the individual isotopic peaks. Also, in MALDI, one would like to have sufficient resolution to resolve the contribution of any matrix adduct peaks. The relative abundance of such adducts is unpredictable, and they may shift the apparent centroid of the unresolved cluster to higher mass in an unpredictable fashion. The resolution required for species with masses above 10,000 Da is to a first approximate based on the natural width of the molecular ion envelope. For a 25 kDa protein the envelope of isotopes is ca. 11 Da wide at FWHM, corresponding to an apparent resolution of 2300 (25000/11). Having an *instrumental* resolution higher than 2300 but less than 30,000 (see Fig. 4) will not improve the apparent resolution because the minimum width of the unresolved protein molecular ion cluster is determined by the weighted average of the natural abundance of the isotopes present.

Thus, distinguishing protein variants of similar mass depends on a number of factors in addition to the stated resolution of the mass analyzer, including the absolute mass difference between the components and the width and shape of the protein molecular ion "envelopes", which is determined mainly by the mass at which the measurement is being made. The shape and width are also affected by the adduction of small molecular weight ions, such as metal cations or phosphate anions or the photoadduction of matrix molecules often seen in MALDIMS. For example, detection of an N-terminal formyl group (diff. = 28 Da) on a 25 kDa protein would require a resolution of ca. 2000 (FWHM)

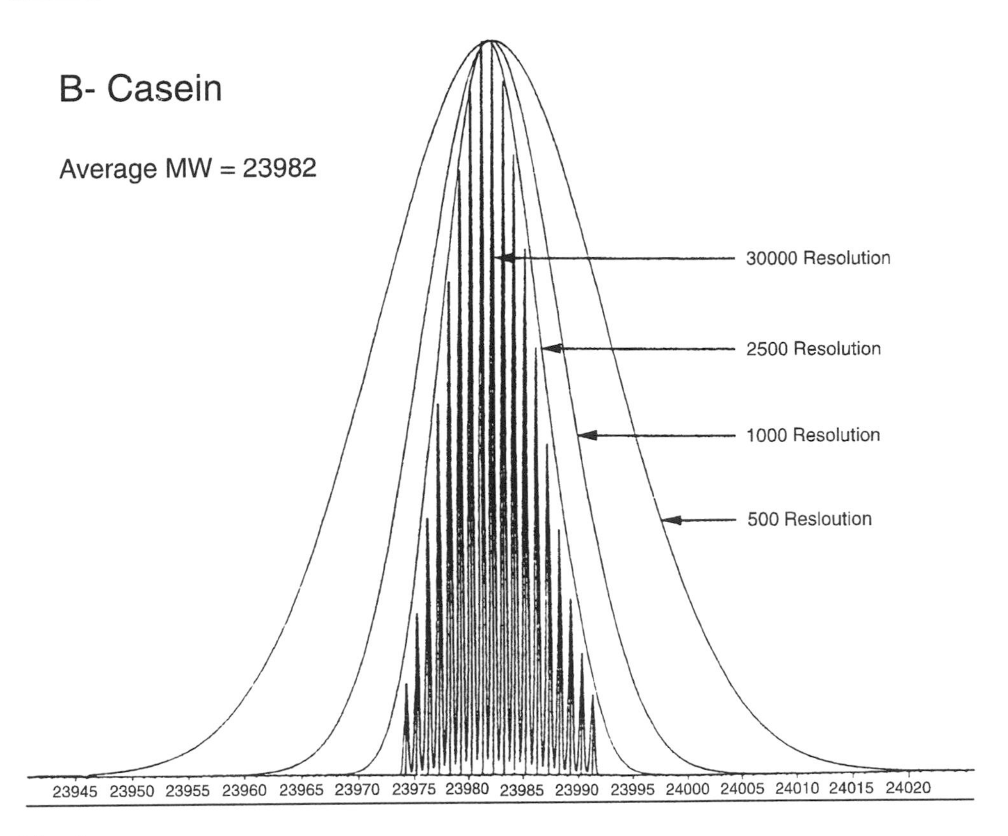

Fig. 4.

($25,000/28_{10\% \text{ valley def.}}$ x 2), which is easily achievable by ESMS on quadrupole mass analyzers. Substantially higher resolution in MALDIMS is typical with a time-of-flight analyzer equipped with an ion mirror or reflectron. Signals due to photochemical adducts of the matrix with the protein are also commonly observed in MALDI. These adducts can give rise to intense signals that, as the MW of the protein increases, fold into the $(M+H)^+$ peak of the protein and skew the observed mass to an artificially high value. These adducts also can interfere with detection of variants.

RESOLUTION OF MULTIPLY-CHARGED IONS

Electrospray ionization mass spectra of peptides or proteins having several basic sites typically show series of peaks arising from a distribution of multiply-charged ions, each having a different m/z value. For a pure compound the charge states of each ion and the corresponding molecular mass can be determined from the spacing of the peaks [9]. However, some peptides may give only one predominant charge state. Furthermore, the spectrum of a mixture may be too complex to identify ion series that belong to each individual component. In such cases it is possible to determine the charge state of each ion if the mass spectrometer has sufficient resolving power, $m/\Delta m$, to separate the stable isotope

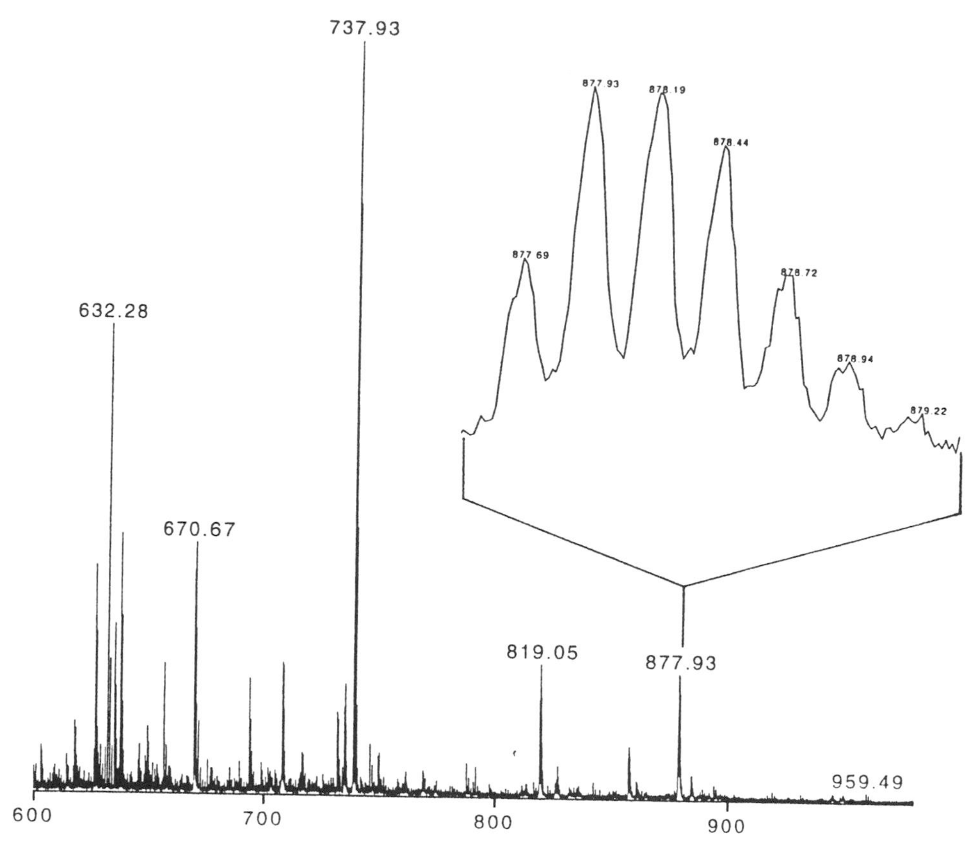

Fig. 5 The partial mass spectrum of an unseparated tryptic digest of 400 fmol of the glycoprotein fetuin. In the inset the peak identified as m/z 877.93 is expanded to reveal a cluster of isotopic components, separated by 0.25 m/z units, thus the charge state for these ions is 4.

profiles. Figure 5 gives an example from the mass spectrum of an unseparated digest of fetuin. This spectrum was recorded at a resolving power of approximately 6,500 FWHM. Expanding the mass scale for the peak identified as m/z 877.93 reveals a cluster of resolved isotope peaks that differ from each other by 0.25 m/z units. As these correspond to isotopic peaks with 1 Da mass differences, the charge state can be calculated from the reciprocal of the peak spacing, i. e., 4. The first peak in this cluster is at m/z 877.69, thus the monoisotopic molecular mass for the peptide giving this MH_4^{4+} ion is given by (877.69 x 4) - 4 = 3506.76.

This spectrum was recorded on a Qq-TOF mass spectrometer, a hybrid instrument in which the relatively high resolving power (up to 12,000 FWHM) derives from the use of an orthogonal acceleration TOF analyzer with a reflectron. This mass resolution is sufficient to allow this procedure to be carried out for peptides and small proteins with an upper mass limit of about 8-10 kDa. Qq-TOF instruments are also capable of tandem mass spectrometry, in which the precursor ions are selected in the quadrupole analyzer, collided with gas molecules, and the ionic fragments are analyzed in the orthogonal TOF. Selection of a multiply-charged precurser ion is likely to give an MS/MS spectrum containing fragments with different charge states. The resolving power of this instrument is the same for fragment ions, thus the charge states can be determined for each peak by inspection of the spacing of each isotope cluster. This information is essential for interpretation of collision-induced dissociation (CID) mass spectra to deduce peptide sequence information.

ACCURATE MASS MEASUREMENTS

A term often encountered in the mass spectrometry literature, is the phrase "accurate mass". Historically, this term refers to measurement of ionic masses to an accuracy of 1-10 parts per million (ppm) . This type of accuracy on small molecules allows the assignment of elemental compositions to the masses measured. Recently, it has also been used to refer to the measurement of masses of large molecules, such as proteins, with an accuracy as low as 1 part per thousand. To avoid confusion, it would be better to restrict the use of the term accurate mass to measurements accurate to 1-10 ppm. Further information on this subject is contained in the references listed below [7, 8].

REFERENCES

1. K. Biemann, *Meth. Enzymol.* **1990**, *193*, 295-305.

2. I. Jardine, *Meth. Enzymol.* **1990**, *193*, 441-455.

3. J. A. Yergey, D. Heller, G. Hansen, R. J. Cotter and C. Fenselau, *Anal. Chem.* **1983**, *55*, 353-356.

4. A. Ingendoh, M. Karas, F. Hillenkamp and U. Giessmann, *Int. J. Mass Spectrom. Ion Proc.* **1994**, *131*, 345-354.

5. C. N. McEwen and B. S. Larsen, *Rapid Comm. Mass Spectrom.* **1992**, *6*, 173-178.

6. J. A. Yergey, *Int. J. Mass Spectrom. Ion Proc.* **1983**, *52*, 337-349.

7. A .L. Burlingame, R. K. Boyd and S. J. Gaskell, *Anal. Chem.* **1994**, *66*, 636R.

8. M. L. Gross, *J. Am. Soc. Mass Spectrom.* **1994**, *5*, 57.

9. M. Mann, C. K. Meng and J. B. Fenn, *Anal. Chem.* **1989**, *61*, 1702-1709.

Author Index

Subject Index

A

B

C

D

G

H

I

J, K

L

M

N

O

P

Q

R

S

T

U

V

W

Y

Z